Environmental Quality Management

Environmental Quality Management

Editor:

Quan Cui, Ph.D.

ArclerPress

www.arclerpress.com

Environmental Quality Management

Editor: Quan Cui, Ph.D.

708 3rd Avenue, 6th Floor
New York
NY 10017
United States of America
www.arclerpress.com

ISBN: 978-1-68094-479-2

Library of Congress Control Number: 2016951340

About the Editor

Quan Cui, Ph.D.

Quan Cui obtained her PhD from Beijing Normal University in 2015. She worked as environmental impact assessment engineer for years. Aiming at linking science to policy, her study interest spans from ecosystem modelling to water resources management. She published peer-reviewed articles about using machine learning models in inflow forecasting, CO2 management and watershed ecosystem health assessment.

Contents

List of Contributors

Gary S. Bilotta
School of Environment and Technology, University of Brighton, Brighton BN2 4GJ, UK
Department for Environment, Food and Rural Affairs, London SW1P 3JR, UK

Alice M. Milner
Department for Environment, Food and Rural Affairs, London SW1P 3JR, UK
Department of Geography, University College London, London WC1E 6BT, UK

Ian Boyd
Department for Environment, Food and Rural Affairs, London SW1P 3JR, UK
College Gate, University of St Andrews, Fife KY16 9AJ, UK

Andrew H. Bingham
Air Resources Division, National Park Service, P.O. Box 25287, Denver, CO 80225, USA

M. Francesca Cotrufo
Department of Soil and Crop Sciences and Natural Resources Ecology Laboratory, Colorado State University, 200 West Lake Street, Fort Collins, CO 80523, USA

Sam D. Taylor
School of Environmental Sciences, University of East Anglia, Norwich Research Park, Norwich, NR4 7TJ, UK

Yi He
Tyndall Centre for Climate Change Research, School of Environmental Sciences, University of East Anglia, Norwich Research Park, Norwich, NR4 7TJ, UK

Kevin M. Hiscock
School of Environmental Sciences, University of East Anglia, Norwich Research Park, Norwich, NR4 7TJ, UK

Ulrich Förstner
Institute of Environmental Technology and Energy Economics, University of Technology Hamburg-Harburg, Eissendorfer Street, 21071 Hamburg, Germany.

Henner Hollert
Department of Ecosystem Analysis, Institute for Environmental Research, ABBt – Aachen Biology and Biotechnology, RWTH Aachen University, Worringerweg 1, 52074 Aachen, Germany.

Markus Brinkmann
Department of Ecosystem Analysis, Institute for Environmental Research, ABBt – Aachen Biology and Biotechnology, RWTH Aachen University, Worringerweg 1, 52074 Aachen, Germany.

Kathrin Eichbaum
Department of Ecosystem Analysis, Institute for Environmental Research, ABBt – Aachen Biology and Biotechnology, RWTH Aachen University, Worringerweg 1, 52074 Aachen, Germany.

Roland Weber
POPs Environmental Consulting, Lindenfirststrasse 23, 73527 Schwäbisch Gmünd, Germany.

Wim Salomons
Kromme Elleboog 21, 9751 RB, Haren, Groningen, Netherlands.

Jean-Christophe Comte
University of Aberdeen, School of Geosciences, Aberdeen, UK
Previously at Queen's University Belfast, Groundwater Research Group, Belfast, UK

Rachel Cassidy
Agri-Food and Biosciences Institute, Belfast, UK
Previously at Queen's University Belfast, Groundwater Research Group, Belfast, UK

Joy Obando
Kenyatta University, Department of Geography, Nairobi, Kenya

Nicholas Robins
Previously at Queen's University Belfast, Groundwater Research Group, Belfast, UK
Previously at British Geological Survey, Wallingford, UK

Kassim Ibrahim
University of Comoros, Faculty of Sciences, Moroni, Comoros

Simon Melchioly
University of Dar Es Salaam, Department of Geology, Dar Es Salaam, Tanzania

Ibrahimu Mjemah
Sokoine University of Agriculture, Department of Physical Sciences, Morogoro, Tanzania

Halimu Shauri
Pwani University, Department of Social Sciences, Kilifi, Kenya

Anli Bourhane
University of Reunion Island, Laboratory Geosciences Reunion, Saint-Denis, France
Water Office Reunion, Saint-Denis, France

Ibrahim Mohamed
University of Comoros, Faculty of Sciences, Moroni, Comoros

Christine Noe
University of Dar Es Salaam, Department of Geology, Dar Es Salaam, Tanzania

Beatrice Mwega
Kenyatta University, Department of Geography, Nairobi, Kenya

Mary Makokha
Kenyatta University, Department of Geography, Nairobi, Kenya

Jean-Lambert Join
University of Reunion Island, Laboratory Geosciences Reunion, Saint-Denis, France

Olivier Banton
University of Avignon, Laboratory of Hydrogeology, Avignon, France

Jeffrey Davies
Previously at British Geological Survey, Wallingford, UK
University College London, London, UK

Odeya Cohen
Department of Emergency Medicine, Recanati School for Community Health Professions, Faculty of Health Sciences, Ben-Gurion University of the Negev, Beer-Sheva 84105, Israel
PREPARED Center for Emergency Response Research, Ben-Gurion University of the Negev, Beer-Sheva 84105, Israel

Arkady Bolotin
Department of Public Health, Faculty of Health Science, Ben-Gurion University of the Negev, Beer-Sheva 84105, Israel

Mooli Lahad
Department of Psychology, Tel-Hai Academic College, Galil-Elyon 12208, Israel
The Community Stress Prevention Centre (CSPC), Kiryat-Shmona 11016, Israel

Avishay Goldberg
Department of Emergency Medicine, Recanati School for Community Health Professions, Faculty of Health Sciences, Ben-Gurion University of the Negev, Beer-Sheva 84105, Israel
PREPARED Center for Emergency Response Research, Ben-Gurion University of the Negev, Beer-Sheva 84105, Israel
Department of Health Systems Management, Faculty of Health Sciences & Guilford Glazer Faculty of Business and Management, Ben-Gurion University of the Negev, Beer-Sheva 84105, Israel

Limor Aharonson-Daniel
Department of Emergency Medicine, Recanati School for Community Health Professions, Faculty of Health Sciences, Ben-Gurion University of the Negev, Beer-Sheva 84105, Israel
PREPARED Center for Emergency Response Research, Ben-Gurion University of the Negev, Beer-Sheva 84105, Israel

Johanna Jacomina Heymans
Scottish Association for Marine Science, Scottish Marine Institute, Oban PA37 1QA, UK

Marta Coll
Institut de Recherche pour le Développement UMR MARBEC (MARine Biodiverity Exploitation & Conservation), Avenue Jean Monnet, BP 171, 34203 Sète Cedex, France

Institute of Marine Science (ICM-CSIC), Passeig Marítim de la Barceloneta, n 37-49, 08003, Barcelona, Spain

Jason S. Link
NOAA Fisheries, Woods Hole,MA 02543, USA

Steven Mackinson
Centre for Environment, Fisheries and Aquaculture Science (Cefas), Pakefield Road, Lowestoft NR33 0HT, UK

Jeroen Steenbeek
Ecopath International Initiative Research Association, Barcelona, Spain

Carl Walters
Institute for the Oceans and Fisheries, University of British Columbia, 2202 Main Mall, Vancouver, BC, Canada

Villy Christensen
Institute for the Oceans and Fisheries, University of British Columbia, 2202 Main Mall, Vancouver, BC, Canada

Ranaivo A. Rasolofoson
Food and Resource Economics, University of Copenhagen, Rolighedsvej 25, 1958 Frederiksberg C, Denmark
School of Environment, Natural Resources and Geography, Bangor University, Thoday Building, Deiniol Road, LL57 2UW Gwynedd, Wales, UK

Paul J. Ferraro
Department of Economics, Andrew Young School of Policy Studies, Georgia State University, Atlanta, GA 30302, USA

Clinton N. Jenkins
School of Environment, Natural Resources and Geography, Bangor University, Thoday Building, Deiniol Road, LL57 2UW Gwynedd, Wales, UK

Julia P.G. Jones
Instituto de Pesquisas Ecológicas, Nazaré Paulista, SP, Brazil

David M. Oliver
Biological & Environmental Sciences, School of Natural Sciences, University of Stirling, Stirling FK9 4LA, UK

Kenneth D.H. Porter
Biological & Environmental Sciences, School of Natural Sciences, University of Stirling, Stirling FK9 4LA, UK

Yakov A. Pachepsky
USDA ARS, Beltsville Agricultural Research Center, Beltsville, MD 20705, USA

Richard W. Muirhead
AgResearch Ltd, Land & Environment, Invermay Research Centre, Private Bag 50034, Mosgiel 9053, New Zealand

Sim M. Reaney
Department of Geography, Durham University, Durham DH1 3LE, UK

Rory Coffey
School of Biosystems Engineering, Agriculture and Food Science Centre, University College Dublin, Belfield, Dublin, Ireland

David Kay
Centre for Research into Environment & Health, Aberystwyth University, Wales SY23 3DB, UK

David G. Milledge
Department of Geography, Durham University, Durham DH1 3LE, UK

Eunmi Hong
USDA ARS, Beltsville Agricultural Research Center, Beltsville, MD 20705, USA

Steven G. Anthony
ADAS Group Ltd, HQ Pendeford House, Pendeford Business Park, Wolverhampton WV9 5AP, UK

Trevor Page
Lancaster Environment Centre, Lancaster University, Lancaster LA1 4YQ, UK

Jack W. Bloodworth
Cranfield Water Science Institute, Cranfield University, Cranfield, Bedfordshire MK43 0AL, UK

Per-Erik Mellander
TEAGASC, Agricultural Catchments Programme, Johnstown Castle, Wexford, Ireland

Patrice E. Carbonneau
Department of Geography, Durham University, Durham DH1 3LE, UK

Scott J. McGrane
Department of Civil & Environmental Engineering, University of Surrey, Guildford, Surrey GU2 7XH, UK

Richard S. Quilliam
Biological & Environmental Sciences, School of Natural Sciences, University of Stirling, Stirling FK9 4LA, UK

Daniel J.C. Skinner
Cranfield University, School of Water, Energy and Environment, Cranfield, Bedfordshire MK43 0AL, UK

Sophie A. Rocks
Cranfield University, School of Water, Energy and Environment, Cranfield, Bedfordshire MK43 0AL, UK

Simon J.T. Pollard
Cranfield University, School of Water, Energy and Environment, Cranfield, Bedfordshire MK43 0AL, UK

R. Paul Bardos
r3 environmental technology UK Ltd, University of Reading Whiteknights Campus, Reading, UK
School of Environment and Technology, University of Brighton, Brighton, UK

Brian D. Bone
Bone Environmental Consultant Ltd., Mickleton, Chipping Campden, UK

Richard Boyle
Homes and Communities Agency, Rivergate, Temple Quay, Bristol, UK

Frank Evans
National Grid Property, National Grid House, Warwick Technology Park, Gallows Hill, Warwick CV34 6DA, UK

Nicola D. Harries
CL:AIRE, 32 Bloomsbury Street, London WC1B 3QJ, UK

Trevor Howard
Environment Agency, Horizon House, Deanery Road, Bristol BS1 5AH, UK

Jonathan W.N. Smith
Shell Global Solutions (UK) Ltd, Lange Kleiweg 40, 2288 GK Rijswijk, The Netherlands
Groundwater Protection & Restoration Group, University of Sheffield, Kroto Research Institute, Sheffield, UK

Byron K. Williams
The Wildlife Society, 5410 Grosvenor Lane, Suite 200, Bethesda, MD 20814, USA

Eleanor D. Brown
Science and Decisions Center, U.S. Geological Survey, 12201 Sunrise Valley Drive, Reston, VA 20192, USA

Magdalena K. Grove
School of Environment and Technology, University of Brighton, Cockcroft Building, Lewes Road, Brighton BN2 4GJ, UK
Aquaread Ltd., Bridge House, Northdown Industrial Estate, Broadstairs, Kent, CT10 3JP, UK

Gary S. Bilotta
School of Environment and Technology, University of Brighton, Cockcroft Building, Lewes Road, Brighton BN2 4GJ, UK

Robert R. Woockman
Department of Civil and Environmental Engineering, The University of Tennessee, 325 John D Tickle Engineering Building, Knoxville, TN 37996, USA

John S. Schwartz
Department of Civil and Environmental Engineering, The University of Tennessee, 325 John D Tickle Engineering Building, Knoxville, TN 37996, USA

David A. Turner
Technology and Society Lab, Empa, Lerchenfeldstrasse 5, 9014 St. Gallen, Switzerland

Ian D. Williams
Centre for Environmental Sciences, Faculty of Engineering and the Environment, University of Southampton, University Road, Southampton, Hampshire SO17 1BJ, United Kingdom

Simon Kemp
Centre for Environmental Sciences, Faculty of Engineering and the Environment, University of Southampton, University Road, Southampton, Hampshire SO17 1BJ, United Kingdom

A.P. Hejnowicz
Environment Department, University of York, Heslington, York YO10 5DD, United Kingdom

M.A. Rudd
Department of Environmental Sciences, Emory College of Arts and Sciences, United States

P.C.L. White
Environment Department, University of York, Heslington, York YO10 5DD, United Kingdom

Letizia Tedesco
Finnish Environment Institute, Marine Research Centre, Helsinki, Finland

Chiara Piroddi
Finnish Environment Institute, Marine Research Centre, Helsinki, Finland
Institute of Marine Science, Spanish Research Council, Barcelona, Spain

Maria Kämäri
Finnish Environment Institute, Marine Research Centre, Helsinki, Finland

Christopher Lynam
CEFAS, Lowestoft, United Kingdom

Eirini Politi· John S. Rowan
Geography, School of Social Sciences, University of Dundee, Dundee DD1 4HN, UK

Mark E.J. Cutler
Geography, School of Social Sciences, University of Dundee, Dundee DD1 4HN, UK

Gemma Dunn
Program on Water Governance, University of British Columbia, 439-2202 Main Mall, Vancouver, BC V6T 1Z4, Canada

Leila Harris
Institute for Resources, Environment, and Sustainability and Program on Water Governance, University of British Columbia, 421-2202 Main Mall, Vancouver, BC V6T 1Z4, Canada

Christina Cook
Geography Department, Hebrew University of Jerusalem, Mount Scopus, Jerusalem, 91905, Israel

Natalie Prystajecky
Department of Pathology and Laboratory Medicine, University of British Columbia, G227-2211 Wesbrook Mall Vancouver, Vancouver, BC V6T 2B5, Canada

Junguo Liu
School of Nature Conservation, Beijing Forestry University, Qinghua East Road 35, Haidian District, 100083 Beijing, China

Qingying Liu
School of Nature Conservation, Beijing Forestry University, Qinghua East Road 35, Haidian District, 100083 Beijing, China

Hong Yang
Eawag, Swiss Federal Institute of Aquatic Science and Technology, Ueberlandstrasse 133, 8600 Duebendorf, Switzerland
Department of Environmental Sciences, University of Basel, Petersgraben 35, 4001 Basel, Switzerland

Jeffery S. Horsburgh
Department of Civil and Environmental Engineering, Utah Water Research Laboratory, Utah State University, 8200 Old Main Hill, Logan, UT 84322-8200, USA

Stephanie L. Reeder
Utah Water Research Laboratory, Utah State University, Logan, UT, USA

Amber Spackman Jones
Utah Water Research Laboratory, Utah State University, Logan, UT, USA

Jacob Meline
Utah Water Research Laboratory, Utah State University, Logan, UT, USA

Preface

The environment plays a crucial role in people's physical, mental and social well-being. The complex relationships between environmental factors and human health, taking into account multiple pathways and interactions, should be seen in a broader spatial, socio-economic and cultural context. The degradation of the environment, through air pollution, noise, chemicals, poor quality water and loss of natural areas, combined with lifestyle changes, may be contributing to substantial increases in rates of obesity, diabetes, diseases of the cardiovascular and nervous systems and cancer — all of which are major public health problems for Europe's population. Reproductive and mental health problems are also on the rise. Asthma, allergies, and some types of cancer related to environmental pressures are of particular concern for children. Better understanding of differences in the social distribution of environmental quality can be helpful for policy.

Environmental quality management means to manage human behaviors to ensure that the environmental quality meets the specific environmental standards. It encompasses a wide range of procedures from pollution control to sustainable development goals. This book includes in-depth research articles about environmental policy making, pollution control and management, modelling of industrial management practice effect on environmental quality, environmental code and policy analysis, environmental resource management, ecosystem resilience, environment risk assessment, geospatial technologies of environmental changes and marine models used to forecast seawater quality et. al. The environment directly affects health status and plays a major role in quality of life, years of healthy life lived, and health disparities. Poor air quality is linked to premature death, cancer, and long-term damage to respiratory and cardiovascular systems. Secondhand smoke containing toxic and cancer-causing chemicals contributes to heart disease and lung cancer in nonsmoking adults. Globally, nearly 25% of all deaths and the total disease burden can be attributed to environmental factors. Poor environmental quality has its greatest impact on people whose health status is already at risk. Safe air, land, and water are fundamental to a healthy community environment. An environment free of hazards, such as secondhand smoke, carbon monoxide, allergens, lead, and toxic chemicals, helps prevent disease and other health problems. Implementing and enforcing environmental standards and regulations, monitoring pollution levels and human exposures, building environments that support healthy lifestyles, and considering the risks of pollution in decision-making can improve health and quality of life for all. Poor air quality contributes to cancers, cardiovascular disease, asthma, and other illnesses. Poor water quality can lead to gastrointestinal illness and a range of other conditions, including neurological problems and cancer. Some chemicals in and around homes and workplaces can contribute to acute poisonings and other toxic effects. The built environment (such as schools, parks, greenways, and transportation systems) affects both individual health and environmental quality.

Editor
Quan Cui, Ph.D.

INTRODUCTION

Environmental quality management means to manage human behaviors to ensure that the environmental quality meets the specific environmental standards. It encompasses a wide range of procedures from pollution control to sustainable development goals. Environmental Quality Management services include air quality, water management, clean-up of contaminated sites. Environmental Quality Standard is a value, generally defined by regulation, which specifies the maximum permissible concentration of a potentially hazardous chemical in an environmental sample, generally of air or water. Environmental sanitation is a set of actions geared towards improving the quality of the environment and reducing the amount of disease. By doing so, the hope is that living conditions will improve and health problems will decrease. The management of water, solid waste, and industrial waste, as well as the topic of pollution and noise control, all fall under the umbrella of environmental sanitation.

The water supply can directly impact pollution and the spread of disease. The most common sources of water include groundwater and surface water; the main sources of groundwater are usually wells and springs, while surface water comes from fresh water sources like lakes and rivers. Both require treatment before being consumed because various chemicals, particles, and biological contaminants, like bacteria, can enter the water.

Many developed countries have water treatment plants where drinking water is cleaned of dirt and particles, disinfected, and stored until needed. This water may be run through pipes directly to homes and business. Other methods of treating water include boiling it or treating it with chemicals to kill harmful bacteria.

In addition to other sources, water can become contaminated from household trash and human waste. If waste is not disposed of properly, it can mix with water in the ground or with surface water that might be collected for drinking and cooking. When people drink tainted water, it can spread disease or even cause new outbreaks.

The environmental conditions of a given area may be affected by waste management, the process used to dispose of garbage. How waste is disposed of varies based on living conditions and the accepted standard of living in a geographical area. While some communities provide wastewater treatment and trash collection, others do not, which reduces the ability to control the well-being of the environment and its people. When waste is not removed and treated properly, pollution may lead to the spread of disease; when proper disposal and treatment methods are followed, disease and pollution can usually be reduced.

Environmental Management Plan

Environmental management plan is required for formulation, implementation and monitoring of environmental protection measures during and after commissioning of projects. The plans should indicate the details as to how various measures have been or are proposed to be taken including cost components as may be required.

Cost of measures for environmental safeguards should be treated as an integral component of the project cost and environmental aspects should be taken into account at various stages of the projects:

- Conceptualization: preliminary environmental assessment
- Planning: detailed studies of environmental impacts and design of safeguards
- Execution: implementation of environmental safety measures
- Operation: monitoring of effectiveness of built-in safeguards

The management plans should be necessarily based on considerations of resource conservation and pollution abatement, some of which are:

- Liquid Effluents
- Air Pollution
- Solid Wastes
- Noise and Vibration
- Occupational Safety and Health
- Prevention, maintenance and operation of Environment Control Systems
- House-Keeping
- Human Settlements
- Transport Systems
- Recovery - reuse of waste products
- Vegetal Cover
- Disaster Planning
- Environment Management Cell

Liquid Effluents

- Effluents from the industrial plants should be treated well to the standards as prescribed by the Central/State Water Pollution Control Boards.

- Soil permeability studies should be made prior to effluents being discharged into holding tanks or impoundments and steps taken to prevent percolation and ground water contamination.

- Special precautions should be taken regarding flight patterns of birds in the area. Effluents containing toxic compounds, oil and grease have been known to cause extensive death of migratory birds. Location of plants should be prohibited in such type of sensitive areas.

- Deep well burial of toxic effluents should not be resorted to as it can result in re-surfacing and ground water contamination. Re-surfacing has been known to cause extensive damage to crop and live stocks.

- In all cases, efforts should be made for re-use of water and its conservation.

Air Pollution

- The emission levels of pollutants from the different stacks, should conform to the polluting control standards prescribed by Central or State Boards.

- Adequate control equipment should be installed for minimizing the emission of pollutants from the various stacks.

- In-plant control measures should be taken to contain the fugitive emissions.

- Infrastructural facilities should be provided for monitoring the stack emissions and measuring the ambient air quality including micro-meteorological data (wherever required) in the area.

- Proper stack height as prescribed by the Central/State Pollution Control Boards should be provided for better dispersion of pollutants over a wider area to minimize the effect of pollution.

- Community buildings and townships should be built up-wind of plant with one-half to one kilometer greenbelt in addition to physiographical barrier.

Solid Wastes

- The site for waste disposal should be checked to verify permeability so that no contaminants percolate into the ground water or river/lake.

- Waste disposal areas should be planned down-wind of villages and townships.

- Reactive materials should be disposed of by immobilizing the reactive materials with suitable additives.

- The pattern of filling disposal site should be planned to create better landscape and be approved by appropriate agency and the appropriately pretreated solid wastes should be disposed according to the approved plan.

- Intensive programs of tree plantation on disposal areas should be undertaken.

Noise and Vibration

Adequate measures should be taken for control of noise and vibrations in the industry.

Occupational Safety and Health

Proper precautionary measures for adopting occupational safety and health standards should be taken.

Prevention, maintenance and operation of Environment Control Systems

- Adequate safety precautions should be taken during preventive maintenance and shut down of the control systems.

- A system of inter-locking with the production equipment should be implemented where highly toxic compounds are involved.

Housekeeping

Proper house-keeping and cleanliness should be maintained both inside and outside of the industry.

Human Settlements

- Residential colonies should be located away from the solid and liquid waste dumping areas. Meteorological and environmental conditions should be studied properly before selecting the site for residential areas in order to avoid air pollution problems.
- Persons who are displaced or have lost agricultural lands as a result of locating the industries in the area, should be properly rehabilitated.

Transport Systems

- Proper parking places should be provided for the trucks and other vehicles by the industries to avoid any congestion or blocking of roads.
- Siting of industries on the highways should be avoided as it may add to more road accidents because of substantial increase in the movements of heavy vehicles and unauthorized shops and settlements coming up around the industrial complex.
- Spillage of chemicals/substances on roads inside the plant may lead to accidents. Proper road safety signs both inside and outside the plant should be displayed for avoiding road accidents.

Recovery reuse of waste products

Efforts should be made to recycle or recover the waste materials to the extent possible. The treated liquid effluents can be conveniently and safely used for irrigation of lands, plants and fields for growing non-edible crops.

Vegetal Cover

Industries should plant trees and ensure vegetal cover in their premises. This is particularly advisable for those industries having more than 10 acres of land.

Disaster Planning

Proper disaster planning should be done to meet any emergency situation arising due to fire, explosion, sudden leakage of gas etc. Firefighting equipment and other safety appliances should be kept ready for use during disaster/emergency situation including natural calamities like earthquake/flood.

Environment Management Cell

Each industry should identify within its setup a Department/Section/Cell with trained personnel to take up the model responsibility of environmental management as required for planning and implementation of the projects.

CHAPTER 1

SYSTEMATIC REVIEWS TO INFORM ENVIRONMENTAL POLICIES

Gary S. Bilotta, Alice M. Milner , Ian Boyd

Keywords: Systematic review, Literature review, Environmental science, Environmental policy

INTRODUCTION

Environmental research varies in its methodological quality, [1] degree of bias, [2] and relevance to policy. Using this heterogeneous, and sometimes polarised, research to inform environmental policies can be a challenging task, which at present is often first approached through the use of narrative literature reviews (Boyd, 2013). It is recognised that these types of literature reviews are vulnerable to author bias, which can occur when the review authors intentionally or unintentionally select or emphasise research according to their own opinions, prejudices or commercial interests (Higgins and Green, 2011). Furthermore, narrative literature reviews rarely consider, in a reproducible and meaningful manner, the methodological quality, degree of bias, and therefore reliability of the primary studies that are cited. These features of narrative literature reviews could lead to ill-informed environmental policies.

In evidence-based policy-making in the healthcare field, systematic review (SR) processes are used in order to tackle these issues, helping to present a comprehensive, policy-neutral, transparent and reproducible synthesis of the evidence. These SR processes are exemplified by the activities of the Cochrane Collaboration (http://www.cochrane.org/); an international network of more than 31,000 researchers and practitioners (a mix of volunteers and paid staff who are affiliated to the organisation), from over 120 countries, who work to help healthcare practitioners, policy-makers, patients, their advocates and carers, make well-informed decisions about healthcare, by preparing, updating, and promoting the accessibility of SRs on the effectiveness of healthcare interventions. The Cochrane Collaboration have published over 5000 SRs so far, all of which are freely available online in the *Cochrane Database of Systematic Reviews*, which is part of *The Cochrane Library* (http://www.cochrane.org/cochrane-reviews/about-cochrane-library).

There is a common belief outside of healthcare, however, that SRs intrinsically adopt a biomedical model that is of relevance only to medicine, for example only capable of using randomised controlled trials (RCTs) and only capable of answering certain types of questions (Petticrew, 2001). As demonstrated in this article, this belief is unjustified. The practices of the Cochrane Collaboration have spurred the development of another international initiative; the Campbell Collaboration (http://campbell.gse.upenn.edu), who prepare, maintain, and disseminate SRs on the effectiveness of social and behavioural interventions in education, social welfare, and crime and justice (Davies and Boruch, 2001). More recently, these practices have spurred the founding of the Collaboration for Environmental Evidence – CEE (http://www.environmentalevidence.org/); an open community of scientists and managers who, from their initial centres in Australia, South Africa, Sweden and the UK, have started to prepare SRs on environmental topics. Nevertheless, at present many environmental researchers, practitioners and policy-makers are typically less familiar with exactly what a SR involves, and often have major misconceptions about their history and purpose (Petticrew, 2001). The aim of this article is to: (1) summarise the process of conducting a SR, using the Cochrane Collaboration's exemplary methodology as an example (http://handbook.cochrane.org/), (2) explain the rationale behind each stage of the process, and (3) examine the prospects and challenges of using SRs to inform environmental policies.

Citation: Gary S. Bilotta, Alice M. Milner, Ian Boyd, Systematic reviews to inform environmental policies, Environmental Science & Policy, Volume 42, October 2014, Pages 67-77, ISSN 1462-9011, http://dx.doi.org/10.1016/j.envsci.2014.05.010.

THE PROCESS OF CONDUCTING A COCHRANE SYSTEMATIC REVIEW

The key stages of producing a Cochrane systematic review (CSR), as described in the Cochrane Handbook (http://handbook.cochrane.org/), are illustrated in Fig. 1 and are summarised and compared to traditional literature reviews in Table 1:

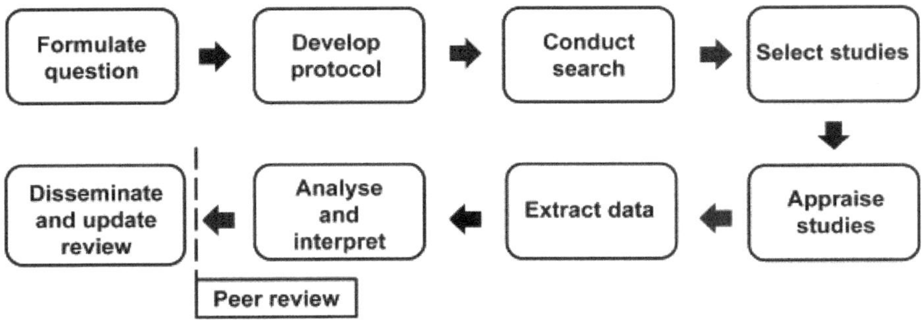

Figure 1: Key stages of conducting a CSR.

Table 1: A comparison of the features of CSRs (http://handbook.cochrane.org/) and traditional narrative literature reviews.

Stage	Cochrane systematic reviews	Narrative reviews
Formulating a question	Start with clear question to be answered or hypothesis to be tested (relating to specific populations, interventions, comparisons and outcomes)	May start with clear question to be answered, but they more often involve general discussion of the subject with no stated hypothesis
Developing a protocol	Protocol to be used is established and documented in advance, prior to knowledge of the available studies. This reduces the impact of review authors' bias, promotes transparency of methods and processes, reduces the potential for duplication, allows peer review of the planned methods, and enables easy maintenance of reviews in the light of new findings	Do not normally follow a pre-published protocol
Conducting the search	Strive to locate all relevant published and unpublished studies to limit the impact of publication and other biases	Do not usually attempt to locate all relevant literature. Often focus on published studies only
Selecting studies	Involve explicit description of what types of studies are to be included to limit selection bias on behalf of author(s)	Usually do not describe why certain studies are included and others are excluded
Appraising studies	Examine in a systematic and unbiased manner, the methods used in the primary studies, and investigate potential biases in those studies and sources of heterogeneity between study results	Often do not consider differences in study methods or study quality
Extracting data for analysis	Data analyses may be narrative, such as a structured summary and discussion of the studies' characteristics and findings, or quantitative, that is involving statistical meta-analysis, supported by Cochrane's review writing software	Typically limited to narrative analyses
Interpreting the synthesis	Base their conclusions on those studies which are most methodologically sound, and present a policy-neutral summary of the body of evidence, supported by Cochrane's review writing software	Often do not differentiate between methodologically sound and unsound studies. Sometimes present a policy-aligned summary of the body of evidence.
Disseminating and maintaining the review	Written by more than one author and peer-reviewed by a number of experts. They are then published online and are sometimes also co-published in peer-reviewed healthcare journals. They must remain free for dissemination in any and all media. Authors are committed to maintaining and updating these reviews, at least every two years, on the *Cochrane Library*	Written by one or more authors, and are sometimes peer-reviewed by experts, but are not always available as open access articles, and are not updated on a central database in the light of new findings

Adapted from Petticrew (2001).

The rationale behind each stage of a Cochrane systematic review

Formulating a question

As with any research, the first and most important decision in preparing a CSR is to determine its focus (O'Connor et al., 2011). This is best done by clearly framing the questions the review seeks to answer. Well-formulated questions will guide many aspects of the review process, including determining eligibility criteria, searching for studies, collecting data from included studies, and presenting findings (Jackson, 1980,Cooper, 1984 and Hedges, 1994). In CSRs, questions are stated broadly as review 'Objectives', and specified in detail as 'Criteria for considering studies for this review' (O'Connor et al., 2011). A statement of the objectives typically begins with a precise statement of the primary objective, normally in the format of a single sentence. For example, for CSRs this may take the form: 'To assess the effects of [*treatment, intervention or comparison*] for [*health problem*] in [*types of people, disease or problem and setting if specified*]'. This might be followed by one or more secondary objectives, relating to different participant groups, different comparisons of interventions or different outcome measures (O'Connor et al., 2011). As this example suggests, the detailed specification of the review question requires consideration of several key components (Richardson et al., 1995 and Counsell, 1997), including the types of populations (or participants), types of interventions and comparisons, and the types of outcomes that are of interest (*PICO* – *P*articipants, *I*nterventions, *C*omparisons and *O*utcomes) (O'Connor et al., 2011). As well as focussing review conduct, the contents of these sections are used by readers in their initial assessments of whether the review is likely to be directly relevant to the issues they face (O'Connor et al., 2011).

Systematic reviews are likely to be more relevant to the end-user and of higher quality if the initial questions and the protocols (see Section 2.1.3) are informed by advice from stakeholders with a range of experiences, in terms of both the topic and the methodology (Khan et al., 2001, Rees et al., 2004 and Thomas et al., 2004). The contribution of consulted stakeholders during the development of a review question and protocol, should be documented in the Acknowledgements section of the protocol or review (Green and Higgins, 2011). Titles for CSRs are agreed by and registered with Cochrane Review Groups, who then oversee the process from publishing the protocol to publishing the final review.

Application to environmental research

It is not difficult to see how the specified PICO components of CSR questions could be adapted for SRs on environmental topics. Human participants may be replaced with specified animal or plant populations, habitats, ecosystems or members of society; healthcare interventions may be replaced with environmental management options such as the use of different agricultural techniques or different plant/animal disease control measures; comparisons between healthcare interventions (e.g. comparing the effectiveness of a pharmaceutical drug against a placebo), may be replaced with environmental studies using before-and-after approaches or interventions versus control experiments; human health outcomes may be replaced with metrics of animal, plant, or ecosystem health or productivity, or even social outcomes. Indeed, the CEE (2013)guidelines for conducting SRs on environmental topics, recommend adopting the PICO approach to question formulation. Policy-based questions may be a starting point to guide SRs on environmental topics, though those carrying out reviews need to be satisfied that the policy development process that has been the source of questions is robust. There may be reasons for challenging policy-makers to understand the rationale for certain questions before embarking upon a particular SR.

Developing a protocol

Preparing a CSR is complex and involves many judgements; in order to minimize the potential for bias in the review process, these judgements should be made in ways that do not depend on the findings of the studies included in the review (Green and Higgins, 2011). Since CSRs are by their nature retrospective, it is important that the methods to be used should be established and documented in advance (Green and Higgins, 2011). Publication of a protocol for a review prior to knowledge of the available studies reduces the impact of review authors' bias, promotes transparency of methods and processes, reduces the potential for duplication, and allows for stakeholder engagement/review of the planned methods (Light and Pillemer, 1984).

While the intention is that reviews will adhere to the published protocol, changes in a review protocol are sometimes necessary (Green and Higgins, 2011). It is important, however, that these changes should not be made on the basis of how they affect the outcome of the research study (Green and Higgins, 2011). Post hoc decisions made when the impact on the results of the research is known, such as excluding selected studies from a SR, are highly susceptible to bias and should be avoided (Green and Higgins, 2011).

In the case of CSRs, protocols are published before the completed SR in the *Cochrane Database of Systematic Reviews* (*CDSR*) (Green and Higgins, 2011). Changes in the protocol are documented and reported in the 'Differences between protocol and review' section of the completed review, and sensitivity analyses exploring the impact of deviations from the protocol are undertaken when possible (Green and Higgins, 2011).

Application to environmental research

It is not difficult to imagine how authors might prepare a Cochrane-style protocol for a SR on an environmental topic. Indeed, the CEE (2013) provide a template for developing, registering and publishing a SR protocol. Their latest guidance can be found at www.environmentalevidence.org/Instructionsforauthors.html. For examples of almost 100 protocols produced at the time of writing this article, visit the CEE *Library* (www.environmentalevidence.org/Library.htm). For examples of reviews that are in progress, visit: http://www.environmentalevidence.org/Reviewsinprogress.html. The protocols produced so far by the CEE, vary in their detail and level of specification, partly reflecting the fact that environmental researchers are, at present, less familiar with the SR process. It would be advantageous for the CEE and wider environmental community to build expertise in protocol registration. Ideally, those who oversee the process should be specialists in this skill, but not have a detailed knowledge of the evidence available to answer the question. This will help to eliminate potential bias resulting from experts directing the review in a certain direction, based on their existing knowledge of the topic.

Conducting the search

Literature searches for CSRs aim to be as extensive as possible to ensure that as many as possible of the relevant studies are included in the review (Lefebvre et al., 2011). It is, however, necessary to strike a balance between striving for comprehensiveness and maintaining relevance when developing a search strategy (Lefebvre et al., 2011). Increasing the comprehensiveness or sensitivity of a search,[3] will reduce its precision,[4]and will retrieve more non-relevant articles (Lefebvre et al., 2011).

The Cochrane Handbook provides detailed guidance on developing a search strategy for a CSR. The search strategy for a CSR (search terms to be used, databases to be searched, etc.) is described in its review protocol, though searching can be an iterative process in which the terms that are used are modified, based on what has already been retrieved (Lefebvre et al., 2011). There are diminishing returns for search efforts; after a certain stage, each additional unit of time invested in searching returns fewer references that are relevant to the review (Lefebvre et al., 2011). Consequently there comes a point where the rewards of further searching may not be worth the effort required to identify the additional references (Lefebvre et al., 2011). The decision as to how much to invest in the search process depends on the question the review addresses and the resources that are available. Lefebvre et al. (2011) suggest that at a conservatively estimated reading rate of two abstracts per minute, the results of a database search can be 'scan-read' at the rate of 120 per hour; so the high yield and low precision associated with CSR searching is not as daunting as it might at first appear.

In CSRs the full final search strategies used for each database searched are included in an appendix of the CSR, so all search strategies should be saved, and notes taken of the number of records retrieved for each database searched (Lefebvre et al., 2011). This ensures that the search procedure is transparent, auditable, and reproducible.

Application to environmental research

The literature search processes used in CSRs were originally designed with a focus on evidence derived from RCTs examining the effectiveness of healthcare treatments. At present, non-randomised studies (NRSs) are a more common study methodology in environmental science.[5] These types of studies could be more time-consuming to search for during the literature search and selection stage of a SR on environmental topics. For example, when a CSR aims to include RCTs only, various approaches are available to restrict the search strategy to RCTs, including:

I. search for previous reviews of the review question,

II. use resources such as CENTRAL or Cochrane Review Group-specific registers that are 'rich' in RCTs,

III. use methodological filters and indexing fields, such as publication type in MEDLINE, to limit searches to studies that are likely to be RCTs, and

IV. search trial registers.

However, to restrict the search to particular NRS designs is more difficult; study design labels are not used consistently by authors and are not indexed reliably by journals or bibliographic databases. Search results thus often contain large numbers of irrelevant citations and abstracts often do not provide adequate detail about NRS design. Therefore, unlike the situation when reviewing studies from RCTs, it may be necessary to obtain and read many full reports in order to identify eligible studies. This challenge does not prevent NRSs from being included in SRs, Indeed, the Cochrane Collaboration have produced a number of reviews which include evidence from NRSs, including qualitative studies and economic data. The literature search and study eligibility assessments may, however, take longer to complete than they would with a review based solely on RCTs.

Authors, publishers and hosts of bibliographic databases could all contribute to improving the reporting and indexing of environmental NRSs so that they are easier to search and check against eligibility criteria in the future. In evidence-based medicine, this standardisation of reporting and indexing of research began in 1996 when an international group of epidemiologists, statisticians, clinical trialists, and medical editors, some of whom were involved with establishing the Cochrane Collaboration, published the CONSORT statement; a checklist of items to be addressed in a report of the findings of RCTs (Turner et al., 2013). CONSORT has twice been revised and updated over time, and the impact has been noted as one of the major milestones on health research methods over the last century (Gabriel and Normand, 2012). Environmental science would benefit from developing an environmental equivalent of the CONSORT statement for NRSs.

In the meantime, the CEE (2013) provide detailed guidance on how best to conduct literature searches for environmental studies.

Selecting the eligible studies

The findings of a SR depend critically on decisions relating to which studies are included, and on decisions relating to which data from these studies are presented and analysed (Higgins and Deeks, 2011). The methods used for these decisions must be transparent, and they should be chosen to minimize biases and human error. A CSR is a review of*studies* that meet pre-specified criteria for inclusion in the review (Higgins and Deeks, 2011). Since each of the studies discovered from the literature search stage may have been reported in several articles, abstracts or other reports, a comprehensive search for studies for the review may identify many *reports* from potentially relevant studies (Higgins and Deeks, 2011). Two distinct processes are therefore used to determine which studies can be included in the review: one is to link together multiple reports of the same study; and the other is to use the information available in the various reports to determine which studies are eligible for inclusion (Higgins and Deeks, 2011). Although sometimes there is a single report for each study, it should never be assumed that this is the case as this could introduce substantial biases if studies are inadvertently included more than once in a meta-analysis (Tramèr et al., 1997).

A typical process for selecting studies for inclusion in a CSR is as follows: (I) merge search results using reference management software, and remove duplicate records of the same report. (II) Examine titles and abstracts to remove obviously irrelevant reports (authors should generally be over-inclusive at this stage). (III) Retrieve full text of the potentially relevant reports. (IV) Link together multiple reports of the same study. (V) Examine full-text reports for compliance of studies with eligibility criteria. (VI) Correspond with investigators, where appropriate, to clarify study eligibility (it may be appropriate to request further information, such as missing results, at the same time). (VII) Make final decisions on study inclusion and proceed to data collection (Higgins and Deeks, 2011).

Decisions about which studies to include in a review are among the most influential decisions that are made in the review process (Higgins and Deeks, 2011). However, they involve judgement – to help ensure that these judgements are reproducible in CSRs, more than one author repeats the process independently, and this is overseen by the Cochrane Review Group. Using at least two authors, searching independently, reduces the possibility that relevant reports will be discarded (Edwards et al., 2002).

Experts in a particular area frequently have pre-formed opinions that can bias their assessments of both the relevance and validity of articles (Cooper and Ribble, 1989 and Oxman and Guyatt, 1993). Thus while it is important that at least one author is knowledgeable in the area under review, it is an advantage to have a second author who is not a content expert (Higgins and Deeks, 2011). Disagreements about whether a study should be included can generally be resolved by auditable discussion (Higgins and Deeks, 2011). Often the cause of disagreement is a simple oversight on the part of one of the review authors, but when the disagreement is due to a difference in interpretation, this may require auditable arbitration by another reviewer (Higgins and Deeks, 2011).

Application to environmental research

Study selection methods used in CSRs could be used, with little or no amendment, in SRs on environmental topics. The CEE (2013) suggest that it is good practice at the beginning of the abstract relevance assessment stage for two reviewers to undertake the same process on a random sub-sample of articles from the original list (the recommended sample is a minimum of 50 articles or 10% up to a maximum of 200 references). To check for consistency in the interpretation of the selection criteria, reviewer relevance decisions can be compared by performing a kappa analysis, which adjusts the proportion of records for which there was agreement, by the amount of agreement expected by chance alone. A kappa rating of 'substantial' (>0.5) is recommended to pass the assessment.

For the latest guidance on how to conduct the study selection stage for environmental SRs, visit www.environmentalevidence.org/Instructionsforauthors.html. Inconsistencies among reviewers at this stage of the SR, could change the outcome of the review. Owing to the importance of decisions made at this stage, it would be sensible for the CEE to enforce the independent assessment of reproducibility of this stage, rather than just recommending it as good practice. Likewise, it would be worthwhile enforcing that at least one of the members responsible for judgements of study eligibility, is a non-expert on the review topic.

Appraising the selected studies

This stage of a CSR is designed to ensure that the review authors are cognisant of the potential biases within primary studies and of how such biases could impact review results and subsequent conclusions (Higgins et al., 2011a). Cochrane SRs assess the methodological quality of primary studies through use of an objective system developed by the Grades of Recommendation, Assessment, Development and Evaluation (GRADE) Working Group (GRADE Working Group, 2004, Schünemann et al., 2006,Guyatt et al., 2008a and Guyatt et al., 2008b). This approach is now used by the World Health Organisation (WHO) and the UK National Institute for Health and Care Excellence (NICE) among 20 other bodies internationally. The GRADE approach specifies four levels of quality (High, Moderate, Low, and Very Low). The highest quality rating is for evidence from RCTs. Review authors can, however, downgrade evidence from RCTs to moderate, low, or even very low quality evidence, depending on the presence of the five factors, including limitations in the design and implementation of available studies suggesting high risk of bias; indirectness of evidence; unexplained heterogeneity or inconsistency of results; imprecision of results; and high probability of publication bias (Higgins et al., 2011a). Review authors will generally grade evidence from sound observational studies as low quality. If however, such studies yield large effects and there is no obvious risk of bias explaining those effects, review authors may rate the evidence as moderate or, if the effect is large enough, even high quality (Higgins et al., 2011a). The very low quality level includes, but is not limited to, studies with critical problems and unsystematic clinical observations (e.g. case series or case reports) (Higgins et al., 2011a).

The risk of bias within studies is assessed through the use of the Cochrane Collaboration's Risk of Bias Tool (Higgins et al., 2011a), which helps review authors to identify potential biases and support judgements of the degree to which they may have influenced the findings of each study. As part of the SR process, authors record their judgments of the risk of bias and provide evidence for each potential source of bias. Through the combined use of GRADE and the Risk of Bias Tool, review authors are guided in their evaluation of each of the included studies. This increases the transparency and auditability of evidence appraisal stage, reducing the potential for authors' bias to influence the conclusions of the review, while helping the authors to discover the consistencies and account for the variability in similar appearing studies through accounting for potential biases in the primary research (Cooper and Hedges, 1994). At least two authors assess the risk of bias within primary studies, and these authors are double-blinded before agreeing a final assessment.

While the Cochrane Collaboration prefer to use evidence derived from RCTs, owing to the lower potential for bias (when conducted according to best practice), the group have produced many SRs based on evidence derived from NRSs. Some review authors have tried to develop, optimise and 'validate' search strategies for NRS (Wieland and Dickersin, 2005, Fraser et al., 2006, Furlan et al., 2006, Golder et al., 2006a and Golder et al., 2006b), and there is a dedicated NRS Methods Group who provide guidance to support authors who are considering including NRS in CSRs (Higgins and Green, 2011). In principle, the assessment of the risk of bias in NRSs is exactly the same as it is for RCTs, but the Collaboration advise that review authors must pay extra attention to the weaknesses of the designs that have been used (such as noting their potential to ascertain causality); the execution of the studies through careful consideration of their risk of bias, especially the potential for selection bias and confounding factors to which all NRSs are suspect; and the potential for reporting biases, including selective reporting of outcomes owing to the lack of study registration systems (Higgins and Green, 2011).

Deeks et al. (2003) noted that there were at least 194 existing tools (scales which score the studies based on a number of weighted criteria; and checklists which assess studies against criteria without producing a score), that could be or have been used to assess methodological quality of NRSs. Until relatively recently, CSRs used a variety of these tools (Lundh and Gøtzsche, 2008). The Cochrane Collaboration's current recommended tool for assessing risk of bias, however is neither a scale nor a checklist. It is a domain-based evaluation, in which critical assessments are made separately for different domains of bias (or aspects of research design).[6] These domains were selected on the basis of empirical evidence linking them to biased findings (Wood et al., 2008 and Gluud, 2006). The approach was developed between 2005 and 2007 by a working group of methodologists, editors and review authors. Because it is impossible to know the extent of bias (or even the true risk of bias) in a given study, the possibility of validating any proposed tool is limited (Higgins and Green, 2011). Experiences in application of the first (2008) version of the tool led to some criticisms concerning its ease of use and reliability (i.e. inter-rater consistency) (Hartling et al., 2009 and Hartling et al., 2012). In response to these criticisms, a revised version was published in 2011, and a working group has since been established to continue development of the Risk of Bias Tool, with version 2.0 due to be released in 2014 (Turner et al., 2013). It is also accepted that issues of study design arise when using the tool to assess risk of bias in NRSs, and therefore an ongoing Cochrane Methods Innovation Fund project will lead to the release of a new version of the tool for assessing NRSs (Turner et al., 2013). Nevertheless, the use of alternative scales for assessing quality or risk of bias is explicitly discouraged in CSRs. The Collaboration argue that while scales offer appealing simplicity; theoretical (Greenland and O'Rourke, 2001), and empirical evidence (Juni et al., 1999), suggests that their associations with intervention effect estimates are inconsistent and unpredictable (Balk et al., 2002, Emerson et al., 1990 and Schulz et al., 1995). Furthermore, calculating a summary score inevitably involves assigning 'weights' to different items in the scale, in ways that are difficult to justify (Higgins et al., 2011b).

Application to environmental research

Practitioners conducting SRs on environmental issues should consider testing, modifying, and adopting the quality assessment tools and risk of bias tools developed by the Cochrane Collaboration; capitalising on more than twenty year's of theoretical and empirical research that has been invested in these tools, while attempting to enhance the ability of the tools to discriminate between different levels of quality within NRSs, and widening the possible sources of bias considered in the Risk of Bias Tool to make it more relevant to environmental research.

That said, while NRSs are the current modus operandi for environmental research, RCTs could and should be used more often as a means of addressing research hypotheses. The RCT is widely regarded as the design of choice for the assessment of the effectiveness of interventions in healthcare, and this is such for a reason. The main benefit of the RCT is the use of a randomisation procedure that, when properly implemented, ensures that the allocation of any study unit to one intervention or another cannot be predicted. The randomisation process makes the comparison groups equal with respect to both known and unknown prognostic factors at baseline, apart from chance bias (D'Agostino and Kwan, 1995). RCTs also tend to benefit from so-called 'inherited properties', which generally mark them out as higher quality studies (Deeks et al., 2003). These properties include the fact that they are prospective studies, with written protocols specifying, and thus standardising, important aspects of study unit enrolment, interventions, observation and analysis (Abel and Koch, 1999). RCTs are also more likely to employ specific measures to reduce or remove bias, such as blinded outcome assessment (Deeks et al., 2003).

Ultimately the strength of the findings of a SR are determined by the quality and risk of bias in the primary studies cited. If environmental researchers were to use RCTs more widely as a research method (where appropriate and feasible), then it is likely that the environmental community would be able to increase the impact of their SRs. In the meantime, the environmental community will have to make use of the evidence currently available, which as mentioned above is often derived from NRSs. The choice of which quality appraisal tool to use to assess these studies is critical, as it has been empirically demonstrated that the use of different quality scales for the assessment of the same studies results in different estimates of quality (Moher et al., 1998 and Juni et al., 1999).

At present, the CEE (2013) does not place restrictions on the use of existing checklists or critical appraisal tools as a basis for study appraisal, but requires that authors either explain why they used the chosen method as is (no modification, because not considered to be needed, and why), or adapted the method for their SR (in which case the decisions made must be stated and justified). They suggest that review-specific a priori assessment criteria for appraising the quality of methodology should be included in the SR protocol, and that two or more assessors should be used for study appraisal. The environmental community should attempt to determine the most suitable tool for quality appraisal in environmental SRs. This would simplify the SR process and reduce the potential for authors to select a quality appraisal tool that emphasises research that meets their own opinions, prejudices or commercial interests.

Extracting data for analysis and interpretation

Analyses within CSRs may be narrative, such as a structured summary and discussion of the studies' characteristics and findings, or quantitative, that is involving statistical analysis (Deeks et al., 2011). Meta-analysis – the statistical combination of results from two or more separate studies – is the most commonly used statistical technique. Cochrane review writing software (RevMan) can perform a variety of meta-analyses, although it is stressed that meta-analysis is not appropriate in all CSRs (Deeks et al., 2011).

In CSRs the analysis plan follows from the scientific aim of the review. Reviews have different types of aims, and may therefore contain different approaches to analysis (Deeks et al., 2011). The most straightforward CSR assembles studies that make one particular comparison between two treatment options. Meta-analysis and related techniques can be used if there is a consistent outcome measure to (I) establish whether there is evidence of an effect, (II) estimate the size of the effect and the uncertainty surrounding that size, and (III) investigate whether the effect is consistent across studies.

Some reviews may have a broader focus than a single comparison. The first is where the intention is to identify and collate studies of numerous interventions for the same disease or condition. The second, related aim is that of identifying a 'best' intervention. Such reviews may include multiple comparisons and meta-analyses between all possible pairs of treatments, and require care when it comes to planning analyses (Deeks et al., 2011).

Occasionally review comparisons have particularly wide scopes that make the use of meta-analysis problematic. When reviews contain very diverse studies a meta-analysis might be useful to answer the overall question of whether there is evidence that a particular intervention can work. But use of meta-analysis to describe the size of effect may not be meaningful if the implementations are so diverse that an effect estimate cannot be interpreted in any specific context (Deeks et al., 2011).

An aim of some CSRs is to investigate the relationship between the size of an effect and some characteristic(s) of the studies (Deeks et al., 2011). This is uncommon as a primary aim in CSRs, but may be a secondary aim.

Application to environmental research

Similar considerations influence decisions on whether the synthesis should be narrative or quantitative in SRs on environmental topics. For the latest CEE guidance on this stage of conducting a SR on environmental topics, visit: www.environmentalevidence.org/Instructionsforauthors.html. An example of an environmental SR with an entirely narrative synthesis (Davies et al., 2006) and a narrative synthesis that complements a quantitative synthesis (Bowler et al., 2010) are available in the CEE Library (http://www.environmentalevidence.org/Reviews.html). Open access data is now becoming more prevalent in environmental research, and this will increase the possibilities for SRs with quantitative meta-analyses in the future.

Disseminating and updating the review

CSRs are written by more than one author and are peer-reviewed by a number of experts (Green and Higgins, 2011). They are then published online and are sometimes also co-published in peer-reviewed healthcare journals (Green and Higgins, 2011). For the Cochrane Collaboration, there is one essential condition of co-publication: CSRs must remain free for dissemination in any and all media, without restriction from any of them (Green and Higgins, 2011). Since evidence on a given subject is generally dynamic and continually evolving, incorporating additional studies as they become available can change the results of a SR (Chalmers and Haynes, 1994). Therefore, SRs that are not maintained run the risk of becoming out of date and even misleading. An important feature of CSRs is that review authors are committed not only to preparing SRs of evidence, but also to maintaining (and updating) these reviews on a regular basis (at least every two years) on the *Cochrane Library* (Green and Higgins, 2011).

Application to environmental research

The environmental equivalent to the Cochrane Collaboration (i.e. the CEE) was founded in 2008 (Pullin and Knight, 2013), and an online open-access library of SRs was created in 2012 to enable widespread dissemination. This is currently a relatively small but growing library of SRs on environmental topics. Policies need to be established to ensure that these reviews are maintained and updated regularly. If society is to move towards better-informed environmental policies for a sustainable global environment and the conservation of biodiversity, these SR activities need to be accelerated urgently (Pullin and Knight, 2013).

CONCLUSIONS

Systematic reviews are powerful tools that aim to provide comprehensive and reproducible summaries of evidence to guide policy decisions. They employ a range of methods that are designed to reduce the influence of author bias while considering bias in the primary studies. This article summarised how to conduct SRs according to best practice in the healthcare field, and explained the rationale for each stage of conducting a SR. It has demonstrated that existing CSR methods can be, and already have been, used in SRs on environmental topics. For example, to date the CEE have published over 60 SRs, with a further 30 SRs in progress. These SRs, which are all available from the CEE Library, cover a range of topics including pure environmental science questions such as 'What is the evidence for glacial shrinkage across the Himalayas?' (Miller et al., 2013), applied environmental management topics such as 'Evaluating the biological effectiveness of fully and partially protected marine areas' (Sciberras et al., 2013), and human–environment interaction questions such as 'What is the evidence that scarcity and shocks in freshwater resources causes conflict instead of collaboration?' (Johnson et al., 2011). We suggest, however that the process of conducting a SR on an environmental topic could be improved through several adaptations of both the SR process, and the manner in which environmental research is conducted, reported and indexed in the future. The literature search stage could be expedited by producing a statement of guidelines to standardise the reporting and indexing of environmental studies, equivalent to the CONSORT statement in healthcare. The consistency of the study appraisal stage of SRs could be augmented by refining the existing quality assessment tools used in the healthcare field, enhancing their ability to discriminate quality and risk of bias in NRSs. Ultimately, the strength of evidence within SRs on environmental topics could be improved through more widespread use of RCTs as a research method, owing to their inherently lower risk of bias when conducted according to best practice.

Society may be on the cusp of an evidence revolution in environmental management but it will take new contributors and investment to ensure that this has impact. The CEE recently proposed a five-year programme to build capacity for the conduct and use of SRs in the environment sector (Pullin and Knight, 2013). This programme aims to: (1) increase the commissioning and use of SR in evidence-based policy; (2) develop the capacity of the global environmental research community to conduct SRs; and (3) develop the capacity of CEE to co-ordinate and promote the conduct of SRs in the environmental sector. All of these plans require a big effort on the part of those already active in CEE, but also provide opportunities for others to join in and contribute to the growth of the network (Pullin and Knight, 2013).

ACKNOWLEDGEMENTS

This work arises from Natural Environment Research Council funding (GrantNE/L00836X/1) in collaboration with the United Kingdom's Department for Environment, Food and Rural Affairs.

REFERENCES

1. Abel, U., Koch, A., 1999. The role of randomization in clinical studies: myths and beliefs. J. Clin. Epidemiol. 52, 487–497.
2. Balk, E.M., Bonis, P.A.L., Moskowitz, H., Schmid, C.H., Ioannidis, J.P.A., Wang, C., Lau, J., 2002. Correlation of quality measures with estimates of treatment effect in meta-analyses of randomized controlled trials. J. Am. Med. Assoc. 287 (22) 2973–2982.
3. Berlin, J.A., Rennie, D., 1999. Measuring the quality of trials: the quality of quality scales. J. Am. Med. Assoc. 282 (11) 1083– 1085.
4. Bowler, D., Buyung-Ali, L., Knight, T., Pullin, A.S., 2010. How effective is 'greening' of urban areas in reducing human exposure to ground level ozone concentrations, UV exposure and the 'urban heat island effect'? Environ. Evid., www.environmentalevidence.org/SR41.html.
5. Boyd, I., 2013. Research: a standard for policy-relevant science. Nature 501, 159–160. Collaboration for Environmental Evidence, 2013. Guidelines for Systematic Review and Evidence Synthesis in Environmental Management. Version 4.2. Environmental Evidence. , www.environmentalevidence.org/Documents/Guidelines/ Guidelines4.2.pdf.
6. Chalmers, I., Haynes, B., 1994. Reporting, updating, and correcting systematic reviews of the effects of health care. Br. Med. J. 309 (6958) 862–865.
7. Cooper, H., Ribble, R.G., 1989. Influences on the outcome of literature searches for integrative research reviews. Knowl.: Creation Diffus. Util. 10, 179–201.
8. Cooper, H.M., 1984. The problem formulation stage. In: Cooper, H.M. (Ed.), Integrating Research: A Guide for Literature Reviews. Sage Publications, Newbury Park, CA.
9. Cooper, H., Hedges, L. (Eds.), 1994. The Handbook of Research Synthesis. Russell Sage Foundation, New York.
10. Counsell, C., 1997. Formulating questions and locating primary studies for inclusion in systematic reviews. Ann. Intern. Med. 127 (5) 380–387.
11. D'Agostino, R.B., Kwan, H., 1995. Measuring effectiveness: what to expect without a randomized control group. Med. Care 33 (4) 95–105.
12. Davies, P., Boruch, R., 2001. The Campbell Collaboration: does for public policy what Cochrane does for health. Br. Med. J. 323 (7308) 294–295.
13. Davies, Z.G., Tyler, C., Stewart, G.B., Pullin, A.S., 2006. Are Current Management Recommendations for Conserving Saproxylic Invertebrates Effective? CEE review 05-011 (SR17). Collaboration for Environmental Evidence. , www.environmentalevidence.org/SR17.html.
14. Deeks, J.J., Dinnes, J., D'amico, R., Sowden, A.J., Sakarovitch, C., Song, F., Petticrew, M., Altman, D.J., 2003. Evaluating nonrandomized intervention studies. Health Technol. Assess. 7 (27) 1–179.
15. Deeks, J.J., Higgins, J.P.T., Altman, D.G., 2011. Analysing data and undertaking meta-analyses. In: Higgins, J.P.T., Green, S. (Eds.), Cochrane Handbook for Systematic Reviews of Interventions Version 5.1.0 (updated March 2011). The Cochrane Collaboration, 2011. (Chapter 9).
16. Edwards, P., Clarke, M., DiGuiseppi, C., Pratap, S., Roberts, I., Wentz, R., 2002. Identification of randomized controlled trials in systematic reviews: accuracy and reliability of screening records. Stat. Med. 21 (11) 1635–1640.
17. Emerson, J.D., Burdick, E., Hoaglin, D.C., Mosteller, F., Chalmers, T.C., 1990. An empirical study of the possible relation of treatment differences to quality scores in controlled randomized clinical trials. Control. Clin. Trials 11 (5) 339– 352.
18. Fraser, C., Murray, A., Burr, J., 2006. Identifying observational studies of surgical interventions in MEDLINE and EMBASE. Br. Med. Counc. Med. Res. Methodol. 6, 41.
19. Furlan, A.D., Irvin, E., Bombardier, C., 2006. Limited search strategies were effective in finding relevant nonrandomized studies. J. Clin. Epidemiol. 59 (12) 1303–1311.
20. Gabriel, S.E., Normand, S.L., 2012. Getting the methods right – the foundation of patient-centered outcomes research. N. Engl. J. Med. 367 (9) 787–790.
21. Gluud, L.L., 2006. Bias in clinical intervention research. Am. J. Epidemiol. 163 (6) 493–501.
22. Golder, S., Loke, Y., McIntosh, H.M., 2006a. Room for improvement? A survey of the methods used in systematic reviews of adverse effects. BMC Med. Res. Methodol. 6, 3.
23. Golder, S., McIntosh, H.M., Duffy, S., Glanville, J., 2006b. Centre for Reviews and Dissemination and UK Cochrane Centre Search Filters Design Group,

Developing efficient search strategies to identify reports of adverse effects in MEDLINE and EMBASE. Health Inform. Libr. J. 23, 3–12.

24. GRADE Working Group, 2004. Grading quality of evidence and strength of recommendations. Br. Med. J. 328, 1490– 1494.

25. Green, S., Higgins, J.P.T., 2011. Preparing a Cochrane review. In: Higgins, J.P.T., Green, S. (Eds.), Cochrane Handbook for Systematic Reviews of Interventions Version 5.1.0 (updated March 2011). The Cochrane Collaboration, 2011. (Chapter 2).

26. Greenland, S., O'Rourke, K., 2001. On the bias produced by quality scores in meta-analysis, and a hierarchical view of proposed solutions. Biostatistics 2, 463–471.

27. Guyatt, G.H., Oxman, A.D., Kunz, R., Vist, G.E., Falck-Ytter, Y., Schu¨nemann, H.J., 2008a. What is 'quality of evidence' and why is it important to clinicians? Br. Med. J. 336 (7651) 995– 998.

28. Guyatt, G.H., Oxman, A.D., Vist, G.E., Kunz, R., Falck-Ytter, Y., Alonso-Coello, P., Schu¨nemann, H.J., 2008b. GRADE: an emerging consensus on rating quality of evidence and strength of recommendations. Br. Med. J. 336 (7650) 924–926.

29. Hartling, L., Ospina, M., Liang, Y., Dryden, D.M., Hooton, N., Seida, J.K., Klassen, T.P., 2009. Risk of bias versus quality assessment of randomised controlled trials: cross sectional study. Br. Med. J. 339, b4012.

30. Hartling, L., Hamm, M.P., Milne, A., Vandermeer, B., Santaguida, P.L., Ansari, M., Tsertsvadze, A., Hempel, S., Shekelle, P., Dryden, D.M., 2012. Testing the Risk of Bias tool showed low reliability between individual reviewers and across consensus assessments of reviewer pairs. J. Clin. Epidemiol. 66 (9) 973–981.

31. Hedges, L.V., 1994. Statistical considerations. In: Cooper, H., Hedges, L.V. (Eds.), The Handbook of Research Synthesis. Russell Sage Foundation, New York, NY.

32. Higgins, J.P.T., Altman, D.G., Sterne, J.A.C., 2011a. Assessing risk of bias in included studies. In: Higgins, J.P.T.,Green, S. (Eds.),Cochrane

33. Handbook for Systematic Reviews of Interventions Version 5.1.0 (updated March 2011). The Cochrane Collaboration, 2011. (Chapter 8).

34. Higgins, J.P.T., Altman, D.G., Gøtzsche, P.C., Ju¨ni, P., Moher, D., Oxman, A.D., Savovic´, J., Schulz, K.F., Weeks, L., Sterne, J.A.C., 2011b. The Cochrane Collaboration's tool for assessing risk of bias in randomised trials. Br. Med. J. 343, d5928.

35. Higgins, J.P.T., Deeks, J.J., 2011. Selecting studies and collecting data. In: Higgins, J.P.T., Green, S. (Eds.), Cochrane Handbook for Systematic Reviews of Interventions Version 5.1.0 (updated March 2011). The Cochrane Collaboration, 2011. (Chapter 7).

36. Higgins, J.P.T., Green, S., 2011. Cochrane Handbook for Systematic Reviews of Interventions Version 5.1.0 [updated March 2011]. The Cochrane Collaboration, 2011. Available from www.cochrane-handbook.org.

37. Jackson, G.B., 1980. Methods for integrative reviews. Rev. Educ. Res. 50, 438–460.

38. Johnson, V., Fitzpatrick, I., Floyd, R., Simms, A., 2011. What is the Evidence that Scarcity and Shocks in Freshwater Resources Cause Conflict Instead of Promoting Collaboration? CEE Review 10-010. Collaboration for Environmental Evidence. , In: www.environmentalevidence.org/SR 10010.html.

39. Juni, P., Witschi, A., Bloch, R., Egger, M., 1999. The hazards of scoring the quality of clinical trials for meta-analysis. J. Am. Med. Assoc. 282 (11) 1054–1060.

40. Khan, K.S., ter Riet, G., Glanville, J., Sowden, A.J., Kleijnen, J., 2001. Undertaking Systematic Reviews of Research on

41. Effectiveness: CRD's Guidance for those Carrying Out or Commissioning Reviews (CRD Report Number 4), 2nd edition. NHS Centre for Reviews and Dissemination, University of York, York, UK.

42. Lefebvre, C., Manheimer, E., Glanville, J., 2011. Searching for studies. In: Higgins, J.P.T., Green, S. (Eds.), Cochrane

43. Handbook for Systematic Reviews of Interventions. Version 5.1.0 (updated March 2011). The Cochrane Collaboration, 2011. (Chapter 6).

44. Light, R.J., Pillemer, D.B., 1984. Summing Up: The Science of Reviewing Research. Harvard University Press, Cambridge, MA.

45. Lundh, A., Gøtzsche, P.C., 2008. Recommendations by Cochrane Review Groups for assessment of the risk of bias in studies. Br. Med. Counc. Med. Res. Methodol. 8, 22.

46. Miller, J., Rees, G., Warnaars, T., Young, G., Shrestha, A., Collins, D., 2013. What is the Evidence About Glacier Melt Across the

47. Himalayas. CEE Review 10-008. Collaboration for Environmental Evidence. , In: www.environmentalevidence.org/SR 10008.html.

48. Moher, D., Pham, B., Jones, A., Cook, D.J., Jadad, A.R., 1998. Does quality of reports of randomised trials affect estimates of intervention efficacy reported in meta-analyses? Lancet 352 (9128) 609–613.

49. O'Connor, D., Green, S., Higgins, J.P.T., 2011. Defining the review question and developing criteria for including studies. In: Higgins, J.P.T., Green, S. (Eds.), Cochrane Handbook of Systematic Reviews of Intervention. Version 5.1.0 (updated March 2011). The Cochrane Collaboration, 2011. (Chapter 5).

50. Oxman, A.D., Guyatt, G.H., 1993. The science of reviewing research. Ann. N. Y. Acad. Sci. 703, 125–133.

51. Petticrew, M., 2001. Systematic reviews from astronomy to zoology: myths and misconceptions. Br. Med. J. 322 (7278) 98–101.

52. Pullin, A.S., Knight, T., 2013. Time to build capacity for evidence synthesis in environmental management. Environ. Evid. 2 (1) 21–22.

53. Rees, R., Kavanagh, J., Burchett, H., Shepherd, J., Brunton, G., Harden, A., Thomas, S., Oakley, A., 2004. HIV Health

54. Promotion and Men who have Sex with Men, (MSM): A Systematic Review of Research Relevant to the Development and Implementation of Effective and Appropriate

55. Interventions. EPPI-Centre, Social Science Research Unit, Institute of Education, University of London, London, UK.

56. Richardson, W.S., Wilson, M.S., Nishikawa, J., Hayward, R.S.A., 1995. The well-built clinical question: a key to evidence based decisions. ACP J. Club A12–A13.

57. Schulz, K.F., Chalmers, I., Hayes, R.J., Altman, D.G., 1995. Empirical evidence of bias, dimensions of methodological quality associated with estimates of treatment effects in controlled trials. J. Am. Med. Assoc. 273 (5) 408–412.

58. Schu¨nemann, H.J., Jaeschke, R., Cook, D.J., Bria, W.F., El-Solh, A.A., Ernst, A., Fahy, B.F., Gould, M.K., Horan, K.L., Krishnan, J.A., Manthous, C.A., Maurer, J.R., McNicholas, W.T., Oxman, A.D., Rubenfeld, G., Turino, G.M., Guyatt, G., 2006. An official ATS statement: grading the quality of evidence and strength of recommendations in ATS guidelines and recommendations. Am. J. Respir. Crit. Care Med. 174 (5) 605– 614.

59. Sciberras, M., Jenkins, S., Kaiser, M., Hawkins, S., Pullin, A., 2013. Evaluating the biological effectiveness of fully and partially protected marine areas. Environ. Evid. 2 (4), in: http:// www.environmentalevidencejournal.org/content/2/1/4.

60. Thomas, B.H., Ciliska, D., Dobbins, M., Micucci, S., 2004. A process for systematically reviewing the literature: providing the research evidence for public health nursing interventions. Worldviews Evid.-Based Nurs. 1, 165–184.

61. Trame` r, M.R., Reynolds, D.J., Moore, R.A., McQuay, H.J., 1997. Impact of covert duplicate publication on meta-analysis: a case study. Br. Med. J. 315 (7109) 635–640.

62. Turner, L., Boutron, I., Hro´bjartsson, A., Altman, D.G., Moher, D., 2013. The evolution of assessing bias in Cochrane systematic reviews of interventions: celebrating methodological contributions of the Cochrane Collaboration. Syst. Rev. 2 (1) 79.

63. West, S., King, V., Carey, T.S., Lohr, K.N., McKoy, M.D.N., Sutton, S.F., Lux, L., 2002. Systems to Rate the Strength of Scientific Evidence. In: Evidence Report/Technology Assessment No. 47 (Prepared by the Research Triangle Institute-University of North Carolina Evidence-based Practice Center under Contract No. 290-97-0011). AHRQ Publication No. 02-E016. Agency for Healthcare Research and Quality, Rockville, MD.

64. Wieland, S., Dickersin, K., 2005. Selective exposure reporting and Medline indexing limited the search sensitivity for observational studies of the adverse effects of oral contraceptives. J. Clin. Epidemiol. 58, 560–567.

65. Wood, L., Egger, M., Gluud, L.L., Schulz, K.F., Juni, P., Altman, D.G., Gluud, C., Martin, R.M., Wood, A.J.G., Sterne, J.A.C., 2008. Empirical evidence of bias in treatment effect estimates in controlled trials with different interventions and outcomes: meta-epidemiological study. Br. Med. J. 336 (7644) 601–605.

CHAPTER 2

ORGANIC NITROGEN STORAGE IN MINERAL SOIL: IMPLICATIONS FOR POLICY AND MANAGEMENT

Andrew H. Bingham, M. Francesca Cotrufo

Keywords: SOM, Sequestration, Factors, Policy decisions, Environmental management

INTRODUCTION

Nitrogen (N) is an essential nutrient which limits productivity in most terrestrial ecosystems (Chapin et al., 2012 and LeBauer and Treseder, 2008). Anthropogenic activity has doubled the rate of terrestrial N fixation and has caused many regions to experience significant impacts from N enrichment such as eutrophication, acidification and loss of biodiversity (Canfield et al., 2010). Soil is the largest pool of fixed and biologically available N; decades of research have improved our understanding of the pathways N moves through in the soil environment, the nature of soil organic N (SON) and its implication for carbon (C) sequestration, work that has been synthetized in important reviews (Knicker, 2011 and van Groenigen et al., 2015). Because most anthropogenic N additions end up in the soil (Johnson and Turner, 2014), a thorough understanding of the processes and factors governing its removal from cycling and sequestration is crucial for informing policy and management decisions, yet this information remains scattered throughout the literature.

Despite the fact that mineral N fixed on clays can make a significant contribution to sub-surface soils N stocks (Stevenson, 1986), the bulk of the soil N stock is in organic matter (OM) forms (Schulten and Schnitzer, 1998). Analytical advances in the past few decades have altered our understanding of soil organic matter (SOM) chemistry and resulted in a paradigm shift of the mechanisms explaining its long-term persistence (Lehmann and Kleber, 2015 and Schmidt et al., 2011b). For many years it was believed that chemical recalcitrance, i.e., the inherent resistance to degradation, was responsible for the longevity of organic matter in soils. Classically, two mechanisms were thought to foster SOM persistence: 1) selective preservation of structurally complex compounds, such as lignin, suberin and cutin, due to their resistance to microbial breakdown, and 2) condensation reactions forming intricate, irregular organic compounds generally referred to as humus (Kögel-Knaber, 1993, Sollins et al., 1996 and Stevenson, 1994). However, empirical evidence of humification has never been found, and advanced chemical analyses of the persistent organic matter pool in soil demonstrated that it is primarily made of chemically labile structures that are easily degraded, with low relative abundance of aromatic groups (Kleber et al., 2011 and references therein). Thus, currently the persistence of organic matter in soil is believed to be an ecosystem property (Schmidt et al., 2011b), controlled by microbial inhibition, physical protection and/or chemical stabilization (Von Lützow et al., 2006).

Many excellent reviews are available outlining the changed understanding of the persistence of SOM (Deb and Shukla, 2011, Kögel-Knabner et al., 2008, Lehmann and Kleber, 2015, Marschner et al., 2008, Nannipieri and Paul, 2009, Sollins et al., 2007 and Von Lützow et al., 2006). To date though, there has not been a comprehensive examination of the implications of this new SOM stabilization paradigm for long-term N sequestration in naturally functioning ecosystems. Because of the importance of N sequestration to ecosystem functioning, as well as to policy considerations such as critical loads and N-saturation, an examination of the factors controlling N sequestration within this revised SOM paradigm is crucial. In this review we will first synthesize the current understanding of the nature of the soil N pool and of the factors controlling its long-term storage in temperate region soils. We will then outline the most important factors (Table 1) such as substrate quality and quantity, mineral composition, and aggregate

Citation: Andrew H. Bingham, M. Francesca Cotrufo, Organic nitrogen storage in mineral soil: Implications for policy and management, Science of The Total Environment, Volumes 551–552, 1 May 2016, Pages 116-126, ISSN 0048-9697, http://dx.doi.org/10.1016/j.scitotenv.2016.02.020.

stability that govern the currently understood mechanisms of N persistence: microbial processing, chemical protection and physical protection from microbial degraders (Fig. 1). Finally, we will examine how recent changes in paradigm may affect policy and management actions undertaken to protect and remediate natural areas experiencing increased N deposition. We will examine consequences for N-saturation under the framework put forward by Lovett and Goodale where the capacity of the soil to process N is seen as the bottleneck leading to saturation and we will address implications for critical loads, mitigation, and restoration policy. We will also elucidate how this new understanding could alter land manager actions and considerations when faced with excess soil N.

Table 1: Factors important to long-term stabilization of N in soils.

Factor	Importance
Substrate	
C and N chemistry	Affects efficiency and quantity of microbial processing and provides energy for microbial activity
Microbial processing	
Soil C and N status	Affects efficiency and quantity of microbial processing as well as type of output (NH4 vs. microbial residues)
Community structure	Affects efficiency and quantity of microbial processing and production of sequesterable OM
Edaphic and environmental	Conditions that favor greater microbial activity will allow greater production of microbially processed OM
Adsorption	
Mineral and protein properties	Affects charge and number of adsorption sites
pH	Lower pH favors stronger bond types
Antecedent OM	Affects number of available adsorption sites, strength of bond
Spatial separation	
Microaggregate formation	Creation of small pore spaces excludes microbes and limits diffusion of water and nutrients
Soil position	Reduced diffusion of water and nutrients at depth limits microbial activity

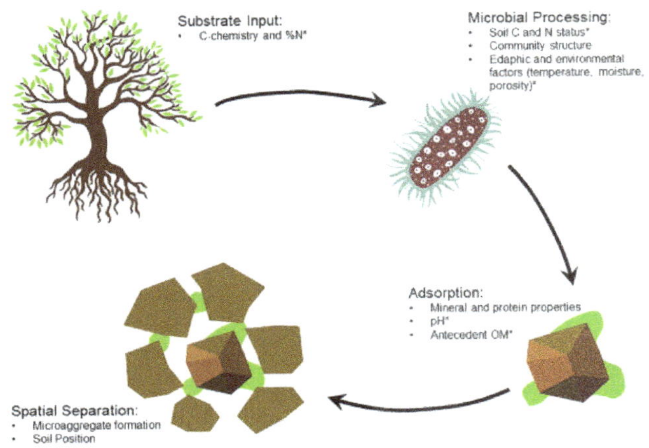

Figure 1: Nitrogen stabilization. Conceptual diagram of the processes involved in N sequestration and the important factors controlling them. Asterisk (*) denotes factor having important policy and/or management implications.

NITROGEN IN SOILS

Nitrogen exists in both inorganic (primarily as ammonium [NH_4^+] and nitrate [NO_3^-]) and organic forms within the soil. Mineral N is introduced to the soil as a product of microbial fixation or through atmospheric deposition, primarily as the result of anthropogenic fossil fuel combustion and emissions from agricultural practices, especially confined animal feedlots, though N inputs from rock weathering may be present as well (Houlton and Morford, 2015 and Thamdrup, 2012). Yet, it has long being recognized that most often, the primary form of N in soil is organic, with Jodidi (1911) noting over a century ago that "practically all of the N occurring in soils is of an organic nature". In fact, organic compounds can comprise up to 95% of the N in some soils (Knicker, 2011, Nannipieri and Paul, 2009 and Rillig et al., 2007), with average pool sizes of about 725 g N m^{-2}, compared to typical pool sizes of 6 g N m^{-2} for mineral N forms (Sylvia et al., 2005). Amides and amines make up the majority of the organic N pool, and aromatic N compounds, while present, contribute a relatively minor share (Leinweber et al., 2013). Most of this organic N, about 400 Tg $year^{-1}$ globally (Chapin et al., 2012), enters the soil as particulate organic matter (POM) or dissolved organic matter through above-ground plant litter and root litter, or in the form of pyrogenic organic matter after fire (Knicker, 2011). Dissolved organic nitrogen (DON) enters the mineral soil through root exudates and leaching of soluble components of plant litter in the organic (O) horizon (Knicker, 2011 and Qualls and Haines, 1991) and concentrations and fluxes of DON entering the mineral soil from the O horizon generally decrease quickly with depth (Rosenqvist et al., 2010). Measurements at the Hubbard Brook Experimental Forest in New Hampshire, a California mudflow chronosequence and European sites have found that especially in ecosystems with leaf litter having a higher %N, above-ground inputs can be an important source of DON, and may provide from half to more than double the input of N to the mineral soil compared to below-ground inputs in some environments (Dittman et al., 2007, Sleutel et al., 2009 and Uselman et al., 2012). The amount of DON in litter leachates varies linearly with the litter N concentration (Soong et al., 2015). In the field, leachate from pine litter was observed to have less DON than that from oak or mixed litter (Silveira et al., 2011). Sleutel et al. (2009) observed that DON comprised 28–46% of dissolved N-fluxes into the mineral soil in deciduous forests in high N deposition environments, compared with < 20% as DON in coniferous forests, and these patterns have been noted in other studies (Currie et al., 1996). Nitrogen is also introduced to the soil through root litter and exudates. In some cases, these inputs may be as or more important than leaf litter inputs for DON (Schmidt et al., 2011a, Uselman et al., 2009 and Uselman et al., 2012). Root N introduced to mineral soil layers is likely to already be in a form that is able to be rapidly processed by microbes and transformed into low molecular weight forms of N (Sanaullah et al., 2011).

Once in the soil, SON may cycle between microbial biomass and residues, adsorption and desorption from soil mineral particles, and dissolution and precipitation from the soil solution before it is held in the soil matrix (Knicker, 2011 and Marín-Spiotta et al., 2014). While bioturbation can be important, especially in upper soil horizons, model results show that liquid phase transport enables most of the movement of SOM (Braakhekke et al., 2011). The ratio of microbially-derived N compounds compared to plant-derived compounds increases with soil depth, while the C:N ratio and DON concentration decrease (Nannipieri and Paul, 2009, Ros et al., 2009 and Schmidt et al., 2011b). This indicates that plant-derived compounds dominate in the upper soil layers while degraded and processed OM consisting primarily of microbial metabolites migrate as dissolved organic matter (DOM) to lower soil horizons (Kaiser and Kalbitz, 2012).

In fire affected ecosystems, organic N may also enter the soil in the form of pyrogenic organic nitrogen (Py N). Py N is believed to contribute largely to the heterocyclic N forms found in soils and to be the only SON pool preserved for long time spans by inherent recalcitrance (Knicker, 2011). In fact, Py N appears to be locked away from the internal N cycling and is minimally transformed by microbes, causing N limitation in frequently burned sites (Soong and Cotrufo, 2015).

Microbial Processing

Nitrogen that is stored for centuries to millennia is predominantly composed of low molecular weight, labile molecules. These compounds can derive directly from plants, as in the case of simple sugars and amino acids, or they may be derived secondarily from microbial processing such as exudation or cell death (Rillig et al., 2007 and references therein). Highly labile compounds such as free amino acids do not exist for long in the soil and are taken up rapidly, within hours or even concurrent with production in some cases (Farrell et al., 2011); because nearly all N sequestered for long periods has undergone microbial processing (possibly many times over) before it is removed from circulation, microorganisms exert an important influence in regulating the amount of N suitable for long term storage (Knicker, 2011 and Sollins et al., 2007). Microbial community composition, as well as environmental factors such as soil pH, moisture and temperature, exerts important influences on microbial activity and the breakdown of proteins within soils (Sylvia et al., 2005).

The proportion of N in substrate is often used as a proxy for its quality and lability. The N content of SOM has long been recognized as an important control over decomposition (Grandy et al., 2009). Observations that leaf litter with a higher %N also had a higher DON concentration (Soong et al., 2015 and Uselman et al., 2012) suggests that a higher amount of N suitable for storage may be produced from higher quality OM. Thomas et al. (2012) used soils from a wide range of biomes to find that the quality of substrate inputs is a main driver of SOM composition. In a crop amendment study, St Luce et al. (2014)found greater incorporation of N into microbial biomass and the mineral N pool with low C:N faba bean residue than with high C:N wheat residue additions. Gillespie et al. (2014) expanded this view from plant residue additions to any type of N inputs as they observed an increase of N-containing microbial byproducts incorporated into the fine fraction following fertilization with mineral N or manure.

High throughput of these higher quality materials can result in a greater mineral-stabilization of SOM, resulting in what would have been earlier considered a paradox: that greater amounts of labile material result in greater amounts of N retained in the soil; a pattern noted by Giardina et al. (2001) in a litter comparison experiment. Cotrufo et al. (2013) have proposed a framework whereby increased substrate quality leads to higher amounts of microbially processed organic matter and thus potential for stored N. According to this framework, litter with more labile forms of C as well as higher %N both lead to increased microbial substrate use efficiency, resulting in a greater production of microbial residues. Evidence for this has been found by Hatton et al. (2015) who used isotopic tracers in a comparison of root and needle litter to find that more labile needle litter preferentially accumulated in mineral associated SOM fractions composed of microbial byproducts than did less labile root material. In litterbag experiments, Hobara et al. (2014) found higher amounts of N in mineral soils and greater amounts of labile, microbial byproducts (amino acids and amino sugars of bacterial origin) from litter with higher initial %N and of higher initial quality. Microbial communities having high substrate use efficiency have also been modeled to produce greater amounts of N-rich necromass, a primary precursor of stable N in soils (Kaiser et al., 2015).

Other evidence indicates that edaphic factors and microbial community dynamics may be as or more important than litter quality in determining the amount of microbial N processing and their substrate use efficiency in soils (Manzoni et al., 2012 and Sinsabaugh et al., 2013). Kaiser et al. (2014) observed that microbial community composition may be able to buffer compositional differences of OM and maintain a constant level of substrate use efficiency despite different C:N ratios and substrate nutrient content. Other research indicates that microbial community composition can facilitate the accumulation of N-rich microbial remains and N-retention to a greater degree than litter C:N composition (Kaiser et al., 2015). Similarly, in mesocosm experiments, soil characteristics were found to be more important than litter quality for C and N transformation rates (Delgado-Baquerizo et al., 2015). Agricultural research has also failed to find differences in long-term soil N pools due to variations in litter quality (Chivenge et al., 2011 and Fonte, 2009), adding to growing evidence of a decoupling between litter quality and long-term SOM stabilization, and the importance of litter quantity and C-content over quality (Carillo et al., 2016 and Castellano et al., 2012). Clearly more research is required to ascertain the effects of substrate chemistry versus edaphic and microbial community composition on OM turnover and production of compounds which can contribute to SON storage.

Existing stocks of accessible C and N in soil also affect the processing of substrate by microbes. At high availability, nitrogen acts to suppress degradation and microbial growth (Frey et al., 2014). In soils with abundant N, amino acids are taken up by microbes primarily for their C-content and N is excreted as NH_4^+ to the soil following deamination within the cell (Farrell et al., 2014 and Tahovská et al., 2013). Although NH_4^+concentrations and mineralization rates may be high, long-term storage is generally suppressed as NH_4^+ is taken up by plants and microbes or converted to NO_3^- that can be lost from the system; not much N is in organic forms suitable for long-term storage. In contrast, N-limited soils with a higher C:N ratio generally favor an organic-N nutrient economy where N exists primarily in organic form and is rapidly assimilated by plants and microbes which compete against one another (Geisseler et al., 2010, Nannipieri and Paul, 2009 and Schimel and Bennett, 2004). In these environments, organic compounds such as amino acids and sugars are taken up directly and are utilized as a source for both C as well as N, which is incorporated into cell biomass (Geisseler et al., 2010). Instead of being excreted to the soil as excess NH_4^+, N is returned to the soil solution through exudation or cell death and lysis in the form of low molecular weight N compounds and cell wall fragments suitable for long-term storage. As long as C remains abundant relative to N, microbial biomass will be considered a stoichiometric sink for N and creation of low molecular weight compounds suitable for N storage will be enhanced (Kopáček et al., 2013 and Tahovská et al., 2013). The amount and chemical composition of C in the soil also influences microbial transformations, providing an energy source for microbial activity and increasing the rate of OM cycling, increasing the amount of low molecular weight materials (Gleixner, 2013). Another factor to consider is that when N is limiting and fresh inputs of labile material are not available, microbes will attack more resistant N sources, transforming more difficult to decompose N compounds into labile microbial byproducts enriched in N compared to the original SOM (Fontaine et al., 2011 and Gleixner, 2013). However, due to the heterogeneity of soils, C and N status is likely to vary widely within short distances, making the interplay between C and N quality and quantity difficult to generalize. These factors play prominent roles in determining the fate of N and C in soils and better understanding of these interactions will pay dividends.

Adsorption

Nitrogen-containing compounds can adsorb directly onto mineral surfaces, retarding transport of these molecules within the soil and reducing susceptibility to oxidative attack (by blocking enzyme attachment, Kögel-Knabner et al., 2008). Adsorption capacity and strength can vary across soil environments and landscapes because of different binding mechanisms, and mineral and protein properties. Proteins adsorb to most surfaces over a wide range of pH and this attachment happens very quickly (Scott and Rothstein, 2014,Sollins et al., 2007 and Yu et al., 2013). Adsorption can occur through several processes that may occur simultaneously on different parts of the N-containing molecule. Polyvalent cation bridges and especially ligand exchange are considered strong binding mechanisms while weaker interactions such as van der Waals forces or H-bonding can also occur (Sanderman and Maddern, 2014 and Von Lützow et al., 2006). Several factors are important to the adsorption capacity of a soil including mineral and protein properties, the pH of the solution and the presence of antecedent OM (Von Lützow et al., 2006 and Yu et al., 2013).

The size and structure of the protein affects adsorption to mineral surfaces. Larger molecules may offer several adsorption sites for binding, enhancing the strength of the connection to the mineral, as well as limiting the number of active sites available to enzymes (Von Lützow et al., 2006). On the other hand, larger proteins tend to be more highly folded and may offer fewer sites for adsorption, and may be too large to fit within the interlamellar areas of clay minerals (Yu et al., 2013). Charge properties of proteins affect the type and strength of bonding, and even whether the protein is attracted or repelled by the mineral surface (Yu et al., 2013).

The presence of clays in soils is important to N storage, though recent research has added nuance to the process by which OM is retained. Phyllosilicate minerals play an important role in OM retention and are often associated with higher levels of N retention (Deb and Shukla, 2011 and Grandy et al., 2009). The high amount of surface area of clay particles leads to a larger number of charged sites and a greater number of binding sites for microbial byproducts. The overall negative charge, small pore spaces and expandable nature of some clay minerals can further increase this capacity (Rillig et al., 2007 and Von Lützow et al., 2006). Non-expandable minerals such as kaolinite or quartz experience weak bonding to SOM, while expandable layer phyllosilicates such as montmorillonite feature extensive internal surfaces available for adsorption (Von Lützow et al., 2006 and Yu et al., 2013). These small spaces, along with the small pore spaces associated with clay mineral particles facilitates multiple bondings with OM, resulting in stronger retention and enhanced adsorption capacity (Kögel-Knabner et al., 2008).Vogel et al. (2014) were able to discern that OM was not attached uniformly to the exterior of clay particles; edges and rough surfaces acted as nuclei for OM attachment. Alternatively, Wei et al. (2014) found that OM decomposition was higher with increasing clay content, likely due to the higher bacterial biomass facilitated by a larger amount of pore space and increased substrate supply. Recent work by Vogel et al. (2015) has also found the lack of a relationship between surface area and OM accumulation and also implicated higher microbial biomass as an important factor in OM and N sequestration. These conflicting results show that clay content alone may not be useful as a proxy for long-term N storage potential.

The composition of the minerals plays an important role. Poorly crystalline Fe-oxides and Al-silicates exhibit properties which lead to increased adsorption of OM and N, especially in subsoils and at lower pH, and may provide better adsorption capacity than phyllosilicate clays (Dippold et al., 2014, Kaiser and Zech, 2000 and Sleutel et al., 2009). In a study of 41 mineral horizons across Canada, Kothawala and Moore (2009) found that the amount of poorly crystalline Fe and Al controlled the adsorption of DON whileDümig et al. (2012) found Fe oxihydroxides dominated the accumulation of OM on mineral surfaces, and the weaker cation bonding led clay minerals to play a minor role. This was also seen by Kuiluweit et al. (2012) who observed an accumulation of amide N from fungi dominated environments on Fe (hydr)oxides. As depth increases, the importance of sorption over aggregate formation as a retention mechanism increases, and the presence of Fe and Al-oxides as an important factor may also increase (Moni et al., 2010 and Rumpel et al., 2012). Because the abundance of Fe and Al minerals generally increases with depth, strong correlations between Fe-oxide content and OM retention have been observed in subsoils (Deb and Shukla, 2011 and Kögel-Knabner et al., 2008).

The pH of the DON solution can be one of the most important factors in adsorption of proteins to minerals by affecting the surface charge of the mineral and the degree of ionization of the protein (Yu et al., 2013). The prevalence of ligand exchange, a strong bonding mechanism, increases as pH decreases, and in acidic conditions most OM is associated with Fe-oxides and Al-silicates (Kleber et al., 2005, Kögel-Knabner et al., 2008 and Von Lützow et al., 2006). Ligand exchange is not possible on clay particles without Fe or Al-oxide coatings, and low pH results in both the clay particle and protein having positive charge, so adsorption takes place through cation exchange (Yu et al., 2013). At neutral pH adsorption shifts from Al or Fe-oxides or cation exchange on clays to primarily electrostatic attraction between the protein and clay particle, a weaker bond (Yu et al., 2013). At basic pH, protein molecules and clay particles become more strongly negatively charged. Electrostatic attraction is weakened and although cation bridging becomes more prevalent, overall adsorption is decreased due to a decrease in the number of available adsorption sites (Yu et al., 2013).

The presence of previously sorbed N plays a role in adsorption as well, and N-containing DOM adsorption on mineral surfaces has been shown to be related to antecedent SOM (Deb and Shukla, 2011). Kothawala and Moore (2009) observed a negative relationship between soil N and DON adsorption which may be due to increased saturation of protective adsorption sites (Von Lützow et al., 2006). Preferential adsorption of N-containing compounds has been observed through increased concentration of N-rich products at the contact point with mineral surfaces (Bonnard et al., 2012 and Hatton et al., 2012b). This agrees with the framework suggesting a layered accumulation of OM on mineral surfaces in which N-rich compounds are favored to comprise the inner layer adjacent to the mineral surface, put forward by Kleber et al. (2007), but see Kaiser and Zech, 2000). This layering, with more strongly bonded, N-rich compounds on the inside allows outer compounds to be easily displaced by inputs of fresh OM, leading to decreased N adsorption capacity when attachment sites are saturated (Scott and Rothstein, 2014).

Sorption alone will only protect the portion of the molecule which is directly attached to the mineral particle and which blocks enzyme active sites (Von Lützow et al., 2006). Unattached portions of the protein will remain bioavailable. Dippold et al. (2014) used an incubation experiment to demonstrate that a small amount of strongly sorbed amino acids accumulated on mineral surfaces but that a remarkably large proportion of the sorbed material was still microbially available and was degraded. Vogel et al. (2015)found that clay minerals with a higher surface area and thus greater number of adsorption sites actually sequestered a smaller amount of N, questioning the importance of the direct adsorption of proteins to mineral surfaces as the dominant sequestration mechanism. Microbial processing of strongly sorbed OM into more easily desorbed products also diminishes the importance of adsorption in long-term protection of OM (Kaiser and Kalbitz, 2012). Current views are mixed and studies to better quantify adsorption and desorption fluxes are needed to better evaluate the importance of this mechanism. Although adsorption appears to be important for protecting a smaller amount of N that is attached directly to the mineral surface, on the whole, its primary importance comes from its ability to retard the movement of N within the soil (Vogel et al., 2014 and Vogel et al., 2015). This allows other retention mechanisms, such as microaggregate formation and spatial separation from microbes to take place.

Spatial Inaccessibility

Occlusion within an aggregate or spatial separation from microbial decomposers effectively preserves N in soil over long periods of time (Von Lützow et al., 2006). Spatial inaccessibility may be the primary driver of N storage by physically protecting OM from microbes and their enzymes through occlusion within an aggregate or within pores which are too small for microbes or their exoenzymes to access, (Marschner et al., 2008). This protection is evidenced by the observation that occluded SOM decomposes at a slower rate and exists in higher concentrations than free SOM (Marschner et al., 2008 and Nichols and Halvorson, 2013).

The size of aggregates can be a key predictor of the relative amount of N retained in OM. Microaggregates (< 250 μm) often have a lower C:N ratio than larger size aggregates, reflecting the increased contribution of microbial metabolites protected in this fraction (Gleixner, 2013, Hatton et al., 2012a, Nannipieri and Paul, 2009 and Pronk et al., 2013). The current prevailing view of aggregate formation begins with the introduction of plant litter to the soil, which is progressively broken down by the soil micro and macrobiota (Hatton et al., 2012a). The smaller compounds produced by this microbial processing bind with mineral surfaces, creating high density, highly stable clay and silt-size microaggregates (< 20 μm; Hatton et al., 2012a, Jastrow, 1996, Rillig et al., 2007, Six et al., 2000 and Von Lützow et al., 2006). These microaggregates can then combine to form larger microaggregates (20–250 μm), with recent studies highlighting the role of microbial necromass in this process (Schurig et al., 2013). Investigations of late hint at the important role cell envelope fragments and polysaccharides can play as a bridge between particles, and these materials may comprise most of the OM attached to mineral surfaces (Dungait et al., 2012, Hatton et al., 2014, Miltner et al., 2012 and Six and Paustian, 2014). Schurig et al. (2013) found that as soil developed following glacial retreat, accumulation of microbial residues on mineral surfaces was likely facilitated by the matrix formed by cell envelope fragments, which eventually coalesced to form microaggregates.

According to current models of aggregate development, the binding mechanisms of microaggregates are generally considered persistent and highly stable, if not permanent, allowing for occlusion to contribute significantly to N storage (Nichols and Halvorson, 2013, Six et al., 2004 and Wilson et al., 2009). Almost 50% of total soil OM has been found to be stored in these small units (Virto et al., 2008). Further aggregation of microaggregates into macroaggregates can occur, especially through the containment of roots and fungal hyphae (Wilson et al., 2009). Stability of these macroaggregates is much more transient however, and unless SOM is continually replenished, aggregate stability will decline (Dungait et al., 2012 and Six et al., 2004). Recent research indicates that microbial cell type (i.e. fungi, actinobacteria, Gram-positive or Gram-negative bacteria) does not influence the amount of microbial necromass which is stabilized in soil (Throckmorton et al., 2015).

Physical separation may be especially important in preserving OM in deeper soils where diffusion of nutrients and water to microbial B14decomposers is more limited (Schmidt et al., 2011b and Scott and Rothstein, 2014) although the more stable subsoil environment does provide more favorable climatic conditions for microbes than the relatively more hostile surface soils that experience a greater range of temperature and moisture regimes (Dungait et al., 2012). Microbial substrate limitation due to reduced diffusion can also be enhanced in dense fractions containing high percentages of clay and silt and in subsoils (Marschner et al., 2008). Reduced microbial activity will reduce the amount of N-containing compounds degraded by microbes, but it will also reduce the amount of microbial byproducts produced which form the main constituents of N that is stored for long periods. Access to the N-containing compounds must be minimized in order for long-term persistence and storage to occur.

POLICY AND MANAGEMENT IMPLICATIONS

The current understanding of sequestered N as composed of chemically labile microbial products, stabilized by adsorption on mineral surfaces and occlusion in aggregate's structure, may require updated management approaches for several aspects of N in soils. It may also have implications for assessing the effectiveness of ecological restoration practices as well as mitigation strategies for reducing anthropogenic N inputs through policy instruments such as critical loads.

Nitrogen Saturation

Improved appreciation of the factors important for long-term N storage can have implications for ecosystem N saturation and NO_3^- leaching. The influential framework for N saturation put forth by Aber et al. (1998) postulated the orderly transition of an ecosystem subject to enhanced N deposition through stages where net primary production, N-mineralization and foliar-N increase with N additions until the system becomes N saturated, at which point net primary production and mineralization decrease, while nitrification and NO_3^- leaching take hold. This framework, however, has been refuted by observations that in many cases NO_3^- leaching is one of the first pathways to respond to additional N and that N levels exceeding the critical load do not automatically lead to leaching (Lovett and Goodale, 2011 and Thimonier et al., 2010). This led Lovett and Goodale (2011) to propose that N saturation must be considered in terms of capacity (the amount of N which can be retained in soil or biological stocks) as wells as kinetics (the rate at which N can be processed).

As we discuss below, the first step in the retention of added N is microbial processing driven by C availability, a driver of kinetic saturation. As the authors note, evidence for capacity saturation on the other hand is weak in most undisturbed temperate ecosystems. Many ecosystems with a variety of vegetation and soils have demonstrated an ability to retain most additional N, with exceptions being areas having obviously low capacity levels such as thin alpine soils (Fenn et al., 2003, Johnson and Turner, 2014 and Sogn et al., 1999). Given these observations we can conclude that capacity saturation is a state that is rarely reached and there is little evidence of an overall capacity for the retention of N. Upon examination at the fine scale or in the short term though, soils may achieve saturation of adsorption or aggregation sites for N in discrete compartments, depending mainly on mineral composition, location within the soil profile and amount of incoming N and C. Adsorption to mineral surfaces is the first step in long-term N storage and although the amount of available adsorption sites will vary with the mineral composition of the soil, it is finite, and generally increases with depth (Guggenberger and Kaiser, 2003). This is also true for SOC, and forms the basis of the C-saturation model put forth by Six et al. (2002) and observed by Stewart et al. (2007), highlighting the interconnectedness of N and C dynamics in soils. The saturation of adsorption sites, especially in upper soil horizons where N initially enters the soil, will result in N-compounds bonding weakly to outer OM in the kinetic zone described byKleber et al. (2007) and observed by Dümig et al. (2012), where hydrophilic (N-rich) molecules are bonded to mineral surfaces but more weakly retained on accreted OM (Bonnard et al., 2012). Fresh inputs of OM will have a higher C:N ratio, closer to that of the original plant material from which it is derived. This more highly hydrophobic material will then displace these weakly-sorbed, outer compounds and allow them to migrate to deeper soil levels where mineral adsorption site capacity is unlikely to be saturated, as illustrated in Fig. 2 (Kaiser and Zech, 2000 and Scott and Rothstein, 2014). This mechanism may be responsible for the observation by Castellano et al. (2012) that N-retention decreases as the amount of mineral-associated OM increases, as well as further explain the commonly observed decrease in soil C:N ratio with depth (e.g., Denef et al., 2013 and Rumpel et al., 2012). Scaling these capacity considerations from fine scale, homogeneous soil volumes to larger scale or ecosystem wide generalizations is problematic though and the fate of these compounds (i.e., whether they are retained or lost from the system) that move to lower soil horizons is not fully understood. Thus, although capacity saturation may occur in small, discrete areas, its importance to long-term ecosystem N retention is likely to be minimal due to its limited occurrence in most systems. The capacity of a soil to process N is increasingly being recognized as the bottleneck that leads to N saturation; this kinetic saturation appears to be driven by an imbalance of N inputs over C inputs, but factors that influence long-term N storage may play a role as well (Kopáček et al., 2013 and Schimel and Bennett, 2004). Pulses of N released during the initial stages of litter breakdown are retained on adsorption sites which allows for attack by microbial degraders (Hatton et al., 2012a). This high N environment, as well as environments with high anthropogenic additions, favors bacterial over fungal communities (Zak et al., 2011). As described earlier, processing by microorganisms represents a crucial step in the creation of sequesterable N so increased processing by bacteria, which have a higher turnover rate and lower C:N ratio than fungi, will increase the production of microbial residues such as cell wall fragments suitable for long-term storage. However, this will be true only up to a point, which may come rapidly as C is consumed. Without a coincident increase in C, microbes will begin to utilize peptides and other small, organic N molecules for C (energy) and excrete excess N as NH_4^+ rather than as organic N products or detritus (Farrell et al., 2012). This NH_4^+ may subsequently be transformed to NO_3^- and leached from the system in stream water, leading to an interesting conclusion where environments that are able to retain higher amounts of N in the short-term through adsorption may actually reduce the amount of N that is stored in the long-term through an earlier transition to an N saturated state, as discussed above. This pulse dynamic was observed by Lewis et al. (2014) who noted that a greater amount of experimentally added N was rapidly transferred to storage in soils with larger O horizons (greater amount of C for microbial activity). Thus management actions to that increase the C status of soils such as protection of old growth areas, the planting of vegetation that produces large quantities of high quality litter, or the eradication of invasive earthworms that consume litter-C may increase the long-term retention of N in soils. In addition, recent work points to the importance of mycorrhizal fungi in mediating N saturation. When adsorption and supply of N in soil is high, plants decrease their C allocation to mycorrhizal fungi (Bahr et al., 2013). This reduction in C increases microbial reliance on peptides for C and decreases the microbial stoichiometric sink strength, which will decrease long-term N storage (Högberg et al., 2014b).

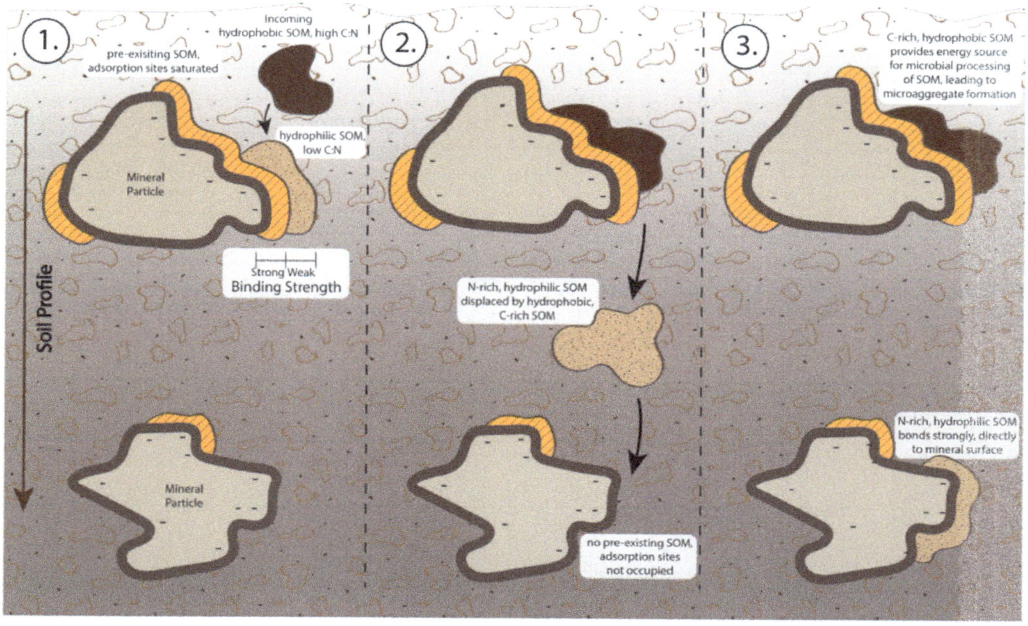

Figure 2: Nitrogen migration. Schematic representation of the process by which N would be preferentially sequestered at depth as observed by Scott and Rothstein (2014). (1) Incoming, C-rich, hydrophobic SOM displaces N-rich, hydrophilic SOM that is weakly bound to existing SOM when adsorption sites are saturated. (2) This N-rich, hydrophilic material is displaced and migrates lower in the soil profile to where (3) mineral adsorption sites are available.

Critical Loads

Although N addition to ecosystems is a natural process, anthropogenic additions, especially from fossil fuel combustion and agricultural emissions from livestock operations, have significantly enhanced rates of N deposition in many areas above naturally occurring levels (Canfield et al., 2010). In order to address these increased inputs a policy instrument called critical loads was developed. Pioneered in Europe in the 1980's, they have been used successfully as a means of mitigating the harmful effects of N deposition (Holmberg et al., 2013). A critical load is defined as the amount of deposition below which no significant effects to the ecosystem are thought to occur according to current knowledge, and is meant to inform the amount of N an ecosystem may endure before unwanted effects become manifest (Nilsson and Grennfelt, 1988).

A common method for generating a critical load is the calculation of a simple mass balance equation, for example,

$$CL_{nut}(\mathrm{N}) = N_i + N_u + N_{de} + N_{le(acc)} \quad (1)$$

where *CLnut*(N) is the critical load of nutrient nitrogen, *Ni* is the long term net immobilization of N in soil, *Nu* is the net removal of N in harvested vegetation and animals, *Nde* is the flux of N to the atmosphere via denitrification and $Nle_{(acc)}$ is the acceptable level of leaching loss of N through the root zone (UBA, 2004). This widely used model describes steady state conditions among other simplifying assumptions. Because of these simplifications, using biologically valid values for the terms becomes critical to obtaining a useful result.

The durable immobilization (*Ni*) term is one of the least well documented of the simple mass balance equation inputs but is crucial to determining an accurate critical load. In this context, *Ni* is the long-term (decades to centuries or longer) accumulation of N in the root zone. In the critical load context, *Ni* represents the amount of N which is retained in a naturally functioning ecosystem without inputs of anthropogenic N, and where the soil C:N ratio does not change. This is to be distinguished from long-term N storage, which represents the total amount of N which is able to be stored in all soil profiles, including enhanced accumulation from man-made sources. Despite this difference, the factors governing the transfer of N into the long-term storage pool should be identical. Protocols for the simple mass balance equation call for using values of 1.0 or 3.5 (revised) kg N ha^{-1} year^{-1} for the *Ni* term, however, estimates of soil N accumulation in North America have ranged from 1.7 kg N ha^{-1} year^{-1} (Turk and Graham, 2009) to as high as 25 kg N ha^{-1} year^{-1} (Bedison et al., 2010) for forest ecosystems (Table 2). Because few studies are specifically designed to measure long-term N accumulation in soils though, often only indirect calculations (such as estimating *Ni* from observed C accumulation and soil C:N ratios) are possible for obtaining estimates for the *Ni* term, and this wide range of estimates creates a large uncertainty in the modeled critical load. An increased understanding of the factors governing N storage will improve the accuracy of values for*Ni* and thus critical load estimates.

Table 2: Nitrogen accumulation rates. Summary of long-term soil N accumulation rates for North American ecosystems. Units are kg N ha^{-1} year^{-1}.

Study	Ecosystem	N accumulation Rate	Notes
Bedison et al. (2010)	Northern Hardwood Forest	25.0	Based on C accumulation rate and C:N ratio over 74years, whole soil profile to 20cm.
Bedison et al. (2010)	Northern Hardwood Forest	11.0	Based on C accumulation rate and C:N ratio over 22years, whole soil profile to 20cm.
Clark and Johnson (2011)	Northern Hardwood Forest	13.0	Based on sites formerly subject to plowed agriculture.
Foote and Grogan (2010)	Northern Hardwood Forest	6.3–9.5	Based on C accumulation rate and C:N ratio.
Gaudinski et al. (2000)	Northern Hardwood Forest	4.0–15.0	Based on C accumulation rate and C:N ratio.
Hooker and Compton (2003)	Northern Hardwood Forest	No significant change	
Goodale and Aber (2001)	Northern Hardwood Forest	20.0	Based on N accumulation in previously burned stands, mineral soil to 10cm.
Kaye et al. (2003)	High Latitude Coniferous	1.0	
Sollins et al. (1983)	Western Coniferous	4.2–5.1	Based on C accumulation and C:N ratio on 1200+ year mudflow, whole soil profile to 70cm.
Turk and Graham (2009)	Western Coniferous	1.7	Mineral soil, but high variability
Giesen et al. (2008)	Western Coniferous	11.3	Old stands following severe fire-average age 550years
Egli et al. (2012)	Alpine	<0.1	Over 20kyears since soil formation

Because N immobilization is now recognized to be driven by ecosystem properties such as edaphic qualities and microbial activity rather than the chemical characteristics of N compounds, more precise and targeted critical loads for N saturation can be developed. Currently, critical loads for N saturation are commonly focused on NO_3^- leaching based on the Aber et al. (1998) framework, which, as discussed in Section 3.1, is not necessarily an indicator of ecosystem N saturation. Using the new paradigm of SON composition and persistence in soils it is possible to identify areas where soil properties and climatic factors are likely to foster lower N processing and decreased stabilization of N. These areas will have a decreased resilience to N saturation and a lower critical load for kinetic N saturation, and may be prioritized for efforts to reduce N deposition.

Mitigation, Restoration, and Other Considerations

The changed paradigm of N storage in soils also has implications for judging the effectiveness of mitigation strategies. The belief that the long term storage of N in soil was due to chemical recalcitrance and condensation reactions meant that for the most part, the persistent, sequestered pool of N in the soil could be removed only through disturbance or via transport through the soil and subsequent leaching. Although a gradient of intrinsic recalcitrance within soils exists, the new understanding that most N is in the form of small, highly labile molecules means that unless well protected from microbial access through protection within the soil matrix, most N can become available again once elevated inputs cease. This can lead to rapid losses of SON where soil factors do not favor sequestration, and may lead to losses of N from the ecosystem through nitrification (because NO_3^- is more mobile within soils) and denitrification, especially when C is limiting. Conversely, since most N in soils is labile, ecosystems with conditions that favor persistence of N may exhibit signs of elevated N for a long period of time after inputs are reduced as the accumulated N is mined. This will be especially true in soils with fewer factors favoring long-term N storage. Ectomycorrhiza have been shown to recover relatively quickly following cessation of N loading and these same fungi are known to mine soil for N when easily accessible inputs do not meet their demand (Hobbie et al., 2013 and Högberg et al., 2010). This extraction of stored N may mask the efficacy of mitigation efforts until the system returns to its previously N-limited state.Högberg et al. (2014a) found elevated levels of foliar N even two decades following reduction of N inputs while grasslands studied by Isbell et al. (2013) failed to recover biodiversity losses decades after N-enrichment ended.

Because adsorption is strongly dependent on pH, mitigation measures that increase the pH of the soil, such as controls on sulfate emissions and subsequent deposition, may change the nature or capacity of N adsorption to minerals (Yu et al., 2013). An increase in pH would change the favored adsorption mechanism from ligand exchange with Al and Fe-oxides or cation bridging with clays to electrostatic attraction on layer silicates, a weaker bond, although properties of the protein are important as well (Rillig et al., 2007 and Von Lützow et al., 2006). This may increase desorption and accessibility of the molecules to enzymatic attack. Empirically, the response of DON to pH is uncertain though and more research is required to elucidate the magnitude of this effect on long-term N storage (Ros et al., 2009).

Climate change and associated increases in floods, droughts and fires will alter N dynamics in soils, but directions and magnitudes are hard to predict since they will depend on the interplay between warming and drying of the soil. Results of studies thus far have been inconsistent and may be highly contingent on site and species specifics (Auyeung et al., 2013). Meta-analysis of warming effects on N dynamics found increased cycling of mineral N, giving credence to studies which have found that warming stimulates N mineralization (Bai et al., 2013). However, meta-analysis also found only a small increase in total soil N content, and no change to microbial N immobilization or microbial N was detected, leaving effects on long-term N storage uncertain and in need of further research (Bai et al., 2013).

Finally, identification of the factors important to long-term N storage should enable better prediction of ecosystems that are able to cope with increased N additions, while better informing biogeochemical models. Using ^{15}N as a tracer, Hatton et al. (2012a) found that litter-derived N followed the same pathway through microbes and macroaggregates into microaggregates where it was protected, despite differing soils, bedrock and OM composition. This suggests that the pathways through which N travels on its way to long-term storage, may be similar for different ecosystems, but edaphic factors such as the presence of Fe and Al-oxides, pH and C and N status, and biotic factors such as substrate quality and microbial community composition may regulate the flux of N to (and less importantly the capacity of) long-term storage sites. It was observed by Lewis et al. (2014) that forest soils with greater C content (such as old growth forests) rapidly integrate greater amounts of N into long-term storage than forest soils with lower C contents. Similarly, Goodale et al. (2015) observed decreased ^{15}N retention in soils during the summer when microbes may have been C-limited compared to spring when ^{15}N was rapidly transferred to stable, heavy-fraction materials, and autumn when C-limitation was alleviated by fresh litter inputs. These observations strongly indicate that factors that facilitate the sequestration of N may also increase resilience to increased N deposition, an important management consideration in the midst of serious anthropogenic alterations to ecosystems through N deposition.

Although afforestation of previously disturbed areas will gradually increase N sequestration, the legacy of lost C from these systems means that priority should be given to the protection of old-growth areas that are naturally more resilient to N additions, where years of C accrual has the potential to rapidly sequester much larger amounts of incoming N (Lewis et al., 2014). An improved understanding of the factors important for N persistence can also guide restoration efforts to places where conditions are such that smaller restoration efforts will have a greater impact. For example, soils with a high potential for N persistence but currently exhibiting low microbial activity could be prioritized for revegetation (which would increase moisture levels in soil and promote microbial activity) over areas with less favorable soils characteristics.

CONCLUSION

The long-term retention of N in ecosystems is important for many reasons. Although many factors are at play, the overall efficiency with which N is sequestered can contribute to whether an ecosystem is generally N or C-limited, and whether it reaches its kinetic capacity to process N, leading to N saturation. The revision of the view of N in soil, from the belief of sequestered N as being chemically recalcitrant and able to resist the attacks of microbes, to a view where in order to persist it must be physically separated from them, has required a reevaluation of the factors that govern N retention. This changed understanding has implications for judging the effectiveness of mitigation measures as well as for our understanding of soil N and C stocks and cycling. For example, because most N stored in soils, the efficacy of mitigation efforts may be masked while accumulated N is processed. Another consequence of this new understanding is that since most N in the long-term storage pool is labile but protected from degradation by inaccessibility, simply measuring the amount of labile N in a soil may overestimate the amount of N in the bioavailable pool and misrepresent the N status of the ecosystem (Darrouzet-Nardi and Weintraub, 2014).

Estimates of the amount of N sequestered in soils have received little attention. Even in Europe where critical loads are used in a regulatory manner, estimates of the amount of N that has accumulated in the long-term lack a consensus; ecosystems in the United States have received even less attention (Duarte et al., 2013). It is increasingly evident that N-saturation is the result of a finite capacity to process N rather than retain it. A better understanding of the factors (Table 1) that govern long-term N storage will improve our understanding of N and C cycling and their effects on ecosystem structure and function, and will improve our understanding of saturation processes. It will improve critical load estimates derived from simple mass balance models to help determine acceptable levels of N inputs to ecosystems, and better inform policy makers and land managers when developing strategies for protection, mitigation and restoration of areas experiencing elevated N inputs (Lovett and Goodale, 2011). Policies that protect soil C, increase microbial activity or raise pH may acquire a higher priority during restoration or mitigation planning. Building on current knowledge, we have outlined the factors that are important for the incorporation of N to long-term storage (Fig. 1), and outlined a mechanism for N sequestration at depth (Fig. 2), yet much work is needed to detail these processes, refine the role of microbes, and examine how the heterogeneous nature of soils, climate and vegetation interact to control this flux. More research is also needed to better quantify adsorption and desorption fluxes, as well as how disturbance, particularly from a changing climate, will affect long-term N storage. Current levels of anthropogenic N inputs have had lasting effects on ecosystems (Thimonier et al., 2010); a better understanding of long-term N storage and the factors that are important to it will help us determine just how lasting those effects could be and help us be better stewards of our environment.

ACKNOWLEDGEMENTS

The authors are grateful to Dan Binkley and Jill Baron for their generosity of time and knowledge, without which this review would not have been possible. We would like to thank Tamara Blett and John Vimont of the National Park Service and Mark Fenn of the US Forest Service for catalyzing this project.

REFERENCES

1. Aber, J., McDowell, W., Nadelhoffer, K., Magill, A., Bernston, G., Kamakea, M., McNulty, S., Currie, W., Rustad, L., Fernandez, I., 1998. Nitrogen saturation in temperate forest ecosystems. Bioscience 921–934.
2. Auyeung, D.S., Suseela, V., Dukes, J.S., 2013. Warming and drought reduce temperature sensitivity of nitrogen transformations. Glob. Chang. Biol. 19, 662–676.
3. Bahr, A., Ellström, M., Akelsson, C., Ekblad, A., Mikusinska, A., Wallander, H., 2013. Growth of ectomycorrhizal fungal mycelium along a Norway spruce forest nitrogen deposition gradient and its effect on nitrogen leakage. Soil Biol. Biochem. 59, 38–48.
4. Bai, E., Li, S., Xu, W., Li, W., Dai, W., Jiang, P., 2013. A meta-analysis of experimental warming effects on terrestrial nitrogen pools and dynamics. New Phytol. 199, 441–451.
5. Bedison, J.E., Johnson, A.H., Willig, S.A., 2010. A comparison of soil organic matter content in 1932, 1984, and 2005/6 in forests of the Adirondak Mountains, New York. Soil Sci. Soc. Am. J. 74, 658–662.
6. Bonnard, P., Basile-Doelsch, I., Balesdent, J., Masion, A., Borschneck, D., Arrouys, A., 2012. Organic matter content and features related to associated mineral fractions in an acid, loamy soil. Eur. J. Soil Sci. 63, 625–636.
7. Braakhekke, M.C., Beer, C., Hoosbeek, M.R., Reichstein, M., Kruijt, B., Schrumpf, M., Kabat, P., 2011. SOMPROF: a vertically explicit soil organic matter model. Ecol. Model. 222, 1712–1730.
8. Canfield, D.E., Glazer, A.N., Falkowski, P.G., 2010. The evolution and future of earth's nitrogen cycle. Science 330, 192–196.
9. Carillo, Y., Ball, B.A., Molina, M., 2016. Stoichiometric linkages between plant litter, trophic interactions and nitrogen mineralization across the litter–soil interface. Soil Biol. Biochem. 92, 102–110.
10. Castellano, M.J., Kaye, J.P., Lin, H., Schmidt, J.P., 2012. Linking carbon saturation concepts to nitrogen saturation and retention. Ecosystems 15, 175–187.
11. Chapin, I.I.I.F.S., Matson, P.A., Vitousek, P.M., 2012. Principles of Terrestrial Ecosystem Ecology. Springer-Verlag, New York.
12. Chivenge, P., Vanlauwe, B., Gentile, R., Six, J., 2011. Organic resource quality influences short-term aggregate dynamics and soil organic carbon and nitrogen accumulation. Soil Biol. Biochem. 43, 657–666.
13. Clark, J.D., Johnson, A.H., 2011. Carbon and nitrogen accumulation in post-agricultural forest soils of western New England. Soil Sci. Soc. Am. J. 75, 1530–1542.
14. Cotrufo, M.F., Wallenstein, M.D., Boot, C.M., Denef, K., Paul, E., 2013. The Microbial Efficiency-Matrix Stabilization (MEMS) framework integrates plant litter decomposition with soil organic matter stabilization: do labile plant inputs form stabile soil organic matter? Glob. Chang. Biol. 19, 988–995.
15. Currie, W.S., Aber, J.D., McDowell, W.H., Boone, R.D., Magill, A.H., 1996. Vertical transport of dissolved organic C and N under long-term N amendments in pine and hardwood forests. Biogeochemistry 35, 471–505.
16. Darrouzet-Nardi, A., Weintraub, M.N., 2014. Evidence for spatially inaccessible labile N from a comparison of soil core extractions and soil pore water lysimetry. Soil Biol. Biochem. 73, 22–32.
17. Deb, S.K., Shukla, M.K., 2011. A review of dissolved organic matter transport processes affecting soil and environmental quality. J. Environment. Analytic Toxicol 1.
18. Delgado-Baquerizo, M., Gacia-Palacios, G., Milla, R., Gallardo, A., Maestre, F.T., 2015. Soil characteristics determine soil carbon and nitrogen availability during leaf litter decomposition regardless of litter quality. Soil Biol. Biochem. 81, 134–142.
19. Denef, K., Del Galdo, I., Venturi, A., Cotrufo, M.F., 2013. Soil carbon and nitrogen stocks and distribution across organic matter fractions in European soils under different land use. Open Journal of Soil Science 3, 297–313.
20. Dippold, M., Biryukov, M., Kuzyakov, Y., 2014. Sorption affects amino acid pathways in soil: implications from position-specific labeling of alanine. Soil Biol. Biochem. 72, 180–192.
21. Dittman, J.A., Driscoll, C.T., Groffman, P.M., Fahey, T.J., 2007. Dynamics of nitrogen and dissolved organic carbon at the Hubbard Brook Experimental Forest. Ecology 88, 1153–1166.
22. Duarte, N., Pardo, L.H., Robin-Abbott, M.J., 2013. Susceptibility of forests in the Northeastern USA to nitrogen and sulfur deposition: critical load exceedance and forest health. Water Air Soil Pollut. 224, 1–21.

23. Dümig, A., Häusler, W., Steffens, M., Kögel-Knabner, I., 2012. Clay fractions from a soil chronosequence after glacier retreat reveal the initial evolution of organo-mineral associations. Geochim. Cosmochim. Ac. 85, 1–18.

24. Dungait, J.A.J., Hopkins, D.W., Gregory, A.S., Whitmore, A.P., 2012. Soil organic matter turnover is governed by accessibility not recalcitrance. Glob. Chang. Biol. 18, 1781–1796.

25. Egli, M., Favilli, F., Krebs, R., Pichler, B., Dahms, D., 2012. Soil organic carbon and nitrogen accumulation rates in cold and alpine environments over 1 Ma. Geoderma 183-184, 109–123.

26. Farrell, M., Hill, P.W., Farrar, J., Bardgett, R.D., Jones, D.L., 2011. Seasonal variation in soluble soil carbon and nitrogen across a grassland productivity gradient. Soil Biol. Biochem. 43, 835–844.

27. Farrell, M., Hill, P.W., Farrar, J., DeLuca, T.H., Roberts, P., Kielland, K., Dahlgren, R., Murphy, D.V., Hobbs, P.J., Bardgett, R.D., Jones, D.L., 2012. Oligopeptides represent a preferred source of organic N uptake: a global phenomenon? Ecosystems 16, 133–145.

28. Farrell, M., Prendergast-Miller, M., Jones, D.L., Hill, P.W., 2014. Soil microbial organic nitrogen uptake is regulated by carbon availability. Soil Biol. Biochem. 77, 261–267.

29. Fenn, M.E., Baron, J.S., Allen, E.B., Rueth, H.M., Nydick, K.R., Geiser, L., Bowman, W.D., Sickman, J.O., Meixner, T., Johnson, D., Neitlich, P., 2003. Ecological effects of nitrogen deposition in the Western United States. Bioscience 53, 404–420.

30. Fontaine, S., Henault, C., Aamor, A., Bdioui, N., Bloor, J.M.G., Maire, V., Mary, B., Revaillot, S., Maron, P.A., 2011. Fungi mediate long term sequestration of carbon and nitrogen in soil through their priming effect. Soil Biol. Biochem. 43, 86–96.

31. Fonte, S.J., 2009. Fertilizer and residue quality effects on organic matter stabilization in soil aggregates. Soil Sci. Soc. Am. J. 73, 961–966.

32. Foote, R.L., Grogan, P., 2010. Soil carbon accumulation during temperate forest succession on abandoned low productivity agricultural lands. Ecosystems 13, 795–812.

33. Frey, S.D., Ollinger, S., Nadelhoffer, K., Bowden, R., Brzostek, E., Burton, A., Caldwell, B.A., Crow, S., Goodale, C.L., Grandy, A.S., Finzi, A., Kramer, M.G., Lajtha, K., LeMoine, J., Martin, M., McDowell, W.H., Minocha, R., Sadowsky, J.J., Templer, P.H., Wickings, K., 2014. Chronic nitrogen additions suppress decomposition and sequester soil carbon in temperate forests. Biogeochemistry 121, 305–316.

34. Gaudinski, J.B., Tumbore, S.E., Davidson, E.A., Zheng, S., 2000. Soil carbon in a temperate forest: radiocarbon-based estimates of residence times, sequestration rates and partitioning fluxes. Biogeochemistry 51, 33–69.

35. Geisseler, D., Howarth, W.R., Joergensen, R.G., Ludwig, B., 2010. Pathways of nitrogen utilization by soil microorganisms — a review. Soil Biol. Biochem. 42, 2058–2067.

36. Giardina, C.P., Ryan, M.G., Hubbard, R.M., Binkley, D., 2001. Tree species and soil textural controls on carbon and nitrogen mineralization rates. Soil Sci. Soc. Am. J. 65, 1272–1279.

37. Giesen, T.W., Perakis, S.S., Cromack, K. Jr, 2008. Four centuries of soil carbon and nitrogen change after stand-replacing fire in a forest landscape in the western Cascade Range of Oregon. Can. J. For. Res. 38, 2455–2464.

38. Gillespie, A.W., Diochon, A., Ma, B.L., Morrison, M.J., Kellman, L., Walley, F.L., Regier, T.Z., Chevrier, D., Dynes, J.J., Gregorich, E.G., 2014. Nitrogen input quality changes the biochemical composition of soil organic matter stabilized in the fine fraction: a longterm study. Biogeochemistry 117, 337–350.

39. Gleixner, G., 2013. Soil organic matter dynamics: a biological perspective derived from the use of compound-specific isotopes studies. Ecol. Res. 28, 683–695.

40. Goodale, C.L., Aber, J.D., 2001. The long-term effects of land-use history on nitrogen cycling in northern hardwood forests. Ecol. Appl. 11, 253–267.

41. Goodale, C.L., Fredriksen, G., Weiss, M.S., McCalley, C.K., Sparks, J.P., Thomas, S.A., 2015. Soil processes drive seasonal variation in retention of 15N tracers in a deciduous forest catchment. Ecology 96, 2653–2668.

42. Grandy, A.S., Strickland, M.S., Lauber, C.L., Bradford, M.A., Fierer, N., 2009. The influence of microbial communities, management, and soil texture on soil organic matter chemistry. Geoderma 150, 278–286.

43. Guggenberger, G., Kaiser, K., 2003. Dissolved organic matter in soil: challenging the paradigm of sorptive preservation. Geoderma 113, 293–310.

44. Hatton, P.J., Kleber, M., Zeller, B., Moni, C., Plante, A.F., Townsend, K., Gelhaye, L., Lajtha, K.,

45. Derrien, D., 2012a. Transfer of litter-derived N to soil mineral–organic associations: evidence from decadal 15N tracer experiments. Org. Geochem. 42, 1489–1501.

46. Hatton, P.J., Remusat, L., Zeller, B., Derrien, D., 2012b. A multi-scale approach to determine accurate elemental and isotopic ratios by nano-scale secondary ion mass spectrometry imaging. Rapid Commun. Mass Spectrom. 26, 1363–1371.

47. Hatton, P.J., Bodé, S., Angeli, N., Boeckx, P., Zeller, B., Boiry, S., Gelhaye, L., Derrien, D., 2014. Assimilation and accumulation of C by fungi and bacteria attached to soil density fractions. Soil Biol. Biochem. 79, 132–139.

48. Hatton, P.J., Castanha, C., Torn, M.S., Bird, J.A., 2015. Litter type control on soil C and N stabilization dynamics in a temperate forest. Glob. Chang. Biol. 21, 1358–1367.

49. Hobara, S., Osono, T., Hirose, D., Noro, K., Hirota, M., Brenner, R., 2014. The roles of microorganisms in litter decomposition and soil formation. Biogeochemistry 118, 471–486.

50. Hobbie, E.A., Ouimette, A.P., Schuur, E.A.G., Kierstead, D., Trappe, J.M., Bendiksen, K.,

51. Ohenoja, E., 2013. Radiocarbon evidence for the mining of organic nitrogen from soil by mycorrhizal fungi. Biogeochemistry 114, 381–389.

52. Högberg, P., Johannisson, C., Yarwood, S., Calleson,

I., Nashölm, T., Myrold, D.D., Högberg, M.N., 2010. Recovery of ectomycorrhiza after 'nitrogen saturation' of a conifer forest. New Phytol. 189, 515–525.

53. Högberg, M.N., Yarwood, S.A., Myrold, D.D., 2014a. Fungal but not bacterial soil communities recover after termination of decadal nitrogen additions to boreal forest. Soil Biol. Biochem. 72, 35–43.

54. Högberg, M.N., Blaško, R., Holm Bach, L., Hasselquist, N.J., Egnell, G., Näsholm, T., Högberg,

55. P., 2014b. The return of an experimentally N-saturated boreal forest to an N-limited state: observations on the soil microbial community structure, biotic N retention capacity and gross N mineralisation. Plant Soil 381, 1–16.

56. Holmberg, M., Vuorenmaa, J., Posch, M., Forsius, M., Lundin, L., Kleemola, S., Augustaitis, A., Beudert, B., de Wit, H.A., Dirnböck, T., Evans, C.D., Frey, J., Grandin, U.,

57. Indriksone, I., Krám, P., Pompei, E., Schulte-Bisping, H., Srybny, A., Váňa, M., 2013. Relationship between critical load exceedances and empirical impact indicators at integrated monitoring sites across Europe. Ecol. Indic. 24, 256–265.

58. Hooker, T.D., Compton, J.A., 2003. Forest ecosystem carbon and nitrogen accumulation during the first century after agricultural abandonment. Ecol. Appl. 13, 299–313.

59. Houlton, B.Z., Morford, S.L., 2015. A new synthesis for terrestrial nitrogen inputs. Soil 1, 381–397.

60. Isbell, F., Tilman, D., Polasky, S., Binder, S., Hawthorne, P., 2013. Low biodiversity state persists two decades after cessation of nutrient enrichment. Ecol. Lett. 16, 454–460.

61. Jastrow, J.D., 1996. Soil aggregate formation and the accrual of particulate and mineralassociated organic matter. Soil Biol. Biochem. 28, 665–676.

62. Jodidi, S.I., 1911. The chemical nature of the organic nitrogen in the soil. J. Am. Chem. Soc. 33, 1226–1241.

63. Johnson, D.W., Turner, J., 2014. Nitrogen budgets of forest ecosystems: a review. For. Ecol. Manag. 318, 370–379.

64. Kaiser, K., Kalbitz, K., 2012. Cycling downwards — dissolved organic matter in soils. Soil Biol. Biochem. 52, 29–32.

65. Kaiser, K., Zech, W., 2000. Sorption of dissolved organic nitrogen by acid subsoil horizons and individual mineral phases. Eur. J. Soil Sci. 51, 403–411.

66. Kaiser, C., Franklin, O., Dieckmann, U., Richter, A., 2014. Microbial community dynamics alleviate stoichiometric constraints during litter decay. Ecol. Lett. 17, 680–690.

67. Kaiser, C., Franklin, O., Richter, A., Dieckmann, U., 2015. Social dynamics within decomposer communities lead to nitrogen retention and organic matter build-up in soils. Nat. Commun. 6.

68. Kaye, J.P., Binkley, D., Rhoades, C., 2003. Stable soil nitrogen accumulation and flexible organic matter stoichiometry during primary floodplain succession. Biogeochemistry 63, 1–22.

69. Kleber, M., Mikutta, R., Torn, M.S., Jahn, R., 2005. Poorly crystalline mineral phases protect organic matter in acid subsoil horizons. Eur. J. Soil Sci. 56, 717–725.

70. Kleber, M., Sollins, P., Sutton, R., 2007. A conceptual model of organo-mineral interactions in soils: self-assembly of organic molecular fragments into zonal structures on mineral surfaces. Biogeochemistry 85, 9–24.

71. Kleber, M., Nico, P.S., Plante, A., Filleys, T., Kramer, M., Swanston, C., Sollins, P., 2011. Old and stable soil organic matter is not necessarily chemically recalcitrant: implications for modeling concepts and temperature sensitivity. Glob. Chang. Biol. 17, 1097–1107.

72. Knicker, H., 2011. Soil organic N — an under-rated player for C sequestration in soils? Soil Biol. Biochem. 43, 1118–1129.

73. Kögel-Knaber, I., 1993. Biodegradation and humification processes in forest soils. In: Bollag, J.-M., Stotsky, G. (Eds.), Soil Biochemistry Vol 8. Marcel Dekker Inc., New York, pp. 101–135.

74. Kögel-Knabner, I., Guggenberger, G., Kleber, M., Kandeler, E., Kalbitz, K., Scheu, S.,

75. Eusterhues, K., Leinweber, P., 2008. Organo-mineral associations in temperate soils: integrating biology, mineralogy, and organic matter chemistry. J. Plant Nutr. Soil Sc. 171, 61–82.

76. Kopáček, J., Cosby, B.J., Evans, C.D., Hruška, J., Moldan, F., Oulehle, F., Šantrůčková, H., Tahovská, K., Wright, R.F., 2013. Nitrogen, organic carbon and sulphur cycling in terrestrial ecosystems: linking nitrogen saturation to carbon limitation of soil microbial processes. Biogeochemistry 115, 33–51.

77. Kothawala, D.N., Moore, T.R., 2009. Adsorption of dissolved nitrogen by forest mineral soils. Can. J. For. Res. 39, 2381–2390.

78. Kuiluweit, M., Bougoure, J.J., Zeglin, L.H., Myrold, D.D., Weber, P.K., Pett-Ridge, J., Kleber, M., Nico, P.S., 2012. Nano-scale investigation of the association of microbial nitrogen residues with iron (hydr)oxides in a forest soil O-horizon. Geochim Cosmochim Ac 95, 213–226.

79. LeBauer, D.S., Treseder, K.K., 2008. Nitrogen limitation of net primary productivity in terrestrial ecosystems is globally distributed. Ecology 89, 371–379.

80. Lehmann, J., Kleber, M., 2015. The contentious nature of soil organic matter. Nature 528, 60–68.

81. Leinweber, P., Kruse, J., Baum, C., Arcand›r, M., 2013. Analytical Techniques. Adv. Agron. 119, 83.

82. Lewis, B.L., Castellano, M.J., Kaye, J.P., 2014. Forest succession, soil carbon accumulation, and rapid nitrogen storage in poorly remineralized soil organic matter. Ecology 95, 2687–2693.

83. Lovett, G.M., Goodale, C.L., 2011. A new conceptual model of nitrogen saturation based on experimental nitrogen addition to an oak forest. Ecosystems 14, 615–631.

84. Manzoni, S., Taylor, P., Richter, A., Porporato, A., Ågren,

G.I., 2012. Environmental and stoichiometric controls on microbial carbon-use efficiency in soils. New Phytol. 196, 79–91.

85. Marín-Spiotta, E., Gruley, K.E., Crawford, J., Atkinson, E.E., Miesel, J.R., Greene, S., CardonaCorrea, C., Spencer, R.G.M., 2014. Paradigm shifts in soil organic matter research affect interpretations of aquatic carbon cycling: transcending disciplinary and ecosystem boundaries. Biogeochemistry 117, 279–297.

86. Marschner, B., Bradowski, S., Dreves, A., Gleixner, G., Gude, A., Grootes, P.M., Hamer, U., Heim, A., Jandi, G., Ji, R., Kaiser, K., Kalbitz, K., Kramer, C., Leinweber, P., Rethemeyer, J., Schäffer, A., Schmidt, M.W.I., Schwark, L., Wiesenberg, G.L.B., 2008. How relevant is recalcitrance for the stabilization of organic matter in soils? J Plant Nutr Soil Sc 171, 91–110.

87. Miltner, A., Bombach, P., Schmidt-Brücken, B., Kästner, M., 2012. SOM genesis: microbial biomass as a significant source. Biogeochemistry 111, 41–55.

88. Moni, C., Rumpel, C., Virto, I., Chabbi, A., Chenu, C., 2010. Relative importance of sorption versus aggregation for organic matter storage in subsoil horizons of two contrasting soils. Eur. J. Soil Sci. 61, 958–969.

89. Nannipieri, P., Paul, E., 2009. The chemical and functional characterization of soil N and its biotic components. Soil Biol. Biochem. 41, 2357–2369.

90. Nichols, K.A., Halvorson, J.J., 2013. Roles of biology, chemistry, and physics in soil macroaggregate formation and stabilization. Open Agricultural Journal 7, 107–117.

91. Nilsson, J., Grennfelt, P., 1988. Critical Loads for Sulphur and Nitrogen: Report from a Workshop Held at Skokloster, Sweden, 19–24 March, 1988. Nordic Council of Ministers, Copenhagen, Denmark.

92. Pronk, G.J., Heister, K., Kögel-Knabner, I., 2013. Is turnover and development of organic matter controlled by mineral composition? Soil Biol. Biochem. 67, 235–244.

93. Qualls, R.G., Haines, B.L., 1991. Fluxes of dissolved organic nutrients and humic substances in a deciduous forest. Ecology 72, 254–266.

94. Rillig, M.C., Caldwell, B.A., Wösten, H.A.B., Sollins, P., 2007. Role of proteins in soil carbon and nitrogen storage: controls on persistence. Biogeochemistry 85, 25–44.

95. Ros, G.H., Hoffland, E., van Kessel, C., Temminghoff, E.J.M., 2009. Extractable and dissolved soil organic nitrogen — a quantitative assessment. Soil Biol. Biochem. 41, 1029–1039.

96. Rosenqvist, L., Kleja, D.B., Johansson, M.B., 2010. Concentrations and fluxes of dissolved organic carbon and nitrogen in a Picea abies chronosequence on former arable land in Sweden. For. Ecol. Manag. 259, 275–285.

97. Rumpel, C., Rodriguez-Rodriguez, A., González-Pérez, J.A., Arbelo, C., Chabbi, A., Nunan, N., González-Vila, F.J., 2012. Contrasting composition of free and mineral-bound organic matter in top- and subsoil horizons of andosols. Biol. Fertil. Soils 48, 401–411.

98. Sanaullah, M., Chabbi, A., Leifeld, J., Bardoux, G., Billou, D., Rumpel, C., 2011. Decomposition and stabilization of root litter in top- and subsoil horizons: what is the difference? Plant Soil 338, 127–141.

99. Sanderman, J., Maddern, T., 2014. Similar composition but differential stability of mineral retained organic matter across four classes of clay minerals. Biogeochemistry 121, 409–424.

100. Schimel, J.P., Bennett, J., 2004. Nitrogen mineralization: challenges of a changing paradigm. Ecology 85, 591–602.

101. Schmidt, B.H.M., Kalbitz, K., Braun, S., Fuß, R., McDowell, W.H., Matzner, E., 2011a. Microbial immobilization and mineralization of dissolved organic nitrogen from forest floors. Soil Biol. Biochem. 43, 1742–1745.

102. Schmidt, M.W.I., Torn, M.S., Abiven, S., Dittmar, T., Guggenberger, G., Janssens, I.A., Kleber, M., Kögel-Knabner, I., Lehmann, J., Manning, D.A.C., Nannipieri, P., Rasse, D.P., Weiner, S., Trumbore, S.E., 2011b. Persistence of soil organic matter as an ecosystem property. Nature 478, 49–56.

103. Schulten, H.R., Schnitzer, M., 1998. The chemistry of soil organic nitrogen: a review. Biol. Fertil. Soils 26, 1–15.

104. Schurig, C., Smittenberg, R.H., Berger, J., Kraft, F., Woche, S.K., Goebel, M.O., Heipieper, H.J., Miltner, A., Kaestner, M., 2013. Microbial cell-envelope fragments and the formation of soil organic matter: a case study from a glacier forefield. Biogeochemistry 113, 595–612.

105. Scott, E., Rothstein, D.E., 2014. The dynamic exchange of dissolved organic matter percolating through six diverse soils. Soil Biol. Biochem. 69, 83–92.

106. Silveira, M.L., Reddy, K.R., Comerford, N.B., 2011. Litter decomposition and soluble carbon, nitrogen, and phosphorus release in a forest ecosystem. Open Journal of Soil Science 1, 86–96.

107. Sinsabaugh, R.L., Manzoni, S., Moorhead, D.L., Richter, A., 2013. Carbon use efficiency of microbial communities: stoichiometry, methodology and modelling. Ecol. Lett. 16, 930–939.

108. Six, J., Paustian, K., 2014. Aggregate-associated soil organic matter as an ecosystem property and a measurement tool. Soil Biol. Biochem. 68, A4–A9.

109. Six, J., Elliott, E.T., Paustian, K., 2000. Soil macroaggregate turnover and microaggregate formation: a mechanism for C sequestration under no-tillage agriculture. Soil Biol. Biochem. 32, 2099–2103.

110. Six, J., Conant, R.J., Paul, E.A., Paustian, K., 2002. Stabilization mechanisms of soil organic matter: implications for C-saturation of soils. Plant Soil 241, 155–176.

111. Six, J., Bossuyt, H., Degryze, S., Denef, K., 2004. A history of research on the link between (micro) aggregates, soil biota, and soil organic matter dynamics. Soil Tillage Res. 79, 7–31.

112. Sleutel, S., Vandenbruwane, J., De Schrijver, A., Wuyts, K., Moeskops, B., Verheyen, K., De Neve, S., 2009. Patterns of dissolved organic carbon and nitrogen fluxes

in deciduous and coniferous forests under historic high nitrogen deposition. Biogeosciences 6, 2743–2758.

113. Sogn, T.A., Stuanes, A.O., Abrahamsen, G., 1999. The capacity of forest soil to absorb anthropogenic N. Ambio 28, 346–349.

114. Sollins, P., Spycher, G., Topik, C., 1983. Processes of soil organic matter accretion at a mud- flow chronosequence, Mt Shasta, California. Ecology 64, 1273–1282.

115. Sollins, P., Homan, P., Caldwell, B.A., 1996. Stabilization and destabilization of soil organic matter: mechanisms and controls. Geoderma 74, 65–105.

116. Sollins, P., Swanston, C., Kramer, M., 2007. Stabilization and destabilization of soil organic matter — a new focus. Biogeochemistry 85, 1–7.

117. Soong, J., Cotrufo, M.F., 2015. Annual burning of a tallgrass prairie inhibits C and N cycling in soil, increasing recalcitrant pyrogenic organic matter storage while reducing N availability. Glob. Chang. Biol. 21, 2321–2333.

118. Soong, J., Parton, W., Calderon, F., Campbell, N., 2015. A new conceptual model on the fate and controls of fresh and pyrolized plant litter decomposition. Biogeochemistry 124, 27–44.

119. St Luce, M., Whalen, J.K., Ziadi, N., Zebarth, B.J., Chantigny, M.H., 2014. Labile organic nitrogen transformations in clay and sandy-loam soils amended with 15N-labelled faba bean and wheat residues. Soil Biol. Biochem. 68, 208–218.

120. Stevenson, F.J., 1986. Cycles of soil: Carbon, Nitrogen, Phosphorus, Sulfur, Micronutrients. Wiley, New York.

121. Stevenson, F.J., 1994. Humus Chemistry: Genesis, Composition, Reactions. John Wiley & Sons, New York.

122. Stewart, C.E., Paustian, K., Conant, R.T., Plante, A.F., Six, J., 2007. Soil carbon saturation: concept, evidence and evaluation. Biogeochemistry 86, 19–31.

123. Sylvia, D.M., Fuhrmann, J.J., Hartel, P.G., Zuberer, D.A., 2005. Principles and Applications of Soil Microbiology. second ed. Pearson Prentice Hall, Upper Saddle River, New Jersey.

124. Tahovská, K., Kaňa, J., Bárta, J., Oulehle, F., Richter, A., Šantrůčková, H., 2013. Microbial N immobilization is of great importance in acidified mountain spruce forest soils. Soil Biol. Biochem. 59, 58–71.

125. Thamdrup, B., 2012. New pathways and processes in the global nitrogen cycle. Annu. Rev. Ecol. Evol. S. 43, 407–428.

126. Thimonier, A., Pannatier, E.G., Schmitt, M., Waldner, P., Walthert, L., Schleppi, P., Dobbertin, M., Kräuchi, N., 2010. Does exceeding the critical loads for nitrogen alter nitrate leaching, the nutrient status of trees and their crown condition at Swiss Long-term Forest Ecosystem Research (LWF) sites? Eur. J. For. Res. 129, 443–461.

127. Thomas, D.C., Zak, D.R., Filley, T.R., 2012. Chronic N deposition does not apparently alter the biochemical composition of forest floor and soil organic matter. Soil Biol. Biochem. 54, 7–13.

128. Throckmorton, H.M., Bird, J.A., Monte, N., Doane, T., Firestone, M.K., Horwath, W.R., 2015. The soil matrix increases microbial C stabilization in temperate and tropical forest soils. Biogeochemistry 122, 35–45.

129. Turk, J.K., Graham, R.C., 2009. Soil carbon and nitrogen accumulation in a forested debris flow chronosequence, California. Soil Sci. Soc. Am. J. 73, 1504–1509. UBA (ed): Manual on Methodologies and Criteria for Modeling and Mapping Critical Loads and Levels, and air Pollution Effects, Risks, and Trends. German Federal Environmental Agency, Berlin, Germany, ca. 190 pp.; constantly updated version at www.icpmapping.org; 2004.

130. Uselman, S.M., Qualls, R.G., Lilienfein, J., 2009. Production of total potentially soluble organic C, N, and P across an ecosystem chronosequence: root versus leaf litter. Ecosystems 12, 240–260.

131. Uselman, S.M., Qualls, R.G., Lilienfein, J., 2012. Quality of soluble organic C, N, and P produced by different types and species of litter: root litter versus leaf litter. Soil Biol. Biochem. 54, 57–67.

132. van Groenigen, J.W., Huygens, D., Boeckx, P., Kuyper, ThW, Lubbers, I.M., Rütting, T., Groffman, P.M., 2015. The soil N cycle: new insights and key challenges. Soil 1, 235–256.

133. Virto, I., Barré, P., Chenu, C., 2008. Microaggregation and organic matter storage at the siltsize scale. Geoderma 146, 326–335.

134. Vogel, C., Mueller, C.W., Höschen, C., Bruegger, F., Heister, K., Schultz, S., Schloter, M., Kögel-Knabner, I., 2014. Submicron structures provide preferential spots for carbon and nitrogen sequestration in soils. Nat. Commun. 5.

135. Vogel, C., Heister, K., Buegger, F., Tanuwidjaja, I., Haug, S., Schloter, M., Kögel-Knabner, I., 2015. Clay mineral composition modifies decomposition and sequestration of organic carbon and nitrogen in fine soil fractions. Biol. Fertil. Soils 1–16.

136. Von Lützow, M., Kögel-Knabner, I., Ekschmitt, K., Matzner, E., Guggenberger, G., Marschner, B., Flessa, H., 2006. Stabilization of organic matter in temperate soils: mechanisms and their relevance under different soil conditions — a review. Eur. J. Soil Sci. 57, 426–445.

137. Wei, H., Guenet, B., Vicca, S., Nunan, N., Asard, H., AbdElgawad, H., Shen, W., Janssens, I.A., 2014. High clay content accelerates the decomposition of fresh organic matter in artificial soils. Soil Biol. Biochem. 77, 100–108.

138. Wilson, G.W.T., Rice, C.W., Rillig, M.C., Springer, A., Hartnett, D.C., 2009. Soil aggregation and carbon sequestration are tightly correlated with the abundance of arbuscular mycorrhizal fungi: results from long-term field experiments. Ecol. Lett. 12, 452–461.

139. Yu, W.H., Li, N., Tong, D.S., Zhou, C.H., Lin, C.X.C., Xu, C.Y., 2013. Adsorption of proteins and nucleic acids on clay minerals and their interactions: a review. Appl. Clay Sci. 80, 443–452.

140. Zak, D.R., Pregitzer, K.S., Burton, A.J., Edwards, I.P., Kellner, H., 2011. Microbial response to a changing environment: implications for the future functioning of terrestrial ecosystems. Fungal Ecol. 4, 386–395.

CHAPTER 3 MODELLING THE IMPACTS OF AGRICULTURAL MANAGEMENT PRACTICES ON RIVER WATER QUALITY

Sam D. Taylor, Yi He, Kevin M. Hiscock

Keywords: Catchment management, Catchment modelling, diffuse water pollution, Mitigation scenarios, SWAT, Water quality

INTRODUCTION

Agricultural diffuse water pollution remains a notable global pressure on surface water and groundwater quality (Carpenter et al., 1998, Vörösmarty et al., 2010 and European Environment Agency, 2012), and trends suggest that agricultural expansion will continue to exacerbate those pressures well into the 21st Century (Tilman et al., 2001). Legislation has been introduced in many parts of the world to protect water bodies from agricultural diffuse water pollution and to improve water quality, including the Nitrates Directive and Water Framework Directive (WFD) in Europe (Council of the European Union, 1991 and Council of the European Union, 2000), and the Clean Water Act in the United States (United States Environmental Protection Agency, 2002). The WFD seeks to improve or maintain water quality through the establishment of River Basin Management Plans (RBMPs) and the development of Programmes of Measures (PoMs), which can be implemented to ensure that each water body within a river basin district achieves good ecological and chemical status (Council of the European Union, 2000). Member states committed to achieving this status by 2015 but many water bodies were not expected to meet the necessary water quality standards before this deadline (European Environment Agency, 2012). According to Solheim et al. (2012), 56% of rivers, 44% of lakes, 67% of transitional waters and 49% of coastal waters that have been classified in Europe do not achieve a good ecological status or potential and 6% of rivers, 2% of lakes, 10% of transitional waters, 4% of coastal waters and 25% of groundwater bodies by surface area are of a poor chemical status. Agricultural diffuse water pollution is cited as a significant pressure in 40% of rivers and coastal water bodies and one-third of lakes and transitional water bodies. Such poor water quality has consequences for the health of aquatic ecosystems, biodiversity, human health, the use of water in industry and agriculture and as a resource for public water supply and recreation (Carr and Neary, 2008).

In Europe, agricultural diffuse water pollution contributes 50–80% of the total nitrogen load and approximately 50% of the total phosphorus load in surface water bodies (European Environment Agency, 2005 and Kronvang et al., 2009). In the United Kingdom (UK) specifically, agricultural diffuse water pollution is estimated to be responsible for 61% of the total nitrogen load and 28% of the total phosphorus load experienced within surface water bodies (Hunt et al., 2004 and White and Hammond, 2007). Nutrient enrichment within surface waters due to the oversupply of phosphorus and nitrogen in agriculture increases the risk of eutrophication (Richardson and Jørgensen, 1996, Withers and Lord, 2002 and Carr and Neary, 2008). While phosphorus pollution has implications for ecosystem health, nitrate pollution also has implications for the supply of water and human health (Withers and Lord, 2002). To protect human health, water is considered to be unfit for human consumption under the Drinking Water Directive applied within Europe if it contains a nitrate concentration above 50 mg L^{-1}(equivalent to 11.3 mg NO_3-N L^{-1}) (Council of the European Union, 1998), but many surface water and groundwater bodies within the UK contain concentrations of nitrate that approach or exceed this limit (European Environment Agency, 2012).

To develop PoMs that can be implemented under the WFD, authorities responsible for establishing RBMPs must be able to assess the effectiveness of potential mitigation options. Given the limited resources available to monitor and quantify the impacts of mitigation options in-field, and the need to provide timely evidence to inform policy, water quality models which

Citation: Sam D. Taylor, Yi He, Kevin M. Hiscock, Modelling the impacts of agricultural management practices on river water quality, Journal of Environmental Management, Volume 180, 15 September 2016, Pages 147-163, ISSN 0301-4797, http://dx.doi.org/10.1016/j.jenvman.2016.05.002.

can quantify the impacts of mitigation options on nutrient losses have been increasingly applied as Decision Support Tools (DSTs) within Decision Support Systems (Collins and McGonigle, 2008 and Volk et al., 2008). This approach can be used to develop targeted mitigation plans, identify critical source areas and times, assess the cost-effectiveness of mitigation options, identify pollution swapping and involve stakeholders in the development of suitable management plans (Bouraoui and Grizzetti, 2014). Effective dialogue and engagement between stakeholders and scientific experts is essential to ensure that the PoMs are appropriate, cost-effective and sustainable and to maximise the effectiveness of the mitigation practices that are introduced (Van Ast, 2000 and Gerrits and Edelenbos, 2004).

The Benchmark Models for the Water Framework Directive project established a set of criteria to assess which models have the potential to assist in the implementation of the WFD (Saloranta et al., 2003). As part of this project, the suitability of the Soil and Water Assessment Tool (SWAT) water quality model for assessing the impacts of mitigation options proposed to meet WFD targets on water quality was examined by Bärlund et al. (2007). Rode et al. (2008) and Volk et al. (2009) also applied SWAT to examine the potential for changes in catchment management to ensure that water bodies achieve WFD targets. SWAT has been widely and successfully applied to assess the impacts of agricultural mitigation options on water quality and can therefore be considered to be an appropriate DST for assisting authorities in managing catchments to achieve statutory water quality targets (e.g. Santhi et al., 2006, Hu et al., 2007, Ullrich and Volk, 2009, Lam et al., 2011, Moriasi et al., 2011, Glavan et al., 2012, Aouissi et al., 2014, Boithias et al., 2014 and Santhi et al., 2014). Examples of mitigation options that have been modelled include buffer strips, nutrient management plans, alternative tillage techniques, alternative crop rotations and changes in land use.

In this study, based in the River Wensum catchment in Eastern England (Fig. 1), the availability of a high-quality, high-frequency dataset of water quality enabled the performance of SWAT in simulating multiple pollutants at a daily time-step to be assessed. SWAT was also used to investigate the impacts of agricultural mitigation options on long-term water quality at a daily resolution and to assess the uncertainties of the predicted impacts of mitigation options on water quality. The unique water quality dataset applied within this study is derived from continuous monitoring at a 30-min temporal resolution. Such a monitoring strategy reduces the uncertainty associated with estimates of in-stream nutrient loads relative to datasets derived from fewer samples collected at longer time intervals and ensures that the model applied within this investigation has been robustly calibrated. This lower uncertainty allows the model to be applied with a higher degree of confidence, creating a more effective and reliable DST.

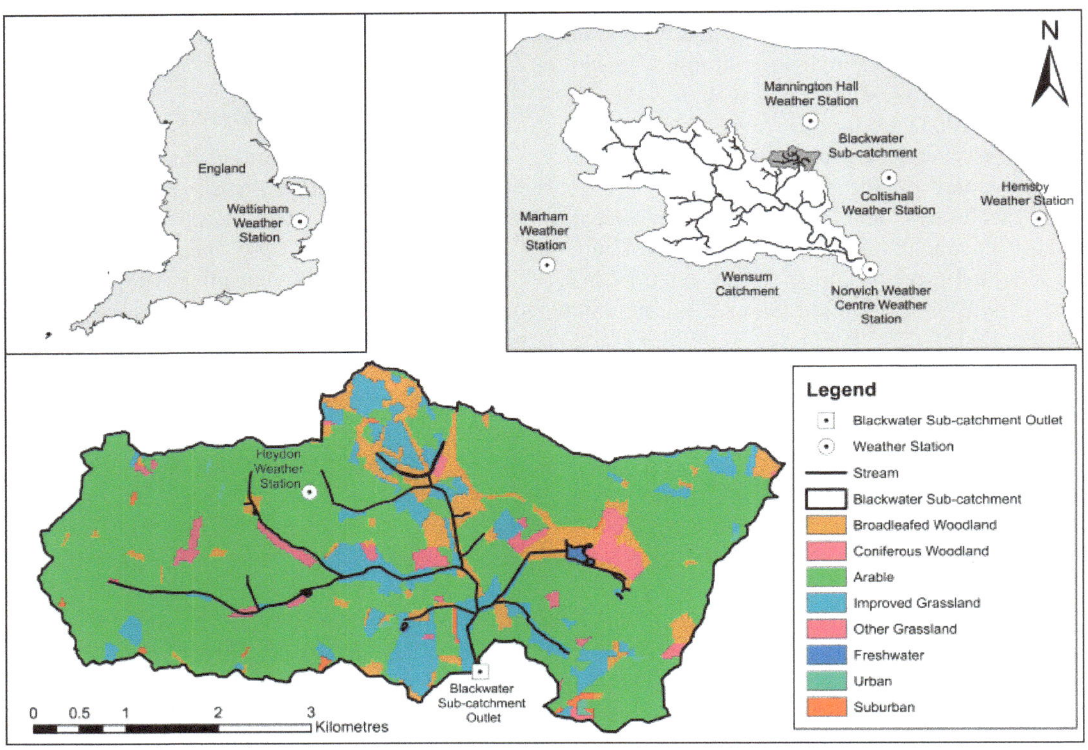

Figure 1: A map of the location and land cover of the Blackwater sub-catchment in relation to the River Wensum catchment within England. The locations of the weather stations used in this investigation and the outlet of the sub-catchment are also shown. Based upon LCM2007 © NERC (CEH) 2011. Contains Ordnance Survey data © Crown Copyright 2007. © third party licensors.

There is no standard or universally accepted metric applied to assess model performance but Moriasi et al. (2007) suggested that models should achieve a Nash-Sutcliffe Efficiency (NSE) coefficient of greater than 0.5 for flow, nitrogen and total phosphorus at a monthly time-step for performance to be considered satisfactory. If we consider this performance criterion to apply at all time-steps, over half of the 115 SWAT hydrological assessments and 37 SWAT pollutant loss studies summarised by Gassman et al. (2007), achieved this level of model performance, but some studies reported poor results for all variables particularly at a daily time-step and it is in this context that we consider the performance of SWAT within the River Wensum catchment.

Since 2010, the River Wensum catchment has been the focus of the Wensum Demonstration Test Catchment (DTC) Project which aims to provide evidence to test the hypothesis that it is economically feasible to reduce agricultural diffuse water pollution through the introduction of agricultural mitigation practices whilst maintaining agricultural productivity (Wensum Alliance, 2014). The Blackwater sub-catchment has been selected as a pilot area where the effects of changes in management will be investigated and is considered to be representative of the rest of the River Wensum catchment. To identify the mitigation options that are most relevant for the River Wensum catchment, there has been close cooperation and engagement between local land owners, farm managers, environmental organisations, government agencies and scientific experts. With knowledge gained from these stakeholders, the aim of this investigation is to apply SWAT to the Blackwater sub-catchment to quantify the long-term impacts of potential changes to agricultural practices on water quality, to assess the uncertainties of those predictions and to identify mitigation options that have the potential to be applied within similar arable catchments to improve water quality. This is one of the first studies to quantify the impacts of agricultural mitigation options, both singly and in combination, on long-term water quality for nitrate and total phosphorus at a daily time-step, in addition to providing an estimate of the uncertainties of those impacts.

In the remaining parts of this paper, a brief review of the study area, the datasets used and the methodology adopted in applying SWAT to the Blackwater sub-catchment is provided. A detailed summary of the mitigation options that were selected and modelled is also supplied. The results of model calibration and validation and the impacts of each agricultural measure on water quality, both singly and in combination, are also presented and discussed. Finally, conclusions and a summary of findings are provided.

MATERIALS AND METHODS

Study Area

The River Wensum has a total catchment area of 675 km^2 and is designated a Special Area of Conservation (SAC), a Drinking Water Protected Area and 71 km of the riparian zone are designated as a Site of Special Scientific Interest (SSSI) (Natural England, 1993, English Nature, 2005 and Environment Agency, 2009). The importance of the River Wensum has also been recognised by the UK Biodiversity Action Plan, which designates the river as a priority chalk river habitat (Biodiversity Reporting and Information Group, 2007). The catchment has a temperate maritime climate and had a mean annual rainfall of 714 mm and an annual rainfall range of 542.6–878.8 mm during 1981–2010 (Met Office, 2014).

This study focuses on the Blackwater River, a tributary of the Wensum, which drains an area of 19.6 km^2 (Fig. 1). The characteristics of the Blackwater sub-catchment are typical of the wider River Wensum catchment and other catchments found in Eastern England. The topography of the sub-catchment is relatively subdued, with elevation ranging from 28 to 70 m above sea level, and 95% of the sub-catchment area has a slope of 5% or less. Streamflow within the Blackwater sub-catchment is derived from groundwater flow, lateral flow in the soil zone, surface runoff and contributions from an extensive tile drain network (Howson, 2012). During periods of low rainfall, streamflow is sustained by baseflow, with a baseflow index similar to that of the Wensum catchment as a whole equal to 0.80 (Outram et al., 2014). At the outlet of the Blackwater sub-catchment during the period from 1 December 2011 to 30 June 2014, 30-min resolution data recorded a daily mean discharge of 0.112 m^3 s^{-1} and daily mean concentrations of 6.16 mg NO_3-N L^{-1} and 0.089 mg P L^{-1} for nitrate and total phosphorus, respectively. Cretaceous Chalk deposits underlay the majority of the sub-catchment, with some Pleistocene Crag deposits on the south-eastern edge of the sub-catchment boundary (Hiscock, 1993). The bedrock geology is overlain by superficial deposits of Quaternary glacial origin composed of boulder clay, sands and gravel that attain a thickness of greater than 20 m (Hiscock, 1993 and Hiscock et al., 1996).

The SWAT Model and Inputs

SWAT is a semi-distributed and physically based water quality model that operates at a continuous time-step (Arnold et al., 2012). The model is designed to simulate the effects of changes in management practices on surface water and groundwater hydrology, diffuse pollution and sediment erosion within catchments. Within SWAT, a catchment is divided into multiple sub-catchments which are then further divided into Hydrologic Response Units (HRUs) that consist of homogeneous land use, slope and soil characteristics (Arnold et al., 2012). Physical processes in SWAT are split into two phases: (i) the land-based phase; and (ii) the channel-based phase (Neitsch et al., 2011). The former includes climate, hydrology, plant growth, erosion, nutrient cycles, pesticides and management practices. The latter routes water, sediment, nutrients and pesticides through the channel network. Input variables define physical properties within the model and parameters are used to define and perform management practices. The model simulates all of the key physical processes found within the Blackwater sub-catchment and is therefore considered to be a suitable model to apply. In order to construct a SWAT model of the Blackwater sub-catchment, ArcSWAT version 2012.10.0.14 was applied (Texas A&M University, 2015). The methodology applied to construct the model is available for reference in Winchell et al. (2013). Readers are referred to Neitsch et al. (2011) for a detailed review of the physical processes modelled within SWAT and Arnold et al. (2014) for a detailed overview of the model input requirements and outputs.Gassman et al. (2007) provide a detailed summary of over 250 previous publications relating to SWAT. Krysanova and Arnold, 2008 and Douglas-Mankin et al., 2010 andTuppad et al. (2011) review the historical development and applications of the model andArnold et al. (2012) present an overview of a methodology that can be adopted when applying the model. The model is subject to ongoing development and future landscape unit and grid-based versions will allow a more detailed spatial representation of catchment practices to be implemented within SWAT (Arnold et al., 2010, Bosch et al., 2010, Bonumá et al., 2014 and Rathjens et al., 2015).

Catchment Agricultural Practices

Data from the Agricultural Census conducted by The Department for Environment, Food and Rural Affairs (Defra) was obtained for the River Wensum catchment for the period 1993–2010 in a 2 km grid square format. Data for the Blackwater sub-catchment was used to identify those crops commonly grown within the sub-catchment (Fig. 2) and to identify an appropriate crop rotation plan to implement within the SWAT model of the sub-catchment (Department for Environment Food and Rural Affairs (Defra), 2016 and EDINA, 2014). Based on this analysis, it was found that the most commonly grown crops within the catchment were wheat, barley, oilseed rape, spring beans and sugar beet. The Salle Estate, which is located in the Blackwater sub-catchment, manages 2000 ha of arable land and operates a seven-year crop-rotation that includes those crop types identified in the agricultural census data (Salle Farms Ltd, 2014). Listed in order of cultivation, the seven-year crop-rotation operated within the sub-catchment and applied within the SWAT model consists of winter barley, winter oilseed rape, winter wheat, sugar beet, spring barley, spring beans and winter wheat (Table 1). The rotation was initiated at different starting points within the rotation based on crop-type and was distributed randomly within the model because actual crop distributions within the sub-catchment were unknown. The Defra RB209 Fertiliser Manual was used to identify appropriate fertiliser application rates for each crop included in the crop-rotation (Defra, 2010a). The timings of planting, harvesting, field tillage and fertiliser application were determined from UK Agriculture (2014) for all crops except sugar beet where the source used was British Sugar (2014).

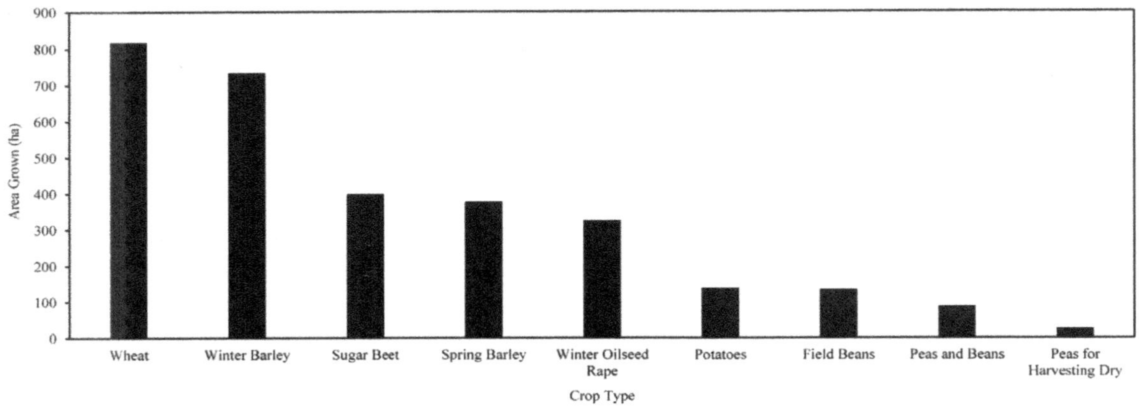

Figure 2: The area of each crop type grown within the Blackwater sub-catchment according to the 2010 Agricultural Census conducted by the Department for Environment, Food and Rural Affairs (Department for Environment Food and Rural Affairs (Defra), 2016 and EDINA, 2014).

Table 1: The seven year crop-rotation scheme and management operations applied within the SWAT model of the Blackwater sub-catchment.

Year	Month	Day	Management operation	Description
1	9	15	Tillage	Generic fall ploughing operation
1	9	30	Tillage	Roterra harrow tillage operation
1	10	1	Cultivation	Plant winter barley
2	3	1	Fertiliser application	Apply 40 kg ha^{-1} elemental nitrogen
2	3	1	Fertiliser application	Apply 60 kg ha^{-1} phosphate
2	4	1	Fertiliser application	Apply 70 kg ha^{-1} elemental nitrogen
2	7	31	Harvest	Harvest winter barley
2	8	15	Tillage	Generic fall ploughing operation
2	8	31	Tillage	Roterra harrow tillage operation
2	9	1	Cultivation	Plant winter oilseed rape
3	3	1	Fertiliser application	Apply 60 kg ha^{-1} elemental nitrogen
3	3	1	Fertiliser application	Apply 50 kg ha^{-1} phosphate
3	4	1	Fertiliser application	Apply 60 kg ha^{-1} elemental nitrogen
3	7	31	Harvest	Harvest winter oilseed rape
3	9	15	Tillage	Generic fall ploughing operation
3	9	30	Tillage	Roterra harrow tillage operation
3	10	1	Cultivation	Plant winter wheat
4	3	1	Fertiliser application	Apply 40 kg ha^{-1} elemental nitrogen
4	3	1	Fertiliser application	Apply 60 kg ha^{-1} phosphate

4	5	1	Fertiliser application	Apply 120 kg ha^{-1} elemental nitrogen
4	8	31	Harvest	Harvest winter wheat
4	9	15	Tillage	Generic fall ploughing operation
5	3	17	Fertiliser application	Apply 50 kg phosphate
5	3	31	Tillage	Roterra harrow tillage operation
5	4	1	Cultivation	Planting sugar beet
5	4	1	Fertiliser application	Apply 40 kg ha^{-1} elemental nitrogen
5	5	1	Fertiliser application	Apply 40 kg ha^{-1} elemental nitrogen
5	10	31	Harvest	Harvest sugar beet
5	11	15	Tillage	Generic fall ploughing operation
6	1	31	Tillage	Roterra harrow tillage operation
6	2	1	Cultivation	Plant spring barley
6	4	1	Fertiliser application	Apply 70 kg ha^{-1} elemental nitrogen
6	4	1	Fertiliser application	Apply 45 kg ha^{-1} phosphate
6	8	31	Harvest	Harvest spring barley
6	11	15	Tillage	Generic fall ploughing operation
7	1	31	Fertiliser application	Apply 40 kg ha^{-1} phosphate
7	1	31	Tillage	Roterra harrow tillage operation
7	2	1	Cultivation	Plant spring beans
7	8	31	Harvest	Harvest spring beans
7	9	15	Tillage	Generic fall ploughing operation
7	9	30	Tillage	Roterra harrow tillage operation
7	10	1	Cultivation	Plant winter wheat
8	3	1	Fertiliser application	Apply 40 kg ha^{-1} elemental nitrogen
8	3	1	Fertiliser application	Apply 60 kg ha^{-1} phosphate
8	5	1	Fertiliser application	Apply 120 kg ha^{-1} elemental nitrogen
8	8	31	Harvest	Harvest winter wheat

To assess the impacts of mitigation options on agricultural diffuse water pollution and water quality within the Blackwater sub-catchment, a variety of mitigation options have been introduced on the Salle Estate as part of the Wensum DTC Project (Lovett et al., 2015). The mitigation options include the introduction of a cover crop during the autumn and winter months which is intended to protect soils from erosion when they would otherwise be bare, to reduce the leaching of nutrients from soils during wet winter months and, when destroyed, to act as a 'green manure', slowly releasing nutrients to the surrounding soil for subsequent crops (Rubæk et al., 2011). The use of strip tillage to establish autumn and spring-sown crops, with the intention of reducing sediment and nutrient loss in surface runoff, has been introduced as an additional mitigation option in some pilot areas of the sub-catchment.

Meteorological Data

The meteorological inputs required to perform simulations within SWAT include daily observations of precipitation, mean wind speed, maximum and minimum temperature, solar radiation and mean relative humidity (Arnold et al., 2014). If no observations are available, SWAT includes a weather generator which has the capacity to generate estimates of meteorological variables. Observations of meteorological variables recorded from January 1980 to June 2014 were obtained from UK Met Office Integrated Data Archive System (MIDAS) Land and Marine Surface Stations Data for application within the model (Met Office, 2012). Observations of daily minimum and maximum temperature, wind speed and relative humidity were obtained from the MIDAS weather station located at Marham (MIDAS Station ID: 409), which is sited approximately 40 km to the south-west of the Blackwater sub-catchment. Observations of daily sunshine hours recorded at Marham weather station were used to estimate a daily record of incident solar radiation for the sub-catchment. Where observations of daily sunshine hours are missing from the Marham record, observations recorded at the nearby MIDAS weather stations located at Coltishall (MIDAS Station ID: 429), Norwich Weather Centre (MIDAS Station ID: 408), Hemsby (MIDAS Station ID: 433) and Wattisham (MIDAS Station ID: 440), selected in order of their proximity to the sub-catchment and the availability of data, were used to interpolate the missing data. Observations of daily precipitation were obtained from the MIDAS weather station located at Heydon (MIDAS Station ID: 4807) (Fig. 1). Where observations of precipitation are missing from the Heydon record, observations recorded at the nearest MIDAS weather station, located at Mannington Hall (MIDAS Station ID: 24219), were used to interpolate the missing data using the nearest-neighbour technique.

Water Quality Data

As part of the Wensum DTC Project, automated equipment including a pressure transducer housed in a stilling well, a Nitratax Plus SC sensor and a Phosphax Sigma analyser, have been used to continuously monitor river stage, nitrate and total phosphorus concentrations, respectively, at 30-min intervals at the outlet of the Blackwater sub-catchment since April 2011 (Fig. 1). Quality assurance and quality control procedures, including the comparison of high-frequency data to laboratory analysed spot samples, were conducted to ensure the validity of data included in this study. Flow gauging using an electromagnetic open channel flow meter was conducted on 16 occasions during high, moderate and low flow events which, in combination with observations of river stage from the pressure transducers, was used to develop a power law stage-discharge rating curve which was applied to estimate daily mean discharge, nitrate load and total phosphorus load exported from the sub-catchment during the period 1 December 2011 to 30 June 2014. These estimates were applied within this study to perform model sensitivity analysis, calibration and validation. To identify the importance of any relationship between sediment transport and total phosphorus concentrations within the sub-catchment, 467 in-stream grab samples collected at the outlet of the Blackwater sub-catchment during the period October 2010 to March 2015 were used to develop a log-log regression model and conduct a linear regression *t*-test to test the hypothesis that the relationship between the concentration of total suspended solids and the concentration of total phosphorus was significant.

Geographical Datasets

The digital terrain model applied within this study has a resolution of 5 m and was obtained from the NEXTMap British Digital Terrain Model Dataset (Itermap Technologies, 2007). Land cover within the study area was identified from the Land Cover Map 2007 (LCM2007) raster dataset which has a resolution of 25 m and divides land cover into 23 distinct classes based on the Broad Habitats defined within the UK Biodiversity Action Plan (Morton et al., 2011). According to LCM2007, land cover within the Blackwater sub-catchment is largely arable with 86.05% of the land area utilised for agricultural purposes (Morton et al., 2011). The dominance of the arable farming industry within the sub-catchment is reflected by the fact that 74.22% of the land area is utilised for growing crops and 11.83% as grazing pasture. Woodland, other areas of grassland and heathland, urban areas and surface water bodies including wetland environments account for the remaining area.

A map of soil types within the sub-catchment was derived from the National Soil Map (NATMAP) vector dataset which displays the spatial occurrence of 300 distinct Soil Associations throughout England and Wales (Cranfield University, 2014a). Each Soil Association is composed of multiple Soil Series and possesses distinct properties. According to NATMAP, five different Soil Associations are present within the Blackwater sub-catchment. Burlingham 1, Wick 2 and Wick 3 cover 83.72% of the sub-catchment and are composed of loamy soils, Beccles 1 covers 16.17% of the sub-catchment and is composed of loamy over clayey soils and Isleham 2 covers 0.11% of the sub-catchment and is composed of sandy soils (Cranfield University, 2014b). The properties of each Soil Association, as required by SWAT, have been determined from the Horizon Fundamentals, Horizon Hydraulics, NSI Textures and NSI Profile datasets (Cranfield University, 2014c and Cranfield University, 2014d). The properties required by SWAT for each layer of each soil type include the depth of soil layer, moist bulk density, available water capacity, saturated hydraulic conductivity, sand, silt, clay and organic carbon content, maximum rooting depth within the soil profile, the fraction of porosity from which anions are excluded, moist albedo of the soil surface and erodibility (Arnold et al., 2014).

Model Calibration and Validation

In order to conduct a sensitivity analysis and to perform model calibration and validation, the Sequential Uncertainty Fitting version 2 (SUFI-2) optimisation algorithm (Abbaspour et al., 2004 and Abbaspour et al., 2007) was applied within the SWAT Calibration and Uncertainty Program (SWAT-CUP) version 5.1.6.2 (Abbaspour, 2014). SUFI-2 is based on the concept of equifinality, which posits that multiple models (i.e. multiple parameter sets) provide equally acceptable predictions and as such, parameter values are treated as uncertain (Beven, 1993 and Beven and Freer, 2001). Model parameters selected for calibration were first assigned an initial global uncertainty range within SWAT-CUP (Table 2). Sensitivity analysis was then performed to identify those parameters that model outputs were sensitive to. In general, a parameter should be included in calibration if sensitivity analysis identifies that there is a 95% probability that the sensitivity of a variable to a particular parameter is significant. Only sensitive parameters were included in the calibration of the model at a daily time-step against observations of discharge and nitrate and total phosphorus loads recorded at the outlet of the Blackwater sub-catchment. Using the sensitive parameters, five iterations of 1000 simulations were performed to calibrate the model. The parameter ranges were updated after each iteration, as identified by the SUFI-2 optimisation algorithm, until prediction uncertainty and model performance was considered satisfactory. The model was applied at a daily time-step during the period from 1 December 2011 to 30 June 2014, of which 1 December 2011 to 31 March 2013 and 1 April 2013 to 30 June 2014 were used as calibration and validation time periods, respectively. An initial warm-up period of four years was applied during calibration and validation to ensure that the model achieved a steady-state and to eliminate any initial bias. Validation involved evaluating model performance against observations recorded outside of the calibration time-period and was utilised as an additional test of model performance.

Table 2: The model parameters identified as significant by the sensitivity analysis and the initial and final calibrated ranges of each parameter.

Parameter	Description	Initial range	Final range
ALPHA_BF	Baseflow recession constant (1/day)	0–1	0.16–0.5
GW_DELAY	Groundwater delay time (days)	0–500	420–490
CH_N2	Manning's roughness coefficient for the main channel	0–0.3	0.03–0.081
CH_K2	Effective hydraulic conductivity of main channel alluvium (mm hr^{-1})	0–100	28–55
ALPHA_BNK	Baseflow recession constant for bank storage (1/day)	0–1	0.73–0.96
GW_REVAP	Groundwater evaporation coefficient	0.02–0.2	0.03–0.1
SURLAG	Surface runoff lag coefficient	1–24	1–4.18
REVAPMN	Threshold depth of water in the shallow aquifer required for the movement of water from the shallow aquifer to the unsaturated zone to occur (mm)	0–500	66–200
OV_N	Manning's roughness coefficient for overland flow	−0.2 to 0.2[a]	−0.035 to 0.087 [a]
CN2 AGRL	Runoff curve number for agricultural land	−0.2 to 0.2[a]	−0.15 to −0.05[a]
CN2 FRSD	Runoff curve number for deciduous forest	−0.2 to 0.2[a]	−0.13 to 0.093 [a]
CN2 PAST	Runoff curve number for pasture land	−0.2 to 0.2[a]	−0.23 to −0.082[a]
SOL_AWC	Available water capacity of soil layer (mm H_2O/mm soil)	−0.2 to 0.2[a]	0.16–0.39 [a]
SOL_Z	The depth from the soil surface to the bottom of soil layer (mm)	−0.2 to 0.2[a]	−0.041 to 0.028 [a]
DDRAIN	Depth to the sub-surface drain (mm)	900–1100	1060–1130
CDN	Denitrification exponential rate coefficient	0–0.1	0.033–0.059
ANION_EXCL	Fraction of void space from which anions are excluded	0.5–0.75	0.68–0.76
SDNCO	Fraction of field capacity above which denitrification takes place	0.9–1	0.94–0.96
SOL_NO3	Initial nitrate concentration in the soil layer (ppm)	0–100	69–96
SOL_SOLP	Initial soluble phosphorus concentration in the soil layer (ppm)	0–100	36–70
GWSOLP	Concentration of soluble phosphorus in groundwater (ppm)	0–0.25	0.06–0.19
SOL_BD	Moist bulk density of soil layer (g cm^{-3})	−0.2 to 0.2[a]	−0.25 to −0.054[a]
RCN	Concentration of nitrogen in rainfall (mg L^{-1})	0–15	3.7–7
CMN	Rate factor for mineralisation of active organic nutrients in humus	0.001–0.003	0.0017–0.0023
NPERCO	Nitrate percolation coefficient	0–1	0.21–0.47
CH_ERODMO	The level of resistance to channel erosion	0–1	0.83–0.96
HLIFE_NGW	Half-life of nitrate in groundwater (days)	0–200	130–200
PHOSKD	Phosphorus soil partitioning coefficient (m^3 Mg^{-1})	100–200	150–180
TDRAIN	Time to drain soil to field capacity (hours)	0–72	46–64
ESCO	Soil evaporation compensation factor	0–1	0.86–1
SHALLST_N	Initial concentration of nitrate in shallow aquifer (ppm)	0–1000	130–310
ERORGP	Phosphorus enrichment ratio	0–0.1	0.0017–0.03

A A relative change which has been applied to the original value of the parameter where the value is multiplied by 1 plus a number from within the defined range.

Objective Functions

Moriasi et al. (2007) recommend that three quantitative statistics are used as objective functions to evaluate model performance, including the Nash-Sutcliffe Efficiency (NSE) coefficient, percentage bias (PBIAS) and the ratio of the root mean square error to the standard deviation of the measured data (RSR). Each of these statistical measures is defined below.

Nash-Sutcliffe Efficiency Coefficient

The Nash-Sutcliffe Efficiency (NSE) coefficient proposed by Nash and Sutcliffe (1970) is defined by Equation (1).

$$NSE = 1 - \frac{\sum_{i=1}^{n}\left(Y_i^{obs} - Y_i^{sim}\right)^2}{\sum_{i=1}^{n}\left(Y_i^{obs} - \overline{Y^{obs}}\right)^2} \qquad (1)$$

where: n is the total number of observations, Y_i^{obs} is the value of the observed variable at the ith time-step, Y_i^{sim} is the value of the simulated variable at the ith time-step and $\overline{Y^{obs}}$ is the mean value of the measured data considered.

NSE is a normalised statistic that describes the degree of the 'goodness-of-fit' between model predictions and observations and can vary between $-\infty$ and 1, where a value of 1 represents a perfect fit. An NSE value of between 0 and 1 is generally recognised as acceptable model performance, whilst a value of less than 0 indicates that the mean of the measured data is a better predictor of a variable compared to the model and indicates unsatisfactory model performance.

Percent Bias

Percent bias (PBIAS) is described as the average tendency of simulated data to overestimate or underestimate a variable relative to observations and is defined by Equation (2). The optimum value of PBIAS is zero, indicating perfect agreement between model simulations and observations. A negative PBIAS value indicates overestimation and a positive value indicates underestimation.

$$PBIAS = \frac{\sum_{i=1}^{n}\left(Y_i^{obs} - Y_i^{sim}\right) * 100}{\sum_{i=1}^{n}\left(Y_i^{obs}\right)} \qquad (2)$$

Ratio of the Root Mean Square Error to the Standard Deviation of the Measured Data (RSR)

RSR is described as the ratio of the Root Mean Square Error (RMSE) to the standard deviation (STDEV) of observed data and is defined by Equation (3) (Moriasi et al., 2007).

$$RSR = \frac{RMSE}{STDEV_{obs}} = \frac{\left[\sqrt{\sum_{i=1}^{n}\left(Y_i^{obs} - Y_i^{sim}\right)^2}\right]}{\left[\sqrt{\sum_{i=1}^{n}\left(Y_i^{obs} - \overline{Y^{obs}}\right)^2}\right]} \qquad (3)$$

RSR can vary from an optimum value of zero, indicating that there is no error between measured and simulated data, up to large positive values (Moriasi et al., 2007). A small RSR indicates a good model performance.

Model Performance Criteria

Moriasi et al. (2007) suggest that for a model to be considered to perform satisfactorily in simulating discharge, nitrate and total phosphorus loads at a monthly time-step, it must achieve a NSE of >0.5, a RSR of 0.7 and a PBIAS of ±25% for discharge and a NSE of >0.5, a RSR of 0.7 and a PBIAS of ±70% for nitrate and total phosphorus loads.

Mitigation Scenarios

As part of the Wensum DTC Project, stakeholders, including farmers and farm-advisers, were consulted to identify and select potential agricultural mitigation options that can be applied within the Blackwater sub-catchment to improve water quality. The Farm Scale Optimisation of Pollutant Emission Reductions (FARMSCOPER) tool, described in detail by Zhang et al. (2012) and Gooday et al. (2014), was also applied to the sub-catchment to evaluate the impacts of potential mitigation options. FARMSCOPER is a spreadsheet-based DST which can identify the impacts of mitigation options on losses of multiple pollutants at the farm scale and assess the costs of each mitigation option (ADAS, 2015). Input requirements include mean annual precipitation, soil type and general farm type, based on the robust farm types classification scheme used by the UK Government (ADAS, 2015 and Department for Environment Food and Rural Affairs (Defra), 2010b). More detailed livestock and cropping information can be included if required. Since application within this project, the tool has undergone considerable development and it can now evaluate the impacts of mitigation options on biodiversity, energy and water use and can be applied at catchment and national scales (ADAS, 2015). The options identified as being suitable by stakeholders and the results provided by FARMSCOPER were broadly similar and were selected for evaluation in this study (see Table 3).

Table 3: The agricultural measures scenarios applied within the SWAT model of the Blackwater sub-catchment.

Number	Name	Description
S0	Control scenario	Baseline scenario representing current conditions and practices
S1	Buffer strip (2 m)	Establishment of 2 m wide buffer strip on arable land
S2	Buffer strip (6 m)	Establishment of 6 m wide buffer strip on arable land
S3	Conservation tillage	A reduced tillage practice compared to the control scenario
S4	Zero tillage	No field tillage and the direct drilling of crops
S5	No tile drains	Removal or blockage of field drainage systems from all arable land
S6	Red clover cover crop	Introduction of a red clover cover crop to the crop rotation scheme

S7	Combined scenario	Buffer strip (6 m) (S2) and red clover cover crop (S6) scenarios combined

The control scenario (S0) is considered to represent current conditions and practices within the catchment and is used as the baseline scenario against which all other mitigation scenarios are assessed. Under scenario S0, a generic ploughing operation (primary tillage) is conducted on agricultural land within the model prior to establishing a crop. Primary tillage involves the aggressive mixing of surface materials and a mixing or burying of crop residues, pesticides and fertilisers leaving a rough soil surface. Primary tillage is followed by a further pulverisation of surface materials (secondary tillage) with a harrow (the Roterra harrow in the SWAT model). Secondary tillage involves a less aggressive mixing of soils, and pulverises soils into a finer material, removing air pockets and preparing the seedbed for cultivation (see Table 4). Such a detailed regime of tillage practice is not often conducted in SWAT. Under scenario S0, tile drains are included on all areas of arable land. Sandy soils (i.e. Isleham 2) where tile drains would otherwise have been excluded are not under arable land use anywhere within the catchment.

Table 4: The mixing depth and efficiency of each tillage technique applied within the model.

Tillage technique	Mixing depth (mm)	Mixing efficiency (fraction)
Generic ploughing operation	150	0.95
Conservation tillage	100	0.25
Roterra harrow	5	0.80

Scenarios S1 and S2 involve the introduction of buffer strips of 2 m and 6 m width, respectively, to areas of arable land within the sub-catchment. Scenario S1 represents a compulsory practice required under cross compliance rules in order to qualify for payments under Common Agricultural Policy schemes (Defra, 2015). Scenario S2 represents a voluntary practice that can be introduced in order to qualify for payments under the Entry Level Stewardship Scheme by achieving good environmental conditions (Natural England, 2014). Scenarios S3 and S4 consider the use of alternative tillage practices within the sub-catchment. Conservation or reduced tillage (S3) involves a less aggressive mixing of soils relative to the control scenario, whereas no tillage (S4) involves the direct drilling of seeds into soils without any cultivation. The mixing depth and mixing efficiency of each tillage technique considered by the SWAT model is provided in Table 4. Scenario S5 involves the removal or blockage of subsurface tile drainage systems from areas of arable land within the sub-catchment in order to simulate the slowing of runoff and solute transport. Under scenario S6, a red clover cover crop was applied within the modelled sub-catchment on two occasions during the crop rotation scheme when arable land would otherwise have been bare prior to the planting of spring crops. The two occasions are between the harvesting of winter wheat and the cultivation of sugar beet from the 1 September to 31 March and between the harvesting of spring barley and the cultivation of spring beans from 1 September to 31 January. Under this scenario, the red clover cover crop is terminated within the model at the end of the growing period and is ploughed back into the field to form a 'green manure'. Finally, to assess the impacts of mitigation options on water quality when introduced in combination, a red clover cover crop (S6) and buffer strips of 6 m width (S2), the two mitigation options that were considered to be most effective at reducing nitrate and total phosphorus losses individually within the Blackwater sub-catchment, respectively, were modelled together under scenario S7. Each mitigation scenario was implemented across all areas of arable land within the sub-catchment.

To quantify the impacts of each mitigation option on long-term water quality, each scenario was run within the SWAT model at a daily time-step for the period 1990–2009, with an initial warm-up period of four years from 1986 to 1989. The period from 1990 to 2009 was used because precipitation during this period reflected full climatic variability, including droughts and wet periods. A total number of 1000 simulations were performed to simulate discharge, and nitrate and total phosphorus loads at a daily time-step under each scenario. This relatively long time period was used in order to consider the response of the sub-catchment to each measure under a variety of conditions over the long term.

RESULTS AND DISCUSSION

Calibration and validation

Sensitivity analysis identified that the parameters listed in Table 2 were required to be included in model calibration. In order to calibrate the model against observations of discharge, and nitrate and total phosphorus loads, five iterations of 1000 simulations were performed. The initial and final calibrated ranges of each parameter are provided in Table 2.

Discharge Simulation

The model performance in simulating daily mean discharge at the outlet of the Blackwater sub-catchment during the calibration and validation time periods is shown in Fig. 3 and Fig. 4. When evaluated at a daily time-step, the model achieved NSE, PBIAS and RSR values of 0.77, −6.0% and 0.48, respectively, during the calibration period and values of 0.68, −24.8% and 0.57, respectively, during the validation period (Table 5). The 95% prediction uncertainty range bracketed 86% and 87% of observed flow data during calibration and validation periods, respectively, indicating that the model achieved a relatively good fit between predictions and observations overall. To evaluate the model performance at a monthly time-step against the performance criteria suggested by Moriasi et al. (2007), daily data were aggregated into monthly time-series. According to those criteria, the model can be considered to perform very well in simulating discharge at both daily and monthly time-steps during the calibration and validation periods (see Table 5). The negative PBIAS values achieved during both time periods indicate that the model tends to overestimate discharge. This overestimation is pronounced during prolonged dry periods in 2013 and 2014 and may indicate a deficiency in simulating base flow during periods of drought.

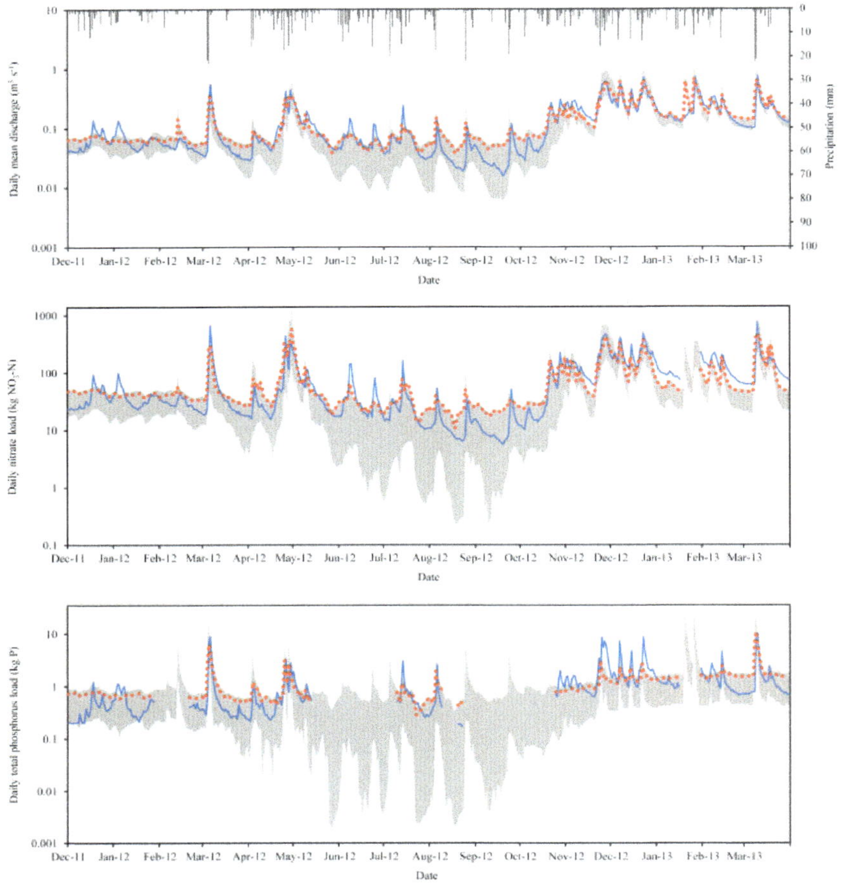

Figure 3: Observed (solid line) and the best simulated (dotted line) daily mean discharge, nitrate and total phosphorus loads recorded at the outlet of the Blackwater sub-catchment during the calibration time period (1 December 2011–31 March 2013). The 95% confidence interval is represented by the hatched area and the daily rainfall amount recorded at Heydon weather station is plotted in the top panel for reference.

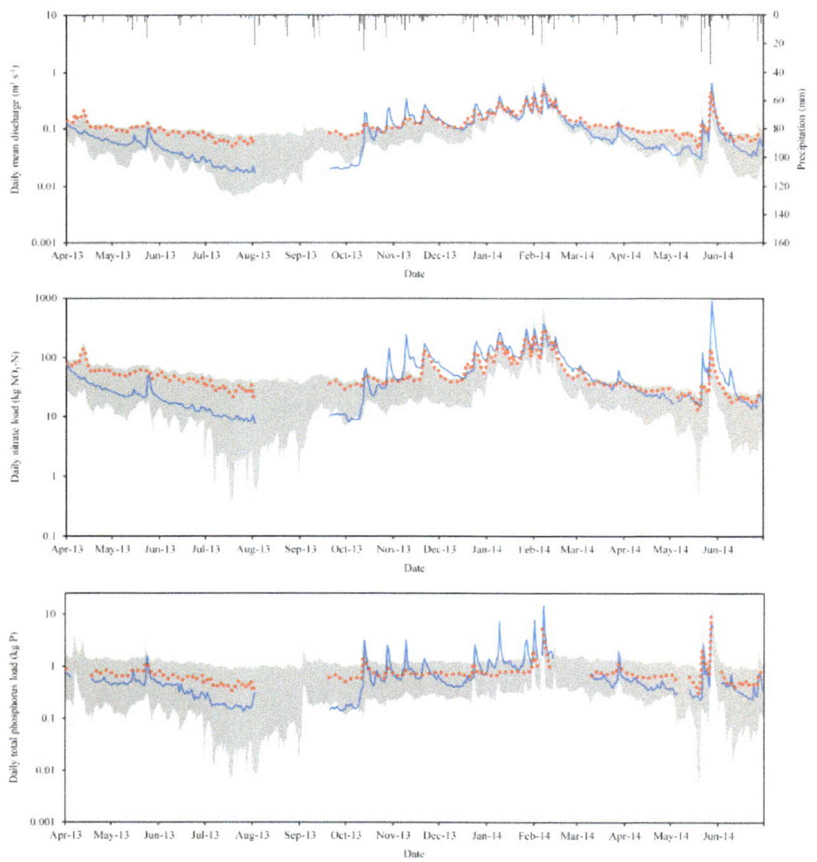

Figure 4: Observed (solid line) and the best simulated (dotted line) daily mean discharge, nitrate and total phosphorus loads

recorded at the outlet of the Blackwater sub-catchment during the validation time period (1 April 2013–30 June 2014). The 95% confidence interval is represented by the hatched area and the daily rainfall amount recorded at Heydon weather station is plotted in the top panel for reference.

Table 5: The statistical performance of the model in simulating mean discharge, nitrate and total phosphorus loads at monthly and daily time-steps at the outlet of the Blackwater sub-catchment during the calibration (1 December 2011–31 March 2013) and validation (1 April 2013–30 June 2014) periods, respectively. NSE is the Nash-Sutcliffe Efficiency coefficient, PBIAS is percentage bias and RSR is the ratio of the root mean square error to the standard deviation of the measured data. The numbers enclosed in brackets are benchmark values suggested by Moriasi et al. (2007).

Variable	NSE	PBIAS (%)	RSR
Daily time-step:			
Calibration:			
Flow	0.77	−6.0	0.48
Nitrate	0.72	5.6	0.53
Total Phosphorus	0.44	0.8	0.75
Validation:			
Flow	0.68	−24.8	0.57
Nitrate	0.46	4.2	0.74
Total Phosphorus	0.36	−2.9	0.80
Monthly time-step:			
Calibration:			
Flow	0.95 (>0.5)	−5.9 (±25)	0.23 (<0.7)
Nitrate	0.86 (>0.5)	5.6 (±70)	0.37 (<0.7)
Total Phosphorus	0.63 (>0.5)	0.8 (±70)	0.61 (<0.7)
Validation:			
Flow	0.92	−15.6	0.28
Nitrate	0.81	−4.7	0.43
Total Phosphorus	0.60	8.5	0.64

Nitrate Simulation

The model performance in simulating daily nitrate loads during the calibration and validation time periods is shown in Fig. 3 and Fig. 4, respectively. When evaluated at a daily time-step, the model achieved NSE, PBIAS and RSR values of 0.72, 5.6% and 0.53, respectively, during the calibration period and values of 0.46, 4.2% and 0.74, respectively, during the validation period (Table 5). The 95% prediction uncertainty range bracketed 76% and 72% of observed nitrate load data during calibration and validation periods, respectively, indicating that the model achieved a relatively good fit between predictions and observations overall. According to the criteria set out in Moriasi et al. (2007), the model performs very well in simulating nitrate loads during the calibration and validation periods if evaluated at a monthly time-step (see Table 5). When evaluated at a daily time-step however, there is a notable decline in model performance during the validation period.

A visual inspection of Fig. 4 indicates that the model generally performs well in simulating nitrate loads during the validation period however there is an observed tendency to underestimate some peaks in nitrate loads. Although the model tends to overestimate discharge in general, it failed to reproduce a number of peaks in discharge (e.g. during March 2012, June–August 2012 and October–December 2013) which appears to translate into an underestimation of nitrate loads. Four factors that may contribute to this deficiency are: (i) rating curve uncertainty under high-flow conditions due to a limited number of flow gauging observations recorded during storm events (McMillan et al., 2010); (ii) difficulties in modelling responses to extreme conditions (Zhang et al., 2014); (iii) difficulties in modelling antecedent conditions within a catchment (Yatheendradas et al., 2008); and (iv) incorrect timing of management practices (e.g. fertiliser application and tillage).

The model also greatly underestimates the mass of nitrate exported from the sub-catchment in response to 35 mm of rainfall recorded at Heydon weather station on 27 May 2014. This is the largest amount of precipitation to have occurred within the sub-catchment on any single day since 2008. During the three consecutive days following this event, nitrate loads observed at the sub-catchment outlet were over 7, 5 and 4 times the mass predicted by the best simulation respectively. It is possible that the response observed within the sub-catchment may result from an incidental loss of nitrate from a farm or from the connection of a previously unconnected nitrate source or so-called legacy stores (Outram et al., 2016) within the system. Such occurrences are difficult to account for within SWAT. If model performance in simulating nitrate loads at a daily time-step during the validation period is evaluated with these three outliers removed, NSE, PBIAS and RSR values of 0.68, −1.43% and 0.56 are achieved, respectively.

According to the criteria set out by Moriasi et al. (2007), the model can be considered to perform very well in simulating nitrate loads at a monthly time-step during the calibration and validation periods (see Table 5). Moriasi et al. (2007) recommend that, in general, the model performance criteria should be less strict when considering a shorter time-step. For the purposes of this investigation, the model is therefore considered to perform adequately in simulating nitrate loads at daily and monthly time-steps.

Total Phosphorus Simulation

The model performance in simulating daily total phosphorus loads during the calibration and validation time periods can be observed in Fig. 3 and Fig. 4, respectively. A visual inspection indicates that the model generally performs well in simulating total phosphorus loads in base flow, however it fails to reproduce a number of peak events during the calibration and validation periods.

The sediment transport component of the SWAT model was not calibrated within this investigation because sediment observations were not available at daily or sub-daily resolutions. 467 stream water samples were, however, collected at the outlet of the Blackwater sub-catchment from October 2010 to March 2015 as part of the Wensum DTC Project and were used to develop a log-log regression model to test the hypothesis that there is a significant relationship between the concentration of total suspended solids and the concentration of total phosphorus (Fig. 5). A linear regression *t*-test found that this relationship has a P-value of >0.001 and is statistically significant. Because of the significance of this relationship and the sensitivity of total phosphorus losses to the transport of sediment during storm events, the lack of high-resolution data means that sediment losses may not be adequately simulated by the model. This observation may account for the apparent deficiency of the model in simulating total phosphorus loads during storm events. Other explanations which may account for the poor performance of the model in reproducing peak total phosphorus events are that: (i) the general representation of fertilizer practice within the model is not sufficiently accurate for total phosphorus at a daily resolution; and (ii) the accumulation of sediment and sediment-associated nutrients within complex tile drainage networks and their subsequent removal during storm events is difficult to reproduce within a generalized model. For example,Kronvang et al. (1997) investigated the transport of sediment and phosphorus in an arable catchment in Denmark and found that the majority of losses occurred during storm events, with subsurface drainage found to be an important pathway.

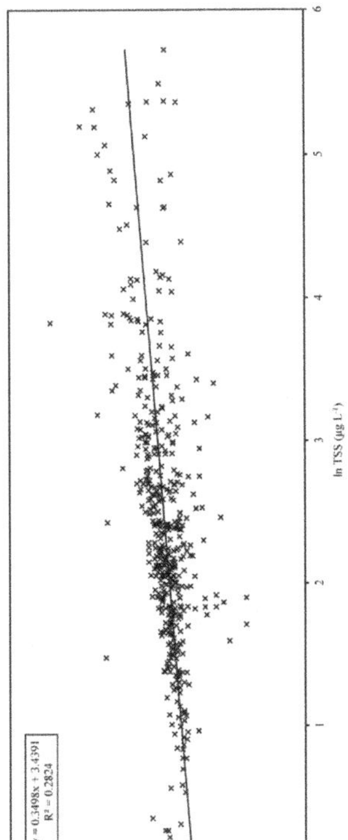

Figure 5: Log-log regression model of the relationship between the concentration of total suspended solids (TSS) and the concentration of total phosphorus (TP) at the outlet of the Blackwater sub-catchment according to stream water samples collected during 1 October 2010–31 March 2015.

Despite the above deficiencies, when evaluated at a daily time-step the model achieved NSE, PBIAS and RSR values of 0.44, 0.8% and 0.75, respectively, during the calibration period and values of 0.36, −2.9% and 0.80, respectively, during the validation period (Table 5). The 95% prediction uncertainty range bracketed 85% and 92% of observed total phosphorus load data during calibration and validation periods, respectively, indicating that the model achieved a relatively good fit between predictions and observations overall. Although the model does not achieve the satisfactory performance criteria suggested by Moriasi et al. (2007) when simulating total phosphorus loads at a daily time-step, the small percentage bias values achieved during the calibration and validation time periods indicate that the model simulates overall total phosphorus loads with reasonable accuracy (Table 5). When evaluated at a monthly time-step, the model performance in simulating total phosphorus loads does achieve the satisfactory performance criteria (Table 5). The priority of this investigation is to achieve good model performance in simulating losses of total phosphorus over the long-term. Given the good performance in this respect, for the purposes of this investigation it is therefore considered that the model performs adequately in simulating total phosphorus loads at both daily and monthly time-steps.

Agricultural Mitigation Options

The satisfactory performance of the model in simulating discharge and nitrate and total phosphorus loads suggests that the model can be applied with high confidence to assess the impacts of agricultural mitigation options on water quality within the Blackwater sub-catchment.

Mitigation Scenario Impacts

Buffer strip scenarios S1 and S2 achieved small reductions in the amount of nitrate lost from the sub-catchment relative to the control scenario (S0) (Fig. 6a). Scenarios S1 and S2 reduced mean annual nitrate losses by 2.3% and 4.6%, respectively, for buffer strips of 2 m and 6 m width. A reduction in the total area of land utilized for agricultural purposes and the reduction in the total amount of fertilizer applied to land within the sub-catchment that results is most likely to be responsible for the reduction in nitrate losses observed under these scenarios. A proportion of the simulated reductions are also likely to result from a reduction in the amount of nitrate lost in surface runoff due to wider buffer strips. In comparison, Glavan et al. (2012) found that introducing buffer strips of 4 m width to arable land and grassland within SWAT reduced losses of total nitrogen by 21.2% and attributed this reduction largely to a drop in the amount of total nitrogen lost in surface runoff. In another study, Lam et al. (2011) found that introducing buffer strips of 10 m width to arable land and pasture land along the main river channel reduced total nitrogen losses by 12.9% and attributed this reduction largely to denitrification within groundwater in the locality of the vegetative buffer. Scenarios S1 and S2 achieved notable reductions in the amount of total phosphorus lost from the sub-catchment relative to the control scenario (S0) (Fig. 6b). Scenarios S1 and S2 reduced mean annual total phosphorus losses by 12.2% and 16.9%, respectively, reflecting an increase in the width of buffer strips from 2 m to 6 m. Increasing the width of buffer strips acts to slow surface runoff, causing more sediment-associated phosphorus to drop out before the runoff enters a stream. In comparison, Glavan et al. (2012) found that introducing buffer strips of 4 m width to arable land and grassland within SWAT reduced losses of total phosphorus by 47.7% and Lam et al. (2011) found that introducing buffer strips of 10 m width to arable land and pastureland along the main river channel reduced total phosphorus losses by 5.3%. Again, it is considered that the effectiveness of buffer strips is dependent on local factors. As evidenced by our study and the findings of others, including Cho et al. (2010), it is clear that the effectiveness of buffer strips varies, depending on local conditions, the width of the buffer strip and the extent of the area to which they are applied. For mean annual losses, the 95% prediction uncertainty range within which 95% of the 1000 model predictions fell, ranged from 2.5 kg NO_3-N ha^{-1} yr^{-1} to 11.5 kg NO_3-N ha^{-1} yr^{-1} and 0.06 kg P ha^{-1} yr^{-1} to 0.28 kg P ha^{-1} yr^{-1} under scenario S1, and from 2.4 kg NO_3-N ha^{-1} yr^{-1} to 11.4 kg NO_3-N ha^{-1} yr^{-1} and 0.05 kg P ha^{-1} yr^{-1} to 0.26 kg P ha^{-1} yr^{-1} under scenario S2 (Fig. 6). Relative to control scenario S0, the lower and upper bounds of the 95% prediction uncertainty range respectively reduced by 5.6% and 2.4% for nitrate and 13.8% and 13% for total phosphorus under scenario S1 and reduced by 7.7% and 3.3% for nitrate and 18.8% and 17.4% for total phosphorus under scenario S2. Although there is some uncertainty associated with model predictions under scenarios S1 and S2, the results indicate a clear reduction in the amount of nitrate and total phosphorus lost from the sub-catchment. This result suggests that buffer strips can be introduced to reduce nitrate and total phosphorus losses over the long-term.

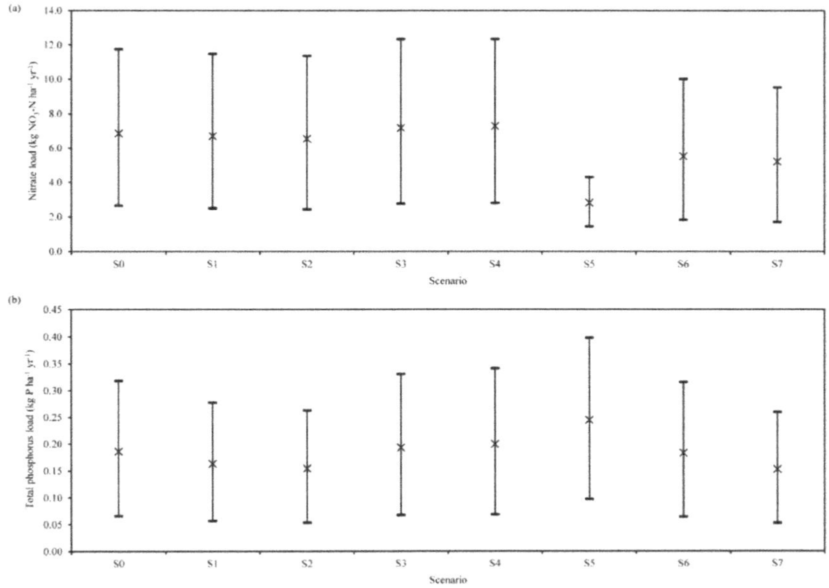

Figure 6: (a) The mean annual nitrate load and (b) the mean annual total phosphorus load exported from the Blackwater sub-catchment during the period 1990–2009 under each mitigation scenario. The upper and lower bounds of the 95% prediction uncertainty range are also shown at the end of each line. The '×' represents the mean value of each scenario.

Alternative tillage scenarios S3 and S4 resulted in small increases in the amount of nitrate and total phosphorus lost from the sub-catchment relative to the control scenario (S0) (Fig. 6). Nitrate losses under scenarios S3 and S4 increased by 4.7% and 6.3%, respectively, and total phosphorus losses increased by 3.8% and 7.2%, respectively. The 95% prediction uncertainty range of mean annual losses ranged from 2.8 kg NO_3-N ha^{-1} yr^{-1} to 12.3 kg NO_3-N ha^{-1} yr^{-1} and 0.07 kg P ha^{-1} yr^{-1} to 0.33 kg P ha^{-1} yr^{-1}under scenario S3, and from 2.8 kg NO_3-N ha^{-1} yr^{-1} to 12.3 kg NO_3-N ha^{-1} yr^{-1} and 0.07 kg P ha^{-1} yr^{-1} to 0.34 kg P ha^{-1} yr^{-1} under scenario S4. Relative to control scenario S0, the lower and upper bounds of the 95% prediction uncertainty range respectively increased by 5.1% and 5% for nitrate and 2.9% and 3.8% for total phosphorus under scenario S3 and increased by 6.2% and 5.0% for nitrate and 4.2% and 7.1% for total phosphorus under scenario S4. Although the 95% uncertainty ranges for losses of nitrate and total phosphorus under scenarios S3 and S4 appear to be relatively large, the upper and lower limits of those ranges depict a small but clear increase in the amount of nitrate and total phosphorus lost from the sub-catchment when alternative tillage practices are introduced. The increase in nitrate and total phosphorus losses was an unexpected result given that alternative tillage systems including conservation tillage and zero tillage have been reported to reduce sediment erosion and losses of total phosphorus and nitrogen (McDowell and McGregor, 1984 and Ulén et al., 2010). Lam et al. (2011)however found that introducing alternative tillage practices within SWAT, including zero-tillage and conservation tillage, did not have a significant impact on total nitrogen and total phosphorus losses and attributed this observation to limited surface runoff and sediment erosion within the catchment (Lam et al., 2010). A number of studies have also reported an increase in the amount of dissolved phosphorus and nitrogen lost from arable fields where reduced tillage systems are implemented for successive years (McDowell and McGregor, 1984 and Ulén et al., 2010). Where plant residues are left undisturbed, the incorporation of fertilizers within soils becomes limited (Ulén et al., 2010) and nutrients accumulate in topsoil (Logan et al., 1991). This practice has the potential to increase the amount of nutrients lost in surface runoff (McDowell and McGregor, 1984 and Ulén et al., 2010) and may account for the small increases in nitrate and total phosphorus losses observed under scenarios S3 and S4. Periodically conducting conventional tillage within a long-term reduced tillage system is recommended by Addiscott and Thomas (2000) in order to redistribute nutrients within the soil subsurface and mitigate this risk.

Scenario S5 involved removing tile drains from the sub-catchment. This measure may not be considered practical or desirable but it is necessary to identify the important pathways of nutrient loss within the sub-catchment. Scenario S5 reduced nitrate losses by 58.9% and increased total phosphorus losses by 31.6%, relative to the control scenario (S0) (Fig. 6). The 95% prediction uncertainty ranges for mean annual losses ranged from 1.4 kg NO_3-N ha^{-1} yr^{-1} to 4.3 kg NO_3-N ha^{-1} yr^{-1} and 0.1 kg P ha^{-1} yr^{-1} to 0.4 kg P ha^{-1} yr^{-1} under scenario S5. Relative to control scenario S0, the lower and upper bounds of the 95% prediction uncertainty range respectively reduced by 45.5% and 63.5% for nitrate and increased by 47.5% and 25.1% for total phosphorus under scenario S5. The result for nitrate indicates that subsurface drainage is a major conduit for nitrate losses from arable land to the river network within the sub-catchment. The large increase in total phosphorus losses results from an increase in surface runoff and soil erosion due to reduced subsurface drainage, and highlights the need to maintain good drainage within arable systems. The 95% confidence interval of the predicted impacts of scenario S5 on nitrate losses within the sub-catchment is also markedly smaller compared to all other scenarios, indicating a higher confidence in model predictions.

Introducing a red clover cover crop to the crop rotation scheme applied within the sub-catchment under scenario S6 reduced nitrate and total phosphorus losses by 19.6% and 1.6%, respectively (Fig. 6). Under scenario S6 the 95% prediction uncertainty range of mean annual losses ranged from 1.8 kg NO_3-N ha^{-1} yr^{-1} to 10.0 kg NO_3-N ha^{-1} yr^{-1} and 0.06 kg P ha^{-1} yr^{-1} to 0.32 kg P ha^{-1} yr^{-1} and, relative to control scenario S0, the lower and upper bounds of the 95% prediction uncertainty range respectively reduced by 30.4% and 14.8% for nitrate and 2.7% and 0.9% for total phosphorus. In comparison,Ullrich and Volk (2009) found that introducing red clover as a cover crop within a SWAT model of the Parthe catchment in central Germany reduced nitrate losses in surface runoff by 63%, relative to a control scenario which involved conservation tillage alone. The large reduction in nitrate loss observed by our study is likely to result from the uptake of nitrate from soils by the cover crop, locking nitrate within organic plant material and preventing it from leaching from soils during wet winter months (Rubæk et al., 2011). The presence of a crop at a time of year when soils would otherwise be bare protects the soil surface and reduces the amount of nutrients lost through wind erosion and surface runoff. The root system of the cover crop also enhances the percolation of water into the soil subsurface, reducing surface runoff and erosion, further reducing nutrient losses. Following the termination of a cover crop, nutrients stored in organic plant material are slowly released to soils through the process of mineralization. The red clover essentially acts as a 'green manure'. The reduction in nitrate losses observed under this scenario and the slow release of nutrients ensure that less nitrogen fertilizer needs to be applied to fields, reducing fertilizer expenditure and improving soil conditions. The magnitude of the reduction in total phosphorus losses is markedly less than that observed for nitrate due to the fact that the uptake of phosphorus by plants is counteracted by the slow desorption of phosphorus from soil particles. This observation limits the potential for cover crops to reduce phosphorus losses, however it is possible to reduce losses of phosphorus through long-term phosphorus mining (Delorme et al., 2000). Mining involves the net removal of nutrients through the harvesting of cover crops, instead of incorporating the organic material of cover crops into soils as a green manure.

Although there is clear uncertainty associated with model predictions for nitrate and total phosphorus losses under each scenario (Fig. 6), the results indicate a clear, if sometimes relatively small, direction of change under each scenario. We can therefore be confident in the impacts of each mitigation option for the management of diffuse pollution, despite the degree of uncertainty that is associated with predictions.

In order to assess which mitigation options have the potential to be applied within the sub-catchment to achieve statutory water quality targets, percent exceedance curves depicting the amount of time any nitrate and total phosphorus concentration is exceeded at the sub-catchment outlet during the period from 1990 to 2009 were developed for each scenario (Fig. 7a and b). With reference to the European Drinking Water Directive, in which water is considered unfit for human consumption if it contains a nitrate concentration above 50 mg L^{-1} (equivalent to 11.3 mg NO_3-N L^{-1}), then under the control scenario (S0), the 50 mg L^{-1} water quality standard is exceeded 0.82% of the time at the sub-catchment outlet, equivalent to 60 days during the period 1990–2009 (Fig. 7a). This risk is reduced to 0.01% of the time or 1 day under scenario S5 in which tile drains are removed from the sub-catchment. Introducing a red clover cover crop to the crop rotation scheme under scenario S6 reduced the amount of time this standard was exceeded to 0.36%, equivalent to 26 days over the 20-year period 1990–2009. Under this scenario, the amount of time that the 50 mg L^{-1} standard was exceeded at the sub-catchment outlet was reduced by over 50% compared to the control scenario, benefiting aquatic ecology and water resource management. Scenarios S1-S4 had a more limited effect on the percent exceedance curves relative to the control scenario (S0) (Fig. 7a). The Diffuse Water Pollution Plan developed for the River Wensum SSSI specifies that for the river to be in a favourable condition, mean annual total phosphorus concentrations must not exceed 0.1 mg L^{-1} at the catchment outlet (Environment Agency, 2010). Under the control scenario (S0), the 0.1 mg L^{-1} target was exceeded 53% of the time at the sub-catchment outlet (Fig. 7b), with the mean annual total phosphorus concentration just below the target at 0.097 mg L^{-1}. This exceedance reduced to 51% and 49% of the time under scenarios S1 and S2, respectively, with 2 m and 6 m wide buffer strips (Fig. 7b). Under scenarios S1 and S2, mean annual total phosphorus concentrations at the sub-catchment outlet were 0.092 mg L^{-1} and 0.091 mg L^{-1}, respectively. Scenario S5, involving the removal of tile drains from arable land, increased the amount of time this target was exceeded to 72% (Fig. 7b). Under this scenario, the mean annual concentration of total phosphorus at the sub-catchment outlet equalled 0.111 mg L^{-1}, exceeding the required target. Scenarios S3, S4 and S6 had a more limited effect on the percent exceedance curves relative to the control scenario (S0) (Fig. 7b). It is clear from the scenarios considered that buffer strips represent the most effective mitigation option that can be applied within an arable catchment to reduce losses of total phosphorus.

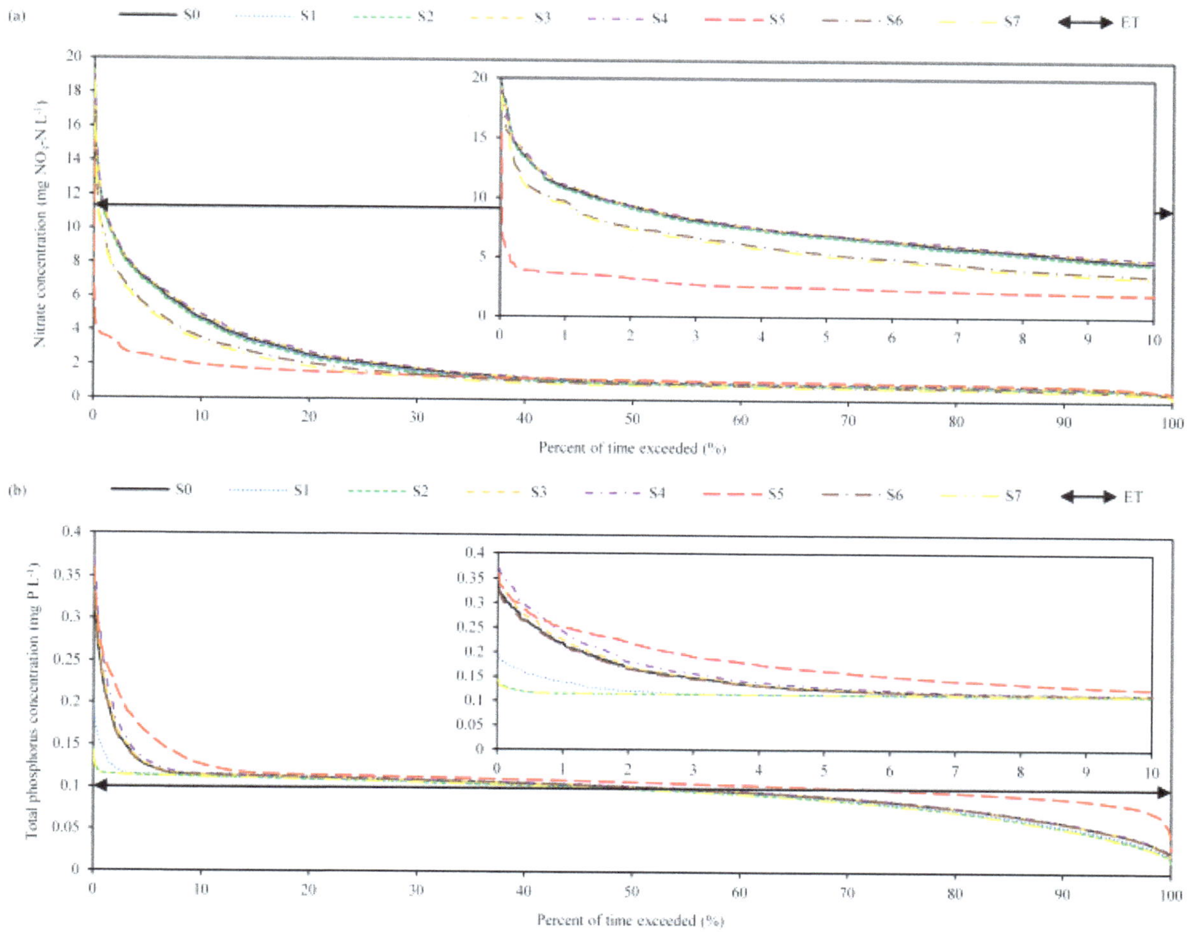

Figure 7: Environmental Targets (ET) and percent exceedance curves for (a) nitrate concentration and (b) total phosphorus concentration as simulated at the outlet of the Blackwater sub-catchment during the period 1990–2009 under each mitigation scenario.

Combined Effectiveness of Mitigation Options

According to the model simulations, the most effective and practical mitigation options considered as part of this investigation in the Blackwater sub-catchment to reduce losses of nitrate and total phosphorus include, respectively, the introduction of a red clover cover crop to the crop-rotation applied within the sub-catchment (scenario S6) and the introduction of buffer strips of 6 m width to areas of arable land (scenario S2). In order to understand the impacts of mitigation options on long-term water quality when introduced to the sub-catchment in combination, these two mitigation options were modelled in combination under scenario S7.

The two mitigation options introduced under scenario S7 reduced nitrate and total phosphorus losses within the sub-catchment by 24.1% and 17.9%, respectively, over the period 1990–2009 (Fig. 6). In comparison, the cumulative impact of these mitigation options, when modelled individually and added together, reduced nitrate and total phosphorus losses over the same period by 24.2% and 18.6%, respectively. This result suggests that the mitigation options considered here simply combine to produce a total effect almost equal the sum of their individual effects. Under scenario S7 the 95% prediction uncertainty range of mean annual losses ranged from 1.7 kg NO_3-N ha^{-1} $yr^{-1}$to 9.5 kg NO_3-N ha^{-1} yr^{-1} and 0.05 kg P ha^{-1} yr^{-1} to 0.26 kg P ha^{-1} yr^{-1} and, relative to control scenario S0, the lower and upper bounds of the 95% prediction uncertainty range respectively reduced by 35.8% and 19% for nitrate and 19.9% and 18.5% for total phosphorus.

The 50 mg L^{-1} drinking water quality standard that applies to nitrate was exceeded 0.34% of the time at the outlet of the Blackwater sub-catchment under scenario S7 (Fig. 7a), equivalent to 25 days during the 1990–2009 period. This result compares to 0.82% of the time or 60 days under the control scenario S0, 0.75% of the time or 55 days under scenario S2 and 0.36% of the time or 26 days under scenario S6. The 0.1 mg L^{-1}water quality target that applies to total phosphorus was exceeded 48.5% of the time at the outlet of the Blackwater sub-catchment during the 1990–2009 period under scenario S7 (Fig. 7b). This result compares to 53.2% of the time under the control scenario S0, 48.6% of the time under scenario S2 and 53.8% of the time under scenario S6. These results further suggest that the combined effect of the mitigation options considered here is nearly equal to the sum of their individual impacts on water quality. Despite this finding, in practice, when choosing mitigation options, it is essential to consider their many potential impacts before introduction in the environment in order to understand the risk of pollution swapping and the potential for unintended environmental consequences (Stevens and Quinton, 2009).

CONCLUSIONS

Water quality models are cost-effective DSTs which can be applied to assess the quantitative impacts of a variety of mitigation options on water quality. Models must be robustly calibrated to achieve this goal, but there is often a scarcity of sufficient data to parameterize and evaluate models. High-frequency water quality monitoring has allowed the successful application of SWAT within this investigation to quantify the impacts of agricultural mitigation options on long-term water quality at a daily resolution in a lowland arable catchment in the UK. The uncertainties of the predicted impacts of each mitigation option on water quality have also been quantified and mitigation options that have the potential to be applied within arable catchments to improve water quality have been identified.

Scenario analysis found that introducing a red clover cover crop to the crop rotation scheme applied within the model reduced nitrate losses by 19.6% and total phosphorus losses by 1.6% over the long-term. This finding suggests that a cover crop can successfully be grown as a 'green manure', improving soil conditions, reducing expenditure on fertilizers and reducing agricultural diffuse water pollution over the long term. The prospect of mining phosphorus through the successive harvesting of cover crops is also considered, but this practice limits the potential for the cover crop to act as a green manure.

Introducing buffer strips of 2 m and 6 m width to arable land was found to be the most effective mitigation options that could be applied to reduce losses of total phosphorus, achieving reductions of 12.2% and 16.9%, respectively, although consideration must be given to the reduction in agricultural productivity that occurs under these scenarios as a result of removing areas of arable land from cultivation.

According to the findings of this investigation, the removal of subsurface tile drainage systems from areas of arable land, albeit not practical in terms of maintaining arable cultivation, represents the single most effective mitigation option that can be adopted to reduce losses of nitrate, achieving a reduction of 58.9%. This measure, however, increased total phosphorus losses by 31.6%, highlighting the need to consider multiple pollutants when evaluating the effectiveness of mitigation options to reduce agricultural diffuse water pollution.

If reductions are to be achieved in both nitrate and total phosphorus losses, the most effective combination of mitigation options that can be applied are a cover crop and buffer strips. When modelled in combination, these two mitigation options were found to have a total impact which was almost equal to the sum of their individual modelled impacts on water quality.

The alternative tillage scenarios applied within the model unexpectedly resulted in small increases in nitrate and total phosphorus losses. This result was attributed to the enrichment of nutrients within topsoil and an increased loss of nutrients in surface runoff. This observation highlights the need to conduct a detailed assessment of the potential impacts of a mitigation option prior to implementation otherwise there is a risk of introducing practices which achieve the opposite of the intended result. This example highlights the benefits provided by water quality models in aiding decision-making and catchment management. The availability of high-frequency water quality data ensures that models can be robustly calibrated. Such techniques can impart a higher degree of confidence to model predictions and, therefore, in the predicted impacts of mitigation options on water quality. This investigation has shown that high-frequency water quality datasets can be applied within SWAT, as an example of one of the many water quality models available, to quantify the long-term impacts of agricultural mitigation options on water quality at a daily resolution and assist in the creation of more effective and reliable DSTs, leading to the development of appropriate diffuse water pollution mitigation plans. Results indicate that there is a relatively large degree of uncertainty associated with model predictions and we would recommend that impact assessments conduct a robust evaluation of prediction uncertainty to improve confidence in model predictions.

ACKNOWLEDGEMENTS

The authors would like to express our sincere thanks to the anonymous reviewers whose insightful comments helped to improve this article. This work was carried out as part of the Wensum Demonstration Test Catchment Project funded by the Department for Environment, Food and Rural Affairs (WQ0212). The authors acknowledge the provision of MIDAS Land and Marine Surface Station Data by the Met Office. S.D.T. acknowledges financial support from an Engineering and Physical Sciences Research Council doctoral studentship [grant number: EP/K503022/1]. The underlying research materials for this study can be accessed by contacting the corresponding author.

REFERENCES

1. Abbaspour, K.C., 2014. SWAT-CUP 2012: SWAT Calibration and Uncertainty Programs - a User Manual. Dübendorf: Eawag and Swiss Federal Institute of Aqutic Science and Technology.
2. Abbaspour, K.C., Johnson, C.A., Van Genuchten, M.T., 2004. Estimating uncertain flow and transport parameters using a sequential uncertainty fitting procedure. Vadose Zone J. 3 (4), 1340e1352.
3. Abbaspour, K.C., Yang, J., Maximov, I., Siber, R., Bogner, K., Mieleitner, J., Zobrist, J., Srinivasan, R., 2007. Modelling hydrology and water quality in the pre-alpine/ alpine Thur watershed using SWAT. J. Hydrol. 333 (2), 413e430.
4. ADAS, 2015. Farmscoper Version 3: Description. ADAS UK Ltd, Wolverhampton. Addiscott, T.M., Thomas, D., 2000. Tillage, mineralization and leaching: phosphate. Soil Tillage Res. 53 (3), 255e273.
5. Aouissi, J., Benabdallah, S., Chabaane, Z.L., Cudennec, C., 2014. Modeling water ^ quality to improve agricultural practices and land management in a Tunisian catchment using the soil and water assessment tool. J. Environ. Qual. 43 (1), 18e25.
6. Arnold, J.G., Allen, P.M., Volk, M., Williams, J.R., Bosch, D.D., 2010. Assessment of different representations of spatial variability on SWAT model performance. Trans. ASABE 53 (5), 1433e1443.
7. Arnold, J.G., Moriasi, D.N., Gassman, P.W., Abbaspour, K.C., White, M.J., Srinivasan, R., Santhi, C., Harmel, R.D., van Griensven, A., Van Liew, M.W., Kannan, N.,

Jha, M.K., 2012. SWAT: model use, calibration, and validation. Trans. ASABE 55 (4), 1491e1508.

8. Arnold, J.G., Kiniry, J.R., Srinivasan, R., Williams, J.R., Haney, E.B., Neitsch, S.L., 2014. Soil and Water Assessment Tool Input/Output File Documentation, Version 2012. Texas Water Resources Institute, Temple.
9. Barlund, I., Kirkkala, T., Malve, O., K € am€ ari, J., 2007. Assessing SWAT model perfor- € mance in the evaluation of management actions for the implementation of the Water Framework Directive in a Finnish catchment. Environ. Model. Softw. 22 (5), 719e724.
10. Beven, K., 1993. Prophecy, reality and uncertainty in distributed hydrological modelling. Adv. Water Resour. 16 (1), 41e51.
11. Beven, K., Freer, J., 2001. Equifinality, data assimilation, and uncertainty estimation in mechanistic modelling of complex environmental systems using the GLUE methodology. J. Hydrol. 249 (1), 11e29.
12. Biodiversity Reporting and Information Group, 2007. Report on the Species and Habitat Review. Department for Environment, Food and Rural Affairs, London.
13. Boithias, L., Srinivasan, R., Sauvage, S., Macary, F., S anchez-Perez, J.M., 2014. Daily nitrate losses: implication on long-term river quality in an intensive agricultural catchment of southwestern France. J. Environ. Qual. 43 (1), 46e54.
14. Bonuma, N.B., Rossi, C.G., Arnold, J.G., Reichert, J.M., Minella, J.P., Allen, P.M., Volk, M., 2014. Simulating landscape sediment transport capacity by using a modified SWAT model. J. Environ. Qual. 43 (1), 55e66.
15. Bosch, D.D., Arnold, J.G., Volk, M., Allen, P.M., 2010. Simulation of a low-gradient coastal plain watershed using the SWAT s. Trans. ASABE 53 (5), 1445e1456.
16. Bouraoui, F., Grizzetti, B., 2014. Modelling mitigation options to reduce diffuse nitrogen water pollution from agriculture. Sci. Total Environ. 468, 1267e1277.
17. British Sugar, 2014. Growing Sugar Beet [online] Available at: http://www. britishsugar.co.uk/Files/Education-Resources/Growing-Sugar-beet/growing_sugar_beet.aspx (Accessed 10.04.14.).
18. Carpenter, S.R., Caraco, N.F., Correll, D.L., Howarth, R.W., Sharpley, A.N., Smith, V.H., 1998. Nonpoint pollution of surface waters with phosphorus and nitrogen. Ecol. Appl. 8 (3), 559e568.
19. Carr, G.M., Neary, J.P., 2008. Water Quality for Ecosystem and Human Health. United Nations Environment Program Global Environment Monitoring System, Burlington.
20. Cho, J., Lowrance, R.R., Bosch, D.D., Strickland, T.C., Her, Y., Vellidis, G., 2010. Effect of watershed subdivision and filter width on SWAT simulation of a coastal plain watershed. J. Am. Water Resour. Assoc. 46 (3), 586e602.
21. Collins, A.L., McGonigle, D.F., 2008. Monitoring and modelling diffuse pollution from agriculture for policy support: UK and European experience. Environ. Sci. Policy 11 (2), 97e101.
22. Council of the European Union, 1991. Council Directive 91/676/EEC of 12 December 1991 concerning the protection of waters against pollution caused by nitrates from agricultural sources. Off. J. Eur. Commun. L 375, 1e8.
23. Council of the European Union, 1998. Council Directive 98/83/EC of 3 November 1998 on the quality of water intended for human consumption. Off. J. Eur. Commun. OJ L 330, 32e54.
24. Council of the European Union, 2000. Directive 2000/60/EC of the European Parliament and of the Council of 23 October 2000 establishing a framework for Community action in the field of water policy. Off. J. Eur. Commun. L 327, 1e73.
25. Cranfield University, 2014a. National Soil Map of England and Wales e NATMAP. Cranfield University [online] Available at: http://www.landis.org.uk/data/natmap.cfm (Accessed 27.05.14.).
26. Cranfield University, 2014b. The Soils Guide. Cranfield University [online] Available at: www.landis.org.uk (Accessed 27.05.14.).
27. Cranfield University, 2014c. Soil Series. Cranfield University [online] Available at: http://www.landis.org.uk/data/series.cfm (Accessed 27.05.14.).
28. Cranfield University, 2014d. National Soils Inventory. Cranfield University [online] Available at: http://www.landis.org.uk/data/nsi.cfm (Accessed 27.05.14.).
29. Delorme, T.A., Angle, J.S., Coale, F.J., Chaney, R.L., 2000. Phytoremediation of phosphorus-enriched soils. Int. J. Phytoremediation 2 (2), 173e181.
30. Department for Environment Food and Rural Affairs (Defra), 2015. The Guide to Cross Compliance in England 2015 (GCCE 2015 V1). Good Agricultural and Environmental Condition (GAEC) 1: Establishment of Buffer Strips along Watercourses [online] Available at: http://adlib.everysite.co.uk/adlib/defra/ content.aspx?id¼1QQUSGMWSS.0INCEX4CZUO4JK (Accessed 24.04.15.).
31. Department for Environment Food and Rural Affairs (Defra), 2016. Structure of the Agricultural Industry [online] Available at: https://www.gov.uk/government/collections/structure-of-the-agricultural-industry (Accessed 22.03.16.).
32. Department for Environment Food and Rural Affairs (Defra), 2010a. Fertiliser Manual (RB209), eighth ed. The Stationary Office, Norwich.
33. Department for Environment Food and Rural Affairs (Defra), 2010b. Definitions of Terms Used in Farm Business Management. Department for Environment Food and Rural Affairs, London.
34. Douglas-Mankin, K.R., Srinivasan, R., Arnold, J.G., 2010. Soil and water assessment tool (SWAT) model: current developments and applications. Trans. ASABE 53 (5), 1423e1431.
35. EDINA, 2014. Agcensus [online] Available at: http://edina.ac.uk/agcensus/ (Accessed 27.05.14.).
36. English Nature, 2005. Citation for Special Area of Conservation (SAC): River Wensum. English Nature, Peterborough.

37. Environment Agency, 2009. Water for Life and Livelihoods. River Basin Management Plan Anglian River Basin District. Annex D: Protected Area Objectives. Environment Agency, Bristol.

38. Environment Agency, 2010. River Wensum SSSI and Potter and Scarning Fens SSSI Diffuse Water Pollution Plan. Environment Agency, Bristol.

39. European Environment Agency, 2005. Source Apportionment of Nitrogen and Phosphorus Inputs into the Aquatic Environment. European Environment Agency, Copenhagen.

40. European Environment Agency, 2012. European Waters e Assessment of Status and Pressures. Publications Office of the European Union, Luxembourg.

41. Gassman, P.W., Reyes, M.R., Green, C.H., Arnold, J.G., 2007. The soil and water assessment tool: historical development, applications, and future research directions. Trans. ASABE 50 (4), 1211e1250.

42. Gerrits, L., Edelenbos, J., 2004. Management of sediments through stakeholder involvement. J. Soils Sediments 4 (4), 239e246.

43. Glavan, M., White, S.M., Holman, I.P., 2012. Water quality targets and maintenance of valued landscape character e experience in the Axe catchment, UK. J. Environ. Manag. 103, 142e153.

44. Gooday, R.D., Anthony, S.G., Chadwick, D.R., Newell-Price, P., Harris, D., Duethmann, D., Fish, R., Collins, A.L., winter, M., 2014. Modelling the costeffectiveness of mitigation methods for multiple pollutants at farm scale. Sci. Total Environ. 468, 1198e1209.

45. Hiscock, K.M., 1993. The influence of pre-Devensian glacial deposits on the hydrogeochemistry of the Chalk aquifer system of north Norfolk, UK. J. Hydrol. 144 (1), 335e369.

46. Hiscock, K.M., Dennis, P.F., Saynor, P.R., Thomas, M.O., 1996. Hydrochemical and stable isotope evidence for the extent and nature of the effective Chalk aquifer of north Norfolk, UK. J. Hydrol. 180 (1), 79e107.

47. Howson, T.J., 2012. Hydrograph Separation Using Stable Isotopes: an Assessment of Flow Contributions in the Blackwater Catchment (M.Sc. thesis). University of East Anglia.

48. Hu, X., McIsaac, G.F., David, M.B., Louwers, C.A.L., 2007. Modeling riverine nitrate export from an East-Central Illinois watershed using SWAT. J. Environ. Qual. 36 (4), 996e1005.

49. Hunt, D.T.E., Dee, A.S., Oakes, D.B., 2004. Updating an Estimate of the Source Apportionment of Nitrogen to Waters in England and Wales. Phase 2 Report to the Department for Environment, Food and Rural Affairs. Department for Environment, Food and Rural Affairs, London.

50. Itermap Technologies, 2007. NEXTMap British Digital Terrain Model Dataset Produced by Intermap. NERC Earth Observation Data Centre [online] Available at: http://catalogue.ceda.ac.uk/uuid/8f6e1598372c058f07b0aeac2442366d (Accessed 19.12.14.).

51. Kronvang, B., Laubel, A., Grant, R., 1997. Suspended sediment and particulate phosphorus transport and delivery pathways in an arable catchment, Gelbaek stream, Denmark. Hydrol. Process. 11 (6), 627e642.

52. Kronvang, B., Behrendt, H., Andersen, H.E., Arheimer, B., Barr, A., Borgvang, S.A., Bouraoui, F., Granlund, K., Grizzetti, B., Groenendijk, P., Schwaiger, E., Hejzlar, J., Hoffmann, L., Johnsson, H., Panagopoulos, Y., Lo Porto, A., Reisser, H., Schoumans, O., Anthony, S., Silgram, M., Venorh, M., Larsen, S.E., 2009. Ensemble modelling of nutrient loads and nutrient load partitioning in 17 European catchments. J. Environ. Monit. 11 (3), 572e583.

53. Krysanova, V., Arnold, J.G., 2008. Advances in ecohydrological modelling with SWAT - a review. Hydrol. Sci. J. 53 (5), 939e947.

54. Lam, Q.D., Schmalz, B., Fohrer, N., 2010. Modelling point and diffuse source pollution of nitrate in a rural lowland catchment using the SWAT model. Agric. Water Manag. 97 (2), 317e325.

55. Lam, Q.D., Schmalz, B., Fohrer, N., 2011. The impact of agricultural best management practices on water quality in a North German lowland catchment. Environ. Monit. Assess. 183 (1e4), 351e379.

56. Logan, T.J., Lal, R., Dick, W.A., 1991. Tillage systems and soil properties in North America. Soil Tillage Res. 20 (2), 241e270.

57. Lovett, A.A., Hiscock, K.M., Outram, F.N., Cooper, R.J., Dugdale, S., Stevenson, J., Sünnenberg, G., Hama-Aziz, Z., Dockerty, T.L., Noble, L., Beamish, J., Hovesen, P., 2015. Experiments with cover crops and cultivation techniques in the Wensum DTC. Aspects Appl. Biol. 129, 85e90.

58. McDowell, L.L., McGregor, K.C., 1984. Plant nutrient losses in runoff from conservation tillage corn. Soil Tillage Res. 4 (1), 79e91.

59. McMillan, H., Freer, J., Pappenberger, F., Krueger, T., Clark, M., 2010. Impacts of uncertain river flow data on rainfall-runoff model calibration and discharge predictions. Hydrol. Process. 24 (10), 1270e1284.

60. Met Office, 2012. Met Office Integrated Data Archive System (MIDAS) Land and Marine Surface Stations Data (1853-current). NCAS British Atmospheric Data Centre [online] Available at: http://catalogue.ceda.ac.uk/uuid/ 220a65615218d5c9cc9e4785a3234bd0 (Accessed 19.12.14.).

61. Met Office, 2014. England 1981e2010 Averages [online] Available at: http://www. metoffice.gov.uk/climate/uk/averages/19812010/areal/england.html (Accessed 27.05.14.).

62. Moriasi, D.N., Arnold, J.G., Van Liew, M.W., Bingner, R.L., Harmel, R.D., Veith, T.L., 2007. Model evaluation guidelines for systematic quantification of accuracy in watershed simulations. Trans. ASABE 50 (3), 885e900.

63. Moriasi, D.N., Steiner, J.L., Arnold, J.G., 2011. Sediment measurement and transport modeling: impact of riparian and filter strip buffers. J. Environ. Qual. 40 (3), 807e814.

64. Morton, D., Rowland, C., Wood, C., Meek, L., Marston, C., Smith, G., Wadsworth, R., Simpson, I.C., 2011. Final

Report for LCM2007 e the New UK Land Cover Map. Centre for Ecology and Hydrology, Wallingford.

65. Nash, J.E., Sutcliffe, J.V., 1970. River flow forecasting through conceptual models part I - a discussion of principles. J. Hydrol. 10 (3), 282e290.
66. Natural England, 1993. River Wensum Site of Special Scientific Interest Citation. 1 Natural England, York.
67. Natural England, 2014. Entry Level Stewardship. Environmental Stewardship Handbook, fourth ed. Natural England, York. Neitsch, S.L., Arnold, J.G., Kiniry, J.R., Williams, J.R., 2011. Soil and Water Assessment Tool Theoretical Documentation, Version 2009. Texas Water Resources Institute, Temple.
68. Outram, F.N., Lloyd, C.E.M., Jonczyk, J., Benskin, C., McW, H., Grant, F., Perks, M.T., Deasy, C., Burke, S.P., Collins, A.L., Freer, J., Haygarth, P.M., Hiscock, K.M., Johnes, P.J., Lovett, A.L., 2014. High-frequency monitoring of nitrogen and phosphorus response in three rural catchments to the end of the 2011e2012 drought in England. Hydrol. Earth Syst. Sci. http://dx.doi.org/10.5194/hess-18- 3429-2014.
69. Outram, F.N., Cooper, R.J., Sünnenberg, G., Hiscock, K.M., Lovett, A.A., 2016. Antecedent conditions, hydrological connectivity and anthropogenic inputs: factors affecting nitrate and phosphorus transfers to agricultural headwater streams. Sci. Total Environ. 545e546, 184e199. http://dx.doi.org/10.1016/j.scitotenv.2015.12.025.
70. Rathjens, H., Oppelt, N., Bosch, D.D., Arnold, J.G., Volk, M., 2015. Development of a grid-based version of the SWAT landscape model. Hydrol. Process. 29 (6), 900e914.
71. Richardson, K., Jørgensen, B.B., 1996. Eutrophication: definition, history and effects. In: Jørgensen, B.B., Richardson, K. (Eds.), Eutrophication in Coastal Marine Ecosystems. American Geophysical Union, Washington DC, pp. 1e19.
72. Rode, M., Klauer, B., Petry, D., Volk, M., Wenk, G., Wagenschein, D., 2008. Integrated nutrient transport modelling with respect to the implementation of the European WFD: the Weisse Elster Case study, Germany. Water SA. 34 (4), 490e496.
73. Rubæk, G.H., Christen, B., Børgesen, C.D., Møller Hansen, E., Heckrath, G.J., Jørgensen, U., van der Salm, C., Schoumans, O.F., Sørensen, P., Thomsen, I.K., 2011. Crop management for mitigation of nutrient losses to the aquatic environment. In: Schoumans, O.F., Chardon, W.J., Bechmann, M., Gascuel-Odoux, C., Hofman, G., Kronvang, B., Litaor, M.I., Lo Porto, A., Newell-Price, P., Rubæk, G.H. (Eds.), Mitigation Options for Reducing Nutrient Emissions from Agriculture. Alterra, Wageningen, pp. 53e62.
74. Salle Farms Ltd, 2014. Crop Rotation [online] Available at: http://www.sallefarms.co. uk/crop-rotation/ (Accessed 27.05.14.).
75. Saloranta, T.M., K€ am€ ari, J., Rekolainen, S., Malve, O., 2003. Benchmark criteria: a tool for selecting appropriate models in the field of water management. Environ. Manag. 32 (3), 322e333.
76. Santhi, C., Srinivasan, R., Arnold, J.G., Williams, J.R., 2006. A modeling approach to evaluate the impacts of water quality management plans implemented in a watershed in Texas. Environ. Model. Softw. 21 (8), 1141e1157.
77. Santhi, C., Kannan, N., White, M., Di Luzio, M., Arnold, J.G., Wang, X., Williams, J.R., 2014. An integrated modeling approach for estimating the water quality benefits of conservation practices at the river basin scale. J. Environ. Qual. 43 (1), 177e198.
78. Solheim, A.L., Austnes, K., Kristensen, P., Peterlin, M., Kodes, V., Collins, R., Semeradov a, S., Künitzer, A., Filippi, R., Prchalova, H., Spiteri, C., Prins, T., 2012. Ecological and Chemical Status and Pressures in European Waters. Thematic Assessment for EEA Water 2012 Report. European Topic Centre on Inland, Coastal and Marine Waters, Prague.
79. Stevens, C.J., Quinton, J.N., 2009. Diffuse pollution swapping in arable agricultural systems. Crit. Rev. Environ. Sci. Technol. 39 (6), 478e520.
80. Texas A&M University, 2015. ArcSWAT [online] Available at: http://swat.tamu.edu/ software/arcswat/ (Accessed 27.03.15.).
81. Tilman, D., Fargione, J., Wolff, B., D'Antonio, C., Dobson, A., Howarth, R.,
82. Schindler, D., Schlesinger, W.H., Simberloff, D., Swackhamer, D., 2001. Forecasting agriculturally driven global environmental change. Science 292 (5515), 281e284.
83. Tuppad, P., Douglas-Mankin, K.R., Lee, T., Srinivasan, R., Arnold, J.G., 2011. Soil and water assessment tool (SWAT) hydrologic/water quality model: extended capability and wider adoption. Trans. ASABE 54 (5), 1677e1684.
84. UK Agriculture, 2014. Crops in the UK [online] Available at: http://www. ukagriculture.com/crops/crops.cfm (Accessed 04.04.14.).
85. Ulen, B., Aronsson, H., Bechmann, M., Krogstad, T., Øygarden, L., Stenberg, M., 2010. Soil tillage methods to control phosphorus loss and potential side-effects: a Scandinavian review. Soil Use Manag. 26 (2), 94e107.
86. Ullrich, A., Volk, M., 2009. Application of the soil and water assessment tool (SWAT) to predict the impact of alternative management practices on water quality and quantity. Agric. Water Manag. 96 (8), 1207e1217.
87. United States Environmental Protection Agency, 2002. Federal Water Pollution Control Act (33 U.S.C. 1251 et seq.). United States Government Publishing Office, Washington, D.C.
88. Van Ast, J.A., 2000. Interactive management of international river basins; experiences in Northern America and western Europe. Phys. Chem. Earth Part B Hydrol. Oceans Atmos. 25 (3), 325e328.
89. Volk, M., Hirschfeld, J., Dehnhardt, A., Schmidt, G., Bohn, C., Liersch, S., Gassman, P.W., 2008. Integrated ecological-economic modelling of water pollution abatement management options in the Upper Ems River Basin. Ecol. Econ. 66 (1), 66e76.

90. Volk, M., Liersch, S., Schmidt, G., 2009. Towards the implementation of the European Water Framework Directive?: Lessons learned from water quality simulations in an agricultural watershed. Land Use Policy 26 (3), 580e588.

91. Vor€ osmarty, C.J., McIntyre, P.B., Gessner, M.O., Dudgeon, D., Prusevich, A., Green, P., € Glidden, S., Bunn, S.E., Sullivan, C.A., Reidy Liermann, C., Davies, P.M., 2010. Global threats to human water security and river biodiversity. Nature 467 (7315), 555e561.

92. Wensum Alliance, 2014. About the River Wensum [online] Available at: http:// www.wensumalliance.org. uk/index.html (Accessed 30.09.14.).

93. White, P.J., Hammond, J.P., 2007. Updating the Estimate of the Sources of Phosphorus in UK Waters. Department for Environment, Food and Rural Affairs, London.

94. Winchell, M., Srinivasan, R., Di Luzio, M., Arnold, J., 2013. ArcSWAT Interface for SWAT2012, User›s Guide. Blackland Research and Extension Center, Temple.

95. Withers, P.J.A., Lord, E.I., 2002. Agricultural nutrient inputs to rivers and groundwaters in the UK: policy, environmental management and research needs. Sci. Total Environ. 282, 9e24.

96. Yatheendradas, S., Wagener, T., Gupta, H., Unkrich, C., Goodrich, D., Schaffner, M., Stewart, A., 2008. Understanding uncertainty in distributed flash flood forecasting for semiarid regions. Water Resour. Res. 44 (5).

97. Zhang, Y., Collins, A.L., Gooday, R.D., 2012. Application of the FARMSCOPER tool for assessing agricultural diffuse pollution mitigation methods across the Hampshire Avon Demonstration Test Catchment, UK. Environ. Sci. Policy 24, 120e131.

98. Zhang, X., Xu, Y.P., Fu, G., 2014. Uncertainties in SWAT extreme flow simulation under climate change. J. Hydrol. 515, 205e222. S.D

CHAPTER 4

DIOXIN IN THE ELBE RIVER BASIN: POLICY AND SCIENCE UNDER THE WATER FRAMEWORK DIRECTIVE 2000–2015 AND TOWARD 2021

Ulrich Förstner, Henner Hollert, Markus Brinkmann, Kathrin Eichbaum, Roland Weber and Wim Salomons

Keywords: Activated carbon, Biota-EQS, Bitterfeld region, Chemical status, Dredged materials, Flood risks, Marine strategy, NGOs, RBC Elbe, Sediment management concept

BACKGROUND

Two prominent objectives of the European Water Framework Directive (WFD 2000 [1]) are the catchment-wide no-deterioration status (article 4) and the reduction of priority pollutants [2]. While the former WFD-principle just entered the public discussion via a ruling of the European Court of Justice from July 1, 2015 on a river deepening project [3], the Actualized River Basin Management Plan of the River Basin Community Elbe from December 22, 2015 [4] states: "without goal-oriented measures for the reduction of primary and secondary pollution sources, the objectives of a good chemical and ecological quality in surface waters until the end of the second WFD management period and a good environmental state according to the European Marine Strategy Framework Directive (MSFD [5]) until 2020 will be strongly endangered." Both examples demonstrate the need for comprehensive and in-depth information from policy and river basin administration with regard to WFD key issues.

The same is valid for an involvement of scientific expertise at critical steps in the WFD implementation process. Here, the River Basin Community Elbe has developed a good understanding of "historical contaminated sediments" [6] and a Sediment Management Concept [7]. On the other hand, the Elbe-typical dioxin problems (mainly originating from federal state area Saxony-Anhalt) were widely ignored by the River Basin Community Elbe, including the representatives from the Free and Hanseatic City of Hamburg [8]. This happened despite the perspective that in the WFD period 2016–2021, PCDD/Fs and dl-PCBs will play a major role: (a) as new priority substances in the list of biota-Environmental Quality Standards (EQS, i.e., effect-based and legally enforcible numerical quality criteria for assessing the chemical status of aquatic systems) and (b) as the most critical substances at the land–sea interface (i.e., between WFD and MSFD) [9].

Our contribution to the discussion regarding the first WFD period at the Elbe River will comprise different scientific aspects, by presenting more detailed biological and toxicological implications of dioxins and a review of new concepts for remediation on dioxin-contaminated sediments. Last, the study should reflect the achievements of consultants and experts in ad hoc groups, mostly in German language, as an invitation to the international scientific community to participate in the next cycles of the Elbe River Basin Management Plan.

Thematic overview along the RBC Elbe background document "pollutants" (2015 [10])

Analyzing the policy of RBC Elbe in the field *"reduction of pollutant loads"* the preferred way will be along the themes of two *background documents* from the ad hoc working group "Pollutants" for the River Basin Management Plan Elbe (German Part, November 11, 2009 [11]) and the Actualized River Basin Management Plan Elbe 2016–2021(December 22, 2015 [4]). This approach includes three *WFD-related EU-directives* on marine strategies [5], "new" priority hazardous substances [12], and flood risks [13] (Table 1, last column).

Citation: Ulrich Förstner, Henner Hollert, Markus Brinkmann, Kathrin Eichbaum, Roland Weber and Wim Salomons, Dioxin in the Elbe river basin: policy and science under the water framework directive 2000–2015 and toward 2021, DOI 10.1186/s12302-016-0075-8.

Table 1: Original statements (in *italics*) in the seven chapters of the RBC Elbe background document "pollutants" [10] and further information in the present work (citations in square brackets refer to this work; last column: reference to WFD-near EU-directives, treated in this work, and key sections of this work)

Background document "reduction of pollutant loads" (RBC Elbe 21.12.2015 [10])	**This work**
1. Introduction, page 5	
The document actualizes the contents of the background paper for the Deduction of Supra-Regional Management Objectives in the German Part of the Elbe River Basin for the Contaminant Focus" (RBC Elbe 2009a [6])	Introduction: Dioxin from Bitterfeld (Box 1)
2. Supra-regional objectives, p 6–7	
"To attain the objectives according to the EG-WFD (2000/60/EG) [4] and EG Marine Strategy Directive 2008/56/EG [5, 13] direct source-related or at least near-source measures are needed in many water bodies of the Elbe catchment"	2008/56/EG MSFD Land-sea interface; Conclusions/outlook
3. Evaluation—chemical status p 8–15	
As a result of an actualization of the assessment from 2013 it has been found that a good chemical status of the Elbe River cannot be met area-wide due to an excess of EQS of mercury (Hg) in biota"(ca. *60 % from re-emissions—soils, sediment,* etc).	2013/39/EU new PHS EQS for dioxins/DLSs (RWTH Aachen et al.)
4. Catchment areas, sources, p 16–20	
"In the Mulde catchment, the organic pollutants HCH and PCDD/Fs are on the top of the agenda. The middle Elbe is a relevant interim reservoir; its stagnant areas (cut-off meanders, harbors, groin fields) can easily be transferred during floodwater events"	Prioritization (Box 3) Elbe R. basin sediment remediation (Table 4)
5. Hitherto activities, p 21–22	
"The Sediment Management Concept of the RBC Elbe [7] should contribute to attain a good chemical and ecological status*; as such it is a basis for the second RBMP Elbe according to WFD and for the achievement of the environmental objectives of MSRL"*	Dioxin stabilization with activated carbon; passive sampling (Box 4)
6. State of implementation, success, p 23–25	
"Investigation on organic pollutants in suspended matter, sediments and floodplains of the Spittelwasser and the Lower Mulde River shows, that the massive fine sediment depots in the Spittelwasser do no more exist today"	Radiometric mapping (Box 2 LAF vs Tauw)
7. Challenges, p 26–27	
The biggest challenges exist with the *very rare flood events* from August 2002 (HHQ[a]: upper Elbe R.), March/April 2006, January 2011 and June 2013 (HHQ: lower Elbe R.). "*Without a targeted stabilization or removal of highly contaminated historical deposits the flood-induced pollutant releases would remain a significant handicap in attaining the objectives of a good chemical (and biological) quality according to WFD"*	2007/60/EC Flood risks, Climate change (Box 5); FloodSearch, DioRAMA (RWTH Aachen et al.)

[a]HHQ: highest observed water level

(1) Introduction

Based on the risk study initially commissioned by the Hamburg Port Authority (HPA [14]), the RBC Elbe presented a first background paper on the aspect chemical contamination (RBC Elbe [6]) for the public discussion of the draft RBMP Elbe, German part (RBC Elbe 2009b [11]). A special conceptual achievement of the RBCs Pollutant Working Group was the early setting of priorities for remedial measures; here, the reduction needs for sediment-bound pollutants were calculated from mass loads in different areas of concern within a river basin (see RBMP Elbe 2009 [11] section 5.1).

(2) Supra-regional objectives

In a controversy between scientists (primarily authors of the HPA-study [14]) and a part of the RBC Elbe adminstration on the relation between pollution sources and river basin wide problem solutions (Box 3 "two versions") the publication of the Sediment Management Concept [7] eventually decided for the preference of the "source-first principle" ("Sediment management concept: prioritization of measures—Box 3: two versions" section). A still open field is the pollutant transfer from the entire Elbe catchment which induces considerable risks for the marine environment and serious restrictions for the handling of sediment in the tidal areas (MSFD [5]). At the end of the present review, the largest deficiencies are stated, which can be attributed to the preoccupation of the responsible RBC Elbe partners related to the dioxin issue ([8]; Conclusions and Outlook, this work).

(3) Evaluation—chemical status

Among the eight newly identified priority substances of Directive 2013/39/EU from August 12, 2013 [12] are PCDD/Fs and dl- PCBs; The modified EQS for the existing list have to be applied beginning on December 22, 2015, and come into force on December 22, 2018 for the new substances [12]. The provisions of the Directive 2013/39/EU had to be transferred into national legislation—German Surface Water Ordinance (OGewV)—by September 14, 2015 [15]. "Directive 2013/39/EC—policy and science for biota-EQS of DLCs" section will deal with EQS for dioxin-like compounds (DLCs) and cell-based bioassays for detection of DLCs [16].

(4) Catchment areas, sources

When preparing remedial measures in the Elbe river basin, the authors of the Sediment Management Concept [7] have focused on the inventory of sediment volumes and their erodibility under different depositional conditions. In a special action within the Actualized RBMP Elbe [4] and related to the Sediment Management Concept of the RBC Elbe [7] the Federal Institute for Hydrology [17], among other studies, conducted a scientific survey on the groin fields along the Elbe River; one of the results was, that the majority of potentially remobilizable pollutant-rich fine-grained sediments occurs in the groin fields located downstream from Elbe-km 350 ("Sediment management concept: prioritization of measures—Box 3: two versions" section in this work).

(5) Hitherto activities

There are still very few actions which can be classified as "measures to reduce the pollution loads" in the closer sense; at best these activities could be described as "establishment of priority measures within an intensive analytical process" ([10] page 21). The proposals in "Dioxin stabilization using activated carbon technologies—Box 4: passive sampling" of this work follow Ortega-Calvo et al. [18], when introducing bioavailability-based concepts at the transition from excavation procedures to regulations of organic chemicals. However, the pretention of the RBC, that the Sediment Management Concept [7] would be a "basis for the achievement of the environmental objectives of the Marine Strategy Framework Directive" ([10] page 22, footnote 3), is not supported by practical activities at the land-sea interface ("Marine strategy framework directive 2008/56/EC—sediments and pollutants" section in this work).

(6) State of implementation, success

The controversial interpretation of the "sediment depots" in the Spittelwasser creek (Fig. 1 in "Radiometric mapping along the Spittelwasser creek—Box 2: Tauw vs. LAF" section) also reflects the different views of science and local administration on the results of consultancy (Box 2 "LAF vs. Tauw [19, 20]"). Sediment cleanup increasingly moves to in situ technologies such as sediment capping, a form of in situ containment, and monitored natural recovery (MNR). With regard to the stabilization of PCDD/Fs, PCBs and other mainly sediment-bound pollutants, sorbent materials such as activated carbon progressed into a proven, reliable technology [21] ("Dioxin stabilization using activated carbon technologies—Box 4: passive sampling" section); a similar development actually happens with passive sampling as a respective assessment method ([22], Box 4). In both fields RBC Elbe has not presented own initiatives, neither in the background document [10] nor in the RBMP [4].

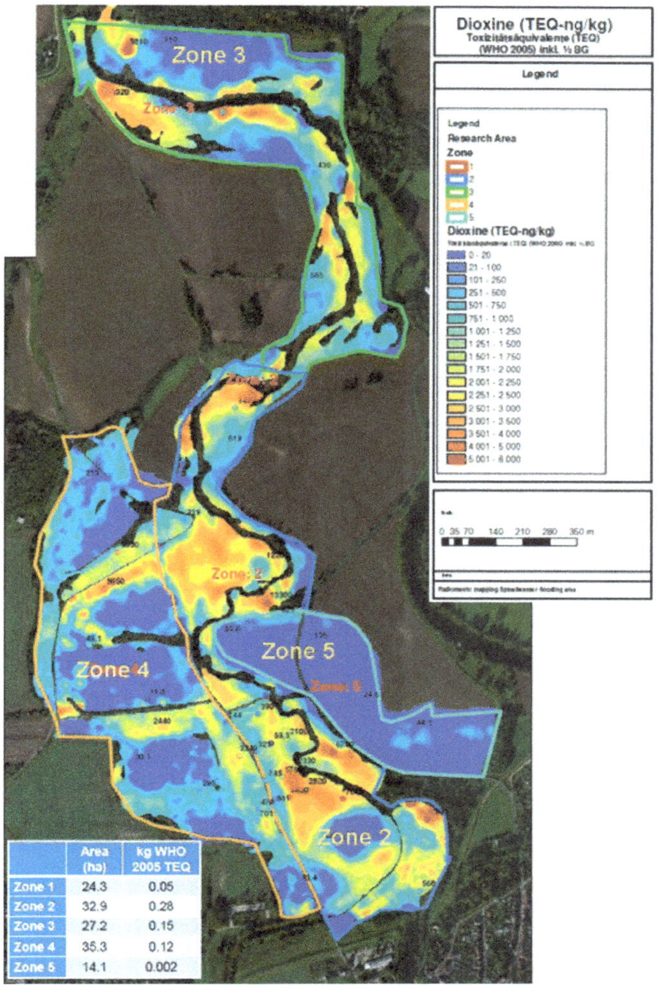

	Area (ha)	kg WHO 2005 TEQ
Zone 1	24.3	0.05
Zone 2	32.9	0.28
Zone 3	27.2	0.15
Zone 4	35.3	0.12
Zone 5	14.1	0.002

Fig1: Radiometric map combining the concentration data [the last seven intervals—*yellow* to *red*—correspond to 2000–6000 ng/kg TEQ] in their areal distribution in Spittelwasser creek bank sediments or fluvisols (Tauw Soil Newsletter from July 2014 [20]) and the information on areal dioxin loads within four different zones of the Spittelwasser site (small *table left below* shows size of the zones [in hectares] and the respective loads [in kg TEQ WHO-2005]; Jacobs et al. 2013 [19]). The map of Jacobs [20] was slightly edited (legend, size)

(7) Challenges

A complex legal approach in support of the chemical quality under the WFD is an assessment and management of flood risks, following Directive 2007/60/EC [13]; the special role of sediment-bound contaminants is due to the exponential increase of solid/pollutant loads with higher water flow velocities [23]. A study in cooperation with RBC Elbe "The flood-water induced remobilization of historical contaminated sediments" (BfG 2014 [17]) calculated the budgets of flood load in the Saale river, which considered both sediment inputs/outputs and the variations of the interim reservoirs. An overview on ecotoxicology in context of sediment mobility and flood risk assessment is presented in "Directive 2007/60/EC—policy and science with special reference to dioxin-like compounds and flood risks" section [24].

BACKGROUND: PCDD/FS AND DL-PCBS IN SEDIMENTS

Many of the current problems regarding hydrophobic substances such as polychlorinated dibenzo-*p*-dioxins and dibenzofurans (PCDD/Fs; often collectively referred to as 'dioxins') and polychlorinated biphenyls (PCBs) under the EU Water Framework relate to sediments mostly originating from historical release and contamination (Nizzetto et al. [25]). PCDD/Fs, PCBs, and other mainly sediment-bound pollutants are migrating together with the sediments downstream and may overtime dominate the pollution load in downstream reaches (Verta et al. [26]). The impacts of sediment-bound pollutants are particularly aggravated during storm events, when these deposits may be mobilized (Wölz et al. [27]; Weber et al. [28]). In addition to increased exposure of aquatic organisms, such as fishes, these sediments can also be transferred to floodplains where they can contaminate the food chain via grazing cattle (Lake et al. [29]; Lake et al. [30], Schulz et al. [31], Kamphues et al. [32]), Weber et al. 2015 [33].

A classic example is the dioxin hot spot in the Spittelwasser creek of the Bitterfeld Chemical Triangle, Germany, the contaminants of which could be traced in sediment samples downstream the Elbe river as far as to the Hamburg harbor area via its congeneric pattern of PCDD and PCDF (Götz et al. [34]).

Sources of dioxins in the Elbe catchment area—Box 1: dioxin from Bitterfeld

The extreme contamination in the Elbe catchment area is largely a legacy from elemental chlorine-based metallurgical production at Bitterfeld and two other sites culminating between 1940 and 1945 [19]. The electrochemical industry in this area was closely connected with aircraft construction, which was based on hydrogen gas as a side product during chlorine alkali electrolysis (Harbodt [41]). In the early 1930s, the booming industrial branch was supplied by the light metal plants at Bitterfeld (Mulde), Aken (Elbe), and Stassfurt (Bode).

In the production process the raw materials, such as magnesium oxide and others, at first reacted with chlorine gas in a conversion furnace; in a second step, the magnesium chloride was transferred to magnesium in a melting electrolytic reaction [19]. The dioxin emissions of such facilities can be extremely high and the only semi-quantified historic emission of a magnesium production to a Norwegian fjord was estimated to 50 to more than 100 kg TEQ (Knutzen and Oehme [42]). According to the Dioxin Toolkit issued by the UNEP (2005), for one ton of magnesium produced using the fused salt electrolysis process, 9 mg I-TEQ are released into the environment via waste water; this represents an estimated 3 kg I-TEQ for the duration of the Second World War at the Bitterfeld site (Umlauf et al. [43]) as a minimum release estimate considering that technology has improved over the 60 years. Based upon characteristic PCDD/F congeners [34, 43], it was at first detected that the extreme contamination of soils and sediments in the 60 km^2 Mulde River and Spittelwasser floodplain (Wilken et al. [44]) was due to emissions from the Bitterfeld plants. More recent analyses by Umlauf et al. [43] suggest, that the sediment samples from the Bode und Saale can also be allocated to the Bitterfeld-Wolfen-Cluster (thermal magnesium and copper production). Whereas in the Saale catchment considerably lower concentrations of PCDD/Fs are found, the much higher suspended load (factor ten, on average) would probably compensate for this difference compared to the Mulde river (Götz et al. [45]).

The dioxin 'hot spot' sediments in the Spittelwasser creek came under special focus during public discussions. Peak concentrations of 140.000 I-TEQ ng/kg PCDD/Fs were measured in a 800-m-long calm water section with a sediment thickness of up to 2 m [46, 47] and approx. 20,000 m^3 fine-grained sediment was estimated in the 1990s (1995 [45], 1997 [46] and 2000 [38]). A rough estimate suggested that just one of these ponds containing 5000 m^{3}with an average of 20,000 ng TEQ per kg sediment could pollute 5 million m^3 of Elbe sediment to 20 ng TEQ/kg ("safe sediment value" [48]); taken the 1990s estimate of 20,000 m^3 hot spot deposits with an average of 20,000 ng TEQ per kg sediment one would calculate a total of 0.4 kg TEQ Dioxin in the ponds of the Spittelwasser creek [14].

It was argued that the Spittelwasser acts as a flood channel of the lower Mulde River when the water discharges in this Elbe tributary exceed a 5-year recurrence flood intensity (HQ_5) (Lindemann [38]). Later, in the SARISK Project (Büttner et al. [49]) the flow trajectories of the spring flood of 2006 were simulated and it was demonstrated that the remobilization of Spittelwasser sediment will start when the Mulde is spreading over the Radegaster Forst at water discharges exceeding 200 m^3/s.

Well then, let mankind wage a slow,

sophisticated war of destruction against this sort of nature.

With sneaking poisons we must try to destroy it.

Novalis, The Novices of Sais

Box 1: Dioxin from Bitterfeld—a common heritage

The Novalis epigram appeared in a publication of Rainer Götz, dioxin expert at the State Environment Agency of Hamburg, who could—in autumn 1989—just speculate on the origin of his PCDD/F findings in the port of Hamburg: probably waste waters from the nine big pulp and paper plants of the German Democratic Republic (GDR) [35]. One year later, on the day of the German reunification, October 3 of 1990, the magazine DER SPIEGEL titled on the situation in Bitterfeld [36]: *"This means revolt; the dioxin values are higher than formerly in Seveso; people apparently do not want to understand the facts."* A first—and so far last!—remediation project against Bitterfeld-sourced dioxin was the collection of so-called "glibberpearls," residues from the production of ion exchangers at the "Chemie AG" [37]; the action of 1994, which was financially supported by the Free and Hanseatic State of Hamburg, provided measurable improvement for the downstream areas of the Elbe river [38].

The findings of the true sources of the large-scale dioxin contamination in the Elbe catchment had no effect on decisions to solve the problem: The feasibility study from 1993 for the sanitation of the Spittelwasser sediments, demanded of the district office Bitterfeld [39], is still kept secret. The dioxin cases of Saxony-Anhalt were obviously not on the shortlist of the sanitation program of the Treuhandanstalt ("Trust agency"); in this action, primarily established to privatize East German enterprises, the remediation budget was cut-down from 100 billion German Mark (~50 billion EURO) to less than 10 % (~4 billion EURO) at the end of the Treuhand-activities [40]. This experience could explain the different "spirit" in Saxony-Anhalt's administration when putting into practice the WFD in the Elbe River Basin community after 2001 [8].

Spittelwasser remediation project (feasibility study 1993)

The sensitive flood situation of the 160 km² lowland area around Bitterfeld and the particular risks from mobile dioxin-rich deposits in the Spittelwasser creek urgently called for immediate remediation measures. Prominent companies participated in the feasibility study 1993 of the District Office of Bitterfeld [39] and after an evaluation of the technical aspects the consultants presented the following proposal for the remediation of the Spittelwasser dioxin hot spots [39]:

"According to available estimation (see above) ~20,000 m³ of sludge with a mean dry substance content of 17 % are deposited in the 3 km long river section under consideration. Based on the properties of the sediments, which—apart from the very high water content—in most cases are characterized by high percentages of fine-grained materials and organic contents, a combination of all reasonable procedures leads to 24 different remediation concepts. For the 12 variants of a shortlist expenses for full solutions can be predicted in the range of 20–30 million German Mark (DM; ~10–15 million EURO). From an ecological, technological and economic view on these variants the consultants prefer the dry recovery and wet separation of the sediments in a sand/gravel and a fine-grain fraction (grain diameter <0.06 mm) with subsequent washing of sand and thermal treatment of the fine fraction (Variant V-6 WV). Expected costs of this variant were 20.9 million DM (~10.5 million EURO)."

The feasibility study from 1993 [39] is still relevant for similar sites of mobile historical contaminated sediments. With the more recent technical developments, e.g., for new hydraulic devices, the loss of contaminated sediments during extraction could be minimized ("Sediment management concept: prioritization of measures—Box 3: two versions" section).

Dioxin concentrations in the Spittelwasser-Mulde-Elbe system 2006/2007

It seems that the systematic sediment survey at the Spittelwasser site has not continued after the year 2000 under Saxony-Anhalt's new Agency for Contaminated Sites (LAF; [8]). Instead, the State Agency for Flood Protection and Water Management of Saxony-Anhalt (LHW [50]) took over the responsibility for the study of dioxin in the waters and soils of the Bitterfeld area as well as in the subsequent Elbe ecosystems originating from secondary sources in the Mulde and Saale catchments. Tables 2, 3 present data on suspended sediments from the official Dioxin Report of the LHW for the Spittelwasser-Mulde-Elbe [50]. Table 2 illustrates, that via the discharge of the Spittelwasser Creek into the Mulde (dioxin level of 750 to approx. 1000 ng I-TEQ/kg in the sediments), a significant increase from 12 ng I-TEQ/kg to 92 ng I-TEQ/kg occurs at the station Mulde/Dessau compared with the upstream Mulde station Bad Düben. A comparison of the data from the Elbe at the stations Dommitzsch and Magdeburg (Table 3) indicates that, due to the influence of the dioxin-rich confluents of Saale and Mulde, there is an increase of the dioxin concentrations in the Elbe River by a factor 7.6 in the year 2006 and by a factor 6.8 in the year 2007 [50].

Table 2: Σ PCDD/F in I-TEQ ng/kg in suspended matter 2006/2007 [50]. Stations Mulde (Bad Düben, Dessau) monthly mixed samples at the automated measurement stations

	Mulde (Bad Düben)		**Spittelwasser**		**Mulde (Dessau)**	
Year	ø	Min/max	ø	Min/max	ø	Min/max
2006	11	10/12	741	445/1052	56.1	16.3/81.8
2007	12	11/13	1032	583/1369	127	96.4/167

Spittelwasser: downstream from Schachtgraben: suspended matter box. Individual samples for mean values 2006/2007: Bad Düben: 2/2; Spittelwasser: 6/3; Dessau: 6/4

Table 3: Samples from automated measurement stations (number of datasets in italics) [50]

	2006	**2007**
Elbe	11.9	8.0
Dommitzsch	*(2)*	*(1)*
Mulde	56.1	126.6
Dessau	*(4)*	*(4)*

Saale Gr Rosenburg	60 *(4)*	94.4 *(4)*
Elbe Magdeburg	91 *(3)*	54.5 *(3)*

The State Agency for Flood Protection and Water Management of Saxony-Anhalt later became involved in the studies on dioxin-polluted sediments in the Saale and Bode rivers [51].

Radiometric mapping along the Spittelwasser creek—Box 2: Tauw vs. LAF

Originally planned as a "hydraulic system analysis of the Bitterfeld area" under the contentious dialog of Saxony-Anhalt's new Agency for Contaminated Sites ([52]; Box 2) it became the most interesting side product of the Tauw-Study "Spittelwasser Pollutant Load Reduction" [19, 20]: The radiometric mapping of contaminants (Sn, HCHs, DDX, and PCDD/PCDF) in the Spittelwasser floodplain, which was performed in collaboration with MEDUSA bv (Groningen, The Netherlands) from 2011 to 2013. The procedure has been successfully used over more than ten years to transform areal data of the natural gamma radiation into maps of sediment and soil structures or—as in the present case—maps of contamination levels (e.g., Van der Graaf et al. [53]).

The transformation step uses the correlation between the pollutant parameter and the respective radionuclide concentration; in the Spittelwasser case, the correlation equation for dioxin was determined from a pilot study as PCDD/Fs (ng TEQ/g) = $2.38 - 0.016*K + 0.103*U$ with R^2 of 0.66 and p value of 0.0028 [17]. In Fig. 1, which combines the information from two sources (Jacobs et al. 2013 [19] and Jacobs 2014 [20]), the concentration data for PCDD/Fs (in ng TEQ/kg) were shown in 18 intervals from less than 20 ng TEQ/kg (this is the upper threshold value for PCDD/Fs of the RBC Elbe [4, 10]) up to 6000 ng TEQ/kg (yellow to red colors >2000 ng TEQ/kg). A first conclusion from the radioactive mapping activity of Tauw [19, 20] should have been that with a more than 100-fold exceedance of the upper threshold value for dioxin at 20 ng TEQ/kg, a sufficient initial suspicion was given for undertaking further actions. Instead, the River Basin Community Elbe declared the problem as being solved (Box 2).

In their expertise, Jacobs et al. [19] calculated an integrated mass load (to 0.3 m sediment depth) of 0.61 kg TEQ for the central Spittelwasser area between Jeßnitz (south-east) and Raguhn (north); this total load is 50 % higher than the estimated load from the former hot spots in the Spittelwasser ponds (last paragraph in "Sources of dioxins in the Elbe catchment area—Box 1: dioxin from Bitterfeld" section). With these data—even as very rough estimates—there are good reasons to consider more detailed studies both for the assessment of priority areas and the selection of appropriate remedial measures (see "Dioxin stabilization using activated carbon technologies—Box 4: passive sampling" section). With regard to areal dioxin load Zone 2 (Fig. 1 [20]), which forms the near-range left and right along the Spittelwasser course, would become the first priority for actions. According to Jacobs et al. [19] Zone 2 was preferably overflown with polluted waters from the formerly highly contaminated Spittelwasser; at regressing floodwater sedimentation of highly contaminated material mainly took place in these low-leveled areas.

Box 2: Consultancy between science and local administration—Tauw vs. LAF

At the end of the first WFD cycle, the Background Document "Pollutants" of the Elbe RBMP [10] called it a "success," when some information from the Tauw Report [19] suggested that the summer flood of 2002 could have eroded the most critical dioxin hot spots from the Spittelwasser river bed. Until now, against better judgement, the picture of a self-cleaning river system remained untouched within the River Basin Community Elbe administration.

The Tauw Report [19] is the outcome of a "contentious dialog," officially installed by Saxony-Anhalt in 2009 [52] after controversial discussions on the "Risk Study" [14] and the source-first principle (Box 3 "two versions"). The report to Saxony-Anhalt's Agency for Contaminated Sites (LAF) dates from October 21, 2013; *in the public on line version of LAF from July 2014 [19] the all-decisive dioxin map—previously presented by Tauw's Patrick Jacobs at a RBC Elbe Workshop on December 17, 2013—was missing:* Did the consultants capitulate in the face of the powerful client and its no-action policy [8]?

Tauw has tried to defend its standards at three occasions: (1) A hint to a promising, yet unpublished neighbor study, (2) a cryptic announcement: "an examination of other measures beyond that will be recommended by the authors," and (3) a later publication of the dioxin map (Fig. 1) in the Tauw newsletter of July 2014 (Jacobs [20]), subsequent to the online release of the final report [19].

SEDIMENT CONCEPTS IN RBMPS

Since the year 2000, any risk assessment in European waters is made by the holistic river basin approach of the Water Framework Directive (WFD) of the European Union [1, 4]. The chemical status of water bodies is to be assessed in terms of compliance with the quality standards (QSs) and under other relevant Community legislation setting environmental quality standards (EQSs) [5]. Initial steps for measures under the WFD were: 2005 first pressure and impact analysis (Article 5), 2006 monitoring programmes to be operational (Article 8) and 2009 establishment of the programme of measures (Article 11).

The development of a WFD concept for historically contaminated sediments

The Scientific Committee on Toxicity, Ecotoxicity and the Environment (CSTEE [54]) concluded from its report "The setting of environmental quality standards for the priority substances included in Annex X of WFD," that "specific quality standards can and should be developed for sediment and biota." The Expert Advisory Forum on Priority Substances and Pollution Control (EAF [55]), while developing a sequence of procedures for the program of measures, proposed the specific source/pathway 'historical pollution from sediments' (S11) for inclusion into an initial 'source screening.'

An initial reference to this type of sediments and pollutants such as dioxins was given from the Elbe river basin by Förstner et al. (2004 [56]); the common characteristics of historically contaminated sediments ("HCSs") and "historical pollutants" ("HPs") is the limited ability for applying proactive measures to reduce their initial entrance into the aquatic environment.

Since then it is clear that the problems associated with both historic issues form a typical internal task of the river basin communities; here, a three-step strategy has been developed by Heise and Förstner (2006 [57]) for the assessment of risks on Rotterdam harbor arising from HCS in the Rhine river basin, by the identification of (i) substances of concern, (ii) areas of concern, and (iii) areas of risk with regard to the probability of polluting the sediments in the downstream reaches. The processes involved are dominated by mechanical re-suspension (Förstner et al. 2007 [58]), i.e., flood events, and this means, with regard to remedial measures, "a targeted stabilization or removal of highly contaminated historical deposits" [10].

On December 24, 2008, the Directive 2008/105/EC of the European Parliament and of the Council on environmental standards in the field of water policy was published [59, 60]; one of the amendments of the original European Water Framework Directive [1] refers to the need to improve the knowledge and data available on sources of priority substances and ways in which pollution occurs in order to identify targeted and effective control options. "Within the framework of the review of Annex X to Directive 2000/60/EC [1], as provided for in Article 16(4) of that Directive, the Commission shall consider inter alia the substances set out in Annex III to this Directive ([59] for possible identification as priority substances or priority hazardous substances. The commission shall report the outcome of its review to the European Parliament and to the Council by 13 January 2011."

The next step in the progress of dioxin issues in aquatic systems was the publication of the Directive 2013/39/EU of the European Parliament and the Council of 12 August 2013 amending Directives 2000/60/EC and 2008/105/EC as regards priority substances in the field of water policy (Anonymous 2013 [12]). The revised EQS should primarily be considered in the River Basin Management Plans for period 2015–2021 [12].

Spittelwasser under the water framework directive

The Spittelwasser area was chosen by the organizers of the international conference ConSoil 2000 for a case comparison and four expert teams from Denmark, Germany, the Netherlands,and the UK were invited. Evaluation of the plan was done by members of the NICOLE (Network for Industrial Contaminated Land) and CLARINET (Contaminated Land Rehabilitation Network) networks [61].

In the study of the German team (Wittmann et al. [62]), a stepwise approach combining monitoring techniques and remediation measures was identified by the environmental authorities to be used for the contaminated floodplain areas [63]. This approach provides for point excavations of critical material and also for the installation of sediment traps. It also includes the promotion of plant growth to stabilize the soils and sediments as well as support evapotranspiration. It has been argued that the design of geotechnical measures will mainly depend on the flow patterns of the water course during flood events. The plan for a pilot or test study on a part of the floodplain area was scheduled for a 4-year implementation period and 15 years for aftercare; it was calculated for initially 2.2 million EUR, not including the costs for sediment traps, excavations, and wetland construction (which would exceed the other costs by one to two orders of magnitude).

When Saxony-Anhalt's new Agency for contaminated sites took over wider responsibilities in 2001, the offer of the Federal Ministry for Education and Research for funding, the German initiative in the Spittelwasser area was not further pursued by the Ministry of Agriculture and Environment in Saxony-Anhalt.

From the involvement into the first Spittelwasser case study of 2000 [61] and from the experience of the various research projects during the following decade—e.g., KORA [64, 65]– the concept "Spittelwasser 2010" could be envisioned as two major steps ([66], Fig. 2):

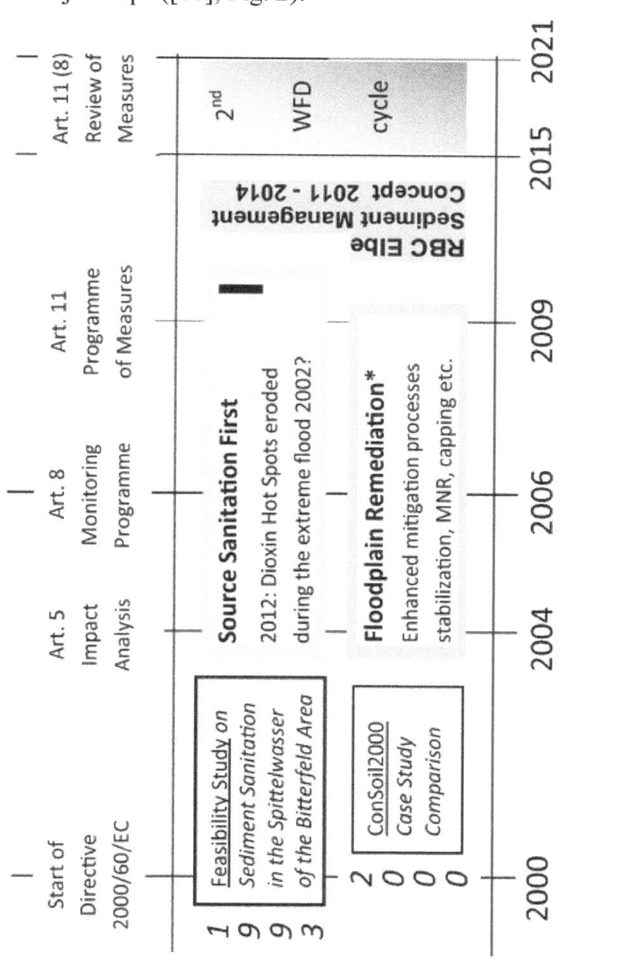

Figure 2: Scheme of the development of an integrated remediation concept for large-scale historical sediment contamination, Spittelwasser in the Bitterfeld District, Germany. *Time scale* is the stepwise implementation of the EU Water Framework Directive (Anonymous 2000a [1]. *Left side* "prerequisites": (i) feasibility study from July 1993 (Anonymous 1993 [39]), (ii) Consoil2000 case study comparison (ConSoil2000 2000b [19, 61]). *In situ processes 2002–2010 (Förstner and Salomons 2010 [66])

1. Source sanitation. Excavation of approximately 20,000 m^3 dioxin hot spot sediment from the Spittelwasser creek ("Spittelwasser remediation project (feasibility study 1993)" section).

2. *Floodplain remediation*, concerning a few tens of km^2 of stable soils ("fluvisols") and erodible channel sediments by new technologies such as monitored natural recovery (MNR), capping, bioremediation, phytoremediation, and embedding into a runoff control system (wetland approach in the widest sense [67]).

Within the common frame of the Large Ecological Project Bitterfeld-Wolfen ("Ökologisches Großprojekt Bitterfeld-Wolfen," ÖGP), the managing State Agency for Contaminated Sites of Saxony-Anhalt (LAF) has spent 230 million EUR in the period from 2001 to 2010 for ground water sanitation, but no substantial responsibility was taken for the sediment issue ([66]; the first Bitterfeld sediment projects in the year 2008 did not even mention the substance group of PCDD/Fs [68, 69].

Sediment management concept: prioritization of measures—Box 3: two versions

The sediment management concept of the River Basin Community Elbe was developed by the Ad hoc Working Group Pollutants/Sediment Management of the RBC Elbe headed by Dr. Peter Heininger (Federal Institute of Hydrology, Koblenz) in the time period of 2011–2013 (see Fig. 2 in "Spittelwasser under the water framework directive" section).

The *quality criteria* of the RBC Elbe for measures on contaminated sediments are based on a differentiation of two groups of Elbe-relevant substances; group 1, forming the regulation level "e" in the RBC Elbe Sediment Management Concept (RBC Elbe 2013 [7]), includes substances, which are explicitly regulated with respect to the protection of human health, e.g., As, Cd, Hg, Pb, HCHs, HCB, benzo(*a*)pyrene (PAH), and PCDD/Fs; other sediment contaminants are listed under the less stringent Group 2.

An initial classification—here for PCDD/Fs—follows the specific criterion (i) "the higher the ranking of the region of origin (here for either "contaminated sites" or "Sediment"; for the latter Bode, Saale or Elbe river) or source type (side structures, lock reservoirs, sedimentation zone, groin fields), respectively, the more urgent is the recommendation"; criterion (ii) is "the number of relevant substances of group 1 per source" and specific criterion (iii) is "the number of relevant substances of group 2 per source." In a fourth step seven general criteria will be applied. The general criteria 1–4 act in the direction of an upgrade; examples are:

1. *Direct source.* The solution of a problem at the source and the elimination of the cause, respectively, had to be preferred (see Box 3: two versions).
2. *Near source.* If the causative source does no longer exist, the solution should be installed as close as possible to the original source ("stairwell cleaning from above").
3. *Resonance 1.* Recommendation would have a positive effect on the other aspects ("hydromorphology" and "shipping").
4. *Resonance 2.* One-shot investment affecting permanently reduced follow-up costs.

The criteria 5–7 indicate a tendency for downgrading the relevance of a certain measure, such as *"degree of difficulty/ requirements for realization"* (No. 5) or *"reliability of predicting the success potential"*, i.e., due to the variability of the system (No. 6). The exclusion criterion *"missing proportionate solution potential"* (No. 7) will only be applied in exceptional cases at very well founded state of knowledge.

The prioritization scheme of the FGG-Elbe has not been applied to a full real situation, but a matrix is given in the Table 6-6 of the Sediment Management Concept (RBC Elbe 2013 [7]) for the selection of recommendations with regard to quality aspects during the sustainable handling of sediments and dredged materials, i.e., for preparing remedial measures in the Elbe river basin. We have posed our focus on dioxin-rich sites from the inventory of sediment volumes and their erodibility under different depositional conditions (Table 4).

Table 4: Criteria for the selection and prioritization of recommendations with regard to quality aspects during the sustainable handling of sediments and dredged materials (after "sediment management concept" of the river basin community Elbe (RBC Elbe 2013, Table 6-6 [7])

Region of origin (source type)	**Substances (health)**	**Direct source**	**Near-source**	**Degree of difficulty**	**Success poten-tial**	**Remediation measures**
Sanitation contaminated sites *within* or *along* rivers/creeks/ditches						
Within **Spittelwas-ser**pre-2002 (1995 - ?)	α-, β-, γ-HCH; Dioxins/Furans	Yes	–	(Medium)	(High)	Excavation, incineration[d]
Along Spittelwasser post-2002 (2013)		–	Yes	Medium	Medium	*Stabiliza-tion* [e] *In situ, AC*
Elimination interim sediment depots						
Saale [a] (side struc-tures)	Hg, Cd, Pb; α-, β-, γ-HCH; benzo(a) pyrene; dioxins/ furans	No	Yes	Medium	??	LHW (2015)[f]
Saale[a] (lock reser-voirs)		No	Yes	Medium	??	LHW (2015)[f]
Lower Bode river [a](sedimentation zones)	Dioxins/furans; Pb	No	Yes	Medium	??	LHW (2015)[f]

Elbe below km 300 (side structures)[b]	Hg, Cd, Pb, As; α-, β-, γ-HCH; HCB, B(a)pyrene; Dioxins/Furans	No	No	Medium	High	*Capping* [g] *Hitzacker/E*
Elbe below km 350 (Groin fields)[c]		No	No	Medium	??	–

In italic letters: own experience with stabilization[e] and active capping[g] technologies

[a]Saale and tributaries. In the navigable section of the Saale river approx. 190,000 tons of fine-grained sediments are deposited (spectrum of pollutants relevant to the river Elbe, incl. dioxins and furans), of which approx. 75 % are classified as remobilizable [(G.E.O.S. 2013 [71]; Wieprecht et al. 2013 [72]). In the Bode River 37,000 tons of fine-grained sediment were found (e.g., weir Stassfurt), 75 % remobilizable, lower Bode river high concentrations of dioxins and furans. The highest concentrations of dioxin—2220 bzw. 6650 ng I-TEQ/kg—were found in deeper layers of core sediments downstream from Stassfurt, where one of the production sites for light metals was located; see introductory section)

[b]Elbe river side structures (harbors, cut-off meanders, bays, blind channels; > 1.000, approx. 50 km^2) comprise a total discharge potential of 20–100 Mio tons; 80 % located in the Elbe river section downstream from km 300 (Heise et al. 2013 [73])

[c]Groin fields. The Inland Elbe River exhibits 6.600 Groin fields which play a role as interim storage for the fine sediment transport, estimated for 1.3 Mio tons along the Elbe; more than 80 % of the muddy, relatively easily remobilizable material is located downstream of Elbe-km 350 (Hillebrand et al. 2014 [74])

[d]Excavation/Incineration. Feasibility Study for the Sanitation of the "Spittelwasser" Sediments (Anonymous 1993 [39]); see "Spittelwasser remediation project (feasibility study 1993)" section

[e]Solidification techniques at the TUHH 1982–2005 (examples [75, 76, 77, 78, 79]

[f]Initial plans for remediation measures were presented at the 23rd Chemical Colloquium of the German Federal Institute of Hydrology in Koblenz, June 11–12, 2015, by Petra Kasimir and Heinz-Jürgen John (Agency for Flood Protection and Water Management of Saxony-Anhalt [51]

[g]Demonstration plant Hitzacker/Elbe has been planned as the final step in a BMBF Research Project on Active Capping (1997–2003, Jacobs and Förstner [80, 81, 82, 83, 84])

The Table shows two different approaches for the remediation of the dioxin pollutants either as hot spots on the bottom of the Spittelwasser creek ("pre-2002-type"; approx. 20,000 m^3 at hotspot, "Spittelwasser remediation project (feasibility study 1993)" section) or as areal deposits along this river's course ("post-2002-type"), according to the findings from the study "Pollution Load Reduction Spittelwasser" of Tauw Consultants [19, 20].

This example reflects the typical development in remediation approaches described in the review on retrospective risk assessment "From bioavailability science to regulation of organic chemicals" by Ortega-Calvo et al. [18], when introducing bioavailability-based concepts at the transition from excavation procedures to regulations of organic chemicals [85]: One important tool was the explanation of bioavailability to regulators using the concepts given in this paper, which made it possible to design new remediation methods. If organic chemicals are immobilized, the flux from the soil to the pore water is low, usually too low for the contaminant to pose risks [...]. Jurisdiction (in the example from Australia) now recognizes that the process of aging can be accelerated via chemically induced immobilization, which results in a rapid decline in bioavailability [...]. After treatment, the bioavailable concentration of the chemical, measured as the concentration in the water phase, remained below the detection limit, and no toxicity for earthworms was observed [...].

Box 3: "Dioxin longitudinal profile 2008"—two versions

A significant step forward for the understanding of transport phenomena and dispersion of dioxin in the Elbe River was made in the "Dioxin Longitudinal Profile 2008" by Umlauf et al. [43]. The findings were: (a) No significant contribution from other PCDD/F sources was observed along a stretch of approximately 400 km (p. 30 in [43]). (b) The concentration levels as well as the downstream profile from 2002 to 2008 were rather similar, indicating minor change of the overall situation since 2002. The similarity of the PCDD/F levels is indicative of a stable interrelationship over the course of a long period of time and this observation provided strong arguments for the "source-first" approach in the study "Assessment of Risks from Particle Bound Substances in the Elbe River Basin" (Heise et al. [14]).

The conclusions of Umlauf et al. [43] in the original—English—version were: "(1) The main dangers with respect to dioxin contamination in the Elbe are high water events occurring in the Spittelwasser-Mulde-Saale system. (2) Consequently, an improvement in the immission situation for the Elbe can only be expected after the corresponding sources have been adequately cleaned up. (3) A reduction in the pollutant loads in the Elbe would have a positive effect on the immission situation in the coastal parts of the North Sea as well". In the German version (Stachel et al. [70]), which was edited by the River Basin Community Elbe, the statements (2) and (3) referring to the source-first principle and to the immission situation of the North Sea were deleted—both would have contradicted to the LAF position in the contentious dialog (see above). Two years later, however, the Sediment Management Concept of the ad hoc Working Group (RCB Elbe [7]) gave the source-first principle the priority among seven general criteria: "The solution of a problem at the source and the elimination of the cause, respectively, have to be preferred."

Dioxin stabilization using activated carbon technologies—Box 4: passive sampling

Since the turn of the century activities at North American sediment cleanup sites increasingly move to in situ technologies such as sediment capping, a form of in situ containment, and monitored natural recovery (MNR), where natural processes are used to mitigate the transfer of particle-bound contaminants into the water phase and/or biota; the latter processes could be supported and enhanced by additives (e-MNR, e.g., [86]). In situtreatment is generally less disruptive and less expensive than traditional sediment cleanup technologies. There continue, however, to be gaps in our knowledge of the fate of contaminants in place, and the effects of in place and ex situ remedial strategies, "which must be filled if management strategies are to be compared and chosen wisely"(Apitz et al. [87]). This means, that "a shift of emphasis is needed toward the use and communication of results from the analyses of multiple lines of evidence," e.g., by examining the potential impacts of large, low-probability events or combination of probabilities (e.g., the 100-year flood and the probability of erosion to a specific depth) on exposure and risk, and the associated uncertainties (Bohlen and Erickson [88]).

Extensive experimental studies and field trials have shown that when applied correctly, in situ treatment via contaminant sequestration and immobilization using a sorbent material such as AC has progressed from an innovative sediment remediation approach to a proven, reliable technology [21]; many cases in the USA and EU [89,90] have demonstrated to decrease the bioavailability of PCBs and PAHs in soils and sediments [91]. Activated carbon reduces pore water concentrations by sequestration of the chemicals through adsorption to the AC surface and within its pore structure (Jonker & Koelmans [92]); in addition, AC has slow kinetics of contaminant desorption, which implies that fluxes of HOCs to the aqueous phase are low, and this limits contaminant mobility in the aquatic environment [90].

Passive sampling has been performed with different systems in which chemicals partition between the dissolved phase and a solid or liquid sampling phase without significantly affecting the soil–water or sediment–water equilibrium [22]. Primary considerations for selecting a passive sampling method (PSM) for a specific application include clear delineation of measurement goals for C_{free}, whether laboratory-based "ex situ" and/or field-based "in situ" application is desired, and ultimately which PSM is best-suited to fulfill the measurement objectives [21] (see Box 4).

Over the past decade, pilot- or full-scale field sediment treatment projects using AC were completed or were underway at more than 25 field sites in the United States, Norway, and the Netherlands; collectively, these field projects (along with numerous laboratory experiments) have demonstrated the efficacy of AC for in situ treatment in a range of contaminated sediment conditions [21]. In general, the effectiveness of AC to reduce bioaccumulation appeared to be species-specific, i.e., bioaccumulation is known to depend on the physiology and behavior of the organisms, which drive factors such as ingestion rate, assimilation efficiency, and elimination (Janssen and Beckingham [21, 96]. *In situ* sediment treatment involves targeted placement of amendments using installation options that fall into two general approaches:

1. Directly applying a thin layer of amendments (which potentially incorporates weighting or binding materials) to surface sediment, with or without initial mixing; and
2. Incorporating amendments into a premixed, blended cover material of clean sand or sediment, which is also applied to the sediment surface [90].

Costs Among the projects mentioned above, field demonstrations in the lower Grasse River (Massena, NY, USA [97] and upper Canal Creek (Aberdeen, MD, USA [98, 99]) included the most comprehensive assessments and available documentation of the longer term efficacy of the in situ AC remediation approach (here mainly on PCB contamination), although similar results have been reported for many of the other field projects; these two sites received the greatest attention in the performance and cost evaluation by Patmont et al. [21] given in Table 5. The general information from these studies is twofold: (1) The costs are in the same order of magnitude for both the excavation/incineration approach of the Spittelwasser feasibility study from 1993 [39] and with the AC in situ stabilization technique, i.e., in the range from 10 to 20 million $US or € for either the hot spot 20,000 m³ or 20 hectares areal deposits; (2) technology development around AC applications involves input of a number of high-ranked scientists from biogeochemical and environmental engineering disciplines.

Table 5: Summary of low and high-range unit costs of AC application (Patmont et al. [21])

Component	Low-range unit cost	High-range unit cost
Activated carbon[a]	$50,000/hectare	$100,000/hectare
Facilitating AC placement using binder/weighting agents[b]	$0/hectare	$70,000/hectare
Facilitating AC placement by blending with Sediment or sand[b]	$0/hectare	$100,000/hectare
Field placement	$30,000/hectare	$200,000/hectare
Long-term moni-toring	$20,000/hectare	$100,000/hectare[c]
Total	$100,000/hectare	$500,000/hectare

Estimated costs for a 4 % AC dose (dry weight basis) over the top 10 cm-sediment layer at a 5-hectare site

[a]Powdered activated carbon (PAC) and/or granular activated carbon (GAC), depending on site-specific designs

[b]To facilitate AC placement, binder or weighing amendment such as SediMite[R] or AquaGate™, or clean sediment or sand (but typically not both) may be required in some applications depending on site-specific conditions and designs

[c]High-end monitoring costs of $100,000 per hectare reflects prior pilot projects and likely overestimates costs for full-scale remedy implementation

Box 4: Passive sampling in the monitoring of dioxin under WFD and US EPA superfund

A recent review in ES&T [93] builds upon the findings of an ICES-workshop on the utility of passive sampling for the risk assessment on contaminated sediments [94] and gives the special focus on the comparison of situations under both the WFD and US EPA Superfund:

In Europe, the strict monitoring requirements laid down in the WFD and its daughter directives impede the implementation of passive sampling for regulatory purposes, whereas in the United States the use of passive sampling in the implementation of remediation processes for contaminated sediments has been encouraged by regulators [95]. The latter is due to the recognition that passive sampling-based C_{free} data, in particular of non-polar organic compounds, provides a better scientific basis for risk assessment, compared with conventional sampling and monitoring procedures. The actual use of passive sampling is limited by the lack of commercial laboratories performing passive sampler deployments and data reporting; in the near future, the scientific community will be crucial in providing guidance on the standardization of passive sampling methods [93].

SCIENTIFIC SUPPORT FOR DIOXIN- AND WFD-RELATED EU-DIRECTIVES

During the implementation of the European Water Framework Directive it became clear that the original legislative system has to be completed by further central "directives," three of them in the aspect "pollutants" and these with special reference to the substance group of "dioxins":

1. The Directive 2013/39/EC from August 24, 2013 reacts on the *"actual legal situation"* (No. 3 in Table 1) at Environmental Quality Standards (EQS) for priority hazardous substances. With the PCDD/Fs a substance group is involved, which has not been regulated so far, but may provide the widest consequences.
2. The Directive 2007/60/EC of October 2007 on the assessment and management of flood risks increasingly deals with the specific *challenges from particle-associated priority pollutants*, such as dioxins (No. 7 in Table 1).
3. Downstream from pollutant sources, possibly until the coastal waters and the North Sea, environmental objectives can only be achieved if the pollutant loads existing in the upstream catchment are reduced or eliminated (Marine Strategy Framework Directive 2008/56/EC of June 2008).

These three directives are highly interconnected. Directive 2013/39/EC is a direct amendment of the WFD, with direct implications for the chances of achieving the "good chemical status" of surface waters. For each Priority Hazardous Substance newly defined by the directive, EQS values, i.e., effect-based numerical quality criteria for assessing the status of an aquatic system, need to be derived and enforced. The implications of including PCDD/Fs and dl-PCBS in the list of Priority Hazardous Substances may have wide consequences of our ability to reach the "good chemical status" in many European catchment areas. "Directive 2013/39/EC—policy and science for biota-EQS of DLCs" section summarizes the current legal situation and novel approaches in the context of Directive 2013/39/EC. Furthermore, it has been stated that it is imperative to closely coordinate measures in context of the Floods Directive, the MSFD and the WFD to be able to reach the individual goals of each of the legislations [100,101]. One particularly illustrative example may be the managed realignment of dikes as a measure to (1) increase the storage capacity of floodwater, which (2) reduces the bed shear stress and the risk of erosion of contaminated sediments while at the same time (3) increasing the structural diversity of the river ecosystem. "Directive 2007/60/EC—policy and science with special reference to dioxin-like compounds and flood risks" section illustrates the implications of Directive 2007/60/EC in policy and science with a special reference to dioxin-like compounds in context of flood events. Last, sediments and suspended particles can be thought of as a transport vector for lipophilic pollutants and thus represent a direct link between WFD and MSFD, since marine systems are the ultimate recipients of the SPM loads of our rivers; this aspect is particularly important in Elbe river basin as discussed in "Marine strategy framework directive 2008/56/EC—sediments and pollutants" section (e.g., [183]).

Directive 2013/39/EC—policy and science for biota-EQS of DLCs

New priority hazardous substances in the field of water policy

The implementation of a community strategy for PCDD/Fs and PCBs was first mentioned expressively in the Second progress report (Annex) (COM(2001)593) from 10.7.2007 (Anonymous 2007 [102]); before, in a Commission Staff Working Document "Impact Assessment," the growing experience was expressed, that "many of the substances of highest concern persist in the environment for a long time; even after their use has been banned and discharge restricted, these substances continue to be found in high concentration in the environment 10, 20, or more years later, and some of them have travelled to remote areas" (Sect. 3 in Anonymous 2006 [103]). In this Working Document, PCDD/Fs and PCBs were not included in the list of priority substances under the WFD as they were considered to be historic pollutants and adequately controlled, "but may be considered for future inclusion depending on new monitoring data" [103]. The Commission Staff Working Document "Impact Assessment" [103] requires from the WFD to submit proposals covering (1) quality standards applicable to the concentrations of the priority substances in surface water, sediments or biota (Article 16(7) WFD) and (2) for priority substances controls for the progressive reduction of discharges, emissions and losses.

Directive 2013/39/EC (from August 24, 2013 [12]) reflects the latest development referring to newly identified priority substances such as PCDD/Fs and dioxin-like PCBs, and there are consequences for the chemical status under the WFD as well as for the adaptation of the German Surface Water Ordinance (OGewV) [15]. Here, the relevance of the biota-EQS criterion for a target achievement is high for mercury, as has been proven from the study of environmental specimen bank organisms (e.g., fish, molluscs, and crustaceans; Wellmitz

[104]). The EQS value of 0.0065 μg TEQ/kg for dioxin and dioxin-like compounds (which is based on the maximum permitted levels in the meat of fish for human consumption defined by regulation 2011/1259/EC) was exceeded by >50 % of the tested organisms in the specimen bank (Mohaupt et al. [105]).

Three EC-Documents refer to methodological aspects: (i) Technical Guidance on the Preparation of an Inventory of Priority and Priority Hazardous Substances [106], (ii) and (iii) regulations for dioxins, dioxin-like PCBs, and non-dioxin-like PCBs in foodstuffs [107, 108]. The following chapter deals with EQS for dioxin-like compounds (DLCs) and cell-based bioassays for detection of DLCs.

EQS for DLCs and cell-based bioassays for detection of DLCs

As mentioned above, the European Water Framework Directive (WFD; EC 2001 [2]), as well as the subsequent directives 2008/105/EC and 2013/39/EU (EC 2008 [59], EU 2013 [13]), defined a total of 45 priority substances. These include PCDD/Fs and dl-PCBs in order to manage and reduce risks based on dioxin-like compounds (DLCs). Sediments can act as long-term sinks for these and possibly other DLCs. Due to their high lipophilicity and persistency these DLCs are known to accumulate in aquatic organisms (e.g., in fish), predominantly via dietary routes of exposure (La Rocca and Mantovani [109], Spagnoli and Skinner [110]) and have been demonstrated to cause a plethora of acute and chronic toxic effects (Mandal [111]). Furthermore, environmental quality standards (EQSs) for the concentrations of these priority substances in the water phase and biota were defined toward a better level of protection in retrospective risk assessment. EQSs, also referred to as action or trigger values, are important tools in sediment assessment frameworks (Apitz and Power [112]) for identifying effects or no effects of sediment-borne contaminants (Wenning and Ingersoll [113]). In 2013, regulation 2013/39/EU [12] entered into force, which established an EQS of 6.5 pg TEQ/gfm for DLCs in biota. Although this represents an important step, it remains unclear how these EQSs for biota translate to EQSs for environmental media, which currently cannot be determined due to the lack of experimental data (EQS dossier 2011 [114]). In particular, the role of sediments, which represent long-term sinks as well as secondary pollution sources, has not yet been sufficiently considered. Since the impact of sediment-borne contaminants on aquatic systems is highly dependent on regional factors, such as erosion risk, sediment particle size distribution, and the structure of the aquatic biocenosis, the directives 2008/105/EC and 2013/39/EU allow the European member states to establish national EQSs for biota and sediments, which can be applied instead of the established water phase EQSs (Carere et al. [115], Maggi et al. [116]).

DLCs share common structural properties. They bind to the cytosolic aryl hydrocarbon receptor (AhR), a ligand-activated transcription factor that regulates the expression of a number of genes. It is believed that the transcriptional responses to AhR activation trigger a major fraction of the documented adverse effects of DLCs (Kawajiri and Fujii-Kuriyama [117], Okey et al. [118]). In this respect, scientists increasingly discuss the role of in vitro bioassays for a biological effect-based assessment in decision making frameworks (Ahlf et al. [119], Besselink et al. [120], den Besten et al. [121]). In vitro bioassays may serve as screening tools for the detection of DLCs in various environmental matrices including foodstuffs (2014/589/EC [108], Eichbaum et al. [122]) because they possess a proven correlation and predictive ability for DLCs. They can support classical, instrumental analysis of individual DLC congeners present in complex mixtures (e.g., sediment or tissue samples) by providing more realistic, ecotoxicologically relevant information. Moreover, they allow for both, the integration of all interactions among DLC congeners and detection of inducers not monitored in compound specific instrumental analyses (Giesy et al. [123], Wernersson et al. [124]). Bioanalytical and instrumental results can be compared by using the approach of toxicity equivalent quotients (TEQs) and biological equivalent quotients (BEQs, Van den Berg et al. [125]).

Since the year 2004, successful implementations of in vitro assays for the screening of DLCs in form of the DR-CALUX assay can be found in the Dutch dredging guideline for coastal sediments, which formerly only included chemical analysis. Here, a biological equivalent quotient (BEQ) signal value of 50 ng BEQ/g dry weight (dw) sediment has been set, which—if exceeded—involves further, detailed investigations (Manz et al. [126]). In German legislation, in vitro assays as semi quantitative methods prior to quantitative instrumental analysis have only been established in the field of food analysis, where BEQs allow for simple yes/no-decisions (2012/252/EU [127]).

Directive 2007/60/EC—policy and science with special reference to dioxin-like compounds and flood risks

In the Elbe river, the summer flood of 2002 was a catastrophic event for a large part of the catchment area (damage of ~25 billion € [BfG 2002 [128]). This case can be seen as the starting point for intensified research on mechanical effects of historical contaminated sediments (HCS, [129, 130, 131]; for example, a comprehensive study on the relocation, dilution and export of metal-polluted sediment in the Saale catchment area was performed by Hanisch et al. [132].

Box 5: Climate change and WFD river basin management—policy and science [137]

With the development of the first river basin management planning under the water framework directive (WFD), which operationally started in 2010, the integration of knowledge about possible climate change impacts on water policy implementation concerns various technical aspects (risk characterization, monitoring, action programs) as well as the evaluation of the "good status" objective's achievements in 2015 (Quevauviller 2011 [137]). The interface between policy and science in this field is dealt in two recent articles: (1) The obvious drawback is that for a single operational framework the scientific basis is divided between two large communities: the disaster risk reduction community and the climate change adaptation community, both of which are bound to different research and operational funding budgets [138]. (2) A review of existing gaps and future research needs based on the findings EU FP7-funded Co-ordination and support action "ClimateWater" [139].

Development of the EU-directive on flood risks

Research projects on flood related hazards including HCS aspects were funded by EU-DG-RTD [133] and BMBF, the latter in the RIMAX (risk management of extreme flood events) coordinated project, e.g., on dry basins and polder for flood retention [134]. During the June flood 2013 at the Elbe river the measurement program "extreme events" of the [RBC Elbe [135] included data of pollutant concentrations and discharges; at Wittenberg and Magdeburg an increase of the dioxin load up to the three to fourfold during this flood event was observed. Contrary to this statement, the draft of the flood risk management plan for the German part of the Elbe river areal unit (RBC Elbe 2014 [136]) just mentions "sediments" once—in the glossary ("the term 'sediment dynamics' comprises transport-, deposition-, and remobilization processes of sediments").

Ecotoxicology in context of sediment mobility and, flood risk assessment

Remobilization of highly contaminated sediments is a key driver for apparent sediment toxicity in aquatic systems (Westrich and Förstner 2007 [140]) and the assessment of sediment stability has been identified as an important and emerging factor that also needs to be considered in the implementation of the WFD (Hollert et al. [141, 142]). Pollutant mobilization from soils and sediments have been described as "time bomb effects" by William Stigliani (1988 [143], 1991 [144]); here, chemical time bombs are defined as "a chain of events resulting in the delayed and sudden occurrence of harmful effects due to the mobilization of chemicals stored in soils and sediments in response to slow alterations of the environment." The re-suspension of sediments certainly possesses some of these characteristics and the frequency and intensity of flood events such as the 500 year flood at the River Elbe in 2002 (Schüttrumpf and Bachmann [145]) are expected to increase in the future because of global climate change [146, 147, 148].

Several experimental laboratory and in-field methods are available to determine the critical bed shear stress for erosion, i.e., the bottom shear stress at which mass erosion of the sediment layer occurs [149, 150, 151, 152, 153,154]. Toxicity testing, however, was mostly conducted in the presence of static sediment layers and is thus only representative of average flow conditions and does not allow prediction of the effects of sediment re-suspension (e.g., [155, 156, 157]). Another approach that was commonly used as a proxy to assess the toxicity of sediments when re-suspended is that of testing sediment elutriates [158, 159]. It should be emphasized, however, that the approach is of relatively limited value for very lipophilic compounds. In particular, experimental research at the intersection between hydrodynamics and ecotoxicology has developed into a promising field that may help to overcome these limitations [160, 161, 162, 163, 164, 165, 166, 167].

The scientific community urgently requires standardized protocols to assess the impact of sediment suspension exposure on biota. This chapter describes a number of recent projects and studies that were initiated in response to these needs.

a) The project framework FloodSearch

In context of the interdisciplinary project framework FloodSearch, which was funded by the German excellence initiative, methods of hydraulic engineering and ecotoxicology were experimentally combined. To this end, exposure experiments with contaminated sediments were conducted in (a) static re-suspension tanks or (b) an annular flume, an experimental facility in which rainbow trout (*Oncorhynchus mykiss*) can be exposed under simulated flood-like conditions (Cofalla et al. [161], Hudjetz et al. [164]. Schüttrumpf et al. [166], Wölz et al. [167], Brinkmann et al. [168, 170]). The investigated sediments were either spiked with a mixture of PAHs or sampled in the field, the latter of which contained different levels of DLC contamination that were aged under natural conditions. A battery of different biomarkers, i.e., measurable biological responses of fish during exposure to the contaminated sediments, was established to verify exposure to and effects of different DLCs. This battery included hepatic activities of the enzymes 7-ethoxyresorufin-*O*-deethylase (EROD), glutathione-*S*-transferase (GST), and catalase (CAT), lipid peroxidation in homogenized liver tissue, as well as the induction of nuclear aberrations (micronuclei) in peripheral erythrocytes. Furthermore, metabolites of PAHs were chemically analyzed in bile liquid of exposed fish to demonstrate uptake and metabolism of these compounds (Kammann [171], Kammann et al. [172]).

Exposure to spiked sediments led to a significantly induced frequency of micronuclei, which correlated well with the concentration of 3-hydroxybenzo[*a*]pyrene in bile liquid, a metabolite of the genotoxic PAH benzo[*a*]pyrene [168]. The enzymatic biomarkers did not indicate any significant alterations in these treatments. In fish exposed to re-suspended natural sediments, significant differences in bile metabolite concentrations as well as in 7-ethoxyresorufin-*O*-deethylase induction were observed compared to control experiments. The biliary concentrations of 1-hydroxypyrene from fish exposed to the three different contamination levels correlated well with the ratio of pyrene concentrations in corresponding bulk sediments and the ratio of particle-bound pyrene in suspended sediments. In contrast, hepatic lipid peroxidation and micronuclei formation represented the different contamination levels less conclusive in these treatments.

Using the newly established interdisciplinary methodology, the studies within the project framework FloodSearch clearly demonstrated that particle-bound DLCs, here predominantly PAHs, from sediments aged under natural conditions may become bioaccessible upon re-suspension during short simulated flood events and are readily absorbed by aquatic organisms such as rainbow trout. The associated short-term effects were clearly documented within the studies, and potential adverse long-term impacts are likely to follow.

b) The DioRAMA project

The DioRAMA project, a cooperation between the Institute for Environmental Research at RWTH Aachen University (Aachen, Germany) and the Department G3 (Biochemistry/Ecotoxicology) of the BfG (Koblenz, Germany), was initiated to promote interdisciplinary research involving chemical and biochemical analyses, ecotoxicology, and risk assessment (Eichbaum et al. [169]). The project was coordinated by the BfG, while experimental investigations were mostly performed at the Institute for Environmental Research. Additional collaborations were established with the University of Saskatchewan (Saskatoon, Canada) and Münster analytical solutions (mas; Münster, Germany). The experimental work within the DioRAMA project was conducted in close exchange with and adjusted to the regulatory practice of the BfG in order to derive useful tools for future application in sediment management. Furthermore, additional collaborations were established with the Thünen Institute of Fisheries Ecology (Hamburg, Germany), the Fraunhofer Institute for Molecular Biology and Applied Ecology IME (Schmallenberg, Germany), the Swiss Federal Institute for Materials Science and Technology (EMPA; Dübendorf, Switzerland), and the Centre for Fish and Wildlife Health (Bern, Switzerland). The DioRAMA project profited largely from the experiences made during the previously introduced projects FloodSearch and FloodSearch II [161, 166, 167].

One major part of the DioRAMA project focused on the establishment and validation of cell-based bioassays for the detection of DLCs in sediments (Eichbaum et al. [16]), while in the second main experiment within the project, the time- and concentration-dependent uptake of DLCs and their associated effects were investigated in rainbow trout (Brinkmann et al. [173]). Exposure experiments were conducted using suspensions of three field-collected sediments from the rivers Rhine and Elbe, which were chosen to represent different levels of contamination. Five serial dilutions of contaminated sediments from the Prossen and Zollelbe sampling sites (both in the Elbe, Germany) were tested and compared with moderately contaminated sediment from Ehrenbreitstein sampling site (in the Rhine, Germany). Fish were exposed to suspensions of these sediment dilutions under semi-static conditions for 90 days. Uptake of particle-bound PCDD/Fs, PCBs, and PAHs was determined by high-resolution gas chromatography and mass spectrometry (HRGC/HRMS) analysis of muscle tissue and high pressure liquid chromatography analysis of bile liquid. Additionally, fish responses to DLCs (EROD activity, micronuclei, and other nuclear aberrations, histopathological, and gross pathological lesions) were investigated.

Analysis of muscle tissue and of bile liquid showed that particle-bound PCDD/Fs, PCBs, and PAHs were readily bioavailable from re-suspended sediments. Uptake of these contaminants and the associated toxicological effects in fish were mostly proportional to their sediment concentrations. The changes in the investigated biomarkers closely reflected the different sediment contamination levels: cytochrome P450 1A mRNA expression and 7-ethoxyresorufin-*O*-deethylase activity in fish livers responded immediately and with high sensitivity, while increased frequencies of micronuclei and other nuclear aberrations, as well as histopathological and gross pathological lesions, were strong indicators of the potential long-term effects of re-suspension events. This study clearly demonstrates that sediment re-suspension can lead to accumulation of PCDD/Fs and PCBs in fish, resulting in potentially adverse toxicological effects. For a sound risk assessment within the implementation of the European Water Framework Directive and related legislation, we propose a strong emphasis on sediment-bound DLCs in the context of integrated river basin management plans. Currently, it is sought for practical implementations of these findings at least for the assessment of dredged materials. In addition to the experimental work conducted within the project, computational models were developed, which will be highly valuable to derive scientifically sound sediment quality standards (Brinkmann et al. [174, 175]).

Marine strategy framework directive 2008/56/EC—sediments and pollutants

The MSFD [12] takes account of land-based as well as fluvial aspects and ensures comparable approaches and methodologies as in the WFD.

Marine strategy framework directive—descriptor 8: contaminants

A commission decision of 1 September 2010 on criteria and methodological standard on good environmental status of marine waters characterized the descriptor 8 "Concentrations of contaminants are the levels not giving rise to pollutions effects," levels of pollution effects on the ecosystem, components concerned, having regard to the selected biological processes and taxonomic groups where a cause/effect relationship has been established and needs to be monitored [176]. The Task Group 8 report "contaminants and pollutions effects" was published in April 2010 under JRC European Commission and ICES [177]. By 15 July 2012, Member States had to prepare the first elements of marine strategies, namely the initial assessment (Article 8), the determination of good environmental status (GES—Article 9), and the establishment of environmental targets and associated indicators (Article 10) and to report them to the Commission by 15 October 2012 [178]. The German report to Article 12 of the MSFD obligations was published on 7 February 2014 [179]. The OSPAR Commission (Convention for the protection of the marine environment of the North-East Atlantic) accompanies the MSFD implementation in several aspects, e.g., with a literature survey on dioxins [180], an advice document on "Good Environmental Status—Descriptor 8: Contaminants" [181] and a "Regional Plan to Improve Adequacy and Coherence of MSFD Implementation 2014–2018 [182]". Wenzel [183] has outlined the significance of the sediment contamination issue between WFD and MSFD with special reference to the Elbe river basin. On the other hand, the draft of the program of measures for the MSFD in the German North Sea and Baltic Sea [184] neglects the aspect sediment contamination and does not mention the dioxin problem in the Elbe; most likely, this is due to the influence of the Federal States Working Group Water and Wastewater (LAWA); deficiencies were observed in their respective contributions to the German versions of the program of measures for the European Water Framework and Flood Risk directives.

Dioxin at the land/sea interface of the Elbe river basin

According to an early dioxin balance for Hamburg [185], in the mid-1990s, approx. 23 g I-TEQ per year were extracted with dredged materials—approx. 300,000 t dry mass—from the Elbe and harbors of the city area, and were deposited on secure sites; approx. 47 g I-TEQ per year were transported in the direction North Sea. Uhlig et al. [186] from analyses of dioxin congener patterns of suspended particulate matter and sediment samples from the Elbe, Spittelwasser, Mulde, Saale, and Schwarze Elster came to the conclusion that 70–82 % of the dioxin contamination of the Elbe sediment in Hamburg can be attributed to sediments from the Mulde, whereby the study took account of both direct transport from the Mulde and also indirect transport via intermediate deposition. For the extreme summer flood of 2002, on the basis of measured SPM deposits on the floodplains of the Elbe, together with analysis results, it was estimated by Stachel et al. [187] that the contamination of the flooded soils was increased to a degree lying between 4.3 and 6.5 g WHO 1998-TEQ. Between 3.1 and 4.6 g WHO 1998-TEQ of PCDD/Fs were transported toward the North Sea over the weir at Geesthacht. Higher levels of dioxins (as compared with a neighboring reference area) were found off-shore in the North Sea South of Helgoland in the area previously used for dumping treated sewage from Hamburg. However, the PCDD/F fingerprints found there point to the Bitterfeld region rather than to impacts from both sewage dumping [43].

CONCLUSIONS AND OUTLOOK

On 1 July 2015, the European Court of Justice (ECJ) rendered its long-awaited judgment interpreting the EU Water Framework Directive 2000/60 in relation to projects such as the deepening of the Weser River in Northern Germany. The ECJ ruled that the environmental objectives of the WFD are not merely objectives for management planning with no link to or impact on individual projects; rather, Member States may not authorize projects which may cause a deterioration of the status of a surface water body unless derogation is granted [3]. Regarding the term "deterioration" the Advocate General Jääskinen recommended a strict interpretation of the WFD with reference to a substance or quality component, without affecting a mandatory classification change [188].

The so far successful initiative of the German NGO "BUND," just in time with the second river basin management plans, confirms the ability of the WFD to adopt concrete measures for both safeguarding a good state and for preventing the deterioration of surface waters. This coincides with the sustainable development strategy, an overriding environmental goal, which is mandatory for all policy sectors and measures within the European Union [189].

From the retrospective to the first WFD phase in the Elbe catchment it seems that "PCDD/Fs and dl-PCB" can be used as a spearhead for a new understanding of problem solutions, where a fundamental change is required in the view on the implementation of the Marine Strategy Framework Directive (below). In the Elbe River basin, following the quasi-ban of dioxin issues from Saxony-Anhalt's environmental agenda in 2009 [9] and the decision of the German Federal States along the North Sea coast to exclude dioxin from their list of "common transitory provisions for handling dredged materials in coastal waters" [190], practical all sanitation activities on dioxin and historical contaminated sediments downstream from the Bode/Saale and Spittelwasser/Mulde were blocked. This has happened despite an ambitious Sediment Management Concept of the River Basin Community Elbe from November 2013 [7] and the inherent responsibilities from WFD-near Directives with respect to the improvement of the chemical quality in surface waters:

1. Under the *Directive 2013/39/EC* [12] the relevance of the biota-EQS criterion for a target achievement of Elbe-typical dioxins and dioxin-like PCBs is particularly high. It is predicted by Mohaupt et al. [105] that in the second River Basin Management Plan, the new and extended EQS-requirements will lead to a failure of the good chemical status in all water bodies; in the third RBMP the EQS values of the eight specific substances will continue to be exceeded, however, with less actual/target status.
2. *Directive 2007/60/EC* [13] on assessment and management of flood risks, relates to the special role of sediment-bound contaminants is due to the exponential increase of solid/pollutant loads with higher water velocities. The assessment of erosion stabilities was one of the major achievements of the Sediment Management Concept of the RBC-Elbe (RBC Elbe 2013 [7]; another was the development of criteria for the prioritization of measures).
3. *The Marine Strategy Framework Directive* (MSFD, 2008/56/EC [5]) is similar to the WFD, in its stepwise implementation mode and with its main objective in the achievement of good environmental status (GES) in the marine environment by 2020. Between the source of a pollutant and its final receptor, the sea, we have the temporal storage in flood plains, lakes, artificial lakes behind dams, deltas, and man-made sedimentation traps (harbors); these are temporary receptors on the temporal scale of years to decades (Salomons [191]). Using scenarios and linking source to receptors has been carried out for *prediction of sediment quality in harbors* (Salomons and Gandrass [192]). In these cases the impact at the receptor could be defined as exceeding the standard for disposal at the sea (Heise et al. [193]); which is sufficient for the harbor manager [191]. For example in the scenario for the Rhine sediment between the barrage Iffezheim and Rotterdam, the hexachlorobenzene (HCB) source can contribute to a failure of the objectives of the WFD in the Rhine Basin [194, 195] and may require additional measures for its control [57]. All of these should be indications for a mandate to the RBC Elbe; however, no activities at the land–sea interface are reported in the Actualized RBMP (RBC Elbe 2015 [4]).

There is a characteristic difference between dioxin and other substances of the Directive 2013/39/EC, many of which can be reduced by technical measures on the basis of product regulations, approvals etc. [196]. For dioxins as historical pollutants, the proactive option is mostly missing, and the aftercare depends on the solidarity among the members of the River Basin Community.[1]

In the costly approval process of the Port of Hamburg for deepening of the Elbe river course [198], involving translocation of additional 1 million m^3/a fine-grained sediment from the tidal Elbe, the pollutant aspect already plays the most critical role [199]. At the end of the day, under the stricter claims from the Marine Strategy Framework Directive compared to the WFD, the only convincing argument would be a significant reduction of the pollutant discharges, with special reference to the Elbe-typical PCDD/Fs, from the catchment area into the North Sea (Additional file 1).

Footnotes

1 As to the RBC Elbe, the Free and Hanseatic City of Hamburg, according to the principles of a coordinated management (WFD Art 4 and Appendix V [4]), has a claim against an upstream riparian, by name Saxony and Saxony-Anhalt, for conducting hydraulic engineering and sanitation measures, which are practically needed to eliminate and reduce specific pollutant inputs from upstream risk areas (Breuer 2008 [197]).

DECLARATIONS

Competing interests

The authors declare that they have no competing interests.

REFERENCES

1. Anonymous (2000) Directive 2000/60/EC of the European Parliament and the Council of 23 October 2000 establishing a framework for Community action in the field of water policy (Water Framework Directive). Official Journal of the EC, L327
2. Anonymous (2001) Directive 2455/2001/EC of the European Parliament and of the Council of 20 November 2001 establishing the list of priority substances in the field of water policy and amending Directive 2000/60/EC. Off J Eur Commun 15:1–5
3. Von Hammerstein F, Lichtblau N (2015) ECJ ruling clarifies "deterioration" under Water Framework Directive. Law-Now—Your World First, 23.07.2015. http://www.cmslaw-now.com/ealerts/2015/07/ecj-ruling-clarifies-deterioration-under-water-framework-directive
4. RBC/FGG Elbe (2016) Aktualisierung des Bewirtschaftungsplans nach § 83 WHG bzw. Artikel 13 der Richtlinie 2000/60/EG für den deutschen Teil der Flussgebietseinheit Elbe für den Zeitraum von 2016 bis 2021. 259 p, FGG Elbe, Magdeburg, 22.11.2015. http://www.fgg-elbe.de/hintergrund-informationen.html (for all RBC/FGG-Elbe Documents)
5. Anonymous (2008) Directive 2008/56/EC of the European Parliament and of the Council of 17 June 2008 establishing a framework for community action in the field of marine environmental policy (Marine Strategy Framework Directive). Off J Eur Union, L 164/19-164/40
6. RBC/FGG Elbe (2009) Hintergrundpapier zur Ableitung der überregionalen Bewirtschaftungsziele für die Oberflächengewässer im deutschen Teil der Flussgebietseinheit Elbe für den Belastungsschwerpunkt Schadstoffe (Abschlussbericht). 29 p, FGG Elbe, 02.04.2009
7. RBC/FGG Elbe (2013) Sedimentmanagementkonzept der FGG Elbe. Vorschläge für eine gute Sedimentmanagementpraxis im deutschen Elbegebiet zur Erreichung überregionaler Handlungsziele. Authors: Heininger P, Blondzik K, Gabriel T, Kasimir P, Keller I, Netzband A, Pfeiffer P, Quick I, Rickert-Niebuhr K, Rohde S, Schwandt D, Schwartz R, Steffen D, Tobian I, Vollmer S, von Seggern D, Wenzel C, Wilke R, Wolf M, 383 p, Magdeburg, 25.11.2013. http://www.fgg-elbe.de/tl_files/Downloads/EG_WRRL/hgi/hgd_bp2-/sedimentmanagementkonzept_fgg-final.pdf
8. Förstner U (2015) Dioxin im Elbegebiet—Landesbehörde gegen den Geist der EU-Wasserrichtlinie. doi:10.13140/RG.2.1.2026.6407, ResearchGate, May 7, 2015; p. 4
9. Förstner U (2015) Editorial: sediments and the eu-water framework directive: revisiting the Elbe 2015 river basin management plan. J Soils Sediments 15(9):1863–1864
10. RBC/FGG Elbe (2015) Hintergrunddokument zur wichtigen Wasserbewirtschaftungsfrage „Reduktion der signifikanten stofflichen Belastungen aus Nähr- und Schadstoffen Teilaspekt Schadstoffe". Ad hoc AG Schadstoffe/Sedimentmanagement der FGG Elbe (Leitung: Dr. Peter Heininger, BfG Koblenz). p 29, Magdeburg, 21.12.2015
11. RBC, FGG Elbe (2009) Bewirtschaftungsplan nach Artikel 13 der Richtlinie 2000/60/EG für den deutschen Teil der Flussgebietseinheit Elbe. FGG Elbe, Magdeburg 11(11):2009
12. Anonymous (2013) Directive 2013/39/EU of the European Parliament and the Council of 12 August 2013 amending Directives 2000/60/EC and 2008/105/EC as regards priority substances in the field of water policy (among the 8 newly identified priority substances are PCDD/Fs and dioxin-like PCBs), Official Journal of the European Union. L 226/1, Brussels 24.8.2013
13. Anonymous (2007) Directive 2007/60/EC of the European Parliament and of the Council of 21 October 2007 on the Assessment and Management of Flood Risks. Official Journal of the European Union L 288/27, Brussels 6.11.2007
14. Heise S, Krüger F, Baborowski M, Stachel B, Götz R, Förstner U (2008) Assessment of risks from particle bound substances in the Elbe river basin (in German). Commissioned by Hamburg Port Authority and River Basin Community (RBC/FGG) Elbe. Hamburg, May 2008, p 349, (Chapter 6: Förstner U "Measures", p 247–341).http://www.tideelbe.de/files/elbestudie_ii_mai_2008_klein.pdf
15. Anonym (2015) Verordnung zum Schutz der Oberflächengewässer (Oberflächengewässerverordnung—OGewV). Entwurf des BMU, Stand: 29. April 2015. http://www.bmub.bund.de/fileadmin/Daten_BMU/Download_PDF/Gesetze/ogewv-entwurf.pdf
16. Eichbaum K, Brinkmann M, Buchinger S, Reifferscheid G, Hecker M, Giesy JP, Engwall M, van Bavel B, Hollert H (2014) In vitro bioassays for detecting dioxin-like activity—Application potentials and limits of detection, a review. Sci Total Environ 487:37–48
17. BfG (2014) Professional Contributions to the Sediment Management Concept Elbe (in German). Mitteilungen Nr. 30. Federal Institute of Hydrology (Bundesanstalt für Gewässerkunde), Koblenz p. 164, ISBN 978-3-940247-10-0, doi:10.5675/BfG_Mitteilungen-_30.2014, URL: http://doi.bafg.de/BfG/2014/BfG_Mitteilungen_30.2014.pdf
18. Ortega-Calvo J-J, Harmsen J, Parsons JR, Semple KT,

Aitken MD, Ajao C, Eadsforth C, Galay Burgos M, Naidu R, Oliver R, Peijnenburg WJGM, Römbke J, Streck G, Versonnen B (2015) From bioavailability science to regulation of organic chemicals. Environ Sci Technol 49:10255–10264

19. Jacobs P, Krautter N, Diesner K (2013) Frachtreduzierung Spittelwasser—Oberflächenwasser-, Sediment- und Auenuntersuchungen; Bewertung und Ableitung von Maßnahmeerfordernissen. Projektleitung: Dr. Patrick Jacobs, Tauw GmbH, Berlin; 21. Oktober 2013. p. 232. Im Auftrag der Landesanstalt für Altlastenfreistellung des Landes Sachsen-Anhalthttp://www.laf-lsa.de/wasserrahmenrichtlinie/

20. Jacobs P (2014) Sediment-bound organic pollutants in the Spittelwasser creek and their impact on the Elbe river basin. Tauw Consultants Berlin, Soil Newsletter July 2014, p. 13. http://www.medusa.online.com/wp-content/uploads/2014/10/Mapping-sediment-bound-organic-pollutants.pdf

21. Patmont C, Ghosh U, LaRosa P, Menzie C, Luthy R, Greenberg M, Cornelissen G, Eek E, Collins J, Hull J, Hjartland T, Glaza E, Bleiler J, Quadrini J (2015) In situ sediment treatment using activated carbon: a demonstrated sediment cleanup technology. Integr Environ Assess Manag 11(2):195–207

22. Ghosh U, Driscoll S, Burgess RM, Jonker MTO, Reible D, Gobas F, Choi Y, Apitz SE, Maruya KA, Gala WR, Mortimer M (2014) Passive sampling methods for contaminated sediments: practical guidance for selection, calibration, and implementation. Integr Environ Assess Manag 10:210–223

23. Hollert H, Dürr M, Haag I, Wölz J, Hilscherova K, Blaha L, Gerbersdorf SU (2007) Influence of hydrodynamics on sediment ecotoxicity. In: Westrich B, Förstner U (eds) Sediment dynamics and pollutant mobility in rivers. Springer, Heidelberg, pp 401–416

24. Brinkmann M, Eichbaum K, Kammann U, Hudjetz S, Cofalla C, Buchinger S, Reifferscheid G, Schüttrumpf H, Preuss T, Hollert H (2014) Physiologically-based toxicokinetic models help identifying the key factors affecting contaminant uptake during flood events. Aquat Toxicol 152:38–46

25. Nizzetto LI (2010) Past, present, and future controls on levels of persistent organic pollutants in the global environment. Environ Sci Technol 44(17):6526–6531

26. Verta M, Kiviranta H, Salo S, Malve O, Korhonen M, Verkasalo PK, Ruokojärvi P, Rossi E, Hanski A, Päätalo K, Vartiainen T (2009) A decision framework for possible remediation of contaminated sediments in the River Kymijoki. Finland. Environ Sci Pollut Res 16(1):95–105

27. Wölz J, Engwall M, Maletz S, Olsmann H, van Bavel B, Kammann U, Klempt M, Weber R, Braunbeck T, Hollert H (2008) Changes in toxicity and dioxin-like activity of suspended particulate matter during flood events at the rivers Neckar and Rhine. Environ Sci Pollut Res 15:536–553

28. Weber R, Gaus C, Tysklind M, Johnston P, Forter M, Hollert H, Heinisch E, Holoubek I, Lloyd-Smith M, Masunaga S, Mocarelli P, Santillo D, Seike N, Symons R, Torres JPM, Verta M, Varbelow G, Vijgen J, Watson A, Costner P, Woelz P, Wycisk P, Zenneg M (2008) Dioxin- and POP-contaminated sites—contemporary and future relevance and challenges. Environ Sci Pollut Res 15:363–393

29. Lake IR, Foxall CD, Fernandes A, Lewis M, White O, Mortimer D, Dowding A, Rose M (2014) The effects of river flooding on dioxin and PCBs in beef. Science Total Environ 491–492:184–191

30. Lake IR, Foxall CD, Fernandes A, Lewis M, Rose M, White O, Lovett AA, White S, Dowding A, Mortimer D (2015) The effects of flooding on dioxin and PCB levels in food produced on industrial river catchments. Environ Inter 77:106–115

31. Schulz A-J, Wiesmüller T, Appuhn H, Stehr D, Severin K, Landmann D, Kamphues J (2005) Dioxin concentration in milk and tissues of cows and sheep related to feed and soil contamination. J Animal Physiol Animal Nutrit 89:72–78

32. Kamphues J, Schulz AJ, Gude K, Bruns-Weller E, Severin K, Appuhn H, Taube V (2011) Investigations concerning the use of dioxin exposed grassland along rivers by food producing ruminants with special regard to food safety. Organohalogen Compd 73:2040–2041

33. Weber R, Hollert H, Kamphues J, Ballschmiter K, Blepp M, Herold C (2015) Analyse und Trendabschätzung der Belastung der Umwelt und von Lebensmitteln mit ausgewählten POPs und Erweiterung des Datenbestandes der POP-Dioxin-Datenbank des Bundes und der Länder mit dem Ziel pfadbezogener Ursachenaufklärung. Umweltforschungsplan des Bundesministeriums für Umwelt, Naturschutz, Bau und Reaktorsicherheit Stoffliche Sicherheit. Forschungskennzahl (UFOPLAN) 3712 65 407 1

34. Götz R, Steiner D, Friesel P, Roch K, Walkow F, Maaß V, Reincke H (1996) Dioxin in the river Elbe—investigations of their origin by multivariate statistical methods. Organohalog Compounds 27:440–443

35. Götz R (1989) Contamination of river sediments by polychlorinated dibenzodioxins (PCDDs) and polychlorinated dibenzofurans (PCDFs) and possible sources [in German]. Vom Wasser 73:457–481

36. Anonym (1990) Das gibt Revolte. Neue Hiobsbotschaften aus Bitterfeld und Umgebung: Die Dioxinwerte liegen höher als einst in Seveso. Die Fakten will niemand so recht wahrhaben. Der Spiegel [Magazine] 40/1990, p. 46–54.http://www.spiegel.de/spiegel/print/d-13500306.html

37. Anonym (1992) Alles durch den Ausguß—Die ostdeutschen Behörden schaffen es nicht, ihre Umweltprobleme unter Kontrolle zu bringen. Der Spiegel 9/1992, p. 52–55 http://www.spiegel.de/spiegel/print/d-13687070.html

38. Lindemann M (2000) Das ökologische Großprojekt Bitterfeld-Wolfen—eine Bilanz aus fachlicher Sicht. Altlasten Spektrum 4(2000):208–214

39. Anonymous (1993) Feasibility Study for the Sanitation of the "Spittelwasser" Sediments. UBS Schwerin and IGB Hamburg on Demand of Landratsamt Bitterfeld,

July 1993 (in German, unpublished)

40. Bonnenberg H (2010) Meine Erfahrungen bei der Treuhandanstalt und ihren Folgegesellschaften. Vortrag anlässlich der Anhörung der Bundestagsfraktion DIE LINKE. Marie Elisabeth Lüders Haus des Bundestags, Berlin, 19. April 2010. 16 S.http://www.treuhandanstalt.com/uploads/media/Treuhandanstalt_DIE_LINKE_01.pdf
41. Harbodt K (2006) History until 1945. In: Friedrich HE, Mordike BL (eds) Magnesium Technology—metallurgy, design data, applications. Berlin: Springer, p. 685
42. Knutzen J, Oehme M (1989) Polychlorinated dibenzofuran (PCDF) and dibenzo-p-dioxin (PCDD) levels in organisms and sediments from the Frierfjord, southern Norway. Chemosphere 19:1897–1909
43. Umlauf G, Stachel B, Mariani G, Götz R (2011) Dioxins and PCBs in solid matter from the River Elbe, its tributaries and the North Sea (longitudinal profile, 2008). EUR—Scientific and Technical Research series—ISSN 1018-5593 (print), p. 118
44. Wilken M, Walkow F, Jager E, Zeschmar-Lahl B (1994) Flooding area and sediment contamination of the river Mulde (Germany) with PCDD/F and other organic pollutants. Chemosphere 29:2237–2252
45. Götz R, Bauer OH, Friesel P, Herrmann T, Jantzen E, Kutzke M, Lauer R, Paepke O, Roch K, Rohweder U, Schwartz R, Sievers S, Stachel B (2007) Vertical profile of PCDD/Fs, dioxin-like PCBs, other PCBs, PAHs, chlorobenzenes, DDX, HCHs, organotin compounds and chlorinated ethers in dated sediment/soil cores from flood-plains of the river Elbe, Germany. Chemosphere 67:592–603
46. Anonym (1997) Aktuelle Bewertung der Gefahrensituation des Spittelwassers I, 8/1997. Staatliches Amt für Umweltschutz Dessau/Wittenberg (unpublished)
47. Anonym (2000) Aktuelle Bewertung der Gefahrensituation des Spittelwassers II, 10/2000. Staatliches Amt für Umweltschutz Dessau/Wittenberg (unpublished)
48. Evers EHG, Laane RWPM, Groenefeld GJJ (1996) Levels, temporal trends and risks of dioxins and related compounds in the Dutch aquatic environment. Organohalogen Compd 28:117–122
49. Büttner O, Böhme M, Rode M (2009) Teilprojekt 3: Hydraulische Modellierung. Abschn 3.8 Überflutung des Salegaster Forsts durch das Hochwasser der Mulde. In: v Tümpling W, Rode M, Böhme M, Gläßer C, Matthies M, Schanze J (eds) SARISK—Entwicklung eines Schadstoffausbreitungsmodells zur stoffbezogenen Risikoanalyse und -bewertung extremer Hochwasserereignisse am Beispiel des Landkreises und der Stadt Bitterfeld. Endbericht des Verbundprojektes. BMBF-Förderkennzeichen PTJ 0330690 A-D, p. 49–80
50. Anonym (2010) Dioxinbericht. Landesbetrieb für Hochwasserschutz und Wasserwirtschaft Sachsen-Anhalt. Gewässerkundlicher Landesdienst (LHW). Nr. 1/2010, 35 p S. http://www.sachsen-anhalt.de/LPSA/fileadmin/Elementbibliothek/Master-Bibliothek-/Landesbetriebe/LHW/neu_PDF5.1/Dokumente_GLD/Dioxinbericht-2007-fertig.pdf
51. Kasimir P, John H-J (2015) Umsetzung des Sedimentmanagementkonzeptes in Sachsen-Anhalt. Presentation at the 23th Chemical Colloquium of the German Federal Institute of Hydrology in Koblenz, June 11–12, 2015
52. Anonym(2009)Kz05.022.0014.Ausschreibungsanzeiger Sachsen-Anhalt vom 29. Mai 2009, p 37/38 Wettbewerblicher Dialog "Frachtreduzierung überwiegend schwebstoffgebundener Schadstoffe der im Grundwasserkörper VM 2-4 gelegenen Fließgewässer mit dem Schwerpunkt Spittelwasser/Schachtgraben". Pilotprojekt Bitterfeld-Wolfen zur Umsetzung der WRRL im Land Sachsen-Anhalt
53. Van der Graaf ER, Koomans RL, Limburg J, de Vries K (2007) In situ radiometric mapping as a proxy of sediment contamination: assessment of the underlying geochemical and –physical principles. Appl Radiation Isotopes 65:619–633
54. CSTEE (2004) Opinion of the Scientific Committee on Toxicity, Ecotoxicity and the Environment (CSTEE) on 'The setting of environmental quality standards for the priority substances included in Annex X of Directive 2000/60/EC in accordance with Article 16 thereof', p. 32 Adopted by the CSTEE during the 43rd plenary meeting of 28 May 2004
55. EAF (2004) Agenda item 5: Emission sources and controls for priority substances (including comments on the proposed quality standards for specific substances). Concept paper from 8 June 2004, p. 77, Expert Advisory Forum on Priority Substances and Pollution Control, 7th EAF Meeting, Brussels, 14–15 June 2004
56. Förstner U, Heise S, Schwartz R, Westrich B, Ahlf W (2004) Historical contaminated sediments and soils at the river basin scale. J Soils Sediments 4(4):247–260
57. Heise S, Förstner U (2006) Risks from historical contaminated sediments in the Rhine basin. Water Air Soil Pollut Focus 6:625–636
58. Förstner U, Flemming H-C, Frimmel FH, Gust G, Jirka GH, von der Kammer F, Larsen O, Markofsky M, Meyer-Reil L-A, Nießner R, Westrich B, Worch E, Zielke W (2007) Sediment- and pollutant-related processes—interdisciplinary approach. In: Westrich B, Förstner U (eds) Sediment Dynamics and Pollutant Mobility in Rivers—An Interdisciplinary Approach. Springer, Heidelberg, pp 15–34
59. Anonymous (2008) Directive 2008/105/EC of the European Parliament and of the Council of 16 December 2008 on quality standards in the field of water policy, amending and subsequently repealing Council Directives 82/176/EEC, 83/513/EEC, 84/156/EEC, 84/491/EEC, 86/280/EEC and amending Directive 2000/60/EC of the European Parliament and of the Council. Official Journal of the European Union, 24.12.2008, L348/84-348/9
60. Förstner U (2009) Sediments and priority substances in river basin—new directive 2008/105/EC; sediment issues in management plans. J Soils Sediments 9(2):89–93

61. ConSoil2000 (2000) Contaminated soil 2000, Case study: comparison of solutions for a large contamination based on different national policies. 7th Int FZK/TNO Conf Contaminated Soil. Leipzig. p. 163

62. Wittmann U, Förstner U, Gier S, Delschen Th, Franzius V, Frauenstein J, Fuchs E, Gläßer W, Lindemann M, Müller R, Meiners G, Wetzel V, Wilke B-M (2000) Case Study Bitterfeld. Contamination of River Basin and Flooding Areas from the Example of the „Spittelwasser"Lowlands at Bitterfeld. German Contribution to Case Study ConSoil2000, 18.09.-22.09.2000 in Leipzig, p. 27

63. Förstner U, Gerth J, Lindemann M, Wittmann U (2001) Managing contaminated sediments—III. In-situ sediment treatment (Spittelwasser case study). J Soils Sediments 1(3)181–187

64. Förstner U, Gerth J, Bley S, Neumann-Hensel H (2007) Methoden der Erfassung, Bewertung und Prognose der intrinsisch/zeitlich verstärkten Schadstoffrückhaltung in kontaminierten Sedimenten (KORA, TV 6). p. 138, Förderkennzeichen 0330519. Laufzeit 01.10.2003 bis 31.12.2006. BMBF, Projektträger Jülich 2007

65. Schwartz R, Gerth J, Neumann-Hensel H, Bley S, Förstner U (2006) Assessment of highly polluted fluvisol in the Spittelwasser floodplain based on national guideline values and MNA-criteria. J Soils Sediments 6(3):145–155

66. Förstner U, Salomons W (2010) Sediment research, management and policy—a decade of JSS. J Soils Sediments 10:1440–1452

67. Haberl R, Grego S, Langergraber G, Kadlec RH, Cicalini A-R, Martins Dias S, Novais JM, Aubert S, Gerth A, Thomas H, Hebner A (2003) Constructed wetlands for the treatment of organic pollutants. J Soils Sediments 3(2):109–124

68. Wisutec (2008) Analyse und Bewertung der technischen Möglichkeiten für eine Problemlösung im Bereich der Oberflächenwässer des Grundwasserkörpers in Bitterfeld. On demand of Großmann Ingenieur Consult GmbH (GICON), 19. June 2008, p. 50, WISUTEC Chemnitz (unpublished)

69. Wisutec (2008) Gewässerunterhaltungsmaßnahme Wiederherstellung des Abflussprofils des Schachtgrabens. Abschlussdokumentation. On demand of P-D ChemiePark Bitterfeld Wolfen GmbH/Landesanstalt für Altlastenfreistellung Sachsen-Anhalt, 05. November 2008, p. 27, WISUTEC Chemnitz (unpublished)

70. Stachel B, Mariani G, Umlauf G, Götz R (2011) Dioxine und PCBs in Feststoffen aus der Elbe, ihren Nebenflüssen und der Nordsee (Längsprofilaufnahme 2008) mit Unterstützung der Flussgebietsgemeinschaft Elbe,FGG Elbe. Redaktion, Frank Krüger

71. G.E.O.S (2013) Bestandsaufnahme belasteter Altsedimente in relevanten Gewässern Sachsen-Anhalts. Phase II. Sedimentbeprobung und Sedimentuntersuchungen. G.E.O.S. Ingenieurgesellschaft mbH, im Auftrag von: Landesbetrieb für Hochwasserschutz und Wasserwirtschaft Sachsen-Anhalt; 30.11.2012; 252 S. http://www.lhw.sachsen-anhalt.de/file-adminBibliothek/Politik_und_Verwaltung/Landesbetriebe/LHW/neu_PDF/5.1/Dokumente_GLD/Sedimentmanagement/Fachgutachten_GEOS/TEIL_II_Bericht_Sedimentuntersuchung.pdf

72. Wieprecht et al (2013) Ermittlung des Remobilisierungspotentials belasteter Altsedimente in ausgewählten Gewässern Sachsen-Anhalts. Institut für Wasser- und Umweltsystemmodellierung (IWS). Lehrstuhl für Wasserbau und Wassermengenwirtschaft der Universität Stuttgart, im Auftrag von: Landesbetrieb für Hochwasserschutz und Wasserwirtschaft Sachsen-Anhalt; Stuttgart, den 11.06.2013; 153 S. http://www.lhw.sachsen-Anhalt.de/fileadmin/Bibliothek/politik_und_Verwaltung/Landesbetriebe/LHW/neu_PDF/5.1/Dokumente_GLD/Sediment-management/Fachgutachten_IWS/Bericht-_Remobilisierungspotential_05-2013.pdf

73. Heise S (2013) Potenzielle Bedeutung der Seitenstrukturen für die Belastung der Elbe—Herangehensweise und erste Ergebnisse. Vom Wasser 111(4):146–147

74. Hillebrand G, Claus E, Heininger P (2014) Die Buhnenfelder der Elbe. In: BfG (ed) Fachbeiträge zum Sedimentmanagementkonzept Elbe. Mitteilungen Nr. 30. Bundesanstalt für Gewässerkunde, Koblenz, p 23–54

75. Wienberg R, Heinze E, Förstner U (1986) Experiments on specific retardation of some organic contaminants by slurry trench materials. In: Assink JW, van den Brink WJ (eds) Contaminated Soil, pp. 848–857. Martinus Nijhoff: Dordrecht

76. Khorasani R, Wienberg R, Förstner U (1988) Solidification, stabilization and encapsulation of organic compounds from remedial actions using inorganic and organic fillers and binding agents. In: Wolf K, van den Brink JW, Colon FJ (eds) Contaminated Soil '88. TNO/BMFT-Conference on Contaminated Soil, pp. 967–976. Kluwer Academic: Dordrecht

77. Salomons W, Förstner U (1988) Environmental Management of Solid Waste: Dredged Material and Mine Tailings. p 390. Springer: Berlin

78. Förstner U (1993) Geochemical concepts in solid waste management. In: Vernet JP (ed) Environmental Contamination. Studies in Environmental Science No. 55, pp 259–292. Elsevier: Amsterdam

79. Förstner U, Gerth J (2001) Natural attenuation—non-destructive processes. In: Stegmann R et al (eds) Treatment of Contaminated Soil. Springer, Berlin, pp 567–586

80. Jacobs PH (2003) Kontaminierte Sedimente und Baggergut: Aktive Barriere-Systeme für die Subaquatische Lagerung und Abdeckung. Dissertation at the University of Technology Hamburg-Harburg, December 3, 2003. p. 198

81. Jacobs PH (2002) A new rechargeable dialysis sampler for monitoring subaqueous in situ sediment caps. Wat Res 36(13):3121–3129

82. Jacobs PH, Förstner U (1999) Concept of subaqueous in situ capping of contaminated sediments with active barrier systems (ABS) using natural and modified zeolites. Wat Res 33(9):2083–2087

83. Jacobs P, Förstner U (2001) Managing contaminated sediments. IV: Subaqueous storage and capping of dredged material. J Soils Sediments 1(4)205–212

84. Jacobs PH, Förstner U (2003) Gewässersedimente und Baggergut. Förstner U, Grathwohl P Ingenieurgeochemie—Natürlicher Abbau und Rückhalt, Stabilisierung von Massenabfällen. Springer, Berlin, pp 330–360

85. Anonymous (2013) Guideline on Ecological Risk Assessment. Schedule B5a; National Environment Protection Council: Canberra, Australia; https://www.comlaw.gov.au/Details/-F2013L00768/97cca2d5-92a7-46a0-80cf-c894d0e913e9

86. Förstner U, Apitz SE (2007) Sediment remediation: U.S. focus on capping and monitored natural recovery. J Soils Sediments 7(6)351–358

87. Apitz SE, Davis JW, Finkelstein K, Hohreiter DL, Hoke R, Jensen RH, Kirtay VJ, Jersac J, Mack EE, Magar V, Moore D, Reible D, Stahl R (2005) Assessing and managing contaminated sediments. Part I: developing an effective investigation and risk evaluation strategy. Integr Environ Assess Manag 1:2–8

88. Bohlen WF, Erickson MJ (2006) Incorporating sediment stability with the management of contaminated sediment sites: a synthesis approach. Integr Environ Assess Manag 2:24–28

89. Ghosh U, Luthy RG, Cornelissen G, Werner D, Menzie CA (2011) In-situ sorbent amendments: a new direction in contaminated sediment management. Environ Sci Technol 45:1163–1168

90. Kupryianchyk D, Rakowska MI, Reible D, Harmsen J, Cornelissen G, van Veggel M, Hale SE, Grotenhuis T, Koelmans AA (2015) Positioning activated carbon amendment technologies in a novel framework for sediment management. Integr Environ Assess Manag 11(2):221–234

91. Harmsen J, Naidu R (2013) Bioavailability as a tool in site management. J Hazard Mater 261:840–846

92. Jonker MTO, Koelmans AA (2002) Sorption of polycyclic aromatic hydrocarbons and polychlorinated biphenyls to soot and soot-like materials in the aqueous environment: mechanistic considerations. Environ Sci Technol 36:3725–3734

93. Booij K, Robinson CD, Burgess RM, Mayer P, Roberts CA, Ahrens L, Allan IJ, Brant J, Jones L, Kraus UR, Larsen MM, Lepom P, Petersen J, Proefrock D, Roose P, Schäfer S, Smedes F, Tixier C, Vorkamp K, Whitehouse P (2016) Passive sampling in regulatory chemical monitoring of nonpolar organic compounds in the aquatic environment. Environ Sci Technol 50(1):3–17

94. ICES (2013) Report of the Workshop on the Application of Passive Sampling and Passive Dosing to Contaminants in Marine Media (WKPSPD); ICES CM 2013/SSGHIE:02; International Council of the Exploration of the Sea: Copenhagen

95. US EPA (2012) Guidelines for using passive samplers to monitor organic contaminants at superfund sediment sites; OSWER Directive 9200.1-110 FS.; United States Environmental Protection Agency, Office of Superfund Remediation and Technology Innovation, and Office of Research and Development: Washington 20460

96. Janssen EML, Beckingham B (2013) Critical review: biological responses to activated carbon amendments in sediment remediation. Environ Sci Technol 47(14):7595–7607

97. Beckingham B, Ghosh U (2011) Field-scale reduction of PCB bioavailability with activated carbon amendment to river sediments. Environ Sci Technol 45:10567–10574

98. Bleiler J, Gardner K, Greenwood S, McCarthy R, Ruiz N (2013) Evaluation of activated-carbon treatment in a PCB-contaminated wetland, in soils/sediments. In: Proc 7th Int Conf Remediation of Contaminated Sediments, 2013, 4–7 February, Dallas (TX) USA. Battelle, C–029

99. Menzie C, Amos B, Kane-Driscoll S, Ghosh U, Gilmour C (2014) Evaluating the efficacy of a low-impact delivery system for in situ treatment of sediments contaminated with methylmercury and other hydrophobic chemicals. ESTCP Environmental Restoration Project ER-200835 http://www.serdp.org/Program-Areas/Environmental-Restoration/Contaminated-Sediments/ER-200835

100. Mostert E, Junier SJ (2009) The European flood risk directive: challenges for research. Hydrol Earth System Sci Disc 6(4):4961–4988

101. Hering D, Borja A, Carstensen J, Carvalho L, Elliott M, Feld CK, Solheim AL (2010) The European water framework directive at the age of 10: a critical review of the achievements with recommendations for the future. Sci Total Environ 408(19):4007–4019

102. Anonymous (2007) Annex to the Communication from the Commission to the Council, the European Parliament and the European Economic and Social Committee, on the implementation of the Community Strategy for dioxins, furans and polychlorinated biphenyls (COM(22001) 593). Second progress report, Brussels, 10.7.2007, p. 15

103. Anonymous (2006) Proposal for a Directive of the European Parliament and of the Council of environmental quality standards in the field of water policy and amending Directive 200/60/EC, Commission Staff Working Document 'Impact Assessment', SEC(2006) 947, Brussels, 17.7.2006, p. 77

104. Wellmitz J (2010) Vergleich der EU-Umweltqualitätsnorm (UQN) für Quecksilber in biologischen Matrizes mit der Belastungssituation in deutschen Oberflächengewässern – Stand der Belastung und Vorschläge für Handlungsoptionen. Fachbericht Umweltbundesamt Berlin Dezember 2010, Fachgebiet II 2.5, p. 26

105. Mohaupt V, Irmer U, Heiss C, Rau F, Schröter-Kermani (o.J.) Stoffliche Herausforderungen für die Wasserwirtschaft—Vorgaben, Grenzwerte, erforderliche Maßnahmen. Umweltbundesamt Berlin, Vortrag beim Ruhrverband. http://www.ruhr-ver-band.de/fileadmin/pdf/wissen/Fachveranstaltungen/Erfurter_Gespraeche/2014/03_mohaupt_DWA_1401.pdf

106. Anonymous (2012) Common implementation strategy for the Water framework directive (2000/60/EC)—Guidance document n° 28—Technical Guidance on the

Preparation of an Inventory of Emissions, Discharges and Losses of Priority and Priority Hazardous Substances. European Commission, Luxemburg

107. Anonymous (2011) Commission regulation (EU) No 1259/2011 of 2 December 2011 amending Regulation (EC) No 1881/2006 as regards maximum levels for dioxins, dioxin-like PCBs and non dioxin-like PCBs in foodstuffs, L 320/18. Official J Eur Union 03(12):2011
108. Anonymous (2014) Commission regulations (EU) No 589/2014 of 2 June 2014 laying down methods of sampling and analysis for the control of levels of dioxin, dioxin-like PCBs and non-dioxin-like PCBs in certain foodstuffs and repealing regulation (EU) No 252/2012, L 164/18. Official J Eur Union 03(06):2014
109. La Rocca C, Mantovani A (2006) From environment to food: the case of PCB. Ann Istituto Superiore di Sanita 42:410
110. Spagnoli JJ, Skinner LC (1977) PCB's in fish from selected waters of New York State. Pesticides monitoring journal 11:69–87
111. Mandal PK (2005) Dioxin: a review of its environmental effects and its aryl hydrocarbon receptor biology. J Comparat Physiol B 175:221–230
112. Apitz S, Power E (2002) From risk assessment to sediment management an international perspective. J Soils Sediments 2:61–66
113. Wenning RJ, Ingersoll CG (2002) Summary of the SETAC Pellston Workshop on Use of Sediment Quality Guidelines and Related Tools for the Assessment of Contaminated Sediments. 17–22 (2002) Fairmont, Montana. Society of Environmental Toxicology and Chemistry (SETAC), Pensacola
114. EQS Dossier (2011) Polychlorinated dibenzo-p-dioxins (PCDDs), polychlorinated dibenzofurans (PCDFs), and dioxin-like polychlorinated biphenyls (dl-PCBs). Prepared by the sub-group on review of the priority substances list (under Working Group E of the Common Implementation Strategy for the Water Framework Directive, p. 35
115. Carere M, Dulio V, Hanke G, Polesello S (2012) Guidance for sediment and biota monitoring under the common implementation strategy for the water framework directive. TrAC Trends Anal Chem 36:15–24
116. Maggi C, Ausili A, Boscolo R, Cacciatore F, Bonometto A, Cornello M, Berto D (2012) Sediment and biota in trend monitoring of contaminants in transitional waters. TrAC Trends Anal Chem 36:82–91
117. Kawajiri K, Fujii-Kuriyama Y (2007) Cytochrome P450 gene regulation and physiological functions mediated by the aryl hydrocarbon receptor. Archiv Biochem Biophys 464:207–212
118. Okey AB, Riddick DS, Harper PA (1994) The Ah receptor: mediator of the toxicity of 2,3,7,8-tetrachlorodibenzo-*p*-dioxin (TCDD) and related compounds. Toxicol Lett 70:1–22
119. Ahlf W, Hollert H, Neumann-Hensel H, Ricking M (2002) A guidance for the assessment and evaluation of sediment quality. A German approach based on ecotoxicological and chemical measurements. J Soils Sediments 2:37–42
120. Besselink HT, Schipper C, Klamer H, Leonards P, Verhaar H, Felzel E, Murk AJ, Thain J, Hosoe K, Schoeters G, Legler J, Brouwer B (2004) Intra- and interlaboratory calibration of the DR CALUX® bioassay for the analysis of dioxins and dioxin-like chemicals in sediments. Environ Toxicol Chem 23:2781–2789
121. den Besten P, de Deckere E, Babut M, Power B, DelValls TA, Zago C, Oen AP, Heise S (2003) Biological effects-based sediment quality in ecological risk assessment for European waters. J Soils Sediments 3:144–162
122. Eichbaum K, Brinkmann M, Buchinger S, Hecker M, Engwall M, Bavel B, Reifferscheid G, Hollert H (2013) The dioRAMA project: assessment of dioxin-like activity in sediments and fish (*Rutilus rutilus*) in support of the ecotoxicological characterization of sediments. J Soils Sediments 13:770–774
123. Giesy JP, Jude DJ, Tillitt DE, Gale RW, Meadows JC, Zajieck JL, Peterman PH, Verbrugge DA, Sanderson JT, Schwartz TR, Tuchman ML (1997) Polychlorinated dibenzo-p-dioxins, dibenzofurans, biphenyls and 2,3,7,8-tetrachlorodibenzo-p-dioxin equivalents in fishes from Saginaw Bay, Michigan. Environ Toxicol Chem 16:713–724
124. Wernersson A-S et al (2015) The European technical report on aquatic effect-based monitoring tools under the water framework directive. Environ Sci Eur 27:7
125. Van den Berg M, Birnbaum L, Denison M, De Vito M, Farland W, Feeley M, Fiedler H, Hakansson H, Hanberg A, Haws L, Rose M, Safe S, Schrenk D, Tohyama C, Tritscher A, Tuomisto J, Tysklind M, Walker N, Peterson RE (2006) The 2005 World Health Organization reevaluation of human and mammalian toxic equivalency factors for dioxins and dioxin-like compounds. Toxicol Sci 93:223–241
126. Manz W, Krebs F, Schipper CA, Den Besten PJ (2007) Status of ecotoxicological assessment of sediment and dredged material in Germany and the Netherlands – Dutch-German Exchange (DGE) on dredged material
127. Anonymous (2012): Commission Regulation (EU) No 252/2012 of 21 March 2012 laying down methods of sampling and analysis for the official control of levels of dioxins, dioxin-like PCBs and non-dioxin-like PCBs in certain foodstuffs and repealing Regulation (EC1883/2006, Official Journal of the European Union
128. BfG (2002) Das Augusthochwasser 2002 im Elbegebiet. Mit einem Sonderprogramm zur Schadstoffbelastung der Elbe-Schwebstoffe während des Hochwassers, p. 27–31. Koblenz, September 2002
129. Franke S, Heinzel N, Specht M, Francke W (2004) AP 3.10: Organische Schadstoffe im Gebiet der Unteren Mulde—Non Target Screening: Analytik von Wasser, Grundwasser und Sediment—Ergebnisse und Konsequenzen. In: Geller W et al. (eds) Schadstoffbelastung nach dem Elbe-Hochwasser 2002. Endbericht des Ad hoc-Verbundprojekts, p. 206–223
130. Schwartz R, Kozerski HP (2004) AP 4.2: Bestimmung des Gefahrenpotentials feinkörniger Buhnenfeldsedimente

für die Wasser- und Schwebstoffqualität der Elbe sowie den Stoffeintrag in Auen. In: Geller W et al. (eds) Schadstoffbelastung nach dem Elbe-Hochwasser 2002. Endbericht des Ad hoc-Verbundprojekts, p. 258–274

131. Pepelnik R, Niedergesäß R, Erbslöh B, Aulinger A, Prange A (2004) AP 3.2: Längsprofiluntersuchungen zur Beurteilung von Auswirkungen des Hochwassers vom August 2002 auf die Wasser- und Sedimentqualität der Elbe. In: Geller W et al. (eds) Schadstoffbelastung nach dem Elbe-Hochwasser 2002. Endbericht des Ad hoc-Verbundprojekts, p. 82–100

132. Hanisch Ch, Zerling L, Junge FW, Czegka W (2005) Verlagerung, Verdünnung und Austrag von schwermetallbelasteten Flusssedimenten im Einzugsgebiet der Saale. Abh der Sächsischen Akademie der Wissenschaften zu Leipzig, Math-Naturw Klasse Band 64, H 1, Leipzig, p. 135

133. Anonymous (2005) Extract of the DG RTD Unit 1.4 Catalogue of Contracts. Topic: Natural Hazards. Flood related EU hazard research projects. Directorate General for Research. Version October 2005.http://ec.europa.eu/research/environment/pdf/ec_flood-_rtd_-pro–ject_catalogue_en.pdf

134. Bardossy A, Färber A, St Schönau, Westrich B, Jancke Th, Schmid G, Wurms S, Stahr K, Gaiser T, Frauenknecht H, Streck Th, Ingwersen J, Kreschnak C, Kern U, Heppelmann H, Förstner U, Jacobs J (2005) RIMAX-Verbundprojekt Entwicklung eines integrativen Bewirtschaftungskonzepts für Trockenbecken und Polder zur Hochwasserrückhaltung. UWSF Z Umweltchem Ökotox 18(1):67–69

135. RBC Elbe (2014) Das Messprogramm Extremereignisse beim Junihochwasser der Elbe 2013—Schadstoffkonzentrationen und –frachten. p. 36. Koordinierung: Bundesanstalt für Gewässerkunde (BfG) Koblenz, Oktober 2014)

136. RBC Elbe (2014) Entwurf des Hochwasserrisikomanagementplans gemäß § 75 WHG bzw. Artikel 7 der Richtlinie 2007/60/EG über die Bewertung und das Management von Hochwasserrisiken für den deutschen Teil der Flussgebietseinheit Elbe. p. 84. FGG-Elbe, Magdeburg, 02. Dezember 2014

137. Quevauviller P (2011) WFD River Basin management planning in the context of climate change adaptation—policy and research trends. European Water 34:19–25

138. Quevauviller P, Gemmer M (2015) EU and international policies for hydrometeorological risks: operational aspects and link to climate action. Adv Climate Change Res 6(74e):79

139. Garnier M, Harper DM, Blaskovicova L, Hancz G, Janauer GA, Jolánkai Z, Lanz E, Porto AL, Mándoki M, Pataki B, Rahuel J-L, Robinson VJ, Stoate C, Tóth E, Jolánkai G (2015) Climate change an European water bodies, a review of existing gaps an future research needs: findings of the climate water project. Environ Manag 56:271–285

140. Westrich B, Förstner U (2007) Sediment dynamics and pollutant mobility in rivers—An interdisciplinary approach. p. 430, Springer: Berlin

141. Hollert H, Heise S, Keiter S, Heininger P, Förstner U (2007) Wasserrahmenrichtlinie—Fortschritte und Defizite. Umweltwiss Schadst Forsch 19:58–70

142. Hollert H, Brinkmann M, Hudjetz S, Cofalla C, Schüttrumpf H (2014) Hochwasser—ein unterschätztes Risiko. Biol unserer Zeit 44:44–51

143. Stigliani WM (1988) Changes in valued "capacities" of soils and sediments as indicators of nonlinear and time-delayed environmental effects. Environ Monitoring Assess 10:245–307

144. Stigliani WM (1991) Chemical time bombs: definition, concepts, and examples. Executive report 16 (CTB basic document). IIASA Laxenburg, Austria, p. 23

145. Schüttrumpf H, Bachmann D (2008): Klimaänderungen und Hochwasserschutz. In: Pinnekamp J (ed), 41. Essener Tagung für Wasser-und Abfallwirtschaft Aachen (Gewässerschutz, Wasser, Abwasser), p 69/1–69/14

146. Hulme M, Jenkins G, Lu X, Turnpenny J, Mitchell TD, Jones RG, Lowe J, Murphy J, Hassell D, Boorman P, McDonald R, Hill S (2002) Climate Change Scenarios for the United Kingdom: The UKCIP02 Scientific Report, Tyndall Centre for Climate Change Research, School of Environmental Sciences, University of East Anglia, Norwich, UK, p. 120

147. Kay AL, Jones RG, Reynard NS (2006) RCM rainfall for UK flood frequency estimation II. Climate change results. J Hydrol 318:163–172

148. Wilby RL, Orr HG, Hedger M, Forrow D, Blackmore M (2006) Risks posed by climate change to the delivery of water framework directive objectives in the UK. Environ Intern 32:1043–1055

149. Haag I, Kern U, Westrich B (2001) Erosion investigation and sediment quality measurements for a comprehensive risk assessment of contaminated aquatic sediments. Sci Total Environ 266:249–257

150. Kern U, Haag I, Holzwarth M, Westrich B (1999) A flume for the determination of depth-dependent erosion stability of aquatic sediments: the SETEG duct system. Wasserwirtschaft 89:72–77

151. Spork V, Ruland P, Cornelisse JM, McLoughlin B (1994) European large installation plan: improving experimental methods for erosion tests, Proc 4th Nearshore and Estuarine Cohesive Sediment Transport Conference INTERCOH. Wallingford, England

152. Spork V, Eisler R, Köngeter J (1998) Optimisation of experimental conditions for annular flumes by LDV measurements. In: Jajawardena AW, Lee JHW, Wang ZY (eds), Proc 7th Intern Conf River Sedimentation, Hong Kong, pp. 329–335

153. Thomsen L, Gust G (2000) Sediment erosion thresholds and characteristics of resuspended aggregates on the western European continental margin. Deep Sea Res Part 1. Oceanogr Res Pap 47:1881–1897

154. Tolhurst TJ, Black KS, Shayler SA, Mather S, Black I, Baker K, Paterson DM (1999) Measuring the in situ erosion shear stress of intertidal sediments with the cohesive strength meter (CSM). Estuar Coast Shelf Sci 49:281–294

155. Aardema MJ, MacGregor JT (2002) Toxicology and genetic toxicology in the new era of "toxicogenomics" impact of "-omics" technologies. Mutat Res 499:13–25

156. Diepens NJ, Arts GHP, Brock TCM, Smidt H, Van Den Brink PJ, Van Den Heuvel-Greve MJ, Koelmans AA (2013) Sediment toxicity testing of organic chemicals in the context of prospective risk assessment: a review. Crit Rev Environ Sci Technol 44:255–302

157. Edge KJ, Dafforn KA, Simpson SL, Ringwood AH, Johnston EL (2015) Resuspended contaminated sediments cause sublethal stress to oysters: a biomarker differentiates total suspended solids and contaminant effects. Environ Toxicol Chem 34:1345–1353

158. Burton GA (1991) Assessing the toxicity of freshwater sediments. Environ Toxicol Chem 10(12):1585–1627

159. Heininger P, Pelzer J, Claus E, Pfitzner S (2003) Results of long-term sediment quality studies on the River Elbe. Acta Hydrochim Hydrobiol 31(4–5):356–367

160. Brinkmann M, Hudjetz S, Keiter S, Seiler TB, Wölz J, Hallare AV, Hollert H, Cofalla C, Roger S, Schüttrumpf H, Gerbersdorf SU (2010) Toxizität und Risk Assessment fluvialer Sedimente und Schwebstoffe: eine kurze Übersicht bisheriger und neuerer Entwicklungen. Umweltwiss Schadst-Forsch 22:651–655

161. Cofalla C, Hudjetz S, Roger S, Brinkmann M, Frings R, Wölz J, Schmidt B, Schäffer A, Kammann U, Hecker M, Hollert H, Schüttrumpf H (2012) A combined hydraulic and toxicological approach to assess re-suspended sediments during simulated flood events—part II: an interdisciplinary experimental methodology. J Soils Sediments 12:429–442

162. Gerbersdorf S, Hollert H, Brinkmann M, Wieprecht S, Schuttrumpf H, Manz W (2011) Anthropogenic pollutants affect ecosystem services of freshwater sediments: the need for a "triad plus x" approach. J Soils Sediments 11:1099–1114

163. Gerbersdorf SU, Cimatoribus C, Class H, Engesser K-H, Helbich S, Hollert H, Lange C, Kranert M, Metzger J, Nowak W, Seiler T-B, Steger K, Steinmetz H, Wieprecht S (2015) Anthropogenic Trace Compounds (ATCs) in aquatic habitats—Research needs on sources, fate, detection and toxicity to ensure timely elimination strategies and risk management. Environ Intern 79:85–105

164. Hudjetz S, Herrmann H, Cofalla C, Brinkmann M, Kammann U, Schäffer A, Schüttrumpf H, Hollert H (2014) An attempt to assess the relevance of flood events—biomarker response of rainbow trout exposed to resuspended natural sediments in an annular flume. Environ Sci Pollut Res 21:13744–13757

165. Oetken M, Stachel B, Pfenninger M, Oehlmann J (2005) Impact of a flood disaster on sediment toxicity in a major river system–the Elbe flood 2002 as a case study. Environ Pollut 134:87–95

166. Schüttrumpf H, Brinkmann M, Cofalla C, Frings R, Gerbersdorf S, Hecker M, Hudjetz S, Kammann U, Lennartz G, Roger S, Schäffer A, Hollert H (2011) A new approach to investigate the interactions between sediment transport and ecotoxicological processes during flood events. Environ Sci Europe 23:39

167. Wölz J, Cofalla C, Hudjetz S, Roger S, Brinkmann M, Schmidt B, Schäffer A, Kammann U, Lennartz G, Hecker M, Schuttrumpf H, Hollert H (2009) In search for the ecological and toxicological relevance of sediment re-mobilisation and transport during flood events. J Soils Sediments 9:1–5

168. Brinkmann M, Hudjetz S, Cofalla C, Roger S, Kammann U, Zhang X, Wiseman S, Giesy J, Hecker M, Schuttrumpf H, Wolz J, Hollert H (2010) A combined hydraulic and toxicological approach to assess re-suspended sediments during simulated flood events: part I—multiple biomarkers in rainbow trout. J Soils Sediments 10:1347–1361

169. Eichbaum K, Brinkmann M, Buchinger S, Hecker M, Engwall M, Bavel B, Reifferscheid G, Hollert H (2013) The dioRAMA project: assessment of dioxin-like activity in sediments and fish (*Rutilus rutilus*) in support of the ecotoxicological characterization of sediments. J Soils Sediments 13:770–774

170. Brinkmann M, Hudjetz S, Kammann U, Hennig M, Kuckelkorn J, Chinoraks M, Cofalla C, Wiseman S, Giesy JP, Schäffer A, Hecker M, Wölz J, Schüttrumpf H, Hollert H (2013) How flood events affect rainbow trout: evidence of a biomarker cascade in rainbow trout after exposure to PAH contaminated sediment suspensions. Aquat Toxicol 128–129:13–24

171. Kammann U (2007) PAH metabolites in bile fluids of dab (*Limanda limanda*) and flounder (*Platichthys flesus*)—spatial distribution and seasonal changes. Environ Sci Pollut Res 14:102–108

172. Kammann U, Brinkmann M, Freese M, Pohlmann J-D, Stoffels S, Hollert H, Hanel R (2013) PAH metabolites, GST and EROD in European eel (*Anguilla anguilla*) as possible indicators for eel habitat quality in German rivers. Environ Sci Pollut Res 21(4):2519–2530

173. Brinkmann M, Eichbaum K, Reininghaus M, Koglin S, Kammann U, Baumann L, Segner H, Zennegg M, Buchinger S, Reifferscheid G, Hollert H (2015) Towards science-based sediment quality standards—effects of field-collected sediments in rainbow trout (*Oncorhynchus mykiss*). Aquat Toxicol 166:50–62

174. Brinkmann M, Eichbaum K, Buchinger S, Reifferscheid G, Bui T, Schäffer A, Hollert H, Preuss TG (2014) Understanding receptor-mediated effects in rainbow trout: in vitro–in vivo extrapolation using physiologically based toxicokinetic models. Environ Sci Technol 48(6):3303–3309

175. Brinkmann M, Freese M, Pohlmann J-D, Kammann U, Preuss TG, Buchinger S, Reifferscheid G, Beiermeister A, Hanel R, Hollert H (2015) A physiologically based toxicokinetic (PBTK) model for moderately hydrophobic organic chemicals in the European eel (*Anguilla anguilla*). Sci Total Environ 536:279–287

176. Anonymous (2010) Commission decision of 1 September 2010 on criteria and methodological standard on good environmental status of marine waters. Document 2010/477/EU, Official J European Union, 02.09.2010, L 132/14

177. Law R, Hanke G, Angelidis M, Batty J, Bignert A, Dachs J, Davies I, Denga Y, Duffek A, Herut B, Hylland K, Lepom P, Leonards P, Mehtonen J, Piha H, Roose P, Tronczynski J, Velikova V, Vethaak D (2010) Marine Strategy Framework Directive, Task Group 8 Report, Contaminants and Pollution Effects. Joint Report, JRC Scientific and Technical Reports, EUR 24335 EN—April 2010, p. 171, Ispra
178. Anonymous (2014) The first phase of implementation of the Marine Strategy Framework Directive (2008/5/EC)—The European Commission's assessment and guidance. Commission Staff Working Document, Commission Report to the Council and the European Parliament, COM(2014) 97 final, Brussels, 20.02.2014
179. Anonymous (2014) Section 8, Descriptor 8 (Contaminants). In: Article 12 Technical Assessment of the MSFD 2012 obligations of Germany. Final version, 7 February 2014, pp. 34–42
180. OSPAR (2007) Background Document on Dioxins, 2007 update. OSPAR (Convention for the Protection of the Marine Environment of the North-East Atlantic) Commission, Hazardous Substances Series, p. 59, ISBN 978-1-905859-47-4
181. OSPAR (2012) MSFD Advice document on Good environmental status—Descriptor 8: Contaminants A living document—Version of 2 March 2012. Approaches to determining good environmental status, setting of environmental targets and selecting indicators for Marine Strategy Framework Directive descriptor 8 Hazardous Substances Series, version of March 2, 2012. p. 26
182. OSPAR (2014) OSPAR Regional Plan to improve adequacy and coherence of MSFD implementation 2014-2018, version 10 December 2014, p. 13
183. Wenzel C (2012) Die EG-Meeresstrategie-Rahmenrichtlinie (MSRL)—Bedeutung für die Bewertung von Sedimentkontaminationen. In: Magdeburger Gewässerschutzseminar 2012 Die Elbe und ihre Sedimente. Hamburg 10./11 Oct 2012, p. 72–76. http://www.ikse.mkol.org/ileadmin/download/MGS/MGS_2012_Tagungsband_Sbornik.pdf
184. Anonymous (2015) Entwurf des MSRL-Maßnahmenprogramms zum Meeresschutz der deutschen Nord- und Ostsee. Bericht der Bundesregierung gemäß §45 h Absatz 1 des Wasserhaushaltsgesetzes. Stand: 31.03.2015. p. 138
185. Anonymous (1995) Dioxin-Bilanz für Hamburg, University of Bayreuth and Trischler & Partner, on demand of the Environmental Agency of the Free and Hanseatic City of Hamburg. Summary, September 1995, p. 9
186. Uhlig S, Simon K, Kunath K (2007) Dioxine und Furane in Elbe, Mulde und Spittelwasser: Statistische Analyse der Kongenerenmuster. Gesellschaft für Qualitätsmanagement und Statistik mbH, Anlage 1 zur HPA-Risikostudie Mai 2008, 60 S. http://www.tideelbe.de/files/anlage_1_-_dioxine_und -furane_in_elbe__mulde_und_spittelwasser.pdf
187. Stachel B, Christoph EH, Götz R, Herrmann T, Krüger F, Lay J, Löffler J, Päpke O, Reincke H, Schröter-Kermani C, Schwartz R, Steeg E, Stehr D, Uhlig S, Umlauf G (2006) Contamination of the alluvial plain, feeding-stuffs and foodstuffs with polychlorinated diebenzo-p-dioxins, polychlorinated dibenzofurans (PCDD/Fs), dioxin-like polychlorinated biphenyls (DL-PCBs) and mercury from the River Elbe in the light of the flood event in August 2002. Sci Total Environ 364:96–112
188. Court of Justice of the European Union (2014) Schlussanträge des Generalanwalts Nilo Jääskinen vom 23. Oktober 2014 in der Rechtssache C-461/13 “Bund für Umwelt und Naturschutz Deutschland e.V. gegen Bundesrepublik Deutschland“. Luxembourg, http://www.eu-koordination.de/PDF/schlussantrag-eugh%20generalanwalt%20weservertiefung.pdf
189. Anonymous (2006) Renewed EU Sustainable Development Strategy. Review of the EU Sustainable Development Strategy (EU SDS). Council of the European Union, 10917 Annex, Brussels 26 June 2006
190. Anonymous (2009/2013) Common transitory provisions for handling dredged materials in coastal waters. August 2009, p. 39 www.htg-baggergut.de/(in German)
191. Salomons W (2008) Linking soils, sediments and catchment ecosystems. J Soils Sediments 8(2):67–68
192. Salomons W, Gandrass J (2001) Dredged material in the port of Rotterdam—Interface between the Rhine catchment and the sea. http://coast.gkss.de/aos/dredged_material
193. Heise S, Förstner U, Westrich B, Jancke T, Karnahl J, Salomons W (2004) Inventory of Historical Contaminated Sediment in Rhine Basin and its Tributaries. Technical University of Hamburg-Harburg and University of Stuttgart, on behalf of the Port of Rotterdam. October 2004, Hamburg, p. 225
194. Schipper C, Schout P (2003) De weg naar implementatie van de Chemie-Toxiciteit-Toets. Werkdocument: RIKZ/2003.036. AKWA/RIKZ 04.005. ISBN: 36934761, p. 82
195. Lyklema T (2005), reprenting the NGOs Stichting Reinwater, Stichting De Noordzee, Waddenvereniging, Gelderse Milieufederatie, Natuur en Milieufederatie Utrecht, Rhein Kolleg. Complaint about Iffezheim. http://www.reinwater.nl/docs/Iffezheim.pdf
196. Vietoris F (2013) Ableitung von Umweltqualitätsnormen (UQN-RL) für das Erreichen von Handlungszielen im Gewässerschutz. In: Bundesanstalt für Gewässerkunde (ed) Bioakkumulation in aquatischen Systemen: Methoden, Monitoring, Bewertung. Kolloq. 6./7. 03.2013 Koblenz. S. 5-10. DOI: 10.5675/BfG_Veranst_2013.7
197. Breuer R (2008) Sedimentmanagement für die Elbe – Rechtsgutachten von Prof. Dr. Rüdiger Breuer, Köhler & Klett Rechtsanwälte Partnerschaft, Köln, zum Strombau- und Sedimentmanagementkonzept, das von Hamburg Port Authority (HPA) und der Wasser- und Schifffahrtsverwaltung (WSV) des Bundes am 01.06.2008 vorgelegt worden ist. 19 S.www.tideelbe.de/files/sedimentmanagement_fuer_die_elbe.pdf
198. Anonym (2008) Strombau- und Sedimentmanagementkonzept für die Tideelbe. Hamburg Port Authority und Wasser- und Schifffahrtsverwaltung

des Bundes. p. 39, June 1, 2008. http://www.hamburg-port-authority.de/de/presse/studien-und-berichte/Documents/SB-SM-Konzept-HPA-WSV.pdf

199. Anonym (2014): Sedimentmanagement Tideelbe—Strategien und Potenziale—Systemstudie II. Ökologische Auswirkungen der Unterbringung von Feinmaterial. Band 1 (2), Endbericht. Im Auftrag des Wasser- und Schifffahrtsamtes Hamburg. Bundesanstalt für Gewässerkunde, Koblenz, BfG-1763, p. 266, doi: 10.5675/BfG-1763

200. Förstner U (2015) Dioxin und Sedimentaltlasten im Elbeeinzugsgebiet unter der WRRL 2000-2015 - Kontroversen zwischen Behörden und Fachwissenschaften. Institut für Umwelttechnik und Energiewirtschaft, Technishe Universität Hamburg-Harburg, p. 104

CHAPTER 5

CHALLENGES IN GROUNDWATER RESOURCE MANAGEMENT

Jean-Christophe Comte, Rachel Cassidy, Joy Obando, Nicholas Robins, Kassim Ibrahim, Simon Melchioly, Ibrahimu Mjemah, Halimu Shauri, Anli Bourhane, Ibrahim Mohamed, Christine Noe ,Beatrice Mwega, Mary Makokha, Jean-Lambert Join, Olivier Banton, Jeffrey Davies

Keywords: Groundwater, Coastal aquifer, Eastern Africa, Environmental change, Governance, Community engagement

GROUNDWATER IN COASTAL REGIONS OF EAST AFRICA

Africa is the continent with the fastest growing populations (United Nations, 2011) with coastal regions projected to experience the highest rates of population growth in coming decades (Vafeidis et al., 2011). At the continental level, East Africa has the second highest rate of population growth, after the Central African region (Ashton and Turton, 2009), while having the lowest renewable freshwater resource of Sub-Saharan Africa (Braun and Xu, 2010). In the Mozambique Channel region, the populations of the Comoros Islands, Kenya and Tanzania have quadrupled in the last 50 years (World Bank, 2014a), twice the global average and growing. Most of the population increase is in urban areas, which are already densely populated and where water resources, particularly groundwater which is often the only source of water of acceptable quality, are already under intense pressure (Steyl and Dennis, 2010; MacDonald et al., 2012; Walraevens et al., 2015). Recent papers have shown the importance of groundwater resources that are under growing pressure in developing regions elsewhere, but crucial for economic development (e.g., Mukherjee et al., 2015 and Watto and Mugera, 2015). Adequate provision of water is central to human health and economic development in these regions where water scarcity is a serious impediment to growth and poses a threat to political stability.

In a demographic context within Sub-Saharan Africa, groundwater resources are heavily relied upon and the focus of strategic development (Ashton and Turton, 2009) because of both the relative resilience of aquifers to anticipated climate change and the widespread contamination of surface water resources. However, as exemplified by findings from previous inter-African reviews of regional and national groundwater management frameworks (e.g. Robins et al., 2006, Adelana et al., 2008, Braune and Xu, 2008 and Braune and Xu, 2010; Knüppe, 2011), groundwater resources, except in countries totally dependent on them, (i) still suffer from an under-evaluation of their importance and significance, (ii) are often managed separately from surface water and (iii) management institutions (governments, communities, NGOs, consultants) are largely fragmented and lack a central strategy. Groundwater information services, i.e. databases and systematic long-term monitoring, are non-existent or of inadequate quality and fragmented. The involvement of stakeholders, including communities, in decision-making processes and resource utilisation is insufficient, under-acknowledged by managers and governments and requires urgent capacity building of all parties, from individual to institutional (e.g. BGR, 2007; Foster et al., 2008 and Braune and Xu, 2010).

Efforts towards the integration of groundwater management within holistic water management frameworks, such as river basin organisations have only emerged recently, mainly through national Integrated Water Resource Management (IWRM) frameworks (Global Water Partnership, 2000 and Global Water Partnership, 2002; World Water Council, 2006) that aim to address these deficiencies and contribute to the Millennium Development Goals (MDGs). Nonetheless, groundwater management appears subsumed under broader policy, legal and institutional frameworks dealing with the management of water resources; hence the integration of groundwater into national policy requires development of adequate cross-sector dialogue within government (Mumma et al., 2011 and Foster et al.,

Citation: Jean-Christophe Comte, Rachel Cassidy, Joy Obando, Nicholas Robins, Kassim Ibrahim, Simon Melchioly, Ibrahimu Mjemah, Halimu Shauri, Anli Bourhane, Ibrahim Mohamed, Christine Noe, Beatrice Mwega, Mary Makokha, Jean-Lambert Join, Olivier Banton, Jeffrey Davies, Challenges in groundwater resource management, Journal of Hydrology: Regional Studies, Volume 5, March 2016, Pages 179-199, ISSN 2214-5818, http://dx.doi.org/10.1016/j.ejrh.2015.12.065.

2012). The management of coastal groundwater poses a further challenge due to its vulnerability to seawater contamination and the specific physical and socio-economic characteristics of the coastal zone. The review ofSteyl and Dennis (2010) is one of the very few works that provides insights on common issues with regards to groundwater management in coastal aquifers of Africa. There is a notable gap in the literature regarding the inter-disciplinary aspects of the management of coastal groundwater resources in Sub-Saharan Africa and the regionally-specific socio-environmental drivers.

In the East African coastal region, demographic change has led to an overall increase in groundwater abstraction, with increased drilling of deep boreholes with higher abstraction rates than traditional dug wells and shallow boreholes. High abstraction rates and concentrated well fields are incompatible with the nature of coastal aquifers. These aquifers are mainly low-lying with shallow water tables and are susceptible to seawater intrusion if not carefully managed, regardless of aquifer productivity or recharge rates (Robins, 2013 and Werner et al., 2013). The abstraction of inland groundwater, by way of contrast is generally only limited by aquifer productivity and available recharge.

The geology of much of the East African coastal aquifers is young (Mesozoic to present), composed of soft or unstable sediments and volcanic deposits, which need supporting during and after drilling to avoid collapse and bloc fall. Though slower and labour intensive to construct, large diameter traditional wells are more suited to such environments; causing less drawdown than boreholes and giving access for clearance of debris (e.g. Bourhane et al., 2015). A larger number of shallow wells widely distributed across an area is effective in minimising the risk of seawater intrusion. Such strategies, however, pose a management challenge in urban areas (Edmunds, 2012) where high densities of abstraction points may be necessary, initial drilling costs are high and risk of contamination is considerable.

East Africa is identified as one of the regions at greatest risk globally from the impacts of climate change (Hinkel et al., 2012). Most coastal areas are low-lying and will experience significant inundation even with modest rises in sea level. The most recent climate change projections (IPCC, 2014a and Cai et al., 2014) anticipate an accelerated rise in global mean sea level in coming decades (Watson et al., 2015) combined with an increase in rainfall during the wet seasons and higher annual temperatures. These may increase potential evapotranspiration and increase seawater intrusion in low-lying coastal settings and islands (Comte et al., 2014). According to the latest report of the Intergovernmental Panel on Climate Change (IPCC, 2014a and IPCC, 2014b), the available information or evidence on projected impacts of climate change on coastal aquifers is globally limited. More specifically, the impact on aquifer recharge in East Africa remains poorly constrained (Döll, 2009) due to a lack of hydrometric monitoring which could be used to inform forward modelling and proactive development of management approaches (Robins et al., 2006, Adelana, 2009 and Steyl and Dennis, 2010).

The combined effects of demographic, socio-economic, climatic and political changes on coastal groundwater resources negatively impact both the availability and sustainability of potable groundwater. Although an adverse impact is expected, the extent of the problem is not well established because of the large number of driving factors involved and the paucity of data with which to test them. Better understanding requires the implementation of integrated multidisciplinary research and expertise (including hydrogeologists, hydrologists, geographers, and socio-environmentalists) in order to improve the knowledge and understanding both of the vulnerability of coastal groundwater and its long-term response to change. This will permit identification of appropriate management strategies to mitigate negative impact on resources. This constitutes a necessary and major step towards the sustainable development and management of water accessibility for people in coastal East Africa.

AIMS AND OBJECTIVES

A suite of investigations was undertaken at three contrasting coastal areas in East Africa; Grande Comore (Ngazidja) Island (11°40′S; 43°20′E) in the Comoros Archipelago, Kilifi County, North of Mombasa, in Kenya (3°37′S; 39°50′E), and Kilwa, a rural area in southeast Tanzania (8°55′S; 39°30′E). These represent three of the lowest-income countries (LIC) (as classified by the World Bank, 2014b) of the East Africa/South-western Indian Ocean zone (Fig. 1) and are exemplars of the physical, social and cultural diversity of the coastal fringes of the region. An important component of the work was testing whether the concerns raised by Steyl and Dennis (2010), regarding the sustainable management of African coastal groundwater resources, apply consistently to the three study areas.

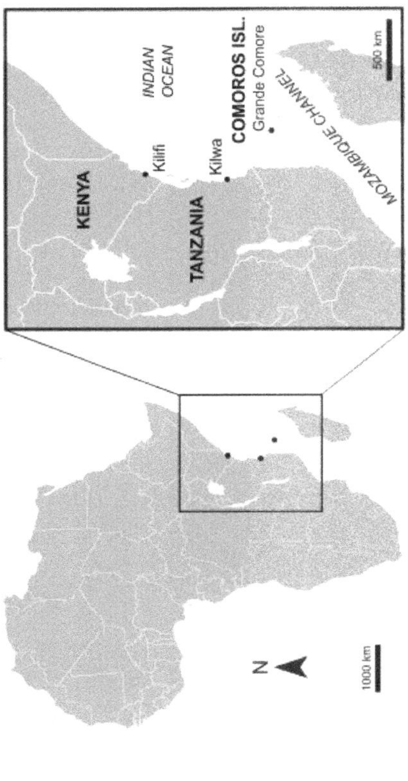

Figure 1: Location of the three study sites in the coastal zones of the Comoros Islands, Kenya and Tanzania.

Three research sites covering the (1) societal, (2) environmental and (3) hydrogeological status of the study areas were established and within each a consistent work programme was applied. The overall research programme aimed to understand the physical and societal aspects of groundwater accessibility, to provide up-to-date regional groundwater databases and identify critical aspects of groundwater access as well as knowledge gaps that will require further longer-term monitoring and research. In this paper we focus on the hydrogeological outcomes of the research, framed within a summary of the principal societal and environmental issues identified in each area. Issues relating groundwater management with governance and institutional frameworks in the study areas would be addressed in detail in another publication (Obando et al., in preparation) focussing specifically on the socio-economic aspects of this research programme.

STUDY AREAS-OVERVIEW

Comoros Islands—Grande Comore (Ngazidja) (see map in Fig. 2 a)

In the Comoros archipelago, the island of Grande Comore (1148 km^2) was selected for the study. As the most populated island in the region, and growing at over 2% per annum (~400,000 inhabitants estimated in 2015), more than half of the population is concentrated within 5 km of the shoreline due to the steep gradients of the elevated and afforested interior of the island. About 65% of the population do not have permanent access to groundwater and primarily harvest rainwater from roofs into tanks, which are only sufficient during the wet season and have issues with bacterial contamination. During the dry season, water use is rationed. For families that can afford it, freshwater is distributed by water trucks delivering from the few fresh abstraction wells on the island. These comprise 54 wells drilled in the volcanic aquifers of the coastal zone which supply 20 localities representing about 35% of the island's population (Mohamed and Othman, 2006). Fewer than 30% of the wells provide groundwater of acceptable quality, i.e. Total Dissolved Solids (TDS) < 1 g L^{-1} (Bourhane et al., 2015) and consequently the local drinking water salinity guideline is usually taken at 3 g L^{-1} instead of 1 g L^{-1} as recommended by the World Health Organisation (WHO, 2003). Wells with higher salinity continue to be used for irrigation, livestock or washing. This situation is not projected to improve in coming decades with anticipated population growth (demand is increasing at about twice the rate of groundwater development) and projected climate change (observed decrease in rainfall; Vincent et al., 2011). The large variations in groundwater salinity are locally responsible for community conflicts with regard to water costs.

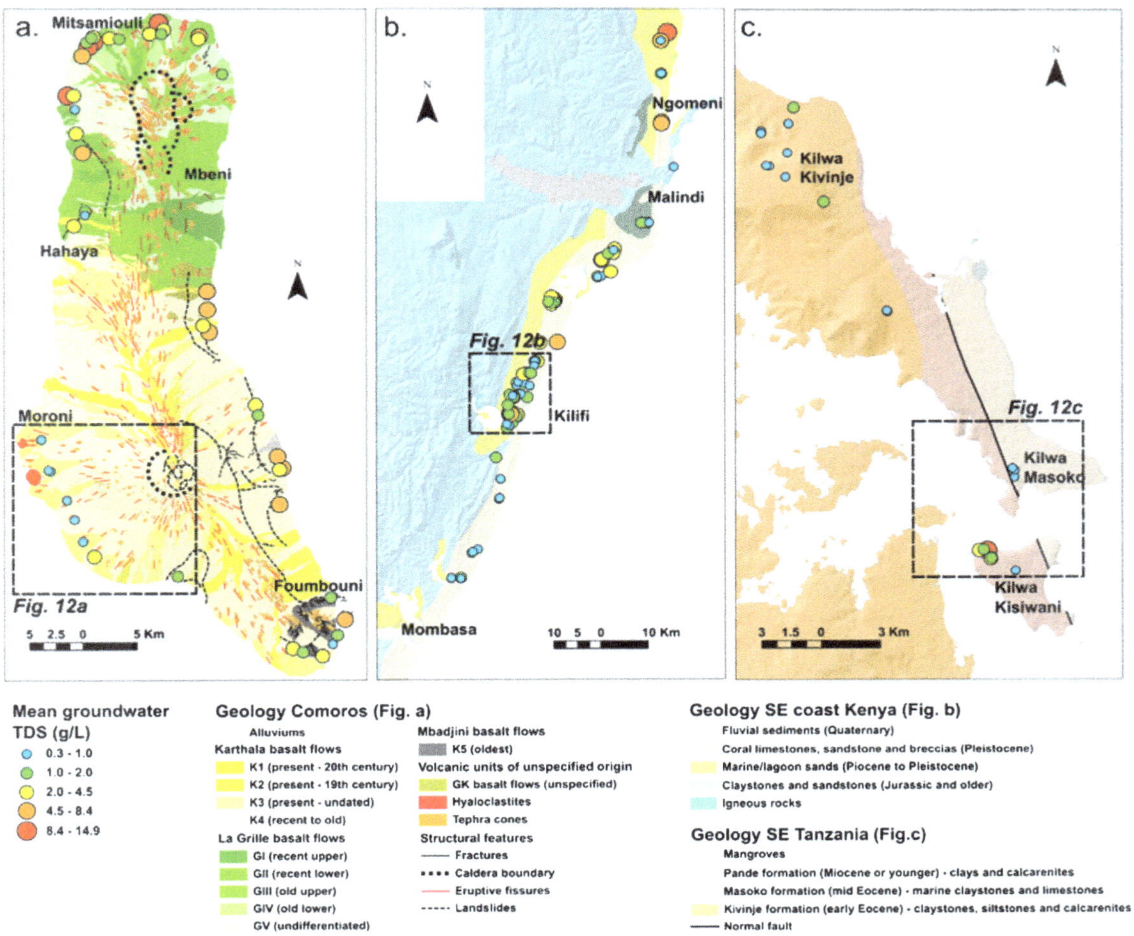

Figure 2: Overview of the study areas, including the main geological units and groundwater salinities of surveyed wells in (a) Grande Comores Island (Comoros), (b) Kilifi region (Kenya) and (c) Kilwa district (Tanzania). See Fig. 1 for the general location on these sites in East Africa. Dashed boxes indicate more detailed areas plotted in Fig. 12.

Kenya–Kilifi County (see map in Fig. 2b)

Kilifi County (13,006 km^2) lies along the Kenyan coast between Malindi to the north and Mombasa to the south, with a total population of 1,109,700 growing at a rate of 3.1% per annum. The area experiences a semi-arid climate, with irregular wet seasons and frequent drought (Onyancha et al., 2010). Although much of the piped water supply in the county is currently supplied from surface water sources, it is expected that reliance on groundwater in the area will increase as the demand for domestic and agricultural water increases. Groundwater yields vary along the coastal zone with poor quality and low yields in areas of Jurassic shales and Pleistocene coral limestones and higher quality and yields in Triassic sandstone and Quaternary sands. Excessive abstraction has already resulted in water contamination and saline intrusion (Musingi et al., 1999 and Munga et al., 2006). Surface water sources in the county are generally inland or to the north where the Athi-sabakia and Tana rivers, which originate in the Kenyan highlands, flow towards the Indian Ocean. The security of these sources is currently threatened by planned large-scale agricultural projects in the Kenyan interior and plans for dams upstream. Most of the water supplied to the densely populated Kilifi township and the linked coastal zone, which is the focus of this study, is piped from these rivers and inland lakes. Daily water demand in the township is estimated to be 200,000 m^{3}against the available 130,000 m^3 (NEMA, 2010) and is leading to pressure and conflict in the area. The contrast between tourist developments and affluent coastal residences and the amenity-poor, densely populated urban zone is pronounced. Many of the existing boreholes yield either saline or brackish water, and the remaining freshwater aquifers are in danger of overexploitation and are susceptible to pollution.

Tanzania—Kilwa district (see map in Fig. 2c)

Kilwa district (19,998 km^2) is a largely rural, low-lying coastal and island environment in south-eastern Tanzania consisting of coral limestone to the east and clays in the west, which dip eastward. Overburden comprises weathered lateritic soils and some Quaternary to Recent deposits of blown sand which are the dominant water bearing formations. The district is predominantly agricultural focussing on crop production such as cashews, simsim and coconut with some livestock rearing. Kilwa Masoko is a growing urban centre, the centre of local administration, with a developing tourism industry. In 2012 it was designated a service port for the gas pipeline installations to the north at Songo Songo island. The population has grown in recent years (currently ~15,000), fuelled by migration from the coastal hinterland. Water demand exceeds supply in the town and across the district. Currently the town is supplied by 4 public supply boreholes at a well field at the centre of the urban area, about 2 km from the coast, supplying approximately 1500 m^3/day. Demand is twice this volume so additional piped water is taken from open springs 15 km to the north at Mpara, with more wells under construction. The islands of Kilwa Kisiwani and Songo Mnara are historic stone-built towns (12thC) now designated a UNESCO world heritage site and a tourist attraction. Only Kilwa Kisiwani (area 17 km^2) is populated, with about 850 people settled in the more elevated (15 m) northwest of the island. The island is composed of coral limestone, overlying marls and clays (Nicholas et al., 2006). Groundwater occurs in shallow lenses, replenished during seasonal rains. Drainage is localised with no rivers or streams. The main employment is subsistence farming and fishing. Water shortages are major issues for the island community. The current sources of water are open shallow wells, the less saline dating from the 12th Century, which were added to by installation of boreholes and hand pumps in 2006 (Marobhe and Songo, 2006). All but one of the boreholes are saline or collapsed and the hand pumps are broken.

METHODOLOGY

Systematic investigations were undertaken in each of the study areas and hydrometric monitoring infrastructure installed at each site. The societal component of the study targeted issues of accessibility facing local communities and stakeholders in each area. An environmental characterisation, covering land use change and examination of existing data sets on sea-level and climate provided insight into the potential factors limiting groundwater availability and sustainability both at present and in the future. The hydrogeological component aimed to establish a quantitative and qualitative background status of coastal groundwater resources.

Societal Constraints on Water Accessibility—Perspectives of Stakeholders and User Groups

Gaining a local perspective on issues affecting access to water was one of the core objectives. Surveys of water users (householders, farmers and industries) and stakeholders (e.g. drilling companies, water managers) provided valuable complementary data to existing demographic observations and census data. Direct interviews with households (a total of 142 households in Kilifi, 154 in Kilwa and 185 in Grande Comore) were undertaken using a stratified random sampling method to select households within local electoral areas. Focused group discussions with community members and health providers were used to identify sources and quality of water supply and public health issues including common waterborne and water-related diseases.

Local stakeholder groups in each target area were established and a Participatory Action Research PAR (Pretty and Vodouhê, 1997) approach used to understand and address the social-economic and environmental factors of water resource management (Van Niekerk and Van Niekerk, 2009 and Mapfumo et al., 2013). This involved working with the selected communities through an inception workshop followed by community workshops and meetings and a final dissemination workshop. The PAR included use of questionnaires, Focused Group Discussion (FGD), in depth interviews and visioning (all in a participatory manner) where the stakeholders (youth groups, women's groups, elders) reflect on the past, look at the current situation and develop a 'vision' of what should be in the future.

The outputs provide an overview of: (a) existing issues with water access in each study area and perceptions of causes; (b) societal issues affecting water resources management; (c) current water resources management strategies and environmental factors affecting the same; (d) past practices and traditional methods of water management; (e) community stakeholder analysis on best practices of water resources management; (f) community perceptions on changing water resources and effects on their livelihoods; and (g) water resources management and environmental health.

Environmental Change—Land Use Change and Climatic Trends

For each study area available secondary data sources on climate, land use and recharge were compiled, and supplemental monitoring implemented where data gaps were identified. An automatic weather station provided rainfall and potential evapotranspiration time-series data at hourly intervals to promote understanding of borehole and well level and electrical conductivity monitoring.

Land cover and use were mapped from available aerial imagery for each area, and validated in the field. Temporal change was evaluated by comparison of the most recent aerial imagery with past imagery of equal resolution. Sources were limited by budget and as such the period varied depending on the availability of imagery for each site. Land use change was mapped and classified within a Geographic Information System (GIS) to provide information on settlement growth/decline and the clearance of land for agriculture, industry and infrastructure.

Long-term trends in temperature and rainfall in the study areas were examined, including the frequency and intensity of rainfall events, and temperature trends. Weather data for available coastal synoptic stations across Kenya, Tanzania and Comoros were acquired from the British Atmospheric Data Centre Global Weather Observation data sets over the period 1984–2014. The sampling time-step for the stations was normally every 3 h but quality and completeness of the data sets varied considerably among sites, with marked improvements from the late 1990s onward.

Hydrogeology and Groundwater Resources

Hydrogeological investigations were carried out to assess the current hydrogeological conditions and the aquifer vulnerability to seawater intrusion across each area.

Comprehensive baseline surveys were undertaken to establish the existing water well infrastructure. Existing datasets were used to map the location of wells and compile their technical characteristics, their use, pumping rate and records of changing salinity. A strong collaboration with local water services was initiated to ensure a continuous exchange and updating of borehole databases.

Baseline geophysical investigations, using 2D Electrical Resistivity Tomography (ERT), were carried out at each site, to characterise the aquifer structure and the current extent of saltwater intrusion. One to two profiles were undertaken at each site, perpendicular to the shore, using boreholes or wells with existing hydrogeological information to inform the interpretation.

Additional well monitoring was carried out to provide an indication of the groundwater variability in both space and time. Spatially, water levels, Electrical Conductivity (EC) and temperature profiles of wells were measured across a sub-set (dependent on density and accessibility) at each site, for a number of campaigns (2 per year corresponding to the dry and wet seasons). Temporally, in 1–3 strategic boreholes/wells at each site, a borehole datalogger was installed to acquire high-resolution temporal data on temperature, EC and water level, to investigate recharge processes and aquifer diffusivity.

RESULTS

Societal—Community and Stakeholder Perspectives

Results from Analysis of Census Data

Analysis of the available census data, from at least 2 decadal censuses, for each site provides an overview of demographic change (Fig. 3) and associated pressures on groundwater infrastructure (Table 1).

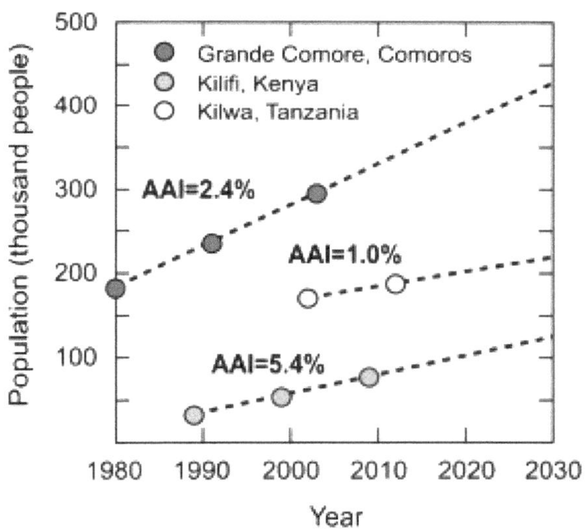

Figure 3: Population changes in the three study areas over the last three decades (from the most recent available census data). AAI = average annual increase; dashed lines are linear projections by 2030.

Table 1: Main source of water used in the three regions as compared to the whole country and change over time.

% (Rounded to upper%)	Remotely piped water	Protected well/ spring	Unprotected well/spring	River/pond/ dam	Vendors (e.g. trucks)	Rainfall tank/Jabias	Other
Tanzania (excl. Zanzibar) 2012	34	20	31	14	2		
Tanzania (excl. Zanzibar) 2002	37	18	29	11	5		
Lindi Region 2012 (Tanzanian coast)	19	14	53	10	4		

Lindi Region 2002 (Tanzanian coast)	13	17	55	15	1		
Kenya 2009	30	36		27	8		1
Kenya 1999	30	33		33	2		3
Kilifi County 2009	57	24		19	1		1
Kilifi County 1999	50	25		24	1		2
Grande Comore 2014	0.0	20a/2b	0.0	2	5	59	18c
Grande Comore 2000	0.0	15a/2b	0.0	3	3	70	10c
Comoros Union 2014	0.0	33a/3b	0.0	3	No data	30	29c
Comoros Union 2000	0.0	20a/2b	0.0	4	No data	40	24c

Kenya values from KNBS.

a UNDP deep wells.

b Private wells.

c Public fountains supplied by groundwater (protected wells).

Demographic Changes

In Kilifi, Kenya, the analysis focused on the township to the north of the estuary and found that population and household densities have grown rapidly. The population in this area more than doubled over the 20 year period, increasing from 255 persons km^{-2} in 1989–535 persons km^{-2} in 2009, and a household density of 54 households km^{-2} in 1989–111 households km^{-2} in 2009.

Population growth in Kilwa district, Tanzania was weaker by contrast; increasing by 10% between 2002 and 2012. The area is still predominantly rural with a population density of 13 people km^{-2}. The urban population increased from 8% to 10% of the total over the decade. In the urban centre of Kilwa Masoko, which has seen some growth in tourism and the development of the port for servicing the gas pipelines to the North.

In Grande Comore (Ngazidja) the population increased by 62% over the 23 years between 1980 and 2003 (last census). The population density was about 250 people km^{-2} in 2003. In 2005 72% of the population was rural, against 76% in 1991. The capital Moroni's population (13.5% of the island's population in 2003) more than doubled between 1980 (17,267 people) and 2003 (40,050 people) and is currently estimated at over 50,000 people. Tourism is as yet embryonic in the Comoros but has potential and is expected to develop in the next decades (Shaaban et al., 2013) due to the many natural and cultural assets.

Water Use and Evolution of Infrastructures

Data on water use (Table 1) were obtained from census records in each country (Kenya, Comoros Union and Tanzania) at the highest resolution available for each study area (respectively, Kilifi County, Grande Comore island and Lindi Province in which Kilwa district is located).

In Kenya, reliance on piped water supply is high in Kilifi with 56.8% of households surveyed in 2009 using piped water over other supplies. There was an increase of 7.1% in piped supplies from 1999 and a corresponding decrease in the use of well or spring supplies. The Kilifi region has no permanent rivers and the piped supplies are abstracted from River Tana and Athi-Sabaki rivers originating from the Kenyan highlands that flow into the Indian Ocean. Currently the supply from inland is inadequate to meet demand and is supplemented by groundwater abstraction, supplying a combined total of 130,000 m^3 daily towards an estimated demand of 200,000 m^3. The decrease in groundwater as a supply (24.5–23.2% between 1999 and 2009) may be linked to the high salinity of coastal aquifers in the area combined with water pollution, environmental degradation and recurrent droughts of wells (Marete, 2006 and Onyancha et al., 2010); which make piped supplies preferable where they can be afforded. Similar patterns were observed in the county during the surveys and FGDs undertaken as part of this study. The use of water vendors and supply trucks has increased by 6.5% in the county over the decade (up to 2009) while the use of water from rivers, ponds and dams directly by households declined by 6.5%.

Piped water is less commonly used in Lindi Region, of which Kilwa is the easternmost district, accounting for 19.1% of use in 2012 (Table 1). This is still low in comparison to the percentage of households with access to piped water at national level (Tanzanian mainland), which was 33.5% in 2002 and reflects the rural character of the area. The increase in piped water use in Lindi from 2002 is attributed to some improvement in infrastructure and a larger urban population in proximity to existing piped networks. Potentially linked to this is the decline in households obtaining water from boreholes, wells and springs from a combined total (unprotected and protected) of 72.1 to 66.9%. Some 79% of all wells/springs in Lindi were considered unprotected in 2012 compared to 76% in 2002. Growth in use of alternative water supplies such as rainwater and supply by water vendors from 0.4% in 2002 to 7.0% in 2012 may reflect the growing urbanisation of the population and increasing demand driving alternatives to traditional well sources as well as a growing perception of water vending as a rewarding business opportunity.

Water supply in Grande Comore Island relies exclusively on local aquifers and on rainfall harvesting. Supply from boreholes and wells more than doubled between 1985 and 2010 as a result of the commissioning of 54 new wells drilled in the early 1980s (UNDP, 1987). Groundwater is currently estimated as supplying just over a third of the island's population with safe water. Alternative water supplies such as rainfall harvesting have decreased as a consequence, from 70% to 59% between 2000 and 2014. Water supply by tanker trucks from operating wells has strongly increased (from 3% in 2000 to 5% in 2014) particularly in the last decade and has become a highly lucrative business (Mohamed, 2012).

Results From Household Surveys, Water Stakeholder and Community Meetings

Further to the available census data, local observations at each of the study areas were undertaken through a combination of household questionnaires, stakeholder and community meetings. For brevity a summary of the key findings is presented here; the detail of the socio-economic work programme will be covered elsewhere.

Water Accessibility

In terms of water access, results from the household survey (Fig. 4) established that less than half of households in Kilifi, Kilwa and Comoros used between 20 and 100 L day^{-1}(mean household size = 6 persons). This falls short of the basic water requirements of 20 L person^{-1} day^{-1} recommended by the World Health Organisation (WHO, 2003). Furthermore, less than a third of households used between 120 and 200 L day^{-1} which is classed as 'reasonable access' but still falls below the 50 L person^{-1} day^{-1} which is considered adequate for domestic use (Gleick, 1996). The primary factors contributing to poor water access across all three areas based on the household surveys and community discussions were: lack of finances to purchase greater quantities or purchase storage/harvesting equipment, insufficient storage capacity, water budgeting/rationing in certain areas or during the dry seasons and a failing or inadequate infrastructure.

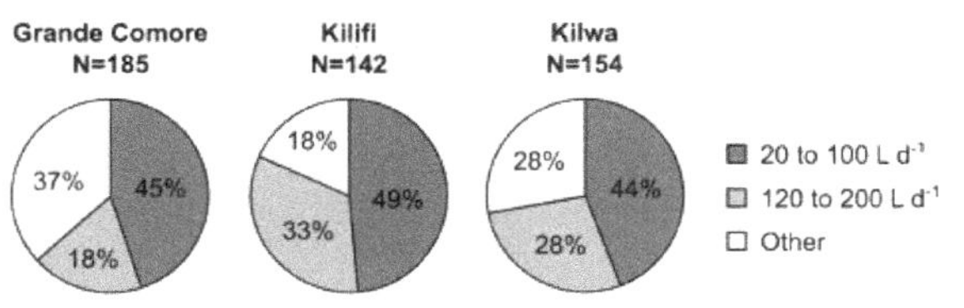

Figure 4: Daily volume of water used per household (mean size of 6 persons) in the three regions. N = number of households surveyed.

Across all three sites the common issues raised in community and stakeholder meetings and discussions were inadequate and intermittent supply, water scarcity, increased salinity, poor quality and high cost. The causes of these issues were attributed to drought (lack of rainfall/changing weather patterns), poor water sector management and governance, sea water intrusion and failing infrastructure. Issues with water quality result primarily from contamination of unprotected wells and poor source protection, with unregulated dumping highlighted as an issue in all areas. In Comoros there is a problem with unregulated use of agricultural fertilizers and sediment erosion due to deforestation.

Participation in Water Resources Management

Across the three study sites the stakeholder and community discussions raised issues with governance and disproportionate resource allocation, highlighting failures at district and national levels. Effective water governance requires active participation of all the stakeholders in management of water resources. This can be determined by the proportion of the local community which actively contributes to water resource decision making through membership of social groups and being involved in leadership roles contributing to the decision making process. Findings from the survey indicate that less than half of the households surveyed were members of social groups relating to water resources (Fig. 5a) or involved in meetings for decision making in relation to water resources (Fig. 5b). Furthermore, less than 25% in Kilifi and Grande Comore and less than 5% in Kilwa had received any training relating to groundwater or surface water management, water quality or water use (Fig. 5c). Additionally, 10–20% of households surveyed identified poor water sector management as a major water problem (Fig. 5d), which needs to be addressed.

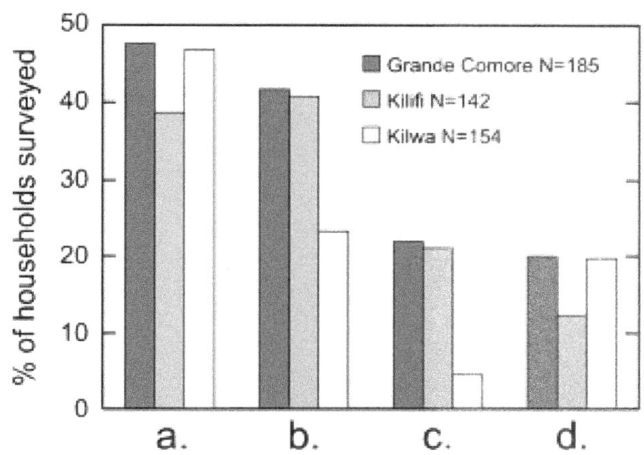

Figure 5: Participation of end-user households in water resources management. (a) Membership of a social group relating to water access; (b) participation in community meetings for decision-making relating to water; (c) water related training received; (d) identification of poor water sector management as a major water problem.

Strategies for Management

In summary it was noted that community members were aware that groundwater resources are being depleted and that salinity is a key water quality issue. The community members are willing and ready to engage in water resources management and to pay for water of good quality.

Several strategies were proposed for sustainable management of existing groundwater resources during the meetings including the implementation of effective water resources management to provide water of good quality that is properly and the provision of appropriate infrastructure for water storage, especially for rainwater harvesting. Furthermore, there should be proper siting of boreholes and wells, with regular maintenance programmes, that should involve women as major stakeholders at micro level and encourage alternative sources of water supply. Traditional knowledge systems should be integrated with modern technologies. The need for awareness and training on water issues is still critical (see Fig. 5c). Micro-finance and social networks are also vital as avenues for capacity building of community for effective management of water resources. From the visioning conducted with the community members the desired state for selected aspects of water resources management was captured in Table 2.

Table 2: Summary of community aspirations based on focussed group discussions on water resource management across the 3 study areas.

Current state	Desired state
Not all community members receive high quality water	All community members receive high quality drinking water
Not all community members have adequate supply of water particularly shortages in dry season	All community members have adequate supply of water in relation to the demand during all seasons
Not all community members can afford to pay for water—prohibitive cost	Affordable water pricing
Waiting too long to fetch water	Sufficient water points per unit of population
Long distances to water source	Water source near dwelling or piped water—enhance connection of water
Community not aware of the water issues, particularly the poor	Participation of community in water governance
Poor water resource management	Effective water resources management

Environmental Change

Regional Climatic Data

Seasonal rainfall varies across the region. The rainy seasons occur during boreal spring and autumn; the long rains occurring in spring and short rains in autumn. The strength of the rainy seasons varies with latitude and changes in circulation patterns over the Pacific and Indian Oceans. Orographic influences are particularly strong in Comoros where volcanic massifs create a topographic high that results in high variability across the island.

From the regional synoptic stations (Fig. 6), average annual rainfall in Mombasa, 20 km south of Kilifi, is 1800 mm (average over 2007–2013) with the majority falling in May and June, and generally some rainfall in all months, though little in January and February (<25 mm in 2013). Seasonal rainfall decreases to the north with an average annual rainfall in Lamu, the most northerly district on the Kenyan coast, of 1700 mm over the 2007–2013 period, for which the record is almost complete. Since the 2005–2008 period, during which Kenya experienced extremes of drought and rainfall (Hastenrath et al., 2010) with 3160 mm recorded in 2008 at Mombasa, annual rainfall has not exceeded 2000 mm, with 2011 particularly dry at 1140 mm. The low rainfall pattern has been linked to variability in both the Indian and Pacific Oceans and to the El Niño-Southern Oscillation (Ummenhofer et al., 2009 and Clark et al., 2003). The effect of prolonged droughts places increases pressure on groundwater resources in the region as surface water dries up.

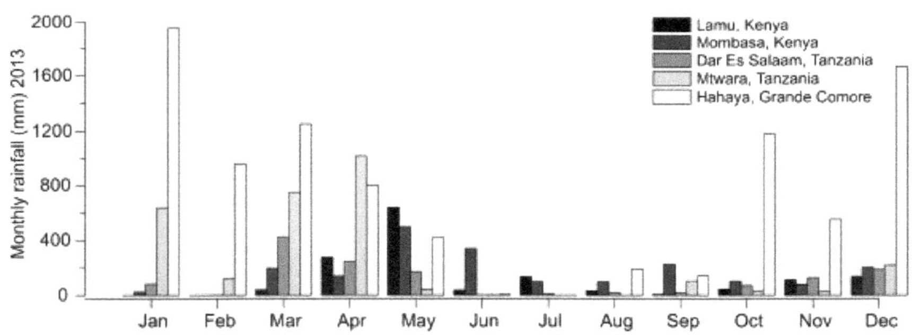

Figure 6: Comparative monthly rainfall totals for 2013 indicating the annual variation in rainfall across seasons from Lamu in northern Kenya to Mtwara in southern Tanzania and Hahaya on Grande Comore island.

To the south at Dar es Salaam in Tanzania, which lies 100 km north of Kilwa, seasonality is more pronounced with clear rainy seasons and almost no rainfall in June, July, August and September and February (43 mm total over these months in 2013 in Dar es Salaam). Average rainfall recorded in Dar es Salaam over the 2007–2013 period was 2570 mm. At Mtwara 170 km south of Kilwa, Tanzania the average annual rainfall was 2430 mm over the 2007–2013 period. In Grande Comore, an average of 4395 mm rainfall was recorded over the same period.

Observations at the weather stations installed at each of the study sites agree well with the patterns at the national synoptic stations overall (Fig. 7). Over the period 25 September 2013–5 July 2014, for which a complete record exists for all three sites, 800 mm of rainfall was recorded in Kilifi (maximum rainfall = 40.8 mm h^{-1}, median rainfall 0.8 mm h^{-1} with rain recorded 5% of time), 1056 mm in Kilwa (maximum rainfall = 46.2 mm h^{-1}, median rainfall 1 mm h^{-1}, with rainfall recorded 4.5% of the time) and 3688 mm in Vouvouni in Grande Comore (maximum rainfall 97.6 mm h^{-1}, median rainfall 1.4 mm h^{-1} and rain recorded 11% of time).

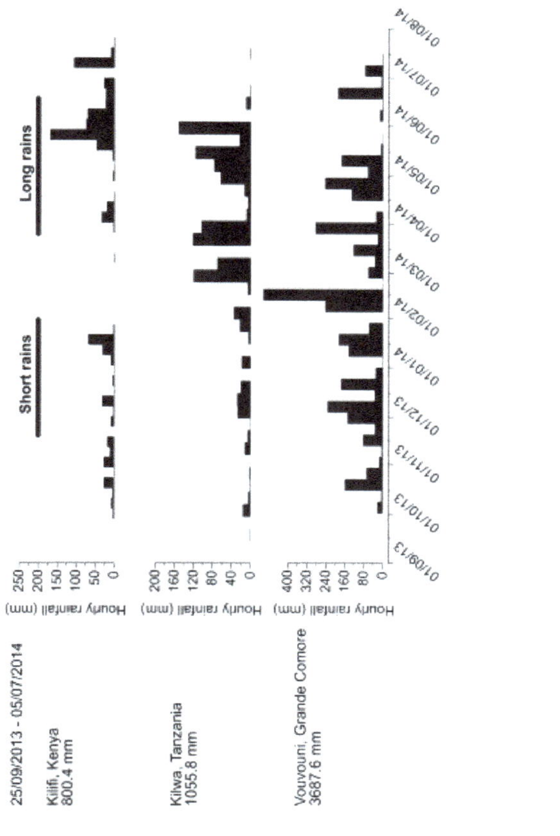

Figure 7: Rainfall over the period 25/09/2013–05/07/2014 at the weather stations at Kilifi, Kenya; Kilwa, Tanzania; and Vouvouni, Grande Comore.

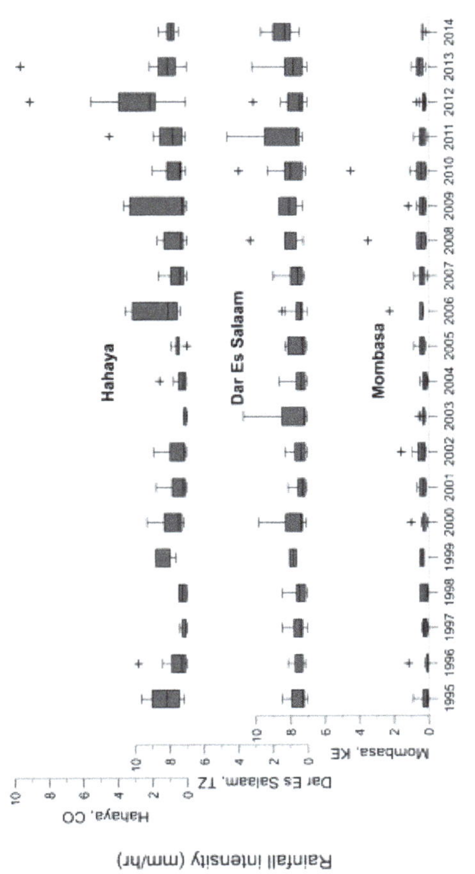

Figure 8: Box whisker plots of monthly rainfall intensities calculated for the closest synoptic stations in each country.

In Comoros, and particularly in Tanzania, which in late April 2014 experienced heavy rainfall and extensive flooding across the north east of the country, the number and magnitude of large rainfall events in recent years was a frequent comment in the social surveys undertaken in the study areas. An analysis of rainfall intensities based on the 3 hourly synoptic station data supports this (Fig. 8) with increasing magnitude events recorded in the last decade at sites (Hahaya, Dar es Salaam and Mombasa) for which the data record is almost complete and where potentially erroneous measurements were filtered based on the Quality Check scores attached to the record. Due to large data gaps and omissions the period prior to 1995 was not included in the analysis. An increase in the extremes of rainfall is apparent from 2005 onward for Hahaya in Comoros and Dar es Salaam, Tanzania. There is less indication of change for Mombasa, which experiences lower annual rainfall.

Many projections exist for the impact of climate change on Global Mean Sea-level (GMSL). The rate of change has been widely debated, with recent work (Watson et al., 2015) which corrected for errors in satellite altimetry data over recent decades, indicating that the actual rate of sea-level rise has accelerated. At a regional level, tidal data (Fig. 9) were obtained from the University of Hawaii Sea Level Center (http://uhslc.soest.hawaii.edu/data/download/rq) for the Western Indian Ocean region (gauges at Mombasa, Kenya; Lamu, Kenya; Zanzibar, Tanzania and Pointe des Galets, Réunion Island). Data gaps are present in almost all data sets but from 1986 to 2012 a comparable section from each time series was extracted and a linear fit applied to data during this period for each station. Not accounting for issues with gauge maintenance and instrument drift a clear positive trend is common to all with estimated annual increases for the gauges ranging from 3 to 9 mm. The potential implications for the coastal communities in this area are clear: accelerated coastal erosion, inundations of low-lying land and resulting inland extension of saltwater intrusion which will impact on the availability of freshwater resources.

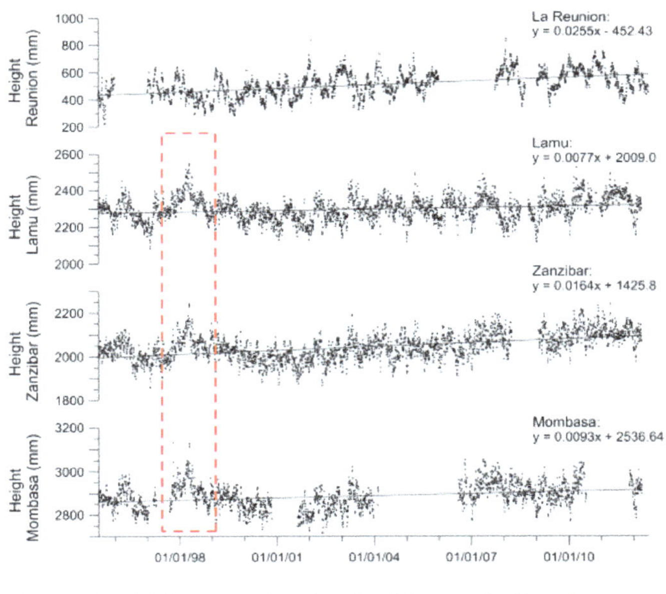

Figure 9: Tidal gauge data for the Western Indian Ocean region proximal to the study area. An upward trend is noted for all sites excepting a peak (red dash box) recorded at Lamu, Zanzibar and Mombasa circa 1998.

Land Use Change

Change in land use across each study area provided an indication of the effects of demographic change and the potential impact on recharge and groundwater quality. Clearance of vegetation for agriculture affects evapotranspiration and recharge and where bare earth remains there is increased potential for soil erosion during intense rainfall events. Increased sediment entering surface waters through erosion affects their use as water sources and with intensification of agriculture comes a risk of contamination from pesticides and herbicides, animal faecal material and chemical fertilisers. Pressure for space in urban areas leads to dispersion and enlargement of the urban periphery. New houses require water and where piped water supplies are not available new wells are necessary or the amount of time spent per day on collecting water increases.

Mapping was undertaken within defined areas at each study area. The focus was on mapping ground cover (extent of vegetation) and urban areas. Buildings were digitised from aerial imagery in order to provide an indication of changing housing and population patterns in the study areas. All buildings within selected areas, including houses, business establishments and public buildings were mapped. Land use and housing density change is presented for Kilwa, Tanzania over the period 2006–2013 (Fig. 10 and Fig. 11). The comparatively rural situation in this study area makes alterations in the landscape easier to map; the quality of imagery available was high.

Figure 10: Land use change in Kilwa Masoko (mainland) and Kisiwani (island), Tanzania between 2006 (left) and 2013 (right).

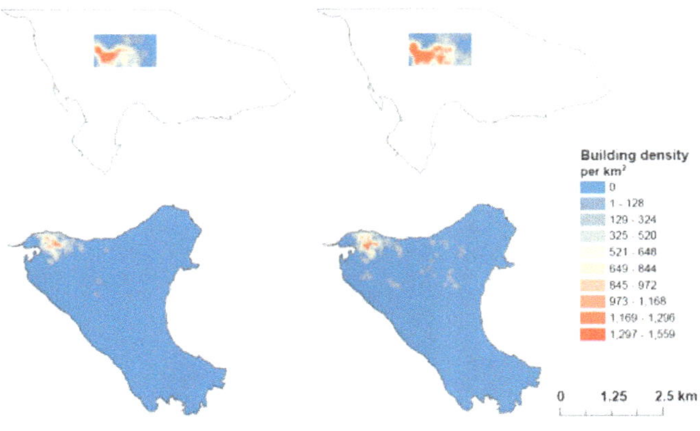

Figure 11: Building density in Kilwa Kisiwani (island) and for a sub-area on the urban periphery of Kilwa Masoko, Tanzania in 2006 (left) and 2013 (right). Density is expressed as number per km^2.

Within dense urban areas, such as Kilifi, the quality of imagery available limited the extent to which change could be mapped. However across the area an increase in land enclosure was noted.

In Kilwa Kisiwani clearance of land for cultivation and to provide fuel has reduced tree cover on the island over the 2006–2013 period (Fig. 10). Areas of formerly dense woodland to the south and east of the village have been thinned and cut for firewood and to provide land for cultivation, with new houses established. Areas with wood cover in the range 80–100%, which made up 35% of the island area in 2006 had reduced to 27% in 2013. The density of housing in the village increased over the period (Fig. 11) and traditional huts were replaced with block built houses. On the mainland, in Kilwa Masoko, the built area increased by 8% over the period, with large areas to the north of the town on both sides of the road cleared for construction. Housing densities increased during the period in formerly low density areas. New housing is largely block-built.

Groundwater Resources

Geological Structure and Spatial Patterns of Groundwater Salinity

The hydrogeological investigations provided baseline information on aquifer characteristics and on groundwater quality across the three study areas (Fig. 2 and Fig. 12).

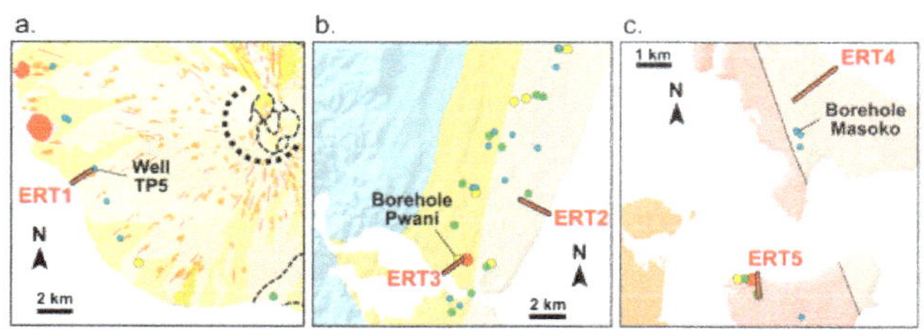

Figure 12: Areas of more detailed hydrogeological investigations, including geophysical profiling (ERT) and high frequency temporal groundwater monitoring in (a) well TP5, Vouvouni area, Grande Comore island; (b) Pwani borehole, Kilifi area and (c) Masoko borehole, Kilwa Masoko. See Fig. 2 for geological legend and borehole colour legend. ERT profiles 1–5 are presented on Fig. 13.

In Grande Comore island (Fig. 2a), aquifers are composed of young volcanic rocks organised in three volcanic massifs of different ages (Bachelery and Coudray, 1993). The Mbadjini massif is the oldest of these (Miocene) and outcrops at limited locations in the south-east of the island. It is characterised by deep weathering, which makes it difficult to distinguish the individual lava flows. The La Grille massif is of intermediate age (Mid-Pleistocene) and outcrops in the north of the island. The degree of weathering varies from moderate to low depending on the chronology of flow emplacement. The Karthala massif is the youngest (Quaternary), and active, volcano on the island characterised by a low degree or absence of weathering of the lava flows.

Water wells on the island are mostly deep (50–100 m), hand-dug, large-diameter wells excavated in the 1980s. Very large spatial variations in groundwater salinities are observed among the wells, which can be related to some extent to geological heterogeneity. In Grande Comore, all wells are located within a few kilometres of the coast due to the much higher topographic gradients compared to hydraulic gradients. This necessitates a large increase in well depths with distance from the coast to reach water, and much higher associated costs. In the Mbadjini massif, groundwater salinity is generally low (<2 g L^{-1}) due to the intensity of weathering resulting in reduced aquifer permeability and, therefore, reduced seawater intrusion due to higher hydraulic gradients. In addition, the weathering of the Mbadjini massif promotes the development of perched aquifers within more recent lava flows that overlie and are disconnected from the basal aquifer which has been subjected to seawater intrusion (Bourhane et al., 2015). In the La Grille and Karthala massifs, water wells have higher salinities (generally above 2 g L^{-1}) except on the western flank of the Karthala massif which experiences higher rainfall. Seawater intrusion in those two massifs is promoted by the high permeabilities of the lava flows and limited weathering; it is reduced by the rainfall intensity notably higher on the western flank of the Karthala volcano due to its exposition to dominant south westerly winds and its high elevation. Important spatial variations of salinities can be noted and attributed to the complexity of imbrication of lava flows of different ages resulting in different degrees of weathering and associated permeabilities.

The Kilifi coast in eastern Kenya (Fig. 2b), is characterised by a 5–15 km wide band of generally highly productive Tertiary and Quaternary sediments overlying less productive Jurassic or older sediments. The Tertiary (Pliocene) and lowest units of Quaternary (Pleistocene) sediments are composed of relatively homogeneous fine-grained marine or lagoon sands. They are overlain by, or for the uppermost units laterally pass into, Pleistocene coral sandstones and limestones displaying variable degrees of karstification. All those formations are covered by Holocene sand dunes along a narrow fringe of about 1 km width along the coastline. Most of the water wells in the coastal zone are shallow (typically <20m) large-diameter wells intersecting late Tertiary and Quaternary aquifers. Groundwater is mainly used for irrigation and as drinking water. Measured salinities display large spatial variations, but remain generally lower than those measured in Grande Comore (2/3 of the surveyed wells in Kilifi region have salinities lower than 2 g L^{-1} and 1/3 lower than the recommended WHO limit of 1 g/L; Fig. 2b). A gradient of increasing salinity also occurs from south to north reflecting the decrease in rainfall northwards in coastal Kenya.

Kilwa district, Tanzania, is composed of Tertiary to possibly Quaternary sediments comprising shale and silt (Nicholas et al., 2006) that are of little hydrogeological interest. However, calcarenite and coral limestone intercalations offer useful yields and are common in the Masoko formation (mid Eocene) that constitutes the western side of Kilwa Masoko peninsula and most of Kilwa Kisiwani island (Fig. 2c). A fault separates the Masoko formation on the west and the more recent (Miocene and younger) sediments of the Pande formation on the east. The moderately productive public boreholes providing drinking water for Kilwa Masoko are drilled in the Pande formation close to this fault. Measured salinities in Kilwa region are generally low, with 70% of wells providing water with less than 1 g/L of salt. This is the case for the majority of boreholes intersecting poorly productive shale and silt aquifers in the Kivinje and Pande formation. Wells drilled in the Masoko formation all occur on Kisiwani Island. They are mostly shallow large diameter hand-dug wells drilled in proximity to habitations on the northwest coast. They have much higher salinities (1–9 g L^{-1}) except for one borehole in the centre of the island.

Fig. 12 shows more detailed, enlarged views of the areas investigated in the three countries. More detailed investigations included geophysical surveys and high frequency temporal monitoring of groundwater in selected wells.

Geophysical Investigations

Electrical Resistivity Tomography (ERT) was applied to all three study areas using similar acquisition methodologies. Several 2D resistivity cross-sections were obtained from selected transects run perpendicular to the coast; results are plotted on Fig. 13. All cross-sections show both the spatial organisation of different types of geological units and the spatial patterns of saltwater distribution in the aquifers. All results highlight the relatively high degree of aquifer salinization associated with the interface between freshwater and saltwater occurring at relatively shallow depths with a low angle of dip away from the coast. Such a low angle is consistent with a low hydraulic gradient resulting from the relatively large permeabilities of the coastal aquifers. Significant differences in mean resistivities are also observed between sites. Both the volcanic aquifers of Grande Comore (ERT1) and coral limestones/sandstones of Kilifi (ERT2) have much higher overall resistivities than in the marine/lagoon sands of Kilifi (ERT3) and clastic sediments of Kilwa (ERT4 and 5), which have lower resistivities overall due to a higher aquifer clay content. Because of these lower mean resistivities, the contrast between freshwater and saltwater is not as clear for those latter sites (ERT3, 4 and 5)

In the Vouvouni area, south west of Grande Comore Island (Fig. 13a), the volcanic aquifer is characterised by a thick (from 10 m at the sea cliffs up to 50 m at 1 km from the shoreline) and highly resistive (>1000 Ω m) unsaturated zone. The transition zone between freshwater-saturated and seawater-saturated basalts (50–1000 Ω m) dips towards the centre of the island at a low angle. Freshwater thickness reaches a maximum of 30 m at 1 km from the coastline. This is consistent with the average low salinity measured in well TP5 (~0.2 g L^{-1}) located at a few hundred meters upslope of the north east end of the section and drilled within about 5–10 m below the water table. The

Vouvouni area has the highest rainfall on Grande Comore (more than three times the average rainfall); therefore the freshwater thickness is less elsewhere on the areas of the island that are also underlain by the same young Karthala basalt flows.

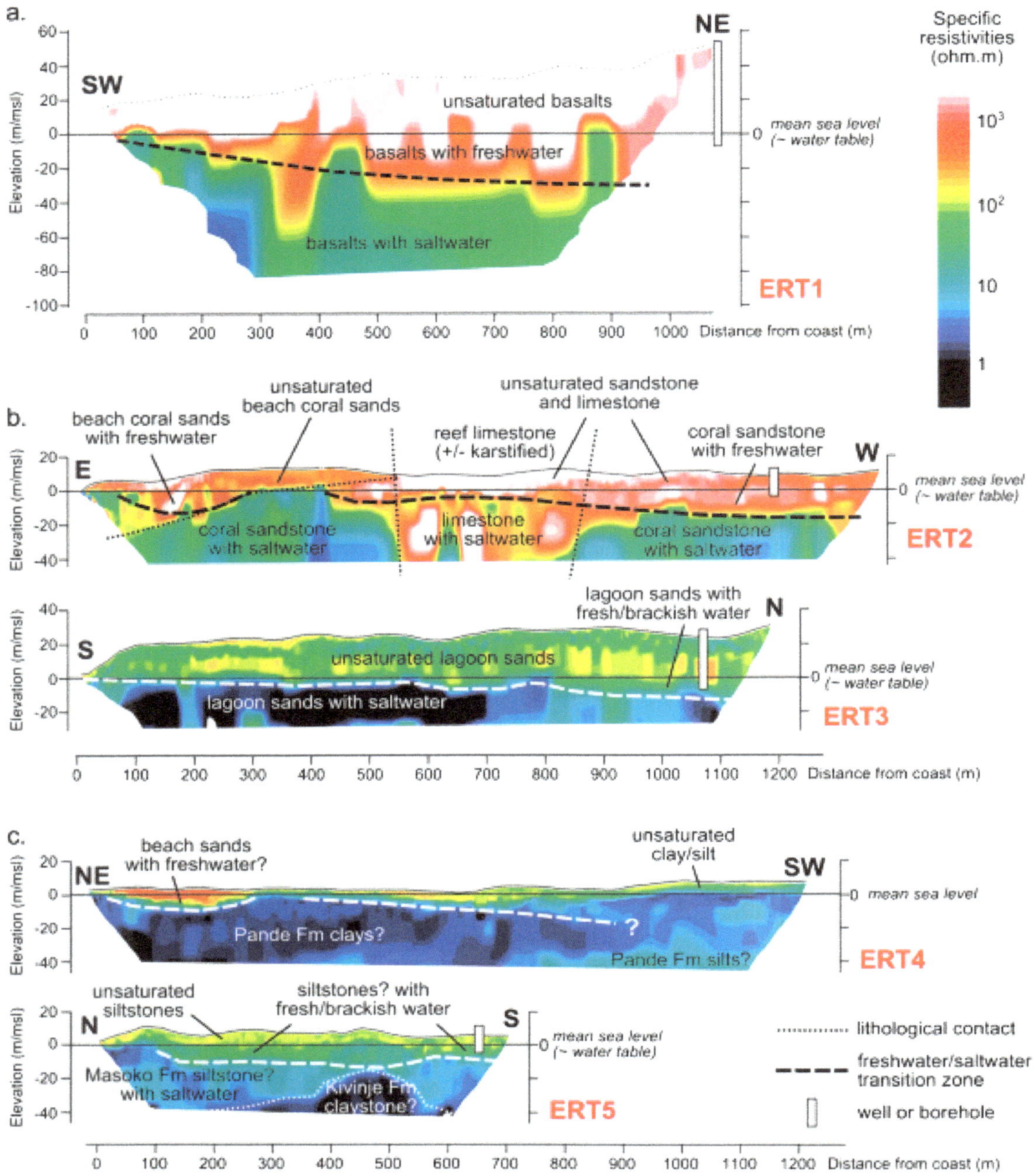

Figure 13: Results of ERT investigations in (a) Grande Comore; (b) Kilifi area and (c) Kilwa Masoko (top) and Kilwa Kisiwani (bottom).

Around Kilifi (Fig. 13b and c) resistivity signatures are different depending on the geological nature of the aquifer. Mean resistivities are generally higher in the Quaternary coastal units of the Pleistocene coral limestones and the Holocene sand dunes (Fig. 13b top) compared to the inland Tertiary marine/ lagoon sands (Fig. 13b bottom). Coastal quaternary formations display a relatively high degree of heterogeneity. The Holocene coral sands (resistivities of ~20–1000 Ω m depending on the salinity of pore water and degree of saturation) overlie the Pleistocene basement with thickness increasing toward the sea (up to 30–40 m thick at the studied location) from about 500 m from the coastline. The Pleistocene basement is composed of variability consolidated coral sandstones and breccias (2–1000 Ω m depending on salinity and saturation) with local occurrences of highly resistive reef limestone (10–10000 Ω m). A reef formation occurs between about 500 and 900 m from the coastline in the cross section of Fig. 13b (top). Strong internal variations in resistivities suggest karstification. Superimposed on the geological structure are different degrees of seawater intrusion.

Within the coastal Holocene sands, a freshwater lens (resistivity of 200–2000 Ω m) occurs reaching up to 20 m thickness at 200 m from the coastline. The maximum width of the lens is about 200–300 m. Seawater intrusion is reaching a greater distance inland in the underlying coral sandstones/ breccias due to their higher permeability. A freshwater wedge (100–1000 Ω m) starts to develop at about 400 m from the coastline with thickness slowly increasing with distance, reaching about 20 m at 1400 m at the end of the ERT profile. Within this formation (500–900 m), the reef limestone displays extremely heterogeneous resistivities suggesting seawater intrusion is taking place within the karstified structures (10–100 Ω m) while the compact limestone remains relatively unaffected due to its low storativity (100–1000 Ω m). Generally, freshwater thickness appears lower within the karstified limestone than within the coral sandstone/breccias. The hydrogeological structure observed in this profile extends laterally, with similar structural patterns within Holocene/Pleistocene formations observed along the coastline north and south of Kilifi town.

The profile ERT3 (Fig. 13b bottom) trends north east from Kilifi creek within the Pleistocene/Holocene marine and lagoon sand units. Aquifer resistivities are much lower, relatively, than those observed in the Quaternary coral units. This is due to a finer granulometry and higher clay content. The unsaturated zone displays resistivities ranging between 10 and 200 Ω m because of horizontal lithological stratification within the sands. Below sea level, the aquifer displays very low resistivities: seawater intruded sands have values generally lower than 10 Ω m and lower than 1 Ω m in the most southern part of the profile close to the creek. There is a thin freshwater/brackish water lens of resistivity values ranging 10–100 Ω m reaching a thickness of only about 15 m at 1100 m from the creek. Presence of fresh/brackish water at this distance is confirmed by a well where groundwater electrical conductivity has been measured at 2500 microS/cm (corresponding to a salinity of about 1.4 g L^{-1}).

In Kilwa, south east Tanzania, the silt-clay dominant lithology of Tertiary sediments is responsible for low resistivities, typically less than 100 Ω m (Fig. 13c). Only coastal sands occurring on a narrow band of about 200 m along the western coastline have higher resistivities of up to 1000 Ω m (ERT4, Fig. 13c top). No clear resistivity contrasts are observable in the Pande Formation (ERT4, Fig. 13c top) except for the transition between the unsaturated (10–100 Ω m) and the saturated zone (<10 Ω m). Because of such low resistivities in the aquifer›s saturated zone, it is difficult to distinguish between freshwater and seawater. However, a general decrease in resistivity towards the coast suggests diffuse saline intrusion. Kilwa Masoko public water boreholes drilled in this formation a few hundred metres south of the south west end of this profile provide freshwater with salinity of about 0.4 g L^{-1}; this suggests that low resistivities are primarily attributed to the clay/silt lithology rather than the saltwater content. This profile also suggests the presence of a thin freshwater lens of 5–10 m thick within or below the coastal sands.

Within the Masoko formation, profile ERT5 (Fig. 10c bottom) is characterised by slightly higher resistivities due probably to a more silty/carbonate lithology. The unsaturated zone clearly shows resistivities ranging 50–200 Ω. In the saturated zone, a fresh/brackish water lens of about 15 m thickness is suggested by slightly lower resistivities of 20–100 Ω m. The underlying seawater-saturated aquifer has resistivities lower than 10 Ω m. At the base of the profile, the even lower resistivities (<1 Ω m) could be attributed to the underlying Kivinje claystone formation, a clay facies within the Masoko formation, both saturated with saltwater. The existence of a well and a borehole at distance of about 600 m, where salinities were measured at 1.2 and 4.3 g L^{-1} is consistent with a brackish nature of the lens rather than freshwater.

Controls of Geology, Distance To Sea and Rainfall on Salinity Distribution in Groundwater

Resistivity profiles clearly reveal a geological control on spatial patterns of seawater distribution in the various coastal aquifers. Other factors such as the distance to the coast and the average rainfall of the area also exert a control on distribution of saltwater within a specific geological unit. Both an increase in distance to the coast and increased rainfall should correspond to a decrease in groundwater salinity. Fig. 14 illustrates such correlations between the combination of distance to coast and amount of rainfall and the salinity observed in wells and boreholes. When all geological types are considered together across the three countries, a poor correlation is observed (Fig. 14a). However, in Grande Comore island (Fig. 14b), strong correlations are observed within the two dominant volcanic units of Karthala and La Grille, respectively with an exponential relationship with R^2 = 0.68 and a power law relationship of R^2 = 0.85. This suggests relative aquifer homogeneity within individual geological units with the magnitude of seawater intrusion predominantly controlled by the aquifer recharge from rainfall and distance to sea.

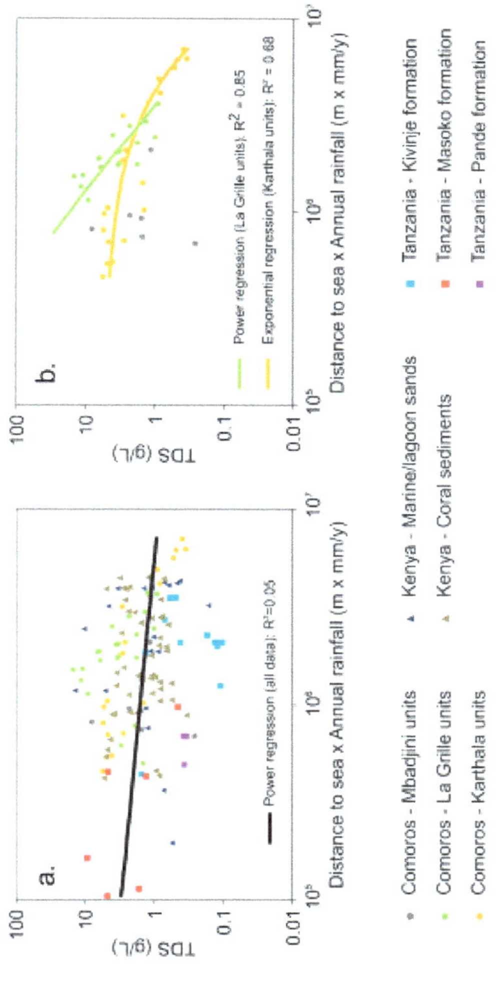

Figure 14: Correlations between borehole/well salinity, geology, distance to coast and rainfall; (a) all data across the three countries; (b) Grande Comore.

In contrast to Grande Comore, the absence of clear correlation for the aquifers of Kilifi and Kilwa can be explained by three main factors: (1) a significant intra-formation aquifer heterogeneity; (2) the large variations of well/borehole depths resulting in large variations of salinity primarily due to the strong vertical salinity gradients in the aquifers; and (3) a large variability in abstraction rates resulting in different degrees of seawater upconing. However, for Kilwa and Kilifi, an analysis of the influence of well depth and pumping rate on salinity is not possible due to lack of available data.

Groundwater Behaviour

Locations of temporally monitored boreholes in Comoros, Kenya and Tanzania are shown on Fig. 12. High frequency groundwater time-series recorded in those boreholes reveal large differences in temporal groundwater behaviour in the three studies sites (Fig. 15). In TP5 well in Vouvouni (Fig. 15a), both groundwater heads and electrical conductivity fluctuations are primarily controlled by semi-diurnal tidal fluctuations. The high water Table during the high tide is systematically accompanied by high electrical conductivity levels. This strong tidal control is promoted by high permeabilities and diffusivities in the volcanic aquifers (Bourhane et al., 2015). The short-term influence of rainfall is not visible, reflecting the large thickness of the unsaturated zone that results in diffuse recharge over time. Seasonal variations of heads and electrical conductivity can be noted with slightly higher water tables and slightly lower electrical conductivity during the wet season as compared to the dry season. TP5 is one of the most intensively pumped wells in Grande Comore. Pumping schedules are designed to minimise drawdowns (about 10 cm, due to high permeability and large diameter of the wells). More interestingly, pumping failures result in a sharp increase in salinity. This is a similar effect to the tidal fluctuationsm—a general rise in the water column results in a rise of more mineralized water from the base of the well. Improving access to groundwater of acceptable quality might require considering different abstraction and management approaches (e.g. widespread low-rate shallow wells or horizontal tunnels). Currently, two programmes funded by the African Development Bank [ADB] and the French Development Agency [AFD], are prospecting for new groundwater resources in Grande Comore. The latter is considering drilling large diameter wells to supply rural communities in both coastal and mountain areas (Join et al., 2013).

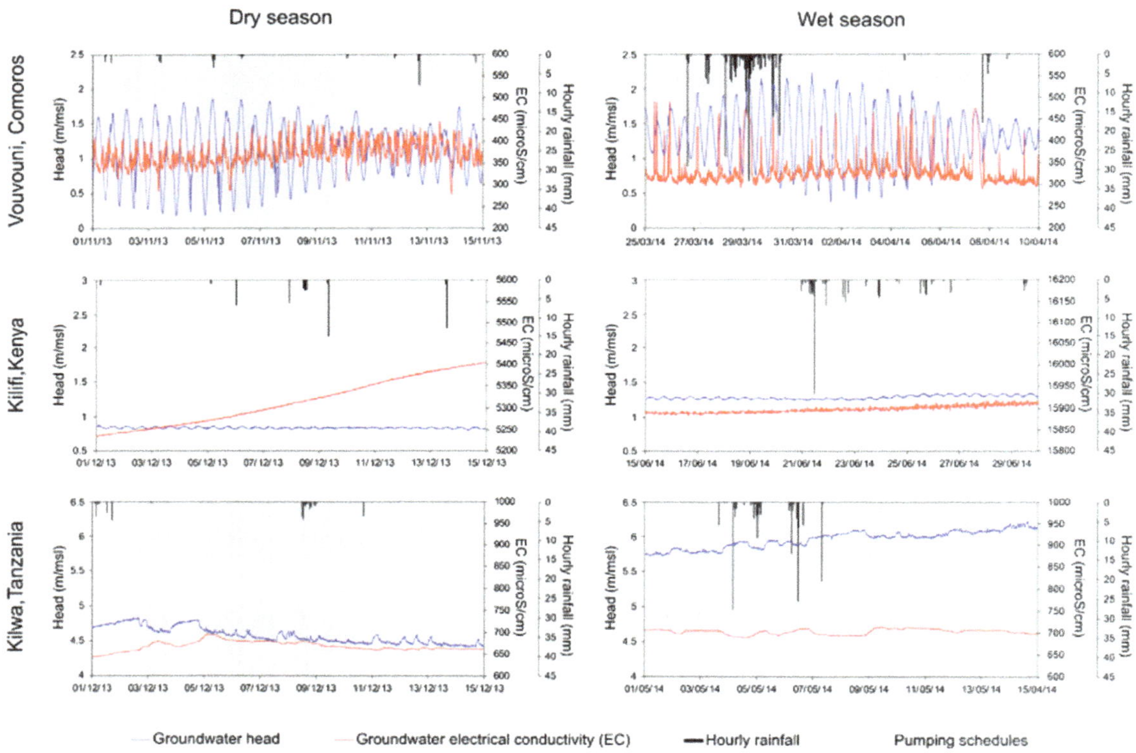

Figure 15: Time series of groundwater heads and electrical conductivities measured in monitored wells (Fig. 12) along with hourly rainfall and pumping schedules for dry and wet seasons.

Pwani University boreholes in Kilifi (Fig. 15b) also show the influence of semi-diurnal tides on groundwater heads and salinity fluctuations, but to a much lower extent than TP5 in Grande Comore. However, hourly rainfall does not show a visible effect on groundwater heads and salinity. As for Grande Comore, the thickness of the unsaturated zone (20–30 m) is likely responsible for diffuse recharge over time during the rainy season, i.e. a more seasonal impact of rainfall than in Grande Comore. This seasonality is seen during the dry season showing a progressive increase in electrical conductivity. Prolonged rainfall is also responsible for a slight rise in the water Table (a few cm) during the wet season. The data logger had to be moved from one borehole to another nearby between the two dry/wet periods because a pump was installed in the original one. The second borehole was previously pumped causing seawater intrusion but pumping was stopped before the datalogger was installed. This explains the much higher EC recorded there during the wet season.

Groundwater records on the Kilwa Masoko boreholes (Fig. 15 bottom) do not show any clear influence from tides. Instead, the influence of pumping is clearly visible on both heads and electrical conductivity: pumping periods are characterised by a drop in head, of about 20 cm, and a rise in electrical conductivity of up to 50 microS/cm. This simultaneous response of both heads and salinity can be attributed to seawater upconing below the boreholes; the drawdown of the water Table is accompanied by a rise in the freshwater/seawater transition zone. Seasonality of rainfall also impacts both heads and salinity, with a progressive decrease in head/increase in salinity during the dry season and increase in head/decrease in salinity during the wet season. Short-term rainfall also shows a slight impact on heads, with highest hourly rains causing a water Table rise of a few cm (e.g. on 4 May 2014).

In summary, the groundwater (heads and salinity) temporal behavior in the volcanic aquifers of Grande Comore appears to be primarily controlled by the tidal fluctuations and secondary by the pumping regime. In the Kilifi marine/lagoon sands, the same parameters appear to be primarily controlled by seasonality and secondly by tides (monitored boreholes were not pumped at the time of monitoring). In the clayey/calcarenite sediments of Kilwa Masoko, they are primarily controlled by the pumping regime and secondly by both short-term rainfall and seasonality.

OVERALL SYNTHESIS, DISCUSSION AND STATE OF UNDERSTANDING

The results of the various investigations carried out in the three countries provide an overview of the status of the groundwater resources and demand for water as well as the challenges faced with regard to expected environmental and societal changes. They also highlight strong similarities as well as notable differences across the three areas. The status of the groundwater resources and water supply is summarised as follows:

- Groundwater resources are important, and often safer sources of domestic and agricultural water across the whole region; used by over 50% of the population in Tanzania and Comoros (including direct supply from groundwater or indirect supply through water truck delivery) and over 20% in Kilifi region, Kenya (in Kilifi about half of the water supply is imported from other catchments and dams on remote major rivers);The observed increase in population of 1.0–5.4% in average annually regards predominantly urban centres and is causing an increase in demand; which is not met by increases in water availability and supply infrastructure;
- There is widespread salinization of wells and boreholes in all areas studied across the region. This is the main limiting factor on the quantity of groundwater availability and was highlighted as a key water quality issue by the community members and stakeholders;
- Fresh groundwater resources in coastal areas systematically occur as thin lenses or wedges; with the freshwater/salt water transition zone is inclined at a low angle so that salt water occurs at relatively shallow depths for a considerable distance inland. This limits the depths to which wells can be bored and the abstraction rates that are possible if fresh water supplies are to be maintained;
- The spatio-temporal variability of groundwater quality is well-correlated with the spatio-temporal distribution of rainfall and geology; in particular, variations in geology are reflected in the temporal response of wells to drivers such as tidal fluctuations (volcanic aquifers of the Comoros), pumping schedules (lower productive sedimentary aquifers of Tanzania), or seasonality (coral aquifers of Kenya);
- Long-term monitoring data for groundwater are lacking across the region and in cases where data are available there are often inconsistencies due to the methodologies used at different times so that comparisons are difficult and trends unclear;
- Well construction strongly influences water quality and long-term sustainability. Traditionally constructed wells are wider and shallower, have better water quality (lower salinity) and are more sustainable (less collapse/failure and more easily refurbished); however unprotected open wells are more likely to have faecal contamination due to dirty collection buckets/ropes (Tanzania and Kenya) and ingress of debris such as animal waste;
- The time-constrained nature of funding and of groundwater development projects are issues, often causing a disconnect where deliverables are required within the short (2–3 years) duration of a project. This results in the prioritisation of delivering a set number of mechanically-drilled production boreholes and rapid speed of installation regardless of the known benefits of hand dug, large diameter wells;
- The importance of local researchers with expertise/experience in coastal hydrogeology and having established links with local water stakeholders is recognised by researchers, stakeholders and communities in order to facilitate literature and data compilation necessary to conduct regional resource assessments.

The environmental drivers (climate and land use) that have potential to affect groundwater resources and their accessibility are:

- An overall change in climate across the region, that is indicated in long-term hydrometric observations and reported by local people who are noticing changes in weather patterns, seasons, dynamics of plants and animal behaviour;
- A clear increase in the extremes of rainfall in the last decade is recorded in Comoros and Tanzania, but is not as clear in Kenya which is drier and less seasonal;
- A clear increasing trend of sea level was recorded over the last two decades ranging from 3 to 9 mm/year across the region and comparable with global estimates for the Indian Ocean region;
- Increasing urbanisation is leading to land clearance for housing and increased demand for wood as a fuel. In areas subject to high intensity rainfall events (Tanzania) this leads to increased potential for runoff and soil

erosion and can reduce infiltration to groundwater;

- Tourism developments and modern plumbed housing place strains on existing water infrastructure through the volume used compared to local housing
- Demand for land in coastal areas is leading to increased enclosure of land (particularly in Kenya) with loss of grazing ground for pastoralists and increased stocking densities on land which is accessible.

The socio-economic drivers and governance affecting groundwater resources and access are:

- Low levels of education, income and limited livelihood characterises the region; all indicate a need for strategies to address poverty in the region;
- Communities feel strongly that it is the responsibility of government to solve water issues and identify shortcomings in governance;
- There are low levels of participation to decision making community meetings and training/education related to water resources management, although community members are keen to engage and voice their opinions; they want to be actively involved in water resource management, identify it as a critical issue and a direct link with daily life;
- Community members are critical of researchers who commonly do not return/share project findings;
- Gender issues and equality are getting increased recognition: women are the main stakeholders at community-level, but rarely involved in discussion, decision and future activities;
- Most users are willing to pay for water, however some resist because water quality is poor (i.e. salty);
- Researchers, managers and communities recognise that technical decisions should be bottom up by local communities and incorporate local knowledge and skills; international/national funding agencies are culpable for perpetuating unadapted and unsustainable groundwater infrastructures;
- Student training is viewed as key for water management as they will be the future decision makers in many areas;
- Particularly in Kenya and Tanzania, which are much larger than the Comoros Islands, the lack of centralisation of basic information on wells and boreholes is pointed out, which causes difficulties in accessing crucial hydrogeological literature and reports.

Overall, stakeholders have recognised the need to understand how the available groundwater resource works so that sustainable management and development plans can be adopted. Also, the investigations as well as the dissemination events strongly pointed to the important role of engagement both with the stakeholder and community and the importance of education as to how coastal groundwater and aquifers work and why they need stakeholder safeguarding.

Recent and specific figures for integrated management of groundwater and coastal aquifers within IWRM frameworks in Sub-Saharan Africa are not yet readily available, though projects relevant to East Africa are in progress (e.g. UNEP, 2012 and Mehta and Movik, 2014). Governance aspects of groundwater management in East Africa lack the mechanism for coordination and for fostering cross-sector linkages. For example, deficiencies have been identified in key areas under existing policy, legal and institutional frameworks in Kenya (Mumma et al., 2011). Given the lack of consideration of groundwater, as pointed out by previous studies (e.g. Braune and Xu, 2010; Knüppe, 2011), one can expect progress to be made on managing coastal groundwater resources in East Africa.

The results of this study provide a detailed regional perspective that complements the previous continent-scale review of coastal aquifers by Steyl and Dennis (2010). It confirms the fundamental importance of coastal groundwater resources for the development of the region and the urgent need for matching the rate of water development with the rate of demographic change, including urban expansion and economic growth. Despite the geological diversity of the region, the sustainable provision of safe groundwater is limited in all areas by both seawater intrusion (rather than aquifer productivity) and the lack of adequate local to national management plans and a policy framework adapted to the specificities of these coastal situations. The positive development of national and inter-African consciousness and leadership with regard to water development mentioned by Steyl and Dennis (2010) must however be accompanied by a realisation of the fundamental importance of implementing coastal groundwater development strategies that are different to those of other physical settings where borehole productivity is the target.

International development projects still fail to provide sustainable groundwater infrastructure and fresh (i.e. non salty) groundwater in coastal regions. Drilling campaigns are often deemed successful once productive boreholes with good quality water are obtained without consideration to future salinisation due to over-pumping. Those strategies, which systematically target a limited number of deep and productive boreholes rather than promoting the development of multiple shallow, large diameter wells, have proven unsuccessful in the long-term. Sustaining the use of large numbers of shallow wells through, for example, modernisation, rehabilitation or deployment of traditional wells however poses significant management challenges requiring a strong collaboration of stakeholders at different levels of society, from communities to providers, consultants, NGOs and regulators. Those collaborative aspects must be embedded or strongly enhanced in the design and implementation of future development projects. In particular, the developing, and lucrative, water trucking business in Kenya and Comoros proves that communities are willing to pay for quality water, which suggests that community/associative management frameworks based on paying for quality water from local wells have potential as a local alternative for water management.

CONCLUSION

The coastal aquifers of East Africa are major water resources for domestic supply, industry and agriculture. They currently require development programmes to support the high rate of population increase in the region, including strong urban development. The region is also experiencing clear trends in long-term environmental change, common to all three countries and probably also shared by many other African coastal countries. Those changes include an increase in rainfall intensity without a clear increase in average rainfall together with significant land clearance and urbanisation. This is expected to have a negative long-term impact on groundwater recharge. Sea level is also showing a consistent rise during the last two decades that may increase coastal erosion and lead to seasonal flooding of low-lying areas, which ultimately will decrease fresh groundwater availability.

Notwithstanding, geological controls across the three areas, groundwater quality from wells and even more from narrow boreholes is broadly poor due to saltwater intrusion, and freshwater availability is low despite highly productive aquifers. Fresh groundwater is often limited to thin freshwater lenses/layers requiring a large number of shallow wells with low abstraction rates. The experience shows that traditional (or modernised) shallow large diameter wells are less saline and are more sustainable than deep boreholes. Observed trends of changes in demography, climate, sea level and land use are not expected to improve the availability of fresh groundwater. Recent and current development projects do not account for hydrogeological requirements for drilling more and shallower wells rather than deep boreholes. Communities as wells as stakeholders are often left with wells having inadequate water quality that are abandoned, opening up business for costly water trucking. Communities are willing to pay for water and to engage in local groundwater management framework provided that groundwater is of appropriate quality.

The recent progress in social consciousness and education on groundwater vulnerability in African coastal regions must be strengthened and rapidly accompanied by the establishment/improvement of national strategies for both groundwater development and management. Those strategies will need to take better account for the specificities of coastal groundwater resources as well as their anticipated changes and will also require enhancing the involvement of communities in the management processes. They also have to be supported by improved knowledge of coastal aquifers and long-term monitoring data, which are lacking in the region.

ACKNOWLEDGMENTS

This research was funded by the UK Natural Environment Research Council NERC, the Department for International Development DFID-UKAID, the Economic and Social Research Council ESRC; UPGRO Grant No NE/L001888/1: towards groundwater security in coastal East Africa. Geophysical data acquisition in the Comoros was funded by the French Development Agency AFD: project GECEAU CKM1068. We acknowledge stakeholders and community members from Grande Comore, Kilwa district and Kilifi Region for their active and fruitful participation in the focussed group discussions and questionnaire surveys. We are grateful to Bonventure Obeka and Stanley Wanjala (MA Sociology students at Pwani University) for assistance throughout the project, to postgraduate and undergraduate students from the Universities of the Comoros, Dar Es Salaam, Pwani and Kenyatta and interns from the Comoros Water Directorate DGEME for assistance in field data acquisition. We thank Isaac Marobhe (University of Dar Es Salaam) for assistance and the loan of field equipment, as well as Peter Karanja Gathu (Kilifi District Water Agency), Magai Kakuru (Kilwa District Water Department), Mahabadi Boinali (Comoros Water Directorate DGEME), Youssouf Ben Abdallah (Comoros Community Representative CIGEO) and Simon Onywere (Kenyatta University) for contributions to the final workshop discussions. We also acknowledge the assistance of the British Institute in Eastern Africa (BIEA) and particularly the late Joseph Mutua for his technical and logistical expertise and untiring assistance through fieldwork in Tanzania and Kenya. We are finally grateful to the Reviewers and Editorial Board of the journal for their valuable comments and suggestions that helped improving the quality of the manuscript.

REFERENCES

1. Adelana, S.M.A., 2009. Monitoring Groundwater Resources in Sub-saharan Africa: Issues and Challenges. IAHS Red Book Publication, pp. 103–113.
2. Adelana, S.M.A., Abiye, T.A., Nkhuwa, D.C.W., Tindimugaya, C., Oga, M.S., 2008. Urban groundwater management and protection in Sub-Saharan Africa. In: Adelana, S.M.A., MacDonald, A.M. (Eds.), Applied Groundwater Studies in Africa, IAH Selected Papers on Hydrogeology 13. CRC Press/Balkema, Leiden, the Netherlands, pp. 1–7.
3. Ashton, P.J., Turton, A.R., 2009. Water and security in Sub-Saharan Africa: emerging concepts and their implications for effective water resource management in the Southern African Region. In: Brauch, H.G., spring, U.O., Grin, J., Mesjasz, C., Kameri-Mbote, P., Behera, N.C., Chourou, B., Krummenacher, H. (Eds.), Facing Global Environmental Change. Springer-Verlag, Berlin, pp. 661–674, http://dx.doi.org/10.1007/978-3-540-68488-6 50.
4. Bachelery, P., Coudray, J., 1993. Carte Géologique Des Comores: Carte Volcanologique De La Grande Comore Au 1/50 000ème Et Carte Volcano-tectonique, Avec Notice Explicative. Ministère Franc¸ ais de la Coopération, 39 p.
5. BGR, Cap-Net, Waternet, WA-Net, 2007. Capacity Building for Groundwater Management in West and Southern Africa. BGR, Hanover, Germany, pp. 65 p.
6. Bourhane, A., Comte, J.-C., Join, J.-L., Ibrahim, K., 2015. Groundwater prospection in Grande Comore Island: joint contribution of geophysical methods, hydrogeological time-series analysis and groundwater modelling. In: Bachelery, P., Lénat, J.-F., Di Muro, A., Michon, L. (Eds.), Active Volcanoes of the Southwest Indian Ocean: Piton de la Fournaise and Karthala. Springer-Verlag, Berlin, pp. 385–401, http://dx.doi.org/10.1007/978-3-642-31395-0 24.
7. Braune, E., Xu, Y., 2008. Groundwater management issues in Southern Africa: an IWRM perspective. Water SA 34 (6), 699–706.
8. Braune, E., Xu, Y., 2010. The role of ground water in sub-Saharan Africa. Groundwater 48, 229–238, http://dx.doi.org/10.1111/j.1745-6584.2009.00557.x.
9. Cai, W., Santoso, A., Wang, G., Weller, E., Wu, L., Ashok, K., Masumoto, Y., Yamagata, T., 2014. Increased frequency of extreme Indian Ocean Dipole events due to greenhouse warming. Nature 510, 254–258, http://

dx.doi.org/10.1038/nature13327.

10. Clark, C.O., Webster, P.J., Cole, J.E., 2003. Interdecadal variability of the relationship between the Indian Ocean zonal mode and east african coastal rainfall anomalies. J. Clim. 16, 548–554, http://dx.doi.org/10.1175/1520-0442(2003)016.

11. Comte, J.-C., Join, J.-L., Banton, O., Nicolini, E., 2014. Modelling the response of fresh groundwater to climate and vegetation changes in coral islands. Hydrogeol. J. 22 (8), 1905–1920, http://dx.doi.org/10.1007/s10040-014-1160-y.

12. Döll, P., 2009. Vulnerability to the impact of climate change on renewable groundwater resources: a global-scale assessment. Environm. Res. Lett. 4 (3), 035006, http://dx.doi.org/10.1088/1748-9326/4/3/035006.

13. Edmunds, M., 2012. Limits to the availability of groundwater in Africa. Environ. Res. Lett. 7, 021003, http://dx.doi.org/10.1088/1748-9326/7/2/021003.

14. Foster, S., Tuinhof, A., Garduno, ˜ H., 2008. Groundwater in sub-Saharan Africa: a strategic overview of developmental issues. In: Adelana, S.M.A., MacDonald, A.M. (Eds.), Applied Groundwater Studies in Africa, IAH Selected Papers on Hydrogeology 13. CRC Press/Balkema, Leiden, the Netherlands, pp. 9–21.

15. Foster, S., Tuinhof, A., van Steenbergen, F., 2012. Managed groundwater development for water-supply security in sub-Saharan Africa: investment priorities. Water SA 38 (3), http://dx.doi.org/10.4314/wsa.v38i3.1.

16. Gleick, P.H., 1996. Basic water requirements for human activities: meeting basic needs. Water Int. 21 (2), 83–92, http://dx.doi.org/10.1080/02508069608686494.

17. Global Water Partnership, 2000. Integrated water resources management, Technical Advisory Committee (TAC) Background Papers, No. 4, 71 p. Available at: http://www.gwp.org/Global/ToolBox/Publications/Background%20papers/04%20Integrated%20Water%20Resources%20Management%20(2000)%20English.pdf. Global Water Partnership, 2002. The Policy Guidance and Operational Tools on IWRM.

18. Hastenrath, S., Polzin, D., Mutai, C., 2010. Diagnosing the droughts and floods in equatorial east Africa during boreal autumn 2005–08. J. Clim. 23 (3), 813–817, http://dx.doi.org/10.1175/2009JCLI3094.1.

19. Hinkel, J., Brown, S., Lars Exner, Nicholls, R.J., Vafeidis, A.T., Kebede, A.S., 2012. Sea-level rise impacts on Africa and the effects of mitigation and adaptation: an application of DIVA. Reg. Environ. Change 12 (1), 207–224, http://dx.doi.org/10.1007/s10113-011-0249-2.

20. IPCC, 2014a. Climate Change 2014a: Impacts, Adaptation, and Vulnerability. Part B: Regional Aspects. Contribution of Working Group II to the Fifth Assessment Report of the Intergovernmental Panel on Climate Change. In: Barros, V.R., Field, C.B., Dokken, D.J., Mastrandrea, M.D., Mach, K.J., Bilir, T.E., Chatterjee, M., Ebi, K.L., Estrada, Y.O., Genova, R.C., Girma, B., Kissel, E.S., Levy, A.N., MacCracken, S., Mastrandrea, P.R., White, L.L. (Eds.). Cambridge University Press Cambridge, United Kingdom and New York, NY, USA, 688 p.

21. IPCC, 2014b. Climate Change 2014b: Impacts, Adaptation, and Vulnerability. Part A: Global and Sectoral Aspects. Contribution of Working Group II to the Fifth Assessment Report of the Intergovernmental Panel on Climate Change. In: Field, C.B., Barros, V.R., Dokken, D.J., Mach, K.J., Mastrandrea, M.D., Bilir, T.E., Chatterjee, M., Ebi, K.L., Estrada, Y.O., Genova, R.C., Girma, B., Kissel, E.S., Levy, A.N., MacCracken, S., Mastrandrea, P.R., White, L.L. (Eds.). Cambridge University Press Cambridge, United Kingdom and New York, NY, USA, p. 1132 p.

22. Join, J.-L., Comte, J.-C., Bourhane, A., Soule, H., 2013. Tests méthodologiques en géophysique en vue de l'implantation de forages d'eau sur l'île de La Grande Comore.P rojet Pilote de gestion du service public de l'eau en milieu rural sur l'île de La Grande Comore. Unpubl. Geophys. Report, Laboratory Geosciences, University of Reunion Island, 64 p.

23. Knüppe, K., 2011. The challenges facing sustainable and adaptive groundwater management in South Africa. Water SA 37 (1), 67–79.

24. MacDonald, A.M., Bonsor, H.C., Dochartaigh, B.É.Ó, Taylor, R.G., 2012. Quantitative maps of groundwater resources in Africa. Environ. Res. Lett. 7 (2), 024009, http://dx.doi.org/10.1088/1748-9326/7/2/024009.

25. Mapfumo, P., Adjei-Nsiah, S., Mtambanengwe, F., Chikowo, R., Giller, K.E., 2013. Participatory action research (PAR) as an entry point for supporting climate change adaptation by smallholder farmers in Africa. Environ. Dev. 5, 6–22.

26. Marete, J.M., 2006. An assessment of water resources and integrated water resources management (IWRM) strategy in Bahari sub-basin, Kilifi District. In: Unpubl. MSc Thesis. University of Nairobi.

27. Marobhe, I., Songo M., 2006. Report on the study of groundwater resources at Kilwa Kisiwani, Kilwa district, and Lindi region, Unpubl. Geol. Report, Dept. of Geology, University of Dar Es Salaam.

28. Mehta, L., Movik, S., 2014. Flows and practices: integrated Water Resource Management (IWRM) in African contexts, IDS Working Paper, 2014, No.438, 34 p.

29. Mohamed, I., 2012. L'eau en Grande Comore: étude de cas d'un petit espace insulaire abondamment arrosé mais en situation de pénurie d'eau. In: PhD Thesis. University of Reunion Island, 353 p.

30. Mohamed, S.H., Othman, S.A., 2006. Etude causale sur la disponibilité de la ressource en eau et la sécurité de l'approvisionnement en eau potable aux Comores, Unpubl. Report, UNDP, Moroni, 42 p.

31. Mukherjee, A., Saha, D., Harvey, C.F., Taylor, R.G., Ahmed, K.M., Bhanja, S.N., 2015. Groundwater systems of the Indian sub-continent. J. Hydrol.: Reg. Stud. 4 (A), 1–14, http://dx.doi.org/10.1016/j.ejrh.2015.03.005.

32. Mumma, A., Lane, M., Kairu, E., Tuinhof, A., Hirji, R., June 2011. Kenya groundwater governance case study. In: Water Papers. Water and Sanitation Program (WSP), The World Bank, Washington, D.C, pp. 2011, available online at URL: http://www.worldbank.org/water.

33. Munga, D., Mwangi, S., Ong'anda, H., Kitheka, J.U., Mwaguni, S.M., Mdoe, F., Barongo, J., Massa, H.S., Opello, G., 2006. Vulnerability and pollution of groundwater in Kisauni, Mombasa, Kenya. In: Xu, Y., Usher, B. (Eds.), Groundwater pollution in Africa. Taylor and Francis (Balkema), the Netherlands, pp. 213–228.

34. Musingi, J.K., Kithiia, S.M., Wambua, B.N., 1999. The urban growth of Mombasa coastal town and its implication for surface and groundwater resources. In: Ellis, J.B. (Ed.), Impacts of Urban Growth on Surface Water and Groundwater Quality. Proceedings of an International Symposium Held during IUGG 99, the XXII General Assembly of the International Union of Geodesy and Geophysics. At Birmingham, UK 18–30 July 1999. IAHS Press (No. 259).

35. NEMA, 2010. Integrated Coastal Zone Management Action Plan for Kenya, 2011–2015. National Environment Management Authority, Nairobi, Kenya, pp. 76 p.

36. Nicholas, C.J., Pearson, P.N., Bown, P.R., Jones, T.D., Huber, B.T., Karega, A., Lees, J.A., McMillan, I.K., O'Halloran, A., Singano, J.M., Wade, B.S., 2006.

37. Stratigraphy and sedimentology of the Upper Cretaceous to Paleogene Kilwa Group, southern coastal Tanzania. J. Afr. Earth Sci. 45 (4), 431–466, http://dx.doi.org/10.1016/j.jafrearsci.2006.04.003.

38. Onyancha, C., Khaemba, A., Sabuni, B., 2010. Aquifer storage and recovery and surface basins for a greener Kilifi District. Nile Water Sci. Eng. J. 3 (3), 26–33.

39. Pretty, J.N., Vodouhê, S.D., 1997. Using rapid or participatory rural appraisal. In: Swanson, B. (Ed.), Improving Agricultural Extension. FAO, Rome, pp. 47–55.

40. Robins, N.S., Davies, J., Farr, J.L., Calow, R.C., 2006. The changing role of hydrogeology in semi-arid southern and eastern Africa. Hydrogeol. J. 14 (8), 1483–1492, http://dx.doi.org/10.1007/s10040-006-0056-x.

41. Robins, N.S., 2013. A review of small island hydrogeology: progress (and setbacks) during the recent past. Q. J. Eng. Geol. Hydrogeol. 46 (2), 157–165, http://dx.doi.org/10.1144/qjegh2012-063.

42. Shaaban, I.A., Ramzy, Y.H., Sharabassy, A.A., 2013. Tourism as a tool for economic development in poor countries: the case of Comoro Islands. Afr. J. Bus. Econ. Res. 8 (1), 127–145.

43. Steyl, G., Dennis, I., 2010. Review of coastal-area aquifers in Africa. Hydrogeol. J. 18 (1), 217–225, http://dx.doi.org/10.1007/s10040-009-0545-9.

44. Ummenhofer, C.C., Sen Gupta, A., England, M.H., Reason, C.J.C., 2009. Contributions of Indian Ocean sea surface temperatures to enhanced East African rainfall. J. Clim. 22, 993–1013, http://dx.doi.org/10.1175/2008JCLI2493.1.

45. UNEP, 2012. The UN-Water Status Report on the Application of Integrated Approaches to Water Resources Management. UNEP Division of Communications and Public Information, Nairobi, Kenya, 119 p., ISBN: 978-92-807-3264-1.

46. United Nations, 2011. World Population Prospects: The 2010 Revision. UN Department of Economic and Social Affairs, Population Division, New York.

47. United Nations Development Plan UNDP, 1987. Recherche et mise en valeur des eaux, Comores: perspectives de mises en valeur des eaux souterraines pour l'alimentation en eau des agglomerations de l'ile de Ngazidja, Unpubl. Techn. Report No. DP/UN/COI-79-005, DP/UN/COI-86-001, New York, 60 p.

48. Vafeidis, A., Neumann, B., Zimmerman, J., Nicholls, R.J., 2011. MR9: analysis of land area and population in the low-elevation coastal zone (LECZ). GB, Foresight, Government Office for Science, London.

49. Van Niekerk, L., Van Niekerk, D., 2009. Participatory action research: addressing social vulnerability of rural women through income-generating activities. J. Disaster Risk Stud. 2 (2), 127–144.

50. Vincent, L.A., Aguilar, E., Saindou, M., Hassane, A.F., Jumaux, G., Roy, D., Booneeady, P., Virasami, R., Randriamarolaza, L.Y.A., Faniriantsoa, F.R., Amelie, V., Seeward, H., Montfraix, B., 2011. Observed trends in indices of daily and extreme temperature and precipitation for the countries of the western Indian Ocean. J. Geophys. Res.: Atmos. 116 (D10), 1961–2008, http://dx.doi.org/10.1029/2010JD015303 (1984–2012).

51. Walraevens, K., Mjemah, I.C., Mtoni, Y., Van Camp, M., 2015. Sources of salinity and urban pollution in the Quaternary sand aquifers of Dar es Salaam, Tanzania. J. Afr. Earth Sci. 102, 149–165, http://dx.doi.org/10.1016/j.jafrearsci.2014.11.003.

52. Watson, C.S., White, N.J., Church, J.A., King, M.A., Burgette, R.J., Legresy, B., 2015. Unabated global mean sea-level rise over the satellite altimeter era. Nat. Clim. Change 5, 565–568, http://dx.doi.org/10.1038/nclimate2635.

53. Watto, M.A., Mugera, A.W., 2015. Econometric estimation of groundwater irrigation efficiency of cotton cultivation farms in Pakistan. J. Hydrol.: Reg. Stud. 4 (A), 193–211, http://dx.doi.org/10.1016/j.ejrh.2014.11.001.

54. Werner, A.D., Bakker, M., Post, V.E.A., Vandenbohede, A., Lu, C., Ataie-Ashtiani, B., Simmons, C.T., Barry, D.A., 2013. Seawater intrusion processes, investigation and management: recent advances and future challenges. Adv. Water Res. 51, 3–26, http://dx.doi.org/10.1016/j.advwatres.2012.03.004.

55. World Bank, 2014. Climate change knowledge portal, http://sdwebx.worldbank.org/climateportal/index.cfm, consulted 10/06/2014.

56. World Bank, 2014. Low income economies classification, http://data.worldbank.org/about/country-classifications, consulted 10/06/2014.

57. World Health Organisation WHO, 2003. Total dissolved solids in drinking-water. Background document for preparation of WHO Guidelines for drinking-water quality no. WHO/SDE/WSH/03.04/16, WHO, Geneva, 8 p.

58. World Water Council, 2006. Water resources development in Africa, Africa Regional Document, 4th World Water Forum, Mexico 2006.

CHAPTER 6

INCREASING SENSITIVITY OF RESULTS BY USING QUANTILE REGRESSION ANALYSIS FOR EXPLORING COMMUNITY RESILIENCE

Odeya Cohen, Arkady Bolotin, Mooli Lahad, Avishay Goldberg, Limor Aharonson-Daniel

Keywords: Community resilience, Resilience, Emergency preparedness, Emergency response plan, Quantile regression, CCRAM

INTRODUCTION

Community resilience (CR) offers a conceptual framework for gauging a community's capacity for coping with environmental changes and emergency situations (Adger, 2006,Cutter et al., 2014 and Bonanno et al., 2015). Community resilience is perceived as a core element of sustainable lifestyle (Magis, 2010 and Wilson, 2012), and indeed it can both help shorten response time during emergencies and mitigate the community's reaction to crises by facilitating purposeful and collective action on the part of its members (Pfefferbaum et al., 2013). As such, CR is increasingly recognized as an important concept in the field of emergency preparedness and response (Aldunce et al., 2015), Community resilience has been explored by a wide variety of specialists, including psychologists (Ungar, 2011 and Norris et al., 2008), geographers (Cutter et al., 2008,Cutter et al., 2010 and Cutter et al., 2014), public health professionals (Castleden et al., 2011 and Poortinga, 2011), engineers (Zoback, 2014), governance specialists (Wilson, 2013), and globalization process managers (Wilson, 2012). This multidisciplinary background is responsible for the fact that the term has diverse definitions and has generated more theoretical models than strong empirical evidence (Engle, 2011,Chandra et al., 2010 and Cutter et al., 2008). The conjoint community resilience assessment measure (CCRAM) is recognized as a valid tool for assessing community resilience by household sampling (Leykin et al., 2013,Cohen et al., 2013 and Cutter, 2015). CCRAM encompasses the various factors, or components, of a community's resilience that are the basis for planning interventions. The factors were identified through a statistical process, yet they are also anchored in the professional literature surrounding the concept of community resilience (Leykin et al., 2013 and Cohen et al., 2013). So far, studies using CCRAM have applied common descriptive or basic statistical regression analyses. The objective of this study is to demonstrate a statistical approach, which is novel to the measurement of community resilience data. Quantile regression, has been used previously in various areas of ecology research (e.g. Cade et al., 1999 and Magzamen et al., 2015) and this is, to the best of our knowledge, its first application in the field of social ecological resilience context.

Quantile regression complements the estimation of conventional mean models by providing a more sensitive insight into the distribution of the dependent variable (Magzamen et al., 2015). As presented by Fitzenberger and Wilke (2015), conventional modeling has focused on the mean of the dependent variable, quantile regression can detect whether the partial effect of a regressor on the quantile is the same for all quantiles or differs across quantiles.

Citation: Odeya Cohen, Arkady Bolotin, Mooli Lahad, Avishay Goldberg, Limor Aharonson-Daniel, Increasing sensitivity of results by using quantile regression analysis for exploring community resilience, Ecological Indicators, Volume 66, July 2016, Pages 497-502, ISSN 1470-160X, http://dx.doi.org/10.1016/j.ecolind.2016.02.012.

MATERIAL AND METHODS

Study Instrument

The CCRAM tool (Leykin et al., 2013) is a self-reported questionnaire with 28 items, 21 of which have been grouped into five factors: leadership, collective efficacy, preparedness, place attachment, and social trust. The mean scores of each the factors portray a community resilience profile (range from 1 to 5). The overall CCRAM score is the average of the scores of 21 survey items with an equal weight. In addition, a single item enquired about perceived community resilience (also on a five point Likert scale) after defining resilience as "the ability to quickly return to routine after an emergency event".

Data Collection

The study was conducted from 2011 to 2014 in Israel. It included 23 small and rural communities (up to 10,000 inhabitants) and six medium-size cities (up to 50,000 inhabitants), with 28% average sampling ratio of households. Data were collected by using door to door surveys at randomly selected addresses and by distributing electronic questionnaires in small settlements with a complete electronic mailing list using Qualtrics (www.qualtrics.com), a web-based survey software. The study was approved by the institutional review board of the Faculty of Health Sciences at Ben-Gurion University of the Negev. Participants gave their written informed consent to take part in the study.

Statistical Analysis

The reliability of both the CCRAM score and its factors were examined using Cronbach's alpha for internal validity. Pearson and Spearman correlation coefficients were then calculated and used to examine the association between CCRAM factors and various background variables. For comparing means of CCRAM scores between sub-groups of participants, an independent t test and analysis of variance (ANOVA) followed by post hoc tests were employed.

Two regression models were used to find the association between dependent variable: perceived community resilience as defined by the single item, and the socio-demographic covariates: gender, age, faith, community type, reported income level, and previous involvement in emergency situations. CCRAM factors were added to the regression equation. The analysis of the two models, examined the regression coefficients of independent variables, along with 95% confidence intervals (CI).

For the logistic regression it was necessary to dichotomize the dependent variable, thus it was recoded into two levels: low (1–2) versus high (4–5), while the intermediate level (3) was omitted from the analysis.

For a non-parametric method of statistical analysis, we exploited quantile regression models, specifically, their most common form – median regression (Gould, 1992 and Hao and Naiman, 2007). Median regression finds a line through the data that minimizes the sum of the absolute residuals rather than the sum of the squares of the residuals, as in ordinary regression.

Both regression models (the ordinary and the quantile) describe the central tendency of the data, of which the mean is one measure and the median another. Since mean, and therefore ordinary linear regression, are sensitive to outliers, the ordinary regression might produce results that do not reflect the central tendency well.

The analysis examined the regression coefficients across quantiles 10, 25, 50, 75 and 90 of the dependent variable. Inter-quantile regression analysis estimated the differences between regression coefficients of independent variables between quantile 10 and quantile 90. Data were analyzed using the Statistical Package for the Social Sciences (SPSS) version 21, and STATA software version 12.1.

RESULTS

Participants

This study was conducted from August 2011 to June 2014. The study included 3152 adults (mean age 41.3, range 18–94, SD = 14.9, median age – 40 years), 1758 (55.8%) from small towns, and 1384 (43.9%) from medium-size cities in Israel. Most (56.1%) participants were women (n = 1767), 70.6% in a permanent relationship (n = 2226). Nearly half (49.1%) had an academic education (n = 1547). Jews comprised 59.7% of participants (n = 1882) and 45.4% were secular (n = 1432). Income: 38.7% reported their income as average (n = 1219), while 24.3% reported their income as less than average (n = 767). Responders reported living in their community for an average of 24.62 years (range 1–92 years, SD = 16.2). 40.7% of the responders noted that they had previously been involved in emergency situations (n = 1284). Major study population characteristics are described in Table 1.

Table 1: Study population distribution with mean CCRAM and mean perceived CR score.

	n	%	Mean CCRAM Score	p-Value (t-test or ANOVA)	Mean perceived CR score	p-Value (t-test or ANOVA)
Total	3152	100	3.24		3.33	
Gender						
Female	1767	56.1	3.25	0.872	3.31	0.228
Male	1332	42.3	3.24		3.36	
Faith						
Jewish	1882	59.7	3.51	<0.001	3.68	<0.001

Non-Jewish	1257	39.9	2.86		2.83	
Community type						
Small towns	1758	55.8	3.39	<0.001	3.52	<0.001
Medium-size cities	1384	43.9	3.06		3.08	
Income						
About average	1219	38.7	3.23	<0.001	3.25	<0.001
Less	767	24.3	3.08		3.16	
More	1037	32.9	3.36		3.53	
Previous involvement in emergency situation						
No	1667	52.9	3.17	<0.001	3.26	0.025
Yes	1284	40.7	3.29		3.35	

The Conjoint Community Resiliency Assessment Measure (CCRAM)

The reliability of the CCRAM score as given by Cronbach's α was 0.94, while that of the factors was as follows: leadership ($\alpha = 0.9$), collective efficacy ($\alpha = 0.87$), preparedness ($\alpha = 0.81$), place attachment ($\alpha = 0.75$), and social trust ($\alpha = 0.78$). All are considered to be satisfactory. The mean CCRAM score were 3.24 (SD = 0.80), with no differences between genders (see Table 1; Table A.1). With reference to the participants' demographic background, Jewish responders were found to have a significantly higher mean CCRAM score than non-Jewish responders ($p < 0.001$). This difference was found to be significant in all CCRAM factors, but the widest gap between Jewish and non-Jewish scores was found in the collective efficacy factor ($p < 0.001$). The CCRAM score was strongly correlated with perceived community resilience as described by a single item ($r = 0.604$, $p < 0.001$) and had a weak positive correlation with age ($r = 0.216, p < 0.001$).

Logistic Regression Model

A logistic regression was performed ($p < 0.001$). After dichotomizing the dependent variable as described above, it included two levels: low (n = 609) versus high (n = 1384), while the intermediate level (n = 1037) was excluded from this analysis.

Table 2 presents the results of the regression. In this analysis the significant socio-demographic covariates were: being of non-Jewish faith, living in medium-size cities versus small towns, and having less than average income. Regarding the CCRAM factors, leadership, collective efficacy and place attachment were found to have a significant positive association with perceived community resilience.

Table 2: Factors associated with perceived community resilience, a logistic regression.

	Odds ratio (OR)	p	95% C.I. for OR	
			Lower	Upper
Gender				
Female	1			
Male	1.288	.100	.953	1.743
Age	.999	.808	.988	1.009
Faith				
Jewish	1			
Non-Jewish	.278	<0.001	.202	.383
Community type				
Small towns	1			
Medium-size cities	.690	.026	.498	.957
Income				
About average	1			
Less	.608	.008	.421	.877
More	1.224	.270	.855	1.753
Previous involvement in emergency situation				
No	1			
Yes	1.256	.150	.921	1.713
CCRAM factors				
Leadership	1.708	<0.001	1.331	2.192
Collective efficacy	2.188	<0.001	1.641	2.916
Preparedness	1.094	.457	.863	1.387
Place attachment	1.527	<0.001	1.266	1.843
Social Trust	1.199	.094	.970	1.482

Quantile Regression Model

A quantile regression model (Gould, 1992 and Hao and Naiman, 2007) was used to analyze the association between demographic covariates and CCRAM factors, as well as perceived community resilience. Fig. 1 presents the significant socio-demographic covariates in the different quantiles of perceived community resilience. Having an income below average, was found to be significant in the lower (Q10, Q25) resilience quantiles. Having an income above average was found to be positively significant in Q75. Age (per year) and previous involvement in emergency situations were found to be positively associated with perceived community resilience in the lowest quantile. A difference between low and high (Q10, Q25, Q50, Q75 versus Q90) was noted in regard to size of community. The association between size of community and perceived community resilience was found to be negative (residents of larger towns had lower community resilience), up to quantile 75, but reversed in the highest quantile.

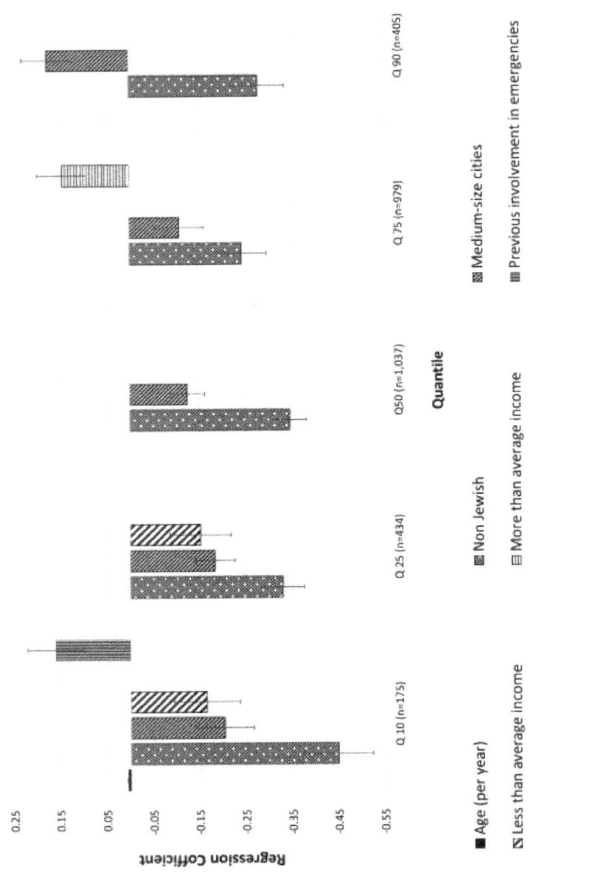

Figure 1: Significant positive and negative socio-demographic co-variates across the quantiles of perceived community resilience.

The complete results for the regression coefficients of the independent variables, along with the 95% confidence intervals (CI) for quantiles 10, 25, 50, 75 and 90, are presented in Table A.2.

Community Resilience Factors

Fig. 2 presents the significant factors according to quantile. In the results derived from the regression equations among quantiles 25 and 50, all the CCRAM factors were found to be associated significantly with perceived community resilience. However, in quantile 10 place attachment and social trust factors were not significant, while in 90 quantile only collective efficacy and place attachment factors were significant. The collective efficacy factor yielded the highest association in all quantiles (RC range 0.336–0.282), but in quantile 10 no significant difference was found between the regression coefficients of the leadership and collective efficacy factors. The leadership factor was found to be significantly different as between quantile 10 and quantile 90 (RC = −0.18, $p < 0.001$, 95% CI −0.265 to −0.0952).

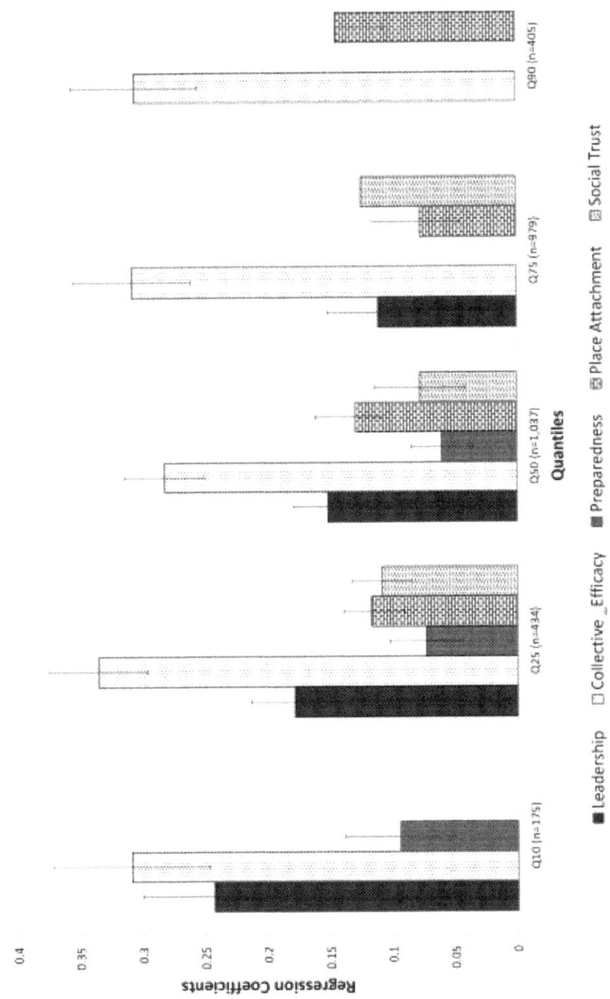

Figure 2: Regression coefficients of significant community resilience factors in quantiles of perceived community resilience.

A summary of the differences between the significant results of the logistic and quantile regression models is presented in Table 3.

Table 3: Significant results and trends of logistic and quantile regression models.

Variable	Logistic regression	Q10	Q25	Q50	Q75	Q90
Gender						
Female						
Male						
Age		+*				
Faith						
Jewish						
Non Jewish	–***	–***	–***	–***	–***	–***
Community type						
Small towns						
Medium-size cities	–*	–**	–***	–***	–*	+**
Income						

About average						
Below average	−**	−*	−**			
Above average					+**	
Previous involvement in emergencies						
No						
Yes		+**				
CCRAM factors						
Leadership	+***	+***	+***	+***	+**	
Collective efficacy	+***	+***	+***	+***	+***	+***
Preparedness		+*	+*	+*		
Place attachment	+***		+***	+***	+*	+***
Social trust			+***	+**	+**	

*$p < 0.05$.

**$p < 0.010$.

***$p < 0.001$.

Trends: − = negative association, + = positive association.

DISCUSSION

Measuring the factors associated with a community's ability to cope with change or disturbance can be a challenging task (Castleden et al., 2011 and Chandra et al., 2010). This manuscript presents an innovative approach based on a 'bottom-up' assessment which takes into account the attitudes of the community's members. This empirical study explores the factors that contribute to perceived community resilience at five levels. The variable perceived community resilience is taken as an indicator of overall community resilience, thereby facilitating the introduction of community resiliency factors as independent variables into the regression equation (Leykin et al., 2013 and Cohen et al., 2013).

In addition to presenting the statistical approach, this paper also demonstrates the wealth of information that can be derived by quantile regression analysis as compared with more traditional models, such as logistic regression.

The use of quantile regression has been more widespread in the context of economic aspects of vulnerability studies (Nordhaus, 2010 and Maloney et al., 2004) and less on psychological aspects of vulnerability studies (Bühler et al., 2012).

One of the earliest quantile regression models, presented by Koenker and Bassett in the late 1970s (1978) in the field of economy, estimated functional relations between variables for all portions of a probability distribution (Cade and Noon, 2003). While this approach has recently gained currency, we have not encountered any record of it being applied in the social context of community resilience in the face of change. In the present study, the application of quantile regression allowed us to examine and compare the impact of various factors on perceived community resilience among sub-populations with various levels of community resilience.

The use of quantile regression makes it possible to critically examine, verify or disprove accepted truths. For example: Community resilience studies perceived economic resources as a significant factor (e.g. Cutter et al., 2014). Our results show that 'less than average income' is negatively correlated with community resilience in the logistic regression model, but the sensitivity of the quantile regression model, enabled to see that in fact, it had a negative significant association only among the lower levels of perceived community resilience (Q10–Q25). Nevertheless, 'more than average income' was not identified as significant by logistic regression, while quantile regression revealed this variable to be positively associated with perceived community resilience in quantile 75.

Previous involvement in an emergency situation was found to have a positive effect in the lower quantile (Q10), contrary to previous studies (Cohen et al., 2013 and Leykin et al., 2013) that used conventional statistical approaches, and to the finding of the logistic regression of the current study which also found it with no significant association.

These findings are of relevance when planning focused intervention plans as it helps identify the most vulnerable population that can benefit most from engagement.

In regard to the CCRAM factors, there were important differences between the two regression models. The preparedness and trust factors were not found to be significant factor associated with perceived community resilience in the logistic regression model, yet quantile regression revealed these factors to be significant, though only among the lower quantiles (Q10–Q50). The impact of each CCRAM factor on perceived community resilience was examined based on the 95% CI of the logistic regression model, but no significant differences were found between the coefficients of the factors. Quantile regression, on the other hand, indicated that the collective efficacy factor had the highest association with perceived community resilience in all groups except quantile 10. As demonstrated in Fig. 2, the most important factor associated with perceived community resilience is collective efficacy. Since it has been documented that a crisis is usually followed immediately by a rise in mutual support and solidarity (Hawdon et al., 2012), intervention plans can be devised, based on the assumption that the population will be willing to help. Collective efficacy may therefore not only contribute to the initial strength and resilience of the society, but it may also help in supporting and enhancing recovery measures.

A community is a multifaceted configuration made up of various elements (individuals) and the connections between them (Yang and Leskovec, 2015). Connections vary in strength and significance, and together they weld this formation into a structure. A community is therefore challenging to study. Research into such a complex entity, with its diverse traits and characteristics, calls for a multidimensional approach.

Bonanno et al. (2015) suggest the use of average-level data on adjustment in studies of adversity. Their approach focuses on the event, rather than on individual reactions to the event, and seeks to characterize differences (on average) between exposed groups and non-exposed groups, or to some other comparative baseline. Their assumption is that the statistical average represents the normal or modal responses to adversity in a community, and therefore that by analyzing only a summary statistic of the dependent variable (such as the mean), we can only see a narrow and incomplete picture (Fitzenberger and Wilke, 2015 and Bonanno et al., 2015).

In this paper we demonstrate that by slicing the population into discrete and more homogeneous slices created by quantiles, we can provide more accurate insights into the community. Departing from the agreed compromise of using the mean in order to describe a group opens up a wide spectrum of opportunities, not only for descriptive or analytic statistics, but also for the practical study of community resilience (and other community characteristics as well). The ability to explore each centile or quantile will facilitate identification not only of persons who need more assistance, but also of those who can help their peers. On the other hand, it will also help us discern what factors exist across the community's sub-groups and which factor is capable of enhancing all of the sub-groups. One limitation of this study is the fact that it was cross-sectional, thus we could identify association but not causality. Future longitudinal studies, particularly those following an intervention or occurrence that influences community resilience, will shed more light on the causes of change identified in the different quantiles. Another limitation is the imperfection of information regarding the response rate, due to the fact that electronic mailing lists were used to approach some of the study population and we are unable to detect mail which did not reach its destination or was not opened.

CONCLUSION

In this study the use of quantile regression analysis for modeling community resilience was shown to yield more sensitive results than traditional regression models, potentially contributing to a deeper understanding of the issues at hand. We recommend that quantile regression be routinely incorporated into the study of community behavior. Quantile regression analysis can play an important role in the development of tailored, more effective response plans for communities facing change and crisis.

ACKNOWLEDGEMENT

The authors would like to thank Mr. Dmitry Leykin and Ms. Aviva Goral for their assistance with data analysis. We would like to express our sincere gratitude to Prof. Friger from the department of Public Health in the faculty of Health Sciences at BGU for exposing us to the potential of using quantile regression analysis in community resilience data and for being the ultimate teacher for Biostatistics. The Conjoint Community Resiliency Assessment Measure (CCRAM) was developed through the group work of a multidisciplinary group of researchers from seven academic institutions in Israel, as detailed in Leykin et al. (2013). We thank them all.

REFERENCES

1. Adger, W.N., 2006. Vulnerability. Global Environ. Change 16 (3), 268–281.
2. Aldunce, P., Beilin, R., Howden, M., Handmer, J., 2015. Resilience for disaster risk management in a changing climate: practitioners' frames and practices. Global Environ. Change 30, 1–11.
3. Bonanno, G.A., Romero, S.A., Klein, S.I., 2015. The temporal elements of psychological resilience: an integrative framework for the study of individuals, families, and communities. Psychol. Inq. 26, 139–169.
4. Bühler, M.D., De Torres Curth, M., Garibaldi, L.A., 2012. Demography and socioeconomic vulnerability influence fire occurrence in Bariloche (Argentina). Landsc. Urban Plan. 110, 64–73.
5. Cade, B.S., Noon, B.R., 2003. A gentle introduction to quantile regression for ecologists. Front. Ecol. Environ. 1 (8), 412–420.
6. Cade, B.S., Terrell, J.W., Schroeder, R.L., 1999. Estimating effects of limiting factors with regression quantiles. Ecology 80, 311–323.
7. Castleden, M., McKee, M., Murray,V., Leonardi, G., 2011. Resilience thinking in health protection. J. Public Health Oxf. Engl. 33, 369–377.
8. Chandra, A., Acosta, J., Meredith, L.S., Sanches, K., Stern, S., Uscher-pines, L., et al., 2010, February. Understanding Community Resilience in the Context of National Health Security A Literature Review, Retrieved from: http://www.prgs.edu/ content/dam/rand/pubs/ working papers/2010/RAND WR737.pdf.
9. Cohen, O., Leykin, D., Lahad, M., Goldberg, A., Aharonson-Daniel, L., 2013. The conjoint community resiliency assessment measure as a baseline for profiling and predicting community resilience for emergencies. Technol. Forecast. Soc.Change 80 (9), 1732–1741.
10. Cutter, S.L., 2015. The landscape of disaster resilience indicators in the USA. Nat. Hazards, 1–18.
11. Cutter, S.L., Ash, K.D., Emrich, C.T., 2014. The geographies of community disaster resilience. Global Environ. Change 29, 65–77.
12. Cutter, S.L., Burton, C., Emrich, C.T., 2010. Disaster resilience indicators for benchmarking baseline conditions. J. Homel. Secur. Emerg. Manag. 7 (1), 1–22.
13. Cutter, S.L., Barnes, L., Berry, M., Burton, C., Evans, E., Tate, E., et al., 2008. A place-based model for understanding community resilience to natural disasters. Global Environ. Change 18, 598–606.
14. Engle, N.L., 2011. Adaptive capacity and its assessment. Global Environ. Change 21, 647–656.
15. Fitzenberger, B., Wilke, R.A., 2015. Quantile Regression Methods. Emerging Trends in the Social and Behavioral Sciences: An Interdisciplinary, Searchable, and Linkable Resource. pp. 1–18.
16. Gould, W.W., 1992. Quantile regression with bootstrapped standard errors. Stata Tech. Bull. 9, 19–21.
17. Hao, L., Naiman, D.Q., 2007. Quantile Regression. Sage, Thousand Oaks, CA.
18. Hawdon, J., Räsänen, P., Oksanen, A., Ryan, J., 2012. Social solidarity and wellbeing after critical incidents: three cases of mass shootings. J. Crit. Incid. Anal. 3, 2–25.
19. Koenker, R., Bassett, R.G., 1978. Regression quantiles. Econometrica: J. Econ. Soc., 33–50.
20. Leykin, D., Lahad, M., Cohen, O., Goldberg, A., Aharonson-Daniel, L., 2013. Conjoint community resiliency assessment measure-28/10 items (CCRAM28 and CCRAM10): a self-report tool for assessing

community resilience. Am. J. Community Psychol. 52 (3–4), 313–323.

21. Magis, K., 2010. Community resilience: an indicator of social sustainability. Soc. Nat. Resour. 23, 401–416.
22. Magzamen, S., Amato, M.S., Imm, P., Havlena, J.A., Coons, M.J., Anderson, H.A., Moore, C.F., 2015. Quantile regression in environmental health: early life lead exposure and end-of-grade exams. Environ. Res. 137, 108–119.
23. Maloney, W.F., Cunningham, W.V., Bosch, M., 2004. The distribution of income shocks during crises: an application of quantile analysis to Mexico, 1992–95. World Bank Econ. Rev. 18, 155–174.
24. Nordhaus, W.D., 2010. The economics of hurricanes and implications of global warming. Clim. Change Econ. 1 (1), 1–20.
25. Norris, F.H., Stevens, S.P., Pfefferbaum, B., Wyche, K.F., Pfefferbaum, R.L., 2008. Community resilience as a metaphor, theory, set of capacities, and strategy for disaster readiness. Am. J. Community Psychol. 41, 127–150.
26. Pfefferbaum, R.L., Pfefferbaum, B.,Van Horn, R.L., Klomp, R.W., Norris, F.H., Reissman, D.B., 2013. The communities advancing resilience toolkit (CART): an intervention to build community resilience to disasters. J. Public Health Manag. Pract. 19 (3), 250–258.
27. Poortinga, W., 2011. Community resilience and health:the role of bonding, bridging, and linking aspects of social capital. Health Place 18, 286–295.
28. Ungar, M., 2011. The social ecology of resilience. Addressing contextual and cultural ambiguity of a nascent construct. Am. J. Orthopsychiatr. 81, 1–17.
29. Wilson, G.A., 2013. Community resilience, policy corridors and the policy challenge. Land Use Policy 31, 298–310.
30. Wilson, G.A., 2012. Community resilience, globalization, and transitional pathways of decision-making. Geoforum 43 (6), 1218–1231.
31. Yang, J., Leskovec, J., 2015. Defining and evaluating network communities based on ground-truth. Knowl. Inform. Syst. 42 (1), 181–213.
32. Zoback, M.L., 2014. Epicenters of resilience. Science 346 (6207), 283.

CHAPTER 7

BEST PRACTICE IN ECOPATH WITH ECOSIM FOOD-WEB MODELS FOR ECOSYSTEM-BASED MANAGEMENT

Johanna Jacomina Heymans, Marta Coll, Jason S. Link, Steven Mackinson, Jeroen Steenbeek, Carl Walter, Villy Christensen

Keywords: Ecopath with Ecosim, Ecological network analysis, Ecosystem modelling, Ecosystem-based management, Monte Carlo, Time series fitting

INTRODUCTION

Ecopath with Ecosim (EwE) is a modelling complex that has been used to create mass balanced models of marine and aquatic ecosystems since the 1980s, when the first Ecopath model of the French Frigate Shoals was created by Jeff Polovina (1984). The software and techniques have been improved to include methods of comparing ecosystems using Ecological Network Analysis (Christensen and Pauly, 1992), to model dynamic changes using Ecosim (Walters et al., 1997), to model spatial changes using Ecospace (Walters et al., 1999) and to model both temporal and spatial dynamics using the spatial-temporal-framework and the habitat foraging capacity model (Christensen et al., 2014).

EwE is the most applied tool for modelling marine and aquatic ecosystems globally, with over 400 models published to date (Colléter et al., 2013). Ecopath was recognized as one of NOAA's top ten scientific breakthroughs (http://celebrating200years.noaa.gov/breakthroughs/ecopath/welcome.html) in its 200 year history, and the 30th anniversary of the approach was recently celebrated with an international conference (Steenbeek et al., 2014 and Coll et al., 2015). The EwE approach is able to address many of the questions asked by managers on marine policy issues such as natural variability and monitoring, management measures, ecosystem goods and services, "Good Environmental Status" targets for the EU Marine Strategy Framework Directive, and environmental change and climate adaptation (Steenbeek et al., 2014 and Hyder et al., 2015). It has also been partially re-coded in different programming languages (Steenbeek et al., 2015) including Fortran (Akoglu et al., 2015), Matlab (Kearney et al., 2013), and R (Lucey et al., 2014). However, of the 105 models used for comparative analyses by Heymans et al. (2014) few have been fitted to historical data using Ecosim (27/105) and even fewer (13/105) were actually calibrated by fitting to time series data (Table S1 in Heymans et al., 2014). Additionally, very few models have actually been used in a management context. A noteworthy exception is the model of the North Sea (Mackinson et al., 2009b), which has recently been used to establish a "key run" (see Section 8 below for a description of key runs) for the ICES Working Group on Multispecies Assessment Methods, WGSAM http://www.ices.dk/community/groups/Pages/WGSAM.aspx (ICES, 2013), and used to evaluate EU Commission proposals for multi-annual plans in the North Sea (STECF, 2015). Models have also been used in the Gulf of Alaska and Bering Sea (Aydin et al., 2007 and Gaichas et al., 2011) and the Western Scotian Shelf (Araújo and Bundy, 2012), where those outputs informed the management process by evaluating previously unexplained mortality.

Because of its ease of use, both the construction and comparison of EwE models can be easily misused by inexperienced users. Ecopath models are easy to construct, with very few automatic checks, but instead requiring the understanding and application of ecological knowledge by users. Ecological Network Analysis (ENA) comparisons

Citation: Johanna Jacomina Heymans, Marta Coll, Jason S. Link, Steven Mackinson, Jeroen Steenbeek, Carl Walters, Villy Christensen, Best practice in Ecopath with Ecosim food-web models for ecosystem-based management, Ecological Modelling, Volume 331, 10 July 2016, Pages 173-184, ISSN 0304-3800, http://dx.doi.org/10.1016/j.ecolmodel.2015.12.007.

are similarly easily undertaken without checking the thermodynamic and mathematical rules behind the analyses (but this is true for all ecological analyses). EwE models, once balanced (whether sensibly or not) can then be used to describe dynamics in the ecosystem without forced calibrations, which again can be used for management without proper verification and validation. This easy access, with limited quality control, may have contributed to the limited utilization of EwE models for management applications.

In order for EwE models to be accepted as being rigorous and consistent enough to be used for management, guidelines are needed to establish best practices in creating and using the models. These guidelines need to take into consideration the thermodynamics and ecological rules available to users, recommended approaches to balance an Ecopath model, the best tools to fit models to time series data, and how to evaluate uncertainty. In addition, we need best practices for comparing ecosystem models of different ecosystems or different models of the same ecosystem via standardized indicators, descriptors, etc. Here, we provide an overview of the best practices for creating Ecopath models, fitting Ecosim models to time series, evaluating uncertainty, and comparing models. The contents herein can serve as guidelines for testing and reviewing EwE models, especially if they are to be used for management purposes.

WHY AND HOW TO CREATE AN ECOPATH MODEL

The FAO have suggested best practices for developing models for an ecosystem approach to fisheries, including considerations for model aggregation, spatial consideration, etc. However, in this paper we are providing the best practices specifically with respect to creating, balancing, fitting and using a EwE model, after one has followed the more generalized best practice guidelines of FAO (2008).

Why this Model?

The choice of modelling technique should always depend on the policy or research question that is to be addressed. The model has to be constructed with the policy or research questions clearly formulated, which will lead to using the correct modelling framework. The FAO report on models used for an ecosystem approach to fisheries (Plagányi, 2007) gives a flow chart that summarizes the models explained and where they are best used. This is a very handy way of making sure the model one uses is able to answer the question being asked. However, it is important that the newest descriptors of the techniques are used when making that decision, as the 2007 report describes the capacity of modelling techniques at that time. According to the report, EwE can be selected to: identify and quantify major energy flows in an ecosystem, describe the ecosystem resources and their interactions among species, evaluate the ecosystem effects of fishing or environmental changes, explore management policy options by incorporating economic, social and ecological considerations of fisheries, evaluate the placement and impact of marine protected areas, or predict the bioaccumulation of persistent pollutants. EwE models are also useful for testing ecosystem theories on resilience, stability and regime shifts (Pérez-España and Arreguín-Sánchez, 2001,Tomczak et al., 2013, Arreguín-Sánchez and Ruiz-Barreiro, 2014 and Heymans and Tomczak, 2015).

Creating an Ecopath Model

A EwE model must represent the main species and trophic levels that are present in the modelled ecosystem and are of relevance for the policy or research question that is to be addressed. The time frame and spatial extent of the EwE model depends on the questions that are to be addressed, as well as data availability. The modelled ecosystem should, as a rule, include the whole habitat area of the main species of concern. If for instance the role of a migratory species on a smaller ecosystem needs to be studied, the biomass of that migratory species would have to be forced in the model. The model will not be able to predict its biomass without information of dynamics outwith the system. The baseline Ecopath models are usually based on average parameters for a given baseline year, and thus typically average over seasonal changes in the ecosystem, although daily (Orr, 2013) and seasonal (Heymans et al., 2002) models have also been developed. To use a EwE model for ecosystem-based management, time series fitting is required to incorporate density-dependence, which best is done by evaluating how the EwE model can reproduce historical dynamics. It follows that sufficiently long time series are required to provide contrast, and the baseline Ecopath model should thus be based on the earliest time where data is available. Temporal extents have ranged among EwE models from 10 to >50 years, depending on the degree of available data. Alternatively, a model can be constructed for the year where most data is available, and then adapted to an earlier timeframe where time series data started, as was done in several examples such as for Central North Pacific (Cox et al., 2002), the Aleutian Islands (Heymans, 2005), Newfoundland (Pitcher and Heymans, 2002), the North Sea (Mackinson and Daskalov, 2007), and the north-western Mediterranean (Coll et al., 2008). It is important, however, to consider the choice of time frame in relation to changes in ecosystem structure that should be accounted for in the dynamic behaviour of the ecosystem model. For example the behaviour might be different in one climate regime than in another. In some cases it is possible to model this change, such as was done in the Baltic Sea (Tomczak et al., 2013), the Gulf of Alaska (Heymans et al., 2007), and the northern Benguela (Heymans and Tomczak, 2016). However, it might not be possible to model this behavioural change, which will severely limit the given model's long-term predictive capabilities.

EwE models must contain at least one detritus group and one consumer group, while there in practice is no upper limits to the number of groups that can be included. Primary producers, while most often included, are not required. As an example, in deep sea models (Tecchio et al., 2013) primary producers may not be present and more than 1 detritus group might be needed to describe the system adequately, for example incorporating marine snow. Often, a combination of functionality and lack of data will require aggregation of several species into 'functional groups' to describe the ecosystem. Functional groups can be individual species or groups of species that perform a similar function in the ecosystem, i.e. have approximately the same growth rates, consumption rates, diets, habitats, and predators. They should be based on species that occupy similar niches, rather than of similar taxonomic groups. All species in a functional group should ideally show similar biomass and catch trajectories over time. In addition, functional groups might be used when a specific policy question requires parsimony. For example it might not be necessary to describe the benthic functional groups in great detail when addressing top predators such as marine mammals. Instead, representation of benthos groups might be aggregated in a simplified way. In most fished ecosystems, fisheries discards may play an important role, and therefore discards might need to be included as a separate detritus group that are eaten by species such as seabirds, crabs and mammals before the discards are broken down to detritus. If more than one detritus group is defined the detritus fate needs to be described in the model. The detritus fate from all functional groups needs to be allocated to an appropriate detritus group otherwise there will be instability when moving towards modelling temporal dynamics.

Top predators are important to constrain the parameters of other consumers in the ecosystem, and therefore it is especially important to have biomass estimates for the top predators in the ecosystem. It is also important to split species into "multi-stanza" groups where it is necessary to capture ontogenetic diet shifts and/or different exploitation patterns. If species are split into multi-stanza groups, estimates are needed of the age (in months) of the split between stanzas (e.g. adult and juvenile), the total mortality per stanza, the von Bertalanffy *K* parameter, and the estimate of weight at maturity as a fraction of weight at infinity (W_{mat}/W_{inf}) (relative weight at maturity, determines size-fecundity relationship). The parameter W_{mat}/W_{inf} is important because it influences the degree of compensation in recruitment (i.e. how productive the juvenile stanzas are at low spawning stock biomass), which has a direct influence on recovery and depletion rates in time dynamics (Ecosim) and estimates of F_{msy}. The reason for this is that when W_{mat}/W_{inf} is small, fish mature at early ages, and it is quite possible for the 1st and/or 2nd "juvenile" stanza(s) to produce enough eggs to sustain recruitment even when the "adult" *F* is very high. This is a classic prediction from equilibrium theory, i.e. that F_{msy} can be infinite if size at first capture is enough larger than size at maturity for animals to have replaced themselves before becoming vulnerable. The default setting for W_{mat}/W_{inf} (0.09) is probably low in many cases, and a better value from the Beverton–Holt invariants would be 0.22. For each stanza, estimates of diet and predation, catches and discards also need to be included.

For each consumer functional group, consumption (Q) = production (P) + respiration (R) + egestion (E) and production a function of the natural and fishing mortality of the ecosystem, so that:

$$P_i = Y_i + B_i \cdot M2_i + E_i + BA_i + P_i \cdot (1 - EE_i) \qquad (1)$$

where *Pi* is the total production rate of group *i*, *Yi* is the total fishery catch rate of *i*, *M2i* is the total predation rate for *i*, *Bi* the biomass, *Ei* the net migration rate (emigration–immigration), *BAi* is the biomass accumulation rate for *i*. *EE* is the ecotrophic efficiency of the group, and indicates the proportion of the production or total mortality (where $P/B = Z$) that is actually explained in the model. 1 − *EE* is the proportion of the production that remains not explained, i.e. the 'other mortality' rate for *i* (Christensen et al., 2008). Normally, biomass, *P*/*B* and *Q*/*B*, in addition to diets and catches, are inputs of the model, while *EE* is an output. However, in some cases when data on biomass are not available, *EE* values are set to allow the model to calculate missing parameters. When using *EE* as input parameters, one should consider:

- Top predators that are not fished or predated upon will have an *EE* of 0 – indicating that none of the production of that group is consumed or fished in the ecosystem, thus a biomass estimate for such groups is always required.
- *EE* values for primary producers are also often significantly less than 1, with only 10% of macrophytes such as kelp being consumed in some ecosystems (Mann, 1988) and phytoplankton in bloom only ~50% is utilized outside of the microbial loop (Azam et al., 1983).
- Most other functional groups should have *EE* values approaching, but lower than 1 if all of their production is utilized within the ecosystem that is modelled.

It is, however, not advisable to use *EE* values to estimate parameters of functional groups that are placed very high or very low in the food web due to the low constraints of the algorithm to properly parameterize those groups: setting *EE* to calculate the biomass of top predators is generally not possible, as they are not consumed within the system and therefore EwE will be unable to calculate the biomass of top predators needed in the system. Similarly, setting *EE* to calculate the primary producer parameters, if necessary, needs to consider that *EE* is generally only high (close to 1) in clear-water systems such as coral reefs or highly oligotrophic systems such as open seas, while systems that have strong seasonal patterns in productivity, such as spring blooms, tend to have *EE*'s of 0.5 or lower.

For each functional group, estimates are required for biomass, production/biomass (*P*/*B*), consumption/biomass (*Q*/*B*) and diet, while estimates of catches and discards are needed for fished groups. The excretion/egestion rate, for which the default is 20% can vary between functional groups, and will typically be higher for groups that predominantly feed on phytoplankton. An excretion rate of 40% is thus often more realistic for zooplankton – to evaluate this, it is important to check the respiration/biomass ratio for a group, and compare this to values from physiological studies. The issue with using the default excretion rate often appears when the estimated *EE* for detritus (which is defined as detritus required/detritus produced) exceeds unity: Increasing the excretion rate results in more detritus being produced.

The microbial loop is important, but not always of importance for a given policy question. Where it is, it is important to consider how to model the loop explicitly, and where it is not, it can often be considered as part of the detritus/phytoplankton part of a model. It often appears in models where the primary production is entered that it is insufficient to meet the demands of the consumers. Where models, however, have explicitly represented the microbial loop, e.g. North Sea (Mackinson and Daskalov, 2007), a considerable part of the production that needs to be passed on to the higher predators can be derived through microbial pathways. When including the microbial loop, estimates of primary production are rarely exceeded. In fact, a lot of primary production is not utilized directly by zooplankton, but settles to the bottom and enters the food web through detrital pathways. Estimates of biomass for each functional group need to reflect the biomass of the group in the whole model area, not just the area within the ecosystem that they occupy. Thus, for species that only occupy part of the area the biomass need to be pro-rated by area. Biomass estimates are calculated in weight/area, e.g. t km^{-2} and can be calculated through standard surveying or stock assessment methods (Hilborn and Walters, 1992 and Walters and Martell, 2004).

Diet estimates for functional groups are obtained from stomach content analyses and should reflect the best available data on diets of predators by weight, not frequency of occurrence or numbers (Stergiou and Karpouzi, 2002). Alternatively, they can also be obtained from stable isotopic analyses (Ramsvatn, 2013 and Deehr et al., 2014) using Bayesian isotopic mixing models. Where functional groups are used, weighted averages using consumption rates or biomass proportions of species specific diets should be used. In some cases, species might consume food outside the ecosystem (i.e. birds that migrate from sea – land, etc.) and as such this should be reflected by an import to the diet of the species. Often the best diet estimates are not available for the time frame of the initial Ecopath model, in which case it is prudent to create a balanced Ecopath model for the year where most diet data is available as a first step, before adapting that model for the timeframe needed, by assuming that changes in prey biomass could reflect change in the diet while consumer preferences would stay similar over time as was done for the Aleutian Islands (Heymans, 2005), Newfoundland (Pitcher and Heymans, 2002) and the North Sea (Mackinson and Daskalov, 2007). In addition, it is important to be aware of the issues of cannibalism, which might impact the estimation of biomass. A rule of thumb is <5% cannibalism in one group, as this causes computational problems when estimating biomass, but also improves the representation of the groups (Christensen et al., 2008). If cannibalism is a problem it is as a rule best to split the group into a multi-stanza group.

Estimates of *P/B* and *Q/B* can be obtained from laboratory experiments. However, if these are not available readily available empirical equations can be used. The *P/B* ratio is related to the turnover rate of a functional group and is equal to the total mortality (Z, $year^{-1}$), where is equal to fishing mortality (F, $year^{-1}$) plus natural mortality (M, $year^{-1}$), all expressed as instantaneous annual rates. F is easily obtained from stock assessments or calculated as:

$$F = \frac{\text{Catch}}{\text{Biomass}} \qquad (2)$$

And for fish, M may be estimated from Pauly's empirical equations (Pauly, 1980):

$$\log M = -0.0066 - 0.279 \cdot \log L_{\infty} + 0.6543 \cdot \log K + 0.4634 \cdot \log T, \text{ or} \qquad (3)$$

$$\log M = -0.2107 - 0.0824 \cdot \log W_{\infty} + 0.6757 \log K + 0.4627 \log T \qquad (4)$$

where L_{∞} and W_{∞} are the length (total length in cm) and wet weight (g) at infinity for the population, K is the von Bertalanffy growth parameter ($year^{-1}$), and T(C) is the mean annual water temperature at which the population is maintained.

Similarly, if not estimated directly from experiments, at least for fish the *Q/B* ratio ($year^{-1}$) may be estimated from the empirical ratio of Palomares and Pauly (1998):

$$\log\frac{Q}{B} = 7.964 - 0.204 \cdot \log W_{\infty} - 1.965 \cdot T' + 0.083 \cdot A + 0.532 \cdot h + 0.398 \cdot d \qquad (5)$$

where T' is the mean annual habitat temperature for the population expressed as 1000/(T + 273.1), A is the aspect ratio of the tail, $h = 1$ when the fish is a herbivore and 0 when not, and $d = 1$ when the fish is a detritivore and 0 when not. Alternatively,Palomares and Pauly (1998) estimated the *Q/B* ratio as:

$$\log\frac{Q}{B} = 5.847 - 0.28 \cdot \log Z - 0.152 \cdot \log W_{\infty} + 1.360 \cdot T' + 0.062 \cdot A + 0.51 \cdot h + 0.39 \cdot d \qquad (6)$$

where Z is the total mortality. The aspect ratio of the tail is a measure of the swimming and metabolic activity of the fish expressed as:

$$A = \frac{h^2}{s} \qquad (7)$$

where h is the height (mm) of the caudal fin, and s the surface the area (mm^2) of the fin, extending to the narrowest part of the caudal peduncle (Palomares and Pauly, 1998). For invertebrates, the best estimates for natural mortality and *Q/B* ratios are obtained from the work of Thomas Brey, 1999, Brey, 2010 and Brey, 2012. These can be estimated using the formulas available at http://thomas-brey.de/science/virtualhandbook, or from the allometric relationships found in Peters (1983).

For seabirds the consumption rate (DR or daily rate of fish consumed in g) can be estimated from Nilsson and Nilsson (1976) as:

$$\log DR = -0.293 + 0.85 \cdot \log W \qquad (8)$$

where W is the mean body weight of the species (g). While for marine mammals the consumption rate (DR or daily rate, in kg) can be estimated from Innes et al. (1987) as:

$$DR = 0.1 \cdot W0.8 \qquad (9)$$

where W is the mean body weight of the species expressed in kg.

PREBAL: RULES OF THUMB

Once the input data to a model has been collected, it is imperative that the underlying assumptions are tested. This is described by Link (2010), who proposes a set of pre-balance (PREBAL) diagnostics including checking the slopes of biomass ratios, vital rates, total production, etc. based on trophic levels. These diagnostics are based on general ecological and fisheries principles, and should be checked before a model is balanced. The PREBAL criteria include some "rules of thumb" such as the assumption that biomass estimates should span 5–7 orders of magnitude, with >7 indicating too many taxonomic/ age-structured groups, and <5 indicating that the model might be too focussed on specific trophic levels (Link, 2010). In addition, the slope of the biomass (on a log scale) should decline by 5–10% across all the taxa arrayed by trophic level (Link, 2010), and groups that are above or below the slope-line should be checked for data integrity. See Link (2010) for further information. An example of some PREBAL diagnostics for a newly published model of the West Coast of Scotland

(Alexander et al., 2015) is given in Fig. 1. These easy to plot diagnostics show that biomass (Fig. 1a) estimates of harbour seals, cetaceans, seabirds, poor cod, lobster and most of the juvenile stanzas (haddock, whiting and cod) might potentially be underestimated in this model, while that of crustaceans, mackerel, blue whiting and large demersals, might be over-estimated. As with biomass, vital rates of predators should be lower than those of their prey (Link, 2010). In the West Coast Scotland model, annual *P*/*B*ratios (Fig. 1b) for cetaceans, lobster, edible crabs and scallops seem low, while those of adult whiting, epifauna and phytoplankton are high. Similarly, annual *Q*/*B* ratios (Fig. 1c) of seals, cetaceans, seabirds, cephalopods, epifauna, infauna and small zooplankton seem very high, while that of crustaceans, horse mackerel, edible crab, rays, large demersals and monkfish, seem low. When creating a new model, these estimates should be checked and although it was undertaken to some extent in the West Coast of Scotland model (Alexander et al., 2015), further analyses and re-parameterization might still be required, especially for the top predators. PREBAL has not been used extensively although examples are now forthcoming (Lassalle et al., 2014 and Zetina-Rejón et al., 2015).

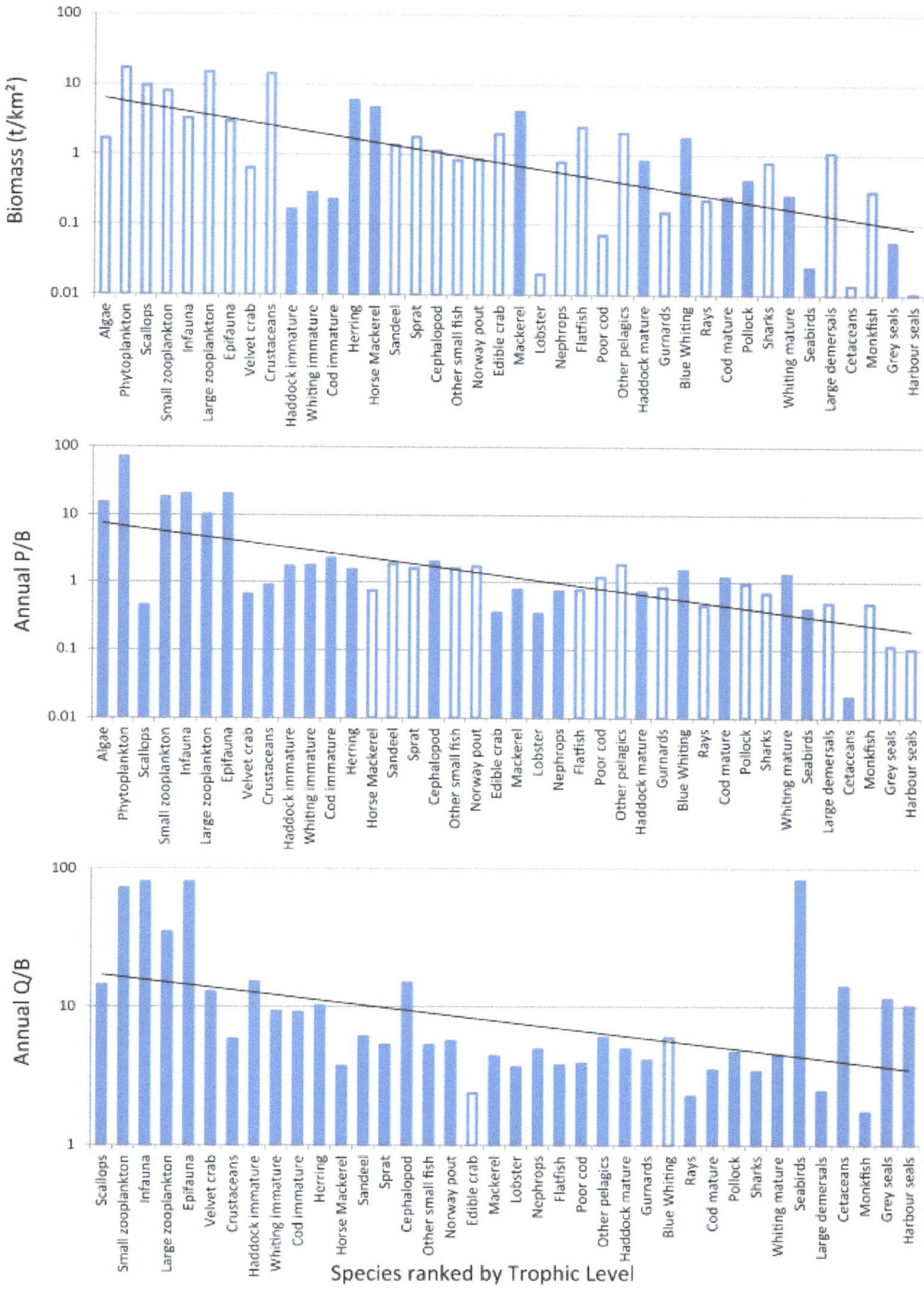

Figure 1: PREBAL of the West Coast of Scotland model (Alexander et al., 2015) plotting (a) biomass estimates (t/km²), (b) production/biomass ratio (per year), and (c) consumption/biomass (per year) on a log scale vs species ranked by trophic level, from lowest to highest trophic level. Solid bars indicate input to the model and white bars indicate model estimated parameters.

A further diagnostic, the *P*/*Q* ratio or the gross food conversion efficiency, indicates that a group cannot produce more than a fraction of what it has eaten, based on the 2nd law of thermodynamics (Link, 2010). In fact, the *P*/*Q* ratio should mostly be between 0.1 and 0.3, as described in Darwall et al. (2010). Prior to final balancing, we recommend that these rules of thumb should be checked. A new PREBAL diagnostic algorithm is currently being developed for release in a new version of EwE (Fig. 2). This diagnostic is only meant to be a first check, and cannot undertake the more in depth analyses of the ratios of small pelagics to top predators, etc. as suggested in Link (2010). These more in depth analyses should be done before the model topology is decided and before the above mentioned diagnostics are run.

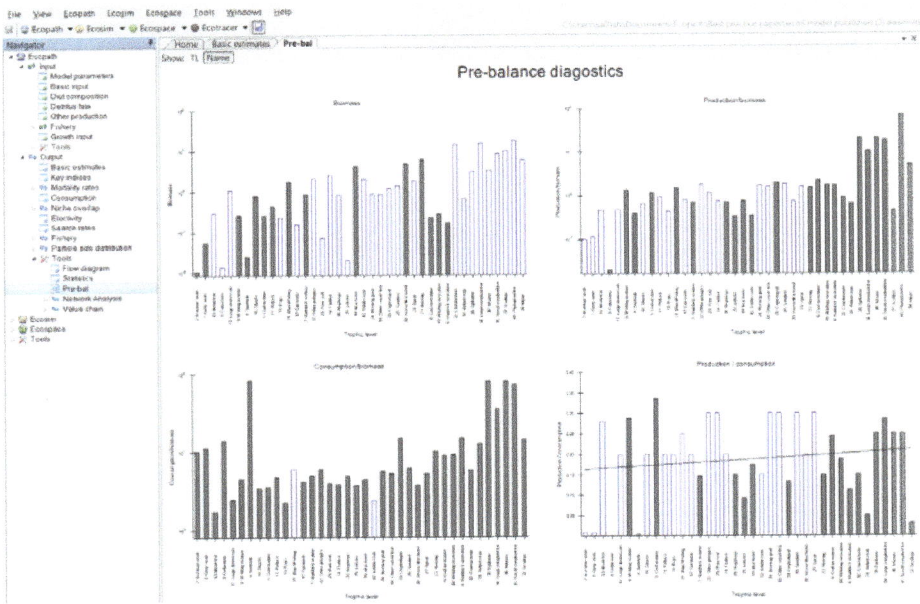

Figure 2: PREBAL diagnostics available for in Ecopath with Ecosim version 5 as given for the West Coast of Scotland.

BALANCING ECOPATH MODELS

Once the Ecopath parameters have been entered into the software and the underlying assumptions tested with the PREBAL approach, the model requires balancing to maintain the laws of thermodynamics. When balancing the model it is imperative that thermodynamic and ecological rules are followed. Darwall et al. (2010) describes the ecological and thermodynamic rules in an easy to read box (Box 1Darwall et al., 2010). Once the model has been balanced it might be useful to recheck the PREBAL estimations, to check for incompatible vital rates (*P*/*B*, *Q*/*B*, etc).

Box 1: Ecological and thermodynamic rules for balancing Ecopath models (from Darwall et al., 2010).

An ecologically and thermodynamically balanced Ecopath model requires a series of logical constraints:

- ***EE*** < 1.0. Ecotrophic Efficiency (*EE*) is a measure of the proportion of production that is utilized by the next trophic level through direct predation or fishing. The value for *EE* (often a calculated output of Ecopath) can never exceed 1.0 as it is not possible for more production to be passed on to the next trophic level than was originally produced—unless the population is in decline. As a guideline an*EE* value near to 1.0 is expected when the main part of production is consumed by predators or taken by the fishery. A value near to 0.0 is expected for a group, such as an apex predator, which suffers no predation and is not exploited by a fishery.
- < GE < 0.3. Gross food conversion efficiency (*GE*) in general has a value between 0.1 and 0.3. Values greater than 0.5 are not often found but may be encountered in groups such as bacteria or in specially bred farmed fish.
- Net Efficiency > ***GE***. Net Efficiency is the value for food conversion after accounting for unassimilated food (*U*) for which the Ecopath default value is 20%, which is most applicable for finfish (Winberg, 1956). Thus Net Efficiency = $P/(P + R)$. *GE* is P/Q, and $Q = P + R + U$. It is therefore clear that *GE*can never exceed Net Efficiency.
- Respiration/Assimilation Biomass (***RA***/***AS***) < 1.0. The proportion of biomass lost through respiration cannot exceed the biomass of food assimilated. As a guideline, *K* selected species, which are expected to invest a relatively small proportion of energy intake in somatic and gonadal tissue production are expected to have *RA*/*AS* ratios close to 1.0. In contrast, *r*-selected species are more likely to invest a greater proportion of energy intake into growth and reproduction resulting in an *RA*/*AS* ratio well below 1.0.
- Respiration/Biomass (***RA/B***) indicates the "metabolic activity level" of a group. *RA/B* ratios are expected to be within 1–10 $year^{-1}$ for fish and may be as high as 50–100 $year^{-1}$ for groups with higher turnover such as copepods. The default value for the proportion of unassimilated food (20%) may be changed to better reflect the *RA/B* ratio value expected of the group in question.

- Production/Respiration (***P/RA***) < 1.0. This ratio effectively expresses the fate of assimilated food. Odum (1969) stated that *P/RA*, which is typically less than 1, approaches 1 as the system matures. However, Christensen and Pauly (1993)comparing 41 Ecopath models found that *P/RA* ranged from 0.8 to 3.2. The high ratio values were thought to have arisen because of the omission of bacterial activity that led to an underestimation of respiration.

It is important to annotate Ecopath input data with appropriate references and to describe the origin of the data used to create the model in the model pedigree (Christensen and Walters, 2004 and Morissette, 2007), which describes the precision of the data and sets confidence intervals to be used with this data if undertaking Monte-Carlo simulations. These estimates can be picked up by the Monte-Carlo routine to test the efficacy of the input data on the dynamic simulations in Ecosim and can be used to calculate confidence intervals in ecological indicators. Model descriptions are often provided in technical reports, where data and methods used for the model are described in detail. A good example is given in Mackinson and Daskalov (2007).

COMPARING ECOPATH MODELS

Over the past 30 years, the construction and comparison of Ecopath models and the use of Ecological Network Analysis (ENA) indices to compare ecosystem models has been the mainstay of many MSc and PhD theses, as well as for a wide range of projects. This started prior to the main use of Ecopath with the work of Baird et al. (1991), who compared six marine ecosystems using ENA approaches, followed by the estimates of maturity from Christensen (1995) based on the work of Odum (1969). However, only when an Ecopath model has been properly checked with PREBAL and balanced should any model comparisons be attempted. As stated above, model comparisons based on ecosystem summary statistics and ENA indices are widespread in the literature. Of the 105 models used for meta-analyses in Heymans et al. (2014), 39 models used ENA indices for comparison (see Table S1 in that reference); however very few of these followed the PREBAL or balancing rules given above.

When comparing Ecopath or ENA models it is important to realize that the topology (number compartments, distribution of species, links between species, etc.) of a model is key since some indicators are dependent of the model structure (Pinnegar et al., 2005). For example, in a model where the primary producers and primary consumers are well defined (i.e. a system where the microbial loop is separated from the detritus Mackinson and Daskalov, 2007), but the top predators are aggregated, the total energy recycled through the ecosystem (Finn cycling index, Finn, 1976) would be much higher than a model with the same number of functional groups where all the primary producers are aggregated, the primary consumers are aggregated, and the top predators disaggregated as much less energy can flow between mid-trophic level and top predators that have a much lower biomass. As suggested by Pinnegar et al. (2005), ecosystem indices are very different in ecosystems that are aggregated differently, i.e. if one places emphasis on the top predators vs. mid-level species, the ecosystem indices might be very different (see Figure 3 in Pinnegar et al., 2005).

A study on the comparison of ecosystem indicators from Ecopath models found eight indicators that were robust to model construction, including the number of living groups, number of linkages or total number of functional groups (Heymans et al., 2014). These indicators are the ratios of primary production (*PP*), consumption (*Q*) and export (*Ex*) to total systems throughput (*TST*), the total biomass of the community (*TBCo*), the total systems throughput, relative ascendency (*A*/*C*), relative overhead (*O*/*C*) and redundancy (*R*) (for definitions see Heymans et al., 2014). When comparing models of different ecosystems these indicators would be the most conservative to use, as they are mainly standardized by *TST*. Most other indicators would more likely indicate differences in model construction. Ecosystem models that have been constructed with similar topologies (e.g. with or without microbial loop defined, similar top predators specifications or similar multiple stanzas for one species, etc.), and are constructed for similar areas can be used for comparison of other indicators. This has been done by Heymans et al. (2007) for ecosystems in the Gulf of Alaska, for upwelling systems (Shannon et al., 2003, Heymans et al., 2004 and Moloney et al., 2005) and for the northeast US large marine ecosystem (Link et al., 2008). Similarly, changes in ecosystems over time can be described by these indicators if ecosystem topologies stay constant in Ecosim (Coll et al., 2009). In addition ecosystem traits such as ecosystem depth, size, and location would naturally show differences in ecological indicators, and therefore should not be used for comparison (Heymans et al., 2014), or should at least be accounted for in any such contrast. They can be used as co-variates in a statistical analysis, for example.

Before looking at parameterization of Ecosim, it is important to reflect that the parameters that define a species' productivity which are specified in Ecopath also play an important role on the models dynamic predictions. For example, a species that has a low biomass and a high *P*/*B* ratio will have the same production as one with a high biomass and low*P*/*B* ratio. The impact of a low *P*/*B* ratio on the dynamics of that species is, however, very different to that of a higher *P*/*B* ratio. As an example, in a West Coast of Scotland model (Alexander et al., 2015), the initial biomass for grey seals (6204 t), and *P*/*B* ratio (0.114 year^{-1}) estimates a production of 707 t year^{-1}, while an order of magnitude more biomass (62,040 t) and a *P*/*B* ratio of 0.0114 year^{-1} will give a similar production. Thus in Ecopath the model will be balanced (even though the lower *P*/*B* is unrealistic), but in Ecosim the dynamics will be very different, and time series fitting is important to evaluate which of such parameter settings are most likely. The key to this is setting the density-dependent vulnerability parameters in Ecosim, setting the predatory prey dynamics between top down or bottom-up control between predators and prey (Ahrens et al., 2012). In Fig. 3, the estimated biomass of grey seals on the West Coast of Scotland is plotted against the input data used to fit that model as per Figure 4a of Alexander et al. (2015). If an order of magnitude mistake was made, the resulting biomass trajectory is shown in blue (on the second *y*-axis). It is clear that although the model would have balanced the predicted biomass trajectory (given the same vulnerability parameters, etc.) the trajectory would have been quite different. If the wrong Ecopath biomass and *P*/*B* ratios were used, the estimated vulnerabilities, etc. would have also been different when fitting to time series.

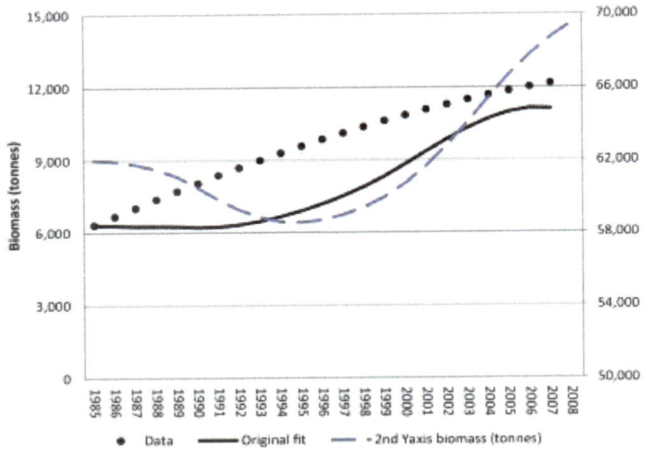

Figure 3: Adapted from Figure 4a in Alexander et al. (2015) where grey seal biomass data (tonnes) and original fit, refers to data and model fit from that paper and 2nd *y*-axis biomass refers to an order of magnitude difference in biomass (tonnes).

TIME SERIES FITTING IN ECOSIM

In order for a model to predict changes in response to temporal changes in fishing and environmental drives, its historical temporal dynamics should first be calibrated. This process compares model predictions to reference data to evaluate its performance in terms the fit to data, and importantly the credibility of its behaviour. The process of fitting the model to time series data is the part of the calibration process which estimates predator-prey parameters that influence degree of density dependence and thus, the rates of change in the biomass of each group. The second part of the calibration process is evaluating whether the parameterization lead to credible behaviour, for example in its predictions of sustainable fishing rates. Taking a pattern oriented approach to assessing the modelling skill (Grimm et al., 2005 and Kramer-Schadt et al., 2007) is the best approach, seeking patterns in both biomass trends and sustainable fishing rate estimates that are reasonable (more on this later).

Special Note: Exploitation Rate '*F*' and Instantaneous *F*

The fishing mortality (F, year^{-1}) in Ecosim is essentially the instantaneous annual F as traditionally defined in fisheries, i.e. by the relation $C = F*B_{bar}$, where C is the annual catch and B_{bar} is the mean (integral) of biomass over the annual period. Ecosim only converts these F's to discrete rates in the multi-stanza monthly age-structure updates, where monthly survival rates are calculated as $\exp(-Z/12)$ and Z is the sum of F's over fisheries plus natural mortality rate components. In Ecosim, catch per year is always the sum of predicted monthly catches, where catch for each month is approximated by$F*B_{month}$, i.e. F times the biomass at the start of the month. In other words, there is no need to convert to discrete time exploitation rates when looking at Ecosim F values.

Furthermore, if a species is to be split into multiple stanzas, the parameterization of the multi-stanza groups should be done with care such that the degree of compensation in recruitment at low stock biomass leads to predictions of F_{msy} (fishing mortality at maximum sustainable yield, MSY) that are at least consistent with published ranges ofF_{msy}. This check needs to be undertaken as part of the fitting process in Ecosim, and should lead to plausible Beverton–Holt type stock-recruitment functions (Walters and Martell, 2004). Otherwise there is a danger that the vulnerabilities estimated for multi-stanza groups during time series fitting can lead to overly optimistic estimation of how productive the groups are at low stock biomass, and thus vastly inflated estimates of F_{msy}(see Walters et al., 2005 and Mackinson et al., 2009b). The 'calibration' of the stock recruitment relationship of multi-stanza groups is essential for analyses seeking to address questions about MSY, but is equally important to all Ecosim work.

To fit an Ecosim model time series, data is required to: (a) affect a change in the model, and (b) to compare the modelled output to reference data using statistical diagnostics (Shannon et al., 2004, Coll et al., 2008 and Mackinson, 2014). Time series data is imported into Ecosim through a comma separated value (csv) file. Data used to affect a change in the model include forced biomass, time forcing data (e.g. on primary production, or for instance temperature trends that can affect the search rate, vulnerability or feeding arena of a predator), effort by gear type, fishing mortality by group, total mortality by group, forced total mortality by group and forced catches. Data used for statistical comparative purposes include relative or absolute biomass, total mortality, catches or average weight (for species that are defined by multi-stanzas). Diet compositions have also been used to fit models to data. The forcing data affects a change in the model that is described in biomass and catch estimates. These estimates are then compared with the reference observational data using a statistical measure of goodness of fit (Christensen and Walters, 2004). The quality of the time series data is important, and needs to be documented properly if the fitted model is to be used for management purposes.

The time series data can be weighted to represent the reliability of the data by giving it a weight > 0, where 0 indicates that it will not be used to calculate the *SS* (Christensen et al., 2008). The weighting factors can be set to reflect the reliability of the time series, e.g. by using the confidence intervals from survey estimates or from retrospective analysis of assessment models. Alternatively, the weights can be used to place more emphasis on target species for modelling. For instance, a model that is focused on cod in the Baltic may increase the weight that is placed on cod biomass. One method that can be used to determine the weighting factors is to perform a signal to noise ratio assessment for each series. The signal can be fitted using "LOcally weighted Scatterplot Smoothing" (LOESS) with the degree of smoothing required given by the optimal span determined from the bias-corrected Akaike information criterion (*AICc*, Akaike, 1974) following the method ofHurvich and Tsai (1998). The noise can be determined from the variance of the model residuals (var_{res}) and weights for EwE can be determined from the inverse of the variance ($1/var_{res}$).

When fitting a model to time series data, the best practice is to use the statistical hypothesis testing method as described by Mackinson et al. (2009a) and refined inTomczak et al. (2012), Mackinson (2014) and Alexander et al. (2015). This procedure is based on the estimation both of the sum of squares (*SS*) and the Akaike Information Index (*AIC*), which penalizes for fitting too many parameters based on the number of time series available for estimating the *SS* (Akaike, 1974). The *AIC* is defined as:

$$AIC = n \cdot \log(minSS/n) + 2K \quad (10)$$

where n is the number of observations, or time series values, i.e. the number of series used multiplied by the number of years for each parameter, min*SS* is the minimum sum of squares calculated by the algorithm, and K is the number of parameters estimated.

In this instance the *AICc* (Burnham and Anderson, 2004) is used, to address small sample size. The *AICc* is defined as:

$$AICc = AIC + 2K \cdot (K-1)/(n-K-1) \quad (11)$$

The fitting procedure is based on testing statistical hypotheses in the context of fitting the model, related to the impact of fishing, changes in predator-prey dynamics, also called the vulnerability settings (Christensen and Walters, 2000), possible changes in primary production through the estimation of a forcing function on primary production specifically, or all of the above. The hypothesis that obtains the lowest *AIC* is therefore the hypothesis that obtains the best fit of the model to the data while using the least number of parameters to do so. The hypotheses to be tested include: (1) Baseline (model run without fishing and with no changes to vulnerabilities, or primary production anomaly); (2) Baseline and changes in vulnerabilities; (3) Baseline and changes in primary production anomalies; (4) Baseline and changes in vulnerabilities and primary production anomalies; (5) Fishing, thus including fishing but excluding any fitting to time series for vulnerabilities or primary production anomaly searches; (6) Fishing and changes in vulnerabilities; (7) Fishing and changes in primary production anomalies; and (8) Fishing and changes in vulnerabilities and primary production anomalies.

The number of parameters that can be estimated (i.e. the number of changes in vulnerabilities, or changes in primary production anomalies, or changes in both) depend on the number of time series available to calculate the *SS*. The number of possible parameters that can be estimated is often significantly more than the data available to constrain the model, thus it is imperative to test the parameters that affect the most detectable change (e.g. Mackinson, 2014). To do so, an algorithm is used to calculate which of the vulnerability parameters creates the largest change of the *SS*, and the most sensitive vulnerability parameters are used. In addition, the number of primary production anomalies must be ≥ 2 as a spline point of 1 reverts to estimating a primary production estimate for each year of the simulation. It is also advisable not to estimate more primary production anomaly spline points than the number of time series you have to constrain your model. The *AIC* assumes that all data points are independent. In most cases fisheries data are not independent, thus as a conservative rule of thumb the total number of parameters estimated should therefore be $K - 1$. Thus, for example in the West Coast of Scotland model (Alexander et al., 2015), there were 48 time series of catch or biomass used for estimating the *SS*, and therefore 47 parameters were estimated, either as vulnerabilities or primary production spline points (up to the number of years in the time series). Thus in this case 2–23 spline points or 47 vulnerabilities could be tested, for a possible 1756 combinations. Alexander et al. (2015) tested each 5th combination (i.e. 5, 10, 15 vulnerabilities, 5, 10 15 spline points, etc.) to reduce the repetitive nature of this task. However, an automated routine to run through these options would be preferable and is currently being programmed as a plug-in to the EwE desktop software (Steenbeek et al., 2015, Scott et al., submitted). It is expected that this routine will be released with the next version of the software. A full description of the methodology for fitting a model is given in the Ecopath wiki: Fit to time series (http://sources.ecopath.org/trac/Ecopath/wiki/EwEugFitToTimeSeries).

Special Note: The Model Fits the Data, but is its Behaviour Credible?

It's not uncommon that the estimated vulnerability parameters that provide the best fit of model predictions to observation data can lead to unrealistically high estimates of F_{msy}. When Ecosim estimates low vulnerabilities, this implies high productivity at low biomass and potentially high estimates of F_{msy} that would be not be considered either feasible, or credible based on other assessments and history of F rates that have led to stock depletions in the past. This presents a dilemma to the modeller – pretty fit or credible model behaviour? Taking a pattern oriented approach, the best method is to find the balance between credible fit and behaviour (see for example, Mackinson, 2014). As already noted, particular attention needs to be paid to multi-stanza groups where the connection between the vulnerability estimated for each stanza's (as well as their other parameters mentioned earlier) have a strong influence on the productivity and estimates of sustainable fishing rates.

ADDRESSING UNCERTAINTY IN INPUT DATA

A rapid assessment of the effect of uncertainty in Ecopath input parameters on Ecosim simulations can be done by running Monte Carlo simulations in Ecosim. For example, in the West Coast of Scotland model (Alexander et al., 2015), 20 Monte Carlo runs based on a coefficient of variation (CV) of 0.1 around the input parameters for biomass, *P*/*B*,*Q*/*B*, *EE*, etc. gives twenty very different outcomes for grey seals (Fig. 4). The best statistical fit (lowest sum of squares) trajectory for grey seals is much higher than the best estimate based on the *AICc* with grey seal biomass being overestimated by the model. However, this Monte Carlo routine provides a good idea of the range of outputs available based on the uncertainty surrounding the input data, as seen by the 5th and 95th percentile values plotted (Fig. 4). The CV of input parameters can be set in the Monte Carlo routine or can be obtained from the quality of the input data given in the pedigree routine. The Monte Carlo routine also runs in a new "ecological indicators" plug in that will be release in the new version of software, enabling to assess the impact of input uncertainty in some output results of Ecopath and of Ecosim (Coll and Steenbeek, 2014).

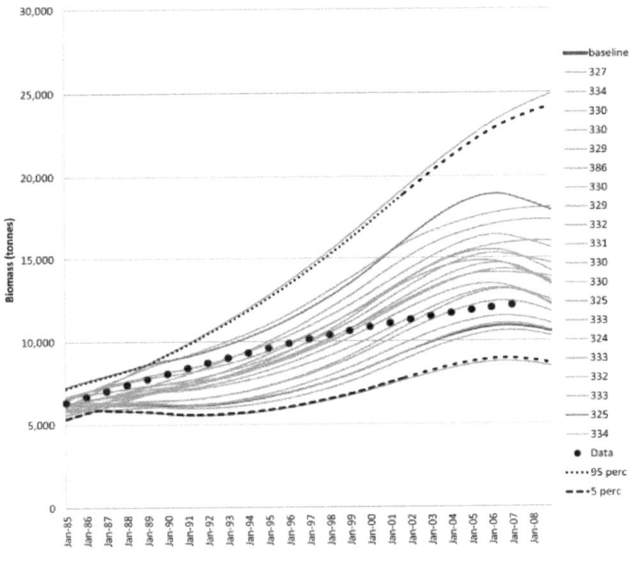

Figure 4: Twenty Monte Carlo simulations of grey seal biomass with best fitted West Coast of Scotland model (Alexander et al., 2015). Thick blue line is final best fitted model (sum of squares of 327), thin blue line is model with least sum of squares (325), grey lines are the other 19 fits with legend values representing the final sum of squares for each run and dots are the data. The 95% percentile and 5% percentile values are also plotted. (For interpretation of the references to color in figure legend, the reader is referred to the web version of the article.)

EwE users have recently developed several sophisticated routines for considering model parameter uncertainty, enabling representation of multiple possible representations of the modelled ecosystem and assessment of the impact of this performance model simulation (e.g. Gaichas et al., 2012, Prato et al., 2014, Guesnet et al., 2015 and Platts and Mackinson, 2015). A new (soon to be released) plugin enables users to create alternative Ecopath models with associated Ecosim parameters set, and evaluate the performance of management strategies when this uncertainty is accounted for (Platts and Mackinson, 2015).

QUALITY ASSURANCE FOR EWE MODEL APPLICATIONS TO MANAGEMENT: KEY RUNS

Only once a model has been evaluated for the quality of the input data through PREBAL and other diagnostics, the balance of the model has been evaluated through testing the thermodynamic and ecological rules, the model has been calibrated to the appropriate time series data, and the effect of parameter uncertainty examined should the model be considered for use in management purposes.

In Europe, the concept of a 'Key run' model (ICES, 2012 and ICES, 2013) is being used as part of quality assurance procedures (ICES, 2013) for assessing models being (or intended to be) used in provision of advice for management. The term Ecopath 'Key-run' originated from earlier ICES multispecies working groups (ICES, 2009) as a way to facilitate a common understanding among modellers using multispecies models. These have typically been used for multispecies models and EwE models. A 'key-run' refers to a model parameterization and output that is accepted as a standard, reviewed and published by ICES Working Group on Multispecies Assessment Methods (WGSAM, http://www.ices.dk/community/groups/Pages/WGSAM.aspx), and thus serves as a quality assured source for scientific input to ICES advice. WGSAM described what this standard should look like for different models as part of their terms of reference in 2013. In general terms, a key run describes and makes accessible the model outputs for a case study of a particular ecosystem. It contains consistent outputs, together with documentation and information on inputs. ICES (2012) defines that the prime purposes of a key run include:

(a) Demonstrating the utility of a particular model formulation in a controlled environment and thereby building confidence that this formulation is appropriate to use in providing advice.

(b) Assisting with the development of multi-model approaches by providing a "standard" setup to aid understanding of different model frameworks, and a worked example of the results that can potentially emerge.

Key runs are typically run every three years, or alternatively, when a substantive change is made to the model parameters, when sufficient new data becomes available, or when the previous key-run is deemed out of date. Thus, they are run regularly, but not at a high frequency (i.e. annually) to avoid any spurious changes attributed to simply just another year of data. A detailed example of the North Sea EwE model 'key run' is presented in Annex 5 of ICES (2012) and the forthcoming ICES report (ICES, 2016). The foundation of a EwE 'key run' is a published and time series fitted Ecosim model. Ongoing quality assurance work in ecosystem models in ICES WGSAM is helping to define protocols and give examples. In the future, there are aims to define and provide examples of a 'key run' spatial model – where a calibrated Ecospace representation is developed using the time series fitted Ecosim model and the spatial-temporal data framework (Steenbeek et al., 2013).

CONCLUSIONS

Ecosystem-based management (EBM) has now been written into policy in Australia, Europe, the USA and other countries. The requirement for executing EBM has led to a large number of ecosystem models being developed. During the past 30 years EwE has been continuously updated and extended, and this paper is a natural extension of codifying it as an important tool for ecosystem-based management. EwE models are easy to construct and manipulate, but we need to raise the standards of use for these models in a management context. The same applies when using these models to explore ecosystem theory. To use ecosystem models, one needs to have confidence in their ability to be executed correctly. To be executed correctly, one needs to understand the uncertainty in the input data, and be able to assess the confidence in the model outputs. This means taking accounting of model parameter uncertainty and undertaking specific performance evaluations. We describe a suite of best practices that should contribute to helping ensure that EwE models are able to be used for ecosystem-based management. We trust that what we have provided here is a useful step in that direction. These best practices will help with the uptake and utilization of model outputs and increase the credibility of EwE models for wider application. EwE models cannot, and should not be used as the only tool for

EBM. In many cases the quality of the data is just not sufficient or the questions being asked are not sufficiently related to predator-prey dynamics underlying this model. However, in many instances where EwE is an appropriate model, a clear expectation of reviewers, managers and stakeholders should be that the models presented have met the standards described above. We see EwE as being one of a suite of tools useful for the development of multi-model ensemble predictions of ecosystem responses to changes, where we seek to establish robust results from a suite of models (Fulton and Smith, 2004).

We realize that these guidelines for best practice may be challenging for some applications, and we sympathize with systems where data quality and quantity is insufficient. In these instances, using EwE models to describe the most critical data gaps, to design research strategies, as part of adaptive management and to test uncertainties in unknown parameters are still valid and useful. However, without best practice and without a clear understanding of the associated uncertainty, it should be expected that management interventions may not yield the desired outcomes.

ACKNOWLEDGEMENTS

The authors wish to thank Dr Karen Alexander and Dr Natalia Serpetti for their thoughtful comments on this manuscript. JJH was supported by the Natural Environment Research Council and Department for Environment, Food and Rural Affairs under the project MERP: grant number NE/L003279/1, Marine Ecosystems Research Programme. SM acknowledges support from Defra contract M1228 'Fizzyfish' and EU FP7 GAP2, Grant agreement 266544. MC was partially funded by the European Commission through the Marie Curie Career Integration Grant Fellowships to the BIOWEB project and by the Spanish National Program Ramon y Cajal. VC acknowledges support through an NSERC Discovery Grant RGPIN-2014-05782.

REFERENCES

1. Ahrens, R.N.M., Walters, C.J., Christensen, V., 2012. Foraging arena theory. Fish Fish. 13, 41–59.
2. Akaike,H., 1974.Anew look atthe statisticalmodelidentification.IEEE Trans.Autom. Control 19, 716–723.
3. Akoglu, E., Libralato, S., Salihoglu, B., Oguz, T., Solidoro, C., 2015. EwE-F 1.0: an implementation of Ecopath with Ecosim in Fortran 95/2003 for coupling. Geosci. Model Dev. Discuss. 8, 1511–1537.
4. Alexander, K.A., Heymans, J.J., MaGill, S., Tomczak, M., Holmes, S.,Wilding, T.A., 2015. Investigating the recent decline in gadoid stocks in the west of Scotland shelf ecosystem using a food-web model. ICES J. Mar. Sci. 72, 436–449.
5. Araújo, J.N., Bundy, A., 2012. Effects of environmental change, fisheries and trophodynamics on the ecosystem ofthe western Scotian Shelf, Canada. Mar. Ecol. Prog. Ser. 464, 51–67.
6. Arreguín-Sánchez, F., Ruiz-Barreiro, T.M., 2014. Approaching a functional measure of vulnerability in marine ecosystems. Ecol. Indic. 45, 130–138.
7. Aydin, K., Gaichas, S., Ortiz, I., Kinzey, D., Friday, N.,2007. A comparison of the Bering Sea, Gulf of Alaska, and Aleutian Islands Large Marine Ecosystems through food web modeling. In: NOAA Technical Memorandum NMFS-AFSC-178. NOAA.
8. Azam, F., Fenchel, T., Field, J.G., Gray, J.S., Meyer-Reil, L.A., Thingstad, F., 1983. The ecological role of water-column microbes in the sea. Mar. Ecol. Prog. Ser. 10, 257–263.
9. Baird, D., McGlade, J.M., Ulanowicz, R.E., 1991. The comparative ecology of six marine ecosystems. Philos. Trans. R. Soc. Lond. B 333, 15–29.
10. Brey, T., 1999. A collection of empirical relations for use in ecological modelling. Naga ICLARM Q. 22, 24–28.
11. Brey, T., 2010. An empirical model for estimating aquatic invertebrate respiration. Methods Ecol. Evol. 1, 92–101.
12. Brey, T., 2012. A multi-parameter artificial neural network model to estimate macrobenthic invertebrate productivity and production. Limnol. Oceanogr.: Methods 10, 581–589.
13. Burnham, K.P., Anderson, D.R., 2004. Multimodel inference: understanding AIC and BIC in model selection. Sociol. Methods Res. 33, 261–304.
14. Christensen, V., 1995. Ecosystem maturity – towards quantification. Ecol. Model. 77, 3–32.
15. Christensen, V., Pauly, D., 1992. ECOPATH II – a software for balancing steadystate ecosystem models and calculating network characteristics. Ecol. Model. 61, 169–185.
16. Christensen, V., Pauly, D.,1993. Trophic models of aquatic ecosystems. In: ICLARM Conference Proceedings 26. 1st edition.InternationalCenter for LivingResources Management, Manila, Philippines.
17. Christensen, V., Walters, C., 2000. Ecopath with ecosim: methods, capabilities and limitations. Pages 79–105. In: Pauly, D., Pitcher, T.J. (Eds.), Methods for Evaluating the Impacts of Fisheries on North Atlantic Ecosystems. Fisheries Centre, University of British Columbia, Vancouver, BC.
18. Christensen, V., Walters, C.J., 2004. Ecopath with Ecosim: methods, capabilities and limitations. Ecol. Model. 172, 109–139.
19. Christensen, V., Walters, C.J., Pauly, D., Forrest, R., 2008. Ecopath with Ecosim Version 6. User Guide. Fisheries Centre, University of British Columbia, Vancouver, Canada.
20. Christensen, V., Coll, M., Steenbeek, J., Buszowski, J., Chagaris, D., Walters, C.J., 2014. Representing variable habitat quality in a spatial food web model. Ecosystems, 1–16.
21. Coll, M., Steenbeek, J., 2014. New software plug-in to calculate biodiversity and conservation-based indicators from EwE food web models. Pages 67–68. In: Steenbeek, J., Piroddi, C., Coll, M., Heymans, J.J., Villasante, S., Christensen, V. (Eds.), Ecopath 30 Years Conference Proceedings: Extended Abstracts. Fisheries Centre Research Reports 22(3). Fisheries Centre, University of British Columbia, Vancouver.
22. Coll, M., Palomera, I., Tudela, S., Dowd, M., 2008.

Food-web dynamics in the South Catalan Sea ecosystem (NW Mediterranean) for 1978–2003. Ecol. Model. 217, 95–116.

23. Coll, M., Palomera, I., Tudela, S., 2009. Decadal changes in a NW Mediterranean Sea food web in relation to fishing exploitation. Ecol. Model. 220, 2088–2102.

24. Coll, M., Akoglu, E., Arreguín-Sánchez, F., Fulton, E.A., Gascuel, D., Heymans, J.J., Libralato, S., Mackinson, S., Palomera, I., Piroddi, C., Shannon, L.J., Steenbeek, J., Villasante, S., Christensen, V., 2015. Modelling dynamic ecosystems: venturing beyond boundaries with the Ecopath approach. Rev. Fish Biol. Fish. 25, 413–424.

25. Colléter, M., Valls, A., Guitton, J., Morissette, J., Arreguín-Sánchez, F., Christensen, V., Gascuel, D., Pauly, D., 2013. EcoBase: a repository solution to gather and communicate information from EwE models. Fisheries Centre Research Reports. UBC Fisheries Centre.

26. Cox, S.P., Essington, T.E., Kitchell, J.F., Martell, S.J.D., Walters, C.J., Boggs, C., Kaplan, I., 2002. Reconstructing ecosystem dynamics in the central Pacific Ocean, 1952–1998. II. A preliminary assessment of the trophic impacts of fishing and effects on tuna dynamics. Can. J. Fish. Aquat. Sci. 59, 1736–1747.

27. Darwall, W.R.T., Allison, E.H., Turner, G.F., Irvine, K., 2010. Lake of flies, or lake of fish? A trophic model of Lake Malawi. Ecol. Model. 221, 713–727.

28. Deehr, R.A., Luczkovich, J.J., Hart, K.J., Clough, L.M., Johnson, B.J., Johnson, J.C., 2014. Using stable isotope analysis to validate effective trophic levels from Ecopath models of areas closed and open to shrimp trawling in Core Sound, NC, USA. Ecol. Model. 282, 1–17.

29. FAO, 2008. Fisheries Management. The Ecosystem Approach To Fisheries. Best Practices in Ecosystem Modelling to Informing an Ecosystem Approach to Fisheries. FAO.

30. Finn, J.T., 1976. Measures of ecosystem structure and function derived from analysis of flows. J. Theor. Biol. 56, 363–380.

31. Fulton, E.A., Smith, A.D.M., 2004. Lessons learnt from a comparison of three ecosystem models for Port Phillip Bay, Australia. Afr. J. Mar. Sci. 26, 219–243.

32. Gaichas, S.K., Aydin, K.Y., Francis, R.C., 2011. What drives dynamics in the Gulf of Alaska? Integrated hypotheses of species, fishing, and climate relationships using ecosystem modeling. Can. J. Fish. Aquat. Sci. 68, 1553–1578.

33. Gaichas, S.K., Odell, G.,Aydin,K.Y., Francis,R.C., 2012. Beyond the defaults:functional response parameter space and ecosystem-level fishing thresholds in dynamic food web model simulations. Can. J. Fish. Aquat. Sci. 69, 2077–2094.

34. Grimm, V., Revilla, E., Berger, U., Jeltsch, F., Mooij, W.M., Railsback, S.F., Thulke, H.- H., Weiner, J., Wiegand, T., DeAngelis, D.L., 2005. Pattern-oriented modeling of agent-based complex systems: lessons from ecology. Science 310, 987–991.

35. Guesnet, V., Lassalle, G., Chaalali, A., Kearney, K., Saint-Béat, B., Karimi, B., Grami, B., Tecchio, S., Niquil, N., Lobry, J., 2015. Incorporating food-web parameter uncertainty into Ecopath-derived ecological network indicators. Ecol. Model. 313, 29–40.

36. Heymans, J.J., 2005. Ecosystem model of the Western and Central Aleutian Islands in 1963, 1979 and 1991. Pages 8–82. In: Guénette, S., Christensen, V. (Eds.), The Steller sea lion decline: models and data ofthe Northeast Pacific, Fisheries Centre Research Reports 13(1).

37. Heymans, J.J., Tomczak, M.T., 2016. Regime shifts in the Northern Benguela ecosystem: challenges for management. Ecol. Model. 331, 151–159.

38. Heymans, J.J., Ulanowicz, R.E., Bondavalli, C., 2002. Network analysis of the South Florida Everglades graminoid marshes and comparison with nearby cypress ecosystems. Ecol. Model. 149, 5–23.

39. Heymans, J.J., Shannon, L.J., Jarre, A., 2004. Changes in the northern Benguela ecosystem over three decades: 1970s, 1980s and 1990s. Ecol. Model. 172, 175–195.

40. Heymans, J.J., Guénette, S., Christensen, V., 2007. Evaluating network analysis indicators of ecosystem status in the Gulf of Alaska. Ecosystems 10, 488–502.

41. Heymans, J.J., Coll, M., Libralato, S., Morissette, L., Christensen, V., 2014. Global patterns in ecological indicators of marine food webs: a modelling approach. PLOS ONE 9, e95845.

42. Hilborn, R., Walters, C.J., 1992. Quantitative Fisheries Stock Assessment: Choice, Dynamics and Uncertainty. Kluwer Academic Publishers, Boston.

43. Hurvich, C.M., Tsai, C.-L., 1998. A crossvalidatory AIC for hard wavelet thresholding in spatially adaptive function estimation. Biometrika 85, 701–710.

44. Hyder, K., Rossberg, A.G., Allen, J.I., Austen, M.C., Barciela, R.M., Blanchard, J., Burrows, M.T., Defriez, E., Dorrington, T., Edwards, K., Garcia-Carreras, B., Heath, M.R., Hembury, D.J., Heymans, J.J., Holt, J., Houle, J.E., Jennings, S., Mackinson, S., McPike, R., Mee, L., Mills, D.K., Montgomery, C., Pearson, D., Pinnegar, J.K., Pollicino, M., Popova, E.E., Rae, L., Rogers, S.I., Speirs, D., Spence, M., Thorpe, R., Turner, R.K., van der Molen, J., Yool, A., Paterson, D.M., 2015. Making modelling count – increasing the contribution of shelf-seas community and ecosystem models to policy development and management. Mar. Policy 61, 291–302.

45. ICES, 2009. Report of the Working Group on Multispecies Assessment Methods (WGSAM), 5–9 October 2009, ICES Headquarters, Copenhagen, ICES CM 2009/RMC:10 ICES CM 2009/RMC:10, ICES, Copenhagen.

46. ICES, 2012. Report of the Working Group on Multispecies Assessment Methods (WGSAM), 10–14 October 2011, Woods Hole, USA, ICES CM 2011/SSGSUE:10 ICES CM 2011/SSGSUE:10, ICES, Copenhagen.

47. ICES, 2013. Interim Report of the Working Group on Multispecies Assessment Methods (WGSAM), 21–25 October 2013, Stockholm, Sweden, ICES CM 2013/SSGSUE:10 ICES CM 2013/SSGSUE:10, ICES, Copenhagen.

48. ICES, 2016. Interim Report ofthe Working Group on Multispecies Assessment Methods (WGSAM), 9–13 November 2015, Woods hole, USA, ES CM 2016/ SSGSUE ICES CM 2016/SSGSUE, ICES, Copenhagen.

49. Innes, S., Lavigne, D.M., Earle, W.M., Kovacs, K.M., 1987. Feeding rates of seals and whales. J. Anim. Ecol. 56, 115–130.

50. Kearney, K.A., Stock, C., Sarmiento, J.L., 2013. Amplification and attenuation of increased primary production in a marine food web. Mar. Ecol. Prog. Ser. 491, 1–14.

51. Kramer-Schadt, S., Revilla, E., Wiegand, T., Grimm, V., 2007. Patterns for parameters in simulation models. Ecol. Model. 204, 553–556.

52. Lassalle, G., Bourdaud, P., Saint-Béat, B., Rochette, S., Niquil, N., 2014. A toolbox to evaluate data reliability for whole-ecosystem models: application on the Bay of Biscay continental shelf food-web model. Ecol. Model. 285, 13–21.

53. Link, J.S., 2010. Adding rigor to ecological network models by evaluating a set of pre-balance diagnostics: a plea for PREBAL. Ecol. Model. 221, 1582–1593.

54. Link, J., Overholtz, W., O'Reilly, J., Green, J., Dow, D., Palka, D., Legault, C., Vitaliano, J., Guida, V., Fogarty, M., Brodziak, J., Methratta, L., Stockhausen, W., Col, L., Griswold, C., 2008. The Northeast US continental shelf Energy Modeling and Analysis exercise (EMAX): ecological network model development and basic ecosystem metrics. Contributions from Advances in Marine Ecosystem Modelling Research II 23–26 June 2008, Plymouth, UK. J. Mar. Syst. 74, 453–474.

55. Lucey, S.M., Aydin, K.Y., Gaichas, S.K., Fogarty, M.J., Hyun, S.-Y., Cadrin, S.X., 2014. Improving the EBFM toolbox with an alternative open source mass balance model. Pages 204–205. In: Steenbeek, J., Piroddi, C., Coll, M., Heymans, J.J., Villasante, S., Christensen, V. (Eds.), Ecopath 30 years conference proceedings: Extended Abstracts, Fisheries Centre Research Reports 22(3). Fisheries Centre UBC, Vancouver.

56. Mackinson, S., 2014. Combined analyses reveal environmentally driven changes in the North Sea ecosystem and raise questions regarding what makes an ecosystem model's performance credible? Can. J. Fish. Aquat. Sci. 71, 31–46.

57. Mackinson, S., Daskalov, G., 2007. An ecosystem model of the North Sea for use in research supporting the ecosystem approach to fisheries management: description and parameterisation. Cefas Technical Report Cefas Science Series Technical

58. Report 142. CEFAS, Lowestoft.

59. Mackinson, S., Daskalov, G., Heymans, J.J., Neira, S., Arancibia, H., Zetina-Rejón, M., Jiang, H., Cheng, H.Q., Coll, M., Arreguin-Sanchez, F., Keeble, K., Shannon, L., 2009a. Which forcing factors fit? Using ecosystem models to investigate the relative influence of fishing and changes in primary productivity on the dynamics of marine ecosystems. Ecol. Model. 220, 2972–2987.

60. Mackinson, S., Deas, B., Beveridge, D., Casey, J., 2009b. Mixed-fishery or ecosystem conundrum? Multispecies considerations inform thinking on long-term management of North Sea demersal stocks. Can. J. Fish. Aquat. Sci. 66, 1107–1129.

61. Mann, K.H., 1988. Production and use of detritus in various freshwater, estuarine, and coastal marine ecosystems. Limnol. Oceanogr. 33, 910–930.

62. Moloney, C.L., Jarre, A., Arancibia, H., Bozec, Y.-M., Neira, S., Roux, J.-P., Shannon, L.J., 2005. Comparing the Benguela and Humboldt marine upwelling ecosystems with indicators derived from inter-calibrated models. ICES J. Mar. Sci. 62, 493–502.

63. Morissette, L., 2007. Complexity, Cost and Quality of Ecosystem Models and Their Impact on Resilience: A Comparative Analysis, with Emphasis on Marine Mammals and the Gulf of St. Lawrence. University of British Columbia, Vancouver.

64. Nilsson, S.G., Nilsson, I.N., 1976. Numbers, Food Consumption, and Fish Predation by Birds in Lake Möckeln, Southern Sweden. Ornis Scand. 7, 61–70.

65. Odum, E.P., 1969. The strategy of ecosystem development. Science 164, 262–270. Orr, K.K., 2013. Predicting the Ecosystem Effects of Harvesing Beach-cast Kelp for Biofuel. University of Aberdeen, Oban.

66. Palomares, M.L.D., Pauly, D., 1998. Predicting food consumption of fish populations as functions of mortality, food type, morphometrics, temperature and salinity. Mar. Freshw. Res. 49, 447–453.

67. Pauly, D., 1980. On the interrelationships between natural mortality, growth parameters, and mean environmental temperature in 175 fish stocks. Journal du Conseil international pour l'Exploration de la Mer 39, 175–192.

68. Pérez-Espana, ˜ H., Arreguín-Sánchez, F., 2001. An inverse relationship between stability and maturity in models of aquatic ecosystems. Ecol. Model. 145, 189–196.

69. Peters, R.H., 1983. The Ecological Implications of Body Size, 1st ed. Cambridge University Press, Cambridge.

70. Pinnegar, J.K., Blanchard, J.L., Mackinson, S., Scott, R.D., Duplisea, D.E., 2005. Aggregation and removal of weak-links in food-web models: system stability and recovery from disturbance. Ecol. Model. 184, 229–248.

71. Pitcher, T.J., Heymans, J.J., 2002. Ecosystem models of Newfoundland for the time periods 1995, 1985, 1900 and 1450. Fisheries Centre Research Reports 10(5).

72. Plagányi, É.E., 2007. Models for an ecosystem approach to fisheries. FAO Fisheries Technical Paper No. 477. FAO Fisheries Technical Paper No. 477. 477. FAO, Rome.

73. Platts, M., Mackinson, S., 2015. Assessing the Effect of Model Parameter Uncertainty on the Performance Management Strategies in Ecosim.

74. Polovina, J.J., 1984. Model of a coral reef ecosystem I. The ECOPATH model and its application to French Frigate Shoals. Coral Reefs 3, 1–11.

75. Prato, G., Gascuel, D., Valls, A., Francour, P., 2014. Towards a balance between complexity and feasibility in food-web models of Mediterranean coastal ecosystems: addressing uncertainty while accounting for data

collection constraints. Fish. Centre Res. Rep. 22, 70–71.

76. Ramsvatn, S., 2013. Investigating Coastal Ecosystem Structure and Dynamics using Ecopath Mass-Balance Modelling and Stable Isotope Data. University of Tromso, Tromso.

77. Scott, E., Serpetti, N., Steenbeek, J., Heymans, J.J., 2015. A Stepwise Fitting Procedure for Automated Fitting of Ecopath with Ecosim Models. SoftwareX, under review.

78. Shannon, L.J., Moloney, C.L., Jarre, A., Field, J.G., 2003. Trophic flow in the southern Benguela during the 1980s and 1990s. Contributions from Advances in Marine Ecosystem Modelling Research II 23–26 June 2008, Plymouth, UK, 39. J. Mar. Syst., 83–116.

79. Shannon, L.J., Christensen, V., Walters, C.J., 2004. Modelling stock dynamics in the southern Benguela ecosystem for the period 1978–2002. Afr. J. Mar. Sci. 26, 179–196.

80. STECF, 2015. Multiannual Management Plans (North Sea), Working Graoup 16–20th March 2015, Ispra, Italy. STECF EWG 15 02.

81. Steenbeek, J., Coll, M., Gurney, L., Mélin, F., Hoepffner, N., Buszowski, J., Christensen, V., 2013. Bridging the gap between ecosystem modeling tools and geographic information systems: driving a food web model with external spatial–temporal data. Ecol. Model. 263, 139–151.

82. Steenbeek, J., Piroddi, C., Coll, M., Heymans, J.J., Villasante, S., Christensen, V., 2014. Ecopath 30 Years Conference Proceedings: Extended Abstracts. Fisheries Centre Research Reports.

83. Steenbeek, J., Buszowski, J., Christensen, V., Akoglu, E., Aydin, K., Ellis, N., Felinto, D., Guitton, J., Lucey, S., Kearney, K., Mackinson, S., Pan, M., Platts, M., Walters, C., 2015. Ecopath with Ecosim as a model-building toolbox: source code capabilities, extensions, and variations. Ecol. Model., http://dx.doi.org/10.1016/j.ecolmodel.2015.06.031.

84. Stergiou, K.I., Karpouzi, V.S., 2002. Feeding habits and trophic levels of Mediterranean fish. Rev. Fish Biol. Fish. 11, 217–254.

85. Tecchio, S., Coll, M., Christensen, V., Company, J.B., Ramírez-llodra, E., Sardà, F., 2013. Food web structure and vulnerability of a deep-sea ecosystem in the NW Mediterranean Sea. Deep Sea Res. Part I: Oceanogr. Res. Pap. 75, 1–15.

86. Tomczak, M.T., Niiranen, S., Hjerne, O., Blenckner, T., 2012. Ecosystem flow dynamics in the Baltic Proper—using a multi-trophic dataset as a basis for food–web modelling. Ecol. Model. 230, 123–147.

87. Tomczak, M.T., Heymans, J.J., Yletyinen, J., Niiranen, S., Otto, S.A., Blenckner, T., 2013. Ecological network indicators of ecosystem status and change in the Baltic Sea. PLOS ONE 8, e75439.

88. Walters, C.J., Martell, S.J.D., 2004. Fisheries Ecology and Management. Princeton University Press, Princeton.

89. Walters, C., Christensen, V., Pauly, D., 1997. Structuring dynamic models of exploited ecosystems from trophic mass-balance assessments. Rev. Fish Biol. Fish. 7, 139–172.

90. Walters, C., Pauly, D., Christensen,V., 1999. Ecospace: predicting of mesoscale spatial patterns in trophic relationships of exploited ecosystems, with emphasis on the impacts of marine protected areas. Ecosystems 2, 539–554.

91. Walters, C.J., Christensen, V., Martell, S.J., Kitchell, J.F., 2005. Possible ecosystem impacts of applying MSY policies from single-species assessment. ICES J. Mar. Sci./Journal du Conseil 62, 558–568.

92. Winberg, G.G., 1956. Rate of Metabolism and Food Requirements of Fishes.

93. Zetina-Rejón, M.J., Cabrera-Neri, E., López-Ibarra, G.A., Arcos-Huitrón, N.E., Christensen, V., 2015. Trophic modeling of the continental shelf ecosystem outside of Tabasco, Mexico: a network and modularity analysis. Ecol. Model. 313, 314–324.

CHAPTER 8

EFFECTIVENESS OF COMMUNITY FOREST MANAGEMENT AT REDUCING DEFORESTATION

Ranaivo A. Rasolofoson,Paul J. Ferraro, Clinton N. Jenkins Julia P.G. Jones

Keywords: Community Forest Management, Deforestation, Impact evaluation, Matching, Madagascar, Payments for Ecosystem Services

INTRODUCTION

The major role of tropical forests in biodiversity and climate change has led the world to search for effective ways to slow deforestation. Many approaches have come in and out of fashion. Strictly protected areas, which prohibit most human activities, were popular in the early days of conservation and remain so today. As an alternative to strict protected areas, Community Forest Management (CFM) emerged in the late 1980s (Hutton et al., 2005). By virtue of involving local forest users in management, CFM is promoted as having the potential to benefit both forests and local livelihoods (Behera, 2009). This potential, however, has been questioned (Behera, 2009) and its evidence base has been found to be weak (Bowler et al., 2012). Although Payments for Ecosystem Services (PES) have become the most recent fashion in efforts to reduce deforestation, CFM remains an important part of the forest management toolkit in many developing countries (Blaikie, 2006). It is also promoted as a means by which PES schemes can be implemented. High quality studies evaluating the effectiveness of CFM are therefore important for shaping future development and investment in approaches to reduce deforestation. We aim to provide robust evidence on effectiveness of CFM at reducing deforestation.

Citation: Ranaivo A. Rasolofoson, Paul J. Ferraro, Clinton N. Jenkins, Julia P.G. Jones, Effectiveness of Community Forest Management at reducing deforestation, Biological Conservation, Volume 184, April 2015, Pages 271-277, ISSN 0006-3207, http://dx.doi.org/10.1016/j.biocon.2015.01.027.

Studies investigating the effectiveness of conservation interventions often fail to adequately control for confounding factors that affect both the assignment of interventions and the outcomes of interest (Bowler et al., 2012, Ferraro and Pattanayak, 2006 and Joppa and Pfaff, 2010). Recent studies investigating the effectiveness of protected areas at reducing deforestation have made progress in controlling for confounding factors by the use of statistical matching (e.g. Andam et al., 2013, Andam et al., 2008, Carranza et al., 2014 and Ferraro et al., 2013). Matching selects comparison areas that have pre-intervention baseline values of confounding factors most similar to intervention area values, and thus makes it possible to control for these confounding factors (Joppa and Pfaff, 2011). However, we know of only one study (Somanathan et al., 2009) that has used matching to investigate the effectiveness of CFM at reducing deforestation.

A significant challenge for evaluating the effectiveness of CFM is the large variation in forest management practices and designs within the approach, both among and within countries (Lund et al., 2009). In terms of practices, examples of this variation range from cases where the community has a good understanding of their rights and responsibilities to cases where CFM exists on paper only (Benjamin, 2008 and Lund et al., 2009). An example of design variation is that some CFMs allow communities to benefit from commercial use of forest resources within their managed forests while others do not (Persha et al., 2011). Failure to consider this variation compromises the potential for learning about design and implementation factors that promote CFM effectiveness.

The Malagasy government legislated CFM in the late 1990s (Raik, 2007) to reduce deforestation and protect the significant part of the world's biodiversity that is endemic to Madagascar (Le Saout et al., 2013). The number of CFM units increased rapidly and continues to grow (Aubert et al., 2013). Many publications review the institutional and political aspects of Madagascar's forest decentralization process (Pollini et al., 2014,Pollini and Lassoie, 2011, Raik and Decker, 2007, Rives et al., 2013 and Urech et al., 2013), but only a few focus on empirically estimating the performance of CFM in terms of conservation outcomes (CIRAD, 2013, Sommerville et al., 2010 and Toillier et al., 2011). None adequately control for factors that may confound impact estimates.

Using statistical matching to control for factors that confound impact estimates, we investigate the effectiveness of Madagascar CFM at reducing deforestation between 2000 and 2010. To our knowledge, this is the first national scale study of performance of CFM at delivering conservation outcomes. First, we assess the overall effectiveness of Madagascar's forest decentralization policy at reducing deforestation by looking at all CFM units across the country. Second, we distinguish and study effectiveness in a subsample of CFM units where we have information to suggest that CFM was implemented on the ground. Finally, we differentiate between CFM that allows and does not allow commercial use of forest resources and study effectiveness conditional on whether CFM permits or prohibits commercial use. Note that we do not consider other important potential outcomes from CFM including impacts, positive or negative, on human welfare.

METHODS

Study Areas

In Madagascar, the transfer of forest management to local communities involves three main steps; the creation of a local forest management group, adoption of forest rules, and signed contract between the local forest management group, the state forest department and possibly the municipality where the forests are located (Aubert et al., 2013 and Pollini and Lassoie, 2011). In our study, CFM refers to forests managed by communities that achieved these three steps.

Our study covers CFM established between 2000 and 2005 (Fig. 1). Because 2010 is the end of the period of our analyses, selecting CFM established between 2000 and 2005 allows observing at least five years of deforestation impacts post CFM establishment. There is no national database containing current information on all CFM units. We therefore gathered information from multiple sources including organizations involved in implementation; namely, Direction Générale des Forêts, Office National pour l'Environnement, Asity, Fanamby, Durrell Wildlife Conservation Trust, Conservation International, Wildlife Conservation Society, and World Wide Fund for Nature.

Malagasy CFM varies in their implementation quality. Some were established with little input from local communities (Rives et al., 2013), and others received little or no external support (Hockley and Andriamarovololona, 2007). It is very difficult to get information of the implementation quality of the individual CFM unit. We used whether a CFM unit passed the forest department evaluation that is undertaken three years after the contract (Pollini and Lassoie, 2011) as an indicator of whether the project was indeed implemented. While not an ideal indicator, it does at least suggest the CFM unit has met the basic institutional, socio-economic and environmental criteria of the evaluation. We refer to units that passed the evaluation as CFM units that have information to suggest implementation.

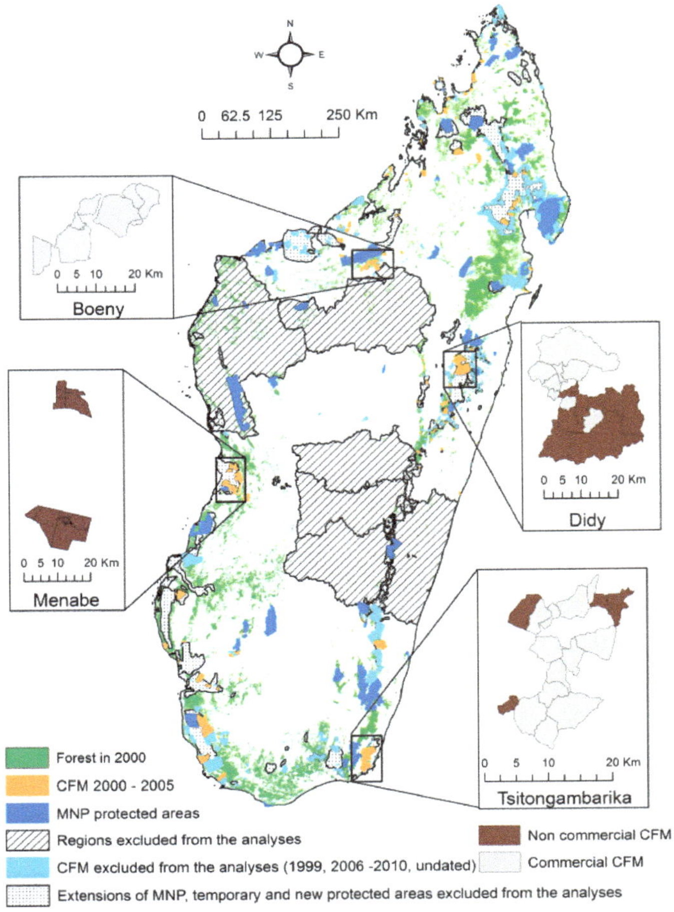

Figure 1: Map showing CFM established between 2000 and 2005, commercial and non-commercial CFM sites and areas excluded from the analyses (Projection: Laborde Madagascar).

CFM implementation in Madagascar varies according to regulations related to commercial use of forest resources. Commercial CFM allows commercial uses and adopts it as a conservation strategy. Non-commercial CFM does not permit commercial uses and follows a pure conservation strategy (Randrianarivelo et al., 2012). Because there are no reliable national data regarding where commercial uses are permitted within CFM, we conducted analyses on commercial and non-commercial CFM for four sites only, where we were able to ascertain information on commercial uses through field visits, interviews with site managers or search of existing literature. The four sites are Didy, Tsitongambarika, Menabe and Boeny (Fig. 1). All CFM units that we considered in these four sites had passed the forest department evaluation. Table 1 presents the number of CFM units, the area of land and natural forest covered by each type of CFM considered in our analyses.

Table 1: Number of units and dimension of different types of Community Forest Management (CFM).

Types	Study site (scale)	CFM unit	Land area (ha)	Natural forest area in 2000 (ha)
All CFM	Madagascar	231	699,961	308,290
CFM with information suggesting implementation	Madagascar	116	399,861	211,666
Commercial CFM	Didy	8	29,104	23,409
	Tsitongambarika	12	18,089	7214
	Boeny	7	30,920	10,768
Non commercial CFM	Didy	8	40,164	32,757
	Tsitongambarika	4	3757	866
	Menabe Antimena	4	22,042	13,991

Non-CFM areas refer to forests that, up to 2010, were not technically and financially supported by particular organizations and thus were under government control. Since the government has been weak and unable to enforce forest laws, these forests are subject to open access (Raik, 2007 and Urech et al., 2013)

We excluded six out of the 22 administrative regions of Madagascar where we were unable to collect CFM data (Fig. 1). Because we analyzed CFM established between 2000 and 2005, undated CFM and CFM established before 2000 or after 2005 were excluded. We also excluded protected areas managed by Madagascar National Parks. Finally, extensions of protected areas, temporary and new protected areas created since 2003 were excluded. However, any portions of these newly created protected areas that were known to be community managed were considered as CFM (Fig. 1, see Appendix D Table D1 for how CFM, non-CFM and excluded areas fit into official Madagascar forest statuses since 2003).

Matching, Unit of Analysis, Sampling

Conservation interventions like CFM are not randomly assigned. The site characteristics that affect where conservation interventions are assigned also affect deforestation, thus confounding attempts to estimate intervention impacts (Ferraro and Pattanayak, 2006). To control for these confounding factors, some empirical studies have used matching (Andam et al., 2008 and Joppa and Pfaff, 2011). Matching selects comparison areas that are similar to the intervention areas in terms of their values of the confounding factors at the pre-intervention baseline. Thus, one assumes that the outcomes of the comparison group represent, in expectation, the counterfactual outcomes of the intervention sites had they not been exposed to the conservation intervention.

Table 2: Different types of analyses.

Analysis	Intervention	Counterfactual	Estimand
Effectiveness of all CFM	All CFM	Non-CFM	Difference of deforestation between CFM and CFM had there been no intervention
Effectiveness of CFM with information suggesting implementation	CFM with information suggesting implementation	Non-CFM	Difference of deforestation between CFM and CFM had there been no intervention
Effectiveness of commercial CFM	Commercial CFM	Non-CFM	Difference of deforestation between CFM and CFM had there been no intervention
Effectiveness of noncommercial CFM	Noncommercial CFM	Non-CFM	Difference of deforestation between CFM and CFM had there been no intervention
Relative effectiveness of commercial and noncommercial CFM on the types of CFM forests where commercial use has been permitted	Commercial CFM	Noncommercial CFM	Difference of deforestation between actual commercial CFM forests and these forests had commercial use been prohibited
Relative effectiveness of noncommercial and commercial CFM on the types of forests where commercial use has been prohibited	Noncommercial CFM	Commercial CFM	Difference of deforestation between actual noncommercial CFM forests and these forests had commercial use been permitted

The unit of analysis is a forested pixel from the 2000 forest cover baseline (See Appendix A for limitations of using 2000 baseline forest cover and CFM established between 2000 and 2005, and Appendix B for how we deal with potential pseudo-replication in which pixels within a particular CFM are not independent). For each forested pixel at baseline, covariates take the values of each confounding characteristic at that pixel location. For each analysis (Table 2), we selected random forested pixels in intervention areas. Then, we used matching to pair each randomly selected pixel with the most similar pixel in comparison areas in terms of covariates. The outcome variable is whether a pixel remained forested or not in the 2010 land cover. The estimated difference in deforestation between intervention and similar comparison areas represents the impact of the intervention on deforestation for intervention sites or the Average Treatment effect on the Treated (ATT).We used independent samples T-test to compare deforestation in intervention and similar comparison or counterfactual areas. Our study comprises six analyses (Table 2). The first analysis compares all CFM established between 2000 and 2005 to non-CFM. The second compares the CFM that has information that suggests implementation to non-CFM. The four remaining analyses regard commercial and non-commercial CFM. Ferraro et al. (2013) demonstrated that four ATTs are policy relevant for studies involving two types of intervention. The third and fourth analyses compare commercial and non-commercial CFM to non-CFM. The fifth and sixth ones compare commercial to non-commercial CFM, and vice versa. The difference between these two latter analyses rests upon the type of the sampled pixels used in the comparison. The CFM forests where commercial use is permitted may be observably different from the CFM forests where such use is prohibited (in terms of the confounding factors). To understand how deforestation in commercial CFM forests would have been different without commercial use requires that we compare commercial CFM forests to non-commercial CFM forests that are observably similar at baseline (in terms of the confounding factors). So the comparison of commercial to non-commercial CFM uses an intervention group of all the randomly selected commercial CFM pixels and a comparison group of only the best matches of non-commercial CFM pixels. Dissimilar pixels from the non-commercial CFM sample are discarded. The estimate from this comparison represents the average impact of permitting commercial use on the types of CFM forests where commercial use has been permitted. The impact of commercial use may be different on the types of CFM forests where commercial use has been prohibited. To estimate this impact, we formed an intervention group of all the randomly selected non-commercial CFM pixels and a comparison group of only the best matches of commercial CFM pixels.

In all analyses, we used Mahalanobis covariate matching because it better balances covariates than other matching options. We performed exact matching on vegetation zones (eastern humid, western deciduous and southern spiny forests, Appendix E Fig. E1). We executed bias adjustment regression to correct for any remaining post-matching covariate imbalance (Abadie and Imbens, 2006). We used the "matching" package in R (Sekhon, 2011).

We aimed to select sample sizes that balance our interests in achieving high statistical power and reducing computer processing time. Learning from multiple trial analyses, we decided on a sample of around 30,000 pixels for all intervention areas in each analysis. For comparison areas, we sampled around two to four times more pixels (Appendix D Table D2). The larger sample size from comparison areas increases the probability of finding a good match for each intervention pixel.

Forest Cover

We used 2000 and 2010 deforestation data developed by ONE et al. (2013). These are based on images from Landsat TM and Landsat ETM+ and have a resolution of 28.5 m and an accuracy rate close to 90%. Full methods are in Harper et al. (2007).

Covariates or Confounding Baseline Characteristics

Based on Madagascar CFM practitioners' opinion, and CFM and deforestation studies in Madagascar and other tropical countries (Barsimantov and Kendall, 2012, Bowler et al., 2012, Forrest et al., 2008, Gorenflo et al., 2011 and Sussman et al., 1994), we identified pressure and access as potentially confounding factors. To control for these factors, we used measures of agricultural suitability, slope, elevation, distance to recent deforestation (1990–2000), distance to forest edge, distance to a village, distance to an urban center, distance to a road, distance to a cart track, duration of trip to an urban center and population density (see Appendix D Tables D3, D4 for sources of covariate data). Because community characteristics received little consideration in selection of community forests for CFM designation, we did not consider community characteristics as confounding factors but only condition on these site characteristics indicating pressure and access (see Appendix C).

Sensitivity Analysis to Unobservable Bias

While matching can ensure that the distributions of observable covariates are similar between intervention and comparison groups, the groups may still differ in terms of unobserved covariates that affect both deforestation and assignment to intervention. To check the robustness of our estimates of effectiveness to such unobservable covariates, we performed Rosembaum's (2010) sensitivity test. A parameter measures the dissimilarity in the likelihood of receiving intervention between intervention and counterfactual units due to unobservable covariates. In the absence of unobservable differences, takes the value of one 1. The higher the value of , the more dissimilar is the likelihood of receiving intervention for the matched pair due to unobserved variables. The sensitivity analysis consists of increasing the values of and determining a critical at which the estimate of effect of intervention is not significantly different from zero. In other words, we seek to measure how strong an unobservable confounder would have to be in order for the estimated effect not to be significantly different from zero. The higher the value of the critical , the more robust is the estimate of intervention effect to unobservable bias. We carry out sensitivity tests with the "rbounds" package in R (Keele, 2010).

RESULTS

Before matching, CFM pixels are, on average, located closer

to recent deforestation, to a road and to an urban center and are characterized by shorter trip durations to an urban center than non-CFM pixels. Although these patterns suggest CFM is assigned to areas of higher deforestation pressure, CFM is also located on lands less suitable for agriculture and at higher elevation (Appendix D Table D5).

Commercial CFM pixels are, on average, associated with lands more suitable for irrigated rice, closer to a village, a road and an urban center, shorter trip duration to urban center, higher population density, but they are located on lands less suitable for agriculture and on steeper slopes than non-commercial CFM sites (Appendix D Tables D9, D10).

Matching generally improves covariate balance. The mean differences and the mean raw eQQ differences of covariates in intervention and counterfactual areas tend toward zero after matching (Appendix D Tables D5–D10). An exception is suitability for agriculture in the comparisons of commercial and non-commercial CFM, and vice versa. Matching does not improve balance for this factor (Appendix D Tables D9, D10). This is because all suitable lands for agriculture are found only in the non-commercial CFM in Menabe. Thus, there are no matched suitable lands in commercial CFM. We describe potential effects of this imbalance in the discussion.

Between 2000 and 2010, CFM sites had, on average, 0.02% less deforestation than matched non-CFM sites, a statistically insignificant difference (p = 0.99, Fig. 2). When we consider only CFM with information suggesting implementation, CFM had 0.76% less deforestation than matched non-CFM, but still statistically insignificant (p = 0.71). Differentiating CFM by whether commercial uses are allowed, we estimate that commercial CFM experienced 1.83% more deforestation than matched non-CFM (p = 0.16). Non-commercial CFMs reduced deforestation by 2.01% relative to matched non-CFM ($p < 0.001$). When we compare commercial CFM to matched non-commercial CFM, to investigate their relative effectiveness on the types of CFM forests where commercial use has been permitted (i.e., forests on lands more suitable for irrigated rice, closer to a village, a road and an urban center, shorter trip duration to urban center, higher population density), commercial CFM experienced 3.24% more deforestation ($p < 0.001$). Comparing non-commercial CFM to matched commercial CFM, to investigate their relative effectiveness on the types of forests where commercial use has been prohibited (i.e., forests on lands less suitable for irrigated rice, farther to a village, a road and an urban center, longer trip duration to urban center, lower population density), we estimate non-commercial CFM reduced deforestation by 5.59% ($p < 0.001$, Fig. 2).

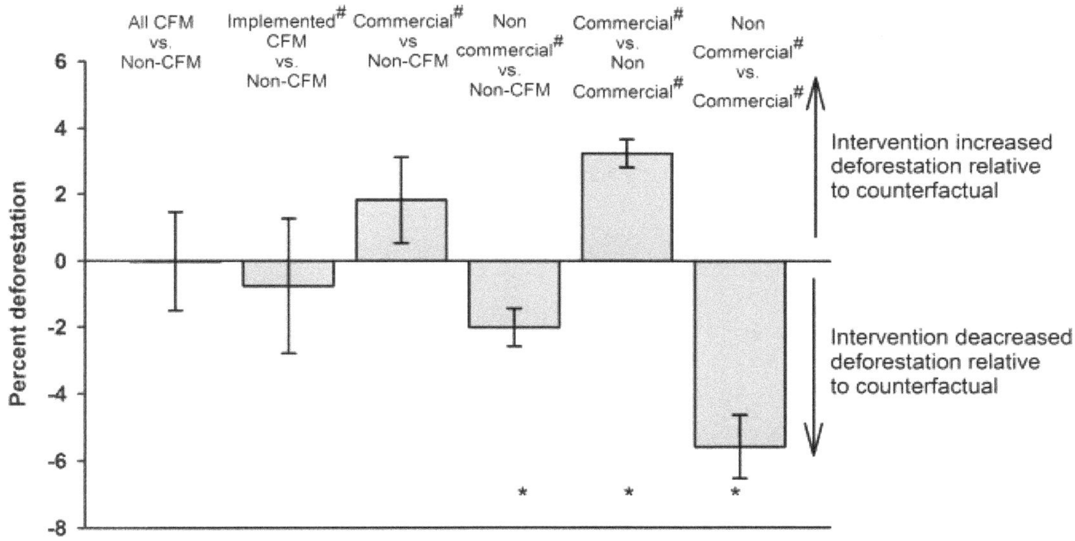

Figure 2: Differences in percent deforestation between intervention and counterfactual (#CFM where we have information to suggest implementation, * significant at $p < 0.001$, error bars: standard errors for post-matching estimates that are calculated using a variance formula that is robust to heteroskedasticity and adjusts the variance estimator for repeated matches among control units (Abadie and Imbens, 2006).

Table 3 presents the results of the sensitivity of our analyses to hidden bias (i.e., an unobservable covariate). For example, where the parameter is 1.38, the estimate of 2.01% remains significantly different from zero at a p value of 0.05 even if an unobservable covariate makes non-commercial CFM pixels 1.38 times more likely to receive intervention than non-CFM pixels. In other words, unobservable covariates need to increase the likelihood of the non-commercial pixels to receiving intervention by a factor greater than 1.38 in order for the impact estimate not to be statistically different from zero.

Table 3: Sensitivity tests to unobservable covariates.

Analysis	Critical at $p = 0.05$
Noncommercial CFM vs. non CFM	1.38
Commercial CFM vs. noncommercial CFM	1.50
Noncommercial CFM vs. commercial CFM	5.85

DISCUSSION

Decentralization of forest management to local communities in Madagascar has not, on average, achieved its forest conservation goal. In terms of deforestation, we cannot detect an effect, on average, of CFM compared to no CFM, even after restricting the sample to only where we have information to suggest CFM implementation on the ground.

Many studies report success and failure of CFM at delivering conservation outcomes (Cox et al., 2010, Pagdee et al., 2006 and Porter-Bolland et al., 2012). However, in a systematic review of CFM performance in developing countries, Bowler et al. (2012)showed that evidence on effectiveness of CFM is weak because of poor study design. They proposed a "gold standard" that would produce quality CFM assessment. We believe that our study meets the "gold standard" as far as is possible (though only for a single potential outcome from CFM – that of reducing deforestation). That is, we use comparator sites, baseline forest cover data, multiple CFM across Madagascar and paired or matched design. We also sample randomly the unit of analysis and allow enough time (5–10 years) for impacts to take place. Finally, we identify and control for confounding factors that may bias impact estimates.

While CFM failed, on average, to reduce deforestation relative to non-CFM, non-commercial CFM appears to have had more success, albeit a small one. Putting all types of CFM in one basket would lead to the single conclusion that CFM is not an effective approach to reduce deforestation, obscuring the positive impact non-commercial CFM appears to have had. This result emphasizes the importance of differentiating among types of CFM in evaluation (Lund et al., 2009). Potential mechanisms through which non-commercial CFM may have had more success are complementary direct payments for conservation. Some non-commercial CFM in our study sites in Didy, Tsitongambarika and Menabe practice a direct payment for conservation scheme to offset restrictions introduced by interventions (e.g. Brimont and Bidaud, 2014 and Sommerville et al., 2010).

The estimated reduction in deforestation from non-commercial CFM is important given that the role of commercial use of forests in conservation is a subject of much debate in theoretical and empirical studies. Some studies argue that by assigning value to forests, commercial use provides means and incentives to local communities to protect forests, while others show that it can trigger the destruction of the resources being commercialized (Agrawal and Chhatre, 2006, Barsimantov and Kendall, 2012 and Persha et al., 2011). Our findings do not support the argument that permitting commercial extraction can enhance the deforestation-reducing impacts of CFM.

Our matching algorithm was unable to remove the pre-matching difference between commercial and non-commercial CFM in terms of agriculture suitability (Appendix D Tables D9, D10). After matching, commercial CFM has lower suitable lands for agriculture (0%) than non-commercial CFM (29%) has. However, knowledge of the direction of the effect of agriculture suitability on deforestation allows us to infer the implications of the post-matching imbalance. Gorenflo et al. (2011) show that lower suitability for agriculture is associated with lower deforestation rates in Madagascar. Thus, the post-matching imbalance should occasion lower deforestation in commercial CFM than non-commercial CFM. Therefore, if matching had balanced the suitability for agriculture between the two types of CFM, commercial CFM performance relative to non-commercial CFM would have appeared even worse because the lower deforestation occasioned by the lower land suitability in commercial CFM would have been erased. Our estimates of impacts for commercial vs. non-commercial CFM, and vice versa are thus conservative.

At the national level, our findings substantiate the rather gloomy pictures of CFM in Madagascar depicted in a number of institutional and policy studies (Pollini et al., 2014,Pollini and Lassoie, 2011, Raik and Decker, 2007, Rives et al., 2013 and Urech et al., 2013). To explain the ineffectiveness of CFM, these studies describe inadequate integration of local participation, resource capture by elites, unfulfilled support promises by different organizations, and lack of capacity of the community and state, among other factors. While these studies point to institutional and policy shortcomings, we advance our understanding of CFM performance by empirically showing that at the national scale, there was no impact in terms of delivering a central objective: reducing deforestation. A recent empirical study (CIRAD, 2013) is of particular interest because it also looked explicitly at the impact of CFM on deforestation and covered part of our study areas, and its results contradict ours. It found that deforestation was significantly less in CFM than in areas without community conservation. It also shows that commercial CFM was more effective at reducing deforestation than non-commercial CFM. The results are not directly comparable to ours because the analyses cover a different time period and are at a different spatial scale, but the CIRAD study should be interpreted with care because they did not adequately control for the biases in confounding factors that we do here. Failure to adequately control for such biases can result in incorrect impact estimates (Andam et al., 2013, Andam et al., 2008 and Joppa and Pfaff, 2011).

Many factors can influence effectiveness of CFM (Agrawal, 2003). We focused on the potential role of commercial use of forest resources (and given our study area, the potential that complementary direct payment for conservation could have in non-commercial CFM). Another potential moderating factor is the amount of resources invested, which may explain the apparently better performance of non-commercial CFM relative to commercial CFM we observe. During our visit to Didy, commercial CFM officials complained about receiving smaller resources relative to their neighboring non-commercial CFM (implemented by different organizations with different funding). However, we lacked the quantitative information on spending to allow this potential moderator of success to be included in the analysis. Data on this moderator in the future will also offer opportunities to extend our study by exploring CFM cost-effectiveness.

We focused on comparing CFM to non-CFM and different types of CFM. Other studies attempt to compare CFM to strictly protected areas (IUCN categories I–IV) to investigate the relative effectiveness of these two different approaches (Bray et al., 2008 and Porter-Bolland et al., 2012). We do not attempt such a comparison because we believe that investigation of the relative effectiveness of strictly protected areas and CFM at reducing deforestation cannot be done credibly for Madagascar with the same robust and rigorous methodology we used here. Matching covariates and baseline forest cover ideally should be measured before intervention (Andam et al., 2008). CFM units in Madagascar were established in or after 2000 while 80% of the strictly protected areas were created before 1970. Thus, strictly protected areas and CFM have different baselines (2000 vs. pre-1970). If these baseline measures are to be used, the start time for the analyses will have to be different for CFM (2000) and strictly protected areas (pre-1970). Therefore, the comparisons of CFM to strictly protected areas, and vice versa, are difficult for Madagascar because the impacts will be estimated for dramatically different time periods. In addition, communities around and managers of strict protected areas and those of CFM have different length of experience in exposure to intervention that may explain the difference of impacts between the two approaches. Such comparisons may yield credible results in specific sites or in other countries, where strictly protected areas and CFM were established around the same time.

In conclusion, we provide robust evidence that CFM and commercial use of forest resources are not guarantees of conservation success. Our findings also suggest that differentiating among types of CFM is important when evaluating effectiveness. By explicitly estimating impacts conditional on the type of CFM, scholars can shed light on the factors that promote effective CFM.

ACKNOWLEDGMENTS

We thank the European Commission for providing funding through FONASO and H. Larsen, J. Gibbons, and two anonymous reviewers for commenting on drafts of this paper. We also thank the following people and organization in Madagascar for providing data: T. Tondrasoa Ramaharitra, A. Lohanivo, Direction Générale des Forêts, A. Ravoahangy and R. Andriamasimanana (ASITY), Conservation International, A. Volahy (DWCT), A. Rasolonjanahary (FANAMBY), H. Rakotomanana (ONE), V. Andriamahery (WCS), A. Andriamanalina and L. Rakotondrasoa (WWF MWIOPO). We also acknowledge the p4ges project (NE/K010220/1).

APPENDIX A. SUPPLEMENTARY MATERIAL

Appendices

Appendix A. Limitations of using 2000 baseline forest cover and CFM established between 2000 and 2005

Ideally, baseline confounding characteristics, particularly baseline forest cover, should be measured at the very time CFM is established (Andam et al., 2008). In this study, baseline forest cover is dated in 2000 while CFM are established from 2000 to 2005. When planners establish CFM, they make decision based on the landscape they are facing. They are likely to establish CFM in location where there are forests at the time of establishment. Thus, if a CFM was established after 2000, it was likely to be forested in that post-2000 year of establishment while its counterfactual, which was forested in the 2000 baseline may already be deforested in the post-2000 year of establishment. The fact that the CFM was still forested in the post-2000 year of establishment could signify that it unobservably has smaller deforestation probability than its counterfactual, which was already cleared. While matching can control for observable bias, it cannot deal with this kind of unobservable bias. That CFM could unobservably have smaller deforestation probability than its counterfactual means that our impact estimates can be considered as upper bounds. That is, if the probability of deforestation of the intervention and its counterfactual were the same, effectiveness of the intervention would be smaller. Where intervention has no significant effect, this will not change the conclusion that it is ineffective. However, where there is significant effect, this means that the effect estimate can be smaller. For the latter, the sensitivity tests to unobservable bias provide means to know how sensitive an estimate is to potential unobservable bias.

Appendix B. Potential pseudo replication

We may have pseudo replication because our units of analysis (pixel) within a management unit (CFM) may not be independent. If we assume that deforestation is affected by pixel-level attributes and pre-treatment management unit-level attributes, but that the pre-treatment management unit attributes are not correlated with treatment assignment, whereas the pixel-level attributes are (and we can observe the relevant confounding pixel-level attributes), then we can estimate impacts by conditioning on the pre-treatment, pixel-level attributes (as we do in our study). But we would have a problem estimating the variance of the estimated impact – we would underestimate the true variance by ignoring pre-treatment management unit effects on deforestation.

Some of the pseudo replication concern is mitigated by our sampling procedure. To address the issue of spatial autocorrelation, we set a minimum distance of 68 meters between sampled pixels (the mean distance between a deforested pixel and the nearest deforested one if deforestation from 2000 to 2010 was randomly distributed throughout the forested areas in 2000; Ebdon, 1985; Mitchell, 2005). But this sampling procedure will not eliminate the pseudo replication problem.

We therefore performed a falsification (placebo) test to provide evidence that our pixel-level matching covariates are sufficiently rich to eliminate effects of the management units in the absence of treatment. To do that, we selected random pixel in CFM areas, matched them with similar pixels in control in terms of pixel-level covariates and compared deforestation in the two areas. But we did the analysis with pre-treatment (1990 – 2000) forest cover data, a period in which no treatment effect should be detectable (because treatment had not yet been assigned). We cannot reject the null hypothesis of zero difference in the deforestation rates in CFM and control sites ($p = 0.11$). That result provides indirect evidence that our pixel-level covariates are sufficiently rich to eliminate effects of the management units in the absence of treatment. In other words, we can assume that the management unit-effects, were they to exist, do not have large impact on estimate of variance.

Appendix C. Selection of sites to CFM in Madagascar

In addition to the physical characteristics used in the matching procedure, there may also be community characteristics that affect which communities a planner selects for being eligible for CFM and which communities self-select to accept CFM. For example, Ostrom (2000) identified seven community attributes that facilitate the emergence of CFM. These attributes are also likely to affect forest management in the absence of formal CFM designation, and thus would be confounding factors. The presence of these attributes is likely to increase the likelihood of CFM designation and reduce deforestation in the absence of CFM (because they promote better forest management). Thus, failing to control for these community characteristics in our analyses will bias our impact estimator upward, i.e. our estimates are upper bound estimates of the true effect of CFM.

We argue that such bias would be small, if it exists at all, in the case of Madagascar. Very few CFM designations were initiated by communities or based on community characteristics. Instead, conservation non-governmental organizations (NGOs), pressured by donors, approached communities and rushed to implement many CFM contracts to fit their agenda (Pollini et al., 2014; Pollini and Lassoie, 2011). For example, many of the existing CFM sites have been established to form a "green belt" to shield the core zones of the newly created protected areas after 2003. Communities around these newly protected areas were approached by conservation NGOs to establish CFM regardless of the community characteristics. The goal was to form the "green belt".

Interviews with a NGO expert in CFM implementation revealed that their first criterion to select CFM sites is pressure. More pressured sites are likely to be selected to CFM. The second criterion is access. Groups of neighboring CFM sites are often implemented together to reduce costs. The only community characteristic is community motivation. However, community motivation is another factor in which access plays a major role. Implementation organizations carry out awareness campaign to boost local motivation. Easier accessed areas closer to implementation organization offices that are often located in larger towns or cities are more likely to receive more intense awareness campaign and thus more likely to be motivated to engage in CFM.

Given that community characteristics receive little consideration in CFM implementation, there is no surprise that only two of Ostrom (2000)'s seven community attributes conducive to CFM have been present in CFM sites established in eastern Madagascar (Urech et al., 2013). Community characteristics have made these sites ineligible for CFM but pressure of the international conservation agenda pushed these sites to be designated CFM sites.

In brief, failure to consider community characteristics as confounding factors is likely to have small effects in our estimates because community characteristics receive little consideration in selection of sites to CFM. The physical characteristics we chose as confounding factors are likely to include factors judged by the international conservation agenda as important to the selection of sites to CFM.

Appendix D. Tables

Table D1: Presence of community forest management areas (CFM), control areas and areas excluded from the analyses in different statuses of forests in Madagascar

Status	CFM	Non-CFM	Excluded
MNP protected areas	No	No	Yes
Extension of MNP protected areas	Yes	No	Yes
Temporary protected areas	Yes	No	Yes
New protected areas	Yes	Yes (if creation of protected area not yet started or at a very early stage)	Yes (if stage of creation of protected area unknown or advanced)
Public domain	Yes	Yes	No

Table D2: Different analyses and sample sizes

Analysis	Sample size (number of pixels)	
	Intervention	Comparison
Effectiveness of all CFM	CFM: 37 679	Non-CFM: 112 709
Effectiveness of CFM with information suggesting implementation	CFM: 30 000	Non-CFM: 112 709
Effectiveness of commercial CFM	Commercial CFM: 30 000	Non-CFM: 120 000
Effectiveness of noncommercial CFM	Noncommercial CFM: 30 000	Non-CFM: 82 407
Relative effectiveness of commercial and noncommercial CFM in commercial CFM setting	Commercial CFM: 30 000	Noncommercial CFM: 53 528*
Relative effectiveness of noncommercial and commercial CFM in noncommercial CFM setting	Noncommercial CFM: 30 000	Commercial CFM: 45 657*

*Because of the imposed minimum distance between sample pixels, it was not possible to sample comparison pixels two to four times more than intervention pixels. Thus, we sampled the maximum number of pixels that could be generated from the comparison pixel population.

Table D3: Baseline characteristics likely to affect both assignment to CFM and rate of deforestation

Confounding variables	Unit	Source
Agricultural suitability	Pixel 9km*9km	IIASA(Fischer et al., 2002) †
Irrigated rice suitability	Pixel 90m*90m	Ramaharitra Tondrasoa, 2012
Elevation	Pixel 90m*90m	SRTM Digital Elevation Model (Shuttle Radar Topography Mission SRTM)
Slope	Pixel 90m*90m	
Distance to recent deforestation (1990 – 2000)	Pixel 90m*90m	ONE, DGF, CI, FTM and MNP (ONE et al., 2013), SRTM Digital Elevation Model (Shuttle Radar Topography Mission SRTM)
Distance to forest edge (2000)	Pixel 90m*90m	
Distance to a village	Pixel 90m*90m	UN OCHA ROSA (UN OCHA ROSA, 2007) , SRTM Digital Elevation Model (Shuttle Radar Topography Mission SRTM)
Distance to an urban center	Pixel 90m*90m	
Distance to a road	Pixel 90m*90m	FTM (Foiben-Taosarintanin´i Madagasikara FTM, 1990), SRTM Digital Elevation Model (Shuttle Radar Topography Mission SRTM)
Distance to a cart track	Pixel 90m*90m	
Duration of trip to urban center	Commune‡	ILO (ILO, 2003)
Population density in 2003	Fokontany‡	Vieilledent et al., 2013

†We have reclassified the agriculture constraints levels of IIASA (Table D4)

‡Administrative levels in Madagascar from the smallest to the largest: Fokontany, Commune, District, Region, Nation

Table D4: Reclassification of the agriculture constraints levels of IIASA (Fischer et al., 2002)

Level of constraints in IIASA	Reclassification
No constraints	None*
Very few constraints	Suitable for agriculture
Few constraints	Suitable for agriculture
Partly with constraints	Suitable for agriculture
Frequently severe constraints	Unsuitable for agriculture
Very frequent severe constraints	Unsuitable for agriculture
Unsuitable for agriculture	Unsuitable for agriculture

*There is no "no constraints" area in Madagascar

Table D5: Covariate balance for all CFM vs. non-CFM

Variable	Mean CFM	Mean non-CFM	Difference of means	Mean raw eQQ difference
Suitable land for agriculture (%)				
- Unmatched	18.801	36.798	-17.997	17.997
- Matched	18.801	18.743	0.058	0.058
Unsuitable land for agriculture (%)				
- Unmatched	78.665	61.730	16.935	16.935
- Matched	78.665	78.723	-0.058	0.058
Suitable land for irrigated rice (%)				
- Unmatched	7.710	6.022	1.688	1.688
- Matched	7.710	7.710	0.000	0.000

Slope (degree)				
- Unmatched	6.947	6.488	0.459	0.463
- Matched	6.947	6.945	0.002	0.147
Elevation (m)				
- Unmatched	504.590	423.180	81.410	98.859
- Matched	504.590	511.040	-6.450	14.741
Distance to recent deforestation (km)				
- Unmatched	2.340	3.746	-1.407	1.407
- Matched	2.340	2.149	0.191	0.215
Distance to forest edge (km)				
- Unmatched	0.643	0.369	0.274	0.283
- Matched	0.643	0.555	0.088	0.089
Distance to a village (km)				
- Unmatched	4.792	4.332	0.461	0.465
- Matched	4.792	4.564	0.229	0.287
Distance to a road (km)				
- Unmatched	10.420	11.648	-1.228	1.258
- Matched	10.420	9.835	0.585	0.934
Distance to cart track (km)				
- Unmatched	2.989	3.298	-0.310	0.310
- Matched	2.989	2.982	0.006	0.108
Distance to an urban center (km)				
- Unmatched	60.427	87.909	-27.482	27.507
- Matched	60.427	63.890	-3.463	5.404
Trip to urban center (hour)				
- Unmatched	25.068	32.743	-7.675	10.022
- Matched	25.068	25.273	-0.205	2.146
Population density in 2003 (inh./km^2)				
- Unmatched	14.147	14.469	-0.322	1.391
- Matched	14.147	14.132	0.015	1.020

Table D6: Covariate balance for CFM with information suggesting implementation on the ground vs. non-CFM

Variable	Mean CFM	Mean non-CFM	Difference of means	Mean raw eQQ difference
Suitable land for agriculture (%)				
- Unmatched	13.817	36.798	-22.981	22.980
- Matched	13.817	13.790	0.027	0.027
Unsuitable land for agriculture (%)				
- Unmatched	85.017	61.730	23.287	23.287
- Matched	85.017	85.043	-0.026	0.027

Suitable land for irrigated rice (%)				
- Unmatched	8.817	6.022	2.795	2.793
- Matched	8.817	8.817	0.000	0.000
Slope (degree)				
- Unmatched	6.578	6.488	0.090	0.667
- Matched	6.578	6.654	-0.076	0.192
Elevation (m)				
- Unmatched	532.180	423.180	109.000	138.900
- Matched	532.180	540.610	-8.430	21.194
Distance to recent deforestation (km)				
- Unmatched	2.535	3.746	-1.211	1.212
- Matched	2.535	2.309	0.226	0.255
Distance to forest edge (km)				
- Unmatched	0.704	0.369	0.335	0.342
- Matched	0.704	0.607	0.097	0.098
Distance to a village (km)				
- Unmatched	5.181	4.332	0.849	0.853
- Matched	5.181	4.882	0.299	0.359
Distance to a road (km)				
- Unmatched	11.896	11.648	0.248	0.970
- Matched	11.896	11.089	0.807	1.296
Distance to cart track (km)				
- Unmatched	3.007	3.298	-0.291	0.293
- Matched	3.007	3.025	-0.018	0.152
Distance to an urban center (km)				
- Unmatched	59.889	87.909	-28.020	28.020
- Matched	59.889	64.667	-4.778	7.364
Trip to urban center (hour)				
- Unmatched	31.370	32.743	-1.373	7.189
- Matched	31.370	30.543	0.827	3.029
Population density in 2003 (inh./km^2)				
- Unmatched	11.525	14.469	-2.944	3.268
- Matched	11.525	11.782	-0.257	1.035

Table D7: Covariate balance for commercial CFM vs. non-CFM

Variable	Mean commercial CFM	Mean non-CFM	Difference of means	Mean raw eQQ difference
Suitable land for agriculture (%)				
- Unmatched	0.000	27.956	-27.956	27.957
- Matched	0.000	0.000	0.000	0.000

Unsuitable land for agriculture (%)				
- Unmatched	1.000	70.539	29.461	29.463
- Matched	1.000	1.000	0.000	0.000
Suitable land for irrigated rice (%)				
- Unmatched	23.310	7.216	16.094	16.093
- Matched	23.310	23.310	0.000	0.000
Slope (degree)				
- Unmatched	9.702	8.157	1.546	1.550
- Matched	9.702	9.711	-0.009	0.205
Elevation (m)				
- Unmatched	731.570	519.200	212.370	241.450
- Matched	731.570	721.950	9.620	29.690
Distance to recent deforestation (km)				
- Unmatched	1.499	4.412	-2.913	2.921
- Matched	1.499	1.667	-0.168	0.169
Distance to forest edge (km)				
- Unmatched	0.822	0.386	0.437	0.458
- Matched	0.822	0.677	0.145	0.145
Distance to a village (km)				
- Unmatched	3.112	4.542	-1.431	1.431
- Matched	3.112	3.131	-0.019	0.090
Distance to a road (km)				
- Unmatched	7.256	13.118	-5.862	5.959
- Matched	7.256	7.178	0.078	0.494
Distance to cart track (km)				
- Unmatched	2.233	3.691	-1.457	1.457
- Matched	2.233	2.458	-0.224	0.226
Distance to an urban center (km)				
- Unmatched	34.614	71.736	-37.222	37.227
- Matched	34.614	42.464	-7.850	7.883
Trip to urban center (hour)				
- Unmatched	14.704	32.283	-17.579	17.804
- Matched	14.704	13.972	0.732	4.999
Population density in 2003 (inh./km^2)				
- Unmatched	19.281	13.689	5.592	7.658
- Matched	19.281	19.482	-0.201	2.366

Table D8: Covariate balance for non commercial CFM vs. non-CFM

Variable	Mean non commercial CFM	Mean non-CFM	Difference of means	Mean raw eQQ difference

Suitable land for agriculture (%)				
- Unmatched	29.437	27.842	1.595	1.593
- Matched	29.437	29.380	0.057	0.057
Unsuitable land for agriculture (%)				
- Unmatched	70.430	70.743	-0.313	0.313
- Matched	70.430	70.487	-0.057	0.057
Suitable land for irrigated rice (%)				
- Unmatched	14.923	7.168	7.755	2.793
- Matched	14.923	14.923	0.000	0.000
Slope (degree)				
- Unmatched	6.402	8.090	-1.688	1.699
- Matched	6.402	6.455	-0.053	0.236
Elevation (m)				
- Unmatched	731.470	515.330	216.140	279.240
- Matched	731.470	725.580	5.890	45.265
Distance to recent deforestation (km)				
- Unmatched	1.743	4.409	-2.666	2.677
- Matched	1.743	1.663	0.080	0.237
Distance to forest edge (km)				
- Unmatched	1.177	0.387	0.790	0.794
- Matched	1.177	1.044	0.133	0.133
Distance to a village (km)				
- Unmatched	4.657	4.542	0.114	0.587
- Matched	4.657	4.389	0.268	0.315
Distance to a road (km)				
- Unmatched	12.098	13.087	-0.989	1.706
- Matched	12.098	12.481	-0.383	0.726
Distance to cart track (km)				
- Unmatched	2.778	3.700	-0.922	0.992
- Matched	2.778	3.074	-0.296	0.304
Distance to an urban center (km)				
- Unmatched	47.020	71.739	-24.719	25.727
- Matched	47.020	51.129	-4.109	4.751
Trip to urban center (hour)				
- Unmatched	17.212	32.346	-15.134	17.092
- Matched	17.212	17.487	-0.275	6.447
Population density in 2003 (inh./km^2)				
- Unmatched	9.651	13.599	-3.948	6.719
- Matched	9.651	9.559	0.091	2.327

Table D9: Covariate balance for commercial CFM vs. non commercial CFM

Variable	Mean commercial CFM	Mean non commercial CFM	Difference of means	Mean raw eQQ difference
Suitable land for agriculture (%)				
- Unmatched	0.000	29.207	-29.207	29.207
- Matched	0.000	24.620	-24.620	24.966
Unsuitable land for agriculture (%)				
- Unmatched	1.000	70.625	29.375	29.377
- Matched	1.000	75.380	24.620	24.966
Suitable land for irrigated rice (%)				
- Unmatched	23.310	14.779	8.531	8.530
- Matched	23.310	23.107	0.203	0.187
Slope (degree)				
- Unmatched	9.702	6.462	3.240	3.240
- Matched	9.702	9.384	0.318	0.412
Elevation (m)				
- Unmatched	731.570	731.240	0.330	101.980
- Matched	731.570	680.350	51.220	51.522
Distance to recent deforestation (km)				
- Unmatched	1.499	1.729	-0.230	0.236
- Matched	1.499	1.264	0.235	0.252
Distance to forest edge (km)				
- Unmatched	0.822	1.163	-0.340	0.340
- Matched	0.822	0.796	0.026	0.044
Distance to a village (km)				
- Unmatched	3.112	4.617	-1.505	1.505
- Matched	3.112	3.298	-0.186	0.235
Distance to a road (km)				
- Unmatched	7.256	12.139	-4.883	4.964
- Matched	7.256	6.318	0.938	1.061
Distance to cart track (km)				
- Unmatched	2.233	2.763	-0.530	0.530
- Matched	2.233	2.133	0.100	0.176
Distance to an urban center (km)				
- Unmatched	34.614	46.922	-12.308	12.31
- Matched	34.614	44.160	-9.546	9.541
Trip to urban center (hour)				
- Unmatched	14.704	17.205	-2.501	2.806
- Matched	14.704	15.436	-0.732	0.808

Population density in 2003 (inh./km^2)				
- Unmatched	19.281	9.688	9.593	10.297
- Matched	19.281	17.908	1.373	3.613

Table D10: Covariate balance for non commercial CFM vs. commercial CFM

Variable	Mean non commercial CFM	Mean commercial CFM	Difference of means	Mean raw eQQ difference
Suitable land for agriculture (%)				
- Unmatched	29.437	0.000	29.437	29.437
- Matched	29.437	0.000	29.437	28.669
Unsuitable land for agriculture (%)				
- Unmatched	70.430	1.000	-29.570	29.570
- Matched	70.430	1.000	-29.570	28.797
Suitable land for irrigated rice (%)				
- Unmatched	14.923	23.475	-8.552	8.550
- Matched	14.923	14.923	0.000	0.000
Slope (degree)				
- Unmatched	6.402	9.659	-3.257	3.256
- Matched	6.402	6.619	-0.217	0.435
Elevation (m)				
- Unmatched	731.470	515.330	216.140	279.240
- Matched	731.470	725.580	5.890	45.265
Distance to recent deforestation (km)				
- Unmatched	1.743	1.495	0.249	0.249
- Matched	1.743	1.378	0.365	0.371
Distance to forest edge (km)				
- Unmatched	1.177	0.805	0.372	0.372
- Matched	1.177	1.141	0.036	0.057
Distance to a village (km)				
- Unmatched	4.657	3.093	1.563	1.563
- Matched	4.657	4.159	0.497	0.633
Distance to a road (km)				
- Unmatched	12.098	7.193	4.905	4.973
- Matched	12.098	10.837	1.261	1.805
Distance to cart track (km)				
- Unmatched	2.778	2.240	0.538	0.538
- Matched	2.778	2.517	0.261	0.345
Distance to an urban center (km)				
- Unmatched	47.020	34.720	12.300	12.306
- Matched	47.020	42.138	4.882	4.983

Trip to urban center (hour)				
- Unmatched	17.212	14.506	2.706	2.994
- Matched	17.212	17.252	-0.040	0.723
Population density in 2003 (inh./km^2)				
- Unmatched	9.651	19.534	-9.883	10.560
- Matched	9.651	11.437	-1.786	3.252

Appendix E. Types of vegetation in Madagascar

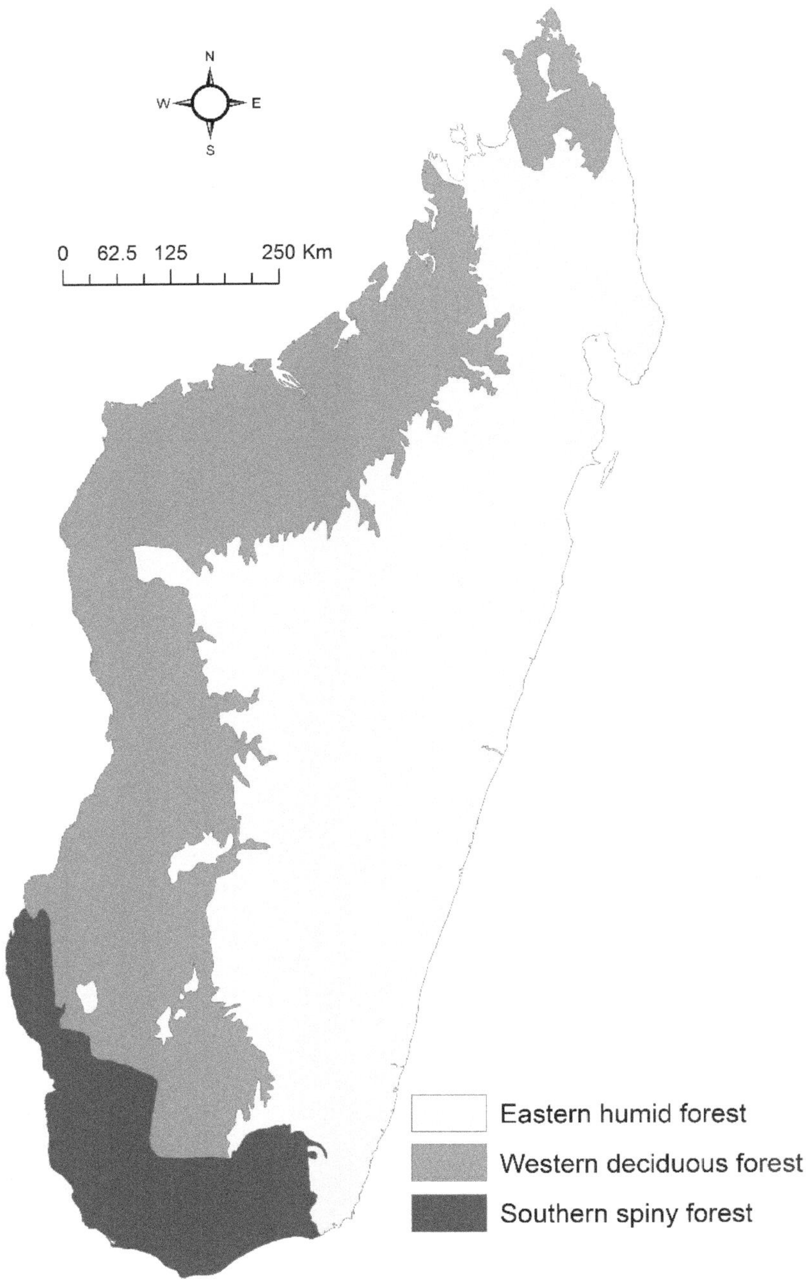

Figure E1. Three major types of vegetation in Madagascar (Sources: Moat and Du Puy, 1997; Schatz and Lescot, 2003; Projection: Laborde Madagascar)

REFERENCES

1. Abadie, A., Imbens, G., 2006. Large sample properties of matching estimators for average treatment effects. Econometrica 74, 235–267.
2. Agrawal, A., 2003. Sustainable governance of common pool resources: context, methods, and politics. Annu. Rev. Anthropol. 32, 243–262.
3. Agrawal, A., Chhatre, A., 2006. Explaining success on the commons: community forest governance in the Indian Himalaya. World Dev. 34, 149–166.
4. Andam, K.S., Ferraro, P.J., Pfaff, A., Sanchez-Azofeifa, G.A., Robalino, J., 2008.
5. Measuring the effectiveness of protected area networks in reducing deforestation. P. Natl. Acad. Sci. USA 105, 16089–16094.
6. Andam, K.S., Ferraro, P.J., Hanauer, M.M., 2013. The effects of protected area systems on ecosystem restoration: a quasi-experimental design to estimate the impact of Costa Rica's protected area system on forest regrowth. Conserv. Lett. 3, 317– 323.
7. Aubert, S., Tchousso, A., Razafiarijaona, J., 2013. Les tenants et les aboutissants juridiques des transferts de gestion des resources naturelles renouvelables. In: Montagne, P., Rives, F. (Eds.), Kajiala. Tattali, Djekabaara. Valoriser les Produits pour Mieux Conserver les Forêts, tome 2. CITE, Antananarivo, Madagascar, pp. 49–67.
8. Barsimantov, J., Kendall, J., 2012. Community forestry, common property, and deforestation in eight Mexican states. J. Environ. Dev. 21, 414–437.
9. Behera, B., 2009. Explaining the performance of state–community joint forest management in India. Ecol. Econ. 69, 177–185.
10. Benjamin, C.E., 2008. Legal pluralism and decentralization: natural resource management in Mali. World Dev. 36, 2255–2276.
11. Blaikie, P., 2006. Is small really beautiful? Community-based natural resource management in Malawi and Botswana. World Dev. 34, 1942–1957.
12. Bowler, D.E., Buyung-Ali, L.M., Healey, J.R., Jones, J.P.G., Knight, T.M., Pullin, A.S., 2012. Does community forest management provide global environmental benefits and improve local welfare? Front. Ecol. Environ. 10, 29–36.
13. Bray, D.B., Duran, E., Ramos, V.H., Mas, J.-F., Velazquez, A., McNab, R.B., Barry, D., Radachowsky, J., 2008. Tropical deforestation, community forests, and protected areas in the Maya forest. Ecol. Soc. 13, 56.
14. Brimont, L., Bidaud, C., 2014. Incentivizing forest conservation. Payments for environmental services and reducing carbon emissions from deforestation. In: Scales, I.R. (Ed.), Conservation and Environmental Management in Madagascar. Routledge, Oxon, UK, pp. 299–319.
15. Carranza, T., Balmford, A., Kapos, V., Manica, A., 2014. Protected area effectiveness in reducing conversion in a rapidly vanishing ecosystem: the Brazilian Cerrado. Conserv. Lett. 7, 216–223.
16. CIRAD Centre de coopération Internationale en Recherche Agronomique pour le Développement, 2013. Assessment of the design elements of a sharing mechanism of benefits from carbon revenues. Madagascar CAZ project. CIRAD, Antananarivo, Madagascar.
17. Cox, M., Arnold, G., Tomás, S.V., 2010. A review of design principles for communitybased natural resource management. Ecol. Soc. 15, 38.
18. Ferraro, P.J., Pattanayak, S.K., 2006. Money for nothing? A call for empirical evaluation of biodiversity conservation investments. PLoS Biol. 4, e105.
19. Ferraro, P.J., Hanauer, M.M., Miteva, D.A., Canavire-Bacarreza, G.J., Pattanayak, S.K., Sims, K.R.E., 2013. More strictly protected areas are not necessarily more protective: evidence from Bolivia, Costa Rica, Indonesia, and Thailand. Environ. Res. Lett. 8, 025011.
20. Forrest, J.L., Sanderson, E.W., Wallace, R., Lazzo, T.M.S., Cerveró, L.H.G., Coppolillo, P., 2008. Patterns of land cover change in and around Madidi National Park, Bolivia. Biotropica 40, 285–294.
21. Gorenflo, L.J., Corson, C., Chomitz, K.M., Harper, G., Honzák, M., Özler, B., 2011. Exploring the association between people and deforestation in Madagascar. In: Cincotta, R.P., Gorenflo, L.J. (Eds.), Human Population. Its Influences on Biological Diversity. Springer, Heidelberg, Germany, pp. 197–221.
22. Harper, G.J., Steininger, M.K., Tucker, C.J., Juhn, D., Hawkins, F., 2007. Fifty years of deforestation and forest fragmentation in Madagascar. Environ. Conserv. 34, 325–333.
23. Hockley, N.J., Andriamarovololona, M.M., 2007. The economics of community forest management in Madagascar: is there a free lunch? An analysis of Transfert de Gestion. USAID United States Agency for International Development, Antananarivo, Madagascar.
24. Hutton, J., Adams, W.M., Murombedzi, J.C., 2005. Back to the barriers? Changing narratives in biodiversity conservation. Forum Dev. Stud. 32, 341–370.
25. Joppa, L.N., Pfaff, A., 2010. Reassessing the forest impacts of protection: the challenge of nonrandom location and a corrective method. Ann. NY Acad. Sci. 1185, 135–149.
26. Joppa, L.N., Pfaff, A., 2011. Global protected area impacts. P. Roy. Soc. B – Biol. Sci. 278, 1633–1638.
27. Keele, L., 2010. An overview of rbounds: an R package for Rosenbaum bounds sensitivity analysis with matched data. <http://CRAN.R-project.org/package= rbounds>.
28. Le Saout, S., Hoffmann, M., Shi, Y., Hughes, A., Bernard, C., Brooks, T.M., Bertzky, B., Butchart, S.H.M., Stuart, S.N., Badman, T., Rodrigues, A.S.L., 2013. Protected areas and effective biodiversity conservation. Science 342, 803–805.
29. Lund, J., Balooni, K., Casse, T., 2009. Change we can believe in? Reviewing studies on the conservation impact of popular participation in forest management. Conserv. Soc. 7, 71–82.
30. ONE Office National pour l'Environnement, DGF Direction Générale des Forêts, CI Conservation

International, FTM Foiben-Taosarintanin'i Madagasikara, MNP Madagascar National Parks, 2013. Evolution de la couverture de forêts naturelles à Madagascar 2005–2010. ONE, DGF, CI, FTM, MNP, Antananarivo, Madagascar.

31. Pagdee, A., Kim, Y., Daugherty, P.J., 2006. What makes community forest management successful: a meta-study from community forests throughout the world. Soc. Natur. Resour. 19, 33–52.

32. Persha, L., Agrawal, A., Chhatre, A., 2011. Social and ecological synergy: local rulemaking, forest livelihoods, and biodiversity conservation. Science 331, 1606–1608.

33. Pollini, J., Lassoie, J.P., 2011. Trapping farmer communities within global environmental regimes: the case of the GELOSE legislation in Madagascar. Soc. Natur. Resour. 24, 814–830.

34. Pollini, J., Hockley, N., Muttenzer, F.D., Ramamonjisoa, B.S., 2014. The transfer of natural resource management rights to local communities. In: Scales, I.R. (Ed.), Conservation and Environmental Management in Madagascar. Routledge, Oxon, UK, pp. 172–192.

35. Porter-Bolland, L., Ellis, E.A., Guariguata, M.R., Ruiz-Mallén, I., Negrete-Yankelevich, S., Reyes-García, V., 2012. Community managed forests and forest protected areas: an assessment of their conservation effectiveness across the tropics. For. Ecol. Manage. 268, 6–17.

36. Raik, D., 2007. Forest management in Madagascar: an historical overview. Madagascar Conserv. Dev. 2, 5–10.

37. Raik, D.B., Decker, D.J., 2007. A multisector framework for assessing communitybased forest management: lessons from Madagascar. Ecol. Soc. 12, 14.

38. Randrianarivelo, G., Montagne, P., Ravelona, M., 2012. Gestion durable et exploitation clandestine de bois d′ oeuvre dans la forêt classée d′ Ambohilero, commune rurale de Didy. In: Montagne, P., Bertrand, A. (Eds.), Kajiala, Tattali, Djekabaara. Valoriser les Produits pour Mieux Conserver les Forêts, tome 1. CITE, Antananarivo, Madagascar, pp. 87–106.

39. Rives, F., Carrière, S.M., Montagne, P., Aubert, S., Sibelet, N., 2013. Forest management devolution: gap between technicians'design and villagers'practices in Madagascar. Environ. Manage. 52, 877–893.

40. Rosembaum, P.R., 2010. Design of Observational Studies. Springer, Heidelberg, Germany.

41. Sekhon, J.S., 2011. Multivariate and propensity score matching. Software with automated balance optimization: the Matching Package for R. J. Stat. Softw. 7, 1–52.

42. Somanathan, E., Prabhakar, R., Singh, B., 2009. Decentralization for cost-effective conservation. P. Natl. Acad. Sci. USA 106, 4143–4147.

43. Sommerville, M., Milner-Gulland, E.J., Rahajaharison, M., Jones, J.P.G., 2010. Impact of a community-based payment for environmental services intervention on forest use in Menabe, Madagascar. Conserv. Biol. 24, 1488–1498.

44. Sussman, R.W., Green, G.M., Sussman, L.K., 1994. Satellite imagery, human ecology, anthropology, and deforestation in Madagascar. Hum. Ecol. 22, 333–354.

45. Toillier, A., Serpantié, G., Hervé, D., Lardon, S., 2011. Livelihood strategies and land use changes in response to conservation: pitfalls of community-based forest management in Madagascar. J. Sustain. For. 30, 20–56.

46. Urech, Z.L., Sorg, J.P., Felber, H.R., 2013. Challenges for community-based forest management in the KoloAla site Manompana. Environ. Manage. 51, 602–615.

CHAPTER 9

MICROBIAL WATER QUALITY

David M. Oliver, Kenneth D.H. Porter, Yakov A. Pachepsky, Richard W. Muirhead, Sim M. Reaney, Rory Coffey, David Kay, David G. Milledge, Eunmi Hong, Steven G. Anthony, Trevor Page, Jack W. Bloodworth, Per-Erik Mellander, Patrice E. Carbonneau, Scott J. McGrane, Richard S. Quilliam

Keywords: Catchment management, diffuse pollution, Faecal indicator organism, Human health, Pathogens

INTRODUCTION

Determining the microbial quality of recreational, drinking, irrigation and shellfish-harvesting waters is important to ensure compliance with health-related standards and associated legislation. Levels of microbial pollution in environmental matrices are often measured by quantifying faecal indicator organisms (FIOs). FIOs (e.g. *Escherichia coli*and intestinal enterococci) play a significant global role in informing regulators and environmental managers of levels of faecal pollution and hygienic status of water resources. For example, in the European Union, FIOs are key parameters monitored within the Bathing Water Directive (BWD) (CEC, 2006a) and Shellfish Waters Directive (CEC, 2006b) to regulate microbial water quality in protected areas. In the USA, FIOs are monitored under the Clean Water Act, and pathogen indicators (predominantly *E. coli*) are the leading cause of watercourse impairment (USEPA, 2015). The microbial quality of irrigation water and implications for food safety is also gaining increased recognition (Pachepsky et al., 2011) particularly in the USA after the Food Safety Modernization Act was signed into law in 2011. In New Zealand, the National Policy Statement for Freshwater Management (NPS-FW, 2014) plays a pivotal role in driving an agenda for delivering better national water quality. The NPS-FW (2014) includes a National Objectives Framework, which sets out water quality standards and "national bottom lines" which are minimum standards that all waters in NZ should comply with. One of these national minimum standards relates to *E. coli* concentrations in all waters.

Historically most research into waterborne disease has focused on human faecal sources such as sewage and storm-water (McBride et al., 2013). However, there is a growing recognition that animal faeces also represent a significant human health risk (Soller et al., 2015, Dufour et al., 2012 and Till et al., 2008). Human faecal sources are generally easier to manage 'point' sources from sewage systems, while agricultural sources are the more difficult to manage 'diffuse' sources (Vant, 2001). Some of the challenge of managing these faecal sources comes from a range of ownership issues. For example, in a large catchment there may be a number of point sources of human faecal discharge to a river but the management of this may be the responsibility of only one or a few municipal authorities or sewerage undertakers that have clear responsibility for water quality management and/or public health. In contrast, sources of animal faeces will be the responsibility of (a) the many farmers who manage domesticated animals and (b) different government (or quasi-government) departments for the management of wild animals (e.g. water fowl, deer, and beavers), neither of which have water quality or public health as a prime responsibility. Catchment regulators implementing, for example, the EU Water Framework Directive (2000) or the US Clean Water Act are therefore faced with a complex system for multiple sources of diverse sanitary significance. To make sense of, and to support decision-making on, multiple sources of faecal material discharges from a range of different activities we need catchment scale models that can fairly apportion water quality risks to individual contaminant sources. The

Citation: David M. Oliver, Kenneth D.H. Porter, Yakov A. Pachepsky, Richard W. Muirhead, Sim M. Reaney, Rory Coffey, David Kay, David G. Milledge, Eunmi Hong, Steven G. Anthony, Trevor Page, Jack W. Bloodworth, Per-Erik Mellander, Patrice E. Carbonneau, Scott J. McGrane, Richard S. Quilliam, Microbial water quality, Science of The Total Environment, Volume 544, 15 February 2016, Pages 39-47, ISSN 0048-9697, http://dx.doi.org/10.1016/j.scitotenv.2015.11.086.

application of catchment models to predict microbial water quality, coupled with direct detection and quantification, plays an important role for guiding decision-making associated with the management of water resources. Recent examples of model applications to inform environmental management of FIOs include the use of catchment screening tools to determine FIO source apportionment at bathing waters according to land use (Kay et al., 2010), development of farm-scale indexing approaches to elucidate FIO risks arising from different farm practices (Muirhead, 2015), and the simulation of *E. coli* persistence in cowpats to help further our understanding of population dynamics of FIOs on grazed pastures (Martinez et al., 2013). However, modelling the fate and transfer of FIOs at different spatial scales poses a considerable challenge to the research and policy community (Oliver et al., 2009, Coffey et al., 2007 and Pachepsky et al., 2006). This is because our level of understanding of FIO behaviour in the environment is still relatively poor compared to other agricultural pollutants such as phosphorus (P) and nitrogen (N) (Oliver et al., 2015 and Kay et al., 2008a). The persistence of FIOs in the environment is highly context specific and the modelling of FIOs in agricultural catchments therefore presents a different set of challenges to P and N. This set of challenges includes their potential to replicate and increase in number under favourable environmental conditions, or experience population die-off at varying rates depending on a range of unfavourable biotic and abiotic factors (Stocker et al., 2015). Thus, it is critical that we try to understand the survival, transfer and export of FIOs better once they are excreted into the wider environment, because improved knowledge and data on the behavioural characteristics of these micro-organisms will enhance our ability to model and predict their interactions with, and responses to, the world around us (e.g. Oliver et al., 2015, Guber et al., 2015, Quilliam et al., 2014, McKergow and Davies-Colley, 2010,Wyer et al., 2010, Kay et al., 2008b and Muirhead et al., 2004). To address the gaps in our current understanding, some of the key research questions under investigation by the research community include:

(i) how do the hydrological pathways that connect FIO sources to water bodies vary in space and time across different catchment types and how does this impact on FIO travel times through the environment?;

(ii) To what extent does the probability of FIO die-off vary for different environmental conditions (in different environmental matrices) around the world?

(iii) How do we integrate FIO behavioural characteristics (e.g. their ability to persist or move in the environment) into model frameworks that are useful for decision-makers?

(iv) How will the export of FIOs from the landscape alter under projected climate change and/or land use change?

Answers to these questions rely on a combination of laboratory, field and modelling approaches and will take time to mature. Furthermore, it is well recognised that the quality of modelled outputs has an implicit reliance on the quality of the field and laboratory data used to populate each model (de Brauwere et al., 2014). With microbial water pollution generating increased attention through, for example, tightening of the BWD standards in Europe (Quilliam et al., 2015) and FIOs being recognised as the leading cause of watercourse impairment in the USA, there is an increasing demand by policy-makers for models to help inform catchment microbial dynamics in order to prioritise efforts for environmental and human-health protection (Coffey et al., 2015). Given the limited evidence-base on which FIO models are built relative to other agricultural pollutants (e.g. nutrients) it is therefore imperative that the end-user expectations of FIO models are appropriately managed by model developers. In response, this commentary highlights four over-arching questions that must be considered as part of good practice prior to the deployment of any modelling approach, irrespective of their geographic application in the world, to predict FIO behaviour in catchment systems. In turn, these questions link to a series of short and longer-term research priorities to enable more effective FIO modelling.

OVER-ARCHING RATIONALE TO GUIDE DEVELOPMENT OF FIO MODELLING IN CATCHMENTS

Models are often used to help answer research or policy-orientated questions. Fig. 1depicts a generic modelling process used to generate answers (and new questions) to environmental problems. We use this generic process as a template to discuss modelling challenges associated with FIOs. The modelling process can be broken down into four core components, namely: 'model purpose', 'model approach', 'data' and 'application'. Individually, each component of the modelling process will accommodate a range of requirements for fundamental research to improve our wider understanding of FIO fate and transfer. However, taken together they provide an integrated and iterative approach to microbial fate and transport model development. The four modelling components, each coupled with a higher level question, are discussed below. We argue that an internationally relevant framework of over-arching questions is both timely and important to steer future laboratory and field-based research in order to underpin and improve our modelling capability in the rapidly developing field of catchment microbial dynamics. Rather than contributing in-depth discussions to an already overcrowded literature of comprehensive FIO and pathogen modelling reviews this commentary aims to provide a more critical focus for future research direction, enabling more effective deployment of models for predicting water quality in agricultural catchments.

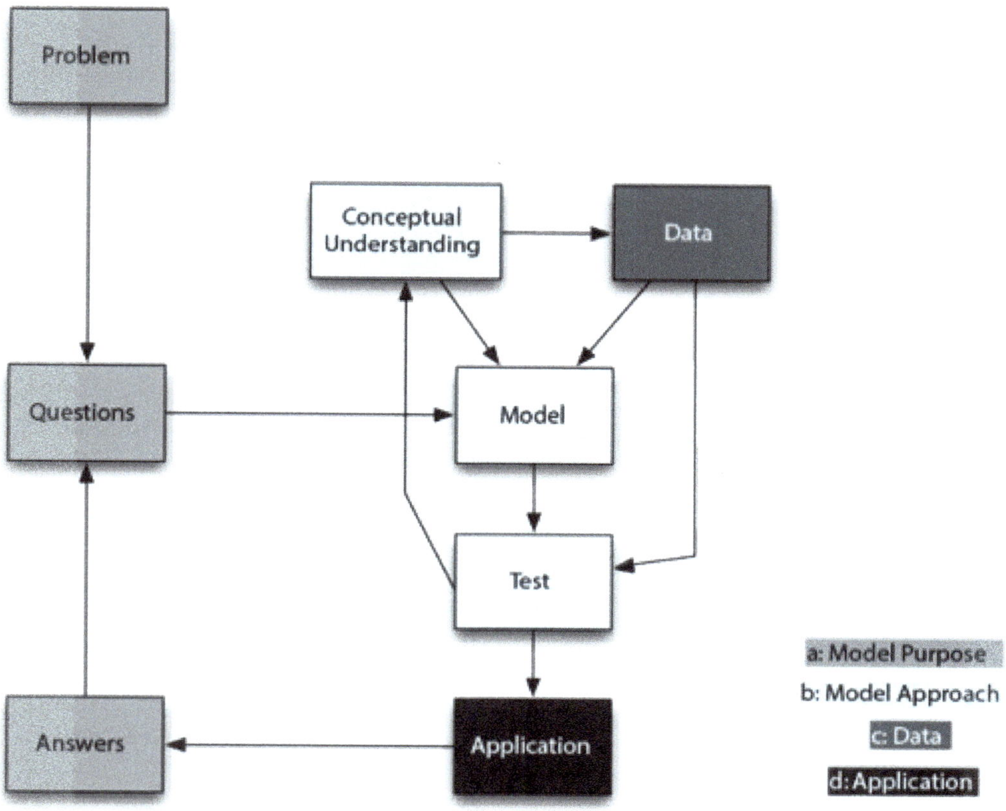

Figure 1: Generic iterative modelling process used to generate answers (and new questions) to environmental problems.

Model Purpose: What is the Question We Want to Answer?

Irrespective of environmental context or pollutant, no model can be developed to predict everything, everywhere, all of the time. Yet, end-users often misunderstand a model's predicative capability, or apply it inappropriately (Dickey-Collas et al., 2014) thus limiting its predictive capability. Both misapplications can be avoided by assessing a model's purpose and clarifying its expected outputs, defining the question of interest then determining whether a particular model can answer it.

The prediction of FIO behaviour in the environment can be undertaken for a number of reasons. For example, to: (i) explore real-time prediction of microbial pollution in recreational waters; (ii) model the dynamics of short term pollution incidents in the aquatic environment; (iii) understand the contributions of FIOs delivered to receiving waters via different hydrological pathways to inform mitigation and management; (iv) enable screening to guide regulators and/or policy-makers in prioritising decisions; (v) predict future scenarios due to climate or land-use change; or (vi) test new hypotheses to understand an environmental system better and advance our knowledge further. Recognising that no single model can inform all of these objectives is crucially important. End-users have a responsibility to ensure that they understand a model's purpose prior to its application. Model developers have a responsibility to be transparent about their model's strengths and limitations and to not promote model acceptance irrespective of end-user needs.

Model purpose and scale are intricately linked, and appreciating the acceptable range of scales at which a model can operate effectively is essential for robust prediction. There are significant questions concerning the extent to which FIO responses and behaviour observed and quantified on the microscale can be extrapolated to reflect behaviour at the catchment scale. This challenge is not restricted to microbial pollution and there are opportunities to learn from other research disciplines (e.g. nutrient management) (Vermeulen et al., 2015). Demand for models capable of predicting pollutant flux in catchment systems is increasing in part due to policy drivers that require assessments of how changes in management might impact on water quality in the future (Kay et al., 2010). Such predictions are complicated by the effects of scale, such as its impact on lag times between management actions and observed water quality improvements (Murphy et al., 2015a). In addition, calibrating FIO model simulations of mitigation effectiveness can be hampered by a lack of good quality FIO data (e.g. losses and exports) at the farm scale (Muirhead, 2015, Coffey et al., 2010 and Coffey et al., 2013). While generic management actions can be incorporated into models at a coarse resolution (e.g. typical manure application periods and grazing durations) it is the subtleties of management actions that are notoriously difficult to capture in model frameworks. This difficulty is largely because management is not static over time and changes in management at the farm business level can lead to discrepancies between on-the-ground actions versus modelled assumptions (e.g. grazing regimes and manure management) (Winter et al., 2011). Model frameworks need flexibility to respond to shifts in management practice (e.g. Bloodworth et al., 2015); the development of novel modelling approaches are now urgently required to better represent the effects of changing management actions on FIO dynamics.

Modelling Approach: What Approach Should Be Used And Why?

The structure of a model is based on the conceptual understanding of the system of interest. Conceptually, the framing of how microbial fate and transfer interact in agricultural systems is improving but the conversion of perceptual and conceptual understanding into procedural models is fraught with challenges (Krueger et al., 2007). One significant barrier is the level of uncertainty in answers to fundamental research questions (Section 1). In the UK and Europe, this is largely due to a legacy of riverine monitoring programmes that have historically focussed on nutrient rather than microbial pollutants, leading to relative scarcity of spatially distributed FIO data in catchment systems (Kay et al., 2007). This is coupled with a greater focus on longer-term monitoring of bathing waters (end-point receptors) as driven by the BWD and risks from sea-bathing. In contrast, countries like New Zealand, the USA and Canada carry out much greater spatial sampling of microbial water quality across catchments, driven by cultural differences in recreational pursuits and interactions with waterbodies e.g. differences in river versus sea-bathing, and therefore arguably have a far better spatial understanding of catchment microbial dynamics relative to the UK (Wilkinson et al., 2011). The amenity value of surface water bodies and coastal waters therefore plays a significant role in governing the availability of existing large-scale FIO datasets.

Whatever level of underpinning knowledge, there are a number of different modelling approaches available for exploring FIO dynamics in catchment systems in response to the purposes outlined in Section 2.1. These can include export coefficient and regression (McGrane et al., 2014, Sinclair et al., 2009, Kay et al., 2008b and Crowther et al., 2002), probabilistic or risk based approaches which represent inputs or parameters as probability distributions, that can be propagated though the model analytically or numerically (e.g. using Monte Carlo simulation) (Muirhead et al., 2011) and mechanistic/process-based models (Iudicello and Chin, 2015, Coffey et al., 2010 and Ferguson et al., 2007). These approaches operate at different temporal resolutions, or time-steps, ranging from sub-daily through to annual or longer timeframes and, as with spatial scale, the temporal scale of operation should help to dictate the best approach for the question of interest. In addition, the transferability of these different approaches across contrasting agricultural systems needs attention, with a particular focus on the challenges posed by inherent differences in catchment characteristics.

The real-time prediction of water quality at bathing areas is a topical modelling challenge in the UK. It has received increased impetus following revisions to the BWD that allow for discounting of a proportion of regulatory samples of poor microbial water quality if information has been provided to the public to warn about a risk of exposure to substandard water quality (e.g. via real-time electronic signage systems). Targeted monitoring has revealed that FIO concentrations exhibit diurnal variability on any given bathing day and that the variation can span across different bathing water classifications (Wyer et al., 2013). This diurnal variability provides clear rationale for the need of models that can dynamically account for FIO behaviour at a sub-daily time-step for informing improved real-time predictions. In Scotland the forecasting of bathing water quality is built around a simple decision-tree approach utilising river discharge and antecedent rainfall information (Stidson et al., 2012). In the USA the software package 'Virtual Beach' develops site-specific statistical models (including multiple linear regression) for the prediction of FIOs at recreational beaches (Frick et al., 2008). For wide-scale adoption these predictive tools need to be developed for each individual bathing site and cannot necessarily be directly transferred to another bathing site without modification.

More sophisticated models, with inherent complexity in the model structure, may not (yet) deliver useful results for such purposes. Complex models often require considerable expertise and resources for computations, data preprocessing, or data post-processing. As a result they are poorly suited to operational applications that require real-time data processing. In addition, users of complex environmental models often fail to deal with uncertainty in their predictions due to the computational expense; however, when they do provide uncertainty estimates these are often ignored or misused by decision-makers (Beven, 2009). Sometimes, a simple model approach can be more effective in guiding decisions or informing knowledge than a model that attempts to capture a comprehensive range of catchment processes and interactions. Simple models also accommodate uncertainty but often users of these approaches have a greater trust in outputs because of increased transparency of the inner workings of the tool.

In moving from a more complex to relatively simple approach it is likely that the original question we are asking of the model needs to be reframed in order to accommodate a different modelling strategy. One simple approach is to develop a risk index where the aim is to provide an indication of the relative contribution of different sources of faecal contamination to the aquatic environment (Muirhead, 2015 and Oliver et al., 2010). These models can take into account many of the underlying processes of source, mobilisation, transport and connectivity to streams but do not explicitly model the processes, and often incorporate expert judgements to inform the risk index (see Fish et al., 2009). This approach can indicate the expected relative change in stream FIO concentration and identify the largest FIO sources, which is useful for prioritising mitigation programs, but generally does not attempt to predict in-stream FIO concentrations (Muirhead, 2015). Advantages of the risk index approach are transferability (providing the sources are consistent) and the ability to easily modify the calculations used as new information/knowledge becomes available.

In a recent review, de Brauwere et al. (2014) commented that regression-based models cannot *replace* mechanistic models for long term assessments of management practices or understanding underlying FIO dynamics. Regression models are not constructed for this purpose thus their application in this context would reflect a case of model misuse. Replacing one model with another is not always necessary or desirable but recognising their different underlying purposes is clearly important. There is even much to be gained from observations of model failure in a particular context. Utilising models is a learning process, valuable insight can be gained from examining why a model fails to match observations, which can in turn lead to the formation of new hypotheses and novel questions. Part of this learning process should ensure that sensitivity and uncertainty analysis is embedded within the modelling approach (Beven, 2015), although to date relatively little attention has been given to these aspects of FIO modelling.

Approaches to modelling can also be strongly shaped by co-operation. There are inefficiencies associated with model proliferation, where multiple groups working on a particular topic develop their own model. Reducing duplication of effort is important if advances in FIO modelling are to be optimised. The need to purchase software licences or the lack of access to source code can prevent other researchers from exploring the power, transferability and potential of different models for their respective FIO datasets. The promotion of collaboration, transparency and open source code would open up model architecture to new ideas from different FIO model developers and accelerate improvements, opportunities for transferability, or new applications of existing model frameworks (Wilkinson et al., 2015). In doing so it is paramount to stress the need for version tracking and quality control protocols in order to capture the reporting of model modifications and data provenance (Vitolo et al., 2015). Without doing so there is an increased risk of models being developed without full audit trails of how they have evolved and without the associated narrative of why they are performing in such a way. The development of FIO models without tracking such changes could negate the benefits of opening up source code to the developer community given the pace at which people (and expertise) move on institutionally. Recent developments in software publishing now allows for model developers to gain recognition for publishing models and source code in the public domain. However, there is a need to support developers to maintain and develop the approaches once funded research projects have been completed.

Data Availability: What Data Do Us Need And What Is Available?

Often our ability to model pollutant response and behaviour in the environment runs ahead of the data that is available to drive or test such models. This is not an issue that is unique to microbial pollution in catchment systems. There is, however, a *relative* lack of data for FIOs when compared with other pollutants monitored in agricultural catchments (Muirhead, 2015). This is partly a legacy of a focus on nutrients in catchment monitoring, together with generally poor central data management of academic, agency and private data resources. Furthermore, monitoring data is designed to determine water quality at a given location, rather than understanding where the contaminants are coming from and therefore, often lacks data at key times and/or locations needed to calibrate models (Wilkinson et al., 2011).

Beyond the immediate issues of spatial and temporal quality of historical datasets and associated sampling capability, there still exist a number of gaps in fundamental understanding of FIO processes in the environment. Progress is being made but issues remain, for example, in determining good quantitative data on wildlife contributions to FIO loading in, and export from, agricultural catchments (Coffey et al., 2015 and Guber et al., 2015), securing a robust library of FIO die-off in fresh faeces of younger livestock (Oliver et al., 2012a), obtaining robust parameters to enable prediction of manure release and removal in site-specific conditions (Blaustein et al., 2015a, Blaustein et al., 2015b and Blaustein et al., 2016), incorporating bed sediment reservoirs of FIOs into existing modelling frameworks (Coffey et al., 2014) and also in understanding the contributions of septic tanks to the impairment of microbial water quality. Septic tank inputs of P to watercourses are now gaining significant attention (Withers et al., 2014) and in many ways this has accelerated the focus of their potential role on FIO contributions to the aquatic environment — helping to 'bridge the gap' between these two aspects of diffuse pollution research. This represents an idea (based on anecdotal association) linking nutrient to FIO research that has had a positive impact. However, it has progressed because of the testing of hypothesised links so that the cross pollination across research disciplines contributed inspiration rather than allowing anecdote to prevail over evidence.

Techniques such as flowpath separation are also attracting recognition in terms of their value for advancing our understanding of FIO transfer dynamics in catchments (Murphy et al., 2015b). However, there remain significant issues with currently available data, and in particular the rates of persistence of FIOs within the multitude of different environmental matrices that they are associated with, e.g. freshwaters of varying turbidity, different soil classifications, and there remain limited field-relevant studies relative to laboratory-based reporting of FIO population dynamics in environmental matrices. Contributions of FIOs from groundwater sources also reflect a relatively unexplored area in the field of catchment microbial dynamics. Certainly the majority of FIO models typically do not consider groundwater contributions as a source of faecal pollution or more specifically the distribution of inputs from groundwater as a source of low FIO input to surface waters via baseflow. Data needs, such as those described above, are fundamental for the advance of FIO modelling and prediction, and the parameterisation of models of FIO behaviour. There needs to be an increased effort to evaluate ranges of FIO parameters during the model parameterisation process; making datasets more transparent and more widely available will help facilitate this.

With data that are available there will remain questions over their transferability to different environmental contexts. Many of these questions could be addressed by providing more detailed meta-data (e.g. weather variables and catchment characteristics) associated with the experimental conditions or monitoring programmes that have generated the data. While the research community may want 'everything' in terms of accompanying data, in reality there are obviously cost and resource constraints that limit wider data collection. Opinions differ on what environmental data should be provided to support FIO modelling, highlighting the need for the research community to debate and identify a 'baseline' of essential supplementary information to facilitate international transferability of FIO data. One basic example is the coupling of FIO concentration and discharge data in hydrological studies; concentration data alone inhibits a wider assessment of catchment issues and exported FIO loads (Pachepsky et al., 2006).This opens up wider debates as to whether arithmetic or geometric mean FIO data should be combined with discharge, and how to compare FIO concentrations determined by different methods (e.g. membrane filtration versus most probable number).

In some countries, nationally available datasets or data derived from farm and fertiliser practice surveys can be explored in further detail to try to redress some of the finer-scale issues of management impacts on FIO dynamics (e.g. Aitken, 2003). Key data that can be extracted from such sources for onward use in FIO modelling include: information on yard management (clean and dirty water separation); frequency of livestock stream fording; and manure management including computed ages of manures and slurries at time of application to land. Access to, and knowledge of, such sources of supplementary data can be enhanced by the involvement of stakeholders and anticipated end-users throughout the entire modelling process, from development to evaluation (Hamilton et al., 2015). Further research is needed to enable successful and reliable integration of mitigation and best management practices (BMPs) into model space but there are large collections of supplementary data that can be explored in order to strengthen FIO modelling for some purposes. However, it is likely that FIO source apportionment or model calibration can be especially sensitive to biased assumptions about practices (e.g. cattle access to streams and duration of manure storage) in particular catchments, based on the down-scaling of regional or national surveys of farm practices. Therefore improvements in farm activity data collection will always be welcome to better inform FIO modelling.

Model Application: How Do We Apply Our Model (Effectively)?

Applying and testing models in landscapes typical of different catchment systems around the world enables an assessment of how transferable a model structure might be beyond its calibration catchment in order to evaluate its global applicability. Transferability of model structure is a favourable attribute but one that presents considerable challenges at the international level, largely as a result of availability of the datasets used to drive models, subtle differences in the classifications used within such datasets across different nations, and given that different people are all asking subtly different questions. Source loads (species, prevalence, concentration etc.), FIO die-off, FIO-sediment interactions and other fate/transport parameters are non-transferable in an environmental context. Values selected are often used as a means of model calibration without any empirical or physical justification. The model parameterisation process needs to be coupled with wider discussion and inclusion of statistical methods (e.g. probabilistic treatment and propagation of parameter uncertainty) to improve model design. This treatment would help to account for specificity of FIO data in model performance evaluation and model calibration.

Awareness and respect for different modelling approaches is also important. Those that apply process-based FIO models (e.g. SWAT and HSPF) generally argue that despite their limitations there is no viable alternative (Pachepsky et al., 2006). However, with questions increasingly focusing on when and where to prioritise spatial targeting of mitigation and management within catchments there is an emerging need for models that take a risk-based approach (e.g. Reaney et al., 2011; Oliver et al., 2010), which highlight in relative (rather than absolute) terms, areas of catchments most conducive for facilitating FIO transfer from land to water. Export coefficient models can do this too but a risk-based approach is the simplest approach to take and prioritisation does not need anything more complex. Something as simple as a risk based model is therefore consistent with the level of uncertainty in the data that underpins it. Although these types of models have not yet been widely adapted for modelling FIOs, risk-based approaches could be parameterised with existing datasets, which is in contrast to the current overreliance on deterministic complex models that often lack the availability of data needed to drive those models effectively (Dean et al., 2009). The selection of useful models for inappropriate applications therefore needs to be more thoroughly guarded against by both the user and developer communities. Legacy effects of 'go-to' industry-standard models being used over long time-periods may be one reason for a mismatch of model application relative to purpose. These 'work-horse' models may be deployed to tackle new and emerging questions that they were simply not designed to answer even with modification or integration of new FIO modelling routines.

Finally, important advances in the science of catchment microbial dynamics often remain inaccessible to those who manage landscape risk on a day-to-day basis. The extent to which models are useful, and that they are used, is driven not only by the rigour of the model but also by the needs of the user, and the availability and accessibility of the model. There are important technical questions around the usability and visualisation of models, and this remains an area of active research (Karpouzoglou et al., 2016 and Rink et al., 2014). Visualisations of environmental risk can powerfully communicate complex environmental risk assessments to decision-makers (Lahr and Kooistra, 2010). Yet many attempts at risk communication are poorly received by end-users (Oliver et al., 2012b), largely because they fail to engage and involve their target audience in the design of such tools (Whitman et al., 2015). Thus, where models or their visualised outputs are intended as tools to help people understand risk, a 'human-centric' approach is required. Studies vary in the point at which model developers seek participation from those that might use the model e.g. from interpretation and modification of visualisation, through model development to definition of the initial research question. These participatory approaches can force model developers to alter both their focus and their ways of working. This shift can be uncomfortable and though calls for participatory modelling are common, examples of its application are far rarer (e.g. Landstrom et al., 2011 and Whitman et al., 2015).

HORIZON SCANNING & FUTURE OPPORTUNITIES

Challenges for the future of FIO modelling in agricultural catchments include, but are not limited to, the need for good quality data on FIO fate and behaviour in the environment. Here we build on the issues identified through our discussion of the four over-arching questions associated with good modelling practice and map future research priorities in terms of short (< 5 year) and longer-term (5-10 year) needs (Table 1). As these priorities are addressed, a pathway to more effective deployment of models for characterising catchment microbial dynamics should begin to emerge.

Table 1: Summary of future short- and longer-term research opportunities for faecal indicator organism (FIO) modelling.

Future research opportunities for FIO modelling	Time-frame
Improve the underpinning evidence-base of FIO fate & behaviour in the environment	<5 years
Strengthen the available data on wildlife/wildfowl FIO sources	<5 years
Increased consideration of the role of groundwater contributions as a source of faecal pollution	<5 years
Better understanding of the differential behaviour of FIOs & specific pathogens	<5 years
Future FIO model performance criteria to include, as a minimum, an uncertainty analysis of modelled predictions	<5 years
Identify a 'baseline' of essential supplementary information to facilitate international transferability of FIO data	<5years
The integration of dynamic management practices and mitigation options into FIO model space within a wider framework of climate and land use change	<10years
Extrapolating FIO fate & transport data from experimental plots to fields and catchments (scaling methodologies)	<10 years
Innovative combinations of FIO modelling and MST for informing on catchment microbial risks	<10 years
Integration of different modelling approaches (including risk-based approaches & QMRA) to promote more holistic modelling of catchment microbial dynamics	<10 years

MST: Microbial Source Tracking; QMRA: Quantitative Microbial Risk Assessment.

Short Term Research Needs (0–5 Years)

Modelling of microbial pollutants in agricultural catchments is only as good as the data that drives the model. Review articles and commentaries often feature repeated pleas for research to 'plug the gaps' in understanding that may begin to sound overused, but for FIO modelling those gaps are surprisingly large (and more so for specific pathogens) and seriously compromise modelling efforts to deliver wider environmental benefits. A case in point is the lack of relevant data concerning wildlife contributions to catchment FIO dynamics. The lack of data on wildlife/wildfowl FIO sources and their potential importance were reported in the literature almost 20 years ago (Weiskel et al., 1996), and while some effort is now being converted into published findings (Guber et al., 2015,Muirhead et al., 2011 and Kiefer et al., 2012), international coverage of useful information remains weak. Uncertainty in FIO loading from wildlife and wildfowl, and behaviour of these FIO populations can increase the uncertainty and undermine model predictions of microbial pollution in agricultural systems. Where FIOs are apportioned to livestock, the agricultural sector will often respond by querying the relative contribution of migratory wildlife or wildfowl. This issue is further exacerbated by the difficulty in calculating faecal loadings since this requires the weight of faeces, its microbial concentration and correct algorithms to calculate the true FIO loads (Muirhead and Cave, 2014). Without being able to understand and evidence these contributions, scepticism over model outputs among some catchment stakeholders will remain high. In parallel, and to help facilitate fundamental data collection such as that described above, a co-ordinated clear voice is needed from the international FIO research community to highlight the relative scarcity of empirical data, and established meta-data standards, across a broad spectrum of environmental conditions.

The pursuit of good quality FIO data does need to be complemented with continued efforts to understand the differential behaviour of FIOs and pathogens, and the associated implications for modelling the efficacy of mitigations designed to reduce microbial watercourse pollution. Thus, while we have focussed our discussion on FIO modelling it is important for resource managers and policy makers to recognise that FIO models do not necessarily inform on specific pathogen behaviour. However, the nexus of FIO modelling and risk assessment is gaining significant momentum and the use of quantitative microbial risk assessment (QMRA) offers another modelling framework with which to estimate human infection risk from exposure to pathogen contaminated waters (Soller et al., 2015).

Benefits of FIO modelling will be further enhanced if the research community promotes a consistent message to model developers calling for any future model performance criteria to include, as a minimum, an uncertainty analysis of modelled predictions. Thus, the pedigree of uncertainty associated with model outputs needs to be transparent and included as standard in the reporting of any FIO modelling. There are mechanisms that can help deliver this drive for greater awareness of model uncertainty. For example, there are more frequent opportunities to promote knowledge exchange (KE) across the science-policy interface enabling scientists and policy-makers to debate the role and impact of model uncertainty. Research Councils in the UK, and in other nations too, are increasingly funding policy-placement opportunities whereby scientists are embedded within government departments thus providing an important route for aiding KE. This should be encouraged in the general field of environmental decision-making using models. In addition, those responsible for funding research could improve model transparency by ensuring that projects with a modelling component fulfil basic uncertainty analysis criteria. Indeed, several academic journals, whose remit includes environmental modelling, now stipulate uncertainty analysis of any modelling as a prerequisite for publication. This template should also be adopted at the funding stage for research projects.

Longer Term Research Needs (5–10 Years)

Increased mutual respect for different modelling philosophies, including those that deviate from a process-based approach (e.g. Odoni and Lane, 2010) and also in promoting a more co-ordinated strategy for the integration of different disciplinary expertise is essential (Hamilton et al., 2015). Convincing model developers that sometimes their own model is not the right tool for the job can be a problem across environmental modelling (Prell et al., 2007). Arguably the biggest risk to modelling frameworks is a shift in the conceptualisation of how the system operates. From an environmental perspective this is perhaps not likely, but the management of agricultural systems is more dynamic and able to accommodate change. Thus, there is a need to ensure that modelling efforts are able to reflect those changes should they occur — this equates to flexibility and futureproofing of modelling approaches and anticipating dynamic agricultural practices within a wider framework of climate and land use change (Coffey et al., 2014). Such efforts are attractive but extremely difficult in practice given the multitude of different changes that must be accounted for. Less flexible modelling approaches will be more at risk of becoming outdated — this is not necessarily a problem if outdated models are recognised rather than continually supported as a legacy of end-user preference. The integration of dynamic management practices and mitigation options into FIO model space represents a clear challenge for model developers. While there is undoubted progress on this challenge it is hindered by the gaps in empirical data and understanding described above.

Efforts to future-proof modelling approaches must equate to an increased effort by the research community to focus on the development of scaling rules for FIO behaviour to enable greater confidence in extrapolation of findings from smaller to larger scales. Extrapolating FIO fate and transport data from experimental plots to fields and catchments will require accounting for increased opportunities for in-field FIO intermittent accumulation, growth and inactivation, spatial variability in overland flow discharge and frequency, etc. Experiments to understand this scaling are difficult and costly, but necessary to increase the value of plot studies. The use of FIO modelling to support the emerging science of microbial source tracking (MST) will also serve to strengthen the toolbox of techniques available to environmental managers for assessing microbial risks in catchment systems. Though MST remains largely qualitative and the subject of much debate with regard to source sensitivity and source specificity (Astrom et al., 2015) it is a technique that is gathering pace and innovative combinations of FIO modelling and MST may hold promise for informing on catchment microbial risks in the future. Likewise, more real-time monitoring techniques are being explored and utilised by researchers to understand temporal variability in faecal contamination of transient surface and groundwater systems (e.g. Sorensen et al., 2015). This presents both an opportunity and a challenge for improved modelling of FIO transfers in the environment.

Given the difficulties of FIO and pathogen data collection in catchments, modelling could be deployed to help guide monitoring programs. Currently, questions of when, where and how to sample are not driven by FIO model calibration, sensitivity analysis, uncertainty evaluation, data assimilation, or other operations that improve modelling reliability and usefulness. For example, storm-flow can export up to 98% of the annual *E. coli* load from agricultural catchments but represent only 6–30% of routine water quality monitoring samples (Muirhead, 2015 and McKergow and Davies-Colley, 2010). Longer term challenges should reflect a move to more integrated approaches to modelling. To realise this ambition requires an appreciation from the regulatory and policy communities that different types of models can be applied in unison to complement rather than 'replace' each other addressing different parts of the larger, multi-scaled problem of integrated catchment management. Integrating different modelling approaches to promote a more holistic approach to catchment modelling is challenging. The development of engaging user interfaces, the incentivising of 'opening-up' the underlying model source code and enabling wider access to multiple model platforms free of licencing issues will require co-operation and collaboration across multiple disciplinary fields and a range of organisations. Most notably, links across psychology, the social and environmental sciences and the environmental modelling community need to be strengthened even further to ensure better human-computer interactive experiences in the modelling process.

CONCLUSION

Modelling microbial fate and transport has been, and will continue to be, helpful in elucidating and integrating knowledge about the intimate relationships between water, soil, FIOs and catchment management. Thus, modelling of catchment microbial dynamics supports our efforts to bridge scales between research and policy making, and plays an important role in helping to understand and address the challenges of protecting food safety and water quality in the face of climate and land use change. Current environmental agendas and increased awareness has put FIO research in focus for many national and international bodies and groups meaning that the time is now right to revisit, revise, and advance factual, conceptual, and technical foundations of FIO fate and transport modelling. Using the recommendations from the over-arching questions posed in this commentary will help to promote better model deployment in the field of catchment microbial dynamics.

ACKNOWLEDGEMENTS

This work was supported by the UK Natural Environment Research Council, as part of the 'PRACTICAL Modelling' project (NE/M005860/1). We are grateful for the constructive comments of the three reviewers and the handling editor.

REFERENCES

1. Aitken, M.N., 2003. Impact of agricultural practices and river catchment characteristics on river and bathing water quality. Water Sci. Technol. 48, 217–224.
2. Astrom, J., TJR, P., Reischer, G.H., Norberg, T., Hermansson, M., 2015. Incorporating expert judgements in utility evaluation of Bacteroidales qPCR assays for microbial source tracking in drinking water source. Environ. Sci. Technol. 49, 1311–1318.
3. Beven, K., 2015. Facets of uncertainty: epistemic uncertainty, non-stationarity, likelihood, hypothesis

testing, and communication. Hydrol. Sci. J. http://dx.doi.org/10.1080/ 02626667.2015.1031761.

4. Beven, K., 2009. Environmental Modelling: An Uncertain Future. Routledge, London.
5. Blaustein, R.A., Pachepsky, Y.A., Shelton, D.R., Hill, R.L., 2015a. Release and removal of microorganisms from land-deposited animal waste and animal manures: a review of data and models. J. Environ. Qual. 44, 1338–1354.
6. Blaustein, R.A., Pachepsky, Y.A., Hill, R.L., Shelton, D.R., 2015b. Solid manure as a source of fecal indicator microorganisms: release under simulated rainfall. Environ. Sci. Technol. 49, 7860–7869.
7. Blaustein, R.A., Pachepsky, Y.A., Hill, R.L., Shelton, D.R., 2016. Rainfall intensity effects on removal of fecal indicator bacteria from solid dairy manure applied over grasscovered soil. Sci. Total Environ. 539, 583–591.
8. Bloodworth, J.W., Holman, I.P., Burgess, P.J., Gillman, S., Frogbrook, Z., Brown, P., 2015. Developing a multi-pollutant conceptual framework for the selection and targeting of interventions in water industry catchment management schemes. J. Environ. Manag. 161, 153–162.
9. CEC, 2006a. Council Directive 2006/7/EC of the European Parliament and of the council of 15th February 2006 concerning the management of bathing water quality and repealing Directive 76/160/EEC. Off. J. Eur. Union L64, 37–51.
10. CEC, 2006b. Council Directive 2006/113/EC of 12 December 2006 on the quality required of shellsh waters (codied version). Off. J. Eur. Union L376, 14–20.
11. Coffey, R.P., Benham, B., Kline, K., Wolfe, M.L., Cummins, E., 2015. Modeling the impacts of climate change and future land use variation on microbial transport. Journal of Water and Climate Change 6, 449–471.
12. Coffey, R., Cummins, E., Bhreathnach, N., Flaherty, V.O., Cormican, M., 2010. Development of a pathogen transport model for Irish catchments using SWAT. Agric. Water Manag. 97, 101–111.
13. Coffey, R., Dorai-Raj, S., O'Flaherty, V., Cormican, M., Cummins, E., 2013. Modeling of pathogen indicator organisms in a small-scale agricultural catchment using SWAT. Human and Ecological Risk Assessment: An International Journal 19, 232–253.
14. Coffey, R., Cummins, E., Cormican, M., Flaherty, V.O., Kelly, S., 2007. Microbial exposure assessment of waterborne pathogens. Human and Ecological Risk Assessment: An International Journal 13, 1313–1351.
15. Coffey, R., Benham, B., Krometis, L.A., Wolfe, M.L., Cummins, E., 2014. Assessing the effects of climate change on waterborne microorganisms: implications for EU and US Water Policy. Human and Ecological Risk Assessment: An International Journal 20 (3), 724–742.
16. Crowther, J., Kay, D., Wyer, M.D., 2002. Faecal-indicator concentrations in waters draining lowland pastoral catchments in the UK: relationships with land use and farming practices. Water Res. 36, 1725–1736.
17. Dean, S., Freer, J., Beven, K., Wade, A.J., Butterfield, D., 2009. Uncertainty assessment of a process-based integrated catchment model of phosphorus. Stoch. Env. Res. Risk A. 23, 991–1010.
18. de Brauwere, A., Ouattara, N.K., Servais, P., 2014. Modeling fecal indicator bacteria concentrations in natural surface waters: a review. Crit. Rev. Environ. Sci. Technol. 44, 2380–2453.
19. Dickey-Collas, M., Payne, M.R., Trenkel, V.M., Nash, R.D.M., 2014. Hazard warning: model misuse ahead. ICES J. Mar. Sci. 71, 2300–2306.
20. Dufour, A., Bartram, J., Bos, R., Gannon, V. (Eds.), 2012. Animal Waste, Water Quality and Human Health: WHO — Emerging Issues in Water and Infectious Disease series. International Water Association, USEPA and WHO, London.
21. Ferguson, C.M., Croke, B.F.W., Beatson, P.J., Ashbolt, N.J., Deere, D.A., 2007. Development of a process-based model to predict pathogen budgets for the Sydney drinking water catchment. Journal of Water & Health 5, 187–208.
22. Fish, R., Winter, M., Oliver, D.M., Chadwick, D.R., Selfa, T., Heathwaite, A.L., Hodgson, C., 2009. Unruly pathogens: eliciting values for environmental risk in the context of heterogeneous expert knowledge. Environ. Sci. Pol. 12, 281–296.
23. Frick, W.E., Ge, Z.F., Zepp, R.G., 2008. Nowcasting and forecasting concentrations of biological contaminants at beaches: a feasibility and case study. Environ. Sci. Technol. 42, 4818–4824.
24. Guber, A.K., Fry, J., Ives, R.L., Rose, J.B., 2015. Escherichia coli survival in, and release from, white-tailed deer feces. Appl. Environ. Microbiol. 81, 1168–1176.
25. Hamilton, S.H., ElSawah, S., Guillaume, J.H.A., Jakeman, A.J., Pierce, S.A., 2015. Integrated assessment and modelling: overview and synthesis of salient dimensions. Environ. Model. Softw. 64, 215–229.
26. Iudicello, J.J., Chin, D.A., 2015. In-stream bacteria modelling as a function of the hydrologic state of a watershed. J. Environ. Eng. 141 (#04014073).
27. Karpouzoglou, T., Zulkafli, Z., Grainger, S., Dewulf, B.W., Hannah, D.M., 2016. Environmental Virtual Observatories (EVOs): prospects for knowledge co-creation and resilience in the information age. Curr. Opin. Environ. Sustain. 18, 40–48.
28. Kay, D., Edwards, A.C., Ferrier, R.C., Francis, C., Kay, C., Rushby, L., Watkins, J., McDonald, A.T., Wyer, M., Crowther, J., Wilkinson, J., 2007. Catchment microbial dynamics: the emergence of a research agenda. Prog. Phys. Geogr. 31, 59–76.
29. Kay, D., Crowther, J., Fewtrell, L., Francis, C.A., Hopkins, M., Kay, C., McDonald, A.T., Stapleton, C.M., Watkins, J., Wilkinson, J., Wyer, M.D., 2008a. Quantification and control of microbial pollution from agriculture: a new policy challenge? Environ. Sci. Pol. 11, 171–184.
30. Kay, D., Crowther, J., Stapleton, C.M., Wyer, M.D., Fewtrell, L., Anthony, S., Bradford, M., Edwards, A.,

Francis, C.A., Hopkins, M., Kay, C., McDonald, A.T., Watkins, J., Wilkinson, J., 2008b. Faecal indicator organism concentrations and catchment export coefficients in the UK. Water Res. 42, 2649–2661.

31. Kay, D., Anthony, S., Crowther, J., Chambers, B.J., Nicholson, F.A., Chadwick, D., Stapleton, C.M., Wyer, M.D., 2010. Microbial water pollution: a screening tool for initial catchment-scale assessment and source apportionment. Sci. Total Environ. 408, 5649–5656.

32. Kiefer, L.A., Shelton, D.R., Pachepsky, Y., Blaustein, R., Santin-Duran, M., 2012. Persistence of Escherichia coli introduced into streambed sediments with goose, deer and bovine animal waste. Lett. Appl. Microbiol. 55, 345–353.

33. Krueger, T., Freer, J., Quinton, J.N., Macleod, C.J.A., 2007. Processes affecting transfer of sediment and colloids, with associated phosphorus, from intensively farmed grasslands: a critical note on modelling of phosphorus transfers. Hydrol. Process. 21, 557–562.

34. Lahr, J., Kooistra, L., 2010. Environmental risk mapping of pollutants: state of the art and communication aspects. Sci. Total Environ. 408, 3899–3907.

35. Landstrom, C., Whatmore, S.J., Lane, S.N., Odoni, N.A., Ward, N., Bradley, S., 2011. Coproducing flood risk knowledge: redistributing expertise in critical 'participatory modelling'. Environ. Plan. A 43, 1617–1633.

36. Martinez, G., Pachepsky, Y.A., Shelton, D.R., Whelan, G., Zepp, R., Molina, M., Panhorst, K., 2013. Using the Q10 model to simulate E. coli survival in cowpats on grazing lands. Environ. Int. 54, 1–10.

37. McBride, G.B., Stott, R., Miller, W., Bambic, D., Wuertz, S., 2013. Discharge-based QMRA for estimation of public health risks from exposure to stormwater-borne pathogen in recreational water in the United States. Water Res. 47 (14), 5282–5297.

38. McGrane, S.J., Tetzlaff, D., Soulsby, C., 2014. Application of a linear regression model to assess the influence of urbanised areas and grazing pastures on the microbiological quality of rural streams. Environ. Monit. Assess. 186, 7141–7155.

39. McKergow, L.A., Davies-Colley, R., 2010. Stormflow dynamics and loads of Escherichia coli in a large mixed land use catchment. Hydrol. Process. 24, 276–289.

40. Muirhead, R.W., 2015. A farm-scale risk-index for reducing fecal contamination of surface waters. J. Environ. Qual. 44, 248–255.

41. Muirhead, R.W., Cave, V.M., 2014. Accounting for groups of animals in QMRA of recreational waters. In: Ames, D.P., Quinn, N.W.T., Rizzoli, A.E. (Eds.), International Environmental Modelling and Software Society (iEMSs), 7th Intl. Congress on Environmental Modelling and Software, San Diego, CA, USA (http://www.iemss.org/society/index. php/iemss-2014-proceedings).

42. Muirhead, R.W., Elliot, A.H., Monaghan, R.M., 2011. A model framework to assess the effect of dairy farms and wild fowl on microbial water quality during base-flow. Water Res. 45, 2863–2874.

43. Muirhead, R.W., Davies-Colley, R.J., Donnison, A.M., Nagels, J.W., 2004. Faecal bacteria yields in artificial flood events: quantifying in-stream stores. Water Res. 38, 1215–1224.

44. Murphy, P.N.C., Mellander, P.-E., Melland, A.R., Buckley, C., Shore, M., Shortle, G., Wall, D.P., Treacy, M., Shine, O., Mechan, S., Jordan, P., 2015a. Variable response to phosphorus mitigation measures across the nutrient transfer continuum in a dairy grassland catchment. Agric. Ecosyst. Environ. 207, 192–202.

45. Murphy, S., Jordan, P., Mellander, P.-E., O'Flaherty, V., 2015b. Quantifying faecal indicator organism hydrological transfer pathways and phases in agricultural catchments. Sci. Total Environ. 520, 286–299.

46. NPS-FW (2014) National policy statement for freshwater management 2014. Issued by NZ Government 2 July 2014. Available from: http://www.mfe.govt.nz/publications/ fresh-water/national-policy-statement-freshwater-management-2014.

47. Odoni, N.A., Lane, S.N., 2010. Knowledge-theoretic models in hydrology. Prog. Phys. Geogr. 34, 151–171.

48. Oliver, D.M., Heathwaite, A.L., Fish, R.D., Chadwick, D.R., Hodgson, C.J., Winter, M., Butler, A.J., 2009. Scale appropriate modelling of diffuse microbial pollution from agriculture. Prog. Phys. Geogr. 33, 358–377.

49. Oliver, D.M., Porter, K.D.H., Heathwaite, A.L., Zhang, T., Quilliam, R.S., 2015. Impact of low intensity summer rainfall on E. coli-discharge event dynamics with reference to sample acquisition and storage. Environ. Monit. Assess. 187, 426.

50. Oliver, D.M., Page, T., Hodgson, C.J., Heathwaite, A.L., Chadwick, D.R., Fish, R.D., Winter, M., 2010. Development and testing of a risk indexing framework to determine field-scale critical source areas of faecal bacteria on grassland. Environ. Model. Softw. 25, 503–512.

51. Oliver, D.M., Page, T., Zhang, T., Heathwaite, A.L., Beven, K., Carter, H., McShane, G., O'Keenan, P., Haygarth, P.M., 2012a. Determining E. coli burden on pasture in a headwater catchment: combined field and modelling approach. Environ. Int. 43, 6–12.

52. Oliver, D.M., Fish, R., Winter, M., Hodgson, C.J., Heathwaite, A.L., Chadwick, D.R., 2012b. Valuing local knowledge as a source of expert data: Farmer engagement and the design of decision support systems. Environ. Model. Softw. 36, 76–85.

53. Pachepsky, Y., Shelton, D.R., McLain, J.E.T., Patel, J., Mandrell, R.E., 2011. Irrigation waters as a source of pathogenic microorganisms in produce: a review. Adv. Agron. 113, 73–138.

54. Pachepsky, Y.A., Sadeghi, A.M., Bradford, S.A., Shelton, D.R., Guber, A.K., Dao, T., 2006. Transport and fate of manure-borne pathogens: modelling perspective. Agric. Water Manag. 86, 81–92.

55. Prell, C., Hubacek, K., Reed, M., Quinn, C., Jin, N., Holden, J., Burt, T., Kirby, M., Sendzimir, J., 2007. If you have a hammer everything looks like a nail: traditional versus participatory model building. Interdiscip. Sci. Rev. 32, 263–282.

56. Quilliam, R.S., Jamieson, J., Oliver, D.M., 2014. Seaweeds and plastic debris can influence the survival of faecal indicator organisms in beach environments. Mar. Pollut. Bull. 84, 201–207.

57. Quilliam, R.S., Kinzelman, J., Brunner, J., Oliver, D.M., 2015. Resolving conflicts in public health protection and ecosystem service provision at designated bathing waters. J. Environ. Manag. 161, 237–242.

58. Reaney, S.M., Lane, S.N., Heathwaite, A.L., Dugdale, L.J., 2011. Risk-based modelling of diffuse land use impacts from rural landscapes upon salmonid fry abundance. Ecol. Model. 222, 1016–1029.

59. Rink, K., Bilke, L., Kolditz, O., 2014. Visualisation strategies for environmental modelling data. Environmental Earth Sciences 72, 3857–3868.

60. Sinclair, A., Hebb, D., Jamieson, R., Gordon, R., Benedict, K., Fuller, K., Stratton, G.W., Madani, A., 2009. Growing season surface water loading of fecal indicator organisms within a rural watershed. Water Res. 43, 1199–1206.

61. Soller, J., Bartrand, T., Ravenscroft, J., Molina, M., Whelan, G., Schoen, M., Ashbolt, N., 2015. Estimated human health risks from recreational exposures to stormwater runoff containing animal faecal material. Environ. Model. Softw. 72, 21–32.

62. Sorensen, J.P.R., Lapworth, D.J., Marchant, B.P., Nkhuwa, D.C.W., Pedley, S., Stuart, M.E., Bellb, R.A., Chirwac, M., Kabikac, J., Liemisae, M., Chibesa, M., 2015. In-situ tryptophan-like fluorescence: a real-time indicator of faecal contamination in drinking water supplies. Water Res. 81, 38–46.

63. Stidson, R.T., Gray, C.A., McPhail, C.D., 2012. Development and use of modelling techniques for real-time bathing water quality predictions. Water & Environment Journal 26, 7–18.

64. Stocker, M.D., Pachepsky, Y.A., Hill, R.L., Shelton, D.R., 2015. Depth-dependent survival of Escherichia coli and enterococci in soil after manure application and simulated rainfall. Appl. Environ. Microbiol. 81, 4801–4808.

65. Till, D., McBride, G., Ball, A., Taylor, K., Pyle, E., 2008. Large-scale freshwater microbiology study: rationale, results and risks. Journal of Water and Health 6 (4), 443–460.

66. USEPA (2015). http://iaspub.epa.gov/waters10/attains_nation_cy.control?p_report_ type=T#causes_303d. Accessed on 5th October 2015.

67. Vant, W.N., 2001. New challenges for the management of plant nutrients and pathogens in the Waikato River, New Zealand. Water Sci. Technol. 43 (5), 137–144.

68. Vermeulen, L.C., Hofstra, N., Kroeze, C., Medema, G., 2015. Advancing waterborne pathogen modelling: lessons from global nutrient export models. Curr. Opin. Environ. Sustain. 14, 109–120.

69. Vitolo, C., Elkhatib, Y., Reusser, D., MacLeod, C.J.A., Buytaert, W., 2015. Web technologies for environmental big data. Environ. Model. Softw. 63, 185–198.

70. Weiskel, P.K., Howes, B.L., Heufelder, G.R., 1996. Coliform contamination of a coastal embayment: sources and transport pathways. Environ. Sci. Technol. 30, 1872–1881.

71. Whitman, G.P., Pain, R., Milledge, D.G., 2015. Going with the flow? Using participatory action research in physical geography. Prog. Phys. Geogr. (in press).

72. Wilkinson, M.E., Mackay, E., Quinn, P.F., Stutter, M., Beven, K.J., MacLeod, C.J.A., Macklin,

73. M.G., Elkhatib, Y., Percy, B., Vitolo, C., Haygarth, P.M., 2015. A cloud based tool for knowledge exchange on local scale flood risk. J. Environ. Manag. 161, 38–50.

74. Wilkinson, R.J., McKergow, L.A., Davies-Colley, R.J., Ballantine, D.J., Young, R.G., 2011.

75. Modelling storm-event E. coli pulses from the Motueka and Sherry Rivers in the Zouth Island, New Zealand. N. Z. J. Mar. Freshw. Res. 45 (3), 369–393.

76. Winter, M., Oliver, D.M., Fish, R., Heathwaite, A.L., Chadwick, D., Hodgson, C., 2011. Catchments, subcatchments and private spaces: scale and process in managing microbial, pollution from source to sea. Environ. Sci. Pol. 14, 315–326.

77. Withers, P.J.A., Jordan, P., May, L., Jarvie, H.P., Deal, N.E., 2014. Do septic tank systems pose a hidden threat to water quality? Front. Ecol. Environ. 12, 123–130.

78. Wyer, M.D., Kay, D., Watkins, J., Davies, C., Kay, C., Thomas, R., Porter, J., Stapleton, C.M., Moore, H., 2010. Evaluating short-term changes in recreational water quality during a hydrograph event using a combination of microbial tracers, environmental microbiology, microbial source tracking and hydrological techniques: a case study in Soutwest Wales, UK. Water Res. 44, 4783–4795.

79. Wyer M, Kay D, Morgan H, Naylor S, Govier P, Clark S, Watkins J, Davies C, Francis C, Osborn H, and Bennett S. (2013). Statistical modelling of faecal indicator organisms at a marine bathing water site: results of an intensive study at Swansea Bay, UK. A report from the Interreg 4a Smart Coasts — Sustainable Communities Project. 67 p. Available from: http://www.smartcoasts.eu/gallery/launch.asp.

CHAPTER 10

ENVIRONMENTAL RISK ASSESSMENTS

Daniel J.C. Skinner, Sophie A. Rocks, Simon J.T. Pollard

Keywords: Uncertainty dimensions, Uncertainty management, Resource allocation, Risk analysis, Elicitation protocol

INTRODUCTION

The primary objective of an environmental risk assessment (ERA) is to estimate the magnitude or level(s) of risk posed to environmental receptors by sources of potential harm (Department for Environment, Food and Rural Affairs [Defra], 2011). Confidence in the risk estimate can be lowered by uncertainties in the ERA, which can weaken, or stall, the basis for risk management. Whilst risk analysts recognise uncertainty should be considered (Funtowicz and Ravetz, 1990, Costanza et al., 1992 and Handmer et al., 2001), many ERAs still fail to identify uncertainties (European Environment Agency, 2007, Hart et al., 2007 and Dale et al., 2008). This paper introduces an approach to aid uncertainty identification and prioritisation within ERAs.

Uncertainty in environmental systems is investigated through its constituent dimensions (Walker et al., 2003); namely, its *location* (where the uncertainty is manifest in the various stages of a risk assessment), its *nature* (due to the incompleteness of knowledge or the inherent variability of natural systems), and its *level* (the severity of the uncertainty, ranging from determinism to ignorance). Ideally, the location of the uncertainty must be known, since without this, no action to reduce it can be implemented. The nature of the uncertainty dictates the extent to which it can be managed; knowledge-based uncertainties can be quantified, reduced, and potentially removed; whilst those that reflect the randomness of natural processes can only be quantified (Kelly and Campbell, 2000). Finally, the level of uncertainty informs selection of an appropriate management technique (Refsgaard et al., 2007). In order to manage uncertainty effectively, it is essential that reasonable attempts to identify all dimensions are made (Walker et al., 2003, Janssen et al., 2003, Refsgaard et al., 2007 and Knol et al., 2009).

The identification of uncertainties relies on risk analysts compiling lists of potential uncertainties with their analyses. These commonly-termed uncertainty typologies aim to define and communicate the important features of uncertainty. When complete, typologies provide reliable characterisations of uncertainties for ERAs. However, the differing abilities and experiences of ERA practitioners results in typologies being used inconsistently (Gillund et al., 2008 and Knol et al., 2009). Some typologies deploy terminology that is contradictory; communicate frequencies and dimensions of uncertainties that can be unclear; and they can deploy source information from limited data sets and that cannot be applied, on an individual basis, to ERAs to characterise the wide range of uncertainties (Skinner et al., 2014a). Critically, an approach for describing uncertainty comprehensively – with specific guidance regarding the identification of uncertainties in ERAs than is found in a typology – is lacking (Sigel et al., 2010). This paper introduces such an approach, based on validated maps of the risk assessment process as a decision system, supported by structured expert elicitations. Here, the approach is illustrated through application to the domain of agricultural pesticide risks to surface water organisms, a subject that has attracted the wide attention of environmental risk assessors.

METHODS

Overview

The methodological process began by creating and validating a template (a risk assessment process map) containing the

Citation: Daniel J.C. Skinner, Sophie A. Rocks, Simon J.T. Pollard, Environmental risk assessments, Science of The Total Environment, Volume 572, 1 December 2016, Pages 23-33, ISSN 0048-9697, http://dx.doi.org/10.1016/j.scitotenv.2016.07.164.

key stages of a generic ERA. For the topic of agricultural pesticide risk to surface water organisms (the risk 'domain'), our 'case study', an evidence base of ERAs was assembled from the prior art from which a dominant risk relationship was selected as the focus. A case study version of the ERA template was created, using information from the gathered ERAs, and validated. Following this, an elicitation was designed using the information within the generic and case study ERA templates, and completed using case study experts selected from their publication of relevant ERAs. The elicitations were combined to describe the levels, natures, and locations of uncertainties within ERAs for the case study. The method is explored in detail below.

Generic ERA template

An examination of peer-reviewed and regulatory literature allowed collation of the important features of ERAs, constructed for this study as a generic ERA template, version 1 – in short, a generalised risk assessment framework comprised of the accepted stages of analysis. Validation of this template was performed in two rounds using the opinions of ERA experts who were asked whether they recognised and could confirm the key stages of an ERA as expressed by the generic process in the template. Experts were sourced using Scopus™, which covers a wide journal range (including academic and industrial trade journals; Falagas et al., 2008), using search terms joined by Boolean connectors '*risk assessment* (in article title) AND *ecological* OR *environmental* OR*human* (in keywords)'. Results were limited to the last 10 years, so to identify experts active in ERA. The location of the experts was not defined, allowing worldwide input. In-built filtering within Scopus™ was used to remove non-relevant literature, with further filtering to ensure the returned sources related to ERAs, performed using the information within titles and abstracts. Filtered records were ordered by decreasing citation count, and the contact details of the first 2000 records exported. This threshold was set to allow for a 50% redundancy in records (due to outdated contact details, duplicates, or retired individuals), and a 5–10% response rate from the remainder (Speirs-Bridge et al., 2010), resulting in 50–100 responses. The email addresses of the corresponding authors were extracted from each record, and duplicate addresses removed.

For the first round of validation, the generic ERA template version 1 was sent to every other contact in the compiled list (i.e. all odd entries) with explanatory information on the research and details of the validation request. Experts were asked to validate the risk assessment phases (e.g. problem formulation), sub-phases (e.g. defining the conceptual model), and individual technical tasks (e.g. identifying sources or stressors, pathways, and receptors) within the generic template. A response time of four weeks was offered, after which the views of the experts were collated, and alterations to the generic ERA template, version 1, made, for suggestions where two or more experts agreed. This exercise yielded the generic ERA template, version 2. The procedure was repeated for a second round of validation, with the generic ERA template, version 2 sent to the remaining experts in the contacts list (i.e. all even entries). Completion of the second validation yielded generic ERA template, version 3, which was then used as the structure for the case study ERA template. To an extent, version 3 of the generic ERA template represents a practitioner consensus on the phases, sub-phases and individual tasks of an environmental or ecological risk assessment.

Case study ERA template

Next, a domain-specific version of the validated generic ERA template version 3 was constructed for the case study. A new evidence base of peer-reviewed ERAs was compiled from Scopus™, using the search term '*risk assessment* (in keywords) and *pesticide* (in keywords) and *water* (in keywords)'. Results were not restricted temporally (e.g. the last 10 years), but assessed for relevance using the procedure described above. The returned ERAs were analysed, with all risk relationships (connections of sources, pathways, and receptors; S-P-Rs; Pollard et al., 2002 and Pollard et al., 2006) recorded. A single S-P-R risk relationship was selected on the basis of the most frequently occurring set of S-P-Rs, and the evidence base updated to contain the corresponding ERAs only. For this case study, the risk relationship explored was the chemical risk (consequence) posed to surface water macroinvertebrates (receptor) as expressed by direct exposure (pathway) to agricultural pesticides (source). Information within the ERAs was used to populate the generic ERA template version 3, so creating a case study ERA template for the specific risk relationship. Following validation, information within the case study ERA template was used to devise an uncertainty-based expert elicitation.

Uncertainty-based expert elicitation

Elicitations were performed (Slottje et al., 2008, United States Environmental Protection Agency, 2009 and Knol et al., 2010) following a seven-step procedure derived from published methodologies (Knol et al., 2010).

Step 1: characterisation of uncertainties

The elicitation used a characterisation of uncertainties based on 171 peer-reviewed ERAs (Skinner et al., 2014a) which consisted of three types of uncertainty within the nature dimension (epistemic, aleatory, and combined) and seven types of uncertainty across the location dimension (data, language, system, variability, extrapolation, model, and decision). Sub-types of location-based uncertainties (Skinner et al., 2014b) were not included. The level of uncertainty was expressed as a range of integers (where zero represents a deterministic understanding of the uncertainty and 10 represents total ignorance to it), consistent with Krayer von Krauss et al. (2004).

Step 2: scope and format of the elicitation

Elicitations deployed a questionnaire implemented in Microsoft Excel 2007 (Microsoft Corporation, Redmond WA) using controls and macros, distributed to experts via email, completed remotely and returned before a set deadline. Experts were approached as individuals.

Step 3: selection of experts

The experts selected to participate were the same as those invited to validate the domain-specific ERA templates, unless they stated a wish not to, and were considered to be subject-matter experts drawn from academia, industry, and regulatory agencies. The credentials of an expert who completed an elicitation were checked to ensure recent and extended involvement in the relevant domain.

Step 4: design of the elicitation protocol

The elicitation was organised according to the phases, groups of tasks, and individual tasks within the validated generic ERA template, all of which were contextualised for the case study ERA template. Experts were asked to assess four aspects for every task within each ERA phase:

1) Using tick-box controls, whether the task was deemed necessary in an ERA of the case study domain, thereby providing an extra validation of the elicitation contents. If deemed unnecessary, experts were able to move to the next task;
2) Using slide bars, the level of the uncertainty associated with performing the task (on a scale of zero to 10, as above);
3) Using tick-box controls, the nature of the uncertainty associated with performing the task (epistemic, aleatory, or combined); and
4) Using tick-box controls, the location(s) of the uncertainty associated with performing the task (data, language, system, variability, extrapolation, model, or decision).

The maximum number of tasks evaluated by an expert during the elicitation was the same as the total number of tasks contained within the validated case study ERA template.

To ensure experts understood uncertainty concepts, they were asked to complete a practice exercise prior to elicitation, based on the introduction of a DNA vaccine into aquaculture (Gillund et al., 2008). The practice task, which consisted of five questions, followed the same format as the main elicitation, so to familiarise the experts with the structure. The answers provided, relating to the level and nature of uncertainty, were compared to the 'control' set from the published elicitation (Gillund et al., 2008) in which the location of uncertainty was not assessed. Experts were considered to understand, and be able to assess the level of uncertainty (i.e. not be overly optimistic or pessimistic when faced with a scenario) if their averaged results were within ± 50% of the control group. With regard to the nature of the uncertainty, experts were expected to agree with the control group to a minimum level of 60% (i.e. three out of five questions). Provided these two criteria were met, their judgements within the corresponding elicitations were deemed valid. A thorough written or verbal communication was held with experts who failed to complete the practice exercise, to satisfy that they understood the associated uncertainty-based concepts.

Step 5: preparation of the elicitation session

The distribution of the case study ERA template to all potential experts, for the purpose of validation, ensured their proper preparation. Both the validated generic ERA and case study ERA templates were provided for experts to view as part of the introductory information within the elicitation system.

Step 6: elicitation of expert judgements

The elicitation was organised into three sections: an introduction, with an elicitation overview and background on uncertainty dimensions and the ERA process; instructions on how to complete the elicitation, including the method used to assess the levels, natures, and locations of uncertainty, as well as the pre-elicitation practice exercise; and the main elicitation, separated into the four ERA phases.

Step 7: possible aggregation and reporting

Due to the stringent selection criteria above, the responses of all experts were considered of equal importance, with equal weighting methods selected as the basis for combining results (Clemen and Winkler, 1999 and Slottje et al., 2008). Specifically, the responses for the levels of uncertainty were aggregated using measures of central tendency, with the natures and locations of uncertainty combined into occurrence rates (the following sub-section contains more information about the types of data gathered).

Data analysis

ERA tasks within the elicitations had two kinds of data associated with them: (i) the level of uncertainty (measured using slide bars) was represented through integer values in the range zero to 10; whilst (ii) the nature and location of uncertainty (measured using tick box controls) were treated as binary values. The data from completed elicitations were extracted to a spreadsheet and assigned an expert identifier. With respect to the level of uncertainty associated with each ERA task, relationships were evaluated using central tendencies, variations from central tendency, and the high-low ranges of responses. Suitable parametric (mean and standard deviation) or non-parametric (median and inter-quartile range) measures were applied, as highlighted in the results below, based on an assessment of their normality. Binary data does not need to be assessed for normality, and was considered to be non-parametric due to the small size of the datasets analysed. All binary data relating to each category within the nature and location dimensions were converted to occurrence rates (%) describing the degree to which that category was selected by experts during the elicitations.

RESULTS

Generic ERA template creation and validation

The generic ERA template, version 1 was created from several academic and grey literature sources (United States Environmental Protection Agency, 1992, United States Environmental Protection Agency, 2003, Suter, 1996, United States Environmental Protection Agency, 1998, Fairman et al., 1998, Department of Environment, Transport and the Regions/Enviro, Department of Health and Ageing, 2002, Landis, 2005, Beer, 2006, Briggs, 2008 and Department for Environment, Food and Rural Affairs, 2011). The 35 experts (21 from academia, 4 from industry, 10 from regulatory agencies) in the first validation provided 111 comments (28 in problem formulation, 22 in exposure assessment, 35 in effects assessment, and 26 in risk characterisation) on the correctness and completeness of this template; 23 of which were implemented (8 in problem formulation, 4 in exposure assessment, 7 in effects assessment, and 4 in risk characterisation. Thirteen experts (9 from academia, 2 from industry, 2 from regulatory agencies) in the second validation provided 37 comments (11 in problem formulation, 10 in exposure assessment, 11 in effects assessment, and 5 in risk characterisation), 3 of which were further implemented (1 in problem formulation, 1 in exposure assessment, 0 in effects assessment, and 1 in risk characterisation. Implementation resulted in generic ERA template, version 3 (Fig. 1); organised by its phases, sub-phases, groups of tasks, and individual tasks (Table 1).

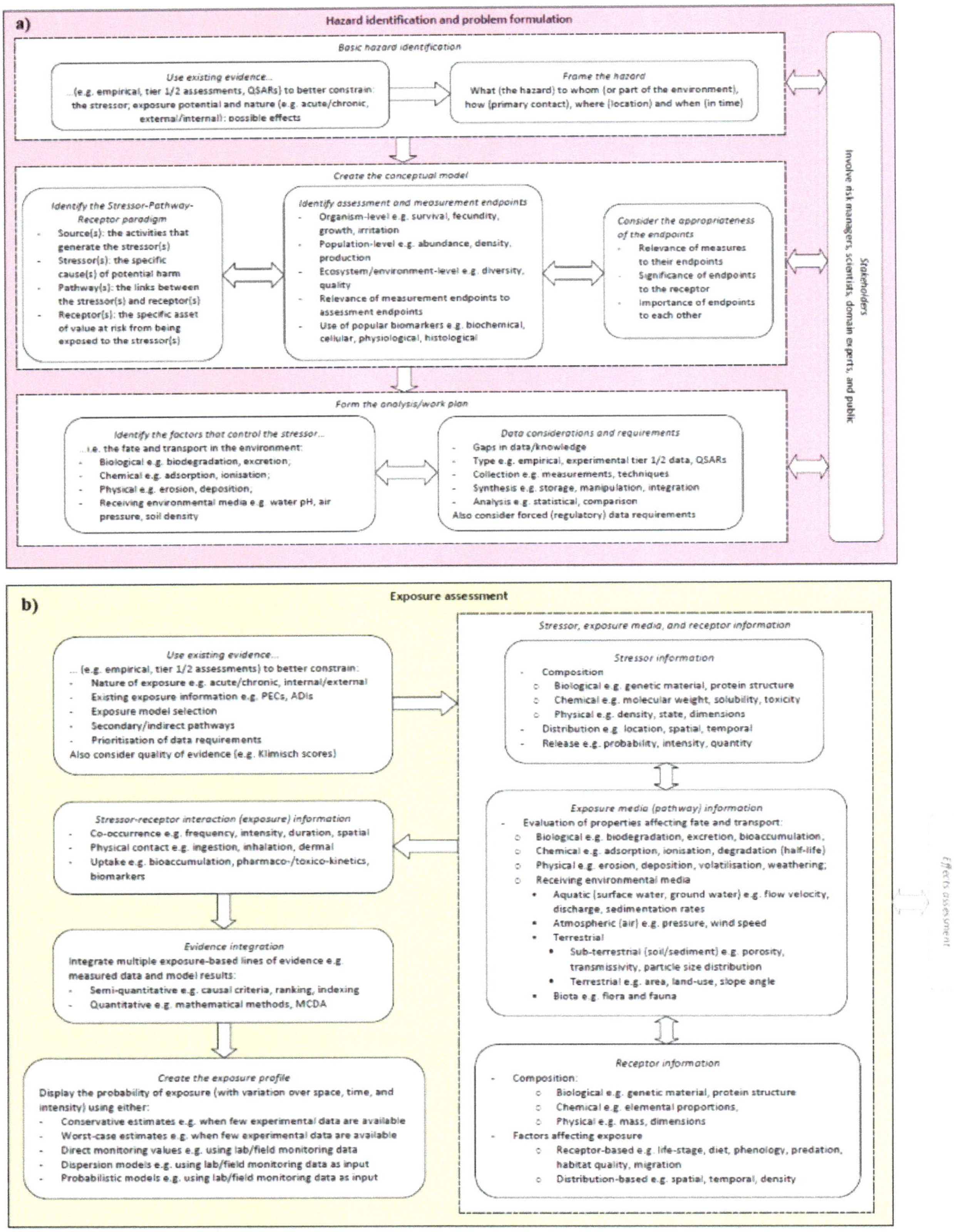

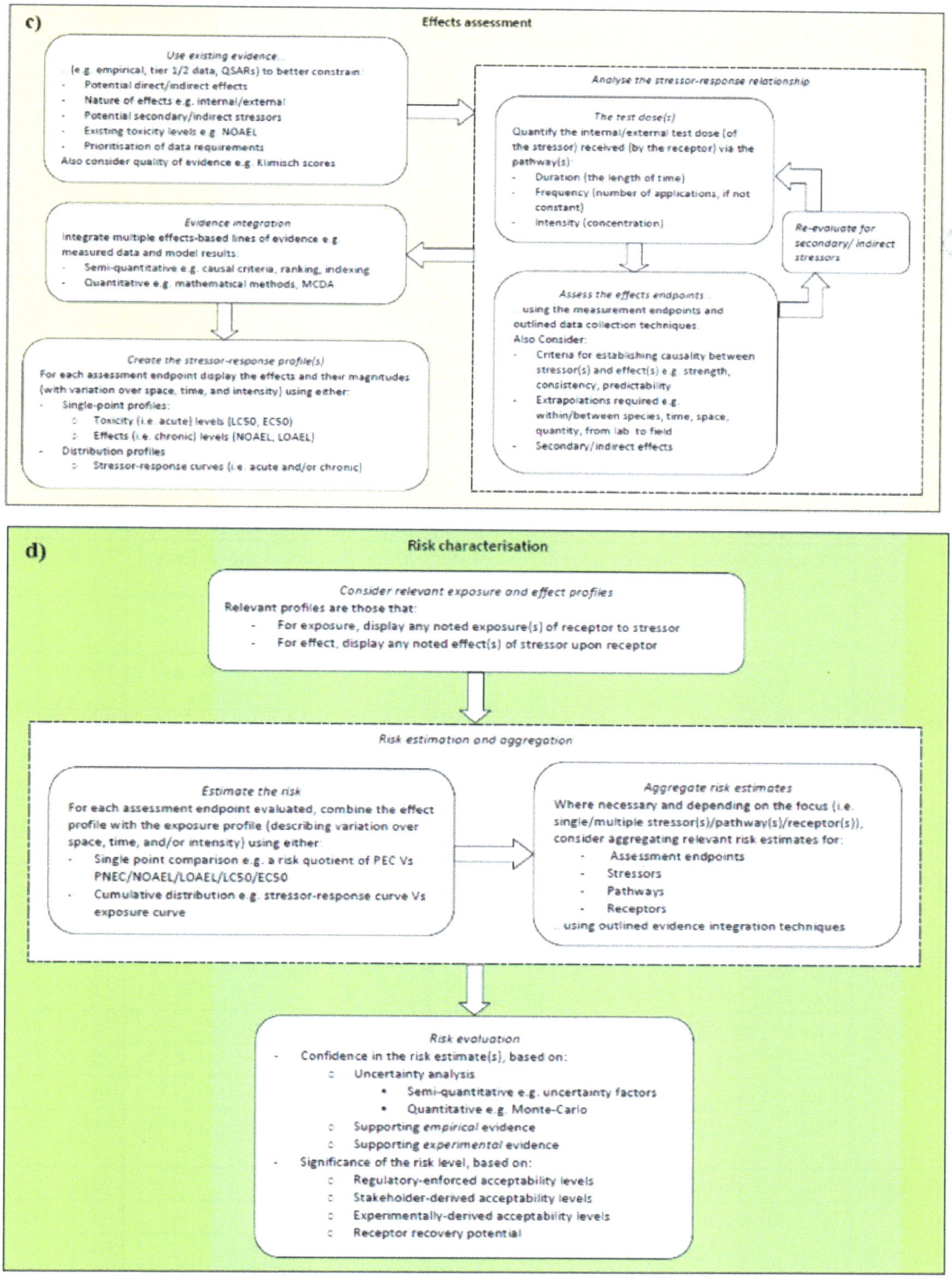

Figure 1: The generic ERA template, version 3, created through the expert validation of versions 1 and 2, describing the important aspects within the phases of: a) hazard identification and problem formulation; b) exposure assessment; c) effects assessment; and d) risk characterisation.

Table 1: Summary of the 105 ERA tasks in the generic ERA template, version 3, organised by ERA phase, sub-phase, and task-group, to be potentially included in the case study expert elicitation exercises (individual tasks are listed in Table S1).

ERA phase	ERA sub-phase	ERA task group	ERA task number
Problem formulation	Preliminary hazard identification	1. Use available evidence to better constrain…	1–4
		2. Framing the hazard	5–9
	Define the conceptual model	3. Identify the S-P-R paradigm, including…	10–13
		4. Choose assessment and measurement endpoints	14–21
		5. Consider the appropriateness of the endpoints	22–24
	Form the analysis/work plan	6. Identify the factors controlling fate and transport of the stressor	25–28
		7. Identify data considerations	29–32
Exposure assessment	Use available evidence to better constrain…	8. (Use available evidence to better constrain…)	33–37
	Stressor, exposure media, and receptor information	9. Collect information about the stressor's composition	38–40
		10. Collect information about the stressor's distribution	41–42
		11. Collect information about the stressor's release	43–45
		12. Collect information about properties affecting fate and transport	46–53
		13. Collect information about the receptor	54–57
	Evaluate stressor-receptor contact	14. Evaluate co-occurrence for…	58–60
		15. Evaluate…	61–62
	Integrate multiple LOEs using…	16. (Integrate multiple LOEs using…)	63–64
	Create the exposure profile(s) using…	17. (Create the exposure profile(s) using…)	65–69
Effects assessment	Use available evidence to better constrain…	18. (Use available evidence to better constrain…)	70–74
	Analyse the stressor-response relationship	19. Determine the test dose for the…	75–77
		20. Assess effect endpoints	78–85
	Integrate multiple LOEs using…	21. (Integrate multiple LOEs using…)	86–87
	Create stressor-response profile using…	22. Single point or distribution methods showing…	88–91
Risk characterisation	Select relevant profiles…	23. (Select relevant profiles…)	92–93
	Estimate and aggregate risk	24. Estimate risk using…	94–95
		25. Aggregate risk estimates for…	96–99
	Evaluate risk levels	26. Assess confidence in the risk levels using…	100–101
		27. Assess the significance of the risk levels using…	102–105

Case study risk relationship (S-P-R)

The literature searches, after filtering and relevance-checking, returned 127 peer-reviewed articles, reduced on the basis of missing information within the articles to a pesticides evidence base of 49 ERA articles. This was analysed for risk relationships, the most frequent of which (n = 5) were 'potential agricultural chemical pesticide risk to surface water macroinvertebrates' and 'potential agricultural chemical pesticide risk to surface water quality'. With dominant parameters described for the source categories (agricultural and chemical, respectively), but less so for the receptor category (macroinvertebrates and water quality, respectively), several receptor parameters were combined (multiple organisms, algae, crustaceans, and macroinvertebrates), establishing a defined risk relationship (n = 13; Cuppen et al., 2000, Mastin and Rodgers, 2000, Palma et al., 2004, van Wijngaarden et al., 2004, Wan et al., 2006, Schuler and Rand, 2008, Siemering et al., 2008, van den Brink et al., 2009, Vryzas et al., 2009,Burgert et al., 2011, Damásio et al., 2011, Guy et al., 2011 and Vryzas et al., 2011) of *'potential agricultural chemical pesticide risk to surface water organisms'*, which was as specific as defensible and selected as the risk relationship for the case study. This allowed the detailed examinations of uncertainties for this case study ERA.

Case study ERA template creation and validation

The generic ERA template version 3 was populated with relevant information from 13 peer-reviewed articles to form the agricultural chemical pesticide risk to surface water organisms ERA template, version 1. Experts (n = 22) provided 119 comments (46 in problem formulation, 37 in exposure assessment, 20 in effects assessment, and 16 in risk characterisation) regarding the correctness and completeness of this template, 32 of which were implemented (14 in problem formulation, 7 in exposure assessment, 8 in effects assessment, and 3 in risk characterisation), enabling the creation of the case study ERA template, version 2. Experts were from academia (n = 15), industry (n = 3), and regulation (n = 4), and resided in Argentina (n = 1), Belgium (n = 1), Canada (n = 4), China (n = 1), Denmark (n = 1), France (n = 2), Netherlands (n = 5), Portugal (n = 1), Serbia and Montenegro (n = 1), Switzerland (n = 1) and the USA (n = 4).

Case study expert elicitation exercise

Nine experts participated in the elicitation, assessing 102 ERA-based tasks (31 in problem formulation, 36 in exposure assessment, 21 in effects assessment, and 14 in risk characterisation) for the levels, natures and locations of uncertainty. Experts were drawn from academia (n = 3), industry (n = 1), and regulation (n = 5), and resided in Canada (n = 2), France (n = 1), Greece (n = 1), Netherlands (n = 2), Spain (n = 1), Switzerland (n = 1) and the UK (n = 1). ERA tasks 16, 36 and 80 from the generic ERA template were not included since they did not appear in case study ERA template, version 2. Data relating to the level dimension were treated as non-Gaussian after assessment of the mean, median, and mode values; central tendency and spread were measured using median values and inter-quartile ranges (IQRs). Shapiro-Wilk normality test (Shapiro and Wilk, 1965) was applied, where significance values for 16 out of the 102 datasets evaluated (i.e. the ERA tasks) fell below the 0.05 significance threshold, confirming a non-Gaussian dataset. The median level of uncertainty across the tasks in problem formulation was 3.0, in the range of statistical uncertainty (Fig. 2a). The group of tasks considering the appropriateness of endpoints had the highest median level of any group in this phase (6.0; group 5). Experts agreed to a moderate extent, with a median IQR of 3.0 across all tasks in this phase. The nature of uncertainty was deemed to consist of a combined epistemic and aleatory category, with a median occurrence rate of 67% across the tasks in this phase (Table 2). However, there were six tasks for which no dominant nature was apportioned, due to occurrence rates of below 50%. Four locations of uncertainty had median occurrence rates of at least 50% in this phase, namely data (67%), system, variability and extrapolation (all 56%; Table 2).

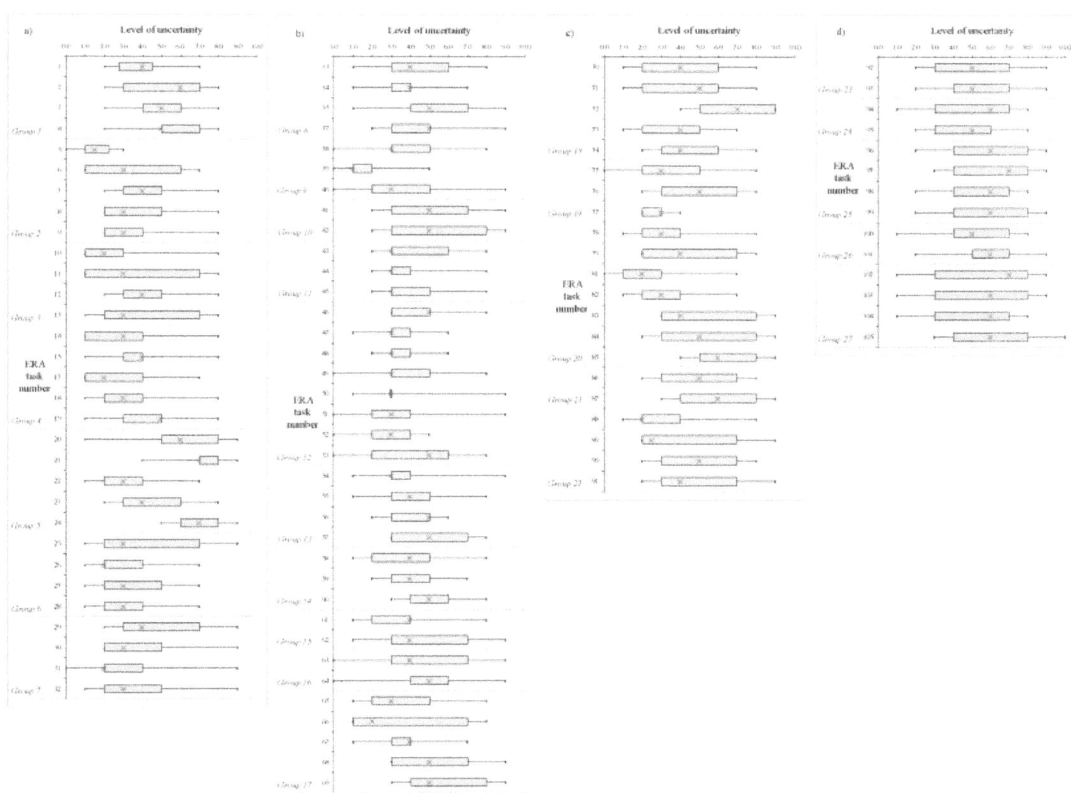

Figure 2: The level of uncertainty communicated by the experts (n = 9) across the 102 assessed ERA tasks, organised into the ERA phases of a) problem formulation, b) exposure assessment, c) effects assessment, and d) risk characterisation, and described

using median values (red crosses), inter-quartile ranges (boxes), and low-high range values (dashed lines) on a 0 (representing determinism) to 10 (representing total ignorance) scale. The ERA tasks are separated into the groups listed in Table 1. (For interpretation of the references to colour in this figure legend, the reader is referred to the web version of this article.)

Table 2: Median occurrence rates (%) for the individual natures and locations of uncertainty provided by experts (n = 9), organised by ERA phase and showing the highest rates (of at least 50%) for the nature (shaded blue) and location (shaded red) of uncertainty associated with each ERA phase.

ERA phase	Nature of uncertainty (%)			Location of uncertainty (%)						
	Epistemic	Aleatory	Combined	Data	Language	System	Variability	Extrapolation	Model	Decision
Problem formulation median	22	11	67	67	11	56	56	56	44	11
Problem formulation mode	22	0	67	67	0	67	56	56	56	11
Exposure assessment median	33	11	56	67	0	3	50	39	22	11
Exposure assessment mode	33	11	44	56	0	33	56	44	22	11
Effects assessment median	22	11	67	67	0	44	56	33	44	0
Effects assessment mode	11	0	67	67	0	56	67	33	56	0
Risk characterisation median	0	11	89	39	0	56	56	78	78	33
Risk characterisation mode	0	11	89	33	0	44	67	78	78	33
Overall median	22	11	67	67	0	44	56	44	44	11
Overall mode	22	11	67	67	0	33	56	56	44	11

The median level of uncertainty in exposure assessment was 4.0, in the range of scenario uncertainty (Fig. 2b). The set of tasks with the highest median level involved collecting information about the distribution of the source term (5.0; group 10), whilst several groups contained similarly low median levels (3.0; groups 9, 11 and 12). Experts showed a high degree of agreement in this phase, with an overall median IQR of 2.0. Half of the 32 assessed ERA tasks in exposure assessment did not have a primary nature of uncertainty associated with them, with the other 16 tasks associated with the combined category (Table 2). The phase median for the latter was 56%, the lowest for the combined category in any phase of the case study. Data uncertainty was the most frequent location based uncertainty, with a median occurrence rate of 67% across the tasks in this phase (Table 2). The high level of uncertainty communicated above for group 10 was shown here to manifest primarily through data-based uncertainty, which had a particularly high median rate of 100%.

Effects assessment had a median level of uncertainty of 4.0, in the range of scenario uncertainty (Fig. 2c). Group 22, involving the creation of stressor-response profiles, contained the lowest median levels of uncertainty (3.3). Experts showed large disagreement about the values of the individual tasks, with a median IQR of 5.0. This phase was again dominated by the combined nature category, with a median occurrence rate of 67% (Table 2). However, task 70, using available evidence at the beginning of the phase to better constrain the potential effects of the source term on the receptor, contained the only example of one of the other two categories occurring more frequently. As for the previous two phases in this case study, effects assessment consisted primarily of data uncertainty, with a median occurrence rate of 67% (Table 2). The uncertainty was also found to exist in the form of variability, but to a lesser extent (56%). Risk characterisation contained the highest median level of uncertainty of the four phases (6.0; Fig. 2d), although was still in the range of scenario uncertainty. Specifically, groups 25, aggregating risk estimates, and 27, assessing the significance of the risk levels, had the highest associated values, of 6.0. The median levels across the 14 tasks all fell within the range of 5.0 to 7.0, but despite this, the median IQR for this phase was 3.0, highlighting the variation in responses. All 14 tasks in this phase were deemed to consist of the combined category, with an overall median rate of 89% (Table 2). The locations of extrapolation and model uncertainty were the most frequent, with median levels of 78% across this phase (Table 2), making risk characterisation the only phase in which data uncertainty was not the primary location-based concern.

The overall case study median level of uncertainty was 4.0, in the range of scenario uncertainty. Experts consistently communicated that the uncertainty seen was both epistemic and aleatory in nature. Data was the highest location-based uncertainty with a median rate of 67%.

DISCUSSION

Uncertainty within an ERA in the case study domain

Case study selection

Agricultural chemical pesticides, including insecticides, herbicides, and fungicides, have proven benefits including improved yield, quality, nutritional value, and cosmetic appearance (Damalas and Eleftherohorinos, 2011). However, the persistence and transport of chemical pesticides can cause wide-ranging adverse effects across different environmental compartments (Chèvre et al., 2006). Regulatory authorities, such as the United States Environmental Protection Agency (US EPA) and the Chemicals Regulation Directorate in the UK, recognise the potential risks of chemicals in pesticide formulations, and oversee strict licensing processes for which environmental risk assessments are key to informing risk management and regulatory control (Schwarzman and Wilson, 2009). In the context of this research, a suitable risk domain is one which: has a large amount of associated empirical evidence; where ERAs are prevalent throughout and where environmental uncertainty is present.

Uncertainty across the phases of an ERA in the case study domain

We are interested in where uncertainties are distributed in ERAs and their extent because decisions on whether to manage risk must rely on risk estimates with which decision-makers have confidence. Decisions about whether to refine the ERA further, or act now based on a knowledge of attending uncertainty, depend on knowing where uncertainties reside in the risk assessment process, how large they are, and whether they are resolvable or not. In turn, this influences the degree of precaution decision-makers are prepared to tolerate. The median levels of uncertainty communicated across the ERA phases for the case study were found to lie in the range of statistical or scenario uncertainty. However, these median values had relatively high IQRs. A high IQR is the result of a stretched distribution resulting from disparate values (Manikandan, 2011), which equates here to disagreement amongst experts. A potential explanation for disagreement is the risk relationship selected as the focus, into which several potential types of agricultural pesticide, multiple environmental pathways, and a multitude of aquatic organisms could still feasibly fit. Experts may be drawing from slightly different interpretations of the risk relationship, basing their responses on past experiences and knowledge as well as on the information presented (Knol et al., 2010); therefore high IQRs may also be the result of disparate experiences. However, they may also result from inherently different attitudes to assessing uncertainty, since the experts involved in this case study provided varied responses to the level-based questions during the practice exercise, with an average variance of 27% to the control group.

Regarding the location of uncertainty, of note is the extrapolation category which returned high median values, especially during risk characterisation (Table 2). Several articles from the pesticides literature discuss the concerns of extrapolating from, for example, predicted no-effect concentrations (PNECs) and no observed adverse effect levels (NOAELs) to expressions of risk (primarily quotients; Palma et al., 2004 and Chèvre et al., 2008), as well as basing PNECs and NOAELs on questionable data during effects assessment (Uricchio et al., 2004). For more specific observations to be made, these patterns must be investigated at the ERA task level.

Uncertainty across the tasks of an ERA in the case study domain

Framing uncertainties in the context of specific ERA tasks not only provides more granularity than at the ERA phase level, but allows for more specific and targeted guidance on the selection and implementation of potential techniques to manage the uncertainty. The ERA tasks with the lowest levels of uncertainty (Table S1, presented in Supplementary Material) predominantly reside within the problem formulation phase; just under 50% of the 20 tasks with the lowest levels were found elsewhere. The sub-phases of defining the conceptual model (with four entries) and forming the analysis/work plan (with five entries) are areas in which experts had most confidence. The lowest levels of uncertainty seen in the case study were 1.0. According to the experts, uncertainty was present in all 102 tasks and should be considered throughout. The natures of these low-level uncertainties cannot be described for eight of the 20 tasks with the lowest levels, since their occurrence rates were below 50% (i.e. they occur less frequently than they occur), similarly for just one of the 20 locations (task 10). Of the locations that do feature, data uncertainty dominates with an occurrence rate of above 50% in 15 of these 20 tasks. These observations demonstrate the difficulty in providing detailed descriptions of uncertainty for tasks where levels and occurrence rates are low.

Of the 20 ERA tasks with the highest median levels of uncertainty (Table S1, presented in Supplementary Material), 50% are found in the risk characterisation phase, despite its smaller size (n = 14 tasks) compared to the other phases, with seven in problem formulation, and three in effects assessment. None of these 'most uncertain' tasks occur in the exposure assessment phase. The two tasks with the highest level of uncertainty appeared in the problem formulation phase, for this case the inclusion of the primary production and nutrient cycling endpoints (task 21), and the evaluation of the relative importance of the endpoints to each other (task 24). The former is difficult to measure and often omitted from ERAs (Barnthouse, 2008), and the latter reflects the large number of endpoints for consideration (n = 7). These early-stage uncertainties are the exceptions, with tasks from risk characterisation more commonplace, such as aggregating risk estimates for multiple stressors (task 97) and pathways (task 98) and assessing the significance of these risk estimates using thresholds derived through experimentation (task 101) or regulation (task 102).

Across the 20 tasks with the highest median levels of uncertainty, the nature of uncertainty was exclusively communicated as being epistemic and aleatory combined, and there was representation from five of the seven locations of uncertainty, with language and decision not appearing. These results show that where uncertainty is present, especially at the 'deep' levels (i.e. recognised ignorance) seen in these tasks, it is essential that all aspects of uncertainty are considered and potentially managed (Walker et al., 2003, Kandlikar et al., 2007, Refsgaard et al., 2007 and Knol et al., 2009).

Increased levels, natures, and locations of uncertainty allow for informative descriptions and management guidance to be provided, but also highlight the extent of the uncertainty for the analyst to manage. The results from the case study may be considered useful for analysts based in this specific research domain, but there are some potentially influential aspects that should be considered.

Potentially influential methodological aspects

The method of creating and validating (risk assessment) process maps as the basis for expert engagement, so to elicit views about risks and uncertainties, is gaining in popularity within the risk research community (Krayer von Krauss et al., 2004, Gillund et al., 2008, Krayer von Krauss et al., 2008, Ravnum et al., 2012, Smita et al., 2012 and Zimmer et al., 2012). Here, every effort was made to ensure the templates were as complete and correct as possible, using amassed evidence, before being distributed to experts for validation. However, in sourcing reference materials, there is the chance of an influential source being overlooked. The majority of validation suggestions corresponded to domain-specific terminology and across the templates the number of alterations made was small.

The scope and format of expert elicitations can be greatly influenced by the resources available to the researcher (Knol et al., 2010). In this case, it was beneficial to conduct remote elicitations. Remotely-executed elicitations have benefits over face-to-face formats, including that they are less expensive, their content and structure can be standardised easily, and experts can complete them at their leisure (United States Environmental Protection Agency, 2009 and Knol et al., 2010). With a remote-elicitation format, experts were approached as individuals rather than as groups; hence, the views of experts were considered to be theirs alone and not that of a collective, as is the case with Delphi methods, for example. The types of experts included was also a consideration. Three types of experts are noted: *generalists*, who have substantial knowledge in a discipline connected to the elicitation, and a solid understanding of the elicitation context itself; *subject-matter experts*, who hold detailed knowledge in the subject of the elicitation and are considered by their peers as an authority in the field; and *normative experts*, who have knowledge or experience that can aid the elicitation process, such as statistical or psychoanalytical skills (Kotra et al., 1996, Loveridge, 2004 and Knol et al., 2010). Experts participating in the elicitation process were required to be subject-matter experts with appropriate levels of domain-expertise, in order for a range of balanced and valid opinions to be communicated (US EPA, 2009).

The phrasing of questions and the language used can significantly affect the responses of experts (Meyer and Booker, 1991). To avoid this influence, all questions (namely the ERA tasks) within the elicitations were worded in a consistent manner, using the validated common terminology within the generic and case study ERA templates. Furthermore, potential bias in responses was reduced by providing the experts with personal and professional anonymity throughout. Suggested participant numbers vary depending on the subject, format and budget of the elicitation, and the availability and willingness of experts to participate (US EPA, 2009). The inclusion of a minimum of five or six experts is considered sufficient to cover the breadth of scientific opinion on a given topic, with little benefit in including additional experts beyond 12 (Clemen and Winkler, 1999 and Cooke and Probst, 2006). This research aimed for participant numbers of between five and 12, in order for results to be considered representative.

Typically, tests such as Shapiro and Wilk (1965) or Kolmogorov-Smirnov (Daniel, 1990) are used to determine whether a distribution can be classified as normal (Gaussian), and by extension as parametric or non-parametric (Rees, 2000). However, these tests can be unreliable when performed on small datasets (of less than five and seven values for Kolmogorov-Smirnov and Shapiro-Wilk, respectively), since outliers can easily skew otherwise consistently-distributed data. Furthermore, they are not designed for datasets with high frequencies of duplicate values, as seen here. In such circumstances, a more appropriate method, adopted in this research, is to compare the mean, median, and mode, where consistently similar values for each denote normally-distributed data (McCluskey and Lalkhen, 2007). Whilst these constraints have a potential to affect the approach described in this research, they should not be considered prohibitive to repeated executions within the case study domain, or elsewhere. Having noted these aspects, it is also appropriate to discuss the potential merits of the approach.

On the identification of uncertainties in ERAs

The primary method for identifying uncertainties in ERAs is application of one or more uncertainty typologies by risk analysts (Morgan et al., 1990, van Asselt and Rotmans, 2002 and Knol et al., 2009). The differing abilities and experiences of ERA practitioners results in these typologies being used inconsistently (Gillund et al., 2008 and Knol et al., 2009), which allows uncertainties in ERAs to go unidentified (European Environment Agency, 2007, Hart et al., 2007 and Dale et al., 2008) and sometimes, arguably, treated in a rather token fashion. Some attempts have been made to turn the traditional typology format (i.e. a list of organised uncertainties and their related definitions) into a more useful tool for uncertainty identification. For example, Walker et al. (2003) introduced an uncertainty matrix, reproduced by Janssen et al. (2003), that contained blank sections the analyst was encouraged to complete using qualitative or quantitative information relevant to the system under study. However, in this approach it was still the sole responsibility of the analyst to locate the defined uncertainties, since no specific system-related guidance was provided. The matrix approach was extended to include aspects of systems in which the different dimensions of uncertainty may have been present (e.g. data uncertainty relating to measures of population exposure; Refsgaard et al., 2007 and Knol et al., 2009). However, these aspects, drawn from the domains of environmental modelling and burden of disease assessments were not representative of the full range of potential aspects that would require consideration in such systems. Furthermore, analysts conducting ERAs were not significantly aided by this guidance, due to the limited cross-over between the research domains, and were in a similar position as when equipped with a typology or blank uncertainty matrix. More guidance was offered by Knol et al. (2009) who suggested their uncertainty typology should be used to:

1) identify sources of uncertainty, by either:

a. analysing each step of the assessment and relating uncertainties from the typology to those steps; or

b. considering the uncertainties in the typology and explaining where in the assessment these uncertainties may occur;

2) prioritise each uncertain element within the assessment according to their relative importance; and

3) select one or more suitable uncertainty management techniques (UMTs) to further analyse the identified uncertainties.

Thus, should the analyst have been capable of using the typology or matrix to identify all relevant uncertainties within the system under study, they would still be required to prioritise those uncertainties – according, for example, to a chosen significance metric, or temporal and financial restrictions – before one or more UMTs could be implemented. Guidance related to the selection of one or more UMTs, as noted by Knol et al. (2009), is sufficiently good. If an analyst requires guidance on managing identified uncertainties, it is reasonable to assume that they also require guidance on identifying those uncertainties in the first place. Up to now, this guidance has been lacking, and uncertainty identification has remained a weakness of the uncertainty management process.

Our method, results and observations equip environmental risk analysts with guidance on the locations, natures and levels of uncertainties that are associated with all aspects of the ERA process, for the case study domain. The observations from this research reduce the reliance on the analyst to identify and prioritise uncertainties before they can be managed. The researchers are in the process of applying the method to more case study domains in order to aggregate results and create a generic system for identifying uncertainties within ERAs. This will make the results and associated observations relevant to a larger proportion of environmental risk analysts, when compared to the subset involved in assessing potential agricultural chemical pesticide risk to surface water organisms.

CONCLUSIONS

A reliable characterisation of potential uncertainties is critical for ERAs. Without this, decisions about refining a risk assessment, or acting on the available evidence now, are difficult to make. In the extreme, paralysis over this dilemma could delay time-critical risk management efforts and exacerbate harm to the environment. However, the typologies for uncertainties currently in use can be implemented inconsistently. There is a requirement for an approach which both comprehensively describes uncertainty and offers specific guidance regarding the identification and prioritisation of uncertainties in ERAs.

(i) This research developed an approach based on the results of nine structured elicitations, for agricultural chemical pesticide risks to surface water organisms, in which subject matter experts validated aspects of ERA templates for three dimensions of uncertainty: level; nature; and location. The elicitations describe the uncertainty associated with 102 distinct tasks across four phases of an ERA in this domain; 31 in problem formulation, 36 in exposure assessment, 21 in effects assessment, and 14 in risk characterisation.

(ii) The risk characterisation phase contained the highest median level of uncertainty of 6.0 (on a scale from deterministic understanding of the uncertainty at 0.0, to total ignorance of the uncertainty at 10.0), which was specifically associated with estimating (tasks 94 and 95), aggregating (tasks 96 to 99) and evaluating (tasks 100 to 105) risk estimates. Exposure assessment and effects assessment contained the joint-second highest median level at 4.0, whilst problem formulation returned the lowest median level of uncertainty at 3.0. The median nature of uncertainty across the 102 ERA tasks was almost exclusively a simultaneous combination of epistemic and aleatory. Regarding the locations in which uncertainty was manifest, data uncertainty was dominant in problem formulation, exposure assessment and effects assessment, and had median occurrence rates of at least 50% in 77 out of 102 tasks, followed by variability (54 out of 102), extrapolation (47), system (43), and model (39), with extrapolation and model the dominant locations during risk characterisation.

(iii) This comprehensive description of uncertainty enables risk analysts based in the case study domain to prioritise ERA phases, tasks, and groups of tasks according to either the highest levels of uncertainty (the level dimension), the potential for the uncertainty to be reduced or only quantified (the nature dimension), or the associated types of location-based uncertainty (the location dimension), which further allows one or more appropriate tools for managing the uncertainty to be selected.

The following are the supplementary data related to this article.

Table S1: The case study ERA tasks organised in order of descending median level of uncertainty, with accompanying ranked occurrence proportions (for median values of at least 50%) for the associated nature and locations of uncertainty (see Section 3.4).

ERA Task #	ERA Phase	ERA sub-phase	ERA task group	ERA task	Level[a]	Nature[b]	Location(s)[c]
21	Problem	Define the conceptual model	4. Choose assessment and measurement endpoints	Ecosystem: primary prod. and nutrient cycling	7.0 (Ig)	Co	1: Sys; 2: Dat, Mod; 3: Var, Ext;
24	Problem	Define the conceptual model	5. Consider the appropriateness of the endpoints	Relative importance of endpoints to each other	7.0 (Ig)	Co	1: Ext; 2: Sys, Mod; 3: Var;
72	Effects	Use available evidence to better constrain…	18. (Use available evidence to better constrain…)	Secondary stressors	7.0 (Ig)	Co	1: Dat; 2: Sys, Mod;
97	Risk	Estimate and aggregate risk	25. Aggregate risk estimates for…	Stressors	7.0 (Ig)	Co	1: Ext; 2: Sys;

102	Risk	Evaluate risk levels	27. Assess the significance of the risk levels using…	Regulatory levels	7.0 (Ig)	Co	1: Ext; 2: Sys;
101	Risk	Evaluate risk levels	26. Assess confidence in the risk levels using…	Experimental evidence	6.0 (Sc)	Co	1: Ext;
20	Problem	Define the conceptual model	4. Choose assessment and measurement endpoints	Population: Biodiversity	6.0 (Sc)	Co	1: Dat; 2: Sys, Mod; 3: Var, Ext;
85	Effects	Analyse the stressor-response relationship	20. Assess effect endpoints	Ecosystem: primary prod. and nutrient cycling	6.0 (Sc)	Co	1: Sys; 2: Dat, Ext; 3: Var;
98	Risk	Estimate and aggregate risk	25. Aggregate risk estimates for…	Pathways	6.0 (Sc)	Co	1: Sys, Mod; 2: Ext;
2	Problem	1. Preliminary hazard identification	1.a Use available evidence to better constrain…	Potential receptors	6.0 (Sc)	Co	1: Dat; 2: Var; 3: Sys;
87	Effects	Integrate multiple LOEs using…	21. (Integrate multiple LOEs using…)	Quantitative methods	6.0 (Sc)	Co	1: Mod; 2: Dat, Var; 3: Sys, Ext;
94	Risk	2. Estimate and aggregate risk	2.a Estimate risk using…	Single-point profiles	6.0 (Sc)	Co	1: Ext, Mod; 2: Sys;
96	Risk	2. Estimate and aggregate risk	2.b Aggregate risk estimates for…	Assessment endpoints	6.0 (Sc)	Co	1: Mod; 2: Sys, Ext;
99	Risk	Estimate and aggregate risk	25. Aggregate risk estimates for…	Receptors	6.0 (Sc)	Co	1: Ext; 2: Sys, Mod; 3: Var;
104	Risk	3. Evaluate risk levels	3.b Assess the significance of the risk levels using…	Experimental levels	6.0 (Sc)	Co	1: Ext, Mod; 2: Var;
105	Risk	3. Evaluate risk levels	3.b Assess the significance of the risk levels using…	Receptor recovery potential	6.0 (Sc)	Co	1: Mod; 2: Sys, Ext; 3: Dat, Var;
103	Risk	3. Evaluate risk levels	3.b Assess the significance of the risk levels using…	Stakeholder levels	6.0 (Sc)	Co	1: Ext; 2: Sys, Mod;
3	Problem	1. Preliminary hazard identification	1.a Use available evidence to better constrain…	Potential exposure	5.0 (Sc)	Co	1: Sys, Var; 2: Dat, Ext; 3: Mod;
4	Problem	1. Preliminary hazard identification	1.a Use available evidence to better constrain…	Potential effects	5.0 (Sc)	Co	1: Var; 2: Dat, Ext; 3: Mod;

19	Problem	2. Define the conceptual model	2.b Choose assessment and measurement endpoints	Population: abundance	5.0 (Sc)	Co	1: Dat; 2:Var, Ext, Mod;
37	Exposure	1. Use available evidence to better constrain…	…	Prioritisation of data	5.0 (Sc)	Co	1: Sys; 2: Dat;
46	Exposure	2. Stressor, exposure media, and receptor information	2.b Collect information about properties affecting fate and transport	Biological	5.0 (Sc)	Co	1: Dat; 2: Var;
56	Exposure	2. Stressor, exposure media, and receptor information	2.c Collect information about the receptor	Spatial distribution	5.0 (Sc)	-	1: Dat; 2: Ext;
60	Exposure	3. Evaluate stressor-receptor contact	3.a Evaluate co-occurrence for…	Intensity of overlap	5.0 (Sc)	Co	1: Var; 2: Dat;
64	Exposure	4. Integrate multiple LOEs using…	…	Quantitative methods	5.0 (Sc)	Co	1: Sys; 2: Dat, Mod; 3: Var;
35	Exposure	1. Use available evidence to better constrain…	…	Model selection	5.0 (Sc)	Co	1: Mod; 2: Sys, Ext; 3: Dat, Var;
93	Risk	1. Select relevant profiles…	…distribution methods showing	For effects	5.0 (Sc)	Co	1: Ext, Mod; 2: Dat, Var;
95	Risk	2. Estimate and aggregate risk	2.a Estimate risk using…	Cumulative distributions	5.0 (Sc)	Co	1: Ext; 2: Var, Mod;
100	Risk	3. Evaluate risk levels	3.a Assess confidence in the risk levels using…	Empirical evidence	5.0 (Sc)	Co	1: Ext; 2: Var;
41	Exposure	2. Stressor, exposure media, and receptor information	2.a.2 Collect information about the stressor's distribution	Spatial	5.0 (Sc)	Co	1: Dat; 2: Var;
53	Exposure	2. Stressor, exposure media, and receptor information	2.b Collect information about properties affecting fate and transport	Environmental media: Aquatic	5.0 (Sc)	Co	1: Var; 2: Dat;
57	Exposure	2. Stressor, exposure media, and receptor information	2.c Collect information about the receptor	Temporal distribution	5.0 (Sc)	Co	1: Dat; 2: Var, Ext;
68	Exposure	5. Create the exposure profile(s) using…	…	Stressor-based models	5.0 (Sc)	Co	1: Mod; 2: Sys; 3: Ext;
69	Exposure	5. Create the exposure profile(s) using…	…	Receptor-based models	5.0 (Sc)	Co	1: Mod; 2: Ext; 3:Sys;
71	Effects	1. Use available evidence to better constrain…	…	Direct/indirect effects	5.0 (Sc)	-	1: Dat; 2: Sys;

76	Effects	2. Analyse the stressor-response relationship	2.a Determine the test dose for the…	Frequency	5.0 (Sc)	Co	1: Dat, Ext, Mod;
86	Effects	3. Integrate multiple LOEs using…	…	Semi-quantitative methods	5.0 (Sc)	Co	1: Sys; 2: Dat, Ext; 3: Var;
90	Effects	4. Create stressor-response profile using…	…single point methods showing…	Effects levels	5.0 (Sc)	Co	1: Dat; 2: Ext; 3: Mod;
92	Risk	1. Select relevant profiles…	…distribution methods showing	For exposure	5.0 (Sc)	Co	1: Mod; 2: Dat, Var;
42	Exposure	2. Stressor, exposure media, and receptor information	2.a.2 Collect information about the stressor's distribution	Temporal	5.0 (Sc)	Co	1: Dat; 2: Var;
84	Effects	Analyse the stressor-response relationship	20. Assess effect endpoints	Ecosystem: Biodiversity	5.0 (Sc)	Co	1: Dat; 2: Sys; 3: Var, Ext, Mod;
15	Problem	2. Define the conceptual model	2.b Choose assessment and measurement endpoints	Organism: behaviour	4.0 (Sc)	Co	1: Dat; 2: Ext;
34	Exposure	1. Use available evidence to better constrain…	…	Exposure levels	4.0 (Sc)	-	1: Dat; Sys, Var, Ext;
67	Exposure	5. Create the exposure profile(s) using…	…	Direct monitoring values	4.0 (Sc)	Co	1: Dat, Var;
1	Problem	1. Preliminary hazard identification	1.a Use available evidence to better constrain…	Potential stressors	4.0 (Sc)	Co	1: Dat; 2: Sys; 3: Var, Ext;
7	Problem	1. Preliminary hazard identification	1.b Framing the hazard	Frame the 'how'	4.0 (Sc)	Co	1: Ext, Mod; 2: Dat, Var;
12	Problem	2. Define the conceptual model	2.a Identify the S-P-R paradigm, including…	The pathway(s)	4.0 (Sc)	Co	1: Mod; 2: Dat;
55	Exposure	2. Stressor, exposure media, and receptor information	2.c Collect information about the receptor	Receptor characteristics	4.0 (Sc)	-	1: Dat;
59	Exposure	3. Evaluate stressor-receptor contact	3.a Evaluate co-occurrence for…	Temporal overlap	4.0 (Sc)	Co	1: Dat; 2: Sys;
61	Exposure	3. Evaluate stressor-receptor contact	3.b. Evaluate…	Nature of contact	4.0 (Sc)	-	1: Dat;
23	Problem	2. Define the conceptual model	2.c Consider the appropriateness of the endpoints	Significance of endpoints to receptor	4.0 (Sc)	Co	1: Var; 2: Sys, Ext; 3: Mod;
33	Exposure	1. Use available evidence to better constrain…	…	Nature of exposure	4.0 (Sc)	Co	1: Dat, Sys;

58	Exposure	3. Evaluate stressor-receptor contact	3.a Evaluate co-occurrence for…	Spatial overlap	4.0 (Sc)	Co	1: Dat; 2: Var;
73	Effects	1. Use available evidence to better constrain…	…	Toxicity levels	4.0 (Sc)	Co	1: Dat, Ext; 2: Sys;
74	Effects	1. Use available evidence to better constrain…	…	Prioritisation of data	4.0 (Sc)	Co	-
29	Problem	3. Form the analysis/work plan	3.b Identify data considerations	Gaps in data	4.0 (Sc)	Co	1: Sys; 2: Dat, Mod;
62	Exposure	3. Evaluate stressor-receptor contact	3.b. Evaluate…	Uptake by receptor	4.0 (Sc)	Co	1: Dat, Var;
63	Exposure	4. Integrate multiple LOEs using…	…	Semi-quantitative methods	4.0 (Sc)	Co	1: Mod; 2: Dat, Sys, Var, Ext;
70	Effects	1. Use available evidence to better constrain…	…	Nature of effects	4.0 (Sc)	-	1: Dat, Sys;
91	Effects	4. Create stressor-response profile using…	…distribution methods showing	Effects levels	4.0 (Sc)	Co	1: Dat, Mod; 2: Var, Ext;
79	Effects	2. Analyse the stressor-response relationship	2.b Assess effect endpoints	Organism: behaviour	4.0 (Sc)	Co	1: Dat, Var;
83	Effects	2. Analyse the stressor-response relationship	2.b Assess effect endpoints	Population: abundance	4.0 (Sc)	Co	1: Sys; 2: Dat, Var, Ext; 3: Mod;
50	Exposure	2. Stressor, exposure media, and receptor information	2.b Collect information about properties affecting fate and transport	Environmental media: biota	3.0 (St)	-	1: Dat; 2: Var;
44	Exposure	2. Stressor, exposure media, and receptor information	2.a.3 Collect information info about the stressor's release	Probability	3.0 (St)	Co	1: Dat; 2: Var;
47	Exposure	2. Stressor, exposure media, and receptor information	2.b Collect information about properties affecting fate and transport	Chemical	3.0 (St)	-	1: Dat;
48	Exposure	2. Stressor, exposure media, and receptor information	2.b Collect information about properties affecting fate and transport	Physical	3.0 (St)	-	1: Dat;
54	Exposure	2. Stressor, exposure media, and receptor information	2.c Collect information about the receptor	Physical composition	3.0 (St)	-	1: Dat;
77	Effects	2. Analyse the stressor-response relationship	2.a Determine the test dose for the…	Intensity	3.0 (St)	Co	1: Dat; 2: Var;

9	Problem	1. Preliminary hazard identification	1.b Framing the hazard	Frame the 'when'	3.0 (St)	Co	1: Dat; 2: Sys, Var, Ext, Mod;
18	Problem	2. Define the conceptual model	2.b Choose assessment and measurement endpoints	Organism: fecundity	3.0 (St)	-	-
22	Problem	2. Define the conceptual model	2.c Consider the appropriateness of the endpoints	Relevance of measures to their endpoints	3.0 (St)	-	1: Dat; 2: Sys;
28	Problem	3. Form the analysis/work plan	3.a Identify the factors controlling fate and transport of the stressor	Environmental media factors	3.0 (St)	Co	1: Dat; 2: Var; 3: Sys;
38	Exposure	2. Stressor, exposure media, and receptor information	2.a.1 Collect information about the stressor's composition	Biological information	3.0 (St)	-	1: Var;
45	Exposure	2. Stressor, exposure media, and receptor information	2.a.3 Collect information info about the stressor's release	Quantity	3.0 (St)	Co	1: Dat;
49	Exposure	2. Stressor, exposure media, and receptor information	2.b Collect information about properties affecting fate and transport	Environmental media: terrestrial	3.0 (St)	Co	1: Dat;
51	Exposure	2. Stressor, exposure media, and receptor information	2.b Collect information about properties affecting fate and transport	Environmental media: Sub-terrestrial	3.0 (St)	-	1: Dat;
52	Exposure	2. Stressor, exposure media, and receptor information	2.b Collect information about properties affecting fate and transport	Environmental media: Atmospheric	3.0 (St)	Co	1: Dat, Var;
78	Effects	2. Analyse the stressor-response relationship	2.b Assess effect endpoints	Organism: development	3.0 (St)	Co	1: Dat, Var;
82	Effects	2. Analyse the stressor-response relationship	2.b Assess effect endpoints	Organism: fecundity	3.0 (St)	Co	1: Var; 2: Dat;
8	Problem	1. Preliminary hazard identification	1.b Framing the hazard	Frame the 'where'	3.0 (St)	Co	1: Dat; 2: Sys, Mod;
14	Problem	2. Define the conceptual model	2.b Choose assessment and measurement endpoints	Organism: development	3.0 (St)	Co	1: Sys; 2: Dat, Ext;
27	Problem	3. Form the analysis/work plan	3.a Identify the factors controlling fate and transport of the stressor	Physical factors	3.0 (St)	Co	1: Dat; 2: Var; 3: Sys;

30	Problem	3. Form the analysis/work plan	3.b Identify data considerations	Types of data required	3.0 (St)	Co	1: Dat; 2: Mod;
32	Problem	3. Form the analysis/work plan	3.b Identify data considerations	Analysis techniques	3.0 (St)	Co	1: Dat;
40	Exposure	2. Stressor, exposure media, and receptor information	2.a.1 Collect information about the stressor's composition	Physical information	3.0 (St)	-	1: Dat, Var;
43	Exposure	2. Stressor, exposure media, and receptor information	2.a.3 Collect information info about the stressor's release	Intensity	3.0 (St)	Co	1: Dat;
65	Exposure	5. Create the exposure profile(s) using…	…	Conservative estimates	3.0 (St)	-	1: Ext;
75	Effects	2. Analyse the stressor-response relationship	2.a Determine the test dose for the…	Duration	3.0 (St)	Co	1: Dat;
6	Problem	1. Preliminary hazard identification	1.b Framing the hazard	Frame the 'whom'	3.0 (St)	Co	1: Dat;
13	Problem	2. Define the conceptual model	2.a Identify the S-P-R paradigm, including…	The receptor(s)	3.0 (St)	Co	1: Dat, Sys;
25	Problem	3. Form the analysis/work plan	3.a Identify the factors controlling fate and transport of the stressor	Biological factors	3.0 (St)	Co	1: Var; 2: Dat;
11	Problem	2. Define the conceptual model	2.a Identify the S-P-R paradigm, including…	The stressor(s)	3.0 (St)	Co	1: Sys; 2: Dat; 3: Mod;
89	Effects	4. Create stressor-response profile using…	…single point methods showing…	Extreme toxicity	2.5 (St)	Co	1: Dat;
10	Problem	2. Define the conceptual model	2.a Identify the S-P-R paradigm, including…	The source(s)	2.0 (St)	-	-
26	Problem	3. Form the analysis/work plan	3.a Identify the factors controlling fate and transport of the stressor	Chemical factors	2.0 (St)	Co	1: Var; 2: Dat;
31	Problem	3. Form the analysis/work plan	3.b Identify data considerations	Collection techniques	2.0 (St)	-	1: Dat;
81	Effects	2. Analyse the stressor-response relationship	2.b Assess effect endpoints	Organism: survival	2.0 (St)	Co	1: Dat, Var;
88	Effects	4. Create stressor-response profile using…	…single point methods showing…	Conservative toxicity	2.0 (St)	Co	1: Dat;

17	Problem	2. Define the conceptual model	2.b Choose assessment and measurement endpoints	Organism: survival	2.0 (St)	-	1: Sys;
66	Exposure	5. Create the exposure profile(s) using…	…	Worst-case estimates	2.0 (St)	-	1: Dat, Ext;
5	Problem	1. Preliminary hazard identification	1.b Framing the hazard	Frame the 'what'	1.5 (St)	-	1: Dat;
39	Exposure	2. Stressor, exposure media, and receptor information	2.a.1 Collect information about the stressor's composition	Chemical information	1.0 (St)	-	1: Dat;

[a] Ig=Recognised ignorance; Sc=Scenario uncertainty; St=Statistical uncertainty. IQRs and high-low ranges are used to rank like values.

[b] Co=Combined.

[c] Dat=Data; Sys=System; Var=Variability; Ext=Extrapolation; Mod=Model. Median occurrence rates are used to rank like values.

ACKNOWLEDGEMENTS

This research was funded jointly by EPSRC/ESRC/NERC/Defra under grantEP/G022682/1. DJCS was funded by Cranfield University's EPSRC Doctoral Training Accounts EP/P503507/1 and EP/P504147/1. The authors thank the anonymous subject-matter experts who participated in this research.

The data supporting this study are openly available from Cranfield University's repository at http://dx.doi.org/10.17862/cranfield.rd.3495935.

REFERENCES

1. Barnthouse, L., 2008. The strengths of the ecological risk assessment process: linking science to decision making. Integr. Environ. Assess. Manag. 4 (3), 299–305.
2. Beer, T., 2006. Ecological risk assessment and quantitative consequence analysis. Hum. Ecol. Risk. Assess. 12 (1), 51–65.
3. Briggs, D.J., 2008. A framework for integrated environmental health impact assessment of systemic risks. J. Environ. Health 7 (61), 10–1186.
4. Burgert, S., Schäfer, R.B., Foit, K., Kattwinkel, M., Metzeling, L., MacEwan, R., Kefford, B.J., Liess, M., 2011. Modelling aquatic exposure and effects of insecticides – application to south-eastern Australia. Sci. Total Environ. 409 (14), 2807–2814.
5. Chèvre, N., Loepfe, C., Singer, H., Stamm, C., Fenner, K., Escher, B.I., 2006. Including mixtures in the determination of water quality criteria for herbicides in surface water. Environ. Sci. Technol. 40 (2), 426–435.
6. Chèvre, N., Edder, P., Ortelli, D., Tatti, E., Erkman, S., Rapin, F., 2008. Risk assessment of herbicide mixtures in a large European lake. Environ. Toxicol. 23, 269–277.
7. Clemen, R., Winkler, R., 1999. Combining probability distributions from experts in risk analysis. Risk Anal. 19, 187–203.
8. Cooke, R.M., Probst, K.N., 2006. Highlights of the Expert Judgment Policy Symposium and Technical Workshop (Conference Summary).
9. Costanza, R., Funtowicz, S., Ravetz, J., 1992. Assessing and communicating data quality in policy-relevant research. Environ. Manag. 16 (1), 121–131.
10. Cuppen, J.G.M., Van Den Brink, P.J., Camps, E., Uil, K.F., Brock, T.C.M., 2000. Impact of the fungicide carbendazim in freshwater microcosms. I. Water quality, breakdown of particulate organic matter and responses of macroinvertebrates. Aquat. Toxicol. 48 (2), 233–250.
11. Dale, V., Biddinger, G., Newman, M., Oris, J., Suter, G., Thompson, T., Armitage, T., Meyer, J., Allen-King, R., Burton, G., et al., 2008. Enhancing the ecological risk assessment process. Integr. Environ. Assess. Manag. 4 (3), 306–313.
12. Damalas, C.A., Eleftherohorinos, I.G., 2011. Pesticide exposure, safety issues, and risk assessment indicators. Int. J. Environ. Res. Public Health 8 (5), 1402–1419.
13. Damásio, J., Barceló, D., Brix, R., Postigo, C., Gros, M., Petrovic, M., Sabater, S., Guasch, H., de Alda, M.L., Barata, C., 2011. Are pharmaceuticals more harmful than other pollutants to aquatic invertebrate species: a hypothesis tested using multi-biomarker and multispecies responses in field collected and transplanted organisms. Chemosphere 85 (10), 1548–1554.
14. Daniel, W.W., 1990. Applied Nonparametric Statistics. second ed. PWS-Kent, Boston.
15. Department for Environment, Food and Rural Affairs, 2011. Guidelines for Environmental Risk Assessment and Management: Green Leaves III. HMSO, London, UK.
16. Department of Environment, Transport and the Regions/ Environment Agency and Institute for Environmental Health, 2000o. Guidelines for Environmental Risk

Assessment and Management: Revised Departmental Guidance. HMSO, London, UK.

17. Department of Health and Ageing, 2002. Environmental Health Risk Assessment: Guidelines for Assessing Human Health Risks from Environmental Hazards. HMSO, Canberra, Australia.
18. European Environment Agency, 2007. Late Lessons from Early Warnings the Precautionary Principle 1896–2000: Implications for Policymaking and Research.
19. Fairman, R., Mead, C., Williams, W., de Medio Ambiente, A., 1998. Environmental Risk Assessment: Approaches, Experiences and Information Sources. Office for Official Publications of the European Communities, Luxembourg (ISBN: 92-9167-080-4).
20. Falagas, M.E., Pitsouni, E.I., Malietzis, G.A., Pappas, G., 2008. Comparison of PubMed, Scopus, Web of Science, and Google Scholar: strengths and weaknesses. Fed. Am. Soc. Exp. Biol. J. 22 (2), 338–342.
21. Funtowicz, S.O., Ravetz, J.R., 1990. Uncertainty and Quality in Science for Policy. 15. Springer Science and Business Media.
22. Gillund, F., Kjulberg, K., Krayer von Krauss, M.P., Myhr, A., 2008. Do uncertainty analyses reveal uncertainties? Using the introduction of DNA vaccines to aquaculture as a case. Sci. Total Environ. 407 (1), 185–196.
23. Guy, M., Singh, L., Mineau, P., 2011. Using field data to assess the effects of pesticides on Crustacea in freshwater aquatic ecosystems and verifying the level of protection provided by water quality guidelines. Integr. Environ. Assess. Manag. 7 (3), 426–436.
24. Handmer, J., Norton, T., Dovers, S., 2001. Ecology, Uncertainty and Policy: Managing Ecosystems for Sustainability. Prentice Hall, Harlow, UK.
25. Hart, A., Roelofs, W., Crocker, J., Murray, A., Boatman, N., Hugo, S., Fitzpatrick, S., Flari, V., 2007. DEFRA Research Contract CPEC38: Quantitative Approaches to the Risk Assessment of GM Crops. Central Science Laboratory, York, UK.
26. Janssen, P.H.M., Petersen, A.C., van der Sluijs, J.P., Risbey, J.S., Ravetz, J.R., 2003. RIVM/MNP Guidance for Uncertainty Assessment and Communication: Quickscan Hints and Actions List. RIVM/MNP Guidance for Uncertainty Assessment and Communication Series 2.
27. Kandlikar, M., Ramachandran, G., Maynard, A., Murdock, B., Toscano, A., 2007. Health risk assessment for nanoparticles: a case for using expert judgment. Nanotechnology and Occupational Health. Springer Netherlands, pp. 137–156.
28. Kelly, E., Campbell, K., 2000. Separating variability and uncertainty in environmental risk assessment – making choices. Hum. Ecol. Risk. Assess. 6 (1), 1–13.
29. Knol, A.B., Petersen, A.C., van der Sluijs, J.P., Lebret, E., 2009. Dealing with uncertainties in environmental burden of disease assessment. Environ. Health 8, 21.
30. Knol, A.B., Slottje, P., van der Sluijs, J.P., Lebret, E., 2010. The use of expert elicitation in environmental health impact assessment: a seven step procedure. Environ. Health 9 (1), 19.
31. Kotra, J.P., Lee, M.P., Eisenberg, N.A., DeWispelare, A.R., 1996. Branch Technical Position on the Use of Expert Elicitation in the High-Level Radioactive Waste Program. Division of Waste Management, Office of Nuclear Material Safety and Safeguards, US Nuclear Regulatory Commission.
32. Krayer von Krauss, M.P., Casman, E.A., Small, M.J., 2004. Elicitation of expert judgments of uncertainty in the risk assessment of herbicide-tolerant oilseed crops. Risk Anal. 24, 1515–1527.
33. Krayer von Krauss, M.P., Kaiser, M., Almaas, V., van der Sluijs, J.P., Klopgrogge, P., 2008. Diagnosing and prioritizing uncertainties according to their relevance for policy: the case of transgene silencing. Sci. Total Environ. 390 (1), 23–34.
34. Landis, W.G., 2005. Population-scale assessment endpoints in ecological risk assessment part 1: reflections of stakeholder values. Integr. Environ. Assess. Manag. 2 (1), 86–91.
35. Loveridge, D., 2004. Experts and foresight: review and experience. Int. J. Foresight Innov. Policy 1, 33–69.
36. Manikandan, S., 2011. Measures of dispersion. J. Pharmacol. Pharmacother. 2, 315–316.
37. Mastin, B.J., Rodgers Jr., J.H., 2000. Toxicity and bioavailability of copper herbicides (clearigate, cutrine-plus, and copper sulfate) to freshwater animals. Arch. Environ. Contam. Toxicol. 39 (4), 445–451.
38. McCluskey, A., Lalkhen, A.G., 2007. Statistics II: central tendency and spread of data. Contin. Educ. Anaesthia, Crit. Care Pain 7 (4), 127–130.
39. Meyer, M.A., Booker, J.M., 1991. Eliciting and Analyzing Expert Judgment: A Practical Guide. Academic Press, London, UK.
40. Morgan, M.G., Henrion, M., Small, M.J., 1990. Uncertainty: A Guide to Dealing with Uncertainty in Quantitative Risk and Policy Analysis. Cambridge University Press, Cambridge, UK.
41. Palma, G., Sanchez, A., Olave, Y., Encina, F., Palma, R., Barra, R., 2004. Pesticide levels in surface waters in an agricultural-forestry basin in Southern Chile. Chemosphere 57 (8), 763–770.
42. Pollard, S.J.T., Yearsley, R., Reynard, N., Meadowcroft, I.C., Duarte-Davidson, R., Duerden, S., 2002. Current directions in the practice of environmental risk assessment in the United Kingdom. Environ. Sci. Technol. 36 (4), 530–538.
43. Pollard, S.J.T., Smith, R., Longhurst, P.J., Eduljee, G., Hall, D., 2006. Recent developments in the application of risk analysis to waste technologies. Environ. Int. 32, 1010–1020.
44. Ravnum, S., Zimmer, K.E., Keune, H., Gutleb, A.C., Murk, A.J., Koppe, J.G., et al., 2012. Policy relevant results from an expert elicitation on the human health risks of decabromodiphenyl ether (decaBDE) and hexabromocyclododecane (HBCD). Environ. Health 11 (Supplement 1), S7. Rees, D.G., 2000. Essential Statistics. fourth ed. CRC Press LLC, USA.

45. Refsgaard, J.C., van der Sluijs, J.P., Hojberg, A., Vanrolleghem, P., 2007. Uncertainty in the environmental modelling process - a framework and guidance. Environ. Model. Softw. 22 (11), 1543–1556.
46. Schuler, L.J., Rand, G.M., 2008. Aquatic risk assessment of herbicides in freshwater ecosystems of south Florida. Arch. Environ. Contam. Toxicol. 54 (4), 571–583.
47. Schwarzman, M.R., Wilson, M.P., 2009. New science for chemicals policy. Science 326, 1065–1066.
48. Shapiro, S.S., Wilk, M.B., 1965. An analysis of variance test for normality (complete samples). Biometrika 52, 591–611.
49. Siemering, G.S., Hayworth, J.D., Greenfield, B.K., 2008. Assessment of potential aquatic herbicide impacts to California aquatic ecosystems. Arch. Environ. Contam. Toxicol. 55 (3), 415–431.
50. Sigel, K., Klauer, B., Pahl-Wostl, C., 2010. Conceptualising uncertainty in environmental decision-making: the example of the EU water framework directive. Ecol. Econ. 69 (3), 502–510.
51. Skinner, D.J.C., Rocks, S.A., Pollard, S.J.T., 2014a. A review of uncertainty in environmental risk: characterising potential natures, locations, and levels. J. Risk Res. 17 (2), 195–219.
52. Skinner, D.J.C., Rocks, S.A., Pollard, S.J.T., Drew, G.H., 2014b. Identifying uncertainty in environmental risk assessments: a novel typology and implications for risk characterisation. Hum. Ecol. Risk. Assess. 20 (3), 607–640.
53. Slottje, P., van der Sluijs, J.P., Knol, A.B., 2008. Expert elicitation: methodological suggestions for its use in environmental health impacts assessments. RIVM Letter Report 630004001/2008. RIVM, Copernicus Institute for Sustainable Development and Innovation, Bilthoven, Netherlands.
54. Smita, S., Gupta, S.K., Bartonova, A., Dusinska, M., Gutleb, A.C., Rahman, Q., 2012. Nanoparticles in the environment: assessment using the causal diagram approach. Environ. Health 11 (Supplement 1), S13.
55. Speirs-Bridge, A., Fidler, F., McBride, M., Flander, L., Cumming, G., Burgman, M.A., 2010. Reducing overconfidence in the interval judgments of experts. Risk Anal. 30, 512–523.
56. Suter II, G.W., 1996. Guide for Developing Conceptual Models for Ecological Risk Assessment. No. ES/ER/TM-186. Oak Ridge National Laboratory, Tennessey, United States.
57. United States Environmental Protection Agency, 1992. Framework for Ecological Risk Assessment. United States Environmental Protection Agency Risk Assessment Forum, Washington, USA.
58. United States Environmental Protection Agency, 1998. Guidelines for Ecological Risk Assessment. United States Environmental Protection Agency Risk Assessment Forum, Washington, USA.
59. United States Environmental Protection Agency, 2003. Framework for Cumulative Risk Assessment. United States Environmental Protection Agency Risk Assessment Forum, Washington, USA.
60. United States Environmental Protection Agency, 2009. Expert Elicitation Task Force White Paper. Science Policy Council, Washington, USA.
61. Uricchio, V.F., Giordano, R., Lopez, N., 2004. A fuzzy knowledge-based decision support system for ground water pollution risk evaluation. J. Environ. Manag. 73 (3), 189–197.
62. van Asselt, M.B.A., Rotmans, J.P., 2002. Uncertainty in integrated assessment modelling. Clim. Chang. 54 (1–2), 75–105.
63. van den Brink, P.J., Crum, S.J.H., Gylstra, R., Bransen, F., Cuppen, J.G.M., Brock, T.C.M., 2009. Effects of a herbicide-insecticide mixture in freshwater microcosms: risk assessment and ecological effect chain. Environ. Pollut. 157 (1), 237–249.
64. van Wijngaarden, R.P.A., Cuppen, J.G.M., Arts, G.H.P., Crum, S.J.H., van Den Hoorn, M.W., van den Brink, P.J., Brock, T.C.M., 2004. Aquatic risk assessment of a realistic exposure to pesticides used in bulb crops: a microcosm study. Environ. Toxicol. Chem. 23 (6), 1479–1498.
65. Vryzas, Z., Vassiliou, G., Alexoudis, C., Papadopoulou-Mourkidou, E., 2009. Spatial and temporal distribution of pesticide residues in surface waters in northeastern Greece. Water Res. 43 (1), 1–10.
66. Vryzas, Z., Alexoudis, C., Vassiliou, G., Galanis, K., Papadopoulou-Mourkidou, E., 2011. Determination and aquatic risk assessment of pesticide residues in riparian drainage canals in northeastern Greece. Ecotoxicol. Environ. Saf. 74 (2), 174–181.
67. Walker, W., Harremoes, P., Rotmans, J., van der Sluijs, J.P., van Asselt, M., Janssen, P., et al., 2003. Defining uncertainty: a conceptual basis for uncertainty management in model-based decision support. Integr. Assess. 4 (1), 5–17.
68. Wan, M.T., Kuo, J.-N., McPherson, B., Pasternak, J., 2006. Triazine and metolachlor herbicide residues in farm areas of the Lower Fraser Valley, British Columbia, Canada. J. Environ. Sci. Health B 41 (6), 855–867.
69. Zimmer, K.E., Gutleb, A.C., Ravnum, S., Krayer von Krauss, M.P., Murk, A.J., Ropstad, E., et al., 2012. Policy relevant results from an expert elicitation on the health risks of phthalates. J. Environ. Health 11 (Supplement 1), S6.

CHAPTER 11

SUSTAINABILITY ASSESSMENT AND MANAGEMENT IN CONTAMINATED LAND PRACTICE

R. Paul Bardosa, Brian D. Bone, Richard Boyle,Frank Evans, Nicola D. Harries,Trevor Howard,Jonathan W.N. Smithh,

Keywords: Sustainable remediation, contaminated land management, Remediation, Option appraisal, Sustainability assessment, SuRF-UK

INTRODUCTION

Land contamination is recognised as a threat to soil and water quality, and to the wider environment (Van-Camp et al., 2004), and it can also pose significant health, environmental and social pressures (Environment Agency, 2009). Land contamination problems are common around the world. For example, Van Liedekerke et al. (2014)estimated that 2.5 million sites are potentially contaminated across Europe. The management of contaminated land imposes substantial economic costs, amounting to billions of pounds worldwide each year. The scale of land-contamination problems, and of the responses to them, makes achieving sustainability in contaminated land remediation an important objective (Bardos et al., 2011a and Bardos et al., 2011b). There is now an active international debate about how best to ensure that land contamination is managed in a sustainable manner (Bardos, 2014). In this context, sustainable remediation is the process of effectively managing contaminated land risks to human health and the environment in a manner that minimises the environmental footprint, maximises societal benefits, and minimises the costs of those remediation activities. Ideally all three outcomes are achieved, but where trade-offs are necessary, the assessment provides a framework to identify and select the best, or most sustainable, remediation solution. The Sustainable Remediation Forum in the UK (SuRF-UK) is an initiative established in 2007 to support more sustainable remediation practice in the UK by providing guidance based on multilateral inputs from different practitioners and stakeholder interests (CL:AIRE, 2010).

This paper describes SuRF-UK's latest guidance on preparing for and defining ('framing') the sustainability assessment and for a simple qualitative 'entry-level' to sustainability assessment in remediation projects (CL:AIRE, 2014a) It also presents suggested 'sustainable management practices' for application across all phases of contaminated land activities from planning and procurement, site investigation through to implementation and verification of remediation works (CL:AIRE, 2014b).

HISTORICAL CONTEXT

For the past decade the prevailing international consensus, at least across much of Europe, has been that risk assessment is the most rational approach for determining remediation need and urgency (CLARINET and NICOLE, 1998, NICOLE and COMMON FORUM, 2013 and Vegter et al., n.d). Risk assessment provides a means of understanding which receptors might be affected, and how severely. It evaluates both the magnitude of any consequence and likelihood of the consequence. On this basis decisions can be made on behalf of society about how to best allocate scarce resources. In many countries risk assessment takes into consideration the proposed use of the site following remediation (Nathanail et al., 2013), so that more sensitive end-uses require more stringent remediation goals than less sensitive uses.

Citation: R. Paul Bardos, Brian D. Bone, Richard Boyle, Frank Evans, Nicola D. Harries, Trevor Howard, Jonathan W.N. Smith, Sustainability assessment and management in contaminated land practice, Science of The Total Environment, Volumes 563–564, 1 September 2016, Pages 755-768, ISSN 0048-9697, http://dx.doi.org/10.1016/j.scitotenv.2015.12.001.

The importance of sustainability in this discussion is manifold, but is related to the effective *delivery* of whatever risk management is necessary to protect human health or the wider environment (Hou and Al-Tabbaa, 2014, Holland et al., 2013 and Plant et al., 2015):

- Some technical means of delivering remediation may be more beneficial than others, or have fewer negative impacts;
- In some cases the use of generic risk management thresholds may lead to an over-design of the remediation leading to unacceptable impacts elsewhere, for example compared with a site specific approach;
- There may be opportunities for synergy, for instance with renewable energy, green building, and waste recycling where remediation processes could deliver multiple benefits (*e.g.* biomass production as well as risk management, from phytoremediation (Licht and Isebrands, 2005);
- The potential negative outcomes of delivering a particular set of remediation goals may suggest reconsideration of the design of site-use originally envisaged;
- There may be opportunities to make sustainability gains by considering remediation as part of an overall land management planning process, for example taking into account a mosaic of land uses or changes in approach to the design and layout of buildings;
- Potentially developments in policy and legislation, combined with the limited availability of public funds, provide a major opportunity to shift the policy focus for contaminated sites from management of costs and liability, to value creation.

It could be argued that suitably professional project designers should already have many of these ideas in mind. However, explicit consideration of sustainability puts these considerations into a systematic structure, and perhaps widens the range of available considerations to allow for a more holistic assessment. Properly executed sustainability assessment also allows for these considerations, their assumptions and their evidence to be more effectively discussed across stakeholder interests, transparently recorded, properly documented and ultimately verified (CL:AIRE, 2010).

Since the mid-1990s a broad range of tools have been developed for or applied to the assessment of the wider impacts, or latterly the sustainability, of remediation measures. One of the first was a system developed by TNO in The Netherlands called STEPS in the early 1990s (Aelmans et al., 1993, van Veen et al., 1997 and Ferdinandy and Weenk, 1999). This evolved into the 'REC' tool from The Netherlands which provides three indices related to risk reduction effectiveness, cost and 'environmental merit' (Beinat et al., 1998) to support choices in remedial method selection. Both were derived from life cycle assessment (LCA) concepts, as was the 'Sinsheim' tool developed in Germany (Bender et al., 1998). REC and other LCA based tools continue to be researched as tools for remediation decision-making (Cappuyns and Kessen, 2013). Around the same time the German Federal Environmental Protection Agency was promoting a semi-quantitative approach to support option appraisal for brownfields development. This was based on three indices calculated on the basis of a prescribed series of categories and weightings, intended to be related to monetary value: 'site potential index', 'exploitation potential index' and 'site value' (Grimski et al., 1998). In 2000 the Environment Agency of England and Wales published a review of approaches for understanding the 'wider environmental value' of remediation which suggested a more qualitative approach (Environment Agency, 2000a). Around the same time the Environment Agency also published approaches for cost benefit analysis and cost effectiveness analysis for remediation (Environment Agency, 1999 and 2000b).

In 2002 the outputs of EU funded CLARINET networking project[1] recognised that sustainability in remediation was an integrated consideration of social, economic and environmental factors (Bardos et al., 2002), in line with the concept of sustainable development advanced by the Brundtland Commission (Brundtland, 1987). The initiation of the various sustainable remediation networks described in Table 1 has led to a proliferation of method development for the application of sustainability assessment to remediation.

Table 1: A timeline of some of the key events in the evolution of the sustainable remediation concept (in part summarised from Bardos et al., 2012, Bardos et al., 2013 and Bardos, 2014).

Year	Event
1961	Terms of reference for the Feasibility Study for the Lower Swansea Valley Regeneration project (one of the first ever restoration projects), specifically couched in terms of social and economic benefit
1979–1985	UK led international NATO Challenges of Modern Society (CCMS) study on contaminated land which included a conclusion that treatment based solutions were environmentally preferable to solutions based on containment or removal to landfill (Smith, 1985)
1978–1983	Major land contamination incidents for example in The Netherlands (Lekkerkerk, 1980–81) and the USA (Love Canal — 1978, Time Beach — 1983) trigger widespread concerns about hazards from land contamination. To see an example of the shock of these events read Eckardt C. Beck in the EPA Journal, January 1979aa
1986–2007	A series of NATO/CCMS Pilot Studies involving various NATO Countries and others (e.g. Australia, Japan, EPAC) focusing on treatment based remediation approachesb
1987	Publication of the Brundtland Report
1995	Publication of the REC method in The Netherlands, the first well established decision framework considering three metrics: risk reduction, cost and 'environmental merit' (NOBIS, 1995a,b)
1998	Publication of the findings of the EU funded network on the scientific basis for risk assessment for contaminated land (the 'CARACAS' project, 1995–1998) which established a broad technical basis for risk based decision-making for contaminated land management (Ferguson et al., 1998).

1999/2000	Publication in the UK of Environment Agency reports of the wider environmental value of remediation, and of cost benefit analysis for remediation (Environment Agency, 1999, 2000a, 2000b)
2002	Publications of the findings of the EU funded 'CLARINET' networkc,1998–2001) which established risk-based land management as the principal rationale for decision-making for contaminated land management (Vegter et al., 2002), sustainability explicitly recognised as a consideration in risk management decision making (Bardos et al., 2002)
2004	Publication of the 'Model Procedures' in the UK which anticipated the importance of wider concerns (such as stakeholder opinions and wider impacts of remediation) in contaminated land management decision-making (Environment Agency and Defra, 2004))
2006	Initiation of the Sustainable Remediation Forum (SURF) in the USA which was the first cross-sectoral stakeholder forum explicitly focused on developing sustainability as a consideration in remediation outcomes
2007	Initiation of the Sustainable Remediation Forum in the UK (SuRF-UK), which although quite independent of SURF, was inspired by it, and continues to work with it.
2008	Publication of the first guidance on 'green remediation' by the US EPA
2009	Initiation of SuRF-Australia and New Zealand (they published a sustainable remediation framework based on SuRF-UK in 2011d)
2010	Publication of SuRF-UK framework for sustainable remediation, which was the first cross sectoral sustainable remediation guidance document accepted by national regulatory organisations (CL:AIRE, 2010)
2010–2013	Establishment of SuRF networks in The Netherlands (2010), Brazil (2010), Canada (2011), Italy (2012), Taiwan (2012) and Colombia (2013).
2011	Publication of a sustainable remediation framework in the USA by SURF (Holland et al., 2011)
2011	Publication of the NICOLE 'road map' for sustainable remediation, the first internationally agreed guidance (NICOLE, 2011)
2011	Interstate Technology & Regulatory Council (ITRC) Report on green and sustainable remediation' (ITRC 2011)
2012	Establishment of 'SURF-International' an informal quarterly forum where many of the SURFs and NICOLE gather in web conference calls to exchange information, experiences and idease
2012	Initiation of ISO/DIS 18504 to draft a descriptive standard for 'Sustainable Remediation'.
2013	Risk-Informed and Sustainable Remediation — Joint Position Statement by NICOLE and COMMON FORUM
2013	ASTM Standard Guide for Integrating Sustainable Objectives into Cleanup, E2876-13
2013	The International Committee on Contaminated Land (an international policy and regulatory network) tabled its first discussions on 'green and sustainable remediation' in Durban, South Africa. www.iccl.ch
2016	Planned publication date for a special issue on sustainable remediation in the Journal of Environmental Management
2016	Anticipated publication of ISO 18504 on Sustainable Remediation

a http://www2.epa.gov/aboutepa/love-canal-tragedy Accessed July 2015.

b NATO/CCMS Pilot Study information and some of the later reports can be found at: www.cluin.org/global Accessed July 2015.

c The Contaminated Land Rehabilitation Network For Environmental Technologies in Europe.

d See http://landandgroundwater.com/page/sustainable-remediation-forum-for-australia-new-zealand-surf-anz accessed July 2015.

Sustainability has also developed in importance as a criterion in decision making for brownfields redevelopment. Indeed explicit consideration of social and economic goals took place as early as 1961 for the feasibility study for the Lower Swansea Valley restoration work (Bromley and Humphrys, 1979 and US EPA — US Environmental Protection Agency, 1992).[2] Its terms of reference were to '*establish the factors which inhibit the social and economic use of land in the Lower Swansea Valley and to suggest ways in which the area should be used in the future*'. [3] While brownfields restoration is a broader context than remediation, there are obvious cross-overs, not least that the management of land contamination is frequently required as part of a brownfields restoration (BenDor et al., 2011 and CABERNET, 2006). A range of sustainability assessment techniques have been developed for brownfields decision making. A qualitative approach was developed in Scotland in the late 1990s (Curran and Hart, 1998). In the mid-2000s the European 'RESCUE' project developed a semi quantitative 'sustainability assessment tool' based on site-specific scores and weightings developed through a stakeholder process to identify sustainable development priorities (RESCUE Consortium, 2005). This approach was endorsed by the European 'REVIT' project in 2007 (REVIT Consortium, 2007). The UK SUBR:IM project (Sustainable Urban Brownfield Regeneration: Integrated Management) also suggested that indicators for sustainability assessment for brownfields restoration should be based on a similar process of identifying priorities (CL:AIRE, 2007b). In Germany the 'SAFIRA' project produced an integrated numerical sustainability and economic assessment model (Schädler et al., 2011) called the MMT tool. This software offers two different indicator-based approaches to assess sustainability. The first one, a set of fixed indicators allows assessing the suitability of different land-use types in regional context of the particular site. This makes use of 15 general and normative sustainability goals developed by (Bleicher and Groß, 2010). The second approach explicitly considers spatial planning options. Here, stakeholders define site-specific problems and relate them to indicators suitable for measuring sustainable development (Morio et al., 2011).

CURRENT STATE OF THE ART FOR SUSTAINABLE REMEDIATION

Across the world there is now a developing consensus about the definitions and concepts underpinning what is broadly described as 'sustainable remediation' (Bardos, 2014 and Rizzo et al., 2016) and a growing technical understanding of how it should be implemented. Table 1 provides a timeline of some of the key developments in the evolution of sustainable remediation. Experiences have been shared in three international sustainable remediation conferences[4] (Copenhagen 2009, Vienna 2012, Ferrara 2014) and a special issue of *Journal of Environmental Management* is planned on the topic in 2016. Many of the organisations and networks involved meet quarterly 'on-line' with secretarial support from CL:AIRE. [5] In addition, members of many of these initiatives, along with other practitioners, have collaborated as experts nominated by their respective national standards bodies under the auspices of the International Standards Organisation (ISO/DIS 18504) to draft a proposed *informative* standard for 'Sustainable Remediation'. The proposed standard aims to draw together an international consensus on the nature and process of identifying sustainable remediation on a site specific basis (ISO 2015). In addition to sustainable remediation, within the US a number of government organisations and networks have been developing a related concept of 'Green Remediation', as defined in US EPA (2008a).

Green remediation is intended to reduce the demand placed on the environment during clean-up actions and to conserve natural resources. Green remediation anticipates that the major decision making elements setting the boundaries for remediation action, including economic and social considerations, have already taken place. 'Green remediation focuses on the environmental footprint of Superfund response actions. The broader realm of site sustainability examines environmental issues but also includes social and economic aspects that are typically addressed by site users and local or regional communities' (US EPA, 2010a). The Interstate Regulatory Council (ITRC) in the USA has combined these concepts as 'green and sustainable remediation' — GSR (ITRC, 2011).

From the perspective of implementing sustainable remediation, analogous to the consideration of risk in contaminated land decision-making, two broad activities can be identified (based on Bardos et al., 2011b and NICOLE, 2011):

- Sustainability management: using sustainability as a criterion in decision-making with a view to optimising outcomes
- Sustainability assessment: the process of gaining an understanding of possible outcomes across all three elements (environmental, social and economic) of sustainable development.

In this structure sustainability assessment is used as a tool to support appropriate, sustainable land management.

Within the USA, ASTM have published standards for both green and sustainable remediation, which consider sustainability assessment (ASTM, 2013) and guidance has been produced by the EPA and ITRC (ITRC, 2011 and US EPA — US Environmental Protection Agency, 2008a) and other public sector organisations (*e.g.* NAVFAC, 2012). These allow significant flexibility in approach rather than being strictly prescriptive. There is a wide range of possible sustainability assessment 'tools' (approaches, methods and software) being used/developed in the USA (Holland et al., 2011). The draft ISO standard (ISO 2015) also allows for a range of possible sustainability assessment tools to be used, while setting out the broad principles by which they should be designed and used. SuRF-UK's guidance is very much in line with this approach of setting broad guidelines for how sustainability assessment should be carried out, without prescribing exactly how it should be done (CL:AIRE, 2010 and 2014a). It seems likely the 'standard' to be complied with will generally be determined by the project funder. The Austrian Environmental Protection Agency has, for example, published a required method for publically funded projects in Austria (Döberl et al., 2013).

Table 2 summarises the broad approaches that SuRF-UK has identified as potentially applicable to supporting sustainable remediation assessments (CL:AIRE, 2010). This table also indicates each approach's coverage of the environmental, economic and social elements of sustainable development; whether techniques are quantitative, semi-quantitative or qualitative; and whether contaminated land management applications are known to exist at present. A number of practical examples are analysed in more detail byBeames et al. (2014).

Table 2: Selected decision support techniques with relevance to sustainable remediation assessments. Taken fromCL:AIRE, 2010. © CL:AIRE, 2015. Reproduced by permission.

	Environment	Economy	Society	Type	CLM
Scoring/ranking systems (including multi-criteria analysis), e.g. Balasubramaniam and Voulvoulis (2005)	Narrow to Wide	Narrow to Wide	Narrow to Wide	Qual	Yes, e.g. Bleicher and Groß, 2010
Carbon footprint ('area'), e.g. Carbon Trust et al. (2008b)	Narrow	–	–	Quan	Yes, Ellis and Hadley (2009)
Carbon balance (flows), e.g. Defra, 2006	Narrow	–	–	Quan	
Cost benefit analysis, e.g. Environment Agency (1999, 2000b)	Narrow to Wide	Narrow to Wide	Narrow to Wide	Quan	Yes, e.g. Environment Agency (2000b)

Cost effectiveness analysis, e.g. Environment Agency (1999)	Narrow to Wide	Narrow to Wide	Narrow to Wide	Qual	Yes, e.g. Environment Agency (1999)
Eco-efficiency	Narrow	–	–	Quan	Yes, e.g. Sorvari et al. (2009)
Ecological footprinta	Narrow	–	–	Quan	Yes, e.g. Ferdos and Rosén (2013)
Energy/intensity efficiency	Narrow	–	–	Quan	Yes, Ellis and Hadley (2009)
Environmental risk assessment	Narrow to Wide	–	–	Qual/Quan	Yes, Environment Agency and Defra (2004)
Human health risk assessment		–	Narrow	Qual/Quan	Yes, Environment Agency and Defra (2004)
Environmental impact assessment [for projects]/Strategic environmental assessment [for policies], DCLG, 2006, ODPM 2005	Narrow to Wide	–	–	Qual	Yes, ODPM 2005, Stroud District Council (2015)
Financial risk assessment	–	Narrow	–	Quan	Yes, Finnamore et al. (2000)
Industrial ecologyb	Narrow to Wide	Narrow to Wide	–	Quan	
Life Cycle Assessment (based)	Narrow to Wide	–	–	Quan	Yes, e.g. Cappuyns and Kessen (2013)
Quality of life assessment (US EPA, 1973)	Wide	Wide	Wide	Qual	
Key: Qual=Qualitative (including semi-quantitative methods) Quan=Quantitative CLM=Contaminated Land Management application examples available Narrow: indicates that relatively few of the SURF-UK indicator categories are considered by the this method Wide: indicates that many of the SURF-UK indicator categories could be considered by the this method •The table describes each technique in terms of its typical coverage of particular aspects of sustainability. For example, a carbon footprint appraisal focuses on a 'narrow' segment of environmental sustainability issues (ignoring for example soil functionality, biodiversity and landscape impacts), whereas all of these aspects could be considered by a cost-benefit analysis, providing it was suitably specified.•A dash (–) means that the technique has no coverage.					

a E.g. http://www.footprintnetwork.org/en/index.php/GFN/page/footprint_basics_overview/ Accessed November 2015.

b http://www.gdrc.org/sustdev/concepts/16-l-eco.html Accessed November 2015.

Both qualitative and semi-quantitative methods, such as use of multicriteria analysis (*e.g.*Harbottle et al., 2008a and Harbottle et al., 2008b) can be structured across a broad range of considerations (Smith and Kerrison, 2013 and Beames et al., 2014). With the*possible* exception of cost benefit analysis or approaches to cost-effectiveness (see below), quantitative methods do not even approach a full scope of sustainability considerations. Carbon footprint based measurements are self-evidently one dimensional, albeit covering one important environmental consideration. Life cycle assessment approaches partially cover environmental considerations, but cannot evaluate impacts on important factors for some contexts such as soil functionality or landscape changes which may be particular assets for the non-built re-use of land. LCA also does not speak to social and economic considerations, although attempts are being to adapt the rigour of life cycle thinking to sustainability assessment (Calcas Consortium, 2009).

Cost benefit assessment describes a process of comparing the likely costs of a project with its benefits and is a form of economic valuation. Where this assessment is based on conversion to strictly monetary terms it is described as cost benefit analysis – CBA (Commonwealth of Australia, 2006). The common feature of formal CBA is an attempt to express all values (costs and benefits) using a single monetary unit, which some see as facilitating clear-cut decisions about overall benefit (Brouwer and van Ek, 2004). Including 'externalities' into the CBA process allows wider benefits and impacts to be considered and costed (*e.g.* Doick et al., 2009). There are two significant limitations for the use of CBA as a means of sustainability assessment: the scope of externalities being considered and the approach taken to valuing externalities. There are a range of technical approaches to the valuation of externalities (Cellini and Kee, 2010 and Defra — Department for Environment, Food and Rural Affairs,). However, all valuation methods are subject to errors and bias based on the assumptions used when they are applied, are complex and are not always seen as reliable by project stakeholders, particularly when they are not economists (Atkinson and Mourato, 2008). Cellini and Kee (2010) suggest some externalities and benefits may be regarded as impossible to value. The other important limitation is scope, which is a pragmatic consideration. The time and data needed to prepare a CBA may make wide ranging and holistic assessment of externalities prohibitively expensive.

Taking a staged or tiered approach, starting with simple qualitative approaches, and moving through to more quantitative methods should the need arise, has advantages in terms of cost and resource efficiency as well as providing a structure that is as inclusive as possible and combines the relative strengths of the methods available (CL:AIRE, 2010, EURODEMO Project Consortium, 2007, NICOLE, 2011 and SURF — Sustainable Remediation Forum, 2009). This is also in line with other suggestions for good practice in sustainability assessment (Pollard et al., 2004 and Therivel, 2004). Therivel (2004)suggests that the appropriate point of trade-off between comprehensiveness, rigour, transparency, user friendliness and costs for sustainability assessment tools should depend on the decision that the tool is informing. Furthermore, the amount of time and effort needed as an input should be proportional to the benefits provided as an output. He suggests two stage tools or processes, with a 'shallow' initial stage which gives a broad-brush analysis of a problem, and a 'deep' focus on those issues that were identified in the first stage as being particularly problematic, contentious or important to the decision making process. Within many European countries the contaminated land sector is very familiar with the use of tiered approaches in risk assessment for similar reasons (Environment Agency and Department for Environment and Food an and Nathanail et al., 2013), so this concept already links well with established practices.

However, there can be strong stakeholder demands for quantitative assessments, particularly where one or more stakeholders have a requirement for cost benefit analysis based decision making, or particular quantified organisational/ corporate sustainability goals, for example related to use of fossil carbon (*e.g.* NAVFAC, 2014; or in a more general sense: HM Treasury, 2011 and JRC-IES, European Commission Joint Research Centre, Institu). There may also be cultural leanings towards qualitative or quantitative methods (Rizzo et al., 2016). A tiered approach, starting with more holistic qualitative appraisal, is still useful in these circumstances, in particular to (a) test the assumptions and relevance of any subsequent quantitative appraisals, and (b) provide a more broad ranging benchmark for comparison with the outcomes of quantitative assessments. Furthermore, Smith and Kerrison (2013) have found that simple sustainability appraisal led to the same remediation option selection as more complex appraisal, and can be used to reliably inform environmental management decisions on relatively simple land contamination projects.

As part of a tiered approach, qualitative sustainability assessment might be an important adjunct for cost benefit assessments, where these are required. The qualitative assessment provides a relatively simple means of identifying which of the many possible sustainability considerations are most greatly differentiated between the remediation or other project options being considered. The qualitative assessment provides both a holistic assessment which might be prohibitively expensive to execute as CBA, and also a rationale for a clear range of externalities which should be considered in the CBA. Additionally, qualitative or semi-quantitative assessments can be used in tandem with CBA to distinguish non-monetisable values between options. Such a combined 'cost-effectiveness' approach to decision has been adopted already for sustainable remediation decision making by public investors in Austria (Döberl et al., 2013).

SURF-UK GUIDANCE

SuRF-UK is coordinated by an independent charity, Contaminated Land: Applications in Real Environments (CL: AIRE). It operates *via* a Steering Group who have overseen a series of meetings and projects. Since 2009 SuRF-UK has produced a wide range of outputs, on the basis of funding and in-kind contributions from a wide range of public and private sector contributors from across the UK. These outputs are listed in Fig. 1 and are freely downloadable from www.claire.co.uk/surfuk.

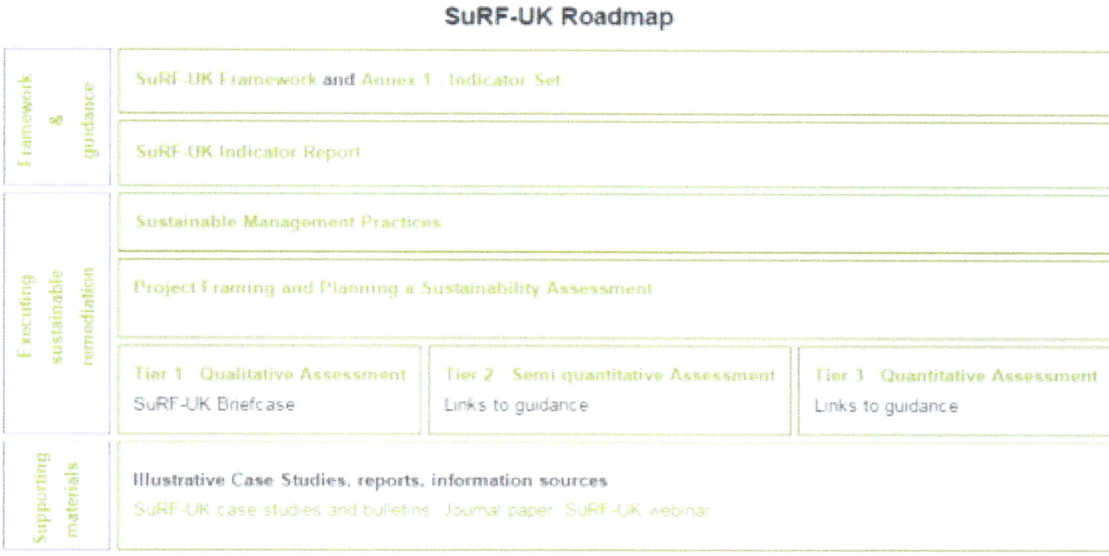

Figure. 1: SuRF-UK Outputs © CL: AIRE 2015, reproduced by permission.

Figure options

There are a range of potential benefits from sustainable remediation practice (CL: AIRE, 2014a):

- Supporting effective risk management
- Generating value by finding optimal solutions for soil and groundwater projects
- Identifying and avoiding project risks
- Demonstrable compliance with government and/or corporate policies and goals for sustainable development
- Providing a positive contribution towards delivery of corporate social responsibility (CSR) programmes, reputation and public relations
- Being a contributor to sustainable development.

SuRF-UK's 'framework for assessing the sustainability of soil and groundwater remediation' (CL:AIRE, 2010), and all of its guidance, has been developed on the basis of consultative processes with a wide range of practitioners drawn across different stakeholder types. The various reports of this process are all available from www.claire.co.uk/surfuk. From the outset SuRF-UK identified the requirement for simple, yet robust, approaches to assist sustainable remediation decision-making as the likely complexity of understanding sustainability was seen as being a potentially significant burden on project management decision-making effort and cost (CL:AIRE, 2007a and CL:AIRE, 2007b). Given that the adoption of sustainable remediation, in the UK at least, is effectively voluntary (CL: AIRE and NICOLE, 2015), significant increases in the cost of decision making disincentivises uptake.

Sustainability was also recognised as being subjective, and assessment approaches as lacking in transparency. Table 3 summarises the constraints on sustainability assessment methods relevant to sustainability management for contaminated sites. Another concern expressed form an early stage of the consultation process was that sustainability might be seen as a substitute criterion for risks to human health and the environment, and so undermine the level of protection from contaminated land management.

Table 3: Limiting factors for sustainability assessment in remediation decision making.[a]

Factor	Relevance and limitations
Subjectivity	Consideration of 'sustainability' in remediation is subjective (Bardos et al., 2011b; Pediaditi et al., 2005; RESCUE, 2005; Sardinha et al., 2013). Sustainable remediation/regeneration frameworks or decision support tools therefore need to provide a flexible, systematic, transparent and recordable process of making these subjective choices.

Stakeholder engagement and transparency	Inclusion of a broad range of stakeholder perspectives is a necessary part of understanding broader value and sustainability (Cundy et al., 2013; Hardi and Zdan, 1997; Pediaditi et al., 2005; Pediaditi et al., 2010; Plant et al., 2015; Rizzo et al., 2016; Sardinha et al., 2013). Practical constraints may mean that engagement activities have to be prioritised (e.g. Bartke and Schwarze, 2015). Simpler (qualitative) approaches seem more likely to be inclusive and engaging for a broader range of stakeholders than more complex processes, especially where assumptions may be 'hidden' in numeric form or the basis for valuations may be contentious (Bardos et al., 2011a).
Scope	In line with the 'Bellagio Principles', sustainability considerations are wide ranging across all three elements of sustainability, environmental, economic and social, see for example the SuRF-UK categories shown in Table 4.
Life cycle thinking	Life cycle thinking (LCT), making use of the disciplines of life cycle assessment, can provide some clear ground rules for effective sustainability assessment to ensure clear goals for the assessment and boundary setting (Bardos et al., 2011b). These disciplines are important parts of the framing of any sustainability assessment, whether it is qualitative, semi-quantitative or quantitative.
Framing an assessment and decision making cost	Assuming that whatever approach adopted has been adequately framed, and is adequately inclusive of stakeholder opinions, then the starting points for qualitative, semi-quantitative or quantitative approaches are broadly equivalent. Incremental costs will tend to be lowest for qualitative approaches as data and calculation effort is less, and consequently highest for quantitative methods. Quantitative methods also require framing and engagement and should not be used as a short cut to circumvent adequate framing and engagement

a Additional detail is available in the Supplementary Information accompanying this paper (See Table S1).

As a consequence SuRF-UK has set out clear underpinning principles for effective sustainable remediation which make explicit the primary role of risk assessment and management in contaminated land decision making (see Table 4) in its sustainable remediation framework. In addition, SuRF explicitly advocates a tiered approach to sustainability assessment to minimise cost and complexity in decision making. SuRF-UK has also provided guidance on identifying which sustainability considerations should be considered to ensure a consistent and holistic approach to determinations of scope, shown in Table 5 (CL:AIRE, 2011). The tiered approach, coupled with the SuRF-UK framework's underpinning principles and its guidance on scope also support engagement, consistency, ensuring suitable quality of evidence and transparency, to provide robust and effective decision-making (CL:AIRE, 2010). The ASTM (2013)standard and the proposed ISO standard (ISO 2015) also recognise the benefits of taking a tiered approach, and the SuRF-UK principles have been adopted in the proposed ISO standard.

Table 4: Key principles associated with sustainable remediation (CL:AIRE, 2010) © *CL:AIRE, 2015. Reproduced by permission.*

Principle 1: Protection of human health and the wider environment.	Remediation [site-specific risk management] should remove unacceptable risks to human health and protect the wider environment now and in the future for the agreed land-use, and give due consideration to the costs, benefits, effectiveness, durability and technical feasibility of available options.
Principle 2: Safe working practices.	Remediation works should be safe for all workers and for local communities, and should minimise impacts on the environment.
Principle 3: Consistent, clear and reproducible evidence-based decision-making.	Sustainable risk-based remediation decisions are made having regard to environmental, social and economic factors, and consider both current and likely future implications. Such sustainable and risk-based remediation solutions maximise the potential benefits achieved.a Where benefits and impacts are aggregated or traded in some way this process should be explained and a clear rationale provided.
Principle 4: Record keeping and transparent reporting.	Remediation decisions, including the assumptions and supporting data used to reach them, should be documented in a clear and easily understood format in order to demonstrate to interested parties that a sustainable (or otherwise) solution has been adopted.

Principle 5: Good governance and stakeholder involvement.	Remediation decisions should be made having regard to the views of stakeholders and following a clear process within which they can participate.
Principle 6: Sound science.	Decisions should be made on the basis of sound science, relevant and accurate data, and clearly explained assumptions, uncertainties and professional judgement. This will ensure that decisions are based upon the best available information and are justifiable and reproducible.

a In certain projects it is recognised that non-optimum remediation decisions may be made because other factors are more influential in optimising the benefit from a wider development scheme. Considering regulatory implications and recording why such a decision was taken should be a minimum requirement for any decision making process.

Table 5: Overarching SuRF-UK Sustainable Remediation Categories (CL:AIRE, 2011) © *CL:AIRE, 2015. Reproduced by permission.*

Environmental	Social	Economic
1 Emissions to air	1 Human health and safety	1 Direct economic costs and benefits
2 Soil and ground conditions	2 Ethics and equity	2 Indirect economic costs and benefits
3 Groundwater and surface water	3 Neighbourhoods and locality	3 Employment and employment capital
4 Ecology	4 Communities and community involvement	4 Induced economic costs and benefits
5 Natural resources and waste	5 Uncertainty and evidence	5 Project lifespan and flexibility

In parallel to providing a framework and guidance for the management and assessment of sustainability in remediation decision making, SuRF UK also recognised that simple operational guidance might significantly improve the sustainability of contaminated land practices. This was partly inspired by the best management practice guidance from the USEPA's 'Green Remediation' programme (US EPA 2008A). Hence overall SuRF-UK identifies two starting points for enabling more sustainable contaminated land management approaches, as part of a tiered process as shown in Fig. 2 (CL:AIRE, 2014b):

- Simple good management practices to mitigate known negative impacts and promote known benefits (sustainability management practices)
- Qualitative sustainability assessment for planning, design and option appraisal.

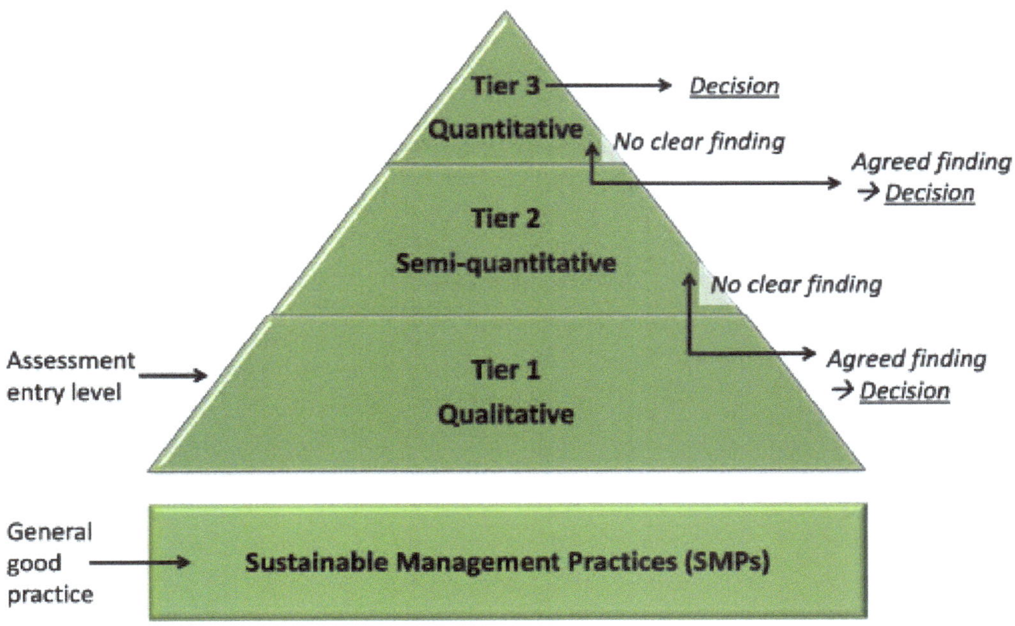

Figure 2: Tiered approach to sustainability assessment © *CL:AIRE 2015, from CL:AIRE,* 2014a, *reproduced by permission.*

There is a broad range of possible sustainability assessment approaches available at least for quantitative and semi-quantitative tools that might be applied to sustainability assessment, as previously described. Within each of the categories there may be a range of proprietary tools and services (*e.g.* Beames et al., 2014 and Ellis and Hadley, 2009). The SuRF-UK consultations indicated a general view that it would be counter-productive to try and reduce these down to a smaller set of SuRF-UK prescriptive methods, and that this exercise would also undermine competition between service providers and disenfranchise them from the (voluntary) SuRF-UK assessment process. SuRF-UK's conclusion is that the values and broad rules applied to sustainability assessment are more important than the exact details of methodology. SuRF-UK has therefore set out a series of broad procedural steps, as illustrated in Fig. 3, to support consistency across all methodologies in the assessment, design, implementation and reporting of sustainable remediation schemes and so establish a reproducible, transparent and robust approach. The application of these principles and procedural stages is specific to each site/project, and SuRF-UK has called this implementation process the 'framing' of the sustainability assessment (CL:AIRE, 2010).

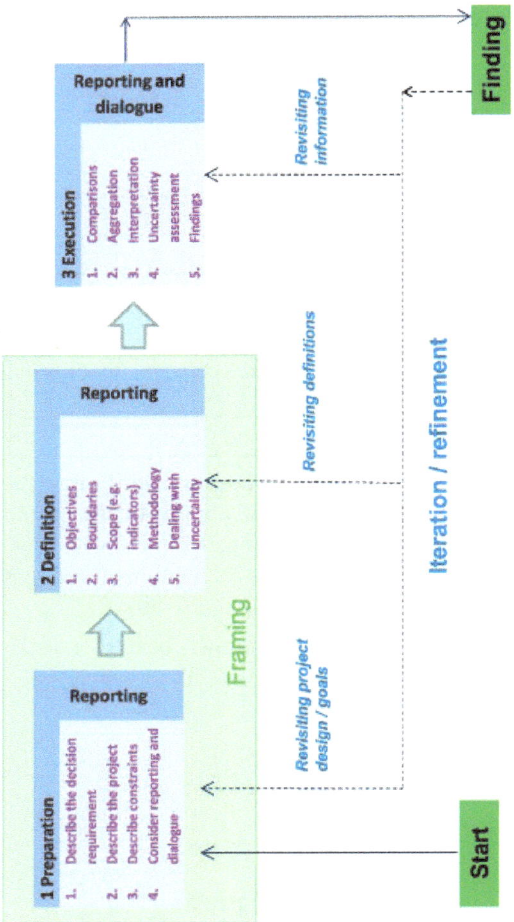

Figure 3: SuRF-UK Approach to Sustainability Assessment (CL:AIRE, 2014a) © *CL:AIRE, 2015. Reproduced by permission.*

Appropriate framing should underpin all sustainability assessments, even if they are only qualitative in nature. SuRF-UK's approach to framing is broadly consistent with the approach taken by the Dutch Sustainable Remediation Forum (SURF Netherlands et al., 2015) and NICOLE (NICOLE, 2011), and has been applied by the SURFs in Australia, New Zealand and other countries (Bardos et al., 2013 and SuRF-ANZ, 2011) as well as by the proposed ISO standard (ISO 2015).

The most recent outputs from SuRF-UK have therefore been concerned with this framing process, sustainable management practices; and also *qualitative* sustainability assessment, as the SuRF-UK consultations had shown that this was relatively poorly supported by published methodologies.

SURF-UK GUIDANCE ON FRAMING A SUSTAINABILITY ASSESSMENT

The overall structural approach to sustainability assessment favoured by SuRF-UK has already been reviewed (Bardos et al., 2011b). The importance of reproducible and transparent structures for sustainability assessment are made very clear in the Bellagio principles, and operational examples of goal setting and setting boundaries are clear from life cycle thinking. SuRF-UK guidance on the 'framing' of sustainability assessment for contaminated land decisions has been designed to be consistent both with the Bellagio Principles and lifecycle thinking.[6] Effective framing provides a more substantive basis for the contaminated land sector to deliver sustainability assessment tools with an appropriate structure and in line with SuRF-UK's sustainable remediation framework principles.

This guidance includes an interactive slide set (in *Adobe PDF* format) supported by a template for a 'log-book' to record decisions (in *Microsoft docx* format), and subsequently a spreadsheet to record assessments (in *Microsoft xlsx* format). These are freely downloadable from www.claire.co.uk/surfuk and are summarised in CL:AIRE (2014a). The guidance was reviewed as it was developed by a series of four practitioner workshops over 2012–13 across the UK, along with the sustainable management practices and qualitative assessment guidance, also reported on the SuRF-UK web site.

The framing process is needed for all tiers of a sustainability assessment process whether qualitative, semi-quantitative or quantitative. Framing includes two groups of activities each with a number of broad steps: the preparation for sustainability assessment followed by the definition of the sustainability assessment approach, as shown in Fig. 3.

- There are four broad steps in preparation for a sustainability assessment: (1) describing the decision requirement, (2) describing the project, (3) describing opportunities and constraints and (4) considering reporting and dialogue. These preparation activities provide the broad frame in which the sustainability assessment must be defined.
- The process of definition considers five issues: (1) objectives, (2) assessment boundaries, (3) scope – sustainable remediation indicators, (4) sustainability assessment methodology, and (5) dealing with uncertainty.

The framing guidance is based on hyperlinked slides that take the user backwards and forwards between slides at different levels of detail according to need. The 'logbook' is intended to assist note-taking by sustainability assessment teams (if required). The aim is to help project managers and sustainability assessors to frame their approach for a sustainability assessment, in several contexts:

- the framing slides as an interactive learning aid
- Use the framing slides as a step by step process of *aide memoire* to develop the sustainability assessment approach
- Use the framing slides to support discussions at meetings
- Use the logbook as a proforma for recording assumptions and findings.

SURF-UK GUIDANCE ON QUALITATIVE SUSTAINABILITY ASSESSMENT

As for the framing guidance, the SuRF-UK 'guidance on

qualitative assessment[7] comprises an interactive slide set (in *Adobe PDF* format) supported by a template for a 'log-book' to record decisions (in *Microsoft docx* format). A spreadsheet tool has been produced which systematically guides users through the key stages of preparation, definition and execution of a qualitative Tier 1 assessment, and encourages transparent documentation of decisions. These are freely downloadable from www.claire.co.uk/surfuk and summarised in CL:AIRE, 2014a.

While Tier 1 is the simplest tier, it still requires that sufficient framing and planning for the assessment has been carried out in advance. Furthermore, while the assessment is qualitative, readily available quantitative information can and should be exploited. The output of the assessment is comprised of simple tables using qualitative categories, such as 'good' or 'neutral' or 'better', or simple rankings (see Fig. 4). If these provide suitably clear differentiation between the options being compared, then more detailed assessment at Tier 2 and 3 may not be needed. The spreadsheet tool was developed to assist sustainability assessment work being undertaken on a commercial basis with multiple stakeholders. During this practical application development it was found that simple rankings (supported by a colour scheme from *green = best* to *red = worst*) provided the most clarity in differentiating options and indicating patterns of behaviours across the sustainability criteria used. While the sustainability assessments themselves are commercial in confidence, Fig. 5 illustrates an example of the framing and qualitative assessment outputs for a fictitious site.

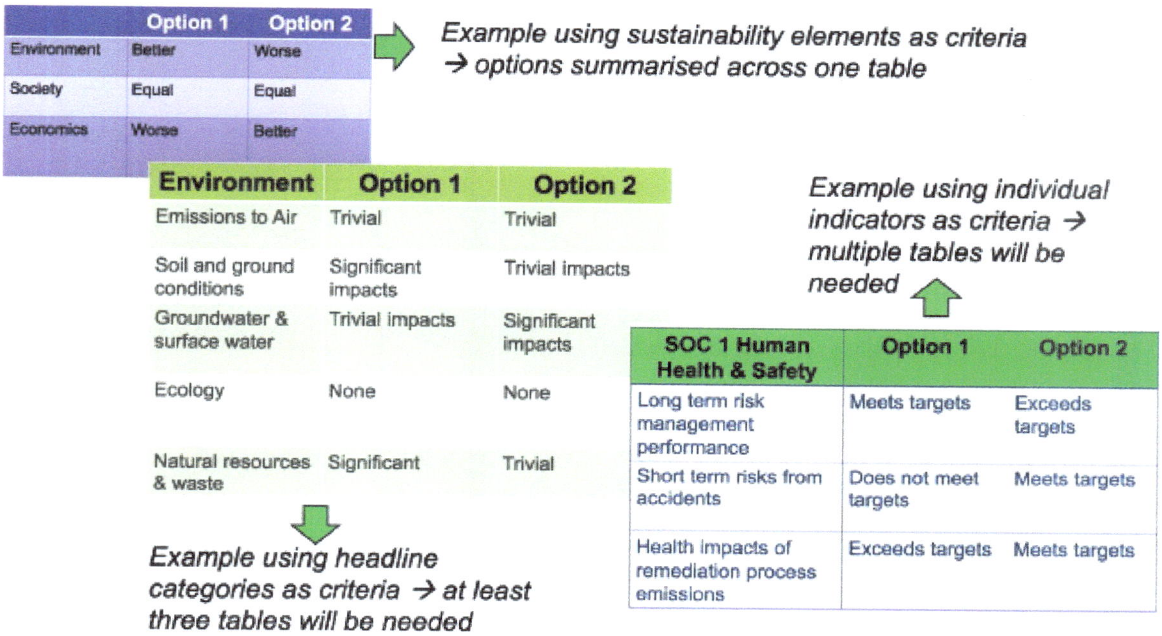

Figure. 4: Illustration of how the selection of *level of detail* for criteria across the SuRF-UK Indicator Set (CL:AIRE, 2011) affect the comparison table outputs © *CL:AIRE 2015, reproduced by permission.*

Assessment criteria	Key Categories	Remediation Options for Ranking					Justification of Scores (refer to 'Execution Supporting' Tab for more detail)	Uncertainty Arising (Description)	Uncertainty Arising (Extent)
		Option A	Option B	Option C	Option D	Option E			
Environmental									
Emissions to air	2*	5	4	3	1	2	CO_2, SOx, NOx dominated by power consumption, off site generation. Option D should be less energy intensive given that there is a significant reduction in operational time. Differences between extent of ozone depleting substances and particulates considered to be of marginal effect overall.	There is a high degree of confidence in the data used. Data used are semi quantitative with emission directly related to power consumption.	Timeframes are worst case consideration
Soil and ground conditions	1*	3	2	1	4	4	D and E achieve no benefit to soils, A & B marginal difference, little improvement, C best	There is a high degree of confidence in the data.	The extent of soil impact is reasonably well defined.
Groundwater and surface water	4*	3	4	2	1	5	Mass removal will have a beneficial effect if applied in the correct location. Based on mass removal, Option D will provide better coverage than Option C and result in betterment of groundwater quality over a larger area.	It is currently uncertain whether mass recovery will be required. Mass recovery estimates are based on knowledge of operation and performance of systems: field trials have been carried out so the extent of effective recovery for Option D is known.	Timeframes are worst case consideration
Ecology		1	1	1	1	1	Based on general disturbance &noise, at ground level and above, impacting bird and other small animals (No significant differences)	There are limited data on ecology at the sites as the driver for risk management is associated with groundwater impact.	The assesment is opinion based, considering limited data set, but also considering risk drivers.
Natural resources and waste	1*	4	3	1	2	5	Option E only has water waste, other options have water waste plus carbon plus PSH and ongoing chemical usage. Time frames of operation accounted for.	There is a high degree of confidence in the data available. Data used are semi quantitative based principally on waste water volumes as those are more significant.	The extent of waste production and natural resource usage is reasonably well defined.

Figure 5: Example qualitative assessment output, © *AECOM 2015, reproduced by kind permission of AECOM.*

SURF-UK GUIDANCE ON SUSTAINABLE MANAGEMENT PRACTICES

The use of good practice by the contaminated land sector has been encouraged in the UK for a few decades, with guidance on environmental risk management first published in 1995 (Department of the Environment, 1995), since superseded in 2000 (DETR, 2000) and 2011 (Defra, 2011). This is supported by a robust range of standards, codes of practice and technical guidance published by authoritative institutions and organisations. Framework documents of note that highlight key principles to be followed and signpost readers to technical advice and guidance include the '*Guiding Principles for Land Contamination*' (Environment Agency, 2010) and the '*Model Procedures for the Management of Land Contamination*' (Environment Agency and Defra, 2004).

The application of best practice was introduced, but not defined, in a wider context for green remediation in Environmental Protection Agency's (USEPA) 'Principles for Greener Cleanups' (2009). This was extended to green and sustainable remediation by the Interstate Technology and Regulatory Council (ITRC, 2011); following the principles introduced in USEPA (2009), but also includes consideration of social elements. The development of SuRF-UK sustainable management practices was informed by these US initiatives. SuRF-UK defines sustainable management practices (SMPs) as 'relatively simple, common sense actions that can be implemented at any stage in a land contamination management project to improve its environmental, social and/or economic performance' (CL:AIRE, 2014a and CL:AIRE, 2014b).

The SuRF-UK SMPs have been linked to the main stages for the management of land contamination (based on two generic management activities and the stages of management of land contamination outlined in Environment Agency and Defra (2004):

- Procurement (ITRC, 2011 and WRAP — Waste and Resources Action Programme, 2003)
- Land use planning (CL:AIRE, 2010 — Stage A assessment)
- Risk assessment (primarily Site Investigation)
- Options Appraisal
- Implementation of remediation — Design
- Implementation of remediation — Construction and Operation
- Implementation of remediation — Verification/Long-term Monitoring and Closure

SMPs can be used to improve the benefits (*e.g.* resource efficiency, cost) or reduce the negative impacts (*e.g.* spillages, complaints) of a project, leading to project 'sustainability gains'. SMPS are intended for use without the need for a formal sustainability assessment. They may also be used where sustainability gains are sought at a programme of work level using generic criteria or standards that can apply to a range of project types. Hence, the use of SMPs is seen by SuRF-UK as very much an entry level activity underpinning whatever additional sustainability based decision making takes place. The use of SMPs to achieve 'sustainability gains' without formal assessment is similar in concept to the 'Level 1' assessment for *Green and Sustainable Remediation*described by ITRC (2011). Fig. 6 describes the implementation process for making use of these SMPs.

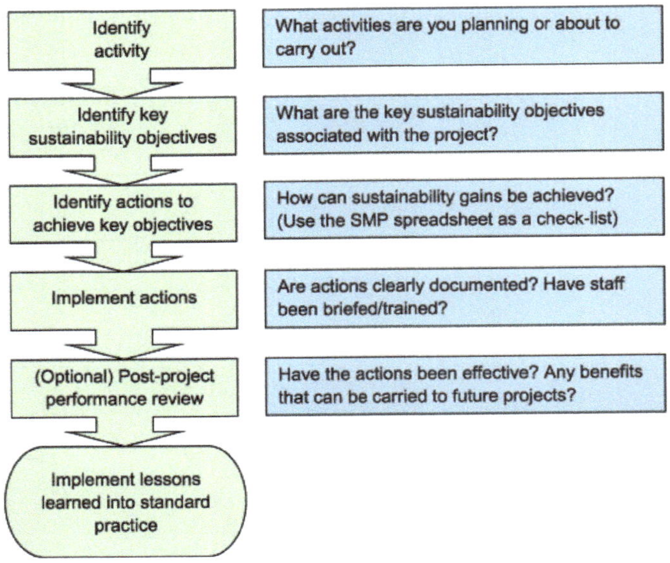

Figure 6: Making use of SuRF-UK Sustainable Management Practices © *CL:AIRE 2015, reproduced by permission.*

SMPs offer a way of changing behaviours or actions to reduce the cost, use of natural resources and/or the negative impact on community or the environment. Actions are mapped against the SuRF-UK indicator categories (see Table 5) to place even simple and low cost actions in a sustainability context. Table 6 sets out selected SMPs identified by SuRF-UK, the source of information, and how they map to SuRF-UK indicator categories. The SMPs themselves were collated from published best practice guidance and codes of practice developed for the land contamination and other sectors in the UK and abroad, as shown for the SMPs in Table 6. The SMPs are mapped against the indicator categories used by SuRF-UK to place simple and low cost actions in a sustainability context. A policy of preventing machinery and transport from being left idling can have multiple sustainability benefits across the environmental, social and economic pillars. Indeed, the Oak Ridge National Laboratory published best practice guidance on idle reduction to improve fuel efficiency, reduce emissions, reduce engine wear and reduce health impacts (ORNL, 2013).

Table 6: Selected sustainable management practices (SMPs) identified by SuRF-UK. Adapted from CL:AIRE (2014b).

SMP	Applicable SuRF-UK indicator category					Source
Adopt a sustainable procurement policy	ALL					WRAP (2003), SPRCa

Implement the Considerate Constructors Scheme Code of Considerate Practice	SOC1	SOC3	ECON2	SOC4	ENV4	http://www.ccscheme.org.uk/index.php/ccs-ltd
Consider non-intrusive surveys to delineate sources and areas affected by contamination	ENV5	SOC5	ECON1			BSI (2011)
Plan to re-use boreholes through each phase of investigation, remediation and long-term monitoring	ENV5	SOC1	SOC3	ECON1		US EPA (2008b,2009)
Consider the use of a mobile laboratory and/or field testing techniques to reduce off-site shipment of samples and improve spatial data	ENV5	SOC3	SOC5	ECON1	ENV1	ITRC (2011)
Identify location of underground services before excavation or drilling	SOC1	SOC3	ECON2			HSE (2014)
Identify drainage systems on-site and design measures to mitigate pollution risks	ENV3	ECON2	SOC3			EA (2012, 2013)
Ensure proper maintenance of vehicles, plant & equipment	ENV1	ENV5	SOC1	SOC3	ECON1	US EPA (2010b); VOSA (2013)
Consider electronic data transfer systems for information exchange between parties, reporting and archiving	ENV5	SOC5	ECON1			ITRC (2011)
Work with local communities to preserve characteristics of the locality	SOC2	SOC4				GLA (2007)

a SPRC — Sustainable Procurement Resource Centre www.sustainable-procurement.org/resources.

SuRF-UK's contention is that SMPs provide practical and generally inexpensive actions that can yield demonstrable 'sustainability gains' for a project. They should be selected where there is a clear benefit in doing so on a project-by-project basis. The SMPs are provided in a *Microsoft Excel* spreadsheet file, downloadable from www.claire.co.uk/surfuk. A report is also available that describes the development of SMPs and instructions for use of the SMP spreadsheet (CL:AIRE, 2014b).

CONCLUSIONS

The wider impacts of remediation have been recognised as important decision criteria in remedy selection for more than 20 years. This consideration was made explicit and extended to recognise the importance of sustainability in risk management decision making by the CLARINET network in 2002. However it is only more recently that the structures (tools, guidance, protocols) for achieving this have begun to emerge; catalysed by the emergence of the sustainable remediation fora in the mid-to-late 2000s. The current international interest in 'sustainable remediation' has resulted in a fairly rapid consensus on descriptions and definitions that are now being crystallised into an ISO standard.

However, approaches to methodologies in detail remain diverse with a range of (mainly) semi-quantitative and quantitative tools developed, or in development. SuRF-UK recognised the need of practitioners to adapt and develop approaches specific to their own needs, but has also identified the need to provide some underpinning guidance ('framing') so that the overall structure of these approaches is consistent, and complies with a set of key principles.

The importance of taking a holistic approach has been underlined by a recent survey undertaken by the Austrian Environmental Protection Agency (reported in Rizzo et al., 2016), which made an analysis of case studies in papers submitted to a number of recent scientific conferences. This found that assessments tended towards 'off the shelf' methodologies such as carbon footprint or life cycle assessments to assess environmental impacts. However a number of key impacts, related for instance to quality of soil, or to the landscape, are not always addressed. This was seen as a sign of the relative immaturity of consultancy practice in sustainable remediation (Döberl and Muller-Grabherr, 2015).

Sustainability assessment is site specific and subjective. There is not, and it is difficult to conceive, an SI unit for sustainability. It depends on the inclusion of a wide range of factors across different stakeholder perspectives. Taking a tiered approach to sustainability assessment offers important advantages, starting from a qualitative assessment and moving through to semi-quantitative and quantitative assessments on an 'as required basis' only. These benefits are (1) the level of effort in decision making is proportionate; (2) a tiered approach supports a more inclusive and transparent approach; (3) the relative strengths and weaknesses of the different tiers are combined in a way that allows both a holistic assessment, and if necessary more detailed assessments for example CBA to be mutually reinforcing; and (4) a tiered approach provides a clear rationale for detailed assessments to be specifically focussed on considerations of high importance. The availability of guidance for 'entry level' qualitative sustainability assessment was found to be lacking and SuRF-UK has therefore provided introductory guidance that is free to access and apply.

It is also clear that there are a number of 'easy wins' that could reduce a range of wider impacts right across the site management process. SuRF-UK has provided a checklist of 'sustainable management practices' to support mitigation of the wider impacts of site management.

SuRF-UK's guidance on framing, qualitative sustainability assessment and sustainable management practices was only released in 2014. The concept of framing has been taken forward by the developing ISO/DIS 18504 standard on Sustainable Remediation.

Across all of the SURF initiatives the major imperatives now are on the collection of case studies of the practical implementation of sustainable remediation concepts in contaminated land management projects, and on the dissemination and training of the broader contaminated land community. This evidence base will grow, and with this tools and approaches can be refined and broader analyses conducted, for example by the academic community. This will lead to better tools and guidance, broader understanding and wider adoption – and ultimately more sustainable development.

ACKNOWLEDGEMENTS

This work was supported by members of the SuRF-UK steering group *via* financial and in-kind contributions. The authors would like to acknowledge the many comments and experiences shared by members of the practitioner community through the development of all of SuRF-UK's outputs since 2007. We would also like to thank Ged Sojka of AECOM for providing Fig. 5, and the kind assistance of Johan van Veen and Frank Swartjes regarding early stage work in The Netherlands.

REFERENCES

1. Aelmans, A.J., Hoekstra, E.J., Leget, A.C., van Veen, H.J., 1993. Sustainable soil use a TNOconcept. TNO Environmental and Energy Research Report. Available from www. eugris.info/displaynewsitem2. asp?NewsID=970; Accessed August 2015.
2. ASTM 2013 Standard Guide for Integrating Sustainable Objectives into Cleanup, E2876- 13, ASTM International, West Conshohocken, PA, 2013, www.astm.org
3. Atkinson, G., Mourato, S., 2008. Environmental cost benefit analysis. Annu rev. Environ. Resourc. 33, 317–344.
4. Balasubramaniam, A., Voulvoulis, N., 2005. The appropriateness of Multicriteria Analysis in environmental decision-making problems. Environ. Technol. 26 (9), 951–962.
5. Bardos, P., 2014. Progress in sustainable remediation. Remediation Journal Winter 2014, 23–32. http://dx.doi.org/10.1002/rem.21412.
6. Bardos, R.P., Lewis, A. J., Nortcliff, S., Mariotti, C., Marot, F. and Sullivan, T. 2002 Review of Decision Support Tools for Contaminated Land Management, and their use in Europe. Final Report. Austrian Federal Environment Agency, 2002 on behalf of CLARINET, Spittelauer Lände 5, A-1090 Wien, Austria. www.commonforum.eu/publications_ clarinet.asp. Accessed July 2015
7. Bardos P., Bakker, L., Slenders, H. and Nathanail, P. 2011a Sustainable remediation. Pp 889–948 in. Swartjes F.A. (Ed.), Book on Contaminated Sites. From Theory Towards Practical Application, Springer Publishers, Dordrecht. ISBN: 978-90-481-9756-9.
8. Bardos, R.P., Bone, B.D., Boyle, R., Ellis, D., Evans, F., Harries, N., Smith, J.W.N., 2011b. Applying sustainable development principles to contaminated land management using the SuRF-UK framework. Remediation Journal Spring 2011, 77–100.
9. Bardos, P., Knight, M. and Humphrey, S. 2012 Sustainable remediation. Environmental Scientist August 2012 pp 46–49. www.ies-uk.org.uk/resources/contaminated-issue
10. Bardos, P., Bakker, L., Darmendrail, D., Harries, N., Holland, K., MacKay, S., Pachon, C., Slenders, H., Smith, G., Smith, J., Wiltshire, L., 2013. Sustainable and green remediation — global update, ThS E3 sustainable use of the subsurface. Proceedings Aquaconsoil 2013 Barcelona, Spain. April 2013. http://dx.doi.org/10.13140/2.1.1812. 2887.
11. Bartke, S., Schwarze, R., 2015. No perfect tools: trade-offs of sustainability principles and user requirements in designing support tools for land-use decisions between green- fields and brownfields. J. Environ. Manag. 153, 11–24.
12. Beames, A., Broekx, S., Lookman, R., Touchant, K., Seuntjens, P., 2014. Sustainability appraisal tools for soil and groundwater remediation: how is the choice of remediation alternative influenced by different sets of sustainability indicators and tool structures? Sci. Total Environ. 470–471, 954–966.
13. Beinat, E., van Drunen, M.A., Janssen, R., Nijboer, M.H., Loolenbrander, J. G. M., Okx, J.P., Schutte, A.R. 1998 The REC Decision Support System for comparing soil remediation alternatives. Phase 2: A methodology based on Risk reduction, Environmental merit and Costs. NOBIS-report 95-1-03, The Netherlands. http://soilpedia.nl/Bikiwiki% 20documenten/NOBIS/Risicoreductie,%20Milieuverdienste%20en%20 Kosten%20(RMK)%20-%2095-1-03/95- 1-03%20rap2.pdf Accessed July 2015
14. Bender, T., Volkwein, S., Battermann, G., Hurtig, H.W., Klopffer, W., Kohler, W., 1998. Life cycle assessment for remedial action techniques: Methodology and Application. In Contaminated Soil'98, Proc 6th Intern FZK/TNO Conference on Contaminated Soil, 17-21 May 1998. Edinburgh, UK. ISBN: 07277 26757, pp. 367–376.

15. BenDor, T.K., Metcalf, S.S., Paich, M., 2011. The dynamics of brownfield redevelopment. Sustainability 2011 (3), 914–936. http://dx.doi.org/10.3390/su3060914.

16. Bleicher, A., Groß, M.S., 2010. Sustainability assessment and the revitalization of contaminated sites: operationalizing sustainable development for local problems. International Journal of Sustainable Development & World Ecology 17 (1), 57–66.

17. Bromley, R.D.F., Humphrys, G. (Eds.), 1979. Dealing With Dereliction the Redevelopment of the Lower Swansea Valley. Pub University of Swansea, Swansea, Wales.

18. Brouwer, R., van Ek, R., 2004. Integrated ecological, economic and social impact assessment of alternative flood control policies in The Netherlands. Ecol. Econ. 50, 1–21.

19. Brundtland, G.H. 1987 Our Common Future. World Commission on Environment and Development. Oxford University Press. www.un-documents.net/our-common-future. pdf Accessed August 2015

20. BSI — British Standards Institution 2011 Investigation of potentially contaminated sites — Code of Practice. British Standards Institution, London. BS 10175:2011.

21. CABERNET 2006 Sustainable brownfield regeneration: CABERNET network report. Eds: Ferber, U., Grimski, D., Millar, K. and Nathanail, P. ISBN 0-9547474-5-3 (University of Nottingham, Nottingham)

22. Calcas Consortium — Co-ordination Action for innovation in Life-Cycle Analysis for Sustainability 2009 D20 Blue Paper on Life Cycle Sustainability Analysis. EC Framework 6 Programme Project no.037075. http://fr1.estis.net/sites/calcas/default.asp Accessed July 2015

23. Cappuyns, V., Kessen, B., 2013. Combining life cycle analysis, human health and financial risk assessment for the evaluation of contaminated site remediation. J. Environ. Plan. Manag. http://dx.doi.org/10.1080/09640568.2013.783460.

24. Carbon Trust, Defra and BSI British Standards 2008b PAS 2050:2008 Specification for the assessment of the life cycle GHG emissions of goods and services. BSI, 389 Chiswick High Road, London W4 4AL, UK. IISBN 978 0 580 50978 0

25. Cellini, S.R. and Kee, J.E. 2010 Cost-Effectiveness and cost–benefit analysis. Pp 493–530 in Handbook of Practical Program Evaluation, 3rd Edition (Eds Wholey, J.S., Hatry, H.P. and Newcomer, K.E.). Wiley. ISBN: 978-0-470-52247-9

26. CL:AIRE (2007a and onwards Meetings, Presentations and Consultations. Listing of activities since 2007. http://www.claire.co.uk/index.php?option=com_content&view= article&id=745:meetings-presentations-and-consultations&catid=963:surfuk&Itemid=78 Accessed July 2015

27. CL:AIRE 2007b. Measuring sustainability: what's in a Number? subr:im Bulletin SUB 4. CL: AIRE, London, UK. www.claire.co.uk Accessed August 2015

28. CL:AIRE 2010. A framework for assessing the sustainability of soil and groundwater remediation. March 2010, CL:AIRE, London, UK. ISBN 978-1-905046-19-5. www.claire.co. uk/surfuk Accessed August 2015

29. CL:AIRE, 2011 The SuRF-UK indicator set for sustainable remediation assessment, CL:AIRE London, UK. ISBN 978-1-905046-1292-5 DoI DOI: 10.13140/RG.2.1.3130.8246 www. claire.co.uk/surfuk Accessed August 2015

30. CL:AIRE 2014a The SuRF-UK Bulletin 4, March 2014, CL:AIRE London, UK. www.claire.co. uk/surfuk Accessed August 2015

31. CL:AIRE 2014b. Sustainable Management Practices for Management of Land Contamination, March 2014, CL:AIRE London, UK. www.claire.co.uk/surfuk Accessed August 2015

32. CL:AIRE and NICOLE 2015 A review of the legal and regulatory basis for sustainable remediation in the European Union and the United Kingdom, CL:AIRE, London, UK. ISBN 978-1-905046-27-0 Accessed August 2015

33. CLARINET and NICOLE 1998 Better decision making now. The use of Risk Assessment and Risk Management for tackling the problems of contaminated land. NICOLE, Netherlands. www.nicole.org/uploadedfiles/1998-the-use-of-risk-assesment-andrisk-management.pdf Accessed July 2015.

34. Commonwealth of Australia 2006 Introduction to Cost-Benefit Analysis and Alternative Evaluation Methodologies January 2006 FINANCIAL MANAGEMENT REFERENCE MATERIAL NO. 5 ISBN 1 921182 02 4 www.finance.gov.au/sites/default/files/Intro_to_ CB_analysis.pdf

35. Cundy, A.B., Bardos, R.P., Church, A., Puschenreiter, M., Friesl-Hanl, M., Müller, I., Neu, S., Mench, M., Witters, N., Vangronsveld, J., 2013. Developing principles of sustainability and stakeholder engagement. J. Environ. Manag. 129, 283–291.

36. Curran, J. and Hart, I. 1998 Sustainable redevelopment: incorporation into project design. pp 201–207. In Land Reclamation Achieving Sustainable Benefits. (Fox, H.R., Moore, H.M., McIntosh, A.D. Eds.) Proc. 4th Intern Conf. Intern. Affiliation of Land Reclamationists. Nottingham, UK, 7–11 Sept 1998. Pub Balkema, ISBN 905809002 7.

37. Department for Communities and Local Government — DCLG 2006 Environmental Impact Assessment: a guide to good practice and procedures. A consultation Paper, DCLG, London, http://webarchive.nationalarchives.gov.uk/20100410180038/http:/ communities.gov.uk/planningandbuilding/planning/sustainabilityenvironmental/environmentalimpactassessment/, Accessed November 2015

38. Department for Environment Food and Rural Affairs — Defra 2006 Carbon balances and energy impacts of the management of UK wastes. Defra R&D Project WR0602 Final Report. ERM and Golder Associates Defra, London, UK. http://randd.defra.gov.uk/ Default.aspx?Menu=Menu&Module=More&Location=None&

Completed=0&- ProjectID=14644 Accessed November 2015

39. Defra — Department for Environment, Food and Rural Affairs 2010 Valuing environmental impacts: practical guidelines for the use of value transfer in policy and project appraisal technical report February 2010. Economics for the Environment Consultancy. https://www.gov.uk/ecosystems-services Accessed October 2013

40. Defra — Department for Environment, Food and Rural Affairs 2011 Guidelines for environmental risk assessment and management. https://www.gov.uk/government/ publications/guidelines-for-environmental-risk-assessment-and-managementgreen-leaves-iii Accessed August 2015

41. DETR — Department of the Environment, Transport and the Regions, Environment Agency and Institute for Health and the Environment, 2000g. Guidelines for Environmental Risk Assessment and Management. The Stationery Office, Norwich, UK.

42. Department of the Environment, 1995. A Guide to Risk Assessment and Risk Management for Environmental Protection. HMSO, London, UK.

43. Döberl, G. andMuller-Grabherr, D. 2015 Recenttrends in the assessment of sustainable remediation — does the tail wag the dog? Page 109. Programme and Book of Abstracts of the 13th International UFZ-Deltares Conference on Sustainable Use and Management of Soil, Sediment and Water Resources 9–12 June 2015, Copenhagen, Denmark. http://www. aquaconsoil.org/assets/acs2015__cg_web.pdf Accessed July 2015

44. Döberl, G., Ortmann, M. and Frühwirth, W. 2013 Introducing a goal-oriented sustainability assessment method to support decision making in contaminated site management Env, Sci. Policy 25 207–217

45. Doick, K.J., Sellers, G., Castan, B.V., Silverthorne, T., 2009. Understanding success in the context of brownfield greening projects: the requirement for outcome evaluation in urban green space success assessment. Urban Forestry and Urban Greening. 8 (3), 163–178.

46. Ellis, R., Hadley, P., 2009. Sustainable remediation white paper—integrating sustainable principles, practices, and metrics into remediation projects. Remediation Journal September 2009, 5–114.

47. Environment Agency 1999 Cost Benefit Analysis for Remediation of Land Contamination, R&D Technical Report P316, prepared by Risk Policy Analysts Ltd. and WS Atkins. Environment Agency, Bristol, UK. ISBN 185705 0371.

48. Environment Agency 2000a Assessing the Wider Environmental Value of Remediating Land Contamination. Bardos, R.P., Nathanail, C.P., and Weenk, A. Environment Agency R&D Technical Report P238. Environment Agency, Bristol, UK. ISBN 185705 0371, DOI: 10.13140/2.1.1255.6486

49. Environment Agency 2000b Costs and Benefits Associated with Remediation of Contaminated Groundwater: Framework for Assessment, R&D Technical Report P279, prepared by Komex Clarke Bond & EFTEC Ltd., Environment Agency, Bristol, UK.

50. Environment Agency 2009 Human Health Toxicological Assessment of Contaminants in Soil. Science Report — Final SC050021/SR2 Environment Agency, Bristol, UK. ISBN: 978-1-84432-858-1

51. Environment Agency 2010 Guiding principles for land contamination. Environment Agency, Bristol, UK. https://www.gov.uk/government/publications/managing-andreducing-land-contamination Accessed August 2015

52. Environment Agency 2012 Working at construction and demolition sites: PPG6 pollution prevention guidelines PMHO0412BWFE-E-E. Environment Agency, Bristol, UK. www. gov.uk/government/uploads/system/uploads/attachment_data/file/290139/ pmho0412bwfe-e-e.pdf Accessed August 2015

53. Environment Agency 2013 Pollution prevention guideline PPG 1. Environment Agency, Bristol, UK. www.gov.uk/government/uploads/system/uploads/attachment_data/ file/290124/LIT_1404_8bdf51.pdf Accessed August 2015

54. Environment Agency and Department for Environment, Food and Rural Affairs 2004 Model procedures for the management of land contamination. Environment Agency R&D Report CLR11. Environment Agency, Bristol, UK. www.gov.uk/government/ uploads/system/uploads/attachment_data/file/297401/scho0804bibr-e-e.pdf Accessed July 2015

55. EURODEMO Project Consortium. 2007. Framework for sustainable land remediation and management. Deliverable reference number: D 5-3. Project no. (GOCE) 003985. Umweltbundesamt, Vienna, Austria.

56. Ferdos, F., Rosén, L., 2013. Quantitative environmental footprints and sustainability evaluation of contaminated land remediation alternatives for two case studies. Remediat. J. 24 (1), 77–98.

57. Ferguson, C., Darmendrail, D., Freier, K., Jensen, B.K., Jensen, J., Kasamas, H., Urzelai, A., Vegter, J., 1998. Risk Assessment for Contaminated Sites in Europe. Volume 1. Scientific Basis. LQM Press, Nottingham.

58. Ferdinandy, M., and Weenk, A. 1999 Milieubeoordeling van verwerkingsketens voor verontreinigde baggerspecie. Eindrapport… RIZA rapport: 97.04. ISBN: 90 369 51 216. www.helpdeskwater.nl/publish/pages/404/posw_2_deel_22.pdf Accessed August 2015

59. Finnamore, J. R., Denton, B. and Nathanail, C.P. 2000 Land contamination — management of financial risk. Report C545, Construction Industry Research and Information Association (CIRIA) publications, London, ISBN 0 86017 545 6

60. GLA — Greater London Authority 2007 Planning for equality and diversity in London. Supplementary Planning Guidance. GLA, London, UK. ISBN 978 1 84781 068 7 www. london.gov.uk/sites/default/files/archives/mayor-strategies-sds-docs-spg-planningfor-equality.pdf Accessed August 2015

61. Grimski, D., Doetsch, P., Rupke, A., 1998. Brownfield versus Greenfield sites under economic and long term environmental considerations. in Proceedings

of the Sixth International FZK/TNO Conference on Contaminated Soil, Edinburgh 17-21 May 1998 vol 1. Published by Thomas Telford, London, pp. Pp 651–660.

62. Harbottle, M.J., Al-Tabbaa, A., Evans, C.W., April 2008a. Sustainability of land remediation, Part 1: overall analysis. Proceedings of the Institution of Civil Engineers Geotechnical Engineering 161 (Issue GE2), 75–92. http// www.icevirtuallibrary.com/content/ article/10.1680/ geng.2008.161.2.75 Accessed August 2015.

63. Harbottle, M.J., Al-Tabbaa, A., C.W, Evans, June 2008b. Sustainability of land remediation, Part 2: impact assessment. Proceedings of the Institution of Civil Engineers Geotechnical Engineering 161 (Issue GE3), 117–127 www.icevirtuallibrary.com/content/ article/10.1680/geng.2008.161.3.117 Accessed August 2015.

64. Hardi, P. and Zdan, T.J. 1997 Assessing sustainable development International Institute for Sustainable Development, 161 Portage Avenue East — 6th Floor, Winnipeg, Manitoba, R3B 0Y4, Canada, ISBN 1-895536-07-3, http://www.iisd.org/pdf/bellagio.pdf Accessed August 2015

65. HM Treasury 2011 The green book. Appraisal and Evaluation in Central Government. Treasury Guidance, TSO, London. www.gov.uk/government/uploads/ system/ uploads/attachment_data/file/220541/green_ book_complete.pdf Accessed August 2015

66. Holland, K., Lewis, R., Tipton, K., Karnis, S., Dona, C., Petrovskis, E., Bull, L., Taege, D., and Hook, C. 2011. Framework for integrating sustainability into remediation projects. Remediation Journal, 21(3):7–38, see also 'Supplement to Framework for Integrating Sustainability into Remediation Projects June 2011' www.sustainableremediation. org/library/ guidance-tools-and-other-resources/metrics-toolbox/ Framework% 20supplement.pdf Accessed July 2015

67. HSE — Health and Safety Executive 2014 Avoiding danger from underground services. HSG47 (Third Edition). Health and Safety Executive. www.hse.gov.uk/ pUbns/ priced/hsg47.pdf Accessed August 2015

68. Holland, K., Karnis, S., Kasner, D.A., Butler, P.B., Hadley, P.W., Nathanail, P., Ryan, J., Smith, L.M., Wice, R., 2013. Integrating remediation and reuse to achieve whole-system sustainability benefits. Remediation Journal Spring 2013, 5–17. http://dx.doi.org/10. 1002/ rem.21345.

69. Hou, D., Al-Tabbaa, A., 2014. Sustainability: a new imperative in contaminated and remediation. Environmental Science and Policy 39, 25–34.

70. International Standards Organisation — ISO 2015 ISO/DIS 18504 Soil quality – Guidance on sustainable remediation http://www.iso.org/iso/home/store/ catalogue_tc/ catalogue_detail.htm?csnumber=62688 Accessed November 2015

71. ITRC — Interstate Technology and Regulatory Council 2011 Green and sustainable remediation: a practical framework. Framework Report GSR-2 ITRC, Washington. www. itrcweb.org/GuidanceDocuments/ GSR-2.pdf Accessed August 2015

72. JRC-IES, European Commission Joint Research Centre, Institute for Environment and Sustainability 2011 Analysis of existing environmental footprint methodologies for products and organizations: recommendations, Rationale, and Alignment. Deliverable 1 to the Administrative Arrangement between DG Environment and Joint Research Centre No. N 070307/2009/552517, including Amendment No 1 from December 2010. http:// ec.europa.eu/environment/eussd/pdf/Deliverable.pdf Accessed August 2015

73. Licht, L.A., Isebrands, J.G., 2005. Linking phytoremediated pollutant removal to biomass economic opportunities. Biomass Bioenergy 28, 203–218.

74. Morio, M., Schädler, S., Finkel, M., Justen, A., Bleicher, A., Bartke, S. and Gross, M. 2011 Megasite management tool suite (MMT) users guide. Version 1.2.0, March 2011. www.ufz.de/index.php?en=19610 Accessed July 2015

75. Nathanail, C. P., Bardos, R.P., Gillett, A., McCaffrey, C., Ogden, R., Scott, D. and Nathanail, J. 2013 International Processes for Identification and Remediation of Contaminated Land. Final Report Project SP1004, Defra, London, UK. http://randd.defra.gov.uk/ Default.aspx?Menu=Menu&Module=More&Location=None&Completed=0&- ProjectID=16289#Description Accessed July 2015

76. NAVFAC — Navy Facilities Engineering Command 2012 Department of the Navy guidance on green and sustainable remediation. User's Guide UG-2093-ENV Rev.1. NAVFAC, Port Hueneme, California 93043–4370, USA. www.navfac.navy.mil/content/dam/ navfac/Specialty%20Centers/Engineering%20and%20 Expeditionary%20Warfare% 20Center/Environmental/ Restoration/er_pdfs/gpr/navfacesc-ev-ug-2093-env-gsr-20120405r1.pdf Accessed August 2015

77. NAVFAC — Navy Facilities Engineering Command 2014 Green and sustainable remediation. ESAT N62583-11-D-0515/TO 0046 NAVFAC, Port Hueneme, California 93043– 4370, USA. www.navfac.navy.mil/content/dam/ navfac/Specialty%20Centers/ Engineering%20and%20 Expeditionary%20Warfare%20Center/Environmental/ Restoration/er_pdfs/g/navfac-ev-fs-gsr-201405v7.pdf Accessed August 2015

78. NICOLE 2011 Road map for sustainable remediation (co-author) NICOLE secretariat, Dutch Sino Business Promotions, Postbus 28249, 3003 KE Rotterdam, The Netherlands. www.nicole.org/pagina/22/Thematic_ Documents.html Accessed August 2015

79. NICOLE and COMMON FORUM 2013 Risk-informed and sustainable remediation. Joint Position Statement by NICOLE and COMMON FORUM, 9 June 2013. www. commonforum.eu/Documents/DOC/ PositionPapers/1177_DDC_FLYER_SR_Joint_ snijlijn_ def_2.pdf Accessed August 2015

80. NOBIS — Netherland Onderzoeksprogramma Biotechnologische In situ Sanering 1995a Risk Reduction, Environmental Merit and Costs. REC-Method, Phase 1. Document 95-1-03, CUR/NOBIS,

81. NOBIS — Netherland Onderzoeksprogramma

Biotechnologische In situ Sanering 1995b Risk Reduction, Environmental Merit and Costs. REC-Method, Phase 2: A methodology based on Risk reduction, Environmental merit and Costs. Document 951-03, CUR/ NOBIS, Gouda, The Netherlands,

82. Office of the Deputy Prime Minister 2005 A Practical Guide to the Strategic Environmental Assessment Directive. ODPM, London. https://www.gov.uk/government/uploads/ system/uploads/attachment_data/file/7657/practicalguidesea.pdf, Accessed November 2005

83. ORNL — Oak Ridge National Laboratory 2013 ORNL operations best practice guide: idle reduction. Sustainable Campus Initiative. ORNL/LTR-2013/102. https://sustainabilityornl.org/documents/ORNL%20Idle%20Reduction%20Guide.pdf Accessed August 2015

84. Pediaditi, K., Doick, K., Moffat, A.J., 2010. Monitoring and evaluation practice for brown- field regeneration to green space initiatives. A meta-evaluation of assessment and monitoring tools. Landsc. Urban Plan. 97, 23–36.

85. Pediaditi, K., Wehrmeyer, W., Chenoweth, J., 2005. Monitoring the sustainability of brownfield redevelopment projects. The redevelopment assessment framework (RAF). Contaminated Land and Reclamation 13 (2), 173–183.

86. Plant, R.A., Chong, J., Lederwasch, A., Prior, J., Asker, S., Boydell, S., 2015. CRC for Contamination Assessment and Remediation of the Environment. Value-based Land Remediation: Improved Decision-making for Contaminated Land, no. CRC CARE Technical Report No. xx, pp. 1–33 (see http://www.uts.edu.au/staff/roelof.plant, unpublished).

87. Pollard, S.J.T., Brookes, A., Earl, N., Lowe, J., Kearney, T.E., Nathanail, C.P., 2004. Integrating decision tools for the sustainable management of land contamination. Sci. Total Environ. 325 (1–3), 15–28.

88. RESCUE Consortium 2005 Best Practice Guidance for Sustainable Brownfield Regeneration. May 2005. Land Quality Press, a Division of Land Quality Management Ltd. ISBN 0-9547474-0-2

89. REVIT Consortium 2007 Working towards more effective and sustainable brownfield revitalisation policies. Ensuring sustainability in Brownfield revitalisation. www. hartvanzuid.nl/downloads/REVIT_booklet.pdf Accessed July 2015

90. Rizzo, E., Bardos, P., Pizzol, L., Critto, A., Giubilato, E., Marcomin, A., et al., 2016. Comparison of international approaches to sustainable remediation. Submitted to Journal Environmental Management (in prep).

91. Sardinha, I.D., Craveiro, D., Milheiras, S., 2013. A sustainability framework for redevelopment of rural brownfields: stakeholder participation at SÃO DOMINGOS mine, Portugal. Journal of Cleaner Production 57, 200–208.

92. Schädler, S., Morio, M., Bartke, S., Rohr-Zänker, R., Finkel, M., 2011. Designing sustainable and economically attractive brownfield revitalization options using an integrated assessment model. J. Environ. Manag. 92, 827–837.

93. Smith, M.A. – ed 1985 Contaminated Land: Reclamation and Treatment Publisher: New York: Plenum Press ISBN: 0306419289. NATO Challenges of Modern Society; v. 8

94. Smith, J.W.N., Kerrison, G., 2013. Benchmarking of decision-support tools used for tiered sustainable remediation appraisal. Water Air Soil Pollut 224 (1706). http://dx.doi.org/ 10.1007/s11270-013-1706-y.

95. Sorvari, J., Antikainen, R., Kosola, M.L., Hokkanen, P., Haavisto, T., 2009. Eco-efficiency in contaminated land management in Finland: barriers and development needs. J. Environ. Manag. 90 (5), 1715–1727.

96. Stroud District Council 2015 Development of potentially contaminated land — guidance for developers, Consultants and Agents. April 2015 (3rd Edition). https://www. stroud.gov.uk/info/contaminated_development.pdf Accessed November 2015

97. SURF — Sustainable Remediation Forum, 2009. Integrating sustainable principles, practices and metrics into remediation projects. In: Ellis, D., Hadley, P. (Eds.), Remediation Journal 19, pp. 5–114 (3).

98. SuRF-ANZ 2011 A framework for assessing the sustainability of soil and groundwater remediation. http://www.surfanz.com.au/pdfs/ SuRF%20Australia%20Framework%20April%2011%20FINAL.pdf Accessed July 2015

99. SURF Netherlands, SKB and Bodembreed Forum 2015 CAHIER Duurzaam saneren en beheren doe je zo! January 2015, SKB, The Netherlands. http://www.surf-nl.com/ publicaties/CAHIER_DuurzaamSaneren_Beheren.pdf Accessed July 2015.

100. Therivel, R. 2004 Sustainable urban environment — metrics, Models and Toolkits: Analysis of sustainability/ social tools, Report to the sue-MoT consortium. 9 June 2004 http://download.sue-mot.org/soctooleval.pdf Accessed July 2015

101. US EPA — US Environmental Protection Agency 1973 The quality of life concept. A potential new tool for decision makers. Available from http://www2.epa.gov/nscep. Accessed November 2015

102. US EPA — US Environmental Protection Agency 1992 Reclamation and redevelopment of contaminated land: Volume II European case studies. EPA/600/SR-92/031

103. US EPA — US Environmental Protection Agency 2008a Green remediation: Incorporating sustainable environmental practices into remediation of contaminated sites EPA/542/

104. R/08/002, USA. https://clu-in.org/download/techdrct/epa-542-f-08-002.pdf Accessed August 2015

105. US EPA — US Environmental Protection Agency 2008b Green Remediation: Best Management

106. Practices for Excavation and Surface Restoration. EPA 542-F-08-012. EPA Office of Superfund Remediation & Technology Innovation. http://epa.gov/tio/download/remed/gr_quick_ref_fs_exc_rest.pdf Accessed August 2015

107. US EPA — US Environmental Protection Agency 2009 Design and installation of monitoring wells. SESDGUID-101-R0. EPA Science & Ecosystem Support

Division. www.epa. gov/region4/sesd/fbqstp/Design-and-Installation-of-Monitoring-Wells.pdf Accessed August 2015

108. US EPA — US Environmental Protection Agency 2010a Superfund Green Remediation Strategy. September 2010. U.S. Environmental Protection Agency, Office of Solid Waste and Emergency Response, Office of Superfund Remediation and Technology Innovation www.epa.gov/superfund/greenremediation/sf-gr-strategy.pdf

109. US EPA — US Environmental Protection Agency 2010b Green Remediation Management Practices: Clean Fuel & Emission Technologies for Site Cleanup. EPA 542-F-10-008. EPA Office of Superfund Remediation & Technology Innovation. www.epa.gov/tio/ download/remed/clean-fuel-emis-gr-fact-sheet.pdf Accessed August 2015

110. Van-Camp L., Bujarrabal, B., Gentile, A-R., Jones, R.J.A., Montanarella, L., Olazabal, C. and Selvaradjou, S-K. 2004. Reports of the technical working groups established under the thematic strategy for soil protection. EUR 21319 EN/1, 872 pp. Office for Official Publications of the European Communities, Luxembourg.

111. Van Liedekerke M., Prokop G., Rabl-Berger S. et al. 2014. Progress in the Management of Contaminated Sites in Europe, JRC Reference Report. European Commission EUR 26376, http://dx.doi.org/http://dx.doi.org/10.2788/4658.10.2788/4658

112. van Veen, H. J. and Weenk, A. and Bardos, R.P. 1997 Selection and setting priorities for remediation options. 5th International FZK/TNO Conference on Contaminated Land. Masastricht, The Netherlands, October 30th to November 3rd 1995. Contaminated Soil'95, Pp 1465–1474, (van den Brink, W.J., Bosman, R. and Arendt, F. Eds) Kluwer Acad Pub., The Netherlands.

113. Vegter, J., Lowe J., & Kasamas, H. (Eds), 2002. Sustainable Management of Contaminated Land: An Overview. Report. Austrian Federal Environment Agency on behalf of CLARINET, Spittelauer Lände 5, A-1090 Wien, Austria. www.commonforum.eu/publications_clarinet.asp.

114. VOSA — Vehicle and Operator Services Agency 2013 HGV Inspection Manual Consolidated Edition. www.gov.uk/government/uploads/system/uploads/attachment_data/file/ 416627/heavy-goods-vehicle-hgv-inspection-manual.pdf, Accessed August 2015

115. WRAP — Waste and Resources Action Programme (2003) Sustainable procurement — making it happen. Available from www.1010global.org/sites/default/files/ sustainable-procurement-making-it-happen.pdf Accessed August 2015

CHAPTER 12

TECHNICAL CHALLENGES IN THE APPLICATION OF ADAPTIVE MANAGEMENT

Byron K. Williams, Eleanor D. Brown

Keywords: Adaptive management, Learning, Nonstationarity, Resilience, Uncertainty, Value of information

INTRODUCTION

Adaptive management, an approach that involves the dual pursuit of management and learning, has been a part of natural resources management for many decades. In natural resources, adaptive management means learning by doing (managing), and altering management activities to reflect new information (learning) (Walters and Holling, 1990). Learning in adaptive management occurs through the practice of management itself, with adjustments to decision making as understanding improves. For many resource management problems, the use of management in an experimental, learning-oriented context is the best, and may be the only, way to gain the understanding needed to manage more effectively (Williams et al., 2007). For many systems such as ecological communities or species populations undergoing rapid change, managers often cannot wait for thorough knowledge. Adaptive management can be used to move forward with a strategy of action that articulates and systematically reduces the multiple aspects of uncertainty that managers face. Though there are of course other approaches to management, some of which may be useful under specific circumstances (Williams et al., 2002), adaptive management serves to fill a need for smart management in the face of uncertainty. Advances in decision analytic theory and methodology since Holling's (1978) andWalters and Hilborn's (1978) early studies of adaptive management of natural resources have led to a proliferation of applications in the conservation of animal and plant species over the last decade. There are by now a great many cases in the literature that document the use of an adaptive approach to resource management. The following examples are by no means exhaustive, but serve to highlight the range of real-world applications and issues that can be addressed through adaptive decision making. For example, studies have involved threatened and endangered species, like Mead's milkweed (*Asclepias meadii*) (Moore et al., 2011a), red-cockaded woodpeckers (*Picoides borealis*) (Moore and Conroy, 2006), red knots (*Calidris canutus rufa*) (McGowan et al., 2011), and endemic fishes in the Tallapoosa River (Irwin and Freeman, 2002). Runge (2011) discussed adaptive management for specific issues such as species recovery and regulatory listing under the U.S. Endangered Species Act. Other studies have focused on particular elements of the decision process, for example, population modeling in wildlife management (reviewed by Lahoz-Monfort et al., 2014) and the central role of monitoring (Lyons et al., 2010) and science–management partnerships (Moore et al., 2011b) in managing habitats. Methodological issues are also at the forefront, such as the role and degree of experimentation in active adaptive management. For example, Parkes et al. (2006) experimented with competing models of control of invasive vertebrates in New Zealand, Runge (2013) examined how optimal decisions could change over the course of a simulated reintroduction of griffon vultures (*Gyps fulvus*), and Rout et al. (2014) considered tradeoffs among prevention, search, and eradication of an invasive species. McCarthy and Possingham (2007) demonstrated optimization of decisions with a case study of native plant revegetation in Australia, and McDonald-Madden et al. (2010) used this method to develop a framework for accelerated learning about populations of the

---

Citation: Byron K. Williams, Eleanor D. Brown, Technical challenges in the application of adaptive management, Biological Conservation, Volume 195, March 2016, Pages 255-263, ISSN 0006-3207, http://dx.doi.org/10.1016/j.biocon.2016.01.012.

Tasmanian devil (*Sarcophilus harrisii*), a species threatened by an emergent fatal infectious disease. Climate change is another major theme, and many researchers have focused on dealing with uncertainties driven by changing climate patterns. For example, adaptive approaches were outlined byConroy et al. (2011) to mitigate impacts on mid-montane bird communities in the southeastern U.S., by Martin et al., 2011a and Martin et al., 2011b to manage fresh water to benefit Florida manatees (*Trichechus manatus latirostris*), and by McDonald-Madden et al. (2011) to determine optimal timing of species translocations.

Nearly all definitions of adaptive management emphasize uncertainty, its integration into a decision making framework, and its reduction through management itself. We use the definition of adaptive management given by Williams et al. (2007): "Adaptive management is a systematic approach for improving resource management by learning from management outcomes." Learning and adaptation is an ongoing process in which learning is used to improve management, and management is used to investigate the resource system (Williams and Brown, 2014). In some cases management interventions can be designed as experimental "treatments." Nonetheless, the ultimate focus is management, and learning is valued for its contribution to better management (Walters, 1986). Adaptive management is distinguished from more common forms of decision making like "trial and error," scenario planning, hedging, and other forms of non-adaptive management that do not focus on uncertainty and its reduction (Williams et al., 2007).

In this paper, we provide an overview of emerging technical issues in adaptive management, bringing together these growing points in its implementation for the first time. Our focus is on the current state of thinking about adaptive management and some important issues facing this approach in the future, rather than on a retrospective of past developments. A number of technical matters in adaptive decision making have recently become apparent, not only in various areas of natural resource management but also in operations research, information theory, systems analysis, and other fields. We hope that by relating these outstanding issues to the practice of adaptive management, their application in biological conservation will be advanced.

We first briefly describe adaptive management processes and their context, including the treatment of uncertainty and its propagation over time. We then use this framework and context to develop a series of emerging issues that we believe pose serious challenges to the approach. These include nonstationarity of long-term environmental trends; the applicability of adaptive management across scales; attributes of models and management alternatives that promote learning; the value of the information produced with adaptive management; management for resilience and sustainability; and institutional (social) learning. Such issues can be relevant to adaptive decision making for many natural resources, including ecological systems, plant and animal species, populations, or habitats, and in this paper we use the generic term "resource" to refer to them.

CONTEXT FOR ADAPTIVE MANAGEMENT

Applications of adaptive management typically involve systems characterized by uncertainty and change over time (Fig. 1). Common features (Williams and Brown, 2014) are: 1) system changes in response to fluctuating environmental conditions and management actions; 2) environmental variation that induces stochasticity in biological and ecological processes, leading to unpredictable system behaviors; 3) periodic and potentially varying management interventions to influence system behaviors either directly or indirectly; and 4) limitation of effective management by uncertainty about the resource system and how it responds, such that reducing this uncertainty can lead to improved management. Most applications involve a systematic approach to management, including problem framing and the use of formal decision analysis; statistical tools such as Bayesian updating (Lee, 1989), Markov processes (Puterman, 1994), and stochastic dynamic programming to optimize decisions (Bertsekas, 1995); and a foundational framework of tightly integrated sequential elements in a decision process that methodically reduces uncertainty (e.g., Williams et al., 2007).

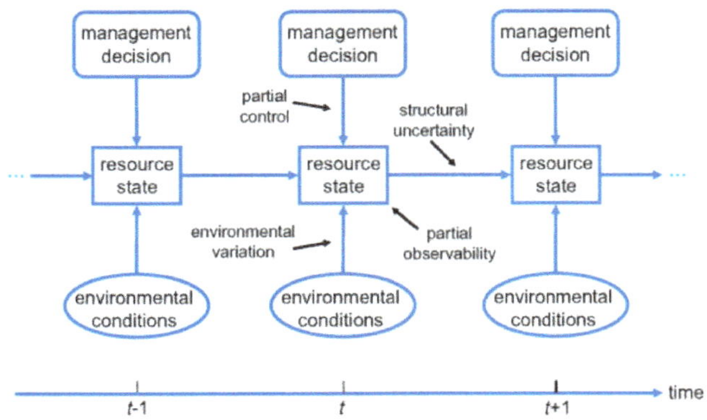

Figure 1: Dynamic resource system, with changes influenced by fluctuating environmental conditions and management actions, with uncertainty factors. Partial control limits the influence of management actions. Environmental variation affects resource system status and dynamics. Partial observability limits the recognition of system status. Structural uncertainty limits the ability to characterize system change.

The fact that management, environmental variation, and resource status are expressed over time provides an opportunity to improve management by learning over the course of the management time frame. A useful notation for this situation denotes the system state at a particular time t by xt,which may represent key resource elements, features, and attributes that change over some time frame. The system is influenced by a conservation action at, chosen from a set of options that are available at time t. Resource dynamics can be represented in discrete time by

$$x_{t+1} = x_t + f(x_t, a_t, e_t), \tag{1}$$

with the resource change $f(xt, at, et)$ from t to $t + 1$ influenced by the resource state xt, the action at, and environmental conditions et at time t.

A key to adaptive management is the recognition and treatment of uncertainty. Here we emphasize four uncertainty factors that affect natural resources, namely environmental variation, partial controllability, partial observability, and structural uncertainty (Williams, 2011a). These uncertainties influence resource management in different ways and at different points in a resource system (Fig. 1), and in combination they can restrict our management ability.

Environmental Variation

Environmental conditions can be viewed as external factors that influence, but are not influenced by, resource conditions and dynamics. Examples include precipitation patterns, temperature regimes, ambient light conditions and other measures, as well as extremes in these conditions. Fluctuations in the environment can be treated as if they vary randomly over time, and interact with landscape changes that co-occur. It often is useful to include unrecognized landscape heterogeneity and unpredictable human impacts on the landscape as a part of "environmental variation."

Environmental conditions combine with demographic stochasticity and other factors to induce random variation in the state transitions in Eq. (1). These transitions can be conveniently represented by a probability $P(xt_{+1} \mid xt, at)$ of transition between successive states, given that action *at* is taken at time *t*. In the following section we primarily use this formulation to describe the change in resource conditions.

Partial Controllability

In many instances the management actions that are actually implemented are not fully determined by the management decisions that are made. Partial controllability refers to the difference between an effect that is intended and the effect that actually occurs. The latter can be characterized conveniently in terms of a distribution of actions that assigns probabilities of their occurrence over a range of potential consequences.

Here we consider a target action *at* in Eq. (1) that is realized only partially, with random variation in the action actually taken. This additional source of stochasticity can be, and often is, combined with environmental variation in the probability structure $P(xt_{+1} \mid xt, at)$ of the transition models. Partial controllability typically increases with geographic scale and ecological complexity, but tends to be less important in localized, smaller-scale projects for which random variation is limited and control can be exercised more directly.

Partial Observability

Partial observability expresses the inability to observe completely the resource system that is being managed. For example, only a part of the area where a fish population occurs can be monitored, and individuals (e.g., plants and animals) often escape detection even in areas that are intensively monitored. As a result, observations are associated with, but not the same as, actual system states. Partial observability obscures the resource status on which effective management depends, which reduces management effectiveness even if environmental variation is minimal and management actions are precisely controlled.

There are several ways partial observability can be addressed. One is to estimate resource status with field data, and then treat the estimate as if it accurately represents resource conditions. Another is to state the uncertainty about resource status explicitly with probabilities for possible resource states, and incorporate them directly into the decision-making process (Williams, 2009). Here we represent partial observability with a distribution *bt* of resource states at each time, where *bt*(*x*) denotes the probability or "belief" that the resource state is *x* at time *t*. This distribution, which often is referred to as a *belief state* (Kaelbling et al., 1998), changes over time as actions are taken, the resource responds, and monitoring data are collected and used to update resource status (Regan et al., 2011, Chadès et al., 2008 and Chadès et al., 2011).

Structural Uncertainty

Structural uncertainty denotes a lack of understanding (or lack of agreement) about the forms and functions of the processes that control resource dynamics. Differing views about how natural processes work and how they respond to management can be framed as hypotheses, which in turn can be embedded in models and used to make testable predictions. The models can be described as above by different transition probability models $Pk(xt_{+1} \mid xt, at)$, where the subscript *k* designates one of several models representing different hypotheses. Structural uncertainty then focuses on the degrees of confidence about the models (and their embedded hypotheses) in representing resource dynamics. The gradual identification of the appropriate model over time is described here as *technical learning*, and it is a key to improved management.

Structural uncertainty is conveniently measured by a distribution *qt* of weights that express relative confidence in the models, with element *qt*(*k*) denoting the degree of confidence in model *k* to represent the resource. The distribution *qt* is often referred to as a *model state*, and it changes over time as resource conditions fluctuate and different management actions are taken. A common mathematical approach for updating the confidence weights is based on Bayes' theorem, which combines the confidence values*qt*(*k*) and resource status xt_{+1} from monitoring data to generate updated confidence values iteratively over time (Lee, 1989):

$$q_{t+1}(k) = \frac{q_t(k)P_k(x_{t+1}|x_t, a_t)}{\overline{P}(x_{t+1}|x_t, a_t, q_t)}$$

with

$$\overline{P}(x_{t+1}|x_t, a_t, q_t) = \sum_k q_t(k)P_k(x_{t+1}|x_t, a_t).$$

Confidence increases for models that make accurate forecasts of resource status, and confidence declines for models that do not make accurate forecasts. Unlike environmental variation (and in some cases partial controllability), which are effectively uncontrolled, structural uncertainty can be reduced with management that targets learning. We note that approaches such as experimentation with analysis of variance statistics, ad hoc metrics that measure distances between and among models (Williams et al., 2002), and other informal procedures also could be used to update model confidence values.

ADAPTIVE DECISION MAKING

A framework for adaptive decision making can be characterized as a two-phase process, which we briefly recapitulate from Williams and Brown (2014) (Fig. 2). A deliberative phase involves the framing of the resource problem in terms of stakeholders, objectives, management alternatives, models (including metrics for the confidence one places on them), and monitoring protocols. An iterative phase uses these elements in an ongoing cycle of *technical learning* about system structure and function, and resource management based on what is learned. Finally, *institutional learning* (a form of social learning [Pahl-Wostl, 2009]) about the decision process is obtained by periodically interrupting the iterative cycle of technical learning to reconsider project objectives, management alternatives, stakeholder engagement, and other elements of the deliberative phase (Fig. 2). The institutional learning cycle complements, but differs from, the cycle of technical learning.

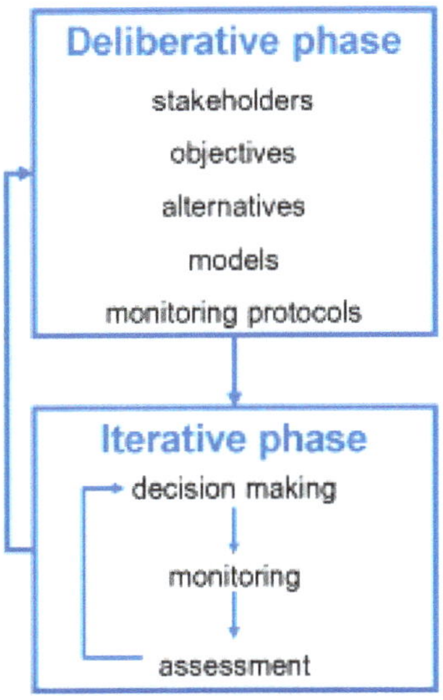

Figure 2: Two-phase learning in adaptive management. The foundation for adaptive management is based on an integrated sequence of elements. Technical learning involves an iterative sequence of decision making, monitoring, and assessment. Institutional learning involves periodic reconsideration of the planning elements of adaptive management. From Williams and Brown (2014).

Value Functions

An assessment of decision options ultimately requires a means to project and value decision outcomes. To be useful in guiding strategy, objectives must be represented by one or more measureable attributes that can be used to evaluate the consequences of management actions (Keeney, 1992 and Arvai et al., 2001). A notation for such a valuation recognizes that actions at any point in time have immediate or short-term effects as well as longer-term effects (Fig. 1). Longer-term effects are represented with the transition models. Short-term effects are often described in terms of the immediate costs and returns associated with an action, and are represented here by time-specific utilities $U(at \mid xt)$ (representing immediate rewards net of any relevant costs) corresponding to a particular action at and resource state xt at time t. If we assume the resource system is fully known and observed, a valuation of management strategy is given in terms of the aggregation of utilities into a *value function* (Williams and Johnson, 2013):

$$V(A_t|x_t) = E\left[\sum_{\tau=t}^{T} U(a_\tau|x_\tau)|x_t\right]. \qquad (2)$$

Here At represents a strategy with state-specific actions over the time frame from t to some terminal time T. Because of the randomness in system responses as captured by the transition probabilities $P(xt_{+1} \mid xt, at)$, the aggregation $\sum_{\tau=t}^{T} U(a_\tau|x_\tau)|x_t$ of utilities over time is also random. A natural way to account for the randomness is to average the value of the aggregated utilities, and hence we use the expected value, where the expectation reflects the transition probability structure. The objective in Eq. (2) can be used to compare potential strategies and select one with high value for any initial resource state.

Several variants of valuation are possible. One extension allows structural uncertainty, which can be captured with different transition models $Pk(xt_{+1} \mid xt, at)$ (and possibly different utilities $Uk(at \mid xt)$). Structural uncertainty is included in the value function as:

$$V(A_t|x_t, q_t) = \sum_k q_t(k)\left\{E\left[\sum_{\tau=t}^{T} U_k(a_\tau|x_\tau)|x_t\right]\right\}. \qquad (3)$$

Eq. (3) is essentially an averaging of model-specific expected values from Eq. (2), based on model state qt. The incorporation of structural uncertainty into decision making transforms the decision framework into a Markovian belief process (Williams, 2011b), thereby increasing substantially the complexity of finding optimal strategies. An example of the use of Markovian belief processes in biological conservation is provided byRunge's (2013) study of optimal decisions for species reintroduction.

Another extension involves the inclusion of partial observability, in which valuation as in Eq. (2) is averaged over the possible resource states at each time:

$$V(A_t|b_t) = E\left[\sum_{\tau=t}^{T}\left\{\sum_x b_\tau(x)U_k(a_\tau|x)\right\}|x_t, b_t\right]. \qquad (4)$$

Eq. (4) represents the averaging of expected values in Eq. (2), based on the belief statebt. The inclusion of partial observability again produces a Markovian belief process (Kaelbling et al., 1998), with commensurate challenges in finding optimal strategies.

Robust Decision Making

The valuation functions for structural uncertainty (Eq. (3)) and partial observability (Eq.(4)) rely on probability structures on which to base the averaging of utilities. But sometimes uncertainty is so severe that it cannot be described probabilistically, so it is not possible to use probability averaging to compute an expected aggregate utility. A candidate approach for severe uncertainty is robust decision making, which focuses not on average aggregate utilities but rather on the production of values that will satisfy a minimum performance criterion over a large extent of model and/or belief states (Williams and Johnson, 2013). For structural uncertainty in particular, robust decision making involves the choice of actions for which the expected utility will be "good enough" over as wide a range of model states as possible (Williams and Johnson, 2013). This shifts the focus from expected utility to the coverage of "good enough" values. The operative question then becomes "what action will allow for the maximum range of model states over which an adequate value is produced?" One analytic approach to this question builds on the work of Ben-Tal and Nemirovski (2002), Ben-Haim (2006), and others on robust optimization theory (though see Sniedovich, 2010 and Sniedovich, 2014and Hayes et al. (2013) for critiques). Applications of robust decision making in natural resources are in their infancy (Williams and Johnson, 2013), and much more work is required to develop both the theory and computing software.

Though the explicit incorporation of learning, i.e., reduction of uncertainty, into decision making opens up new opportunities for improved decision making, it also presents new technical challenges. In what follows we use the framework for adaptive decision making under uncertainty to consider some key challenges facing practitioners as they address the management of systems that are only partially understood, observed, and controlled.

NONSTATIONARY RESOURCE CHANGES

Adaptive management is usually framed in terms of an (often unstated) assumption that the processes influencing resource dynamics are stable over the management time frame, in that patterns of uncontrolled fluctuations change little in their overall direction or range of variation (Williams and Brown, 2012). Approaches to system analysis and control, including the framework typically used in adaptive decision making, have traditionally rested on that assumption of stability. That is, the transition functions (*xt*, *at*, *et*) and transition probability structures $P\ (xt_{+1} \mid xt, at)$ are assumed to be fixed over time, with the recognition that the system can fluctuate randomly because of environmental drivers and other stochastic factors.

An increasingly important need is to extend the framework for adaptive decision making to incorporate nonstationarity, i.e., to allow for decision making when system dynamics do not exhibit stationary patterns as described above. For a great many resource systems, the ecological structures and processes controlling resource dynamics are changing in ways not fully expressed by the standard stationarity assumptions. For example, environmental conditions and the ecological processes influenced by them are exhibiting directional patterns of change. An obvious example is climate change, in which environmental variables such as temperature and precipitation change directionally and cause systemic change in resource dynamics. Large-scale human actions on the landscape can also produce systemic change. A simple way to denote such nonstationarity is to include a temporal index to characterize system transitions. The transition function then becomes *ft*(*xt*, *at*, *et*), where the functional index indicates that the function can change over time, i.e., is nonstationary. In like manner, transition probabilities can be denoted by $P(xt_{+1} \mid xt, at; t)$ to indicate nonstationarity. In terms of adaptive management, the models used to represent uncertainty can be denoted by$Pk(xt_{+1} \mid xt, at; t)$.

Nonstationarity is a newly recognized and serious challenge to adaptive decision making, one for which we need new approaches that go beyond the standard framing of learning-based management. The cycle of learning becomes more difficult when the subjects of investigation – the ecological processes that determine resource change – are themselves changing.

Conservation-oriented examples of an adaptive approach to this problem are provided by Martin et al., 2011a and Martin et al., 2011b in their study of mitigating the effects of sea-level rise on Florida manatees, and Conroy et al. (2011) in their framework for managing forest bird communities in the face of climate change in the Appalachian Mountains. But much more work will need to be done on decision making in general, and adaptive decision making in particular, to address this complex issue.

SPATIAL AND ECOLOGICAL SCALE

One concern in applications of adaptive management is the appropriate scale for decision making. Adaptive management is often associated with big, complex ecosystem-wide applications, such as large-river management (Columbia, Platte, and Missouri Rivers [Quigley and Arbelbide, 1997, Wissmar, and Bisson, 2003, Williams, 2006 and Freeman, 2010]; Glen Canyon Dam on the Colorado River [U.S. Geological Survey, 2008]); continental waterfowl harvest management (Williams and Johnson, 1995 and Williams, 2006); commercial fisheries (Hilborn, 1992 and Conover and Munch, 2002); pest management in forest ecosystems (Shea et al., 2002); and water management (Everglades [Holling et al., 1994]). Ecosystem management at this scale involves economic, social, institutional, and ecological linkages across large landscapes with high degrees of heterogeneity.

However, adaptive decision making as we describe it here applies equally well to local issues, as long as the basic conditions are met (e.g., Williams et al., 2007). Moore et al. (2011b) provide a conservation example involving prairie and wetland habitat restoration on national wildlife refuges. There probably are many more potential applications of adaptive management at local versus larger scales, not only because of the prevalence of such problems but also because they can often be framed more easily, their uncertainties can be identified more readily, stakeholder involvement can be facilitated more directly, and management can often be implemented more easily (McConnaha and Paquet, 1996).

Whatever its scale, the careful framing of a decision problem often requires a combination of both larger and smaller scales than the focal scale of decision making (Holling, 2001), to capture scale-specific factors that influence resource dynamics. Broader and more inclusive scales also are needed when conservation actions conflict with other socio-economic goals. With large spatial scales a problem can arise in linking conservation decisions that are applied locally, and the challenge is to recognize the circumstances when such decisions must be explicitly linked, and when they can be handled independently. Though adaptive management provides a framework and general approach for problems at different scales, these and other scale-specific issues can arise that can make its systematic implementation problematic for complex systems.

MODELS, MANAGEMENT ALTERNATIVES, AND LEARNING

Here we explore the roles of models and management alternatives on the potential for learning and learning rate in adaptive management. Smart decision making with adaptive management is facilitated by learning, which in turn is facilitated by the presence of variation in model performance as well as variation in decision impacts. To aid in the identification of smart management strategies, alternative actions should generate distinctly different predictions. Similarly, learning is promoted when the models under consideration lead to different predictions of resource conditions over time. Put simply, adaptive decision making works best when there is variation in the predictions across models for particular actions, and variation across actions for particular models.

The effect on learning of similar responses can be seen in the Bayesian updating of model weights $qt(k)$ in the model state. Assume that the transition probability structures across the models representing different hypotheses are effectively identical, i.e.,$Pk(xt_{+1} \mid xt, at) = Pk.(xt_{+1} \mid xt, at)$. Given model weights $qt(k)$ at time t, Bayesian updating produces weights at the next time of

$$\begin{aligned} q_{t+1}(k) &= \frac{q_t(k)P_k(x_{t+1}|x_t,a_t)}{\sum_{k'} q_t(k')P_{k'}(x_{t+1}|x_t,a_t)} \\ &= \frac{P_k(x_{t+1}|x_t,a_t)}{P_k(x_{t+1}|x_t,a_t)} \frac{q_t(k)}{\sum_{k'} q_t(k')} \\ &= q_t(k). \end{aligned}$$

Thus, the absence of differences in model performance means there can be no reduction of structural uncertainty, i.e., no learning. Basically, learning as represented by the change in model weights does not occur under these conditions.

A special case of adaptive decision making treats the management alternatives themselves as hypotheses (Williams, 2011a), and examines them using experimental design and hypothesis-testing procedures (Graybill, 1976). Alternatively, when interventions are carried out sequentially one can compare monitoring data against predictions for each alternative to update confidence in the alternatives. One common method is to update confidence weights at each decision point by comparing predicted responses with post-decision monitoring data (Williams et al., 2002), with the change in the weights leading gradually to a recognition of the best intervention. The relevance of Bayesian updating for conservation applications is shown by a few examples, such as optimal control of human disturbance of nesting golden eagles (*Aquila chrysaetos*) (Martin et al., 2009, Martin et al., 2011a and Martin et al., 2011b), dam release prescriptions to protect aquatic habitats (Irwin and Freeman, 2002), and native plant revegetation strategies in Australia (McCarthy and Possingham, 2007).

MONITORING

Monitoring plays a critical role in adaptive management in allowing the comparison of model-based predictions and estimated responses to facilitate learning. In fact, monitoring programs provide data for four key purposes: 1) evaluation of progress toward achieving objectives; 2) determination of resource status, in order to identify appropriate management actions; 3) learning about resource dynamics via the comparison of predictions against survey data; and 4) development and enhancement of models of resource dynamics as needed (Williams et al., 2007).

In consideration of its critical role in any adaptive management approach, monitoring needs to be well-designed and ongoing over the life of an adaptive management project. Some situations that challenge these conditions are the following.

- The frequency of monitoring cannot keep pace with changes in the natural system.
- A design for experimental management and monitoring cannot be developed to test hypotheses, either because understanding of the resource system is too limited or management is too constrained to design a meaningful experiment.
- There is not a firm commitment to funding and institutional support for monitoring for the duration of the learning effort. Though fundamental to adaptive management, monitoring is often the first function to be curtailed by some administrators and managers due to budget pressures.

It may be that with some limited re-orientation and re-designing, an extant monitoring effort can prove useful in meeting the purposes of adaptive decision making. The challenge is to integrate monitoring into the management framework in such a way that it is seen as critical rather than optional. Examples of the crucial role of monitoring in adaptive management applications for conservation have been provided by many authors, such as Conroy et al. (2011), Lyons et al. (2010) and Nichols and Williams (2006).

VALUING INFORMATION

A natural question that arises in adaptive decision making concerns the value of the added information produced. Though the concept of a measurable value for information has been recognized for several decades (Raiffa and Schlaifer, 1961 and Yokota and Thompson, 2004), more recently the "value of information" has begun to surface with increasing frequency in conservation, for example in considering costs and benefits associated with obtaining information from particular sampling techniques, or with resolving particular sources of uncertainty. Conservation examples that incorporate analysis of the value of information include reintroduction of whooping cranes (*Grus Americana*), as discussed by Runge et al. (2011); recovery of the Florida scrub-jay (*Aphelocoma coerulescens*), as discussed by Williams et al. (2011); and management of pink-footed geese (*Anser brachyrhynchus*), as discussed by Williams and Johnson (2015a) and Johnson et al. (2014). In simple terms, the value of information represents a potential increase in value resulting from better information to guide management. If we assume an ongoing if imperfect monitoring effort that informs decision making, the value of information is effectively an indicator of the marginal value of *additional* monitoring.

In the context of structural uncertainty, an intuitive way to understand the value of information is as a comparison of aggregate utility that could be produced if the resource system were fully known, against the aggregate utility produced in the presence of uncertainty (Williams and Johnson, 2015b). The latter expression corresponds to decision making under structural uncertainty, i.e., adaptive management. In that sense the adaptive decision framework serves as a platform for the value of information in iterative decision making. Variations of the value of information include the expected value of partial information, which accounts for the increased value accruing to the elimination of uncertainty from some but not all sources, and the expected value of sample information, which accounts for the gain in value with additional sampling. In both these situations, valuation through adaptive decision making serves as the baseline to assess the improvement of value with reduced uncertainty (Yokota and Thompson, 2004 and Williams and Johnson, 2015a).

SURPRISE, RESILIENCE, AND FLEXIBILITY

Natural resources management is always vulnerable to a "disconnect" between the ecosystem behaviors we expect and those that actually occur (Gunderson, 1999). Surprise can never be eliminated, no matter how learning-based and carefully framed resource management is. Surprise and associated concerns with uncertainty and resilience are central issues in an expansive literature that comes under the rubric of "resilience thinking" (Gunderson and Holling, 2002, Walker and Salt, 2006 and Scheffer, 2009). According to resilience thinking, natural systems are subject not only to reversible short-term change, but also to long-term change that is effectively irreversible. A developing theory of resilience emphasizes "stability basins" for ecological conditions, and thresholds beyond which reversible change within a zone is unlikely once a threshold is crossed. Adaptive management can play an important role in accumulating the knowledge needed to manage for resilience-based objectives over time and improve long-term resource viability (Zellmer and Gunderson, 2009).

One conclusion of resilience thinking is that when management focuses on only one or a few ecosystem attributes, a result is the loss of resilience and an increased vulnerability to unexpected and destructive change (Johnson et al., 2013). Well-known examples include the intensive management of grazing, which can increase the vulnerability of grasslands to drought (Walker and Salt, 2006); and intensive management of commercial fishing, which can lead to the unexpected collapse of a commercial fishery (Walters, 1986). Surprises like these usually are a result of managing in ways that induce stability in targeted ecosystem components in the short term but result in the loss of ecosystem resilience over the long term, and an increase in the vulnerability of the system to extreme and often unwanted changes (Williams and Brown, 2014).

Williams and Brown (2012) identified some steps that can be taken to deal with surprise in the management of ecosystems, as follows.

- Recognize that in any managed ecosystem uncertainty and the potential for surprise are implicit in the scenarios under consideration.
- Incorporate models that are based on broadly differing assumptions, with broadly differing predictions.
- Retain enough management flexibility to adapt to surprise when it occurs.
- Manage the system for sufficient resilience to maintain structure and function when external shocks occur.
- Increase the range of ecosystem conditions, management alternatives, and sources of evidence that are considered.
- Use experimental management and monitoring to learn and manage adaptively.

INSTITUTIONAL LEARNING

Adaptive management involves not only the potential for technical learning about a natural resource system and how it responds to management interventions, it also involves the potential to learn about the decision making process itself. In this sense there are really two learning cycles in adaptive management, with the technical learning cycle nested within a larger cycle of institutional learning (Fig. 3). A typical situation involves multiple iterations of the technical learning cycle during which the institutional framework remains more or less unchanged, followed by a break in technical learning to revisit and potentially restructure some of the institutional elements (Williams and Brown, 2014). Together the two cycles are referred to as double-loop learning (Argyris and Shon, 1978). Pahl-Wostl (2009) expanded this model to include three cycles, by distinguishing socio-political and governance aspects of stakeholder involvement as yet another cycle. In combination the three cycles are held to address three different questions: "Are we doing things right? (Technical learning loop); "Are we doing the right things?" (Process learning loop); and "Who has the rights?" (Socio-political and governance loop) (Johnson et al., 2015). Here we restrict our further discussion to double-loop learning, with governance issues folded into the institutional learning cycle.

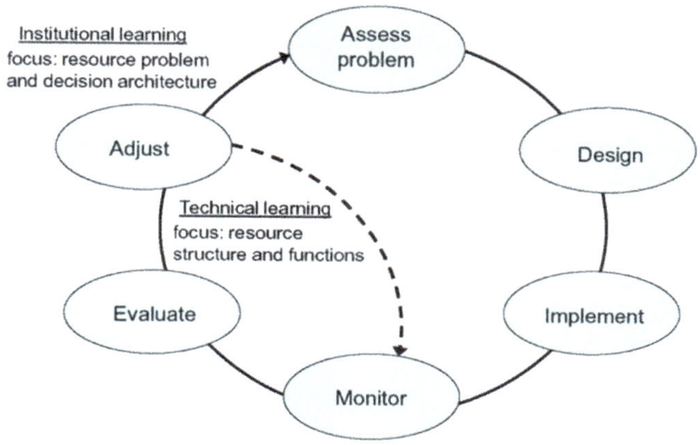

Figure 3: Adaptive management displayed as cycles of technical and institutional learning. The implementation component refers to implementation of a designed process based on problem assessment, which then is used to initiate technical learning. Solid lines represent the complete adaptive cycle, involving deliberative and iterative elements as well as institutional learning. The dashed line indicates feedback defining the technical learning cycle.

Because of the dynamic nature of governance, the need to revisit and adjust the process and governance elements of adaptive management often becomes more pressing as adaptive management proceeds over time. Stakeholder perspectives and values can and do change over time, as previously unanticipated patterns in resource dynamics are exposed and social and cultural values and norms change. Any of these changes can produce a need to adjust objectives, alternatives, and other set-up elements.

Institutional learning sometimes can serve a critical role in helping to overcome disagreements among stakeholders about management objectives, alternatives, and projected management consequences. It is much easier to agree to move forward with a particular management strategy if it is understood that objectives, management alternatives, and the other elements of decision making can be reviewed and renegotiated as new evidence about management performance becomes available. Institutional learning offers an incentive to stakeholders to agree on an initial approach that involves compromise on all sides. In a conservation example, Irwin and Kennedy (2008) found this to be the case during stakeholder negotiations for an initial dam-release prescription to conserve aquatic habitats of fishes endemic to the Tallapoosa River (Irwin and Freeman, 2002). Without such a possibility, on the other hand, negotiation to establish a fixed and permanent approach leaves all parties more entrenched in their positions because they believe the outcome can't be changed.

Challenges in institutional learning include the identification of criteria for when to break out of the technical learning cycle and revisit the decision process elements. If the revisitation is too frequent, the effects of change at the technical and institutional levels become confounded, significantly slowing the rate of learning for both. If it is too infrequent, there is a risk of the loss of commitment of stakeholders as values change, alternatives are marginalized, models cease to perform effectively in predicting system dynamics, objectives lose their relevance, etc.

An additional challenge is how to address potential changes in process elements once they are revisited. One of the biggest challenges in institutional learning involves the identification and updating of objectives, which play a crucial role in driving decisions and evaluating their consequences. Yet it is not uncommon for stakeholders to have very different ideas about what attributes are important and how to measure their importance over time, so that finding common ground is difficult in the first place, and a process by which to reconsider and possibly adjust previously agreed-upon objectives on the basis of experience and performance is an even bigger challenge. There have been a few attempts to formalize a learning process for objectives and other elements (Williams, 2012), but much more needs to be done.

DISCUSSION

Adaptive management recognizes uncertainties about resource processes and their responses to management, and emphasizes the tracking and reduction of uncertainty through management on the basis of what is learned over time. In this paper we have highlighted some of the challenges in the implementation of adaptive management, building on a framework that recognizes both an architecture for decision making and a learning process that folds directly into management. Learning is seen as improvements in technical understanding that aim at biological processes, and institutional improvements that aim at the architectural elements of decision making. We have used this framework to explore a number of practical challenges to adaptive decision making.

In terms of the institutional context of learning, we believe that identifying objectives and getting stakeholders to buy in constitute two of the biggest challenges in adaptive management. Stakeholders often have different ideas about what importance to assign to system attributes. We do not address this issue to any great extent here, and simply note that there is a great wealth of information and advice in the social literature about how to achieve consensus among stakeholders with disparate views. Here we have focused primarily on structural uncertainty, since its reduction is a distinctive point of emphasis in adaptive decision making.

We note that the framework for adaptive management can also accommodate other sources of uncertainty, and in particular partial observability. In fact, there are many similarities in the treatments of partial observability and structural uncertainty (Williams, 2009 and Williams, 2011b). For example, both problems involve the averaging of transitions and utilities across an uncertainty attribute, and both use value functions that incorporate present and expected future utilities. The key differences between them are the nature of the uncertainty attribute, and the way uncertainty is handled in valuation. Each problem is subject to its own form of uncertainty, either about process functions and parameters or about system status. With structural uncertainty the system state is observed, but the process structure is only partially known and must be characterized with time-specific model probabilities. Conversely, with partial observability it is the system structure that is known, but the system state is only partially observed and must be characterized with time-specific state probabilities.

There are other important differences that must be accounted for as well. Under partial observability valuation is conditioned on belief state (Eq. (4)), whereas under structural uncertainty it is conditioned on both resource and model states (Eq. (3)). On the other hand, the inclusion of stochastic observations under partial observability results in a more complicated transition structure than is the case with structural uncertainty. Which of these situations is more difficult to handle no doubt depends on the process structure and stochastic components.

As mentioned earlier, a significant challenge with adaptive management is to integrate multiple sources of uncertainty into decision making, and to assess their interactions. This challenge is made more difficult when one seeks to identify optimal strategies (Williams, 2009), and the challenge is greater still in the presence of nonstationarity. In order to contain costs and maximize value with adaptive decision making, a good deal more thinking will need to go into designing adaptive management projects that successfully deal with multiple sources of uncertainty. A technical assessment of the value of the information may contribute to improved project design, but a low value should not be misinterpreted as meaning that there is little value in using an adaptive approach to management. It should be recognized that low values are often recorded in conservation and natural resources applications (Walters, 1986, Moore and McCarthy, 2010 and Johnson et al., 2014), and this information metric is not the only, and possibly not even the most relevant, measure of value for the decision framework. Among other things, a systematic and structured accounting of the elements of decision making can serve as a mechanism for collaboration and shared decision making, lowering the potential contentiousness and conflict among stakeholders. The value of information can certainly contribute to, but should not obscure, these and other benefits accruing to a structured, adaptive process of decision making.

CONCLUSIONS

It has been a long road in making adaptive management legitimate and useful as a paradigm to guide natural resources decision making, and in particular decision making about biodiversity conservation (Runge, 2011). At this point there is a well-developed framework and process with which to recognize and characterize uncertainties, and reduce them through the use of management itself. But a changing climate, widespread landscape change, accelerating biodiversity loss, and changing cultural values expand the context for resource management and create new problems for the application of an adaptive approach to management.

In many ways these challenges are generic problems for the 21st century, and they apply to the management of many urgent conservation problems. Together they increase the difficulties faced by practitioners in making smart decisions in the face of new problems. Adaptive management provides a context and model for addressing many of these challenges, and in some cases it may be the only way forward for their resolution. However, the adaptive management paradigm will need to be adjusted and expanded to accommodate the new issues that are only now coming into focus.

ACKNOWLEDGMENTS

We thank the USGS Science and Decisions Center for support for BKW during preparation of the manuscript. We are grateful to Fred Johnson for a thoughtful review.

REFERENCES

1. Argyris, C., Shon, D., 1978. Organizational Learning: A Theory of Action Learning. AddisonWesley.
2. Arvai, J.L., Gregory, R., McDaniels, T.L., 2001. Testing a structured decision approach: value-focused thinking for deliberative risk communication. Risk Anal. 21, 1065–1076.
3. Ben-Haim, Y., 2006. Info-gap Decision Theory, Decisions Under Severe Uncertainty. Second ed. Elsevier.
4. Ben-Tal, A., Nemirovski, A., 2002. Robust optimization methodology and applications. Math. Program. B 92, 453–480.
5. Bertsekas, D.P., 1995. Dynamic Programming and Optimal Control vols. I and II. Athena Scientific.
6. Chadès, I., Martin, T.G., Nicol, S., Burgman, M.A., Possingham, H.P., Buckley, Y.M., 2011. General rules for managing and surveying networks of pests, diseases, and endangered species. Proc. Natl. Acad. Sci. 108, 8323–8328.
7. Chadès, I., McDonald-Madden, E., McCarthy, M.A., Wintle, B., Linkie, M., Possingham, H.P., 2008. When to stop managing or surveying cryptic threatened species. Proc. Natl. Acad. Sci. 105, 13936–13940.
8. Conover, D.O., Munch, S.B., 2002. Sustaining fisheries yields over evolutionary time scales. Science 297, 94–96.
9. Conroy, M.J., Runge, M.C., Nichols, J.D., Stodola, K.W., Cooper, R.J., 2011. Conservation in the face of climate change: the roles of alternative models, monitoring, and adaptation in confronting and reducing uncertainty. Biol. Conserv. 144, 1204–1213.
10. Freeman, D.M., 2010. Implementing the Endangered Species Act on the Platte Basin Water Commons. University of Colorado Press.
11. U.S. Geological Survey, 2008. Science Plan for Potential 2008 Experimental High Flow at Glen Canyon Dam. USGS Grand Canyon Monitoring and Research Center.
12. Graybill, F.A., 1976. Theory and Application of the General Linear Model. Duxbury Press.
13. Gunderson, L., 1999. Stepping back: assessing for understanding in complex regional systems. In: Johnson, K.N., Swanson, F., Herring, M., Greene, S. (Eds.), Bioregional Assessments: Science at the Crossroads of Management and Policy. Island Press.
14. Gunderson, L.H., Holling, C.S., 2002. Panarchy: Understanding Transformations in Human and Natural Systems. Island Press.
15. Hayes, K.R., Barry, S.C., Hosak, G.R., Peters, G.W., 2013. Severe uncertainty and info-gap decision theory. Methods Ecol. Evol. 4, 601–611.
16. Hilborn, R., 1992. Can fisheries agencies learn from experience? Fisheries 17, 6–14.
17. Holling, C.S. (Ed.), 1978. Adaptive Environmental Assessment and Management. Wiley.
18. Holling, C.S., 2001. Understanding the complexity of economic, ecological, and social systems. Ecosystems 4, 390–405.
19. Holling, C.S., Gunderson, L.H., Walters, C.J., 1994. The structure and dynamics of the Everglades system: guidelines for ecosystem restoration. In: Davis, S., Ogden, J. (Eds.), Everglades: The Ecosystem and Its Restoration. Lucie Press, St.
20. Irwin, E.R., Freeman, M.C., 2002. Proposal for adaptive management to conserve biotic integrity in a regulated segment of the Tallapoosa River, Alabama, U.S.A. Conserv. Biol. 16, 1212–1222.
21. Irwin, E.R., Kennedy, K.D.M., 2008. Engaging stakeholders for adaptive management using structured decision analysis. Paper presented at the Third Interagency Conference on Research in the Watersheds, Estes Park, CO (http://pubs.usgs.gov/sir/2009/5049/pdf/Irwin.pdf Accessed June 2015).
22. Johnson, F.A., Boomer, G.S., Williams, B.K., Nichols, J.D., Case, D.J., 2015. Multi-level learning in the adaptive management of waterfowl harvests: 20 years and counting. Wildl. Soc. Bull. 39, 9–19.
23. Johnson, F.A., Jensen, G.H., Madsen, J., Williams, B.K., 2014. Uncertainty, robustness, and the value of information in managing an expanding Arctic goose population. Ecol. Model. 273, 186–199.
24. Johnson, F.A., Williams, B.K., Nichols, J.D., 2013. Resilience thinking and a decision-analytic approach to conservation: strange bedfellows or essential partners? Ecol. Soc. 18, 27 (http://www.ecologyandsociety.org/vol18/iss2/art27/ Accessed June 2015).
25. Kaelbling, L.P., Littman, M.L., Cassandra, A.R., 1998. Planning and acting in partially observable stochastic domains. Artif. Intell. 101, 99–134.
26. Keeney, R.L., 1992. Value-based Thinking: A Path to Creative Decision Making. Harvard University Press.
27. Lahoz-Monfort, J.J., Guillera-Arroita, G., Hauser, C.E., 2014. From planning to implementation: explaining connections between adaptive management and population models. Front. Ecol. Evol. 2, 1–5.
28. Lee, P.M., 1989. Bayesian Statistics: An Introduction.

Edward Arnold.

29. Lyons, J.E., Runge, M.C., Laskowski, H.P., Kendall, W.L., 2010. Monitoring in the context of structured decision-making and adaptive management. J. Wildl. Manag. 72, 1683–1692.

30. Martin, J., Fackler, P.L., Nichols, J.D., Lubow, B.C., Eaton, M.J., Runge, M.C., Stith, B.M., Langtimm, C.A., 2011a. Structured decision making as a proactive approach to dealing with sea level rise in Florida. Clim. Chang. 107, 185–202.

31. Martin, J., Fackler, P.L., Nichols, J.D., Runge, M.C., McIntyre, C.L., Lubow, B.C., McCluskie, M.C., Schmutz, J.A., 2011b. An adaptive-management framework for optimal control of hiking near golden eagle nests in Denali National Park. Conserv. Biol. 25, 316–323.

32. Martin, J., McIntyre, C.L., Hines, J.E., Nichols, J.D., Schmutz, J.A., McCluskie, M.C., 2009. Dynamic multistate occupancy models to evaluate hypotheses relevant to conservation of golden eagles in Denali National Park, Alaska. Biol. Conserv. 142, 2726–2731.

33. McCarthy, A.M., Possingham, H.P., 2007. Active adaptive management for conservation. Conserv. Biol. 21, 956–963.

34. McConnaha, W.E., Paquet, P.J., 1996. Adaptive strategies for the management of ecosystems: the Columbia River experience. Am. Fish. Soc. Symp. 16, 410–421.

35. McDonald-Madden, E., Probert, W.J.M., Hauser, C.E., Runge, M.C., Possingham, H.P., Jones, M.E., Moore, J.L., Rout, T.M., Vesk, P.A., Wintle, B.A., 2010. Active adaptive conservation of threatened species in the face of uncertainty. Ecol. Appl. 20, 1476–1489.

36. McDonald-Madden, E., Runge, M.C., Possingham, H.P., Martin, T.G., 2011. Optimal timing for managed relocation of species faced with climate change. Nat. Clim. Chang. 1, 261–265.

37. McGowan, C.P., Smith, D.R., Sweka, J.A., Martin, J., Nichols, J.D., Wong, R., Lyons, J.E., Niles, L.J., Kalasz, K., Brust, J., Klopfer, M., Spear, B., 2011. Multispecies modeling for adaptive management of horseshoe crabs and red knots in the Delaware Bay. Nat. Resour. Model. 24, 117–156.

38. Moore, A.L., McCarthy, M.A., 2010. On valuing information in adaptive-management models. Conserv. Biol. 24, 984–993.

39. Moore, C.T., Conroy, M.J., 2006. Optimal regeneration planning for old-growth forest: addressing scientific uncertainty in endangered species recovery through adaptive management. For. Sci. 52, 155–172.

40. Moore, C.T., Fonnesbeck, C.J., Shea, K., Lah, K.J., McKenzie, P.M., Ball, L.C., Runge, M.C., Alexander, H.M., 2011a. Adaptive decision framework for conservation of a threatened plant. J. Fish Wildl. Manag. 2, 247–261.

41. Moore, C.T., Lonsdorf, E.V., Knutson, M.G., Laskowski, H.P., Lor, S.K., 2011b. Adaptive management in the U.S. National Wildlife Refuge System: science–management partnerships for conservation delivery. J. Environ. Manag. 92, 1395–1402.

42. Nichols, J.D., Williams, B.K., 2006. Monitoring for conservation. Trends Ecol. Evol. 21, 668–673.

43. Pahl-Wostl, C., 2009. A conceptual framework for analysing adaptive capacity and multilevel learning processes in resource governance regimes. Glob. Environ. Chang. 19, 354–365.

44. Parkes, J.P., Robley, A., Forsyth, D.M., Choquenot, D., 2006. Adaptive management experiments in vertebrate pest control in New Zealand and Australia. Wildl. Soc. Bull. 34, 229–236.

45. Puterman, M.L., 1994. Markov Decision Processes: Discrete Stochastic Dynamic Programming. Wiley.

46. Quigley, T.M., Arbelbide, S.J. (Eds.), 1997. An Assessment of Ecosystem Components in the Interior Columbia Basin and Portions of the Klamath and Great Basins, vol. 1General Technical Report PNW-GTR-405. U.S. Department of Agriculture, Forest Service.

47. Raiffa, H., Schlaifer, R.O., 1961. Applied Statistical Decision Theory. Harvard University Press.

48. Regan, T.J., Chadès, I., Possingham, H.P., 2011. Optimal strategies for managing invasive plants in partially observable systems. J. Appl. Ecol. 48, 76–85.

49. Rout, T.M., Moore, J.L., McCarthy, M.A., 2014. Prevent, search or destroy? A partially observable model for invasive species management. J. Appl. Ecol. 51, 555–834.

50. Runge, M.C., 2011. An introduction to adaptive management for threatened and endangered species. J. Fish Wildl. Manag. 2, 220–233.

51. Runge, M.C., 2013. Active adaptive management for introduction of an animal population. J. Wildl. Manag. 77, 1135–1144.

52. Runge, M.C., Converse, S.J., Lyons, J.E., 2011. Which uncertainty? Using expert elicitation and expected value of information to design an adaptive program. Biol. Conserv. 144, 1214–1223.

53. Scheffer, M., 2009. Critical Transitions in Nature and Society. Princeton University Press.

54. Shea, K., Possingham, H.P., Murdoch, W.W., Roush, R., 2002. Active adaptive management in insect pest and weed control: intervention with a plan for learning. Ecol. Appl. 12, 927–936.

55. Sniedovich, M., 2010. A bird's view of info-gap decision theory. J. Risk Finance 11, 268–283.

56. Sniedovich, M., 2014. Response to Burgman and Regan: the elephant in the rhetoric on info-gap decision theory. Ecol. Appl. 24, 229–233.

57. Walker, B., Salt, D., 2006. Resilience Thinking: Sustaining Ecosystems and People in a Changing World. Island Press.

58. Walters, C.J., 1986. Adaptive Management of Renewable Resources. Blackburn Press.

59. Walters, C.J., Hilborn, R., 1978. Ecological optimization and adaptive management. Annu. Rev. Ecol. Syst. 9, 157–188.

60. Walters, C.J., Holling, C.S., 1990. Large-scale

management experiments and learning by doing. Ecology 71, 2060–2068.

61. Williams, B.K., 2006. Adaptive harvest management: where we are, how we got here, and what we have learned thus far. Trans. N. Am. Wildl. Nat. Resour. Conf. 71, 259–274.
62. Williams, B.K., 2009. Markov decision processes in natural resources management: observability and uncertainty. Ecol. Model. 220, 830–840.
63. Williams, B.K., 2011a. Adaptive management of natural resources: framework and issues. J. Environ. Manag. 92, 1346–1353.
64. Williams, B.K., 2011b. Resolving structural uncertainty in natural resources management using POMDP approaches. Ecol. Model. 222, 1092–1102.
65. Williams, B.K., 2012. Reducing uncertainty about objective functions in adaptive management. Ecol. Model. 225, 61–65.
66. Williams, B.K., Brown, E.D., 2012. Adaptive Management: the U.S. Department of the Interior Applications Guide. U.S. Department of the Interior.
67. Williams, B.K., Brown, E.D., 2014. Adaptive management: from more talk to real action. Environ. Manag. 53, 465–479.
68. Williams, B.K., Johnson, F.A., 1995. Adaptive management and the regulation of waterfowl harvests. Wildl. Soc. Bull. 23, 430–436.
69. Williams, B.K., Johnson, F.A., 2013. Confronting dynamics and uncertainty in optimal decision making for conservation. Environ. Res. Lett. 8, 1–16.
70. Williams, B.K., Johnson, F.A., 2015a. Uncertainty and the value of information in natural resource management: technical developments and application to pink-footed geese. Ecol. Evol. 5, 466–474.
71. Williams, B.K., Johnson, F.A., 2015b. Value of information and natural resources decision making. Wildl. Soc. Bull. 39, 488–496.
72. Williams, B.K., Eaton, M., Breininger, D.R., 2011. Adaptive resource management and the value of information. Ecol. Model. 222, 3429–3436.
73. Williams, B.K., Nichols, J.D., Conroy, M.J., 2002. Analysis and Management of Animal Populations. Academic Press.
74. Williams, B.K., Szaro, R.C., Shapiro, C.D., 2007. Adaptive Management: the U.S. Department of the Interior Technical Guide. U.S. Department of the Interior.
75. Wissmar, R.C., Bisson, P.A. (Eds.), 2003. Strategies for Restoring River Ecosystems: Sources of Variability and Uncertainty in Natural and Managed Systems. American Fisheries Society.
76. Yokota, F., Thompson, K.M., 2004. Value of information literature analysis: a review of applications in health risk management. Med. Decis. Mak. 24, 287–298.
77. Zellmer, S., Gunderson, L., 2009. Why resilience may not always be a good thing: lessons in ecosystem restoration from Glen Canyon and the Everglades. Nebr. Law Rev. 7, 894–947.

CHAPTER 13 SUSPENDED SEDIMENT REGIMES IN CONTRASTING REFERENCE-CONDITION FRESHWATER ECOSYSTEMS: IMPLICATIONS FOR WATER QUALITY GUIDELINES AND MANAGEMENT

Magdalena K. Grove, Gary S. Bilotta· Robert R. Woockman, John S. Schwartz

Keywords: Suspended sediment, Suspended solids, Suspended particulate matter, Reference-condition, Water quality, Guidelines

INTRODUCTION

Managing global water resources is one of the greatest challenges of the 21st century (Garrido and Dinar, 2009, Poff, 2009 and Staddon, 2012). Water is a resource that is under growing pressure as the global population rises, and the natural supply, in the form of precipitation, is becoming increasingly variable and uncertain with climate change (Giorgi et al., 2004, Räisänen et al., 2004 and Trenberth et al., 2003). It is therefore essential that water resources are managed sustainably in terms of both their quantity and quality. One of the most commonly attributed causes for the impairment of water quality globally is the presence of excess suspended sediments, ranging from nano-scale particles and colloids to sand-sized sediments (Gray, 2008 and Richter et al., 1997). Suspended sediments (SS) can have a range of detrimental effects on water resources, from aesthetic issues and higher costs of water treatment, to a decline in the fisheries resource and serious ecological degradation (Alabaster and Lloyd, 1982, Bilotta and Brazier, 2008, Cordone and Kelley, 1961, Gammon, 1970, Newcombe and MacDonald, 1991, Owens et al., 2005, Peddicord, 1979, Ryan, 1991 and Wood and Armitage, 1997). Ultimately this can lead to a significant decline in the associated freshwater ecosystem services, estimated to have a global value in excess of $1.7 trillion per annum (Costanza et al., 1998).

In recognition of the potential for SS to cause aquatic degradation, and in an effort to minimise this degradation, government-led environmental organisations from around the world have established water quality guidelines and standards, which state recommended targets for SS (sometimes referred to as suspended solids, and occasionally assessed through proxy measurements such as turbidity) (Bilotta and Brazier, 2008 and Collins et al., 2011). However, at present these guidelines are often blanket values that do not recognise the natural spatial and temporal variations of SS in streams/rivers, and are not well-linked to the biological/ecological impact evidence; therefore ultimately these guidelines may not reflect the specific requirements of the biological communities that they are designed to protect (Bilotta et al., 2012, Collins et al., 2011, Schwartz et al., 2008 and Schwartz et al., 2011). For example, in Europe the Freshwater Fisheries Directive (78/659/EEC) (2006/44/EC) guideline for SS stated that concentrations should not exceed 25 mg L^{-1} in salmonid and cyprinid waters except in exceptional weather conditions or exceptional geographic circumstances (Bilotta and Brazier, 2008 and Collins et al., 2011). This Directive has now been repealed and replaced by the Water Framework Directive

Citation: Magdalena K. Grove, Gary S. Bilotta, Robert R. Woockman, John S. Schwartz, Suspended sediment regimes in contrasting reference-condition freshwater ecosystems: Implications for water quality guidelines and management, Science of The Total Environment, Volume 502, 1 January 2015, Pages 481-492, ISSN 0048-9697, http://dx.doi.org/10.1016/j.scitotenv.2014.09.054.

(2000/60/EC) (2008/105/EC), but there are currently no new guidelines for SS under this Directive.

It is important to recognise that unlike some aquatic pollutants (e.g. pesticides, pharmaceuticals, veterinary medicines), SS is a natural component of freshwater ecosystems, critical to habitat heterogeneity and ecological functioning (Maitland, 2003,Swietlik et al., 2003, Vannote et al., 1980 and Yarnell et al., 2006). There is natural spatial variation in SS conditions. For example, Bilotta et al. (2012) found background concentrations of SS varied by more than 15-fold between 42 temperate ecosystem-types that were in reference condition.[1] Furthermore, these differences could be predicted using data on a number of the natural environmental characteristics associated with each river's catchment, including metrics of climate, catchment geology and topography, and channel hydromorphology. The interpretation of these findings was that differences in the natural environment create unique SS conditions that support unique freshwater communities. If water quality managers wish to protect these unique freshwater communities through establishing and implementing water quality guidelines, then these guidelines should be environment-specific.

There are also natural temporal variations in SS conditions, which may be critical to the ontogeny of certain organisms (e.g. Maitland, 2003), and which should be expected given the temporally variable contributions of SS from channel and non-channel sources. Wood suggest that the principal sources of SS available to a river from channel sources are: (i) river banks; (ii) mid-channel and point bars; (iii) fine bed material; (iv) natural backwaters; (v) fine particles associated with aquatic vegetation; and (vi) other biotic particles including phytoplankton and zooplankton. In addition there may be in-channel generation of SS due to the decomposition of aquatic macrophytes, biofilms and invertebrates. Benthic invertebrate faecal material has been shown to constitute a significant source of fine SS (Ladle and Griffiths, 1980 and Ward et al., 1994). Finally, some aquatic organisms (invertebrates and vertebrates) may also act to re-suspend material stored in the bed and banks of a water body through burrowing, feeding, and breeding behaviours (Harvey et al., 2011, Hassan et al., 2008 and Montgomery et al., 1996). The main non-channel sources of SS supplied to a river are: (i) exposed soils subject to erosion — this material is transported to the channel via surface and subsurface runoff; (ii) mass failures within the catchment, such as landslides and soil creep; (iii) litter fall, principally leaf material from vegetation adjacent to the channel; and (iv) atmospheric deposition, due to aeolian processes and precipitation (Wood and Armitage, 1997).

In catchments where these SS sources have been modified through anthropogenic activities, the resultant modified SS regimes can affect the biological community. However, at present the SS regime required to maintain or restore biological integrity in a given environment has not been defined. It is known that some aquatic organisms are sensitive to changes in SS, in particular, changes to the duration, frequency and timing that a given concentration is experienced (Diehl and Wolfe, 2010, Kerr, 1995, Newcombe and Jensen, 1996, Newcombe and MacDonald, 1991, Reid and Anderson, 1999,Waters, 1995 and Yount and Niemi, 1990), which collectively are referred to as the SS regime (concentration, duration, frequency). The current water quality guidelines for SS, however, through their use of mean annual concentration values or total mean daily loads, fail to recognise the importance of the SS regime, because an observed mean concentration, even if regarded as being compliant with guidelines, could be achieved through an infinite number of concentration scenarios (from highly variable regimes with concentration extremes to relatively stable regimes with little variability about the mean), each scenario potentially having very different biological effects. There is therefore a need for more advanced water quality guidelines that can take into account these natural spatial and temporal variations, so that water quality managers can identify where and when SS pollution is taking place and to what extent. The first step to achieving this is to understand what the natural background SS regimes are in contrasting ecosystems that are in reference-condition.

The aims of this study are to (1) monitor the SS regimes of contrasting reference-condition river ecosystems and to examine the spatial and temporal variations in SS regimes, and based on these findings (2) consider the appropriateness of the current regulatory water quality guidelines and monitoring.

MATERIALS AND METHODS

Field Sites

Field sites were selected from the RIVPACS IV database (May 2011 version) (River Invertebrate Prediction and Classification System — © NERC [CEH] 2006. Database rights NERC [CEH] 2006 all rights reserved). The RIVPACS IV model and database are described in details by Wright et al. (2000) and Clarke et al. (2011), and are only summarised here. The database contains invertebrate, water quality and catchment characteristics data, recorded between 1978 and 2004, from 795 streams and river sites in the UK. All of the sites are in reference condition; defined as having no, or only very minor, anthropogenic alterations to the values of the hydrochemistry and hydromorphology, with biota usually associated with such undisturbed or minimally disturbed conditions. In the development of the RIVPACS model, the sites were divided into similar biological communities, referred to as end groups, based on the invertebrate community composition. Forty-three end group communities were identified in RIVPACS IV. These end groups are a proxy for the wider ecosystems, i.e. specific invertebrate communities occur in specific environments and are associated with specific floral and faunal (vertebrate) communities. In the RIVPACS IV model these end groups were characterised, using Multiple Discriminant Analysis (MDA), in terms of the temporally invariant properties of the environments that they inhabit. The original purpose of the RIVPACS model was for use in assessments of ecological status. The model can be applied to any site wherever these environmental properties can be measured, allowing the user to make a prediction of the expected community composition. The magnitude of deviation between observed and expected community compositions at a site is used as an indication of the ecological status and therefore anthropogenic disturbance.

In this article, the authors used the RIVPACS IV database to identify ecosystems that occur in the most contrasting environments. In order to achieve this, the environmental properties of the 43 end groups were plotted for eight of the twelve environmental variables identified through MDA as being discriminant variables for invertebrate community composition. Latitude and longitude were excluded because they have no direct physical basis and are therefore likely to produce models that are only applicable locally. Alkalinity was excluded because it is a water quality parameter which can be influenced by pollution which could result in an incorrect reference condition model prediction. Air temperature range was excluded because it is highly correlated with mean annual air temperature. The end groups that had the lowest and highest mean values for each of the remaining eight environmental variables were selected as the most contrasting ecosystem types; subsequently the field sites from each of the identified end groups were selected. This was decided by choosing the sites that had the value closest to the mean for that particular environmental variable for that end group. In a few instances, for logistical reasons (proximity/access), the sites that had the second closest value to the mean for the environmental parameter variable were chosen. Using this process, ten field sites were selected to encompass a wide range of environments, varying in their: (1) Climate — mean annual precipitation totals of 450 to 1400 mm and mean annual temperatures ranging from 8.42 to 11.33 °C, (2) Geology — varying from catchments with almost 100% hard igneous rocks to catchments with almost 100% clay or sandstone, and (3) Topography — altitudes at the river›s source varying from 128 to 1216 m above sea level. The stream and river sites also vary in their morphometry with widths ranging from 0.60 to 50 m and depths ranging from 0.08 to 1.90 m (mean of 3 seasonal measurements). As such the findings and model developed from this study will be applicable to a wide range of streams and rivers located in mid-latitude zones. The approximate location of each site can be seen in Fig. 1, which also describes the range of environmental characteristics of each site.

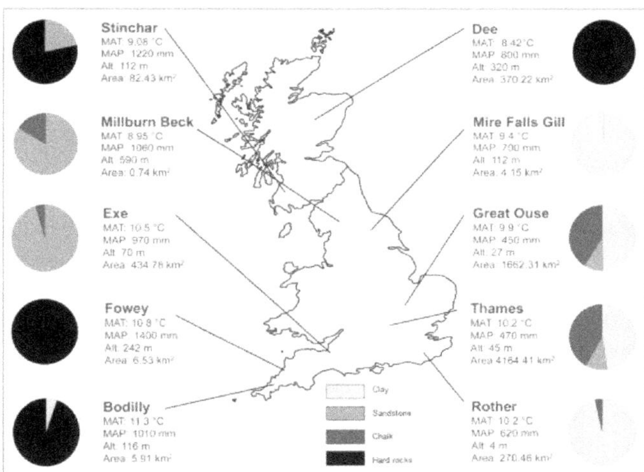

Figure 1: Location of the ten reference-condition stream/river sites, including information on their catchment characteristics: mean annual temperature (MAT), mean annual precipitation (MAP), mean altitude of upstream catchment (Alt), and catchment area (Area). Dominant geology classes, shown in the pie-charts, represent an amalgamation of the 116 Solid Geology classes of the British Geological Survey into 8 River Habitat Survey super-classes. The environmental characteristics of each site, excluding MAP, were sourced from the RIVPACS IV (May 2011 version) database, (Clarke et al., 2011). Mean annual precipitation data were sourced from the UK Meteorological Office (http://www.metoffice.gov.uk/climatechange/science/monitoring/ukcp09/) and derived from 1 km gridded data for the monitoring location during the period 1961–1990.

Sampling, Monitoring and Laboratory Analysis

Each field site was instrumented with a multi-parameter water quality probe (*Aquaread Aquaprobe*) and logger (*Aquaread Aqualogger*), which recorded turbidity — a surrogate measure of SS; every 15 min. These turbidity data were converted to estimated SS concentrations using site-specific turbidity-SS regression equations derived from:

(1) SS data measured through laboratory analysis of water samples that were collected from the sites on a 4–8-week temporal frequency, between May 2011 and May 2013. Water samples were collected manually in 1 L plastic bottles in a depth- and width-integrating manner across each river site. Samples were refrigerated at < 4 °C and stored in the dark prior to analysis in order to avoid degradation. Following Clesceri et al. (1998), samples were analysed in the laboratory for concentrations of SS and volatile organic matter (VOM). This involved filtration of a known volume of sample through a dried, pre-combusted (550 °C for 30 min) and pre-weighed glass fibre filter paper (*Whatman GFF*), with a pore size of 0.70 μm. The filters and their residue were then dried at 105 °C for 60 min and re-weighed. Following this the filters and their residue were ignited in a muffle furnace at 550 °C for 30 min and then re-weighed to determine VOM (VOM equal to the dried particle mass minus the combusted particle mass). This is the standard technique for the determination of VOM (APHA, 1995 and APHA, 1998).

(2) Laboratory-based turbidity-SS calibrations using SS collected from time-integrated samplers between May 2011 and May 2013 that were combined with filtered river water from the site to make up a wide range of known SS concentrations against which to measure turbidity. The time-integrated SS samples were collected using the time-integrating passive trap sampler method described by Phillips et al. (2000). The time-integrated SS samplers were one metre long and consisted of commercially available PE pipes with an outer diameter of 110 mm and a wall thickness of 4.2 mm. They were sealed with a plugged polythene funnel at the inlet and a cap at the outlet. A plastic tube with an inner diameter of 4 mm was passed through the funnel and the cap as inlet and outlet. The samplers were mounted parallel to the riverbed at two upright steel bars driven into the channel bed, with the inlet tube pointing upstream. The greater cross-sectional area of the main cylinder compared to that of the inlet tube reduces the flow velocity within the samplers by a factor of 600 relative to that of the ambient flow. This reduction in flow velocity induces sedimentation of the SS as the water moves through the cylinder towards the outlet tube (Phillips et al., 2000). The time-integrated SS samplers are designed to collect a statistically representative sample under field conditions (Perks et al., 2013, Phillips et al., 2000 and Wildhaber et al., 2012). The SS was extracted from these samplers on a 4–8-week temporal frequency.

The best-fitting regression equations for all of the sites are shown in Fig. 2.

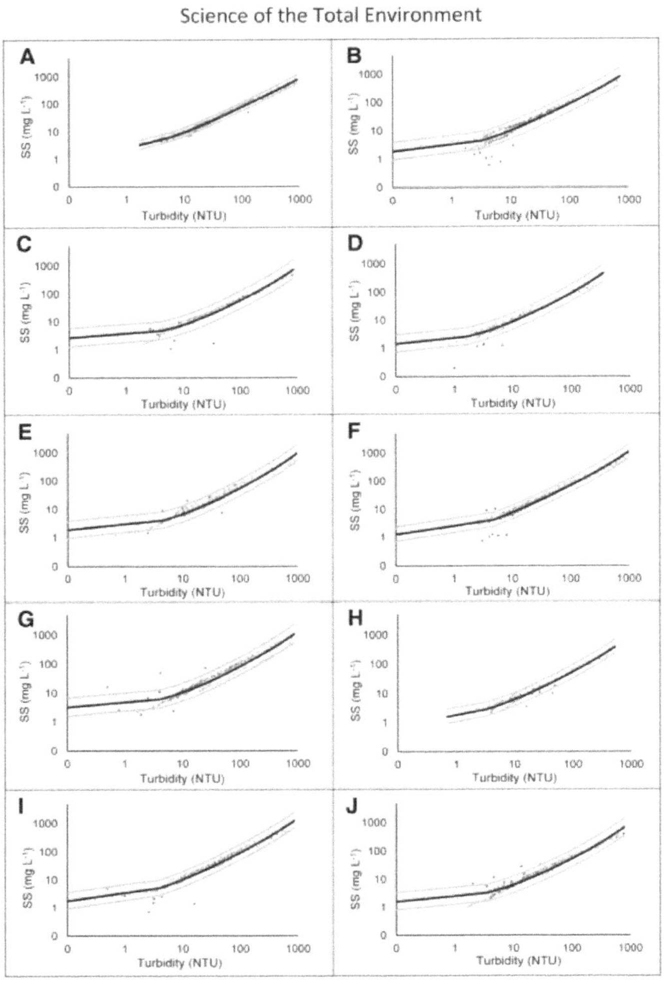

Figure 2: Observed turbidity-SS values (grey crosses for time integrated samples; black diamonds for grab samples) and the predicted values based on regression analysis at each of the study sites (black lines represent the fitted regression value, grey lines represent the upper and lower 95% prediction intervals). Regression equations used to make predictions were carried out using quadratic equations on log-transformed turbidity (NTU) (X-axis) and SS concentration data (Y-axis). All regression equations are statistically significant (p-values $<$ 0.001). (A) Bodilly, $R^2 = 0.98$. (B) Dee, $R^2 = 0.94$. (C) Exe, $R^2 = 0.93$. (D) Fowey, $R^2 = 0.94$. (E) Great Ouse, $R^2 = 0.95$. (F) Millburn Beck, $R^2 = 0.96$. (G) Mire Falls Gill, $R^2 = 0.93$. (H) Rother, $R^2 = 0.95$. (I) Stinchar, $R^2 = 0.96$. (J) Thames, $R^2 = 0.94$.

Data Processing and Analysis

All field data were checked for consistency in turbidity signals by visual inspection as well as against discharge data gathered by the Environment Agency and the Scottish Environment Protection Agency or from the discharge estimations computed from the stage-discharge data collected during the study. In cases where the stream debris or biota covered the turbidity sensors, or if sensors did not respond to changes in stream discharge for prolonged periods of time suggesting equipment failure, the data were removed.

All statistical analysis was conducted using SPSS statistical software (IBM® SPSS® Statistics 20). Prior to the statistical analysis, the SS concentration data sets for all the ten study sites were investigated for normality and homogeneity of variance. In all cases the Kolmogorov–Smirnov test for normality ($p < 0.001$) suggests that the SS concentration data are not normally distributed. The test of homogeneity of variance (Levene Statistic; F (9, 536530) = 7192.1, p = < 0.001) also indicated that the variances are significantly different. Tests of normality and of the homogeneity of variance were also carried out on the transformed SS data (common and natural logarithm, square, square root) with the same outcomes. An examination of the skewness and kurtosis values for the sites also suggests that the SS concentration data are not normally distributed. Based on those findings, non-parametric tests were applied for comparative data analysis.

RESULTS

Spatial Variability

Table 1 presents the summary statistics for the estimated (from herein 'estimated' refers to turbidity-derived) SS concentrations in the ten river sites during the 2011–2013 monitoring period. The raw data are provided in the Supplementary online material. The estimated mean SS concentrations range from 3 mg L^{-1} (Fowey) to 29 mg L^{-1} (Bodilly and Great Ouse). A Kruskal–Wallis and post-hoc tests were conducted to determine if the differences in estimated mean SS concentrations between the sites are statistically significant. Pairwise comparisons were performed using Dunn's (1964) procedure with a Bonferroni correction for multiple comparisons. Statistical significance was accepted at the $p < 0.05$ level for the omnibus test and $p < 0.05$ for the multiple comparisons. Estimated SS concentrations were significantly different between the sites, 2 (9) = 115315.17, $p < 0.001$. Post-hoc analysis revealed statistically significant differences in all comparisons ($p <$ 0.001), except between the following pairs of sites: the River Dee (median = 3.2 mg L^{-1}) and Milburn Beck (median = 3.0 mg L^{-1}) ($p = 0.555$); the River Rother (median = 6.1 mg L^{-1}) and River Stinchar (median = 6.2 mg L^{-1}) ($p = 0.048$); and the River Mire Falls Gill (median = 7.7 mg L^{-1}) and River Exe (median = 6.8 mg L^{-1}) ($p = 0.027$).

Table 1: Estimated mean SS concentrations with lower and upper prediction intervals (based on 95% confidence intervals derived from the turbidity-SS regression), estimated median SS, estimated minimum and maximum SS, and 5th, 30th, 50th and 90th percentiles for estimated SS (mg L^{-1}) in the ten study sites monitored between May 2011 and May 2013. n refers to the number of turbidity readings upon which the estimate is based.

Site	n	Mean	Lower–upper prediction intervals	5th Percentile	30th Percentile	50th Percentile	90th Percentile	Min	Max
				C95	C70	C50 (median)	C10		
Bodilly	59,976	29	20–42	0	11	22	66	0	357

Dee	42,052	6	3–13	0	1	3	14	0	304
Exe	57,840	18	9–39	3	3	7	35	2	591
Fowey	55,489	3	1–6	1	1	1	4	1	460
Great Ouse	51,544	29	15–57	3	3	6	79	2	857
Milburn Beck	62,157	6	3–10	1	1	3	14	1	380
Mire Falls Gill	58,169	15	8–32	3	3	8	27	2	492
Rother	43,331	11	6–19	1	3	6	25	1	407
Stinchar	54,120	15	8–28	1	1	6	32	1	701
Thames	51,862	11	6–22	2	4	7	22	1	234

Table 2 presents the low-resolution, grab sample-derived mean SS concentrations observed in the ten river sites monitored between 2011 and 2013. Observed mean SS concentrations measured between May 2011 and 2013 ranged from 1 mg L^{-1} (Dee) to 17 mg L^{-1} (Bodilly and Mire Falls Gill). Table 2 also compares these observations with the historical low-resolution (monthly grab sample-derived) mean SS concentrations observed in other reference-condition (RIVPACS) sites belonging to the same ecosystem-types or 'end groups'. All of the river sites monitored between 2011 and 2013, except for the River Bodilly, had mean SS concentrations that were within the minimum and maximum mean SS concentrations observed in the sites belonging to the same end-groups. The River Bodilly's mean concentration exceeded the range of mean concentrations observed in the sites belonging to the same end-group. A comparison of the observed (grab-sample) mean SS concentration for each site against the equivalent predicted[2] SS ranges (see Fig. 3), reveals that the model predicted the correct SS range for seven of the sites, and predicted the correct or next closest SS range for the remaining three sites.

Table 2: Comparison of the observed mean SS concentration (mg L^{-1}, based on low-resolution grab sampling) against the mean, minimum and maximum SS concentrations (mg L^{-1}) observed in the sites (n = 638) belonging to the same RIVPACS IV end groups.

River	Observed (grab) mean SS 2011–2013	Mean average SS of sites in the same end group	Min average SS of sites in the same end group	Max average SS of sites in the same end group
Bodilly	17	8	6	9
Dee	1	3	1	6
Exe	13	9	1	24
Fowey	2	4	1	10
Great Ouse	16	19	8	45
Millburn Beck	5	4	1	10
Mire Falls Gill	17	13	5	26
Rother	14	12	4	26
Stinchar	4	4	1	9
Thames	14	19	8	45

Science of the Total Environment

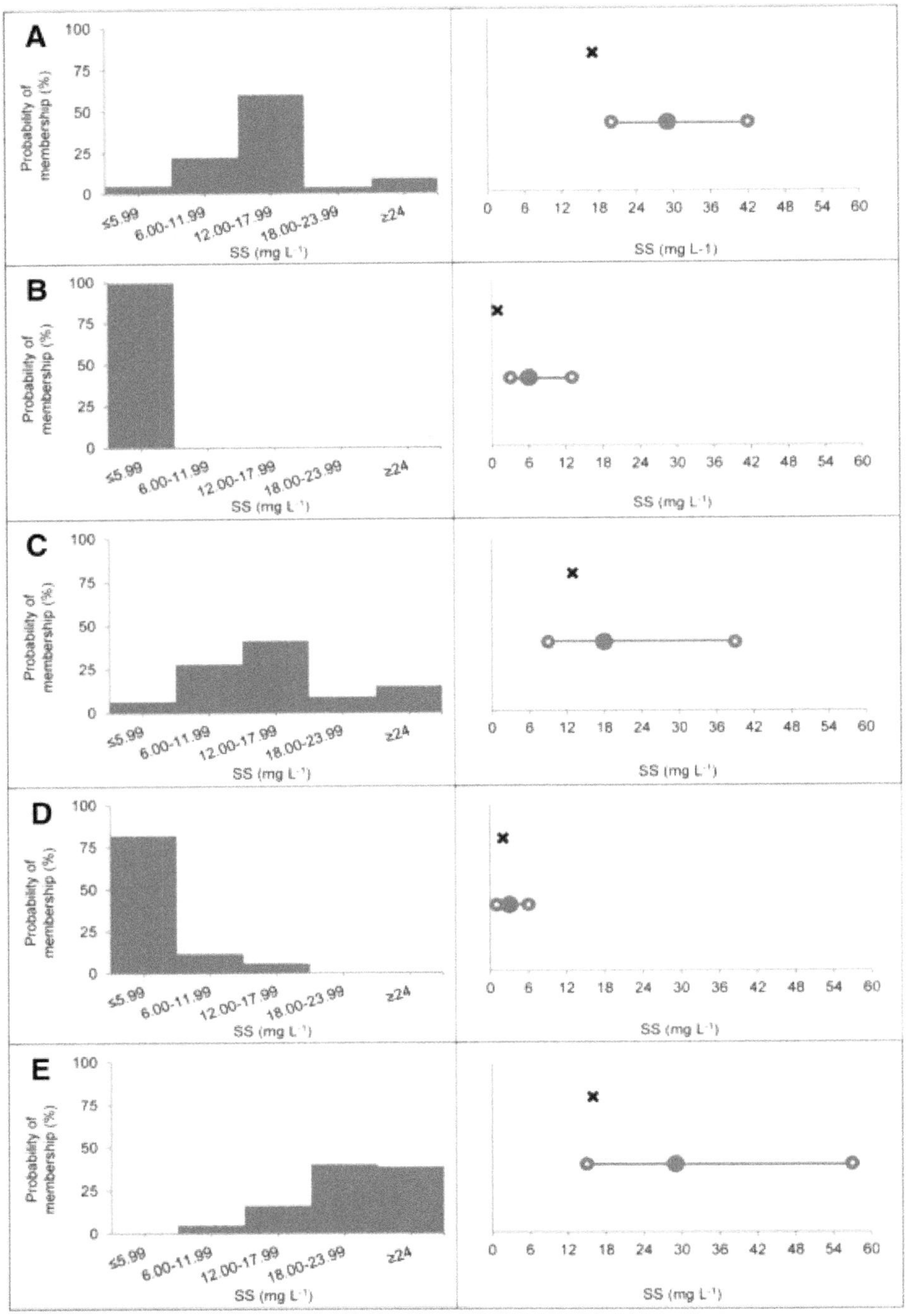

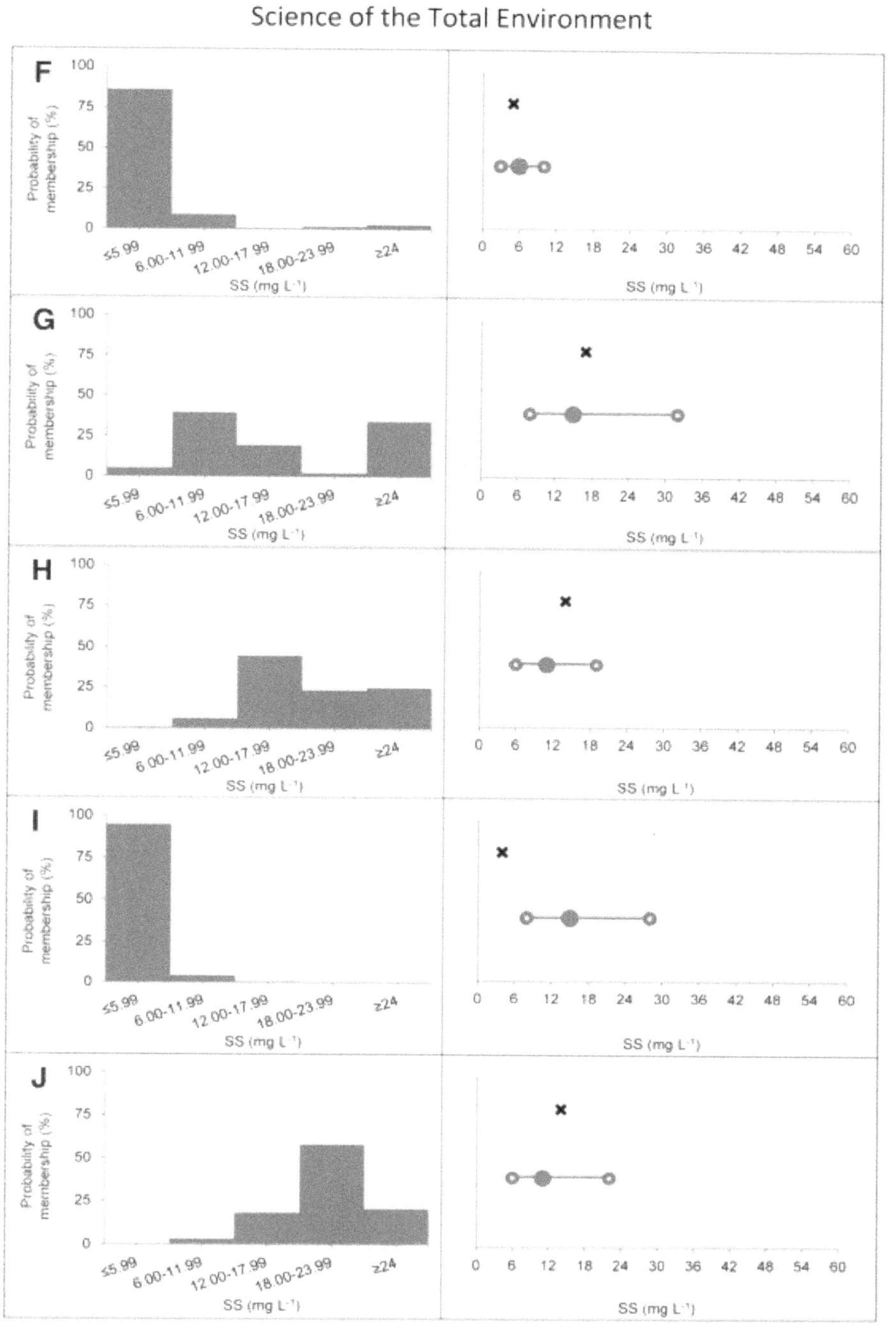

Figure 3: A comparison of the model-predicted SS ranges (bar charts on the left panel) versus the observed (grab sample-derived; illustrated by crosses on the right panel) and estimated (turbidity-derived; illustrated by circles: a closed circle for the central estimate and open circles for the lower and upper prediction intervals) SS concentrations for 2011–2013 monitoring period. (A) Bodilly, (B) Dee, (C) Exe, (D) Fowey, (E) Great Ouse, (F) Millburn Beck, (G) Mire Falls Gill, (H) Rother, (I) Stinchar, (J) Thames.

Temporal Variability

Table 3 presents the estimated mean annual SS concentration observed in each river site for year one (May 2011 to May 2012) and year two (May 2012 to May 2013) of monitoring. These data are also illustrated in Fig. 4 alongside the estimated mean SS concentration for the total monitoring period. A Mann–Whitney U test showed that there are statistically significant differences in the distribution of SS concentrations observed in year one and year two for all the sites ($p < 0.001$ in all cases). All the river sites, except the Dee, Stinchar and Thames, experienced higher estimated mean annual SS concentrations in year two of the study. This difference appears to be partially related to the mean annual discharge. The largest difference in mean annual SS concentration was observed in the River Great Ouse, which experienced a three-fold increase between year one (14 mg L^{-1}) and year two (42 mg L^{-1}) concomitant with roughly three-fold increases in mean annual discharge observed at this site from year one (6.37 m^3 s^{-1}) to year two (19.56 m^3 s^{-1}) (Table 4). The Mann–Whitney U test showed significant differences ($p < 0.05$ in all cases) in annual average discharge observed in year one and year two for all the sites. The relationship between estimated mean annual SS and mean annual discharge is not evident for the Thames site, which experienced remarkably similar estimated mean annual SS concentrations in year one (10.94 mg L^{-1}) and year two (10.86 mg L^{-1}), despite experiencing contrasting mean annual discharges (16.34 $m^3 s^{-1}$and 55.90 $m^3 s^{-1}$).

Table 3: Estimated mean annual SS concentrations in the ten river sites and the percentage differences between the seasons.

SS concentration (mg L^{-1})			
	2011–2012	2012–2013	Percentage difference between 2011–2012 and 2012–2013
	Mean	Mean	
Bodilly	26.70	32.60	22.1
Dee	6.53	6.33	− 3.1
Exe	15.48	21.25	37.3
Fowey	2.21	3.57	61.5
Great Ouse	13.94	42.08	201.9
Milburn Beck	5.64	5.79	2.7
Mire Falls Gill	13.14	17.26	31.4
Rother	8.17	13.02	59.4
Stinchar	17.46	11.76	− 32.6
Thames	10.94	10.86	− 0.7

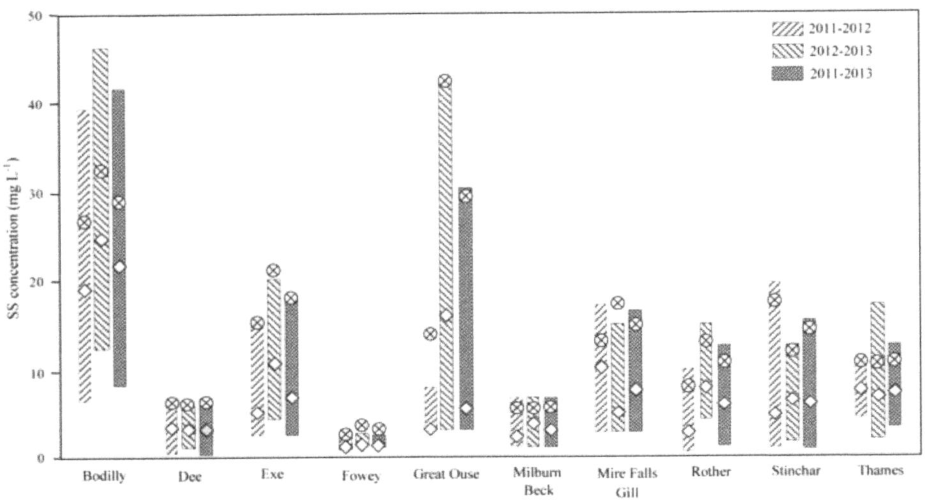

Figure 4: Estimated (turbidity-derived) SS concentrations for the ten study sites based on high-resolution (15-min frequency) monitoring. Data displayed for 2011–2012, 2012–2013 and 2011–2013. Circle symbols represent the mean SS values. Diamond symbols represent the median SS values.

Table 4: Average river discharge for study sites for 2011–2012, 2012–2013, 2011–2013 and the long-term historical average (CEH, 2014).

River	Mean discharge ($m^3 s^{-1}$)			Percentage difference between 2011–2012 and 2012–2013	Historical mean discharge[a] ($m^3 s^{-1}$) and period of record (in parenthesis)
	2011–2012	2012–2013	2011–2013		
Bodilly	0.14	0.30	0.22	114.1	N/A (N/A)
Dee	17.27	13.91	15.61	− 19.4	15.7 (1982–2012)
Exe	8.89	11.85	10.37	33.3	12.7 (1960–2012)
Fowey	0.20	0.34	0.28	68.2	N/A (N/A)

Great Ouse	6.37	19.56	13.18	207.0	11.5 (1972–2011)
Milburn Beck	0.10	0.14	0.12	40.9	N/A (N/A)
Mire Falls Gill	0.05	0.11	0.08	119.3	0.1 (1974–2011)
Rother	1.42	4.42	2.83	211.0	3 (1962–2012)
Stinchar	3.22	2.77	2.99	− 13.9	2.6 (1973–2012)
Thames	16.34	55.90	36.08	242.0	34.5 (1938–2012)

a Data are area-corrected from the National River Flow Archive's supplied by the Centre for Ecology and Hydrology where available (2013).

Table 5: Estimated mean SS concentrations in the study sites based on 15 min resolution and simulated monthly resolution sampling regimes.

		Mean SS concentration ($mg\ L^{-1}$)		% Errors	Absolute difference
		15 min resolution	Monthly resolution	Between mean SS concentration ($mg\ L^{-1}$) sampled every 15 min and monthly	
Bodilly	Mean	29.00	23.07	− 20.5	5.9
	St. Dev.	28.17	18.48		
Dee	Mean	6.42	7.15	11.4	0.7
	St. Dev.	13.17	10.38		
Exe	Mean	18.28	19.17	4.8	0.9
	St. Dev.	40.49	38.32		
Fowey	Mean	3.00	2.99	− 0.3	0.0
	St. Dev.	10.21	3.57		
Great Ouse	Mean	29.38	27.17	− 7.5	2.2
	St. Dev.	61.44	39.14		
Milburn Beck	Mean	5.71	3.25	− 43.0	2.5
	St. Dev.	9.20	3.78		
Mire Falls Gill	Mean	15.34	23.50	53.1	8.2
	St. Dev.	28.00	43.26		
Rother	Mean	10.85	5.66	− 47.8	5.2
	St. Dev.	17.66	4.69		
Stinchar	Mean	14.56	8.34	− 42.7	6.2
	St. Dev.	28.52	10.17		
Thames	Mean	10.90	12.43	14.1	1.5
	St. Dev.	15.01	11.87		

These temporal variations in discharge and SS are natural phenomena that shape biological communities. Therefore ideally (a) water quality guidelines should acknowledge this, and (b) the sampling frequencies used should enable accurate assessments of compliance with these guidelines. In order to assess whether the environment-specific SS prediction model developed by Bilotta et al. (2012) could predict this inter-annual variation, the model input data on mean annual temperature (1961–1990 average) and mean annual precipitation (1961–1990) was replaced with mean annual temperature (°C) and total annual precipitation (mm) measured in the year for which the prediction is to be made at UK Meteorological Office stations closest to each of the ten study sites.[3] (Supplementary online material) illustrates the predicted SS ranges arising from this analysis compared to estimated mean annual SS concentrations (derived from the simulated monthly grab samples described below). As can be seen from, there is good overlap between the predicted SS ranges and the simulated-grab estimated mean annual SS concentrations, for all the sites except the site on the River Rother.

In order to assess the accuracy of the sampling frequency that is currently used in routine regulatory monitoring (Bowes et al., 2009 and Cooper et al., 2003), the high-resolution (15 min frequency) SS datasets were systematically subsampled, selecting one sample per month (selected at 08:00 am on the first day of each month), to calculate the low-resolution mean SS concentrations over the 2011–2013 monitoring period for each site, simulating routine regulatory water quality monitoring. Percentage errors as well as absolute differences were calculated between the high-resolution (15 min frequency) mean SS concentrations and the simulated low-resolution (monthly frequency) mean SS concentrations. The results, presented in Table 5, demonstrate that mean SS concentrations derived from monthly sampling frequencies suffered from errors of up to 53% of values derived from 15-min sampling frequencies, with an average error of 24% for all the ten sites.

DISCUSSION

Spatial Variability

The results presented in Table 1 show that there were over 9-fold differences in estimated mean SS concentrations between the ten sites. Two of the river sites, Bodilly and Great Ouse, had estimated mean SS concentrations (29 mg L^{-1}) that would have exceeded the Freshwater Fisheries Directive (FFD) guideline despite being in reference-condition. The implications of this finding are that the FFD's blanket water quality guideline is likely to have been inappropriate for these specific ecosystem-types as they may be non-compliant regardless of the intensity of land-use. Five of the river sites studied had estimated mean SS concentrations that were less than half of the FFD guideline, with the lowest mean SS observed in the River Fowey (3 mg L^{-1}). Although these river sites would be compliant with the FFD guideline, it is questionable whether the guideline was appropriate for them, as previous research has demonstrated that detrimental impacts can be experienced by some freshwater organisms, of all trophic levels, when exposed to concentrations below 25 mg L^{-1}. For example, Quinn et al. (1992) observed a 40% reduction in the biomass of phytoplankton exposed to SS concentrations of 10 mg L^{-1}. Lloyd (1987) observed a 3–13% reduction in the primary productivity of macrophytes and periphyton exposed to SS concentrations of 8 mg L^{-1}.Rosenberg and Wiens (1978) observed an increase in the amount of drift of benthic invertebrates exposed to SS concentrations of 8 mg L^{-1}. Robertson et al. (2007) observed increased foraging activity in Atlantic salmon when exposed to SS concentrations of 20 mg L^{-1}. Therefore, it is suggested here that some ecosystems may require much lower guideline values in order to be effectively protected.

The degree of spatial variation in estimated mean SS concentrations (3–29 mg L^{-1}) and observed (grab sample-derived) mean SS concentrations (1–17 mg L^{-1}), supports earlier research by Bilotta et al. (2012) that analysed the mean SS concentrations derived from low-resolution (monthly grab sample) monitoring programmes in 638 reference-condition sites in the UK (grouped into 42 ecosystem-types or 'end groups').Bilotta et al. (2012) reported that the mean SS concentrations for the 42 end groups varied between 1.7 and 26 mg L^{-1}. All of the river sites monitored between 2011 and 2013, except for the River Bodilly, had observed (grab sample-derived) mean SS concentrations that were within the minimum and maximum mean SS concentrations observed in the sites belonging to the same end-groups. Bilotta et al. (2012) proposed a model capable of predicting mean (monthly grab sample-derived) SS concentrations to the correct range (6 mg L^{-1} ranges) or the next closest range (in 90% of cases), as a function of a range of site characteristics: mean annual air temperature, mean annual precipitation, mean altitude of upstream catchment, distance from source, slope to source, channel width and depth, the percentage of catchment area comprised of clay, chalk, and hard rock solid geology, and the percentage of the catchment area comprised of blown sand as the surface (drift) material. This model was used to make predictions for the sites monitored in this study. A comparison of the observed (grab-sample) mean SS concentration for each site against the equivalent predicted SS ranges (see Fig. 3), reveals that the model predicted the correct SS range for the seven sites, and predicted the correct or next closest SS range for the remaining three sites.

Temporal Variability

There were statistically significant differences in the distribution of SS concentrations observed in year one and year two of the study for all the sites ($p < 0.001$ in all cases). All river sites, except the Dee, Stinchar and Thames, experienced higher estimated mean annual SS concentrations in year two of the study. This difference appears to be partially related to the mean annual discharge. The largest difference in mean annual SS concentration was observed in the River Great Ouse, which experienced a three-fold increase between year one (14 mg L^{-1}) and year two (42 mg L^{-1}) concomitant with roughly three-fold increases in mean annual discharge observed at this site from year one (6.37 $m^3 s^{-1}$) to year two (19.56 $m^3 s^{-1}$) (Table 4). This inter-annual variability could be predicted reasonably well for all the sites except the River Rother, using the environment-specific model described above, but with modified input data to take into account the mean annual temperature (°C) and total annual precipitation (mm) in the year for which the mean SS prediction was to be made.

The average discharge for the 2011–2013 monitoring period is within 20% of the historical average annual discharge for each site (where data are available). Therefore, although the individual estimated mean annual SS concentrations reported in this paper may represent observations from a drier than average year and a wetter than average year for many sites, the combined mean SS concentration for the whole monitoring period (2011–2013) may be considered to be representative of average hydrological conditions for each site and perhaps therefore average SS conditions for each site. Of course mean values can conceal significant inter-annual differences in the dynamics of both SS and discharge, distorting any relationships between these parameters. For example, although the mean annual SS concentrations for the Thames were similar in year one and year two, the Mann–Whitney *U* test showed that there was a significant difference between the distribution of estimated SS concentrations across the two years.

Sampling Frequency Considerations

The results presented in Table 5 demonstrate that mean SS concentrations derived from monthly sampling frequencies suffer from errors of up to 53% of values derived from 15-min sampling frequencies, with an average error of 24% for all the ten sites. Similar errors have been reported by other researchers (Coynel et al., 2004, Skarbøvik et al., 2012 and Thompson et al., 2014). As the errors vary in both directions there is no simple assumption that less intense sampling underestimates the 'true' values as is sometimes inferred in literature (Coynel et al., 2004 and Thompson et al., 2014). Based on the results of the current study and previous research (Coynel et al., 2004, Skarbøvik et al., 2012 and Thompson et al., 2014), we suggest that water quality managers should place lower confidence in the assessments of compliance with water quality guidelines, when mean annual concentrations are based on monthly-resolution sampling; higher-resolution monitoring of SS would be recommended in order to make more accurate assessments of compliance with water quality guidelines. Thompson et al. (2014)suggest that the optimum monitoring frequency for SS, in terms of accuracy and practicality (e.g. feasible using an autosampler), is likely to be achieved through a 24/7 sampling regime (i.e. a SS sample taken every 7 h to give 24 samples per week). In order to determine the accuracy of this sampling regime in the ten sites studied here, the high-resolution (15-min frequency) SS datasets were systematically subsampled (selecting one sample every 7 h) to calculate a mean SS concentration for each site based on simulated 24/7 sampling. The average error compared to the high-resolution means was just 2%. In addition to enabling a more accurate quantification of the mean SS concentration, high-resolution monitoring of SS also enables water quality managers to gain a better insight into the temporal dynamics of SS (concentration–duration–frequencies) which can be used to predict the impacts on the ecosystem. High-resolution monitoring is more likely to capture the occasional low-frequency high concentration events such as runoff from agricultural land during storms, which may otherwise be missed by low-resolution monitoring but nevertheless can have significant impacts on the aquatic biota. High-resolution monitoring may also help to identify the sources of the pollution based on temporal patterns in concentrations. It must be recognised, however, that: (1) at present there is no environment-specific model that is capable of predicting the high-resolution reference-condition dynamics of SS at a given site in order to aid assessment. At best, the environment-specific model developed by Bilotta et al. (2012) can only predict mean annual concentration ranges. The development of an environment-specific model that is capable of predicting high-resolution dynamics or alternative metrics of exposure (e.g. concentration–duration–frequencies), would require high-resolution monitoring data from studies, similar to this one, collected from a significant number of reference-condition sites. (2) If high resolution monitoring is achieved using a surrogate measure of SS (e.g. turbidity monitoring rather than measuring SS through sampling and analysis), the resultant estimated concentrations will suffer from uncertainties that arise from the turbidity-SS regression analysis. As illustrated in Fig. 3, these uncertainties can be substantial in some cases.

The EU WFD recommends that sampling for the purposes of surveillance monitoring of surface waters should be conducted a minimum of four times a year, but it allows the Member States an option 'to tailor their monitoring frequencies according to the conditions and variability within their own waters' (EC, 2003). The WFD also recognises that sampling programmes will differ greatly between WQ elements, water body types, areas and countries. It also acknowledges that if the quality elements that are to be monitored are temporally variable or spatially discontinuous, more intensive sampling may be required. The observations of this study suggest that SS may be an example of one such temporally variable or spatially discontinuous water quality parameter. In this study the low-resolution sampling error of up to 53% was translated into a maximum absolute error of 8.2 mg L^{-1}, which some might consider to be relatively small? However, it is important to note that this study monitored reference-condition Rivers. The same percentage error on a polluted river with higher mean annual SS concentrations would result in larger absolute errors and potentially significant uncertainties in the assessments of compliance (Skarbøvik et al., 2012).

CONCLUSIONS

Mean SS concentrations, measured between May 2011 and May 2013, in the ten contrasting reference-condition river sites varied up to 9-fold. The degree of spatial variation in estimated mean SS concentrations (3–29 mg L^{-1}), supports earlier research by Bilotta et al. (2012) that analysed the mean SS concentrations derived from low-resolution (monthly grab sample) monitoring programmes in 638 reference-condition sites in the UK (grouped into 42 ecosystem-types or 'end groups'). Bilotta et al. (2012)reported that the mean SS concentrations for the 42 end groups varied between 1.7 and 26 mg L^{-1}. The spatial variation in mean SS could be predicted using the environment-specific SS prediction model developed by Bilotta et al. (2012); which, when applied to the ten study sites, predicted the correct or next closest SS range for all of the sites.

Analysis of temporal variation of SS showed that there are statistically significant differences in the mean annual SS concentrations observed between the two study years. The greatest difference was observed in the Great Ouse river site, which experienced a three-fold increase between year one (14 mg L^{-1}) and year two (42 mg L^{-1}). This difference corresponds well with increases in mean annual discharge observed at this site in year one (6.37 m^3 s^{-1}) and two (19.56 m^3 s^{-1}). These temporal variations in discharge and SS are natural phenomena that shape biological communities and therefore it is important that water quality guidelines acknowledge this. This inter-annual variability in mean annual SS concentration could be predicted reasonably well for all the sites except the River Rother, using the Bilotta et al. (2012) environment-specific SS prediction model, but with modified input data to take into account the mean annual temperature (°C) and total annual precipitation (mm) in the year for which the mean SS prediction is to be made.

This study assessed the accuracy of the sampling frequency that is currently used as part of routine regulatory water quality monitoring. The results demonstrate that mean SS concentrations derived from simulated monthly sampling frequencies suffered from errors of up to 53% compared to 15-min resolution monitoring, with an average error of 24% for all the ten sites. These errors vary in both directions and there is no simple correction that can be applied. Water quality managers should place lower confidence in the assessments of compliance with water quality guidelines, when mean annual concentrations are based on monthly-resolution sampling. The observations of this study suggest that SS is a temporally variable and spatially discontinuous water quality parameter that requires higher-resolution monitoring in order to make an accurate assessment of SS exposure. A 24/7 sampling regime (one sample every 7 h) may be an optimum sampling strategy for SS in terms of accuracy (mean SS concentrations calculated from this regime are, on average, within 2% of the mean SS concentrations calculated from 15-min frequency data) and practicality (samples can be collected by an automated sampler).

The authors are not suggesting that high-resolution SS monitoring should be conducted on a routine basis in all rivers in order to assess compliance with water quality guidelines. This would be prohibitively expensive. Rather, a better system for regulating SS would be to use low-resolution (lower-cost) biological monitoring to determine the ecological status of river sites on a periodic basis (as is current practice across the EU for a number of other water quality parameters) (Turley et al., 2014). At sites where the biological community composition deviates significantly from the reference-condition community composition for that ecosystem-type, high-resolution water quality monitoring could be targeted and used to confirm the likely cause of the biological deviation. If the SS regime differs significantly from that observed in equivalent ecosystems that are in reference condition, then it is likely that SS pollution is occurring and mitigation may be needed. In terms of improving the water quality guidelines for SS, the results of this study demonstrate the need for the more advanced guidelines that take into account natural spatial and temporal variations. One way to achieve this might be to develop environment-specific concentration–duration–frequency guidelines, using high-resolution data from ecosystem-scale observations in reference-condition sites, combined with existing data from microcosm to ecosystem-scale biological/ecological experiments. For the assessments of compliance against modelled guidelines, the resolution of SS monitoring at a given site, must be similar to the resolution of SS data used to develop these guidelines.

ACKNOWLEDGMENTS

The authors would like to acknowledge the funding for this work which was provided by an Engineering and Physical Sciences Research Council CASE Studentship (Grant Reference: EP/1501274/1) which was co-funded by Aquaread Ltd. The authors would also like to acknowledge the following organisations for their contribution to the RIVPACS IV Database (© NERC [CEH] 2006. Database rights NERC [CEH] 2006 all rights reserved) and the WFD119 Project's extension to this database: Centre for Ecology and Hydrology and other Stakeholders/Centre for Ecology and Hydrology, Countryside Council for Wales, Department for Environment, Food and Rural Affairs, English Nature, Environment Agency, Environment and Heritage Service, Freshwater Biological Association, Scotland and Northern Ireland Forum for Environmental Research, Scottish Environment Protection Agency, Scottish Executive, Scottish Natural Heritage, South West Water, Welsh Assembly Government. The authors are extremely grateful for the work of John Davey-Bowker and Michael Dunbar, for their work in compiling the RIVPACS IV database.

REFERENCES

1. Alabaster JS, Lloyd R. Finely divided solids. In: Alabaster JS, Lloyd R, editors. Water Quality Criteria for Freshwater Fish. London: Butterworth Scientific; 1982. p. 1–20.
2. APHA. Standard methods for the examination of water and wastewater. Washington DC, USA: American Public Health Association/American Water Works Association/ Water Environment Federation; 1995.
3. APHA. Standard methods for the examination of water and wastewater. Washington, DC: American Public Health Association; 1998.
4. Bilotta G, Brazier R. Understanding the influence of suspended solids on water quality and aquatic biota. Water Res 2008; 42:2849–61.
5. Bilotta GS, Burnside NG, Cheek L, Dunbar MJ, Grove MK, Harrison C, et al. Developing environment-specific water quality guidelines for suspended particulate matter. Water Res 2012;46:2324–32.
6. Bowes MJ, Smith JT, Neal C. The value of high-resolution nutrient monitoring: a case study of the River Frome, Dorset, UK. J Hydrol 2009; 378:82–96.
7. Clarke R, Davy-Bowker J, Dunbar M, Laize C, Scarlett P, Murphy J. Enhancement of the river invertebrate classification tool. Final Report, Project WFD119; 2011.
8. Clesceri LS, Greenberg AE, Eaton AD. Standard methods for the examination of water and wastewater. Washington, DC.: American Public Health Association; 1998
9. Collins A, Naden P, Sear D, Jones J, Foster ID, Morrow K. Sediment targets for informing river catchment management: international experience and prospects. Hydrol Process 2011; 25:2112–29.
10. Cooper D, Naden P, Smith B, Gannon B. Siltation in rivers. 2: a minimum monitoring strategy for two cSAC rivers. Conserving Natura 2000 Rivers Conservation Techniques Series. Peterborough: English Nature; 2003.
11. Cordone AJ, Kelley DW. The influences of inorganic sediment on the aquatic life of streams. California Department of Fish and Game; 1961.
12. Costanza R, d'Arge R, De Groot R, Farber S, Grasso M, Hannon B, et al. The value of the world's ecosystem services and natural capital. Ecol Econ 1998;25:3–15.
13. Coynel A, Schäfer J, Hurtrez J-E, Dumas J, Etcheber H, Blanc G. Sampling frequency and accuracy of SPM flux estimates in two contrasted drainage basins. Sci Total Environ 2004;330:233–47.
14. Dunn OJ. Multiple comparisons using rank sums. Technometrics 1964;6:241–52.
15. Diehl TH, Wolfe WJ. Suspended-sediment concentration

regimes for two biological reference streams in Middle Tennessee. J Am Water Res Assoc 2010;46:824–37.

16. EC. EC Directive 2000/60/EC of the European Parliament and of the Council of 23 October 2000 of establishing a framework for community action in the field of water policy; 2003. p. 1–160.
17. Gammon JR. Effect of inorganic sediment on stream biota. Vol 18050 DWC 12/70. Washington, DC: USGPO, . USEPA: Water Pollution Control Research Series; 1970.
18. Garrido A, Dinar A. Managing water resources in a time of global change: mountains, valleys and flood plains. London: Routledge; 2009.
19. Giorgi F, Bi X, Pal J. Mean, interannual variability and trends in a regional climate change experiment over Europe. II: climate change scenarios (2071–2100). Climate Dynam 2004;23:839–58.
20. Gray NF. Drinking water quality. Cambridge University Press; 2008.
21. Harvey GL, Moorhouse TP, Clifford NJ, Henshaw AJ, Johnson MF, Macdonald DW, et al. Evaluating the role of invasive aquatic species as drivers of fine sediment-related river management problems: the case of the signal crayfish (Pacifastacus leniusculus). Prog Phys Geogr 2011;35:517–33.
22. Hassan MA, Gottesfeld AS, Montgomery DR, Tunnicliffe JF, Clarke GK, Wynn G, et al. Salmon-driven bed load transport and bed morphology in mountain streams. Geophys Res Lett 2008;35.
23. Kerr SJ. Silt, turbidity and suspended sediments in the aquatic environment: an annotated bibliography and literature review: Ontario Ministry of Natural Resources. Science and Technology Transfer Unit Technical Report TR-008; 1995.
24. Ladle M, Griffiths B. A study on the faeces of some chalk stream invertebrates. Hydrobiologia 1980;74:161–71.
25. Lloyd DS. Turbidity as a water quality standard for salmonid habitats in Alaska. N Am J Fish Manag 1987;7:34–5.
26. Maitland PS. Ecology of the river, brook and Sea Lamprey. Conserving Natura 2000 Rivers Ecology Series No. 5. English Nature, Peterborough. Peterborough: English Nature; 2003.
27. Montgomery DR, Buffington JM, Peterson NP, Schuett-Hames D, Quinn TP. Stream-bed scour, egg burial depths, and the influence of salmonid spawning on bed surface mobility and embryo survival. Can J Fish Aquat Sci 1996;53:1061–70.
28. Newcombe CP, Jensen JO. Channel suspended sediment and fisheries: a synthesis for quantitative assessment of risk and impact. N Am J Fish Manag 1996;16:693–727.
29. Newcombe CP, MacDonald DD. Effects of suspended sediments on aquatic ecosystems. N Am J Fish Manag 1991;11:72–82.
30. Owens P, Batalla R, Collins A, Gomez B, Hicks D, Horowitz A, et al. Fine-grained sediment in river systems: environmental significance and management issues. River Res Appl 2005;21:693–717.
31. Peddicord R. Direct effects of suspended sediments on aquatic organisms. Abstracts of papers of the American Chemical Society. 1155 16th Street NW, Washington DC 20036: American Chemical Society; 1979. p. 38.
32. Perks MT, Warburton J, Bracken L. Critical assessment and validation of a time- integrating fluvial suspended sediment sampler. Hydrol Process 1979.
33. Perks MT, Warburton J, Bracken L. Critical assessment and validation of a time-integrating fluvial suspended sediment sampler. Hydrol Process 2013;28(17):4795–807.
34. Phillips J, Russell M, Walling D. Time-integrated sampling of fluvial suspended sediment: a simple methodology for small catchments. Hydrol Process 2000;14: 2589–602.
35. Poff NL. Managing for variability to sustain freshwater ecosystems. J Water Res Plan Manag 2009;135:1–4.
36. Quinn JM, Davies-Colley RJ, Hickey CW, Vickers ML, Ryan PA. Effects of clay discharge on streams: 2. Benthic invertebrates. Hydrobiologia 1992;248:235–47.
37. Räisänen J, Hansson U, Ullerstig A, Döscher R, Graham L, Jones C, et al. European climate in the late twenty-first century: regional simulations with two driving global models and two forcing scenarios. Climate Dynam 2004;22:13–31.
38. Reid SM, Anderson PG. Effects of sediment released during open-cut pipeline water crossings. Can Water Res J 1999;24:235–51.
39. Richter BD, Braun DP, Mendelson MA, Master LL. Threats to imperiled freshwater fauna. Conserv Biol 1997;11:1081–93.
40. Robertson MJ, Scruton DA, Clarke KD. Seasonal effects of suspended sediment on the behaviours of juvenile Atlantic salmon. Trans Am Fish Soc 2007;136(3):822–8.
41. Rosenberg DM, Wiens AP. Effects of sedimentation on macrobenthic invertebrates on a northern Canadian river. Water Res 1978;12:753–63.
42. Ryan PA. Environmental effects of sediment on New Zealand streams: a review. N Z J Mar Freshw Res 1991;25:207–21.
43. Schwartz JS, Dahle M, Robinson RB. Concentration–frequency–duration curves for stream turbidity: possibilities for use assessing biological impairment. J Am Water Resour Assoc 2008;44(4):879–86.
44. Schwartz JS, Simon A, Klimetz L. Use of fish functional traits to associate in-stream suspended sediment transport metrics with biological impairment. Environ Monit Assess 2011;179:347–69.
45. Skarbøvik E, Stålnacke P, Bogen J, Bønsnes TE. Impact of sampling frequency on mean concentrations and estimated loads of suspended sediment in a Norwegian river: implications for water management. Sci Total Environ 2012;433:462–71.
46. Staddon C. Managing Europe's water resources: twenty-first century challenges. Ashgate Publishing, Ltd.; 2012
47. Swietlik W, Berry W, Gardner T, Hill B, Jha M, Kaufmann P. Developing water quality criteria for suspended and bedded sediments (SABS): potential approaches.

Washington, DC.: US EPA, Office of Water; 2003

48. Thompson J, Cassidy R, Doody DG, Flynn R. Assessing suspended sediment dynamics in relation to ecological thresholds and sampling strategies in two Irish headwater catchments. Sci Total Environ 2014;468–469:345–57.

49. Trenberth KE, Dai A, Rasmussen RM, Parsons DB. The changing character of precipitation. Bull Am Meteorol Soc 2003;84.

50. Turley M, Bilotta GS, Brazier RE, Extence C. Evaluation of a fine sediment biomonitoring tool across a wide range of temperate rivers and streams. Freshw Biol 2014. http://dx.doi.org/10.1111/fwb.12429.

51. Vannote RL, Minshall GW, Cummins KW, Sedell JR, Cushing CE. The river continuum concept. Can J Fish Aquat Sci 1980;37:130–7.

52. Ward G, Ward A, Dahm C, Aumen N. Origin and formation of organic and inorganic particles in aquatic systems. The biology of particles in aquatic systems; 1994. p. 45–73.

53. Waters TF. Sediment in streams: sources, biological effects and control. Am Fish Soc Monogr, Bethesda, Maryland 1995;7:251.

54. Wildhaber YS, Michel C, Burkhardt-Holm P, Banninger D, Alewell C, Van Meerveld I. Measurement of spatial and temporal fine sediment dynamics in a small river. Hydrol Earth Syst Sci 2012;16:1501–15.

55. Wood PJ, Armitage PD. Biological effects of fine sediment in the lotic environment. Environ Manage 1997;21:203–17.

56. Wright JF, Sutcliffe DW, Furse MT. Assessing the biological quality of freshwaters. RIVPACS and other techniques. Ambleside, England: Freshwater Biological Association; 2000.

57. Yarnell SM, Mount JF, Larsen EW. The influence of relative sediment supply on riverine habitat heterogeneity. Geomorphology 2006;80:310–24.

58. Yount JD, Niemi GJ. Recovery of lotic communities and ecosystems from disturbance — a narrative review of case studies. Environ Manage 1990;14:547–69.

CHAPTER 14

COMBINED MATERIAL FLOW ANALYSIS AND LIFE CYCLE ASSESSMENT AS A SUPPORT TOOL FOR SOLID WASTE MANAGEMENT DECISION MAKING

David A. Turner, Ian D. Williams, Simon Kemp

Keywords: Life cycle assessment, Material flow analysis, Solid waste management, Decision support, Greenhouse gas emissions

INTRODUCTION

Currently, around 1.3 billion tonnes of municipal solid waste are generated annually worldwide and generation levels are projected to almost double by 2025, driven by rapid population growth, urbanisation, and socio economic development in developing countries (Hoornweg and Bhada-Tata, 2012). A substantial proportion of this waste material can be viewed as a resource. As demand for natural resources continues to rise, there is increasing pressure on the world's natural resource base, which is having severe environmental consequences (Hertwich et al., 2010). At a global scale, climate change is a serious international concern and the extraction, processing, and use of natural resources contributes directly to climate change through the burning of fossil fuels, whilst the disposal of materials in landfills contributes through emissions of greenhouse gases (GHG). Improving solid waste management (SWM) by recovering value in the form of material and energy resources can contribute towards enhanced resource efficiency and GHG mitigation efforts (UNEP, 2010).

The waste management sector is under increasing pressure to improve its environmental performance. In the European Union (EU), Member States are legally obligated to formulate and implement regional policy instruments to meet the environmental SWM objectives and targets outlined in a broad international legal framework. Article 4(1) of the EU Waste Framework Directive establishes the "waste hierarchy", a five step priority order of waste management comprising, in descending order of priority, prevention, preparation for reuse, recycling, other recovery (e.g. energy from waste), and disposal (EC, 2008). Under the terms of Article 21(1) of the EU Waste Framework Directive, all waste management decisions must be undertaken in line with the waste hierarchy. The Landfill Directive sets a target for member states to reduce the amount of biodegradable municipal solid waste going to landfill to 35% of 1995 levels by 2016 (EC, 1999). A target of achieving 50% recycling of key household waste materials (paper, glass, metals, and plastics) is established in the Packaging and Packaging Waste Directive (EC, 1994). Moreover, at a broader level the EU is committed to reducing its GHG emissions by at least 20% and 40% of 1990 levels by 2020 (EC, 2009) and 2030 (EC, 2014), respectively. Managing resources to maximise environmental sustainability and contribute towards the achievement of national targets set by the EU entails important strategic and investment decisions by local waste managers, who are simultaneously tasked with maintaining a reliable and economical waste removal service to residents under increasing budgetary pressures. To reduce the burden on

Citation: David A. Turner, Ian D. Williams, Simon Kemp, Combined material flow analysis and life cycle assessment as a support tool for solid waste management decision making, Journal of Cleaner Production, Volume 129, 15 August 2016, Pages 234-248, ISSN 0959-6526, http://dx.doi.org/10.1016/j.jclepro.2016.04.077.

local waste managers and promote environmental sustainable practices, there is a need for strong waste policies that guide and enable effective local decisions and actions.

Analytical tools are required to assist local and national governments in evaluating the environmental performance of potential policy measures and local decisions (Turner et al., 2011). Such tools must be capable of handling the increasing complexity of modern 'integrated solid waste management' systems. Modern SWM encompasses a large number of waste treatment technologies, such as incineration, composting, and anaerobic digestion (AD) that are each designed to manage specific waste streams. Many of these technologies provide additional functions, such as secondary materials production and energy production, that necessitate interaction with other sectors, such as manufacturing, agriculture, and energy production (Giugliano et al., 2011). Furthermore, modern SWM systems comprise a global network of facilities, each with distinct technological facets and different levels of operational performance. It is necessary that analytical tools used to support decision making in complex, interdependent systems, such as SWM, adopt a whole system approach that reflects this complexity (Blengini et al., 2012).

In this paper we apply an approach that combines two systems based methodologies, material flow analysis (MFA) and life cycle assessment (LCA), to quantitatively evaluate a complex, municipality-scale SWM system and use scenario analysis to assess the potential effectiveness of different national waste policy measures. The novel contribution of this paper can be summarised as follows:

- Application of a combined MFA and LCA approach to evaluate a complex, multi waste stream SWM system at the meso level.
- Novel use of publically available waste data to comprehensively model waste flows through the system.
- Provision of information to national government regarding the potential effectiveness of waste policy measures.
- Assistance to local government in identifying optimal SWM solutions.

Case Study

Wales is a constituent country of the UK that covers an area of 20,779 km^2 with an estimated population of 3.1 million in 2014 (ONS, 2015). Wales comprises 22 unitary authorities that are individually responsible for arranging waste collection and disposal. The Welsh Government (WG) has introduced a broad and ambitious sustainable development strategy that aims to make sustainable development the core principle of all national and sub national policy and decision making (HMSO, 2015 and Welsh Assembly Government (WAG), 2009). Two key national targets for 2050 have been set: 1) achieving "zero waste" (i.e. eliminating landfilling as far as possible) and 2) reducing national GHG emissions by at least 80% below 1990 levels (HMSO, 2008, Welsh Assembly Government (WAG), 2010a, Welsh Assembly Government (WAG), 2010b and Welsh Assembly Government (WAG), 2011). Furthermore, the WG have also set a target of a 3% reduction in national GHG emissions per year until 2050, to which waste management is required to contribute (WAG, 2010a). The city of Cardiff is the capital of Wales and is located in the south of the country. The city covers an area of 140.3 km^2, of which around 76 km^2 is considered urban, and has a population of approximately 341,100 and a dwelling stock of 135,796. The city has a reported recycling rate of 52.2%, marginally below the national average of 52.3% (StatsWales, 2015b).

The Council operates an alternate weekly kerbside collection service for household residual waste and dry recyclables. Waste materials collected for recycling include paper, card, aluminium cans, steel cans, mixed plastics, and mixed glass. Garden and food waste are each collected separately on a weekly basis from households, whilst an optional absorbent hygiene products collection service operates fortnightly. A bespoke bulky waste collection service is also offered. There are four household waste recycling centres (HWRC) and 14 bring sites (recycling banks) located across the city. The Council also runs services for the collection of wastes from: commercial organisations, street cleaning, fly tipping, and municipal parks/grounds.

METHODOLOGY

A combination of methodologies was applied in this study to quantitatively evaluate the environmental performance of Cardiff's local authority collected waste (LACW) management system and those of alternative systems. LACW comprises all solid waste collected by a local authority. A static MFA approach was applied to assess the mass flows and stocks of LACW into, within, and from the defined systems. MFA is descriptive, systematic approach to assess the metabolism of a defined system and is based on the principle of mass conservation (Brunner and Rechberger, 2004). The freeware STAN v2.5, developed at the Vienna University of Technology, was used to conduct the MFA (Cencic and Rechberger, 2008). STAN allows for visualisation of complex MFA systems, performance of data reconciliation, and consideration of data uncertainties. Mass flow quantities are expressed in terms of mean value and standard deviation, defined as normal distributions. Data reconciliation, based on the method of least squares regression (see Fellner et al., 2011), is performed to enforce mass balance constrains on uncertain, conflicting input data. The extent to which data are reconciled is determined by the initial data uncertainty, which is used as the weighting factor. STAN has been used in many previous MFA studies of SWM systems (e.g. Allegrini et al., 2015, Arena and Di Gregorio, 2014, Boldrin et al., 2011 and Mastellone et al., 2009).

The environmental impacts of the system described by the MFA were assessed using LCA. LCA has been applied extensively to evaluate SWM systems (see Laurent et al., 2014a). The focus of the LCA undertaken in this study was to evaluate the potential climate change impacts of LACW management and to identify the waste streams and waste treatment processes that contribute most significantly to these impacts. The LCA followed the ISO 14040 and 14044 standards for LCA (ISO, 2006a and ISO, 2006b) as far as possible and was performed using the EASETECH software. EASETECH, developed at the Technical University of Denmark, is a LCA model for the assessment of environmental technologies that allows for detailed modelling of complex systems and provides specialist functionality for modelling waste management systems (Clavreul et al., 2014). EASETECH has been used in previous LCA studies of SWM systems (e.g.Butera et al., 2015, Turner et al., 2016 and Yang et al., 2014).

Goal Definition

The goal of this study was threefold: 1) to quantify the mass balance of waste of the existing LACW management system; 2) to quantitatively evaluate the environmental performance of the existing LACW management system with regards to GHG emissions; and 3) to compare the environmental performance of the existing LACW management system with that of alternative systems designed to represent the implementation of national waste policy measures. The purpose was to support regional policy and decision making, i.e. ILCD (International Reference Life Cycle Data System) decision context situation B, meso level decision support (EC et al., 2010). The study was carried out in cooperation with the WG and Natural Resources Wales (NRW), the regulator and principal advisor to the WG, who, along with Cardiff County Council, were the key audience. Additionally, aspects of the study will be of value to other national policy makers and local governments, as well as the wider SWM MFA and LCA communities.

The quantitative assessment of the systems under investigation was carried out using LCA. The LCA followed an "attributional" approach (EC et al., 2010), with allocation avoided by means of system expansion. The assessment included the potential environmental impacts from SWM activities as well as impacts on processes in other systems that are affected by SWM activities, principally the recovery of materials and energy production (Giugliano et al., 2011). The system comprised two subsystems: (1) a foreground system that includes processes directly engaged in the management of the reference flow (here, LACW); and (2) a background system that interacts with the foreground system by supplying or receiving resources, including avoided primary material and energy production (Clift et al., 2000).

Background data were taken from international life cycle inventory (LCI) databases (ecoinvent v.2.2 and ELCD v.2), the UK GHG conversion factors repository (Defra et al., 2013), and literature. Foreground system data were derived from various secondary sources and are described in detail in Section 2.3. Following an attributional approach, average data were generally used to model foreground and background system processes. However, based on the recommendations of the ILCD (EC et al., 2010) and Laurent et al. (2014b) for LCA studies aiming to provide meso level decision support, long term marginal process data were used to reflect the large scale consequences on the energy system from avoided material and energy production. This approach, which is similar to that taken by Rigamonti et al. (2010) and Finnveden et al. (2005), was adopted rather than a "pure" attributional approach as it more accurately reflects the large scale implications of SWM activities on other systems. Specifically, the following assumptions were made concerning material and energy substitution:

- Electricity produced from SWM activities displaced an equivalent amount of electricity generated from combined cycle gas turbines (CCGT), which is considered as being the long term marginal energy source in the UK by the government (DECC, 2014a). Generated electricity is transmitted to the National Grid at an efficiency of 98% (National Grid, 2008).
- Heat produced at facilities fitted with combined heat and power (CHP) units would be used internally due to a lack of established district heating networks in the UK (Hawkey, 2012). Hence, no avoided energy production was associated with heat energy production.
- Secondary products produced from recovered waste materials replace the production of alternative products, including those produced from primary resources. Avoided materials production was modelled based on the average market situation. However, to reflect large scale consequences on the energy system, avoided material production process electricity use was modelled using the long term marginal electricity source, i.e. CCGT. The extent of the displacement was calculated based a substitution ratio, which was calculated as the product of three parameters: (1) recyclability (the amount of a waste material in a recycled product, considering all process material losses); (2) material quality loss (reflects any diminishment in the inherent technical properties of a waste material incurred through reprocessing); and (3) market substitution rate (the actual amount of an alternative product that is replaced at the market). Material losses and material quality losses are defined for each recycled or reused material. Market substitution rates of 1 were used for all product substitution calculations based on the supposition that production of secondary products would not affect the market situation (Briffaerts et al., 2009, Merrild et al., 2012 and Rigamonti et al., 2009).

Scope Definition

The function of the system under investigation is to manage LACW from Cardiff. Hence, the functional unit was defined as the management of the total amount of collected LACW in Cardiff between April 2012 and March 2013. The amount is 168,526 tonnes. LACW is categorised into five primary waste streams (commingled materials, source segregated food waste, source segregated garden waste, source segregated dry recyclables, and residual waste) that are derived from nine waste collection sources (household kerbside collection, bulky waste collection, HWRCs, bring sites, street cleaning, fly tipped materials, non household collections, voluntary sector/community collection, and municipal parks/grounds waste collection). All activities required to manage the LACW from the point of collection to its ultimate disposal, reprocessing, or reuse were considered. Hence, the processes of collection, transportation, intermediate processing (sorting and handling), thermal treatment, biological treatment, material reprocessing and reuse, and landfill were individually quantitatively analysed in terms of their resource consumption and emissions.

System Boundaries

The spatial boundary of the system was defined by the administrative boundary of Cardiff County Council, i.e. only waste generated within this defined spatial boundary was considered in the analysis. The temporal boundary was the April 2012 to March 2013 fiscal year. In this study, the "zero burden assumption" was adopted, whereby the environmental impacts from upstream life cycle stages prior to the collection of LACW were not included (Ekvall et al., 2007). Environmental impacts from capital goods were also excluded, an approach consistent with other LCA studies of waste management systems (Cleary, 2009). Due to difficulties in data collection, environmental burdens from operation of HWRCs were also excluded.

Processes included in the assessed system comprise collection and transport of LACW, waste management facilities (including those for the treatment of residuals, such as incinerator bottom ash or fly ash), and utilisation of secondary products (e.g. digestate or compost). Waste management facilities included in the assessed system were identified from Cardiff's WasteDataFlow[1] Question 100[2] returns for the four quarters of the fiscal year 2012–2013 (WasteDataFlow, 2014), which include details of facilities that handled collected LACW. The Cardiff LACW management system comprised 158 individual facilities located across nine different countries. 139 facilities are located in the UK (41 in Wales and 98 in England), five were located in Europe, and 13 were located in Asia. Facilities was categorised into one of 11 different waste treatment technology types (e.g. AD, incineration, material reprocessing, etc.), based on their description in WasteDataFlow, and were allocated individual identification codes that reflected the facility type and location (domestic or foreign). In addition, 15 generic processes were included to manage collected waste where no waste treatment facility was specified in WasteDataFlow. Details of the facilities (unit processes) included in the assessed system are provided. Facilities were modelled in the MFA and LCA as unit processes.

Impact Coverage

The quantitative potential environmental impact assessment of this study focused on GHG emissions. Hence, climate change was the only potential impact category included in the life cycle impact assessment. This decision was made for two principal reasons:

- Climate change is recognised as being a significant global environmental problem of foremost importance (IPCC, 2014, International Solid Waste Association (ISWA), 2015 and Welsh Assembly Government (WAG), 2010a); and
- Climate change impact has been identified as being a good proxy for the overall environmental impacts of SWM (Defra, 2011).

The GHGs considered in the assessment were carbon dioxide (CO_2) (fossil and biogenic), methane (CH_4), and nitrous oxide (N_2O). Emissions of these GHG represent more than 90% of total GHG emissions from SWM (Bogner et al., 2007).

Scenarios

Four scenarios are compared in this study. The first scenario represents the existing LACW management system, whilst the other scenarios reflect different systems based on national waste policies. Scenarios are compared on the basis of functional equivalence – i.e. it was assumed that waste generation was not influenced by the different scenario assumptions.

S1 *Baseline scenario* Baseline scenario representing the existing LACW management system for Cardiff in April 2012 to March 2013.

S2 *Enhanced food waste capture* Wales' *Municipal sector plan* (WAG, 2011) states that in order for Wales' 70% recycling target for 2025 to be met, it is likely that local authorities will need to separately collect for biological treatment at least 80% of biodegradable waste. To evaluate the potential GHG impacts of enhanced biodegradable waste capture, in this scenario an 80% capture rate of food waste via household kerbside food waste collection is assumed, with food waste diverted from the household residual waste stream. All food waste that is collected is sent to AD, the priority destination for food waste in Wales (WAG, 2010b).

S3 *Enhanced incineration* The *Towards Zero Waste* (WAG, 2010b) waste strategy for Wales sets a target for a maximum level of incineration (with energy recovery) of MSW of 30% by 2025. This scenario was constructed to evaluate the impact of enhanced incineration of LACW. Based on the total quantity of collected LACW for Cardiff in 2012–2013 (168,426 t), the 30% limit set by WG allows for up to 50,530 t of LACW to be incinerated. In this scenario, 50,000 t of residual waste originally collected and sent to landfill is instead sent for incineration.

S4 *Enhanced dry recycling* A target for a minimum level of recycling and biological treatment of municipal solid waste of 70% by 2025 was set by the WG in *Towards Zero Waste* (WAG, 2010b). To evaluate the impact of an enhanced dry recycling strategy, in this scenario optimum household kerbside collection recycling rates (taken from Eunomia et al., 2011), are achieved for key recyclable waste materials (glass, paper, cardboard, steel cans, aluminium cans, and plastics), with materials diverted from the household kerbside collected residual waste stream.

Life Cycle Inventory

Reference Flow Characterisation

A static MFA approach was applied to characterise the mass flows and stocks of LACW into, within, and from the defined system. Due to its complexity and size, the system was divided into four subsystems, representing the management of primary waste streams (residual waste, commingled materials, source segregated dry recyclables, and source segregated food and garden wastes). The import flows into each system were the collected LACW primary waste streams, and the export flows comprised secondary products and emissions. To account for flow multidimensionality, secondary waste types (e.g. paper, glass, food waste) were characterised as fractions of the primary waste stream.

Due to the functional complexity of many processes and the profusion of functionally equivalent processes in the system (i.e. SWM facilities engaged in the same activity), it was necessary to divide many processes into subsystems that contain subprocesses and, in some cases, further subsystems. The four defined systems for the baseline scenario (S1) are, collectively, composed of 548 processes and 1042 flows. Mass flows of waste into the LACW management system and to and between processes were estimated by aggregating Cardiff's WasteDataFlow Question 100 returns for the four quarters of the fiscal year 2012–2013. Where data were not available from WasteDataFlow (e.g. mass flows of wastes and secondary products from treatment facilities and reprocessing), process transfer coefficients were derived from the LCI process inventory models (detailed below). All data used in this study were single value and were inputted into the MFA as mean values and an associated uncertainty, expressed as a percentage. Details of the data reconciliation approach are provided.

Waste Composition

The overall waste composition for the investigated system is presented in Fig. 1. Further details are provided

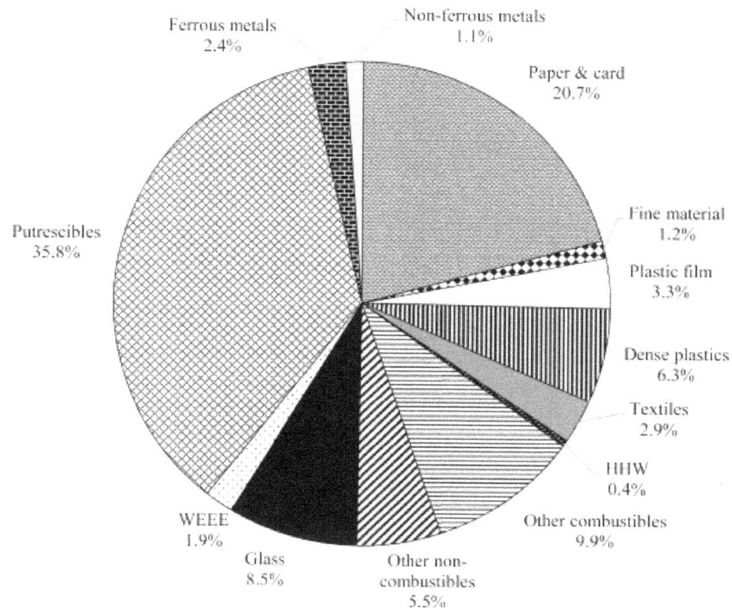

Figure 1: Composition of LACW collected in Cardiff in 2012–2013 by primary material category (% of total LACW input mass flow). LACW, local authority collected waste; WEEE, waste electrical and electronic equipment; HHW, household hazardous waste.

Collection and Transport

Details of the modelling approach for collection and transport are presented.

Transfer Station

Transfer station process electricity and diesel consumptions of 4 kWh/t and 1 l/t, respectively, were assumed, based on measured data from UK based transfer stations reported by Eunomia et al. (2011).

Anaerobic Digestion (Food Waste Only)

The AD of food waste process was modelled based on dry, mesophilic (one stage, one phase) technology, as described by Møller et al. (2009). Details of process inputs and parameter values are presented in Table 1. During waste reception, reject rates of 8%, 50%, and 95% were applied for food waste and fine material, other biodegradable waste, and non biodegradable waste, respectively (Bernstad and la Cour Jansen, 2011). Biogas was assumed to be utilised on site in a CHP engine - the most common biogas utilisation technology in the UK (Horne et al., 2013) - with gross electrical and heat energy generation efficiencies of 31% and 49%, respectively (DECC, 2014b). Heat energy was assumed to be used internally. CHP unit process emissions, produced through incomplete combustion, of 434 mg CH_4/MJ_{biogas} and 1.6 mg N_2O/MJ_{biogas} were assumed (Nielsen et al., 2010). Digestate application was modelled based on the average situation in the UK in 2012, with 91% applied to agricultural land, 6% used in horticulture/gardening, and 3% used as alternative landfill daily cover (Horne et al., 2013). See Section 2.3.8 for details.

Table 1: LCI data for the AD (food waste only) process. Note that data are presented per tonne throughput.

	Unit	Value	Reference
Pre-treatment			
Inputs			
Diesel	l	0.12	Primary data from industrial source.
Electricity	kWh	1.1	Primary data from industrial source.
Anaerobic digestion			
Inputs			
Diesel	kg	1.3	Fisher (2006).
Electricity	kWh	20.6	Fisher (2006).
Process parameters			

Degradation ratio	$\%_{C\_bio\_and}$	70	Gallert and Winter, 1997, Davidsson, 2007 and Yoshida et al., 2012
CH_4 content of biogas	$\%_{biogas}$	63	Smith et al., 2001, Møller et al., 2009 and Banks et al., 2011; EASETECH database.

Composting

Two composting processes were modelled: in vessel composting (IVC) and open windrow composting (OWC). During waste reception, reject rates of 8% and 95% were applied for biodegradable waste and non biodegradable waste, respectively (Bernstad and la Cour Jansen, 2011). Details of process inputs and parameter values are presented in Table 2. The IVC process was assumed to be equipped with a biofilter. Biofilter CH_4 oxidation efficiency was modelled based on the mean of reported values in the literature, whilst the biofilter was assumed to have no effect on N_2O emissions due to a lack of consensus in the literature. A biofilter was not included in the OWC process as it is an open technology. Application of compost produced through IVC and OWC was modelled based on the average situation in the UK in 2012 as follows: for IVC derived compost, 68% to agricultural landspreading, 29% in horticulture/gardening, 2% as alternative landfill daily cover, and 1% for incineration; and for OWC derived compost, 73% to agricultural landspreading, 21% in horticulture/gardening, 5% as alternative landfill daily cover, and 1% for incineration (Horne et al., 2013). See Section 2.3.8 for details.

Table 2: LCI data for the IVC and OWC processes. Note that data are presented per tonne throughput.

	Unit	Value	
		IVC	OWC
Pre-treatment			
Inputs			
Diesel	l	0.12[a]	0.12[a]
Electricity	kWh	1.1[a]	1.1[a]
Aerobic composting			
Inputs			
Diesel	kg	0.26[a]	3[b,c,d]
Electricity	kWh	42.4[a]	0.51[b]
Lubricating oil	l	–	0.023[c]
Process parameters			
C degradation rates			
Paper and card & wood	$\%_{C\ bio}$	35[d]	35[e]
Food waste & fine material	$\%_{C\ bio}$	75[d]	75[e]
Garden waste	$\%_{C\ bio}$	50[d]	50[e]
N degradation rates			
Food waste & fine material	$\%_{N}$	70[e]	70[f]
Garden waste	$\%_{N}$	10[e]	10[f]
Other biodegradable waste	$\%_{N}$	25[e]	25[f]
Biofilter CH_4 oxidation efficiency	$\%_{CH4}$	75[g]	–
Outputs			
Emissions to air			
CH_4	$g/kg_{Degraded\ C}$	2.4[h]	2.1[d]
N_2O	g/kg_{N}	1.8[h,i]	1.5[h]

a Source: Primary data from industrial source.

B zSource: Fisher (2006).

c Source: Boldrin et al. (2009a).

d Source: Andersen et al. (2010).

e Source: Smith et al. (2001).

f Source: EASETECH database.

g Source: Based on the mean of reported values from Dalemo et al., 1997, du Plessis et al., 2003,Streese and Stegmann, 2005, Brown et al., 2008, Boldrin et al., 2009b, Boldrin et al., 2011, Qiang et al., 2011 and Yazdani et al., 2012.

h Source: Boldrin et al. (2009b).

i Source: Amlinger et al. (2008).

Mechanical Biological Treatment

Mechanical biological treatment (MBT) involves a combination of mechanical separation and biological treatment (either bio stabilisation/bio drying, IVC, or AD) (Defra, 2013b). The MBT facility in the investigated system comprised integrated IVC technology. Based on primary data from a UK based MBT facility, process diesel and electricity consumption was assumed to be 0.32 l/t and 7.1 kWh/t, respectively. During mechanical separation, waste is sorted and classified to separate a fine, compostable fraction and two coarse fractions: one for refuse derived fuel (RDF) preparation and the other for disposal in landfill. Ferrous and non ferrous metals are recovered by overband magnets and eddy current separators, respectively, whilst plastics are recovered by near infra red optical separation. Waste transfer coefficients for the mechanical separation stage were derived from facility specific data from Waste Data Flow. For biodegradable waste, 28% is sent for composting, 64% is used as RDF, and 8% is sent for disposal in landfill. 100% of metals and plastics are recovered for recycling. 5% of non biodegradable waste is sent for composting and 95% is sent for disposal in landfill (Bernstad and la Cour Jansen, 2011). The biological treatment stage was modelled based the IVC process model described in Section 2.3.6. The MBT process produces a solid, residue called compost like output. Compost like output application was modelled based on the average situation in the UK in 2012, with 3% applied to agricultural land, 80% used as alternative landfill daily cover, and 17% disposed of in landfill (Horne et al., 2013). See Section 2.3.8 for details.

Application to Land

Application to agricultural land was assumed to require use of a manure spreader, consuming 0.3 l/t and 0.45 l/t of diesel for compost/compost like output and digestate, respectively (Bernstad and la Cour Jansen, 2011 and Boldrin et al., 2009b). The degradation and fate of carbon and nitrogen in solid residues after they have been applied to land was modelled using proxy data from Denmark (Bruun et al., 2006 and Hansen, 2006b) and Europe (Smith et al., 2001), presented in Table 3. Solid residues applied to agricultural land were assumed to partially substitute for commercial fertilisers, with the amounts substituted calculated based on the solid residue nutrient content and a substitution rate. Solid residues used in gardening/horticulture were assumed to substitute for peat (Smith et al., 2001). Details of fertiliser and peat substitution rates applied are presented in Table 3.

Table 3: LCI data for the application to land process. Note that data are presented per tonne throughput.

	Unit	Value	
		Compost/CLO	Digestate
Process parameters			
Carbon bound in soil[a]			
Food waste-derived output	$\%_{C\ bio}$	8.2[b]	8.2[b,c]
Garden waste-derived output	$\%_{C\ bio}$	14[d]	–
Mixed organic waste-derived output[d]	$\%_{C\ bio}$	12[b,d]	12[b,c,d]
Fertiliser substitution rates			
Nitrogen fertiliser	$t/t_{N\ applied}$	0.2[f]	0.48[g]
Phosphorous fertiliser	$t/t_{P\ applied}$	0.5[g]	0.5[g]
Potassium fertiliser	$t/t_{K\ applied}$	0.8[g]	0.8[g]
Peat substitution rate	m^3	1.47[h]	1.47[h]
Outputs			
Emissions to air			
N_2O	g/kg_N	1.4[d]	1.4[d]
$CO_{2\ biogenic}$			
Food waste-derived output	$g/kg_{C\ bio}$	91.8[b]	91.8[b,c]
Garden waste-derived output	$g/kg_{C\ bio}$	86[d]	–
Mixed organic waste-derived output[e]	$g/kg_{C\ bio}$	88[b,d]	88[b,c,d]

a Relevant for the sensitivity analysis only.

b Source: Smith et al. (2001).

c Source: Møller et al. (2009).

d Source: Bruun et al. (2006).

e Based on an assumed composition of 1/3 food waste and 2/3 garden waste.

f Source: Hansen (2006a).

g Source: Evangelisti et al. (2014).

h Based on assumed bulk densities of peat and compost/CLO/digestate of 0.3 t m^{-3} and 0.68 t m^{-3}, respectively (Smith et al., 2001), and an assumed substitution rate of 1 $m^3_{peat}\ m^{-3}_{compost/CLO/digestate}$ (Boldrin et al., 2010).

Material Recovery Facility

Two material recovery facility (MRF) processes were modelled: single stream commingled materials MRF and residual waste MRF. The single stream commingled materials MRF process comprises a series of mechanical separation activities designed to recover the following product streams: mixed container glass, paper, card, high density polyethylene bottles, polyethylene terephthalate bottles, mixed plastics, plastic film, steel cans, and aluminium cans. A residual waste stream is also produced and sent for landfilling. Electricity and diesel consumption of 35 kWh/t and 2 kg/t, respectively, was assumed for the single stream commingled materials MRF process, based on measured data from UK based MRFs reported by Eunomia (2011). Details of product stream material composition are provided. Due to a lack of specific data, electricity consumption for the residual waste MRF process was assumed, based on Pressley et al. (2015), to be approximately 25% more than that of a single stream commingled materials MRF. Hence, an electricity consumption of 44 kWh/t was assumed. Diesel consumption was based on the single stream commingled materials MRF process. The mass of materials recovered at residual waste MRFs was modelled using facility specific data from WasteDataFlow.

Refuse Derived Fuel Production

The RDF production process was modelled based on a theoretical facility with a gas fired drying stage. Electricity consumption of 40 kWh/t of RDF produced was assumed (Burnley et al., 2011). Process RDF yield was based on facility specific data from WasteDataFlow.

Merchant/Exporter

The merchant and exporter processes were modelled using the transfer station process as a proxy.

Incineration

The incineration process was modelled based on an average grate furnace incinerator – the most common incinerator type worldwide (Defra, 2013a) – with dry flue gas scrubbing and de NO_x technology, as described by Boesch et al. (2014). Details of energy and material inputs and waste characteristic transfer coefficients for the incineration process are presented in Table 4 and Table 5, respectively. Thermal energy released during the incineration process is recovered via a boiler and a fully condensing turbine and transformed into electrical energy by a generator at an assumed gross generation efficiency of 22% (European Commission (EC), 2006 and Defra, 2013a). A 2%

electrical energy loss was included to account for part load operation and heat transfer losses (Consonni and Viganò, 2011).

Table 4: Energy and material inputs for the incineration process. Note that data are presented per tonne throughput.

	Unit	Value	Reference
Inputs			
Water	m^3	1	Astrup et al., 2009 and Boesch et al., 2014
Electricity	kWh	142	European Commission (EC), 2006 and Boesch et al., 2014
Heat, heavy fuel oil	MJ	240	Boesch et al. (2014)
Ammonia	kg	0.5	Astrup et al., 2009 and Environment Agency (EA), 2010
Lime	kg	10	EA (2010)
Sodium hydroxide	kg	0.19	Astrup et al., 2009 and Environment Agency (EA), 2010
Activated carbon	kg	0.25	EA (2010)

Table 5: Material characteristic transfer coefficients (%) for the incineration process.

	Bottom ash (%)	Fly ash (%)	Emissions to air (%)	Reference
Ash	87	13	–	EASETECH database
Volatile solids	1	–	–	EASETECH database
Energy	1	–	–	–
Fossil C	–	–	100	EASETECH database
Biogenic C	1	–	99	Boesch et al. (2014)
N	1	–	99	Boesch et al. (2014)
Fe	99	1	–	Koehler et al., 2011 and Boesch et al., 2014
Al	82	18	–	Koehler et al., 2011 and Boesch et al., 2014
Cu	95	5	–	Koehler et al., 2011 and Boesch et al., 2014

Non combustible solid residues (i.e. bottom ash and fly ash) are discharged from the base of the furnace and cooled. Fly ash was assumed to be utilised as a backfilling material, with electricity and diesel consumption of 29 kWh/t and 1.5 l/t, respectively, assumed for the process (Fruergaard et al., 2010). Bottom ash is sent for de scrapping to recover metals for recycling. The de scrapping process was modelled using data fromBosech et al. (2014). 97% of incinerator bottom ash is sent for de scrapping, with the remaining 3% sent to landfill. Recovery rates of 30%, 80%, and 30% were assumed for aluminium, steel, and copper, respectively, with non recovered material sent to landfill. 3 kWh/t of electricity is required for de scrapping process.

Landfill

The non hazardous landfill process (also used as a proxy for inert landfill) was modelled based on an average UK medium sized conventional landfill with gas utilisation. Emissions were modelled over a 100 year time period. Waste material decay rates were taken from IPCC (2006). Energy and material inputs are presented in Table 6. Details of the gas management system and its performance over time are presented in Table 7. Landfill gas recovery efficiencies were averaged over the 100 year time period to better reflect an average tonne of waste deposited, rather than the first mass of waste deposited (see Table 8). Collected landfill gas is utilised in an internal combustion engine to generate electricity, with a gross efficiency of 32% assumed (Patterson et al., 2011). A 1% fugitive CH_4 emissions rate was specified for gas utilisation processes (US EPA, 2011).

Table 6: LCI data for the non-hazardous landfill process. Note that data are presented per tonne throughput.

Inputs	Unit	Quantity	Reference
Diesel	kg	1.8	Hall et al. (2005)
Electricity	kWh	8	Manfredi et al. (2009)
Water	kg	0.00038	Hall et al. (2005)
HDPE (liner)	kg	1	Hall et al. (2005)
Gravel	kg	100	Manfredi et al. (2009)
Steel	kg	0.12	Hall et al. (2005)
Synthetic rubber	kg	0.0011	Hall et al. (2005)
Lubricant	kg	0.0089	Hall et al. (2005)

HDPE, high-density polyethylene.

Table 7: Technical measures and performance associated with landfill gas recovery, utilisation, and oxidation for different time periods.

	Period 1	Period 2	Period 3	Period 4	Period 5
Duration (years)	1	4	15	30	50
CH_4 oxidation (%)	10[a]	10[a]	20[a]	36[b]	36[b]
Gas collected (%)	0[a]	50[a]	75[a]	85[a]	0[c]
Gas management	None	Flare	ICE	ICE	None

ICE, internal combustion engine.

a Source: US EPA (2011).

b Source: Chanton et al. (2009).

C Source: Spokas et al. (2006).

Table 8: Temporally-averaged landfill gas collection efficiencies.

Time period duration (years)	Percentage landfill gas collected	Percentage landfill gas not collected
1	0	100
1	35	65
1	50	50
1	65	35
1	70	30
11	75	25
1	77	23
1	79	21
1	81	19
1	83	17
30	85	15
50	0	100

Source: adapted from US EPA (2011).

Material Reprocessing

Reprocessing of different dry recyclables was modelled using the process models and LCI data presented in Turner et al. (in press).

Reuse

Books and Bric A Brac

Books and bric a brac were each assumed to be resold in charity shops. An electricity input at the charity shop of 357 kWh/t was assumed (James et al., 2011). No material loss or avoided production was assumed.

Paint

No energy or material inputs were included for the reuse of waste paint process. Post reuse, the steel container (200 kg/t) was assumed to be sent for reprocessing. Reused waste paint was assumed to substitute for primary paint production. No material quality loss was assumed. Data for the production of primary paints were adapted for the UK situation from a study undertaken in Abu Dhabi (Nayak and Kumar, 2008). Assuming a paint density of 1 kg/litre, inputs of 33 kWh/t of electricity and 200 kg/t of steel (Saft, 2007) were assumed for the primary paint production process.

Textiles and Footwear

The reuse of textiles and footwear process comprises three stages: 1) sorting; 2) transport to end markets; and 3) reuse. Data for the sorting process were taken from the a UK based textiles recovery plant (EA, 2010). Based on Bartlett et al. (2013), it was assumed that 37% of sorted textiles and footwear are reused domestically, whilst 63% are exported for reuse abroad. Based on the market situation in the year 2008, we assumed that 50% of textiles and footwear exported for reuse abroad are sent to Eastern Europe, 17% to South Asia, and 33% to Sub Saharan Africa (McGill, 2009). Textiles and footwear sent for reuse domestically were assumed to be resold in charity shops (see Section 2.3.15.1). No energy or material inputs were included where textiles and footwear are exported. Reused textiles and footwear were assumed to substitute for the production of textiles from virgin fibres. Due to their comparatively shorter lifetimes and older fashions, market substitution rates of 60%, 75%, and 85% were assumed for textiles and footwear reused in the UK, Eastern Europe, and South Asia and Sub Saharan Africa, respectively (Farrant et al., 2010 and McGill, 2009).

Large Domestic Appliances

Waste electrical and electronic equipment large domestic appliances (LDA) were assumed to be reused domestically. However, based on James et al. (2011), it was assumed that 35% of LDA send for reuse are actually recycled, due to insufficient quality. No energy or material inputs were included for the LDA reuse process. Reused LDA were assumed to substitute for the production of primary LDA. Due to the comparatively shorter lifetimes of reused LDA compared to new LCA, a market substitution rate of 50% was assumed (James et al., 2011).

Life Cycle Impact Assessment

GHG emissions were characterised by Global Warming Potential (GWP) using a 100 year time period (expressed as t CO_2e), with characterisation factors taken from the Intergovernmental Panel on Climate Change (IPCC) Fourth Assessment Report (Forster et al., 2007). The emissions "savings" from biogenic carbon sequestration in landfills or soils (see Christensen et al., 2009) were excluded. This assumption was investigated as part of the sensitivity analysis.

RESULTS AND DISCUSSION

Material Flow Analysis

Comprehensive material flow diagrams for the LACW management systems of each scenario (i.e. the results of the mass balance of waste) are presented. For S1, the baseline scenario, the dominant primary waste stream was residual waste (82,079 t), followed by commingled materials (42,065 t). 63,518 t of waste from the residual waste stream was sent directly to landfill. Of the remaining waste, 18,562 t was sent to one of three residual waste MRFs, with 9,829 t sorted and sent for reprocessing (the remainder was sent to landfill), of which 7,843 t was reprocessed into a secondary product; an overall recycling rate of 42.3% (note that rates varied between facilities). By comparison, all commingled materials were sent to single stream commingled MRFs, with 34,871 t recovered and sent for reprocessing and 27,920 t ultimately reprocessed into a secondary product. This represents an overall single stream commingled MRF recycling rate of 80.0% (note, again, that rates varied between facilities). These results support those of Pressley et al. (2015), who found that recovery rates at single stream commingled MRFs were higher than those at residual waste MRFs. This suggest that policies aiming to increase recycling rates should focus on improving sorting efficiencies at the waste producer level (e.g. households), rather than on recovering materials from the residual waste stream.

Table 9 displays a selection of key indicator results from the MFAs for each scenario. For S1, a reuse/recycling/composting/AD (henceforth, recycling rate) of 49.8% was calculated for the baseline scenario, marginally higher than the 52.8% value reported by Cardiff County Council (StatsWales, 2015b). However, where rejects from reprocessing and biological treatment are included the final (adjusted) recycling rate was estimated as being 45.2%, highlighting the importance of considering downstream reject rates when calculating recycling rates. The recycling rate was highest for S2 (enhanced food waste capture), where high rates of food waste diversion were achieved, whilst the recycling was also higher for S4 (enhanced dry recycling) compared to S1 due to the increased recovery of dry recyclables. For S1, over half of LACW was ultimately landfilled, with 37.7% of LACW sent directly to landfill and 16.4% rejected downstream (Table 9). S3 (enhanced incineration) was found to result in the highest heat and power recovery rate, whilst it was the most successful scenario in terms of diverting material from landfill.

Table 9: Comparison of MFA indicator results for the four scenarios. Note that S1, S2, S3, and S4 refer to Scenarios 1–4, respectively.

Indicator	Description	Scenario			
		S1	S2	S3	S4
Landfill rate (%)[a]	Total tonnage of LACW sent directly to landfill plus total tonnage rejected from other facilities, divided by total tonnage of LACW arisings	54.0	48.0	29.0	52.3
Reuse/recycling/composting/AD rate (%)[a]	Sum of tonnage of LACW sent for re-use, reprocessing, composting, or AD, divided by the total tonnage of LACW arisings	52.8	59.5	52.8	55.1
Heat and power recovery rate (%)[a]	Sum of tonnage of LACW sent for incineration or AD, divided by total tonnage of LACW arisings	8.9	15.4	38.5	8.9
Landfill diversion rate (%)	Sum of tonnage of LACW not sent directly to landfill (i.e. LACW sent to reuse/recycling/composting/AD/thermal treatment), divided by total tonnage of LACW	62.3	68.9	92.0	64.8
Adjusted reuse/recycling/composting/AD rate (%)	Sum of tonnage of LACW sent for reuse, reprocessing, composting, or AD minus sum of tonnage rejected, divided by the total tonnage of LACW arisings	45.2	50.8	44.4	47.0

MFA, material flow analysis; LACW, local authority collected waste; AD, anaerobic digestion.

a Waste management performance indicators for Welsh local authorities.

The results of the MFA suggest that it may not be possible for Wales to achieve its recycling rate target of 70% by 2025 given current waste arisings, even if optimum dry recycling and food waste diversion rates are achieved. Rather, it is likely that significant reductions in residual waste arisings will be required, as well as improvements in recycling rates (Timlett and Williams, 2011). Wales have already made progress in this respect, with a steady rise in the national LACW recycling rate from 37.5% in 2008–2009 to 52.3% in 2012–2013, accompanied by a 9.9% reduction in LACW arisings over the same period (StatsWales, 2015a). However, reductions in total waste arisings have stagnated in recent years. In order for Wales to continue progressing towards its recycling targets, further reductions in waste arisings are likely to be required.

Life Cycle Interpretation

Evaluation of the Existing LACW Management System

Fig. 2 shows the total GHG emissions per scenario and the contribution by primary waste stream. S1, the baseline scenario, was found to result in a net GHG burden (8009 t CO_2e). Note that positive values represent an environmental load (i.e. GHG burden), whilst negative values represent an environmental saving (i.e. GHG benefit). GHG benefits were observed for the management of commingled materials (−9410 t CO_2e), source segregated dry recyclables (−3212 t CO_2e), and source segregated food waste (−2596 t CO_2e). However these benefits were heavily outweighed by the GHG burdens from residual waste management (21,778 t CO_2e).

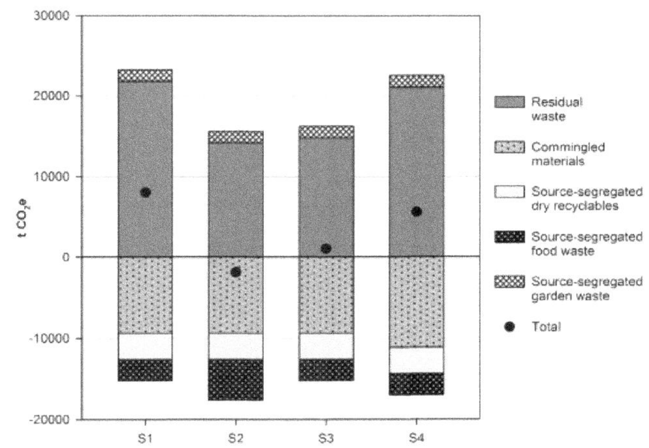

Figure 2: LCA results (GHG emissions, t CO_2e/FU) of different scenarios and the contribution by primary waste stream. Note that S1, S2, S3, and S4 refer to Scenarios 1–4, respectively. LCA, life cycle assessment; GHG, greenhouse gas; FU, functional unit.

Contribution to total primary waste stream GHG impacts by process type for each scenario is shown in Fig. 3. For S1, the dominant contribution in terms of GHG burdens was landfilling of residual waste (25,866 t CO_2e). Material reprocessing & reuse was the only process type that consistently resulted in GHG benefits. For S1, the most significant source of GHG benefits was from reprocessing of commingled materials (−19,219 t CO_2e). The contribution of biological treatment to total GHG impacts was found to be negative for the source segregated food waste stream (i.e. GHG benefit) but positive for the source segregated garden waste stream (i.e. GHG burden). This is due to the different types of biological treatment used to manage each waste stream. Food waste was treated predominately by AD, a process that utilises produced biogas to generate electricity that is exported to the National Grid and substitutes for electricity generated from CCGT, resulting in GHG benefits. By comparison, garden waste was largely treated by composting (either IVC or OWC), through which energy cannot be recovered. This finding correlates with those of Yoshida et al. (2012) andBernstad and la Cour Jansen (2011) who each found that AD of organic waste resulted in a greater GHG benefit compared to composting.

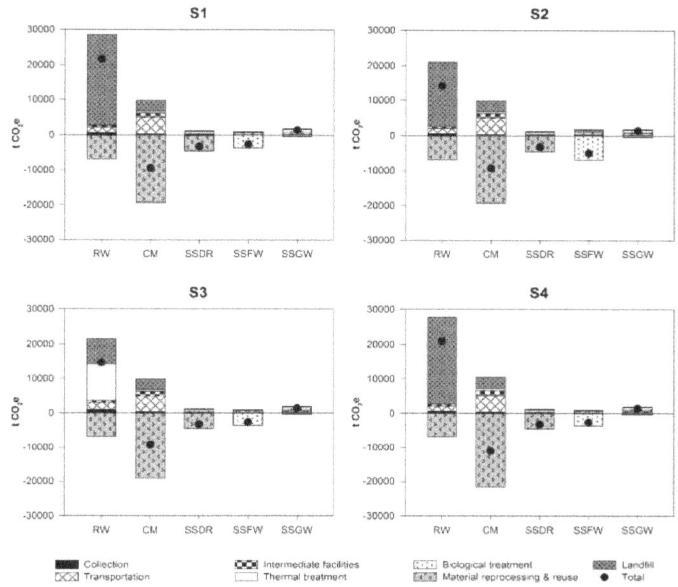

Figure 3: LCA results (GHG emissions, t CO_2e/FU) of different scenarios per primary waste stream and the contribution process type. Note that S1, S2, S3, and S4 refer to Scenarios 1–4, respectively. RW, residual waste; CM, commingled materials; SSDR, source-segregated dry recyclables; SSFW, source-segregated food waste; SSGW, source-segregated garden waste; LCA, life Cycle Assessment; GHG, greenhouse gas; FU, functional unit.

Overall, transport was identified as the third largest contributor to the total GHG impacts for S1 (Fig. 3; 7296 t CO_2e; 10% contribution). However, its relative contribution to the GHG impacts of each primary waste stream management system was highly variable. Whilst the contribution of transport to the GHG impacts from the residual waste (1492 t CO_2e; 4% contribution) and source segregated dry recyclables (160 t CO_2e; 3% contribution) management systems was minor, transport contributed substantially to the GHG impacts from source segregated garden waste management (565 t CO_2e; 24% contribution). This is likely due to the considerable distances that garden waste was transported for treatment. For example, 16,780 t of collected garden waste was transported between 78 and 181 km to an IVC facility, resulting in substantial GHG burdens. The contribution of transport to the total GHG impacts of commingled materials management was also found to be large (4543 t CO_2e; 16% contribution). This is likely due to the large quantity of recyclate that was exported overseas for reprocessing. Of the 42,055 t of commingled materials collected, 11,334 t (predominately paper) was exported to reprocessing facilities in Asia. These results suggests that the calculated GHG impacts from materials recycling in this situation are more greatly affected by transport distance compared to results from previous studies (Larsen et al., 2009, Merrild et al., 2012 and Salhofer et al., 2007). The formulation and implementation of policy measures that promote domestic reprocessing of waste materials may help to reduce the GHG impacts of transport and, consequently, enhance the overall GHG benefits from materials recycling.

Table 10 presents the contributions of wastes from different collection sources to the total GHG impacts for each scenario. For S1, the largest contribution to the total GHG impacts was from household kerbside collected waste, particularly residual waste, which was the most significant source of GHG burdens, and commingled materials, the largest source of GHG benefits. Overall GHG impacts from bulky waste collection, fly tipped materials, the voluntary sector, and municipal parks/grounds waste collection were found to be minor, with each contributing less than 1% to the total GHG impacts. An interesting result from a policy perspective concerns the contribution of street cleaning waste management to the total GHG impacts. Street cleaning waste management was found to contribute 8.8% (751 t CO_2e) of total GHG impacts for S1. Waste from street cleaning comprised both residual waste, the management of which incurred a GHG burden of 2269 t CO_2e, and commingled materials, the management of which resulted in a GHG saving of −1518 t CO_2e. Diversion of street cleaning waste from the residual waste stream by promoting recycling is an area not commonly targeted by policy makers, but could be an effective source of potential GHG benefits.

Table 10: LCA results (GHG emissions, t CO_2e) and contribution analysis ((%t CO_2e=FU) of different scenarios per collection source and primary waste stream. Note that S1, S2, S3, and S4 refer to Scenarios 1–4, respectively.

	Scenario							
	S1		S2		S3		S4	
	t CO_2e	% t CO_2e/FU	t CO_2e	% t CO_2e/FU	t CO_2e	% t CO_2e/FU	t CO_2e	% t CO_2e/FU
Household kerbside collection								
Residual waste	17,457	40.3	9874	25.9	11,645	32.7	16,738	37.8
Commingled materi-als	−7545	17.4	−7545	19.8	−7545	21.2	−9287	21.0
Source-segregated garden waste	1573	3.6	1573	4.1	1573	4.4	1573	3.6
Source-segregated food waste	−2596	6.0	−4954	13.0	−2596	7.3	−2596	5.9
Source-segregated dry recyclables	15	0.0	15	0.0	15	0.0	15	0.0
Total	8904	67.4	−1,0367	63.0	3092	65.7	6443	68.2
Bulky waste collection								
Source-segregated garden waste	2	0.0	2	0.0	2	0.0	2	0.0
Source-segregated dry recyclables	−72	0.2	−72	0.2	−72	0.2	−72	0.2
Total	−70	0.2	−70	0.2	−70	0.2	−70	0.2
HWRCs								
Residual waste	−2084	4.8	−2084	5.5	−1709	4.8	−2084	4.7
Commingled materi-als	−319	0.7	−319	0.8	−319	0.9	−319	0.7
Source-segregated garden waste	192	0.4	192	0.5	192	0.5	192	0.4
Source-segregated dry recyclables	−2560	5.9	−2560	6.7	−2560	7.2	−2560	5.8
Total	−4771	11.9	−4771	13.5	−4396	13.4	−4771	11.6
Bring sites								
Commingled materi-als	−28	0.1	−28	0.1	−28	0.1	−28	0.1
Source-segregated dry recyclables	−595	1.4	−595	1.6	−595	1.7	−595	1.3
Total	−623	1.4	−623	1.6	−623	1.8	−623	1.4

Street cleaning								
Residual waste	2270	5.2	2270	6.0	2017	5.7	2270	5.1
Commingled materials	−1518	3.5	−1518	4.0	−1518	4.3	−1518	3.4
Total	752	8.8	752	10.0	499	9.9	752	8.6
Fly-tipped materials								
Residual waste	208	0.5	208	0.5	237	0.7	208	0.5
Total	208	0.5	208	0.5	237	0.7	208	0.5
Non-household collection								
Residual waste	3926	9.1	3926	10.3	2624	7.4	3926	8.9
Total	3926	9.1	3926	10.3	2624	7.4	3926	8.9
Voluntary sector/community collection								
Source-segregated dry recyclables	0	0.0	0	0.0	0	0.0	0	0.0
Total	0	0.0	0	0.0	0	0.0	0	0.0
Municipal parks/grounds waste collection								
Source-segregated garden waste	−317	0.7	−317	0.8	−317	0.9	−317	0.7
Total	−317	0.7	−317	0.8	−317	0.9	−317	0.7
Total	8008	100.0	−1930	100.0	1046	100.0	5548	100.0

GHG, greenhouse gas; FU, functional unit; LCA, Life Cycle Assessment; HWRC, household waste recycling centre; LACW, local authority collected waste.

Comparison of Scenarios

Compared to S1, each of the three alternative scenarios were found to result in lower overall GHG burdens (Fig. 2). S2 was found to be the best performer with an overall GHG impact of −1930 t CO_2e, the only scenario for which a net GHG benefit was found. The contribution of household kerbside collection residual waste to total GHG impacts was substantially reduced in S2 compared to S1 (Table 10), with a large portion of the GHG burdens of residual waste management replaced by GHG benefits from the management of source separated food waste. This highlights the considerable influence of the organic content of residual waste on total GHG burdens (see Section 3.2.3.2).

After S2, S3 was the next best performer (1046 t CO_2e; Fig. 2). S3 showed the greatest reduction in GHG burdens from landfill compared to S1 (Fig. 3). For S3, the substantial GHG burdens from landfill were replaced by comparatively lower GHG burdens from incineration (Fig. 3), suggesting that incineration represents a better option with regards to GHG impacts compared to landfill. This compares favourably with the findings of previous studies (e.g. Arena et al., 2003, Fernández-Nava et al., 2014 and Massarutto et al., 2011). The GHG benefits of S3 would likely be further enhanced if district heating networks were more widely established in the UK. This would create markets for thermal energy recovered through incineration, leading to further GHG benefits through avoided thermal energy production (Giugliano et al., 2011 and Turconi et al., 2011).

Of the three alternative scenarios, S4 was found to result in the improvement on the results of the baseline scenario, with a total GHG impact of 5548 t CO_2e (Fig. 2). This is likely due to the relatively high dry recyclables sorting efficiencies that are already achieved in Cardiff, particularly for mixed glass, paper, and dense plastics. Considering the results for S2, this finding also suggests that waste policies that target residual waste prevention may be more effective in achieving GHG emissions reductions compared to those that target greater recovery of dry recyclables.

Scenario Analysis

Carbon Sequestration

A sensitivity analysis was performed to investigate the impact of including carbon sequestration on the LCA results through scenario analysis. For each scenario, the amount of carbon sequestered was calculated as the amount of biogenic carbon that, after 100 years, remains in a landfill or, in the case landspreading of organic solid residues from biological treatment, is bound to soil. Fig. 4 shows the total GHG impact for each scenario and the contribution by process type. The inclusion of carbon sequestration had a substantial impact on the calculated results. Overall, net GHG benefits were observed for each of the four scenarios. Landfilling was found to result in overall GHG benefits, effectively rendering landfills a "carbon sink". This contradicts the findings of the UK's national GHG inventory return for GHG emissions from SWM (Salisbury et al., 2015), despite the fact that carbon sequestration is also considered in the national inventory analysis (Hogg et al., 2011).

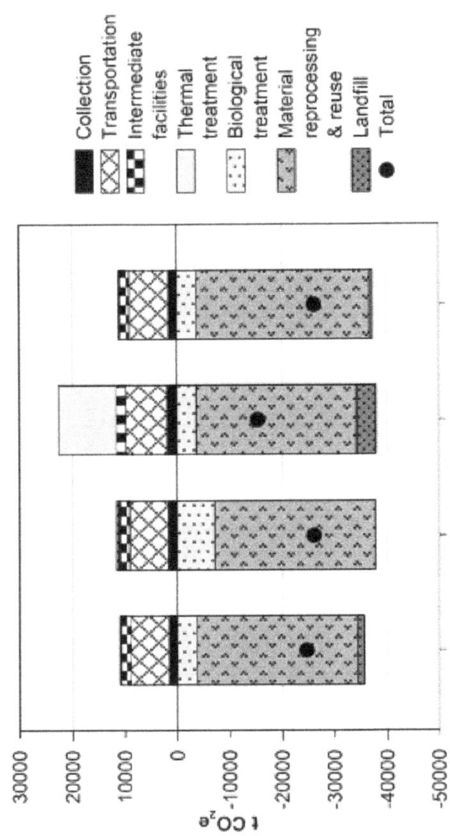

Figure 4: Sensitivity analysis (SA1) results (GHG emissions, t CO_2e/FU) of different scenarios and the contribution by process type – the inclusion of carbon sequestration. Note that SA1.1, SA1.2, SA1.3, and SA1.4 refer to Scenarios 1–4, respectively. GHG, greenhouse gas; FU, functional unit.

The ranking of scenarios in terms of their environmental performance was affected by the choice of including carbon sequestration. Whilst S2 remained the best option and S4 was still found to result in additional GHG benefits compared to S1, the overall GHG burdens for S3 were found to be considerably higher than S1 (Fig. 4). This suggests that when carbon sequestration is included in the calculations, landfilling is generally a better option compared to incineration in terms of GHG impact, which contradicts the results of previous studies (e.g. Arena et al., 2003, Fernández-Nava et al., 2014 and Massarutto et al., 2011). Given the significant effects of including carbon sequestration in the LCA on the calculated results, these authors suggest that further research into carbon degradation rates for waste materials in UK landfills is needed.

Residual Waste Composition

A sensitivity analysis using scenario analysis was performed to test the influence of the modelled organic content of the residual waste stream on the overall LCA results. In this study, residual waste composition data were taken from a national waste composition study (WRAP, 2010) that predates the introduction of separate household food waste collection in Cardiff. The use of these data may, therefore, result in an overestimation of GHG burdens as the total organic content of the input waste may be overestimated. Two scenarios were analysed, with the proportion of organic material in residual waste increased and decreased by 10% compared to the baseline values for each scenario. Compared to the total GHG impact of S1 (8008 t CO_2e), the 10% increase in organic content resulted in a +15.5% variation (result: 9254 t CO_2e), whilst the 10% decrease resulted in a −9.4% variation (result: 6826 t CO_2e). The sensitivity analysis shows that the overall LCA results are highly sensitive to the organic content of residual waste.

CONCLUSIONS AND RECOMMENDATIONS

In this paper we have presented an approach that combines the systematic methodologies of MFA and LCA to quantitatively evaluate the environmental performance of large and complex SWM systems. The approach was applied to evaluate the GHG emissions performance of a local authority SWM system and compare its performance with those of alternative systems to assess the potential effectiveness of different waste policy measures. The results of the MFA show that S2 (enhanced food waste capture) resulted in the highest recycling rate of the investigated scenarios, whilst S3 (enhanced incineration) resulted in the highest rate of waste diversion from landfill. Overall, our results suggest that, in this case, the national government is unlikely to achieve its recycling targets, even if each of the assessed policy measures are implemented optimally. It is likely that in order for the recycling targets to be achieved, these policy measures need to be combined with further policies focused on waste prevention.

The GHG impacts of the existing SWM system were evaluated through LCA, focusing on potential climate change impacts. The results showed that the dominant source of GHG burdens was from residual waste management, principally through landfilling, which was the most significant source of GHG burdens. Material reprocessing and AD were found to be the greatest sources of GHG benefits. GHG burdens from transport were found to be generally minor, except with regards to the management of source segregated garden waste and commingled materials.

Compared to the baseline scenario, each of the three alternative scenarios were found to perform better in terms of GHG impact. Overall, the best performing scenario was S2 (enhanced food waste capture), which was the only scenario that resulted in an overall GHG saving. However, the results of the LCA were found to be highly sensitive to the choice of excluding GHG benefits from carbon sequestration. When carbon sequestration was included, all four scenarios were found to result in net GHG benefits. The order of the four scenarios in terms of overall performance was also found to be affected by the choice, with the GHG burdens for S3 (enhanced incineration) found to be greater than those of the baseline scenario. The results were also found to be sensitive to the organic content of the residual waste stream.

Overall, this paper has demonstrated that the complementary methodologies of MFA and LCA can be used in combination to provide valuable information about the environmental performance of a SWM system. The approach can be applied to assist national governments in appraising existing and possible waste policies and to support local waste managers in identifying environmentally optimal SWM strategies.

ACKNOWLEDGEMENTS

This research was financially supported by the Engineering and

Physical Sciences Research Council (EPSRC; grant number EP/J500537/1). The authors would like to thank Andy Rees and Adrian Jones of the Welsh Government Waste Strategy Branch for their valued input into the project. The EASETECH modelling was undertaken with the support of the Anders Damgaard, to whom, along with Yue Zhang, we are grateful to for their useful comments on this manuscript.

REFERENCES

1. Allegrini, E., Butera, S., Kosson, D.S., Van Zomeren, A., Van der Sloot, H.A., Astrup, T.F., 2015. Life cycle assessment and residue leaching: the importance of parameter, scenario and leaching data selection. Waste Manag. 38, 474e485. http://dx.doi.org/10.1016/j.wasman.2014.12.018.
2. Amlinger, F., Peyr, S., Cuhls, C., 2008. Greenhouse gas emissions from composting and mechanical biological treatment. Waste Manag. Res. 26, 47e60. http:// dx.doi.org/10.1177/0734242X07088432.
3. Andersen, J.K., Boldrin, A., Christensen, T.H., Scheutz, C., 2010. Mass balances and life-cycle inventory for a garden waste windrow composting plant (Aarhus, Denmark). Waste Manag. Res. 28, 1010e1020. http:// dx.doi.org/10.1177/ 0734242X10360216.
4. Arena, U., Di Gregorio, F., 2014. A waste management planning based on substance flow analysis. Resour. Conserv. Recycl. 85, 54e66. http://dx.doi.org/10.1016/ j.resconrec.2013.05.008.
5. Arena, U., Mastellone, M.L., Perugini, F., 2003. The environmental performance of alternative solid waste management options: a life cycle assessment study. Chem. Eng. J. 96, 207e222. http://dx.doi.org/10.1016/j.cej.2003.08.019.
6. Astrup, T., Moller, J., Fruergaard, T., 2009. Incineration and co-combustion of waste: accounting of greenhouse gases and global warming contributions. Waste Manag. Res. 27, 789e799. http://dx.doi.org/10.1177/0734242X09343774.
7. Banks, C.J., Cheshire, M., Heaven, S., Arnold, R., 2011. Anaerobic digestion of sourcesegregated domestic food waste: performance assessment by mass and energy balance. Bioresour. Technol. 102, 612e620. http://dx.doi.org/10.1016/ j.biortech.2010.08.005.
8. Bartlett, C., McGill, I., Willis, P., 2013. Textiles Flow and Market Development Opportunities in the UK, Waste & Resources Action Programme, Banbury, UK. Available from: http://www.wrap.org.uk/content/uk-textile-product-flow-andmarket-development-opportunities.
9. Bernstad, A., la Cour Jansen, J., 2011. A life cycle approach to the management of household food waste e a Swedish full-scale case study. Waste Manag. 31, 1879e1896. http://dx.doi.org/10.1016/j.wasman.2011.02.026.
10. Blengini, G.A., Fantoni, M., Busto, M., Genon, G., Zanetti, M.C., 2012. Participatory approach, acceptability and transparency of waste management LCAs: case studies of Torino and Cuneo. Waste Manag. 32, 1712e1721. http://dx.doi.org/ 10.1016/j.wasman.2012.04.010.
11. Boesch, M.E., Vadenbo, C., Saner, D., Huter, C., Hellweg, S., 2014. An LCA model for waste incineration enhanced with new technologies for metal recovery and application to the case of Switzerland. Waste Manag. 34, 378e389. http:// dx.doi.org/10.1016/j.wasman.2013.10.019.
12. Bogner, J., Abdelrafie Ahmed, M., Diaz, C., Faaij, A., Gao, Q., Hashimoto, S., Mareckova, K., Pipatti, R., Zhang, T., 2007. Waste management. In: Metz, B., Davidson, O.R., Boesch, P.R., Dave, R., Meyer, L.A. (Eds.), Climate Change 2007: Mitigation. Contribution of Working Group III to the Fourth Assessment Report of the International Panel on Climate Change. Cambridge University Press, Cambridge, UK and New York, NY, USA.
13. Boldrin, A., Andersen, J.K., Christensen, T.H., 2009a. Environmental Assessment of Garden Waste Management in Århus Kommune. Technical University of Denmark, Copenhagen, Denmark. Available from: https://www.aarhus.dk/ ~/media/Subsites/AVA/Om-AVA/Bibliotek/Rapporter-og-undersoegelser/ Aarhus-haveaffald-engelsk-rapport.pdf.
14. Boldrin, A., Andersen, J.K., Moller, J., Christensen, T.H., Favoino, E., 2009b. Composting and compost utilization: accounting of greenhouse gases and global warming contributions. Waste Manag. Res. 27, 800e812. http://dx.doi.org/ 10.1177/0734242X09345275.
15. Boldrin, A., Hartling, K.R., Laugen, M., Christensen, T.H., 2010. Environmental inventory modelling of the use of compost and peat in growth media preparation. Resources. Conserv. Recycl. 54, 1250e1260. http:// dx.doi.org/10.1016/ j.resconrec.2010.04.003.
16. Boldrin, A., Andersen, J.K., Christensen, T.H., 2011. Environmental assessment of garden waste management in the Municipality of Aarhus, Denmark. Waste Manag. 31, 1560e1569. http://dx.doi.org/10.1016/j.wasman.2011.01.010.
17. Briffaerts, K., Spirinckx, C., Van der Linden, A., Vrancken, K., 2009. Waste battery treatment options: comparing their environmental performance. Waste Manag. 29, 2321e2331. http://dx.doi.org/10.1016/j.wasman.2009.03.019.
18. Brown, S., Kruger, C., Subler, S., 2008. Greenhouse gas balance for composting operations. J. Environ. Qual. 37, 1396e1410. http://dx.doi.org/10.2134/ jeq2007.0453.
19. Brunner, P.H., Rechberger, H., 2004. Practical Handbook of Material Flow Analysis. Lewis Publishers, Boca Raton, FL, USA.
20. Bruun, S., Hansen, T.L., Christensen, T.H., Magid, J., Jensen, L.S., 2006. Application of processed organic municipal solid waste on agricultural land e a scenario analysis. Environ. Model. Assess. 11, 251e265. http:// dx.doi.org/10.1007/ s10666-005-9028-0.
21. Burnley, S.J., Phillips, R., Coleman, T., Rampling, T., 2011. Energy implications of the thermal recovery of biodegradable municipal waste materials in the United Kingdom. Waste Manag. 31, 1949e1959. http://dx.doi.org/10.1016/ j.wasman.2011.04.015.
22. Butera, S., Christensen, T.H., Astrup, T.F., 2015. Life cycle assessment of construction and demolition waste management. Waste Manag. 44, 196e205. http:// dx.doi.org/10.1016/j.wasman.2015.07.011.

23. Cencic, O., Rechberger, H., 2008. Material flow analysis with software STAN. J. Environ. Eng. Manag. 18, 3e7.

24. Chanton, J., Powelson, D.K., Green, R.B., 2009. Methane oxidation in landfill cover soils - is a 10% default value reasonable? J. Environ. Qual. 38, 654e663. http:// dx.doi.org/10.2134/jeq2008.0221.

25. Christensen, T.H., Gentil, E., Boldrin, A., Larsen, A.W., Weidema, B.P., Hauschild, M., 2009. C balance, carbon dioxide emissions and global warming potentials in LCA-modelling of waste management systems. Waste Manag. Res. 27, 707e715. http://dx.doi.org/10.1177/0734242X08096304.

26. Clavreul, J., Baumeister, H., Christensen, T.H., Damgaard, A., 2014. An environmental assessment system for environmental technologies. Environ. Model. Softw. 60, 18e30. http://dx.doi.org/10.1016/j.envsoft.2014.06.007.

27. Cleary, J., 2009. Life cycle assessments of municipal solid waste management systems: a comparative analysis of selected peer-reviewed literature. Environ. Int. 35, 1256e1266. http://dx.doi.org/10.1016/j.envint.2009.07.009.

28. Clift, R., Doig, A., Finnveden, G., 2000. The application of Life Cycle Assessment to integrated solid waste management. Part 1-methodology. Process Saf. Environ. Prot. 78, 279e287. http://dx.doi.org/10.1205/095758200530790.

29. Consonni, S., Vigano, F., 2011. Material and energy recovery in integrated waste management systems: the potential for energy recovery. Waste Manag. 31, 2074e2084. http://dx.doi.org/10.1016/j.wasman.2011.05.013.

30. Dalemo, M., Sonesson, U., Bjorklund, A., Mingarini, K., Frostell, B., J € onsson, H., € Nybrant, T., Sundqvist, J.O., Thyselius, L., 1997. ORWARE e a simulation model for organic waste handling systems. Part 1: model description. Resour. Conserv. Recycl. 21, 17e37. http://dx.doi.org/10.1016/S0921-3449(97)00020-7.

31. Davidsson, Å., Gruvberger, C., Christensen, T.H., Hansen, T.L., Jansen, J.l.C., 2007. Methane yield in source-sorted organic fraction of municipal solid waste. Waste Manag. 27, 406e414. http://dx.doi.org/10.1016/j.wasman. 2006.02.013.

32. Department of Energy & Climate Change (DECC), 2014a. Valuation of Energy Use and Greenhouse Gas (GHG) Emissions e Background Documentation. DECC, London, UK. Available from: https://www.gov.uk/government/uploads/system/ uploads/attachment_data/file/360044/2014_Background_Documentation_to_DECC_HMT_Supplementary_Appraisal_Guidance.pdf.

33. Department of Energy & Climate Change (DECC), 2014b. What Is Combined Heat and Power? Available from: http://chp.decc.gov.uk/cms/what-is-chp/.

34. Department for Environment Food & Rural Affairs (Defra), 2013a. Incineration of Municipal Solid Waste. Defra, London, UK. Available from: https://www.gov.uk/government/uploads/system/uploads/attachment_data/file/221036/pb13889- incineration-municipal-waste.pdf.

35. Department for Environment Food & Rural Affairs (Defra), 2013b. Mechanical Biological Treatment of Municipal Solid Waste. Defra, London, UK. Available from: https://www.gov.uk/government/uploads/system/uploads/attachment_ data/file/221039/pb13890-treatment-solid-waste.pdf.

36. Department for Environment Food & Rural Affairs (Defra), Ricardo-AEA, Carbon Smart, 2013. Greenhouse Gas Conversion Factor Repository. Available from: http://www.ukconversionfactorscarbonsmart.co.uk/.

37. Department for Environment Food & Rural Affairs (Defra), 2011. Government Review of Waste Policy in England 2011. Defra, London, UK. Available from: https://www.gov.uk/government/publications/government-review-of-wastepolicy-in-england-2011.

38. du Plessis, C., Strauss, J.M., Sebapalo, E.M.T., Riedel, K.-H.J., 2003. Empirical model for methane oxidation using a composted pine bark biofilter. Fuel 82, 1359e1365. http://dx.doi.org/10.1016/S0016-2361(03)00040-1.

39. Ekvall, T., Assefa, G., Bjorklund, A., Eriksson, O., Finnveden, G., 2007. What life-cycle assessment does and does not do in assessments of waste management. Waste Manag. 27, 989e996. http://dx.doi.org/10.1016/j.wasman.2007.02.015.

40. Environment Agency (EA), 2010. Waste and Resources Assessment Tool for the Environment (WRATE) LCA Model, Rotherham, UK.

41. Eunomia Research & Consulting, Resource Futures, HCW Consultants, 2011. Kerbside Collection Options: Wales. Waste & Resources Action Programme, Cardiff, UK. Available from: http://www.eunomia.co.uk/reports-tools/kerbsidecollections-options-wales/.

42. European Commission (EC), 1994. European parliament and council directive 94/62/ EC of 20 December 1994 on packaging and packaging waste. Off. J. Eur. Communities L 365, 10e23. Available from: http://eur-lex.europa.eu/legal-content/ EN/TXT/?uri¼URISERV:l21207.

43. European Commission (EC), 1999. Council directive 1999/31/EC of 26 April 1999 on the landfill of waste. Off. J. Eur. Communities L 182, 1e19. Available from: http:// eur-lex.europa.eu/legal-content/EN/TXT/?uri¼celex:31999L0031.

44. European Commission (EC), 2006. Integrated Pollution Prevention and Control. Reference document on the best available techniques for waste incineration. EC, Luxembourg. Available from: http://eippcb.jrc.ec.europa.eu/reference/BREF/ lcp_bref_0706.pdf.

45. European Commission (EC), 2008. Directive 2008/98/EC of the European parliament and of the Council of 19 November 2008 on waste and repealing certain directives. Off. J. Eur. Union L 312, 3e30. Available from: http://eur-lex.europa.eu/ legal-content/EN/TXT/?uri¼CELEX:32008L0098.

46. European Commission (EC), 2009. Decision No 406/2009/EC of the European Parliament and of the Council of 23 April 2009 on the effort of Member States to reduce their greenhouse gas emissions to meet the Community's greenhouse gas emission reduction commitments up to 2020. Off. J. Eur. Union L 140, 136e148. Available from: http://eur-lex.europa.eu/

LexUriServ/LexUriServ.do? uri¼OJ:L:2009:140:0136:0148:EN:PDF.

47. European Commission (EC), Joint Research Centre (JRC), Institute for Environment and Sustainability (IES), 2010. International Reference Life Cycle Data System (ILCD) Handbook e General Guide for Life Cycle Assessment e Detailed Guidance. Publications Office of the European Union, Luxembourg. http://dx.doi.org/ 10.2788/38479.

48. European Commission (EC), 2014. 2030 framework for Climate and Energy Policies. Available from: http://ec.europa.eu/clima/policies/2030/index_en.htm.

49. Evangelisti, S., Lettieri, P., Borello, D., Clift, R., 2014. Life cycle assessment of energy from waste via anaerobic digestion: a UK case study. Waste Manag. 34, 226e237. http://dx.doi.org/10.1016/j.wasman.2013.09.013.

50. Farrant, L., Olsen, S.I., Wangel, A., 2010. Environmental benefits from reusing clothes. Int. J. Life Cycle Assess. 15, 726e736. http://dx.doi.org/10.1007/s11367- 010-0197-y.

51. Fellner, J., Aschenbrenner, P., Cencic, O., Rechberger, H., 2011. Determination of the biogenic and fossil organic matter content of refuse-derived fuels based on elementary analyses. Fuel 90, 3164e3171. http://dx.doi.org/10.1016/ j.fuel.2011.06.043.

52. Fernandez-Nava, Y., del Río, J., Rodríguez-Iglesias, J., Castrill on, L., Mara n~on, E., 2014. Life Cycle Assessment (LCA) of different municipal solid waste management options: a case study of Asturias (Spain). J. Clean. Prod. 81, 178e189. http:// dx.doi.org/10.1016/j.jclepro.2014.06.008.

53. Finnveden, G., Johnansson, J., Lind, P., Moberg, Å., 2005. Life cycle assessment of energy from solid waste e part 1: general methodology and results. J. Clean. Prod. 13, 213e229. http://dx.doi.org/10.1016/j.jclepro.2004.02.023.

54. Fisher, J.I., 2006. Impact of Energy from Waste and Recycling Policy on UK Greenhouse Gas Emissions. London, UK. Available from: http://randd.defra.gov.uk/Document.aspx?Document¼WR0609_5737_FRP.pdf.

55. Forster, P., Ramaswamy, V., Artaxo, P., Berntsen, T., Betts, R., Fahey, D.W., Haywood, J., Lean, J., Lowe, D.C., Myhre, G., Nganga, J., Prinn, R., Raga, G., Schulz, M., Van Dorland, R., 2007. Changes in atmospheric constituents and in radiative forcing. In: Solomon, S., Qin, D., Manning, M., Chen, Z., Marquis, M., Averyt, K.B., Tignor, M., Miller, H.L. (Eds.), Climate Change 2007. The Physical Science basis. Contribution of Working Group I to the Fourth Assessment Report of the Intergovernmental Panel on Climate Change. Cambridge University Press, Cambridge, UK and New York, NY, USA.

56. Fruergaard, T., Hyks, J., Astrup, T., 2010. Life-cycle assessment of selected management options for air pollution control residues from waste incineration. Sci. Total Environ. 408, 4672e4680. http://dx.doi.org/10.1016/j.scitotenv.2010.05.029.

57. Gallert, C., Winter, J., 1997. Mesophilic and thermophilic anaerobic digestion of source-sorted organic wastes: effect of ammonia on glucose degradation and methane production. Appl. Microb. Biotechnol. 48, 405e410. http://dx.doi.org/ 10.1007/s002530051071.

58. Giugliano, M., Cernuschi, S., Grosso, M., Rigamonti, L., 2011. Material and energy recovery in integrated waste management systems. An evaluation based on life cycle assessment. Waste Manag. 31, 2092e2101. http://dx.doi.org/10.1016/ j.wasman.2011.02.029.

59. Hall, D., Plimmer, B., Taylor, D., 2005. Life Cycle Assessment Landfill Emissions. Golder Associates (UK), Stanton-on-the-Wolds, UK.

60. Hansen, T.L., 2006a. Environmental modelling of use of treated organic waste on agricultural land: a comparison of existing models for life cycle assessment of waste systems. Waste Manag. Res. 24, 141e152. http://dx.doi.org/10.1177/ 0734242X06062485.

61. Hansen, T.L., 2006b. Life cycle modelling of environmental impacts of application of processed organic municipal solid waste on agricultural land (EASEWASTE). Waste Manag. Res. 24, 153e166. http://dx.doi.org/10.1177/0734242X06063053.

62. Hawkey, D.J.C., 2012. District heating in the UK: a technological innovation systems analysis. Environ. Innov. Soc. Transit. 5, 19e32. http://dx.doi.org/10.1016/ j.eist.2012.10.005.

63. Hertwich, E.G., van der Voet, E., Suh, S., Tukker, A., Huijbreghts, M., Kazmierczyk, P., Lenzen, M., McNeely, J., Moriguchi, Y., 2010. Assessing the Environmental Impacts of Consumption and Production: Priority Products and Materials. A Report of the Working Group on the Environmental Impacts of Products and Materials to the International Panel for Sustainable Resource Management, United Nations Environment Programme, Paris, France. Available from: http://www.greeningtheblue.org/sites/default/files/Assessing%20the%20environmental% 20impacts%20of%20consumption%20and%20production.pdf.

64. Her Majesty's Statutory Office (HMSO), 2008. Climate Change Act 2008. HMSO, London, UK. Available from: http://www.legislation.gov.uk/ukpga/2008/27/ contents.

65. Her Majesty's Statutory Office (HMSO), 2015. Well-being of Future Generations (Wales) Act 2015. HMSO, London, UK. Available from: http://www.legislation.gov.uk/anaw/2015/2/contents/enacted.

66. Hogg, D., Ballinger, A., Oonk, H., 2011. Inventory Improvement Project e UK Landfill Methane Emissions Model. Final report to Defra and DECC, Bristol, UK. Available from: http://sciencesearch.defra.gov.uk/Default.aspx?Menu¼Menu& Module¼More&Location¼None&Completed ¼0&ProjectID¼17448.

67. Hoornweg, D., Bhada-Tata, P., 2012. What a Waste. A global review of solid waste management, Washington, DC, USA. Available from: http://hdl.handle.net/10986/17388.

68. Horne, J., Scholes, P., Areikin, E., Brown, B., 2013. A Survey of the UK Organics Recycling Industry in 2012. Banbury, UK. Available from: http://www.wrap.org. uk/sites/files/wrap/ASORI%202012.pdf.

69. International Solid Waste Association (ISWA), 2015.

ISWA Declaration on Climate Change and the Waste and Resource Management Sector. ISWA, Vienna, Austria. Available from: https://www.iswa.org/fileadmin/galleries/Events/GA_ Package_2015/CC_ Declaration_2015.pdf.

70. Intergovernmental Panel on Climate Change (IPCC), 2006. In: Eggleston, H.S., Buendia, L., Miwa, K., Ngara, T., Tanabe, K. (Eds.), 2006 IPCC Guidelines for National Greenhouse Gas Inventories, Prepared by the National Greenhouse Gas

71. Inventories Programme. IPCC, Hayama, Japan. Available from: http://www.ipccnggip.iges.or.jp/public/2006gl/.

72. Intergovernmental Panel on Climate Change (IPCC), 2014. Climate change 2014: synthesis report. In: Pachauri, R.K., Meyer, L. (Eds.), Contribution of Working Groups I, II and III to the Fifth Assessment Report of the Intergovernmental Panel on Climate Change. IPCC, Geneva, Switzerland. Available from: http:// www.ipcc.ch/report/ar5/syr/.

73. International Organization for Standardization (ISO), 2006a. ISO 14044:2006 Environmental Management e Life Cycle Assessment e Requirements and Guidelines. ISO, Geneva, Switzerland. Available from: http://www.iso.org/iso/ home/store/catalogue_tc/catalogue_detail.htm?csnumber¼38498.

74. International Organization for Standardization (ISO), 2006b. ISO 14040:2006 Environmental Management e Life Cycle Assessment e Principles and Framework. ISO, Geneva, Switzerland. Available from: http://www.iso.org/iso/ catalogue_detail?csnumber¼37456.

75. James, K., Maddox, P., Fisher, K., 2011. Benefits of Reuse. Case Study: Electrical Items, Waste & Resources Action Programme, Banbury, UK. Available from: http:// www.wrap.org.uk/sites/files/wrap/Electricals%20 reuse_final.pdf.

76. Koehler, A., Peyer, F., Salzmann, C., Saner, D., 2011. Probabilistic and technologyspecific modeling of emissions from municipal solid-waste incineration. Environ. Sci. Technol. 45, 3487e3495. http://dx.doi.org/10.1021/es1021763.

77. Larsen, A.W., Vrgoc, M., Christensen, T.H., Lieberknecht, P., 2009. Diesel consumption in waste collection and transport and its environmental significance. Waste Manag. Res. 27, 652e659. http://dx.doi.org/10.1177/0734242X08097636.

78. Laurent, A., Bakas, I., Clavreul, J., Bernstad, A., Niero, M., Gentil, E., Hauschild, M.Z., Christensen, T.H., 2014a. Review of LCA studies of solid waste management systemsepart I: lessons learned and perspectives. Waste Manag. 34, 573e588. http://dx.doi.org/10.1016/j.wasman.2013.10.045.

79. Laurent, A., Clavreul, J., Bernstad, A., Bakas, I., Niero, M., Gentil, E., Christensen, T.H., Hauschild, M.Z., 2014b. Review of LCA studies of solid waste management systems e part II: methodological guidance for a better practice. Waste Manag. 34, 589e606. http://dx.doi.org/10.1016/j.wasman.2013.12.004.

80. Manfredi, S., Tonini, D., Christensen, T.H., Scharff, H., 2009. Landfilling of waste: accounting of greenhouse gases and global warming contributions. Waste Manag. Res. 27, 825e836. http://dx.doi.org/10.1177/0734242X09348529.

81. Massarutto, A., de Carli, A., Graffi, M., 2011. Material and energy recovery in integrated waste management systems: a life-cycle costing approach. Waste Manag. 31, 2102e2111. http://dx.doi.org/10.1016/j.wasman.2011.05.017.

82. Mastellone, M.L., Brunner, P.H., Arena, U., 2009. Scenarios of waste management for a waste emergency area. J. Industrial Ecol. 13, 735e757. http://dx.doi.org/ 10.1111/j.1530-9290.2009.00155.x.

83. McGill, M., 2009. Carbon Footprint Analysis of Textile Reuse and Recycling, Centre for Environmental Policy (MSc thesis). Imperial College London, London, UK.

84. Merrild, H., Larsen, A.W., Christensen, T.H., 2012. Assessing recycling versus incineration of key materials in municipal waste: the importance of efficient energy recovery and transport distances. Waste Manag. 32, 1009e1018. http:// dx.doi.org/10.1016/j.wasman.2011.12.025.

85. Møller, J., Boldrin, A., Christensen, T.H., 2009. Anaerobic digestion and digestate use: accounting of greenhouse gases and global warming contribution. Waste Manag. Res. 27, 813e824. http://dx.doi.org/10.1177/0734242X09344876.

86. National Grid, 2008. Investigation into Transmission Losses on UK Electricity Transmission System. National Grid, London, UK. Available from: http://www. nationalgrid.com/NR/rdonlyres/4D65944B-DE42-4FF4-88DF-BC6A81EFA09B/ 26920/ ElectricityTransmissionLossesReport1.pdf.

87. Nayak, G., Kumar, V., 2008. Jotun Paints - Product Life Cycle Assessment. dcarbon8, London, UK. Available from: http://www.thefuturebuild.com/assets/images/ uploads/products/1521/dc8_jotunpaints_ productcarbonfootprint_final_ formasdar1.pdf?/ product-life-cycle-assessment-981475/file-files/ DC8_ JotunPaints_ProductCarbonFootprint_FINAL_ for%2520Masdar.pdf.

88. Nielsen, M., Nielsen, O.-K., Thomsen, M., 2010. Emissions from decentralised CHP= plants 2007-energinet.dk. Environmental project no. 07/1882. Project report 5- emission factors and emission inventory for decentralised CHP production.

89. National Environmental Research Institute, Aarhus University, Aarhus, Denmark. Available from: http://www2.dmu.dk/Pub/FR786.pdf.

90. Office for National Statistics (ONS), 2015. Overview of UK Population, Estimates for England, Wales, Scotland and Northern Ireland Mid-2014-released 25 June 2015. Available from: http://www.ons.gov.uk/ons/rel/pop-estimate/populationestimates-for-ukengland-and-walesescotland-and-northern-ireland/mid- 2014/stye-overview-of-the-uk-population.html.

91. Patterson, T., Esteves, S., Dinsdale, R., Guwy, A., 2011. Life cycle assessment of biogas infrastructure options on a regional scale. Bioresour. Technol. 102, 7313e7323. http://dx.doi.org/10.1016/j.biortech.2011.04.063.

92. Pressley, P.N., Levis, J.W., Damgaard, A., Barlaz, M.A., DeCarolis, J.F., 2015. Analysis of material recovery facilities for use in life-cycle assessment. Waste Manag. 35, 307e317. http://dx.doi.org/10.1016/j.wasman.2014.09.012.

93. Qiang, H., Qiang, Z., Nazim, C., Danny, M., 2011. Biofilter: a promising tool for mitigating methane emission from manure storage. J. Arid Land 3, 61e70. http://dx.doi.org/10.3724/SP.J.1227.2011.00061.

94. Rigamonti, L., Grosso, M., Giugliano, M., 2009. Life cycle assessment for optimizing the level of separated collection in integrated MSW management systems. Waste Manag. 29, 934e944. http://dx.doi.org/10.1016/j.wasman.2008.06.005.

95. Rigamonti, L., Grosso, M., Giugliano, M., 2010. Life cycle assessment of sub-units composing a MSW management system. J. Clean. Prod. 18, 1652e1662. http:// dx.doi.org/10.1016/j.jclepro.2010.06.029.

96. Saft, R., 2007. Life cycle assessment of a pyrolysis/gasification plant for hazardous paint waste. Int. J. Life Cycle Assess. 12, 230e238. http://dx.doi.org/10.1065/lca2007.05.332.

97. Salhofer, S., Schneider, F., Obersteiner, G., 2007. The ecological relevance of transport in waste disposal systems in Western Europe. Waste Manag. 27, S47eS57. http://dx.doi.org/10.1016/j.wasman.2007.02.025.

98. Salisbury, E., Thistlethwaite, G., Young, K., Cardenas, L., Thomas, A., 2015. Greenhouse Gas Inventories for England, Scotland, Wales and Northern Ireland: 1990e2013, London, UK. Available from: http://naei.defra.gov.uk/reports/ reports?report_id¼810.

99. Smith, A., Brown, K., Ogilvie, S., Rushton, K., Bates, J., 2001. Waste Management Options and Climate Change. Final Report to the European Commission. DG Environment, Luxembourg. Available from: http://ec.europa.eu/environment/ waste/studies/pdf/climate_change.pdf.

100. Spokas, K., Bogner, J., Chanton, J., Morcet, M., Aran, C., Graff, C., Golvan, Y., Hebe, I., 2006. Methane mass balance at three landfill sites: what is the efficiency of capture by gas collection systems? Waste Manag. 26, 516e525. http:// dx.doi.org/10.1016/j.wasman.2005.07.021.

101. StatsWales, 2015a. Local Authority Municipal Waste Management, 2008-09 to 2013-14. Available from: https://statswales.wales.gov.uk/Download/File?fileId¼379.

102. StatsWales, 2015b. Reuse/recycling/composting rates by local authority and year. Welsh Government. Available from: http://wales.gov.uk/statistics-andresearch/local-authority-municipal-waste-management/?lang¼en.

103. Streese, J., Stegmann, R., 2005. In: Potentials and Limitations of biofilters for Methane Oxidation, Proceedings Sardinia 2005, Tenth International Waste Management and Landfill Symposium, S. Margherita di Pula, Cagliari, Italy. Available from: http://www.image.unipd.it/cossu/Environmental%20Project% 20Work/Prevous%20years/Articoli%20scientifici/Stegmann%20-%20Biofilters% 20for%20methane%20oxidation.pdf.

104. Timlett, R., Williams, I.D., 2011. The ISB model (infrastructure, service, behaviour): a tool for waste practitioners. Waste Manag. 31, 1381e1392. http://dx.doi.org/ 10.1016/j.wasman.2010.12.010.

105. Turconi, R., Butera, S., Boldrin, A., Grosso, M., Rigamonti, L., Astrup, T., 2011. Life cycle assessment of waste incineration in Denmark and Italy using two LCA models. Waste Manag. Res. 29, 78e90. http://dx.doi.org/10.1177/ 0734242X11417489.

106. Turner, D.A., Kemp, S., Williams, I., 2011. Carbon footprinting in the UK waste management sector. Carbon Manag. 2, 677e690. http://dx.doi.org/10.4155/CMT.11.67.

107. Turner, D.A., Williams, D.I., Kemp, S., 2016. Greenhouse gas emission factors for recycling of source-segregated waste materials. Resour. Conserv. Recycl. 105, 186e197. http://dx.doi.org/10.1016/j.resconrec.2015.10.026.

108. United Nations Environment Programme (UNEP), 2010. Waste and Climate Change: Global Trends and Strategy Framework. UNEP, Osaka, Japan. Available from: http://www.unep.or.jp/ietc/Publications/spc/Waste&ClimateChange/ Waste&ClimateChange.pdf.

109. US EPA, 2011. Background Information Document for Life-cycle Inventory Landfill Process Model. Durham, NC, USA. Available from: http://www4.ncsu.edu/~jwlevis/Landfill-2011.pdf.

110. WasteDataFlow, 2014. Welcome to the WasteDataFlow Website. Available from: http://www.wastedataflow.org/.

111. Welsh Assembly Government (WAG), 2009. One Wales: One Planet. The Sustainable Development Scheme of the Welsh Assembly Government. WAG, Cardiff, UK. Available from: http://www.wales.nhs.uk/sitesplus/documents/829/One% 20Wales-%20One%20Planet%20(2009).pdf.

112. Welsh Assembly Government (WAG), 2010a. Climate Change Strategy for Wales. WAG, Cardiff, UK. Available from: http://gov.wales/topics/ environmentcountryside/climatechange/publications/strategy/?lang¼en.

113. Welsh Assembly Government (WAG), 2010b. Towards Zero Waste: the Overarching Waste Strategy Document for Wales. WAG, Cardiff, UK. Available from: http://gov.wales/topics/environmentcountryside/epq/waste_recycling/zerowaste/? lang¼en.

114. Welsh Assembly Government (WAG), 2011. Municipal Sector Plan. Part 1. Welsh Assembly Government, WAG Cardiff, UK. Available from: http://gov.wales/topics/environmentcountryside/epq/waste_recycling/publication/ municipalsectorplan/?lang¼en.

115. Waste & Resources Action Programme (WRAP), 2010. The Composition of Municipal Solid Waste in Wales. WRAP, Cardiff, UK. Available from: http://www.wrapcymru.org.uk/content/composition-municipal-solid-waste-wales-0.

116. Yang, N., Damgaard, A., Lu, F., Shao, L.M., Brogaard, L.K., He, P.J., 2014. Environmental impact assessment on the construction and operation of municipal solid waste sanitary landfills in developing countries: China case study. Waste Manag. 34, 929e937. http://dx.doi.org/10.1016/j.wasman.2015.07.011.

117. Yazdani, R., Barlaz, M.A., Augenstein, D., Kayhanian, M., Tchobanoglous, G., 2012. Performance evaluation of an anaerobic/aerobic landfill-based digester using yard waste for energy and compost production. Waste Manag. 32, 912e919. http://dx.doi.org/10.1016/j.wasman.2012.01.003.

118. Yoshida, H., Gable, J.J., Park, J.K., 2012. Evaluation of organic waste diversion alternatives for greenhouse gas reduction. Resour. Conserv. Recycl. 60, 1e9. http://dx.doi.org/10.1016/j.resconrec.2011.11.011.

CHAPTER 15

EXPLORING PRIVATE FARM ADVISOR PERSPECTIVES OF AGRI-ENVIRONMENT SCHEMES

A.P. Hejnowicz, M.A. Rudd, P.C.L. White

Keywords: Agri-environmental schemes, Common agricultural policy, Public goods, Biodiversity, Land management, Farmers, Intermediaries

INTRODUCTION

Driven by a range of complex local and global drivers (e.g., globalisation, food security concerns) food production and domestic consumption patterns have undergone rapid transformations (e.g., FAO, 2003, OECD/FAO, 2011, Tscharntke et al., 2012 and Poppy et al., 2014). These changes have been accompanied by significant agricultural intensification and extensification (FAO, 2012, FAO, 2014 and Godfray and Garnett, 2014). Striking a balance between intensification and extensification is a central challenge for modern food production systems (Pretty et al., 2010, Balmford et al., 2012 and Grau et al., 2013). Without balance, environmental risks are high and may include deforestation and forest degradation, loss of biodiversity, soil erosion, decreased water quality, water shortages, increases in greenhouse gas emissions and changes in biogeochemical cycles (e.g., Gibbs et al., 2010, Quinton et al., 2010, Lambin and Meyfroidt, 2011, Lenzen et al., 2012, Mills Busa, 2013 and WWAP, 2014).

In Europe aspects of the agricultural sector have also undergone a degree of intensification (OECD, 2008), with concomitant repercussions for ecosystems, biodiversity and water resources (e.g., Tscharntke et al., 2005, Billeter et al., 2008, Henle et al., 2008, EEA, 2010, Pe'er et al., 2014 and Zanten et al., 2014). The continuing problem European Union (EU) Member States face is trying to maintain thriving and competitive agricultural and forestry sectors whilst also ensuring a secure provision of environmental public goods (Allen and Hart, 2013). In response, to resolve this tension, incentive-based management strategies such as agri-environment schemes (AES) have been introduced throughout the EU (Deal et al., 2012, Lastra-Bravo et al., 2015 and Lefebrve et al., 2015).

Initially optional, the 1992 MacSharry Reform of the Common Agricultural Policy (CAP) made AES a compulsory agricultural measure for all EU Member States; with further consolidation via the Agenda 2000 Reform leading to their provision under Pillar 2 of the CAP (European Commission, 2005 and McCormack, 2012). Essentially, AES operate through voluntary contractual agreements and provide farmers with payments in return for the delivery of environmental public goods and services and/or the adoption of modern environmentally-friendly farming practices (Garrod, 2009, Lastra-Bravo et al., 2015 and Lefebrve et al., 2015). Their implementation is based on the subsidiarity principle, meaning that AES are specially designed to negotiate the particular production-conservation circumstances faced by individual Member States, which they achieve by addressing three intertwined matters, namely: greening farming practices; reducing food production impacts on biodiversity and improving overall countryside management (European Commission, 2005, Smits et al., 2008, European Court of Auditors, 2011, McCormack, 2012, Allen and Hart, 2013 and Burton and Schwarz, 2013). Following their introduction in the UK in 1986 various versions of AES have affected more

Citation: A.P. Hejnowicz, M.A. Rudd, P.C.L. White, Exploring private farm advisor perspectives of agri-environment schemes, Land Use Policy, Volume 55, September 2016, Pages 240-256, ISSN 0264-8377, http://dx.doi.org/10.1016/j.landusepol.2016.04.005.

than 6 million Ha of agricultural land in England alone (Dobbs and Pretty, 2008, Gibbs, 2010 and Tucker, 2010). The most significant recent variant, 'Environmental Stewardship', began in 2005 (Chaplin and Radley, 2010). Its purpose—to offer a fresher, more radical, two-tiered approach to land management characterised as 'broad and shallow' and 'narrow and deep' (Hart, 2010). The 'broad and shallow' tier was designed as a non-competitive and open-access arrangement, while the 'narrow and deep' component was configured as a targeted and competitive option for meeting priority environmental objectives (Boatman et al., 2010). In England, the Entry Level Stewardship (ELS) scheme represents the 'broad and shallow' approach, which also includes organic (OELS) and upland (UELS) variants, while Higher Level Stewardship (HLS) represents the 'narrow and deep' element (Boatman et al., 2010 and Jones et al., 2010; Supporting information Table S1).

So, how effective are AES schemes at meeting their stated environmental goals? At both the European (e.g., Kleijn and Sutherland, 2003 and Kleijn et al., 2011) and UK (e.g.,Whittingham, 2007, Boatman et al., 2008, Defra and Natural England, 2008 and Whittingham, 2011) scale evidence suggests that their ability to provide environmental and conservation benefits have been relatively mixed. In respect of Environmental Stewardship the picture is similarly mixed, with both positive and negative impacts on the supply of environmental benefits identified. In particular, research has tended to focus on the biodiversity impacts of common in-field, margin and boundary options such as crop rotations, hedgerow management, riparian buffer strips and winter stubble regimes on farmland birds (e.g., Davey et al., 2010a, Davey et al., 2010b, Field et al., 2010, Hinsley et al., 2010, Siriwardena, 2010, Baker et al., 2012, Goodwin et al., 2013 and Gruar et al., 2013), and to lesser extents on floristic diversity (e.g., Still and Byfield, 2010 and Morris et al., 2010), insect pollinators (e.g., Fuentes-Montemayor et al., 2011, Critchley et al., 2013, Dunn et al., 2013 and Peyton et al., 2013), natural resource management (e.g., Ramwell and Boatman, 2010), and ecosystem services (e.g., Rollett et al., 2008 and FERA, 2012).

Beyond biodiversity, other analyses have demonstrated that participation in Environmental Stewardship can deliver both human and social capital gains (Mills, 2012), whilst also enhancing local employment and boosting the rural economy (Courtney et al., 2013). Yet, it has also been established that the financial compensation mechanism operated by Environmental Stewardship may promote adverse selection as well as reduce the degree of environmental benefits secured (Fraser, 2009 and Quillérou et al., 2011).

Concerning ourselves with the principal agents involved (e.g., farmers, land managers, independent farm advisors and Natural England) in the implementation of Environmental Stewardship, research has generally favoured addressing the farmer element: focusing primarily on understanding the views of farmers (e.g. FERA, 2013a) and their motivations for engagement in these schemes (e.g. Mills et al., 2013) with little attention paid to intermediaries (e.g., advisors)—particularly independent farm advisors. Yet, drawing on evidence from payment for ecosystem service programmes (PES), a similar mechanism to AES, clearly demonstrates the importance of external advisors – especially as facilitators of agreement processes between participants and contracting authorities –– due to their capacity to provide specialist knowledge and skills (e.g., Ferraro, 2008, Thuy et al., 2010, Lin and Nakamura, 2012, Huber-Stearns et al., 2013, Martin-Ortega et al., 2013, Schomers and Matzdorf, 2013 and Hejnowicz et al., 2014).

In light of this, we posited that examining the farm advisor dimension would represent an important and justified avenue of exploration. By improving our understanding of the views and opinions of farm advisors regarding Environmental Stewardship, it may be possible to identify ways in which to improve the overall implementation and effectiveness of AES: aspects important for achieving conservation objectives, public goods generation and farm business viability. In this research on the English experience, we report results from a survey designed to explore private farm advisors' views regarding their own particular role in the delivery of Environmental Stewardship agreements as well as their opinions concerning farmers, Natural England and other facets of Environmental Stewardship scheme implementation and operationalisation.

Our online survey adopted an exploratory approach, delving into the 'world' of the farm advisor and concentrated on: (i) advisors' views regarding scheme constraints and client motivations and behaviours; (ii) advisors' modes of interaction with their clients and Natural England; (iii) the determinants influencing the content of individual agreements; (iv) mechanisms for balancing client needs and the provision of sufficient levels of environmental public goods, and (v) recommendations for improving the delivery of AES.

It is important to point out that this investigation tells only part of a much larger story. As such, it should be viewed as the starting point, the first stepping stone, to further, more in depth examinations of the farm advisor role which by necessity would need to be triangulate with the views of farmers, land managers and those of Natural England.

BACKGROUND: EVIDENCE TO SUPPORT OUR EXPLORATORY APPROACH

In concentrating on the areas (i–v) we were guided by evidence highlighting key determinants of voluntary incentive scheme operationalisation, implementation and effectiveness (e.g., Martin-Ortega et al., 2013 and Hejnowicz et al., 2014); the general purpose and structure of AES (e.g., European Commission, 2005) and informed by the particular arrangements and specifications of individual Environmental Stewardship schemes (Natural England, 2012a, Natural England, 2012b and Natural England, 2012c). We combined these different strands to develop an Environmental Stewardship framework (Fig. 1) to guide our survey. Reflecting this, the composition of our survey is underpinned by three major foci.

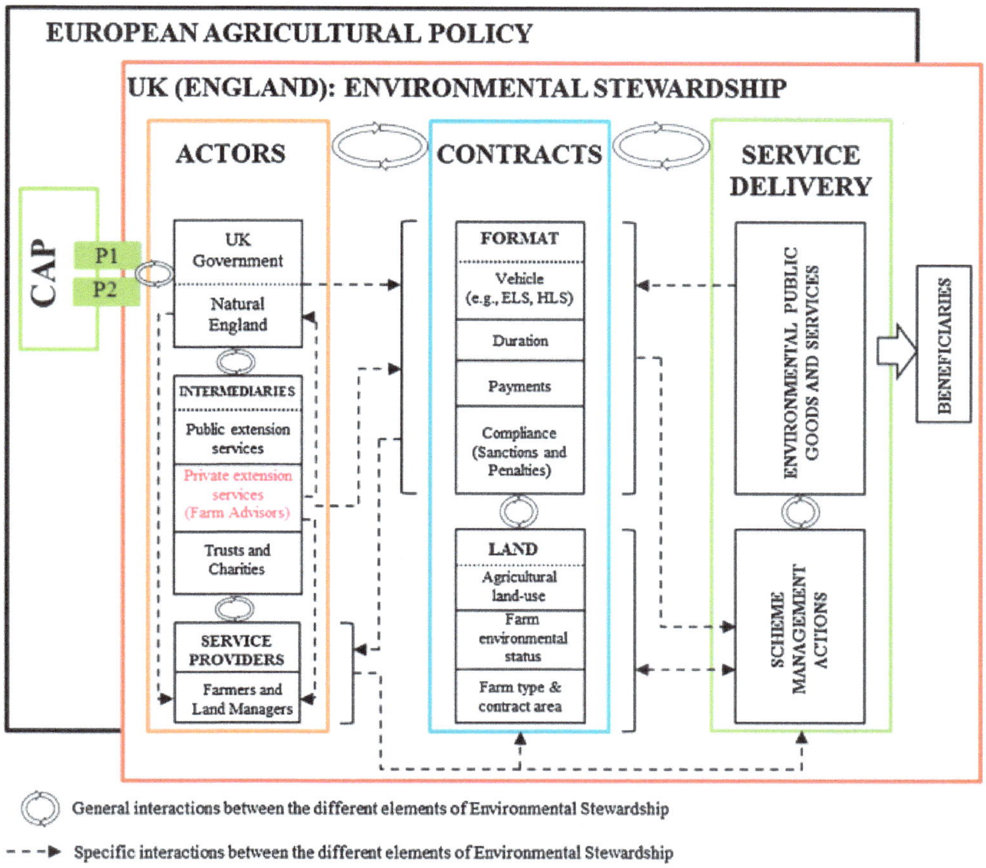

Figure 1: Environmental Stewardship Framework co-opted and adapted from a PES model by Martin-Ortega et al. (2013). The UK/England Stewardship Scheme is placed within the European agricultural policy context expressed through the linkage with the CAP (Pillar 1 and Pillar 2). The Environmental Stewardship programme is divided into three major component parts (actors, contracts and service delivery) with each of these subsequently subdivided into major constituent properties, characteristics and qualities. The framework also emphasises general interactions occurring between components parts as well as highlighting key interactions which are key foci of our survey.

Actors and Their Interactions

Farmers: Motivation, Participation and Knowledge

What motivates participants is important; indeed, it is central to their participation (Lastra-Bravo et al., 2015). Motivations underpinning farmer and land manager decision-making processes have been identified as a complex mosaic of extrinsic and intrinsic values (Siebert et al., 2006), as well as central to delivering effective AES programmes (Mills et al., 2013). The choices farmers make seem to be influenced by a range of external factors like environmental policies; internal drivers such as personal characteristics (e.g., age, education) and farm features, as well as interactive elements related to farm business arrangements and incentive design (Mills et al., 2013). In particular, economic factors related to household income, land tenure, family labour, and farm business structure appear to be particularly influential determinants of participation (Barreiro-Hurlé et al., 2010 and Lastra-Bravo et al., 2015). These represent aspects frequently related to the need to maintain family and farm continuity (Farmar-Bowers and Lane, 2009 and Ingram et al., 2013). The extent to which this mix of farm structural factors and personal farmer characteristics influences AES participation is affected by the likelihood of an agreement producing either major or minor changes to farm operations, and consequently, the potential impacts these changes may have on marginal profits, the size of transaction costs incurred and the level of utility derived from the delivery of environmental goods and services (Barreiro-Hurlé et al., 2010). Overall, as Siebert et al. (2006:319) note:

"There is an intricate interaction of contingencies affected by locality and specific context, such as agronomic, cultural, social and psychological factors, which […] play interwoven roles in each […] specific farm context".

Similarly, alongside motivation, knowledge underpins successful agriculture but the types of knowledge required for engagement in particular land management activities can vary substantially (Winter, 1997). For example, as Ingram notes (2008a:224) in relation to farmers' knowledge of soil management:

"…although farmers are largely knowledgeable many appear to lack the […] knowledge necessary for carrying out more complex […] management practices."

Environmental stewardship more generally has been shown to enhance farmer skills and environmental knowledge, awareness and appreciation (Mills, 2012).

Advisors and Agencies

From this vantage the importance of both public and private advisory extension services and the advice they supply becomes apparent (Lastra-Bravo et al., 2015). The role of external advisors in bridging potential knowledge deficits, acting as a necessary pre-condition for realizing effective voluntary incentive schemes, and delivering successful public policy interventions are widely recognised (Juntti and Potter, 2002, Cooper et al., 2009 and Vesterager and Lindegaard, 2012). Advisors are now required to explain regulatory processes and incentives as well as provide information and training (Vesterager and Lindegaard, 2012). Indeed, the evidence shows that informed farmers are far more likely to participate in AES (Lastra-Bravo et al., 2015). As Radley (2013)states, 'good advisors' are central to the effectiveness of AES. Underscoring this point, private advisors have been shown to positively impact the promotion of both minor and major AES measures as well as influence the willingness of farmers to adopt such measures (Lastra-Bravo et al., 2015). Considered more generally, the evidence indicates that the interactions between the actors (and agencies) involved in voluntary incentive schemes, operating across different institutional levels, as well as the degree of decentralisation and devolution in decision-making, are central to the functionality and effectiveness of these interventions (e.g., Beckmann et al., 2009, Pascual et al., 2010 and Legrand et al., 2013).

Service Delivery

Management Practices and the Provision of Public Goods

The financial rules and decision-making framing AES design are central to how AES achieve environmental outcomes, economic efficiency and widespread uptake (Beckmann et al., 2009). At the heart of AES lies the provision of public goods. The capacity of voluntary management interventions to provide public goods rests on the premise that the measures specified by these initiatives are capable of generating the requisite environmental public goods (i.e. ecosystem services) at scale; and moreover, that participants (e.g., farmers) fully engage in implementing those management practices (Martin-Ortega et al., 2013 and Hejnowicz et al., 2014). Essentially, it is about ensuring schemes are capable of demonstrating additionality, or added value, over and above the business-as-usual case in the absence of any intervention (Ghazoul et al., 2010).

Supplying public goods is an essential function of the CAP; but at the pan-European scale multiple environmental indicators suggest these are currently being undersupplied (Cooper et al., 2009 and Pe'er et al., 2014). With respect to Environmental Stewardship there has been considerable discussion regarding ELS option management uptake and distribution, the inference being that this directly affects their capacity to facilitate the provision of sufficient environmental public goods (e.g., Hodge and Reader, 2010).

Optimal spatial targeting of management options is a key design challenge faced by incentive schemes (Wünscher et al., 2008). The reluctance of many EU/UK farmers to engage in significant environmental management has also been shown to inhibit uptake, although certain activities are guaranteed under cross-compliance (Rollett et al., 2008). Mechanisms to counteract these behavioural dispositions have been developed (Chaplin and Radley, 2010), such as: restricted option choices under a directed ELS regime (Boatman, 2013); focused initiatives like the 'Making Environmental Stewardship More Effective' programme (Blainey, 2013) and the Campaign for the Farmed Environment (Gibbs, 2010). To varying degrees they have established that the delivery of environmental public goods can be enhanced (Boatman, 2013 and Clothier, 2013). Taking the long view, however, there are those who argue that sustaining behavioural change that results in lasting environmental benefits remains a significant challenge (Burton and Paragahawewa, 2011).

Contracts

Agreement Arrangements and Conditions

Connected to issues of uptake and public goods provision are the other core elements of incentive scheme contracts, chiefly, payments, monitoring and compliance (Danielsen et al., 2013, Hejnowicz et al., 2014 and Hejnowicz et al., 2015. The evidence clearly indicates that, for schemes to be effective, the payments entrants receive must be consistent with, and sufficient to cover, the opportunity costs they face through participation (Porras et al., 2013 and Hejnowicz et al., 2014). Consequently, in the case of AES, public agencies need to ensure that payments cover the operational and investment costs, production and profits foregone and private transaction costs incurred by farmers (Falconer, 2000 and Mettepenningen et al., 2009). This is important because, for example, regarding the issue of transaction costs, not only do private transaction costs represent a sizeable proportion of total AES-related costs but fixed transaction costs also act as a major contracting barrier (Falconer, 2000, Ducos et al., 2009 and Mettepenningen et al., 2009). Financial incentives are also key to farmers signing up to longer-term AES agreements, entering schemes that are more prescriptive and joining schemes with significant layers of bureaucracy (Ruto and Garrod, 2009). Hence the significance of payments should not be underestimated, as they directly impact overall income and the feasibility of participation (Wunder et al., 2008) in addition to having fairness and equity ramifications (Borner et al., 2010 and Pascual et al., 2010).

Providing a sufficient incentive also helps to ensure a higher degree of compliance; whilst this is important the other major factor that encourages greater compliance is the provision of contracts with enforceable sanctions and penalties (Ferraro, 2008 and Wunder et al., 2008). Poor enforcement can often undermine scheme performance (Schomers and Matzdorf, 2013). Enforcement only works if there is an adequate monitoring regime in place, and in the case of Environmental Stewardship schemes this is recognised to require considerable improvement (Defra and Natural England, 2008, Boatman et al., 2010, Chaplin and Radley, 2010, Mountford et al., 2013 and Radley, 2013). Collectively, these complex institutional arrangements represent fundamental aspects of the functioning and performance of any voluntary incentive programme (Hejnowicz et al., 2014 and Hejnowicz et al., 2015).

MATERIALS AND METHODS

Data Requirements

Our approach is exploratory. We reasoned the most expedient way to proceed to obtain a broad overview of farm advisors views was to use an online survey. In addition, due to the large number (>900) of Environmental Stewardship advisors this also seemed the most flexible and parsimonious choice. While interviews may provide a far more extensive and nuanced description of farm advisor views, we felt that it was important to gather information from as wide a variety of advisors as possible: given potentially important differences in opinions that might arise due to regional differences in priorities or in differences among advisors who specialized in ELS or HLS schemes. We also anticipated that a broad-based approach would identify specific issues and provide the necessary context to conduct more focused qualitative interviews with farm advisors as well as other agents in the process (e.g., farmers, Natural England advisors) in the future.

Sample

The sample was composed of Natural England registered stewardship advisors whose contact information, which was publically available, we obtained from Natural England's register.[1] After removing duplicate entries the register included information for 958 individual advisors from eight regions of England (North East, North West, East Midlands, West Midlands, East of England, South East, South West, and Yorkshire & Humberside).

Survey Instrument

The survey was constructed with Sawtooth Software's (www.sawtoothsoftware.com) SSI web-based interviewing platform. A pilot survey was tested on 24 stewardship advisors, three from each of eight regions; responses from the pilot informed the design of the final survey. The questions included in the survey explore those issues previously discussed in Section 2. The final survey (Supporting information S2) consisted of 39 questions divided into three parts. Most questions were closed, involving selection of radio buttons (single answers) or checkboxes (multiple answers) but several questions included space for providing extra comments.

More specifically, Part 1 requested advisors' generic background information, including expertise. Part 2 focused on the process of generating a stewardship agreement and comprised sub-sections covering advisors' views regarding: client motivations and knowledge; agreement preparation and constraints; interactions with clients and Natural England; how advisors balance farmer needs and Natural England objectives; and recommendations for improving environmental stewardship agreements in light of future changes to scheme delivery. Part 3 concentrated on the environmental content of agreements, respondents' views regarding Natural England advisor amendments to submitted HLS agreements, and advisors' perceptions of client understanding and acceptance of scheme payment levels and sanctions. Importantly, participants were fully aware of what professional opinions were being asked of them and could decide to opt-out at any stage of the survey.

Survey Implementation

Following standard social science survey protocol (Dilman et al., 2009), Environmental Stewardship advisors were contacted up to five times over a period of five weeks: (8th October to 7 November, 2013, with the survey open from 15 October to 15 November, 2013). Invitation emails contained a unique hypertext link enabling advisors to access the survey directly (Supporting information S3). Of 958 advisors from the register, we assumed that 840 had been contacted after adjusting for 118 non-delivery email notifications.

RESULTS AND DISCUSSION

Advisor Characteristics

Demographics, Experience and Regional Distribution

A total of 354 respondents (42% of the sample) accessed the online survey platform and 251 (29.9%) completed the full survey (Supporting information Table S2). Given the number of completed responses, we can be fairly confident that the views expressed by respondents in our 'survey sample' are broadly representative of those held by the 'population sample'.

Respondents were asked to indicate the regions in which they operated, in terms of commonality: 24.8% of indicated the South West; 20.3% the West Midlands; 17.5% the East of England and Yorkshire-Humberside; 16.3% the East Midlands and the South East; 14.6% the North West; and 8.5% the North East. The majority of respondents (75.3%) operated in one region, with only 14.0%, 5.6%, 2.8% and 2.0% working across two, three, four or five regions respectively.

Relatedly, regions appeared to differ according to their degree of advisor mobility (Supporting information Table S3). In some locations advisors demonstrated a more extensive 'regional working network', operating across a number of different regions, whereas in other cases these 'networks' were more limited. In the East Midlands, for example, advisors exhibited the highest level of mobility with 60.0% working across additional regions. In the South West, on the other hand, just 24.6% of advisors operated outside their own region.

Advisor experience ranged considerably: from one to thirty years. However, the majority of respondents (56.5%) worked in an advisory capacity for 9.6 ± 5.6 yrs, although a large minority (39.04%) were considerably more experienced with 10–20 years of practice. Respondents operating across more regions also appeared to have longer experience as farm advisors, for example, 14.6 ± 2.0 yrs for those working across four and five regions compared to 9.4 ± 0.4 yrs for those working in a single region. Generally respondents indicated that they had knowledge and expertise in relation to two (31.5%) or three (42.2%) Environmental Stewardship schemes, with proficiency in ELS (93.4%) and HLS (82.8%) being the most common (Supporting Information Table S4). There were distinctive patterns of regional expertise (Supporting Information Table S5), probably reflecting the different geographies of these regions as well as the more limited application of some schemes compared to others (e.g., upland versions of ELS compared to standard ELS schemes).

Agreement Formation

Understanding Clients: Farmer Motivations

Identifying the most common motivating factors leading farmers and land managers to engage with Environmental Stewardship, we found that advisors felt both extrinsic and intrinsic values played a motivating role (Table 1). Evidence from previous research indicates that farmers' and land managers' participation in voluntary agri-environment schemes is influenced by a variety of attitudes and values (e.g. Siebert et al., 2006, Cross and Franks, 2007 and Mills et al., 2013). Respondents ascribed a heterogeneity of motivations to the decision-making processes underlying farmer participation, this suggests connections to broader socio-cultural norms, worldviews and goals (Ingram et al., 2013 and Mills et al., 2013)—advancing what Morris and Potter, 1995 term a 'participation spectrum'.

Table 1: Stewardship advisor perceptions of client motivations.

Motivation classificationa and b	Motivating factors (*key themes*)	Primary reason (%,n = 246)	Secondary reason (%,n = 235)	Tertiary reason (%,n = 153)
Extrinsic				
Financial incentives (Cal)	Economics (i.e. income, finance, money, cash, payment, compensation)	64.6	8.9	5.9
Profit maximisation (Cal)	Finance linked to farm viability and management	5.7	–	–
Long-term security and farm viability (Opp)	Continuation of current practices and extension of a prior environmental scheme	4.5	9.8	5.9
Capital investment (Opt)	Use of unproductive marginal land (linkages to finance and profitability)	4.1	12.3	5.9
	Finance linked to recouping monies from modulation	3.7	2.1	1.3
	Income diversification	1.22	2.1	–
	Capital works (e.g. finance, investment and enhancements)	–	6.4	6.5
	General improvement in farm management and operations	–	3.8	7.8
	Increase farm value	–	1.7	1.3
Community image & recognition in wider society (Cat)	Prestige and public perception	–	–	1.9
Regulation (Com)	Cross-compliance	–	0.4	5.9
External non-regulatory obligation*	Peer Pressure	–	0.8	1.9
	Encouragement from Natural England	–	0.4	0.6
Intrinsic				
Personal sense of environmental responsibility and accountability (E)	Environmental benefits (e.g. biodiversity, wildlife, conservation, farm environment)	13.8	39.2	35.3
	Environmental benefits linked to improvements for game and shooting	–	2.1	5.9
	Obligation and responsibility (e.g. moral and environmental aspect)	1.2	3.4	5.9
Commitment and interest in the environment(E)	Personal satisfaction and interest	–	–	5.2

a Motivation classifications (i.e. extrinsic and intrinsic values) and sub-groupings are based on Mills et al. (2013).

b (Cal, Calculating), (Cat, Catalysing), (Com, Compensating), (E, Engaged), (Opp, Opportunistic) and (Opt, Optimising) are categories referring to modes of agri-environment scheme participation identified by Van Herzele et al. (2013). See their paper for category explanations.

* This category is not identified by Mills et al. (2013) but emerged from respondent comments.

Building on the existing literature (e.g., Wilson and Hart, 2000, Wilson and Hart, 2001,Siebert et al., 2006, Cross and Franks, 2007, Defrancesco et al., 2008, Mills et al., 2013 and Lastra-Bravo et al., 2015) our analysis of advisor opinions suggests that extrinsic values such as those related to financial gain, profit maximization, long-term security and capital investment represent the primary motivators encouraging farmer and land manager engagement with environmental stewardship. These observations accord with evidence from UELS agreement holders suggesting that scheme payments act as the principal motivating factors determining participation, with additional agronomic concerns such as the degree of alignment with existing farm practices also influencing engagement (CCRI, 2012). In this regard we identified so-called 'calculating' (i.e., purely financial), 'opportunistic' (i.e., income from existing practices), 'optimizing' (i.e., production potential of marginal land) and, to a lesser extent, 'compensatory' (i.e., regulatory obligation) motivating classifications as the primary drivers underlining farmer participation reinforcing previous analyses (e.g., Pike, 2008, Pike, 2013 and Van Herzele et al., 2013).

Previous studies of AES participants have, for example, highlighted support for environmentally-oriented concepts such as 'wildlife and environment', 'improving the landscape' and 'wildlife conservation benefits' (CCRI, 2012 and FERA, 2013a). Indeed, a positive environmental attitude can be a component of farmers' willingness to engage with AES (Lastra-Bravo et al., 2015). Similarly, respondents felt that intrinsic values were important secondary and tertiary motivators underlying Environmental Stewardship engagement, perhaps in this sense, complementing the more widespread financially-oriented motivations which they proposed that farmers hold: lending credence to the notion that extrinsic and intrinsic values need not be mutually exclusive (Mills et al., 2013). To an extent this may help temper concerns that without significant intrinsic motivations farmers lack the necessary incentive to deliver long-lasting environmental management improvements (Van Herzele et al., 2013).

Farmers' Knowledge and Advisor Advice

Farmers' capacity to undertake on-farm environmental management, in part, relies on their knowledge, skills and understanding of those management requirements, although competencies can vary significantly between individuals (Ingram, 2008a). Consequently, there is a growing recognition that the provision of advice, and the role of advisor, is central for helping farmers negotiate the progressively more complex demands of environmental management and agricultural production (Ingram, 2008b).

The majority of respondents (60.0%) indicated that 50% or more of their clients had a clear notion of the stewardship scheme they wished to enter. However, only 27.6% of respondents agreed that a similar proportion of their clients also understood the intricacies of agreement arrangement (i.e. with particular reference to the application process)—in this case advisors may have been referring to clients that were renewing agreements (FERA, 2013a). Interestingly, respondents' views to both of these questions also demonstrated a degree of regional variation, suggesting the importance of context, although the over-arching picture remained reasonably consistent (Supporting information Table S6). From an advisor perspective, the assertion that famers express strong views regarding the schemes they wish to enter yet demonstrate more limited understanding of scheme-related processes and procedures should not, perhaps, be a surprise—after all one of the reasons famers employ independent advisors is to help navigate the complexity of scheme arrangements (Vesterager and Lindegaard, 2012). For example, as one respondent (id: GRFPA2) remarked:

"Most clients have a general idea of which scheme they would like to enter (i.e. ELS or ELS & HLS) and the majority know the basic options available under ELS & HLS. But I have not met one client yet who has read each handbook (160 pages + in each) from cover to cover before I meet them so they do not realise the complexities involved, particularly in HLS, nor do they always realise the restrictions/requirements involved in the management of some of the options. This always surprises me because I would always want to know the details of something I would be committing to for 10 years especially given the financial and management implications."

In light of this, it is hardly unexpected to find that; overall, respondents regarded their own advice as either 'important' (37.2%) or 'very important' (53.2%) for steering their clients towards the most suitable Environmental Stewardship schemes. Perhaps more surprising was the gender split between advisors, with 63.4% of female respondents regarding their advice as 'very important' compared to 46.3% of their male counterparts. The reason for this apparent gender-based difference is unclear, though it does suggest that the potential role of gender in shaping professional advice needs further exploration.

What we can say, however, is that although we expect respondents to validate their own importance, the evidence seems to suggest that the advice farmers receive can be beneficial. For instance, information from UELS agreement holders points to applicants requiring substantial amounts of advice (CCRI, 2012). Likewise for ELS participants, salient advice has been shown to be essential for enabling farmers to fulfil particular environmental management obligations (Lobley et al., 2010). Thematic analysis of additional open-ended comments not only supports this claim, but further extends it by revealing the wide variety of reasons farm advisors feel their advice is necessary for aiding farmer decision-making (Supporting information Table S7). These reasons include ensuring farmers select the most appropriate type of stewardship scheme and suite of environmental options; as respondent (id: 9V3PZG) clearly stated:

"Farmers are not aware of all the options on offer for each scheme (particularly HLS) and therefore which ones best suit their farm and farming practices. They are also not aware of all the management prescriptions in HLS as these are not in the handbook."

They also reflect the opinion that farmers recognise the need for technical input and appreciate, trust and prefer independent advice; indeed, as one farm advisor (id: 5GHHXD) put it:

"They [that is the farmer] prefer private advice rather than Natural England sponsored/contracted advisors as they rather pay for impartial advice than possibly receive bias advice."

There's also a sense that respondents see their advice has having wider significance for their client's farming system too; a view advanced by respondent (id: ZJGWWD):

"...other factors that 'farmers' need to consider and reflect on including tax implication of dual use and on their single payment, how their business structure is arranged [...], the working and timing of operations at the commencement of the agreement, implications particularly under the HLS regarding the reversion of arable to grass under several options that can affect the capital value of their land..."

Many of these views are reinforced by farmers themselves (CCRI, 2012 and FERA, 2013a).

AGREEMENT PRACTICALITIES

Application Complexity

Respondents noted the variation in application completion times (to the point of submission) between ELS, OELS and UELS schemes, on the one hand, and HLS on the other (Table 2)—illustrating the different labour demands these schemes have. The majority of respondents (88.4%) indicated that ELS applications take 1–3 months to complete, while 51.6% thought that completion times of between 4–9 months were more common for HLS. ELS agreements could sometimes be completed within a few days or weeks because of the efficiency of the ELS online application system. According to open-ended survey responses, major thematic factors contributing to the longer completion times of HLS agreement included: the requirement for detailed surveys (i.e., the Farm Environment Plan [FEP]); farm complexity and holding size; the need for meetings with Natural England (as well as Natural England's capacity and decision-making); draft agreement checks; problems associated with rural land registry mapping, and obtaining historic environment record reports promptly. As one advisor (id: YMEVCN) commented:

"For ELS schemes; the application process is usually very quick – the application forms are available immediately online (or within a week on paper) then they usually take just a few hours to complete. Natural England then process the application to Agreement within 6 to 8 weeks. For HLS; often first contact is made up to or over a year in advance of the start of the Agreement. After this, the FEP must be carried out at the most appropriate time (e.g., mid-late June in species rich hay meadows or March/April on breeding wader ground). It's then necessary to delay the submission of the application until after the FEP has been completed and approved (easily 8 to 12 weeks). It can then be another 8 to 12 weeks from the Agreement to be drawn-up by the HLS Adviser and studied in detail by the applicant before it goes live."

Table 2: Average time taken by respondents to complete Environmental Stewardship agreements (applications).

ES Scheme	(% of Advisors, $n = 250$)					
	1–3 months	4–6 months	7–9 months	10–12 months	1 year +	I don't know
ELS	88.4	4.4	0.0	0.4	0.0	6.8
OELS	57.2	4.8	1.2	0.0	0.0	36.8
U(O)ELS	39.6	4.0	1.6	1.2	0.4	53.2
HLS	20.4	30.4	21.2	8.8	5.6	13.6

The complexity of the application process represents a significant issue in the design and effectiveness of HLS, in particular, the task of undertaking and producing the FEP (Defra and Natural England, 2008). Yet 74.4% of respondents agreed that the FEP was 'important' or 'very important' to the subsequent design of HLS agreements. In addition, 57.6% of respondents indicated that the advice delivered by the FEP grant to farmers and land managers was 'effective' or 'highly effective'. The centrality of FEP to constructing agreements was confirmed in an HLS monitoring programme (FERA, 2013b and Mountford et al., 2013).

Open-ended comments provided a mixture of views with some respondents arguing the value of the FEP, especially its usefulness for: mapping features and selected options; indicating the value of a holding; enabling advisors to advise on marginal areas of the holding and become familiar with the whole farm; and providing the ecological evidence base to support the correct choice of HLS options and indicators. As advocated by respondent (id: RG5J9W) the FEP represents a:

"Very useful exercise for the adviser to get a really good understanding of the farm, what is there and the farming system in place: it gives you the knowledge to sit down with the farmer and talk knowledgeably about their farm and make suggestions, also great to be able to tell them something they don't know and helps to build a good relationship for developing a quality scheme."

On the other hand, some took a more critical stance citing that the FEP was too time-consuming to produce and collected too much unnecessary information (FERA, 2013b). For example, as respondent (id: FJMNC3) notes succinctly:

"I completed a FEP for a farm...where over 75% of the FEP was irrelevant to the plan that was finally agreed. Natural England knows what it wants to focus on – there is no point in wasting time mapping areas that are irrelevant to a future scheme."

Still, others claimed that the FEP was not always read in detail by Natural England project officers; did not properly consider the farm business, and was not used following approval – a view articulated by respondent (id: LDHK8D):

"We get the impression that the FEP is not read in detail by many NE project officers, and that the end stewardship agreement is worked up by verbal discussion rather than by close reference to the FEP – hence I rate the FEP as 'important', but not 'very important'." To some extent these views leave a question mark over whether efforts to simplify FEP methodology and recording have been as successful as originally envisaged (Defra and Natural England, 2008).

Cost and time were identified as the most common principal constraints relating to the preparation and submission of Environmental Stewardship applications (Table 3). Administrative burdens magnify the transaction costs of HLSs, particularly due to labour-intensive activities such as the FEP. Past surveys of both ELS and UELS agreement holders highlighted that the bureaucracy and complexity of schemes are perceived as daunting and can reduce the number of potential applicants (Cross and Franks, 2007,CCRI, 2012 and FERA, 2013a). Reducing scheme complexity and procedures is the most effective means of curtailing transaction costs and increasing AES uptake by farmers (Mettepenningen et al., 2009). In this survey, 62.5% of respondents 'somewhat agreed' or 'highly agreed' that the HLS process needed to be simplified, while a further 80.2% agreed that farmers and land managers perceived the process of applying for environmental stewardship (in particular HLS) as too time consuming and complicated. This resonates with requests to make future Rural Development Programme funding applications far simpler and manageable through, for example, improving available guidance and streamlining application processes and requirements (Defra, 2013).

Table 3: Common constraints (C) identified by respondents in relation to Environmental Stewardship application preparation and submission processes.

Constraints (Emergent Themes)	C1 (%,*n* = 248)	C2 (%,*n* = 211)	C3 (%,*n* = 145)	C4 (%,*n* = 76)	C5 (%,*n* = 40)
Cost (e.g. for clients)	20.6	15.6	4.8	3.9	2.5
Time (e.g. application preparation and submission)	34.7	16.1	6.9	6.6	2.5
Time and cost	6.1	3.8	1.4	0.0	0.0
Costs versus potential rewards of scheme participation	3.2	1.9	2.1	3.9	0.0
Costs associated with implementing scheme options	0.0	4.3	3.5	6.6	2.5
Suitability and alignment with current farming practices (e.g. extent to which farming practices may need to be altered)	8.5	6.6	4.8	5.3	7.5
Farmer preferences, decision-making and expectations of scheme benefits	2.4	1.9	4.1	1.3	0.0
Farmer willingness to engage in the application process and implement scheme requirements	1.6	7.1	6.2	7.9	17.5
Farmer understanding, knowledge and awareness of schemes	0.4	2.4	4.	2.6	2.5
Scheme capacity to be effective and inherent limitations (e.g. environmental option choices and flexibility)	2.4	7.6	11.7	14.5	10.0
Scheme complexity (e.g. in preparation and implementation)	2.4	4.7	14.5	10.5	10.0
Mapping, visualisation tools, information procurement and surveying	5.6	3.3	8.3	2.6	5.0
Natural England	2.8	8.0	8.9	7.9	10.0
Farm limitations and suitability (e.g. size and features)	3.6	3.8	2.8	13.2	0.0
Bureaucracy (e.g. red tape)	0.8	0.0	0.0	1.3	0.0
Scheme funding availability (e.g. in relation to capital works)	0.4	0.5	2.8	2.6	0.0
Other	4.0	9.0	8.9	7.9	27.5

Advisor Roles

Our analysis indicated that respondent interaction (i.e., communication) differed according to agreement type and the corresponding actor. Respondents noted that all Environmental Stewardship schemes involved significant client interaction. However, respondents indicated that ELS, OELS and U(O) ELS schemes involved less client interaction (64.8%–77.6% 'high' to 'very high') compared to HLS schemes (95.5% 'high' to 'very high').

In the context of a 'knowledge exchange encounter' (Ingram, 2008b) studies of advisor-farmer interactions have previously emphasised the power imbalance in that relationship, characterising the advisor as a prospective exploiter; however, following significant privatisation of advisory services farmer demand is potentially reframing that asymmetric power dynamic (Ingram, 2008b). Respondents suggested that interactions between themselves and their clients proceeded based on mutual decision-making (62.0%), or to a lesser extent according to their own requirements (27.0%). Advisors described their interactions with clients mainly in terms of client need (54.0%); with the frequency of interactions being mainly 'sufficient' (40.8%) and/or 'above average compared to other professionals' (28.8%).

The emerging narrative suggests that agreement preparation is not simply a series of box ticking exercises and procedures but also a social process. As Ingram (2008b:414)noted, with respect to the relational interaction between farmers and agronomists:

"...the practice the farmer implements is a negotiated or facilitated outcome between agronomists and farmer rather than a rigid prescriptive practice "'adopted'" by the farmer."

Reinforcing this view, most respondents (80.0%) described the process of agreement preparation as a negotiation between client needs and their expert advice, with clients being relatively flexible on the type of scheme (49.2%) and environmental objectives (56.1%) they would adopt. For example, as one advisor (id: D77SEQ) noted:

"Negotiation is the key to success – one needs to be able to understand the farm, the farmer and then the reasons for being invited and work out a scheme that will achieve success for both parties."

Indeed, the overwhelming majority of advisors (92.0%) indicated that clients demonstrated a high degree of openness towards their advice. This suggests that farmers are a relatively pragmatic group – open to being persuaded on a range of possible recommendations concerning the type and composition of the agreements they enter – and that in this respect advisors can have an important role in guiding farmer-decision making processes. For example, evidence from UELS agreement holders suggests that a fifth of their option uptake is a result of external advice (CCRI, 2012). Likewise, ELS agreement holders have previously stated that advice is:

"...very useful for both option choice and option management" (FERA, 2013a:5).

In comparison with their clients, the levels interaction respondents had with Natural England advisors were substantially lower for ELS, OELS and UELS schemes (8.4% to 11.7% very low to low) but remained similar for HLS (82.0% high to very high). The vast majority (92.7%) of advisors indicated that they knew they could contact Natural England for information and advice if and when required. Their reasons for doing so were many and varied (Table 4); however, thematic analysis indicated that they were commonly related to issues concerning: clarification and guidance on scheme options, appropriateness and suitability; Natural England requirements/expectations for the farm area and application-related issues, details and administrative checks (e.g. help, advice, specific codes, vendor numbers etc.). Overall, additional comments verified that observed differences in the levels of interaction respondents had with clients and Natural England reflected the underlying complexity of the Environmental Stewardship schemes.

Table 4: Common reasons (R) respondents identified for contacting Natural England during agreement preparation.

Emergent Themes	R1 (%,*n* = 237)	R2 (%,*n* = 201)	R3 (%,*n* = 144)	R4 (%,*n* = 59)	R5 (%,*n* = 20)
Clarification and guidance on scheme options, appropriateness and suitability	23.2	16.4	13.2	8.5	20.0
Target options, priorities and features for the local area	6.3	6.0	2.8	1.7	–
Mapping-related issues	8.0	4.0	3.5	5.1	–
Natural England requirements/expectations for the farm area	12.7	6.5	2.8	6.8	5.0
Permission, approval and success of submitted application	3.8	2.0	2.1	–	–
Application-related issues, details and administrative checks (e.g. help, advice, specific codes, vendor numbers etc.)	10.5	7.0	4.9	10.2	10.0
Farm-related features, characteristics and aspects for inclusion in applications	9.3	4.5	5.6	1.7	5.0
Information related to previous applications and current on-farm/neighbouring farm schemes in operation (i.e. practices, compatibility etc.)	4.2	5.0	4.2	–	–
Budgetary availability, flexibility and constraints	2.1	2.5	4.2	1.7	–
Clarification of scheme rules and regulations	2.5	1.5	0.7	1.7	–
FEP/FER/HER related matters	2.5	4.0	8.3	3.4	5.0
Time scale of application submission/deadlines	1.7	3.5	5.6	6.8	5.0
Time scale for management options	0.4	–	–	–	–
Time to meet clients	0.4	0.5	–	1.7	5.0
Information and discussions regarding designated sites	1.3	3.0	1.4	5.1	–
Software/technical issues	0.4	0.5	–	1.7	–
Landowner-related matters (e.g. interest, commitment etc.)	0.4	2.0	4.9	5.1	5.0
Eligibility of land for Stewardship schemes and options	3.8	2.0	1.4	–	–
Field data, measurements and information	1.3	1.5	1.4	3.4	–

Environmental stewardship performance

Public Goods: Promoting Environmental Objectives in Entry Level Stewardship

Delivering public goods implies a degree of spatial optimisation to generate the requisite magnitude and distribution of environmental benefits (Garrod et al., 2012). There has been widespread discussion regarding the effectiveness of ELS option uptake and management activities in relation to realizing environmental benefits and value for money (Defra and Natural England, 2008, Hodge and Reader, 2010 and Jones et al., 2010). In particular, arguments for increased integration of options across the landscape have been proposed (Chaplin and Radley, 2010), with research suggesting that farmers would buy into collaborative AES (Emery and Franks, 2012 and McKenzie et al., 2013).

ELS schemes are designed to achieve five broad environmental objectives: wildlife conservation (WC); landscape quality & character (LQ & C—particularly in relation to water); protection of the historic environment (PHE); natural resource conservation (NRC, particularly in relation to soil) and climate mitigation and adaptation (CM & A). We asked respondents to identify the environmental objectives most frequently met by the agreements they have been involved with (Table 5). Clearly, there are some that appear to be being met more than others. This suggests differences in the way agreements fulfil specific objectives and provide a comprehensive range of environmental benefits (Radley, 2013).

Table 5: Environmental objectives fulfilled by Environmental Stewardship agreements.

Environmental objectives	Presence in stewardship application (%, *n* = 250)	Commonly met environmental objectives	(%)
WC	94.0	WC, LQ & C, PHE	48.0
LQ & C	87.2	WC, LQ & C, NRC	27.6
PHE	67.2	WC, PHE, NRC	11.2
NRC	48.8	LQ & C, PHE, NRC	4.0
CM & A	4.8	–	–

For example, 94.0% of respondents indicated that most ELS schemes in which they have been involved fulfil WC objectives, whereas only 4. 8% identified CM & A as being similarly fulfilled. This trend is also observed at the regional scale. It appears most applications (48.0%) focus on WC, LQ & C and PHE or WC, LQ & C and NRC (27.6%) environmental objectives. This supports previous agreement holder surveys indicating that 'farmland wildlife', 'wildlife conservation benefits' and 'resource protection' were important environmental issues affecting agricultural land (FERA, 2013a). To some extent variations in meeting particular environmental objectives may be a reflection of the outcomes of directed funding streams. For example, significant investments (approximately £200 million) have been channelled towards historic environment conservation activities since 2005 (Natural England, 2014). They may also be a consequence of sectoral variation in agreement and option uptake (Defra and Natural England, 2008).

It is certainly the case that most respondents (87.0%) regard their role as achieving the 'best' agreement for their client. What 'best' actually refers to needs further exploration; however, in this context there is the potential for farmer-advisor co-alignment: either to increase or decrease the environmental ambition of an agreement. One argument raised is that when advisors want to avoid losing the trust of farmers or negatively impacting farmer incomes they may focus on bolstering current on-farm agricultural practices (e.g.,Sutherland et al., 2013). For instance, as respondent (id: 8RAV9H) stated:

"I see my role as trying to fit the scheme with the farming practices/available labour/capabilities & experience of the applicant. My aim is to enhance them all, and to ensure the success of them all."

The majority of respondents (51.4%) stated that 50% or more of the applications they had been involved with included specific environmental objectives based on their priority status, implying that advisors strongly argued for the environmental component and/or that those farmers regarded this as important. As one ardent 'green' thinking advisor (id: TGGE35) commented:

"I try to only work with those people that are truly engaged with the principles of Stewardship. If they are constantly trying to find ways out of doing what is necessary I explain that I can›t help them and leave."

However, for a significant minority (48.59%) this was not the case, suggesting that advisors either failed to effectively promote and push the environmental argument or, that even if a forceful argument was presented; farmers considered other factors to be of greater bearing on their decision-making. A sentiment supported by respondent (id: 725VQ8) who acknowledge that:

"It depends if you are tuned into each other's objectives. I do this with the aim of improving and promoting good environmentally sound farming practice – which has been in serious decline since the 80s. If my clients are similarly inclined, my advice is important and relevant. If they are not of the same opinion, my advice/persuasion falls on deaf ears and is of little importance."

Regarding the latter, although we suggested in Section 4.3.2 that advisors were well placed to influence farmer decision-making – given the level of openness to advice they purportedly demonstrated – respondents identified a range of other significant farm-related factors that strongly affected farmers' decision-making rationale to include/exclude specific management options. The most commonly identified reasons advisors noted were those connected to farm system compatibility. For example, environmental objectives are easy to implement (88.4%); is/are an extension of current farm practices (84.3%), do not significantly impact on the day to day farm routine (84.3%), provide a higher points value (67.1%) and requires few man hours (49.4%). Advisors seem to be indicating that farmers are predisposed towards selecting options that do not significantly influence their farm business (FERA, 2013a), bolstering the idea that the primary predictors of option uptake are agricultural related factors (Hodge and Reader, 2010). Moreover, it supports the generally articulated view of AES that there is a:

"...disjuncture between the policy's supposedly overarching environmental rationale and its realisation in practice through the actions and behaviours of land managers" (Juntti and Potter, 2002:216).

On the other hand, 80.0% of respondents also agreed that in preparing ELS and HLS agreements they needed to balance both client and Natural England needs. However, overall, opinion was split as to whether a stewardship agreement reflected the preferences of their clients or the priorities of Natural England. Nevertheless, when viewed through the lens of gender and expertise differences did emerge. For example, a higher percentage of male respondents (46.3%) compared to female respondents (30.7%) agreed that agreements were more reflective of client preferences than Natural England. In fact; on this issue, the distribution of responses between male and female advisors was significantly different ($H(2) = 7.56$, $df = 1$, $p = 0.006$). However, when considering advisor expertise, those with wider experience of Environmental Stewardship schemes were more inclined to disagree with this position (44.7–54.3%) compared to those with less expertise (30.0–33.8%).

Importantly, 53.2% of respondents acknowledged that there was an inherent conflict between client needs and Natural England priorities, a pattern similarly observed if disaggregated by gender and expertise. This is significant, because it points toward an inherent tension in how agreements meet their statutory obligations whilst acknowledging that, to some degree, they must also co-align with farmer and land manager needs. Although it would require further investigation, it is conceivable that these various decision-making trade-offs may contribute to the skewed pattern of option uptake observed elsewhere (e.g., Boatman et al., 2007, Jones et al., 2010 and Radley, 2013).

To an extent our results add weight to the contention that the voluntary nature of Environmental Stewardship schemes, and their option menus, predispose farmers to undertake only those environmental management activities that would have occurred in their absence (i.e., a lack of additionality); specifically, by subscribing to those practices that fit easily into existing farm activities – leading to the possibility of adverse selection in option choices where the envisaged level of environmental benefits cannot be guaranteed (Hodge and Reader, 2010). For example, research has found that:

"...farmers thought that 61% of features in ELS option would be managed the same if they had not gone into ELS" (FERA, 2013a:9).

And furthermore, that between 21% and 52% of management work, in financial terms, would have occurred 'in the absence of the scheme' (Courtney et al., 2013).

Alteration of Agreements: The Case of HLS

A profusion of advisory services stretching across public and private sectors presents both challenges and opportunities: a pluralistic resource of diverse competencies enriching advice and extending the 'agricultural knowledge system'; yet also providing an opportunity for fragmentation, duplication, and incoherence in policy and delivery; encouraging greater competition between service providers and leading to confusion amongst farmers (Juntti and Potter, 2002 and Sutherland et al., 2013). How these play out in practice can significantly influence the environmental performance of agreements (Sutherland et al., 2013).

It is reasonable to posit that there are different points in the Environmental Stewardship process where the content of agreements can be revised. First, there are those opportunities that arise during agreement preparation, specifically, in relation to the dialogue between farmers, private advisors, and Natural England officers. Second, alterations can be made to applications post-submission when they are being reviewed by Natural England (here we are particularly referring to HLS applications). At this juncture, it is possible that agreements may be altered to favour Natural England's environmental agenda and priorities—potentially shifting the pre-submission content away from more farmer-centric interests.

With this in mind, we asked respondents with experience of HLS agreements ($n = 212$) to comment on the extent to which Natural England advisors made alterations to the environmental content of HLS applications between the original and final approved application. Although some respondents (17.9%) specified that Natural England advisor decisions did not lead to any alterations in environmental content, the majority of respondents indicated that the environmental management composition of applications was either 'somewhat different' (40.1%) or 'moderately different' (38.7%). Notably, from the standpoint of transparency, a majority of respondents (56.3%) declared that Natural England 'very often' or 'always' informed them about changes that had been made:

"I have without exception found the Natural England advisors very helpful and communicative at all stages of the long process it takes to put an application into practice." (Respondent id: WSGZAM)

Listing reasons as to why Natural England advisors altered the environmental content of submitted HLS applications, respondents expressed a range of views with several themes emerging (Table 6). The most frequently cited themes related to changes in Natural England's environmental agenda (23.7%), financial and cost constraints (20.3%), and the calibre of Natural England advisors (11.6%).

Table 6: Emergent themes describing the possible reasons (R) for HLS application modifications.

Emergent themes	R1(%,*n* = 187)	R2(%,*n* = 116)	R3(%,*n* = 64)	R4(%,*n* = 13)	Importance Across Reasons (%)
Changes in Natural England targets and priorities for HLS	26.7	22.4	21.8	0.0	23.7
Budgetary/cost/financial constraints	22.9	18.9	12.5	30.8	20.3
Natural England advisor decision-making, viewpoint and knowledge	9.6	14.7	12.5	7.7	11.6
Other	8.7	12.9	14.1	15.4	11.1
Environmental option suitability	6.9	2.6	7.8	0.0	5.5
Farmers changed their minds	2.7	4.3	9.4	0.0	4.2
Conflict between HLS options and other scheme objectives	4.3	3.4	4.7	0.0	3.9
Farmer and NE negotiated changes	2.1	5.2	4.7	7.7	3.7
HLS prescriptions too burdensome	1.6	5.2	4.7	15.4	3.7
Too little evidence to justify & differences of opinion concerning option inclusion	1.6	4.3	4.7	7.7	3.2
Environmental option eligibility	4.2	0.9	0.0	0.0	2.4
Application mistakes	2.1	2.6	1.6	0.0	2.1
Rarely are changes/adjustments made prior to submission	4.3	0.0	0.0	0.0	2.1
Too many similar options included in HLS applications	1.6	0.9	0.0	7.7	1.3
Inclusion of too many options (overly ambitious application)	0.5	1.7	1.6	7.7	1.3

It is possible, of course, that Natural England modifications to HLS applications do improve their environmental content:

"I work very closely with Natural England and know advisors personally. Although they work to targets they seem to aim for the best 'wildlife' options overall." (Respondent id: HT2RJQ)

Furthermore, a previous assessment of 174HLS agreements indicated that they were generally 'well designed' with 80% of agreements deemed likely to be effective in achieving most outcomes: in each of these cases it is likely that there was considerable Natural England oversight in the delivery of these schemes (Mountford et al., 2013). Conversely, extensive prescriptive revisions to AES have been shown to produce 'excessive uniformity' in habitat management which can undermine biodiversity as well as negatively affect farmer participation (Radley, 2013). The impact on farmers may be potentially quite severe, as respondent (id: 2NUE5U) strongly argued:

"Often the client has already committed to a huge expense to undertake the scheme that they feel they have to accommodate the changes in order to achieve a return. Many are bullied into the changes. This is ineffective as it doesn't take into account if the changes can be managed effectively which can lead to failure in the long term."

The majority of respondents (87.9%) expressed the opinion that they should be included in the discussions leading to changes in finalized HLS agreements. Similarly, 82.2% of respondents disagreed with the notion that it was appropriate for Natural England to make modifications to HLS applications without their input (Supporting Information Table S8). For some advisors involvement is a question of process and procedure:

"If Natural England discuss with advisor then the advisor can understand the rationale and will be able to use advice on other applications. Sometimes the discussions with Natural England can lead to them reverting to original option." (Respondent id: 3US4XH)

Yet, for other respondents, it is about mitigating inefficiencies and the potential straining of relations:

"I've had farmers very upset not understanding the changes and then creating a silly 3 way discussion where it is not easy to know exactly what has been said. It makes sense to have a proper open communication system. The farmer has employed me, why suddenly change to excluding the agent. This varies according to the NE officer." (Respondent id: TFEPJJ)

Nevertheless, Natural England is the statutory authority in charge of implementing Environmental Stewardship and as such oversees the final decisions regarding individual applications—they represent the legitimate institutional authority. It is interesting, therefore, that advisors should feel entitled to have a greater influence over Natural England decision-making. The reasons behind this require further investigation.

Linked to these comments, 60.8% of respondents viewed Natural England's modifications of HLS agreements as not made in the interests of their clients. In fact, 68.8%

of respondents thought that revisions to HLS applications were made to favour Natural England interests (Supporting information Table S8). This point was forcefully conveyed by one advisor (id: MLZEYT) who commented that:

"In my experience they are ALWAYS made to save money and not in the best interests of the client or the environment and, if truth be known, not in the best interests of Natural England either in the long term."

Equally, however, some advisors noted the difficult position Natural England advisors faced:

"Advisors are in the awkward position of trying to achieve Government objectives, habitat and species protection and encouraging land owners to enter the schemes. Advisors work hard to get as much for the money as possible and farmers want as much money as possible." (Respondent id: QK59MK)

Payments, Costs and Income

Financial incentives are a central tool in steering farmers' private interests to provide particular public goods, but to achieve their desired outcome payment levels must account for the opportunity and transaction costs that farmers might incur on entering voluntary schemes (Cooper et al., 2009). Stewardship schemes provide different standard payment amounts. Little consensus emerged in our survey regarding respondents' perspectives on whether clients were satisfied with ELS payments (Supporting information Table S9). There was a sense from some open-ended responses that farmer expectations remain too high, with an attitude of 'maximum gain and minimal cost/impact'. For example, as one advisor remarked:

"They [i.e. farmers] would always be happier with more!" (Respondent id: 7KMC97)

Other advisors observed that payments were sufficient so long as other farming business opportunities were not compromised:

"The £30/ha for ELS is ok provided there is not serious competition for more productive uses." (Respondent id: 4DQMRP)

On the whole there was more general agreement among advisors that clients were satisfied with the standard payment amounts for OELS, UELS and HLS (Supporting information Table S9). Having said that there were also clear differences of opinion, with one farm advisor suggesting that farmers might be overpaid:

"HLS payments are too high for the land based options in the uplands. Extremely large annual payments for delivering very little benefit." (Respondent id: LGQECT)

Whilst in stark contrast, one respondent commented that the present level of HLS payments questions the scheme's viability:

"HLS payments in recent years have been 'trimmed' down to the state that with some schemes the annual return has been so low as to make the scheme non-viable for the farmers." (Respondent id: 7KDH6Y)

However, by and large, with regards to payment satisfaction our results seem to accord with the 66% of UELS agreement holders that considered scheme payments to be 'generous' or 'sufficient' (CCRI, 2012). Ultimately, as respondent (id: P8XVWR) indicates:

"If they [i.e. farmers] were not happy they would not enter an agreement."

Some studies claim farmers view Environmental Stewardship as an income top-up and stabilizer (a form of income security), particularly in instances where the farm business is vulnerable (Mills, 2012). In others, it has been argued that ELS entry may incur modest costs rather than adding to income (Udagawa et al., 2014). Our results suggest that respondents straddle both of these perspectives, as advisors were generally split over whether payments afforded their clients an adequate income stream or not (Supporting information Table S9), although one advisor did comment that:

"Some...farmers look on stewardship payments as income, but I do point out to them that they are required to work for their money – it's not just a free hand-out." (Respondent id: LDHK8D)

However, when specifically reflecting upon the connection between payments and costs (e.g., labour and materials), 56.6% of advisors suggested that clients considered payments insufficient to adequately cover changing input costs versus only 21.1% that thought the contrary. This view was echoed in open-ended comments, with some advisors stating payments did not reflect recent 'escalations in commodity prices' or 'cost increases due to inflation', for example:

"Payments should be index linked. A payment of £30/ha might have been acceptable 5 years ago, but costs have gone up a lot yet scheme incomes have remained the same." (Respondent id: 4A29YS)

Others emphasised that ELS payments were 'not cost effective for arable farmers' and that 'crop values and greening measures will make it harder to encourage renewals', as one advisor outlined:

"Payments are now seen as too low particularly in intensive arable areas where a typical comment is that 'it's hardly worth the hassle'." (Respondent id: 6EATTB)

This is supported by evidence indicating that cereal incomes may be unduly affected by entering ELS schemes (Udagawa et al., 2014), as a respondent noted:

"The £30/ha on ELS was set at a time of low cereal prices and was to compensate for income forgone – in light of much higher prices for crops this aspect needs re-visiting." (Respondent id: 4NGVTB)

While this may be the case other studies have suggested that the percentage of farmers who regard payments as sufficient to cover their costs has grown since 2005 (FERA, 2013a).

Adopting a regional perspective revealed statistically significant heterogeneities in respondents' views regarding scheme payment levels as well as income and input costs (Supporting information Table S10, Figs. S1 and S2). These regional patterns likely reflect individual respondent experiences of specific farm-level socio-economic characteristics as well as the distinctive rural and wider economic circumstances encountered in these locations (see Farm Business Survey).[2] The implication seems to be that payment levels ought to account for these differences, and thus better reflect regional level conditions and farm business circumstances. Overall, advisors' comments suggest that payment levels are a real issue for farmers, in particular, whether the costs incurred actually make entry level schemes unsustainable in the long-term (Udagawa et al., 2014).

Compliance: Penalties and Sanctions

The successful provision of public goods relies on individuals complying with contractual arrangements: this requires agreements to have monitoring and conditionality elements (Hejnowicz et al., 2014). We queried respondents regarding elements of conditionality (i.e., sanctions and penalties) and found that, although a majority, only 51.4% agreed that their clients understood the extent of the penalties that may be applied should they fail to fully comply with agreements (Supporting information Table S9). Notably, a sizeable minority of advisors (33.9%) thought the opposite.

Additional commentary raised a number of issues; for example, some advisors alleged that farmers and land managers, although aware penalties could be applied in cases of non-compliance, were often ignorant of the scope sanctions could take: partly as a consequence of the rules being over-complicated and poorly explained:

"Farmers are often oblivious to the penalties that would be applied for non-compliance. However, there are very few which would knowingly flout the rules and often farmers are found to be non-compliant only as a result of them not being aware of all of the requirements of the complicated scheme (particularly in the case of HLS)." (Respondent id: YMEVCN)

On closer inspection this observation does seem rather puzzling, surely the expectation is that part of the service advisors provide includes spelling out the likely repercussions of non-compliance? If an explanatory deficit exists then surely this rests, in part, on the shoulders of farm advisors themselves? Certainly some advisors made the effort to inform their clients of non-compliance related issues but hinted at the fact that, in their opinion, 'not…many other advisors do this' and also with regards to farmers 'it's amazing how quickly some of them forget!'.

Picking up on the latter point, ignorance and deliberate avoidance have been identified as issues for farmers failing to meet agreement prescriptions, a reality that would seem to support the contention that agreement holders ought to have better access to training in order for them to more closely adhere to agreement conditions (FERA, 2013a). The provision of training has been shown to positively influence farmer behaviour and management activities (Jones et al., 2013).

Training, however, may not be sufficient because, as some respondents disclosed, a number of their clients took a fairly relaxed, even recalcitrant, stance towards compliance issues due to poor monitoring and enforcement:

"Past monitoring of schemes has been poor – I think some farmers believe they are unlikely to be caught breaching ELS/HLS options and may therefore continue existing practices (e.g., supplementary feeding, applying fertilizer) where the options actually forbid this." (Respondent id: QKUYGC)

At the opposite end of the spectrum, however, a number of advisors commented that punitive sanctions actually dissuaded individuals from entering schemes as well as affecting the content of agreements:

"The issue of the current punitive level of sanctions has put off some farmers from going into schemes and has certainly reduced an agreements 'ambition'." (Respondent id: E9DAB2)

Indeed, some respondents suggested that famers regarded sanctions and penalties as the thin-end-of-the-wedge:

"Penalties imposed as a result of inspection are often seen as pedantic and penny pinching for what appears to be minor infringements." (Respondent id: 6EATTB)

Others went further, proposing that sanctions and penalties were inappropriate, poorly formulated and incorrectly realised:

"The main problems with the sanctions and penalties are that they come across as being draconian and in many cases incorrect and based on incorrect information either supplied by the inspector or interpreted by the administration staff." (Respondent id: 97H33F)

and furthermore,

"Rules about penalties are over-complicated and poorly explained. No allowance for intent – issues with extenuating or mitigating circumstances are punished as severely as deliberate non-compliance." (Respondent id: 36E4W9)

Reflecting this view, only 19.9% of respondents agreed that their clients regarded such sanctions as reasonable. Clearly, in the view of advisors, farmers regard the potential sanctions imposed by Natural England as disproportionate, with some suggesting that the fault lies, in part, in the 'uncertainty' and 'inconsistency' with which Natural England tackle these issues.

Perhaps one avenue to help address these compliance issues would be to adopt performance-based payment schemes: here payments are directly linked to the maximization of environmental benefits (Schomers and Matzdorf, 2013). Examples of agricultural results-based payment programmes throughout Europe indicate that they can be successful in generating beneficial ecological, economic and social outcomes (Burton and Schwarz, 2013), and research in England also suggests that payment by results is positively perceived by farmers (Schroeder et al., 2013).

Moving Forwards

A new round of CAP reforms (2014–2020) has been introduced to provide a more streamlined, targeted and greener approach to agricultural production and the rural environment (European Commission, 2014). The degree of inter-pillar transfer from Pillar 1 to Pillar 2 in England, as a consequence, is set to increase from 9% to 15% over the next six years; and Defra has committed to allocate 87% of rural development funds to the environment (Defra, 2014 and Natural England, 2014). These revisions will see Environmental Stewardship programmes eventually phased out and replaced by the New Environmental Land Management Scheme or NELMS for short (Natural England 2013). Although NELMS will be implemented in early 2016, current Environmental Stewardship agreement holders will still be delivering management under the 'older' system. The opportunity therefore exists for further suggestions for refinements to feed into the design and operationalization of NELMS based on lessons learned under the Environmental

Stewardship programme. Respondents were allowed to put forward four recommendations that could enhance Environmental Stewardship uptake and implementation (Table 7). Many recommendations were suggested and thematic analysis identified several broad themes, the most common of which centred on: reorganising Environmental Stewardship delivery (22.5%); simplifying scheme processes and procedures (19.5%); providing more information regarding environmental option management and implementation (14.3%), and improving the targeting of schemes (9.6%). Indeed, the views of respondents also echo those of Lastra-Bravo et al. (2015) who highlight the importance of 'institutional design' and 'stable policy' for aiding farmer engagement with, and adoption of, future agri-environmental schemes.

Table 7: Recommendations (R) for improvements in Environmental Stewardship delivery and implementation.

Emergent themes	R1(%, n = 168)	R2(%, n = 92)	R3(%, n = 56)	R4(%, n = 27)	Overall popularity (%)
Reorganization of Environmental Stewardship delivery (e.g. alternative ways to improve on-the-farm provision)	25.6	16.3	28.6	11.1	22.5
Simplification of Environmental Stewardship processes (e.g. streamline HLS application requirements)	26.2	17.4	7.1	11.1	19.5
Environmental Stewardship scheme options (e.g. degree of flexibility in option implementation)	10.1	21.7	17.9	7.4	14.3
Targeting Environmental Stewardship (e.g. tailoring to meet local environmental needs)	8.3	13.0	8.9	7.4	9.6
Natural England and Natural England Advisors (e.g. knowledge; interaction with farmers)	8.9	9.8	3.6	11.1	8.5
Environmental Stewardship payments (e.g. reassess payments to reflect environmental option requirements and changing labour and input costs)	9.5	6.5	3.6	0.0	7.0
Other	5.4	4.4	7.1	18.5	6.4
Consultation, dialogue and support (e.g. contact with industry; ES support provision for farmers)	4.2	2.2	5.4	11.1	4.4
Agent/advisor training (e.g. ensuring agents are suitability qualified and knowledgeable)	0.6	2.2	1.8	14.8	2.3
Mapping (e.g. improve online mapping tools)	0.6	3.3	1.8	3.7	1.8
Farmer focused (e.g. consideration of farmer viewpoints and operational constraints)	0.0	0.0	8.9	3.7	1.8
Environmental Stewardship scheme reinvention (e.g. establishing and dissolving schemes too frequently)	0.6	0.0	3.6	0.0	0.9
Farmer knowledge (e.g. improving knowledge of ES scheme management)	0.0	2.2	0.0	0.0	0.6
Scheme complementarity (e.g. reduce conflicts between different but co-implemented/managed environmental schemes)	0.0	1.1	1.8	0.0	0.6

Overall, recommendations provided by respondents connected with the themes central to the NEMLS programme (e.g., 'delivering outcomes at a landscape scale'; 'a participative and collaborative approach'; 'outcome focused performance assessment'; 'flexible and adaptable'; 'locally tailored advice and training'; and 'simplification'—see NELMS Project, 2013), as well as reflecting research focusing on the design principles of future agri-environment schemes (FERA, 2013b). Respondent suggestions also aligned with those expressed in recent Defra consultations on CAP reform (Defra, 2014), which included the need to address 'the risk of complexity' and the importance of 'targeting as a means to direct option choice'.

CONCLUSIONS

Our survey has provided an important exploratory assessment of the English experience of Environmental Stewardship viewed through the lens of independent farm advisors: the views of whom are under-represented in the existing literature. In this regard, we have both strengthened and expanded upon the current literature concerning AES, and Environmental Stewardship in particular, by highlighting a broad range of farm advisor views that have not previously been, in this format at least, addressed or assessed.

Our findings indicate that farm advisors exhibit a wide range of expertise and experience, and that in some cases regionality, expertise and gender can play a part in influencing farm advisor perspectives and experience of Environmental

Stewardship schemes. We have also shed some light on the facilitating role farm advisor advice plays in the preparation of Environmental Stewardship agreements, and shown that farm advisors regard what we term the 'knowledge-exchange encounter' as a crucial aspect of this facilitative function.

Initial findings suggest that farm advisors face a difficult balancing act: preparing agreements based around the needs of their clients on the one hand, whilst on the other, ensuring submitted agreements are not at odds with Natural England requirements. In the view of most respondents, there is an inherent tension between farmer and Natural England objectives. This would seem to connect to the finding that, in the eyes of respondents, although farmers display a broad range of extrinsic and intrinsic motivations for engaging in Environmental Stewardship agreements, they are primarily motivated by financially-oriented reasons. And, in addition, they demonstrate a proclivity to make decisions about the environmental management content of agreements based on how closely this aligns with current on-farm practices and the farm business more generally.

Equally, given that famers respect, and are open to, the advice they externally contract, farm advisors may have a decisive role in guiding farmer decision-making processes. Advisors, in this sense, potentially occupy an influential 'soft power' position. This has the prospect of going in one of two directions, either: advisors can encourage farmers to undertake environmentally ambitious agreements that build on intrinsic 'green' motivations or, taking the opposite stance, draw on farmers' extrinsic motivations and produce agreements requiring minimal changes to on-farm practices primarily benefiting existing farm business arrangements.

Allied to these discussions are the observations that many respondents noted, namely, with particular reference to the HLS tier of Environmental Stewardship, that their clients found HLS application processes too burdensome and overly complex, a view they also concurred with, pointing to the need to simplify and streamline the system. Respondents suggested that the complexity of programme arrangements and processes may function as a barrier for farmers and land managers: potentially acting as a contributing factor to decrease the environmental ambition of agreements and increase the likelihood that the management practices incorporated into agreements mirror those of the 'farm system'. Perhaps it is the very labour intensive nature of producing an HLS agreement that provides one explanatory factor for the observation that respondents felt they ought (and were entitled) to be involved in the HLS revision process, a process in which they have no legitimate authority to intervene.

Overall, the narrative we have presented suggests that the knowledge-exchange encounter is not a simple straightforward interaction. Rather, advisers' opinions and comments suggest that there are often important tensions between the goals and agendas of the principal agents involved in preparing, implementing and delivering Environmental Stewardship.

In addition, our survey has also highlighted a feeling among advisors that, for farmers, scheme payments present a real issue, particularly because management interventions can have a considerable impact on overall farm income and; furthermore, may not adequately account for all the costs farmers incur and appropriately reflect regional socio-economic differences. Farm advisors also indicated, and to some extent this may be linked to the issues of scheme complexity we have previously discussed, that there are challenges associated with matters of scheme compliance and sanctions. Among respondents there was a general feeling that a significant minority of their clients were not fully aware of scheme-related penalties and sanctions, and in some cases adopted a fairly relaxed stance towards non-compliance. Such matters pose real issues for how enforcement works and the environmental management effectiveness of schemes, but also, raise important issues regarding proper informed consent (i.e., that agreement holders should be fully informed about, and understand, their contractual obligations at the outset of the process).

Looking ahead, to ensure the success of future AES programmes, teasing out the issues we have started to shed light on in this paper will be necessary, in particular: focusing in more depth on the relationships and tensions existing between farmers, farm advisors and Natural England. This would seem to be the most fertile ground for uncovering those factors determining the overall content, implementation and performance of AES agreements. Ultimately, if NELMS are to fruitfully replace and build on the successes of Environmental Stewardship, as well as avoid any of their pitfalls, then the issues raised by farm advisors in this survey will be important food for thought in developing effective schemes that work in practice. Specifically, by acknowledging the importance of the different agendas and dialogues occurring between farmers, private advisors and Natural England; ensuring that participation and environmental ambition pays and contracts are properly enforced; and that the operation and implementation of schemes is simple, straightforward, easy to put into practice, accommodates farm production and does not alienate potential participants.

ACKNOWLEDGEMENTS

The authors gratefully acknowledge the financial support provided by Natural Environment Research Council (NERC,UK) and the Economic and Social Research Council (ESRC,UK) in enabling this work (Ref: ES/I003851/1). Furthermore, the authors are thankful to those farm advisors that agreed to participate in this study and spent their time completing the online survey. Finally, the authors acknowledge their indebtedness to the two anonymous reviewers who agreed to review the manuscript and whose insightful comments greatly improved the final article.

REFERENCES

1. Allen, B., Hart, K., 2013. Meeting the EU's environmental challenges through the CAP—how do the reforms measure up? In: agri-environment schemes—what have they achieved and where do we go from here? Asp. Appl. Biol. 100, 9–22.
2. Baker, D.J., Freeman, S.N., Grice, P.V., Siriwardena, G.M., 2012. Landscape-scale responses of birds to agri-environment management: a test of the English Environmental Stewardship scheme. J. Appl. Ecol. 49, 871–882.
3. Balmford, A., Green, R., Phalan, B., 2012. What conservationists need to know about farming. Proc. Biol.

Sci. 279, 2714–2724.

4. Barreiro-Hurlé, J., Espinosa-Goded, M., Dupraz, P., 2010. Does intensity of change matter?: Factors affecting adoption of agri-environmental schemes in Spain. J. Environ. Plan. Manage. 53, 891–905.

5. Beckmann, V., Eggers, J., Mettepenningen, E., 2009. Deciding how to decide on agri-environmental schemes: the political economy of subsidiarity, decentralisation and participation in the European Union. J. Environ. Plan. Manage. 52, 689–716.

6. Billeter, R., Liira, J., Bailey, D., Bugter, R., Arens, P., Augenstein, I., et al., 2008. Indicators for biodiversity in agricultural landscapes: a pan-European study. J. Appl. Ecol. 45, 141–150.

7. Blainey, L., 2013. Less management prescription, more outcome focus—making environmental stewardship more effective (MESME) trialling project. Natural England Research Reports No. 047. Natural England. (pp. 1–152).

8. Boatman, N., Jones, N., Garthwaite, D., Bishop, J., Pietravalle, S., Harrington, P., Parry, H., 2007. Evaluation of the operation of environmental stewardship. Final Report, Defra Project No.MA01028 Central Science Laboratory. (pp. 1–9).

9. Boatman, N., Ramwell, C., Parry, H., Bishop, J., Gaskell, P., Mills, J., Dwyer, J., 2008. A review of environmental benefits supplied by agri-environment schemes. Land Use Policy Group, 1–275.

10. Boatman, N.D., Jones, N.E., Gaskell, P., Dwyer, J.C., 2010. Monitoring of agri-environment schemes in the UK. In: agri-environment schemes—what have they achieved and where do we go from here? Asp. Appl. Biol. 100, 9–18.

11. Boatman, N.D., 2013. Evaluating the impacts of limiting free choice in management option selection by Entry Level Stewardship (ELS) applicants. Natural England Commissioned Reports No. 117. pp 89.

12. Borner, J., Wunder, S., Wertz-Kanounnikoff, S., Tito, M.R., Pereira, L., Nascimento, N., 2010. Direct conservation payments in the Brazilian Amazon: scope and equity implications. Ecol. Econ. 69, 1272–1282.

13. Burton, R.J.F., Paragahawewa, U.H., 2011. Creating culturally sustainable agri-environmental schemes. J. Rural Stud. 27, 95–104.

14. Burton, R.J.F., Schwarz, G., 2013. Result-oriented agri-environmental schemes in Europe and their potential for promoting behavioural change. Land Use Policy 30, 628–641.

15. CCRI, 2012. Attitudes to uplands entry level stewardship. Natural England Commissioned Reports No. 091. (pp. 1–139).

16. Chaplin, S.P., Radley, G.P., 2010. Where next for agri-environment schemes, evolution or revolution? In: agri-environment schemes—what have they achieved and where do we go from here? Asp. Appl. Biol. 100, 333–340.

17. Clothier, L., 2013. Campaign for the farmed environment: entry level stewardship option uptake. Defra Agricultural Change and Environment Observatory Research Report No. 32. (pp. 1–13).

18. Cooper T., Hart K., Baldock D., 2009. Provision of public goods through agriculture in the European Union, Report prepared for DG Agriculture and Rural Development, Institute for European Environment Policy, London, 1–396.

19. Courtney, P., Mills, J., Gaskell, P., Chaplin, S., 2013. Investigating the incidental benefits of Environmental Stewardship schemes in England. Land Use Policy 31, 26–37.

20. Critchley, C.N.R., Mole, A.C., Towers, J., Collins, A.L., 2013. Assessing the potential value of riparian buffer strips for biodiversity. In: agri-environment schemes—what have they achieved and where do we go from here? Asp. Appl. Biol. 100, 101–108.

21. Cross, M., Franks, J.R., 2007. Farmer's and advisor's attitudes towards the environmental stewardship scheme. J. Farm Manage. 13, 47–68.

22. Danielsen, F., Adrian, T., Brofeldt, S., 2013. Community monitoring for REDD+: international promises and field realities. Ecol.Soc 18, 41.

23. Davey, C., Vickery, J., Boatman, N., Chamberlain, D., Parry, H., Siriwardena, G., 2010a. Regional variation in the efficacy of Entry Level Stewardship in England. Agric. Ecosyst. Environ. 139, 121–128.

24. Davey, C.M., Vickery, J.A., Boatman, N.D., Chamberlain, D.E., Parry, H.R., Siriwardena, G.M., 2010b. Assessing the impact of Entry Level Stewardship on lowland farmland birds in England. Ibis 152, 459–474.

25. Deal, R.L., Cochran, B., LaRocco, G., 2012. Bundling of ecosystem services to increase forest land value and enhance sustainable forest management. For. Policy Econ. 17, 69–76.

26. Defra and Natural England, 2008. Environmental Stewardship Review of Progress. Department for the Environment, Food and Rural Affairs, London, pp. 1–167.

27. Defra, 2013. Consultation on the Implementation of CAP Reform in England: Summary of Responses and Government Response. Department for the Environment, Food and Rural Affairs, London, pp. 1–102.

28. Defra, 2014. The New Common Agricultural Policy Schemes in England: October 2014 Update. Department for the Environment, Food and Rural Affairs, London, pp. 1–36.

29. Defrancesco, E., Gatto, P., Runge, F., Trestini, S., 2008. Factors affecting farmers participation in agri-environmental measures: a northern Italian perspective. J. Agric. Econ. 59, 114–131.

30. Dilman, D.A., Smyth, J.D., Christian, L.M., 2009. Internet, mail and mixed-mode surveys: the tailored design method, Third Edition. John Wiley & Sons, Inc.

31. Dobbs, T.L., Pretty, J., 2008. Case study of agri-environmental payments: the United Kingdom. Ecol. Econ 65, 765–775.

32. Ducos, G., Dupraz, P., Bonnieux, F., 2009. Agri-environment contract adoption under fixed and variable

compliance costs. J. Environ. Plan. Manage. 52, 669–687.

33. Dunn, J.C., Hartwell, V., Morris, A.J., 2013. Multi-taxa benefits of a targeted single-species agri-environment option. In: agri-environment schemes—what have they achieved and where do we go from here? Asp. Appl. Biol. 100, 137–144.
34. EEA, 2010. The European Environment State and Outlook 2010 Synthesis. Publications Office of the European Union, Copenhagen, pp. 1–228, Available from (http://www.eea.europa.eu/soer/synthesis/synthesis).
35. Emery, S.B., Franks, J.R., 2012. The potential for collaborative agri-environment schemes in England: can a well-designed collaborative approach address farmers' concerns with current schemes? J. Rural Stud. 28, 218–231.
36. European Commission, 2005. Agri-environment Measures: Overview on General Principles, Types of Measures, and Application. European Commission, Directorate General for Agriculture and Rural Development1–24, Available from (http://ec.europa.eu/agriculture/publi/reports/agrienv/rep en.pdf).
37. European Commission, 2014. Overview of CAP Reform 2014–2020. Agricultural Policy Perspectives Brief No. 5. (pp. 1–10), Available from (http://ec.europa.eu/agriculture/policy-perspectives/policy-briefs/05 en.pdf).
38. European Court of Auditors, 2011. Is Agri-environment Support Well Designed and Managed? Special Report No. 7. (pp. 1–82), Available from (http://www.eca.europa.eu/Lists/ECADocuments/SR11 07/SR11 07 EN.PDF).
39. FAO, 2003. World Agriculture: Towards 2015/2030. Earthscan Publications Ltd., pp. 1–444.
40. FAO, 2012. The State of Food and Agriculture 2012. Food and Agriculture Organisation, Rome, pp. 1–182, Available from (http://www.fao.org/docrep/ 017/i3028e/i3028e.pdf).
41. FAO, 2014. The State of Food and Agriculture: Innovation in Family Farming. Food and Agriculture Organisation, Rome, pp. 1–161, Available from (http://www. fao.org/3/a-i4040e.pdf).
42. FERA, 2012. Ecosystem services from environmental stewardship that benefit agricultural production. Natural England Commissioned Reports No. 102. Food and Environment Research Agency and Natural England. (pp. 1–157).
43. FERA, 2013. Monitoring the Impacts of Entry Level Stewardship. Natural England Commissioned Reports No. 133. Food and Environment Research Agency and Natural England. (pp. 1–280).
44. FERA, 2013b. Evidence Requirements to Support the Design of New Agri-environment Schemes—BD5011. Food and Environment Research Agency, pp. 1–201.
45. Falconer, K., 2000. Farm-level constraints on agri-environmental scheme participation: a transactional perspective. J. Rural Stud. 16, 379–394.
46. Farmar-Bowers, Q., Lane, R., 2009. Understanding farmers' strategic decision-making processes and the implication
47. Ferraro, P.J., 2008. Asymmetric information and contract design for payments for environmental services. Ecol. Econ. 65, 810–821.
48. Field, R., Morris, A., Grice, P., Cooke, A., 2010. Evaluating the English higher level stewardship scheme for farmland birds. In: agri-environment schemes—what have they achieved and where do we go from here? Asp. Appl. Biol. 100, 59–68.
49. Fraser, R., 2009. Land heterogeneity, agricultural income forgone and environmental benefit: an assessment of incentive compatibility problems in environmental stewardship schemes. J. Agric. Econ. 60, 190–201.
50. Fuentes-Montemayor, E., Goulson, D., Park, K.J., 2011. The effectiveness of agri-environment schemes for the conservation of farmland moths: assessing the importance of a landscape-scale management approach. J. Appl. Ecol. 48, 532–542.
51. Garrod, G., Ruto, E., Willis, K., Powe, N., 2012. Heterogeneity of preferences for the benefits of Environmental Stewardship: a latent-class approach. Ecol. Econ. 76, 104–111.
52. Garrod, G., 2009. Greening the CAP: how improved design and implementation of agri-environment schemes can enhance the delivery of environmental benefits. J. Environ. Plan. Manage. 52, 581–574.
53. Ghazoul, J., Butler, R.A., Mateo-Vega, J., Koh, L.P., 2010. REDD: a reckoning of environment and development implications. Trends Ecol. Evol. 25, 396–402.
54. Gibbs, H.K., Ruesch, A.S., Achard, F., Clayton, M.K., Holmgren, P., Ramankutty, N., Foley, J.A., 2010. Tropical forests were the primary sources of new agricultural land in the 1980 and 1990. Proc. Natl. Acad. Sci. U. S. A. 107, 16732–16737.
55. Gibbs, C., 2010. The campaign for the farmed environment –a joined up future for agri-environment schemes? In: agri-environment schemes—what have they achieved and where do we go from here? Asp. Appl. Biol. 100, 327–332.
56. Godfray, H.C.J., Garnett, T., 2014. Food security and sustainable intensification. Philos. Trans. R. Soc. B Biol. Sci. 369.
57. Goodwin, C.G., Hugh, N.M.C., Holland, J.M., Leather, S.R., 2013. The influence of environmental stewardship (ES) summer foraging habitat on the territory selection of yellowhammer, Emberiza citronella. In: agri-environment schemes—what have they achieved and where do we go from here? Asp. Appl. Biol. 100, 277–282.
58. Grau, R., Kuemmerle, T., Macchi, L., 2013. Beyond land sparing versus land sharing: environmental heterogeneity, globalization and the balance between agricultural production and nature conservation. Curr. Opin. Environ. Sustain. 5, 477–483.
59. Gruar, D.J., Morris, A.J., Dillion, I.A., 2013. Evaluating the efficacy of winter seed provision by different agri-environment scheme options. In: agri-environment schemes—what have they achieved and where do we go

from here? Asp. Appl. Biol. 100, 259–264.

60. Hart, K., 2010. Different approaches to agri-environment schemes in the EU-27. In: agri-environment schemes—what have they achieved and where do we go from here? Asp. Appl. Biol. 100, 3–7.

61. Hejnowicz, A.P., Raffaelli, D.G., Rudd, M.A., White, P.C.L., 2014. Evaluating the outcomes of payments for ecosystem services programmes using a capital asset framework. Ecosyst. Serv. 9, 83–97.

62. Hejnowicz, A.P., Kennedy, H., Rudd, M.A., Huxham, M.R., 2015. Harnessing the climate mitigation, conservation and poverty alleviation potential of seagrasses: prospects for developing blue carbon initiatives and payment for ecosystem service programmes. Front. Mar. Sci. 2, 32, http://dx.doi.org/10.3389/fmars.2015.00032.

63. Henle, K., Alard, D., Clitherow, J., Cobb, P., Firbank, L., Kull, T., 2008. Identifying and managing the conflicts between agriculture and biodiversity conservation in Europe—a review. Agric. Ecosyst. Environ. 124, 60–71.

64. Hinsley, S.A., Novakowski, M., Heard, M., Bellamy, P.E., Broughton, R.K., Hulmes, S., Peyton, J., Pywell, R.F., 2010. Performance and effectiveness of winter bird food patches established under environmental stewardship: results from the

65. Hillesden experiment. In: agri-environment schemes—what have they achieved and where do we go from here? Asp. Appl. Biol. 100, 151–158.

66. Hodge, I., Reader, M., 2010. The introduction of entry level stewardship in England: extension or dilution in agri-environment policy? Land Use Policy 27, 270–282.

67. Huber-Stearns, H.R., Goldstein, J.H., Duke, E.A., 2013. Intermediary roles and payments for ecosystem services: a typology and programs feasibility application in Panama. Ecosyst. Serv. 6, 104–116.

68. Ingram, J., Gaskell, P., Mills, J., Short, C., 2013. Incorporating agri-environment schemes into farm development pathways: a temporal analysis of farmer motivations. Land Use Policy 31, 267–279.

69. Ingram, J., 2008a. Are farmers in England equipped to meet the knowledge challenge of sustainable soil management? An analysis of farmer and advisor views. J. Environ. Manage. 86, 214–228.

70. Ingram, J., 2008b. Agronomist–farmer knowledge encounters: an analysis of knowledge exchange in the context of best management practices in England. Agric. Hum. Values 25, 405–418.

71. Jones, N.E., Boatman, N.D., Gathwaite, D., 2010. Implementation of environmental stewardship options –additionality and compliance. In: agri-environment schemes—what have they achieved and where do we go from here? Asp. Appl. Biol. 100, 271–278.

72. Jones, G.D., Boatman, N.B., Crowe, A., 2013. Does advice work? Assessing the effectiveness of ETIP. In: environmental management on farmland. Asp. Appl. Biol. 118, 171–177.

73. Juntti, M., Potter, C., 2002. Interpreting and reinterpreting communication, trust and knowledge in the implementation process. Sociol. Ruralis 42, 215–232.

74. Kleijn, D., Sutherland, W.J., 2003. How effective are European agri-environment schemes on conserving and promoting biodiversity? J. App. Ecol. 40, 947–970.

75. Kleijn, D., Rundlöf, M., Scheper, J., Smith, H.G., Tscharntke, T., 2011. Does conservation on farmland contribute to halting the biodiversity decline? Trends Ecol. Evol. 26, 474–481.

76. Lambin, E.F., Meyfroidt, P., 2011. Global land use change economic globalization, and the looming land scarcity. Proc. Natl. Acad. Sci. U. S. A. 108, 3465–3472.

77. Lastra-Bravo, X., Hubbard, C., Garrod, G., Tolón-Becerra, A., 2015. What drives famrers' participation in EU agri-environmental schemes? Results from a qualitative meta-analysis. Environ. Sci. Policy 54, 1–9.

78. Lefebrve, M., Espinosa, M., Gomez y Paloma, S., Paracchini, M.L., Piorr, A., Zasada, I., 2015. Agricultural landscapes as multi-scale public good and the role of the Common Agricultural Policy. J. Environ. Plan. Manage. 58, 2088–2112.

79. Legrand, T., Froger, G., Le Coq, J.F., 2013. Institutional performance of payments for environmental services: an analysis of the Costa Rican program. For. Policy Econ. 37, 115–123.

80. Lenzen, M., Moran, D., Kanemoto, K., Foran, B., Lobefaro, L., Geschke, A., 2012. International trade drives biodiversity threats in developing nations. Nature 486, 109–112.

81. Lin, H., Nakamura, M., 2012. Payments for watershed services: directing incentives for improve lake basin governance. Lakes Reserv.: Res. Manag. 17, 191–206.

82. Lobley, M., Saratsi, E., Winter, M., 2010. Training and advice for agri-environmental management. BOU Proceedings—Lowland Farmland Birds III, 1–5.

83. Martin-Ortega, J., Ojea, E., Roux, C., 2013. Payments for water ecosystem services in Latin America: a literature review and conceptual model. Ecosyst. Serv. 6, 122–132.

84. McCormack, C., 2012. Greening the common agricultural policy. BES Bull. 43 (3), 5–7.

85. McKenzie, A.J., Emery, S.B., Franks, J.R., Whittingham, M.J., 2013. Landscape-scale conservation: collaborative agri-environment schemes could benefit both biodiversity and ecosystem services, but will farmers be willing to participate? J. Appl. Ecol. 50, 1274–1280.

86. Mettepenningen, E., Verspecht, A., Van Huylenbroeck, G., 2009. Measuring private transaction costs of European agri-environmental schemes. J. Environ. Plan. Manage 52, 649–667.

87. Mills Busa, J.H., 2013. Deforestation beyond borders: addressing the disparity between production and consumption of global resources. Conserv. Lett. 6, 192–199.

88. Mills, J., 2012. Exploring the social benefits of agri-environment schemes in England. J. Rural Stud. 28, 612–621.

89. Mills, J., Gaskell, P., Short, C., Boatman, N., Winter, M., 2013. Farmer attitudes and evaluation of outcomes to on-farm environmental management, Report to

Department for Environment, Food and Rural Affairs. CCRI, Gloucester, (pp. 1–222).

90. Morris, C., Potter, C., 1995. ecruiting the new conservationists: farmers' adoption of agri-environmental schemes in the U. K. J. of Rural Studies 11, 51–63.

91. Morris, A.J., Bailey, C.M., Dillon, I.A., Gruar, D.J., Westbury, D.B., 2010. Improving floristically enhanced field margin for wildlife. In: agri-environment schemes—what have they achieved and where do we go from here? Asp. Appl. Biol. 100, 353–357.

92. Mountford, J.O., Cooke, A.I., Radley, G.P., 2013. Higher level stewardship (HLS)—developing a standard method for evaluation of agreements against objectives. In: agri-environment schemes – what have they achieved and where do we go from here? Asp. Appl. Biol. 100, 27–33.

93. Natural England, 2012a. Entry Level Stewardship Scheme. Environmental Stewardship Handbook, fourth edition (NE349). Natural England, pp. 1–180.

94. Natural England, 2012b. Organic Entry Level Stewardship Scheme. Environmental Stewardship Handbook, fourth edition (NE351). Natural England, pp. 1–194.

95. Natural England, 2012c. Higher Level Stewardship Scheme. Environmental Stewardship Handbook, fourth edition (NE350). Natural England, pp. 1–120.

96. Natural England, 2013. Scheme Development Bulletin Edition 3, August 2013. Natural England, pp. 1–8.

97. Natural England, 2014. Scheme Development Bulletin Edition 5, March 2014. Natural England., pp. 1–12.

98. OECD, 2008. Environmental Performance of Agriculture in OECD Countries Since 1990. OECD, Paris, France, pp. 1–9.

99. OECD/FAO, 2011. OECD-FAO Agricultural Outlook 2011–2020. OECD Publishing and FAO, pp. 1–196.

100. Pascual, U., Muradian, R., Rodríguez, L.C., Duraiappah, A., 2010. Exploring the links between equity and efficiency in payments for environmental services: a conceptual approach. Ecol. Econ. 69, 1237–1244.

101. Pe'er, G., Dicks, L.V., Visconti, P., Arlettaz, R., Báldi, A., Benton, T.G., et al., 2014. EU agricultural reform fails on biodiversity. Science 344, 1090–1092.

102. Peyton, J., Heard, M.S., Pywell, R.F., 2013. Testing the benefits of new agri-environment options for pollinating insects. In: agri-environment schemes—what have they achieved and where do we go from here? Asp. Appl. Biol. 100, 291–296.

103. Pike, T., 2008. Understanding behaviours in a farming context. Defra Agricultural Change and Environment Observatory Discussion Paper.

104. Pike, T., 2013. Farmer engagement: an essential policy tool for delivering environmental management on farmland. In: agri-environment schemes—what have they achieved and where do we go from here? Asp. Appl. Biol. 100, 187–191.

105. Poppy, G.M., Jepson, P.C., Pickett, J.A., Birkett, M.A., 2014. Achieving food and environmental security: new approaches to close the gap. Philos. Trans. R. Soc. B Biol. Sci. 369.

106. Porras, I., Aylward, B., Dengel, J., 2013. Sustainable markets monitoring payments for watershed service schemes in developing countries. London, Int. Inst. Environ. Dev., 1–36.

107. Pretty, J., Sutherland, W.J., Ashby, J., Auburn, J., Baulcombe, D., Bell, M., et al., 2010. The top 100 questions of importance to the future of global agriculture. Int. J. Agric. Sustain. 8, 219–236.

108. Quillérou, E., Fraser, R., Fraser, I., 2011. Farmer compensation and its consequences for environmental benefit provision in the higher level stewardship scheme. J. Agric. Econ. 62, 330–339.

109. Quinton, J.N., Govers, G., Van Oost, K., Bardgett, R.D., 2010. The impact of agricultural soil erosion on biogeochemical cycling. Nat. Geosci. 3, 311–314.

110. Radley, G.P., 2013. Lessons for the design of future agri-environment schemes. In: agri-environment schemes—what have they achieved and where do we go from here? Asp. Appl. Biol. 100, 1–8.

111. Ramwell, C.T., Boatman, N.D., 2010. Assessing the impact of environmental stewardship on the protection of water resources. In: agri-environment schemes—what have they achieved and where do we go from here? Asp. Appl. Biol. 100, 175–180.

112. Rollett, A., Haines-Young, R., Potschin, M., Kumar, P., 2008. Delivering environmental services through agri-environment programmes: a scoping study, Report to the Land Use Policy Group, Peterborough, England.

113. Ruto, E., Garrod, G., 2009. Investigating farmers' preferences for the design of agri-environment schemes: a choice experiment approach. J. Environ. Plan. Manage. 52, 631–647.

114. Schomers, S., Matzdorf, B., 2013. Payments for ecosystem services: a review and comparison of developing and industrialized countries. Ecosyst. Serv. 6, 16–30.

115. Schroeder, L., Isselstein, J., Chaplin, S., Peel, S., 2013. Agri-environment schemes: farmers' acceptance and perception of potential payment by results in grassland—a case study in England. Land Use Policy 32, 134–144.

116. Siebert, R., Toogood, M., Knierim, A., 2006. Factors affecting European farmers' participation in biodiversity policies. Sociol. Ruralis 46, 318–340.

117. Siriwardena, G.M., 2010. The importance of spatial and temporal scale for agri-environment scheme delivery. Ibis 152, 515–529.

118. Smits, M.-J., Driessen, P., Glasbergen, P., 2008. Governing agri-environmental schemes: lessons to be learned from the new institutional-economics approach. Environ. Plan. C Gov. Policy 26, 627–643.

119. Still, K.S., Byfield, A.J., 2010. Is environmental stewardship working for rare and threatened plants? In: agri-environment schemes—what have they achieved and where do we go from here? Asp. Appl. Biol. 100, 279–286.

120. Sutherland, L.-A., Mills, J., Ingram, J., Burton, R.J.F., Dwyer, J., Blackstock, K., 2013. Considering the source:

commercialisation and trust in agri-environmental information and advisory services in England. J. Environ. Manage. 118, 96–105.

121. Thuy, P.T., Campbell, R.M., Garnett, S., Aslin, H., Hoang, M.H., 2010. Importance and impacts of intermediary boundary organizations in facilitating payment for environmental services in Vietnam. Environ. Conserv. 37, 64–72.

122. Tscharntke, T., Klein, A.M., Kruess, A., Steffan-Dewenter, I., Thies, C., 2005. Landscape perspectives on agricultural intensification and biodiversity ecosystem service management. Ecol. Lett. 8, 857–874.

123. Tscharntke, T., Clough, Y., Wanger, T.C., Jackson, L., Motzke, I., Perfecto, I., Vandermeer, J., Whitbread, A., 2012. Global food security: biodiversity conservation and the future of agricultural intensification. Biol. Conserv. 151, 53–59.

124. Tucker, R., 2010. Preserving the balance. In: agri-environment schemes—what have they achieved and where do we go from here? Asp. Appl. Biol. 100, 1–2.

125. Udagawa, C., Hodge, I., Reader, M., 2014. Farm level costs of agri-environment measures: the impact of entry level stewardship on cereal farm incomes. J. Agric. Econ. 65, 212–233.

126. Van Herzele, A., Gobin, A., Van Gossum, P., Acosta, L., Waas, T., Dendoncker, N., Henry de Frahan, B., 2013. Effort for money? Farmers' rationale for participation in agri-environment measures with different implementation complexity. J. Environ. Manage. 131, 110–120.

127. Vesterager, J.P., Lindegaard, K., 2012. The role of farm advisors in multifunctional landscapes: a comparative study of three Danish areas, 1995 and 2008. Landsc. Res. 37, 673–702.

128. Wünscher, T., Engel, S., Wunder, S., 2008. Spatial targeting of payments for environmental services: a tool for boosting conservation benefits. Ecol. Econ. 65, 822–833.

129. Whittingham, M.J., 2007. Will agri-environment schemes deliver substantial biodiversity gain, and if not why not? J. Appl. Ecol. 44, 1–5.

130. Whittingham, M.J., 2011. The future of agri-environment schemes: biodiversity gains and ecosystem service delivery? J. Appl. Ecol. 48, 509–513.

131. Wilson, G.A., Hart, K., 2000. Financial imperative or conservation concern? EU farmers' motivation for participation in voluntary agri-environmental schemes. Environ. Plann. A 32, 2161–2185.

132. Wilson, G.A., Hart, K., 2001. Farmer participation in agri-environmental schemes: towards conservation-oriented thinking? Sociologia Ruralis 41, 254–274.

133. Winter, M., 1997. New policies and new skills: agricultural change and technology transfer. Sociologia Ruralis 37, 363–381.

134. Wunder, S., Engel, S., Pagiola, S., 2008. Taking stock: a comparative analysis of payments for environmental services programs in developed and developing countries. Ecol. Econ. 65, 834–852.

135. WWAP (United Nations World Water Assessment Programme), 2014. The united Nations World Water Development Report 2014: Water and Energy. UNESCO, Paris1–230.

136. Zanten, B.T.van., Verburg, P.H., Espinosa, M., Gomez-y-Paloma, S., Galimberti, G., Kantelhardt, J., et al., 2014. European agricultural landscapes, common agricultural policy and ecosystem services: a review. Agron. Sustain. Dev. 34, 309–325.

CHAPTER 16

CAPABILITIES OF BALTIC SEA MODELS TO ASSESS ENVIRONMENTAL STATUS FOR MARINE BIODIVERSITY

Letizia Tedesco, Chiara Piroddi, Maria Kämäri, Christopher Lynam

Keywords: Marine Strategy Framework Directive, Good Environmental Status, Ecosystem models, Baltic Sea, Biodiversity, Marine policy

INTRODUCTION

The Directive 2008/56/EC, known as the Marine Strategy Framework Directive (MSFD), establishes a framework for community action in the field of marine environmental policy[1]. It was formally adopted by the European Union in July 2008. The MSFD outlines a legislative framework for an ecosystem-based approach to the management of human activities that supports the sustainable use of marine goods and services. The overarching goal of the Directive is to achieve Good Environmental Status (GEnS)[1] by 2020 across the European marine environment. The Directive defines GEnS as 'the environmental status of marine waters where these provide ecologically diverse and dynamic oceans and seas which are intrinsically clean, healthy and productive, and the use of the marine environment is at a level that is sustainable, thus safeguarding the potential for use and activities by current and future generations'. With the aim to support its implementation, the MSFD sets out in Annex I 11 qualitative *descriptors* [2] (D1-D11,Table 1), either state or pressure *descriptors*. Later, a Commission decision defines also 29 related *criteria* and 56 related *indicators* [4] that are used in the assessment of the status of the seas. An example of *criteria* and *indicators* defined for biological diversity (D1) is shown in Table 2. With the aim to facilitate the implemention of the MSFD, Borja et al. [5] proposed an operational definition of GEnS, i.e. 'GEnS is achieved when physicochemical and hydrographical conditions are maintained at a level that main structuring components of the ecosystem are present, allowing the functionality of the system to provide resistance and resilience against deleterious effects of human pressures/activities/impacts, maintaining and delivering the ecosystem services that provide societal benefits in a sustainable way'. Despite the fact that several attempts have been made to assess the environmental status of marine waters in an integrative manner e.g. [6], significant gaps still remain for understanding marine ecosystem structures and functions and their response to human pressures e.g. [5]. There are several challenges related to the assessment of GEnS within the MSFD. The assessment of an ecosystem's health requires the setting of adequate reference conditions and/or environmental targets to which data should be compared [7]. The use of robust and appropriate indicators that can assess whether an ecosystem and its services are well maintained and sustainably used is one of the essential steps for the practical implementation of conservation and management policies such as the MSFD [8]. On the other hand, an accurate evaluation requires integrating knowledge across different ecosystem components and linking physical, chemical and biological aspects [9]. To this end, ecological models are a powerful tool for predicting and understanding the consequences of anthropogenic and climate-driven changes in the natural environment e.g. [10].

Citation: Letizia Tedesco, Chiara Piroddi, Maria Kämäri, Christopher Lynam, Capabilities of Baltic Sea models to assess environmental status for marine biodiversity, Marine Policy, Volume 70, August 2016, Pages 1-12, ISSN 0308-597X, http://dx.doi.org/10.1016/j.marpol.2016.04.021.

Table 1: The 11 *descriptors* identified by the Marine Strategy Framework Directive and related number of *criteria*and *indicators*.

#	Type	*Descriptor*	# of *criteria*	# of *indicators*
D1	state	Biological diversity	7	14
D2	pressure	Non-indigenous species	2	3
D3	state	Exploited fish and shellfish	3	8
D4	state	Food webs	3	3
D5	pressure	Human-induced eutrophication	3	8
D6	state	Sea-floor integrity	2	6
D7	pressure	Hydrographical conditions	2	3
D8	pressure	Contaminants	2	3
D9	pressure	Contaminants in fish and seafood	1	2
D10	pressure	Litter	2	4
D11	pressure	Energy and noise	2	2

Table 2: D1 Biological diversity *descriptor* and related *criteria* and *indicators*.

Criteria	*Indicator*
1.1 Species Distribution	1.1.1 Distributional range
	1.1.2 Distributional pattern
	1.1.3 Area covered by the species
1.2 Population size	1.2.1 Population abundance and/or biomass
1.3 Population condition	1.3.1 Population demographic characteristics
	1.3.2 Population genetic structure
1.4 Habitat distribution	1.4.1 Distributional range
	1.4.2 Distributional pattern
1.5 Habitat extent	1.5.1 Habitat area
	1.5.2 Habitat extent
1.6 Habitat condition	1.6.1 Condition of the typical species and communities
	1.6.2 Relative abundance and/or biomass
	1.6.3 Physical, hydrological and chemical conditions
1.7 Ecosystem structure	1.7.1 Composition and relative proportion of ecosystem components

Within this framework, Piroddi et al. [11] assess the most commonly used capabilities of models in five regional European seas (North Sea, Baltic Sea, Mediterranean Sea, Black Sea and Bay of Biscay) to provide information about indicators outlined in the MSFD, particularly on biodiversity-related *descriptors*. They built a catalogue of European models and their derived indicators to assess which models are able to demonstrate the linkages between indicators and ecosystem structure and function, and the impact of pressures on ecosystem state through indicators. A brief summary of the models' catalogue is given in Section 2.1. Thus, Piroddi et al. [11] provide an extensive overview at pan-European scale. As the Baltic Sea is facing several health issues including an enlargement of the eutrophication problem [12] despite the adopted nutrient reduction measures [13], it was found relevant to investigate the Baltic Sea case in more details. To date there has been no evaluation of the capabilities of the ecosystem models of the Baltic Sea to provide information as outlined by the MSFD. This work aims to fill in this knowledge gap by providing a review of the capabilities of nine Baltic Sea ecosystem models to assess the environmental status of marine waters with particular focus on marine biodiversity. Yet, it is acknowledged that this study does not aim to serve as review of all the existing ecosystem models of the Baltic Sea, but instead highlights a process of exploring modelling potential to support this specific European policy. As in Piroddi et al. [11], models were analysed for potentially addressing the MSFD biodiversity-related *descriptors*: biological diversity (D1), non-indigenous species (D2), food webs (D4) and sea-floor integrity (D6). A short description of the characteristics of the Baltic Sea, the main features of the models and the criteria used for deriving indicators and assessing models' capabilities are given in 2.2, 2.3 and 2.4, respectively. The Baltic Sea Model-Derived Indicators (MDI) and their capabilities to inform on biodiversity-related *descriptors* and *pressures* are presented in Section 3.1, whileSection 3.2 gives a more detailed analysis of the capabilities of each of the Baltic Sea models to address, potentially address or not address at all the biodiversity-related*indicators*. Finally, Section 4 highlights the current gaps between the MSFD and the models and suggests the use of different methods and tools as well as the development of new indicators and models to better link ecosystem models to the political framework of the MSFD.

MATERIAL AND METHODS

The Catalogue of European Ecological Models in Brief

This section summarises the methodology Used and the results gained from the analysis of the modelling capabilities of five European regional seas (North Sea, Baltic Sea, Mediterranean Sea, Black Sea and Bay of Biscay) to assess environmental status for marine biodiversity and presented in Piroddi et al. [11].

With the aim of developing new indicators and modelling tools to assess environmental status for marine biodiversity, it is necessary to initially evaluate the capabilities of the state-of-the-art models to do so. The work flow requires a series of sequential steps (Fig. 1). After the identification of the relevant *descriptors* in relation to marine biodiversity (biological diversity (D1) , non-indigenous species (D2) , food webs (D4) and sea-floor integrity (D6) with some relevance for commercial fish (D3) and human-induced eutrophication (D5)), the catalogue of European models that can specifically address these *descriptors* is produced (see the Supplementary material for a detail description of the structure of the catalogue). Every model output is then linked to relevant *descriptors*and related *criteria* and *indicators*, and MDI are then identified. Every MDI is then used for the assessment of its capability to relate to both *descriptors* (Table 1) and *pressures* (Table 3).

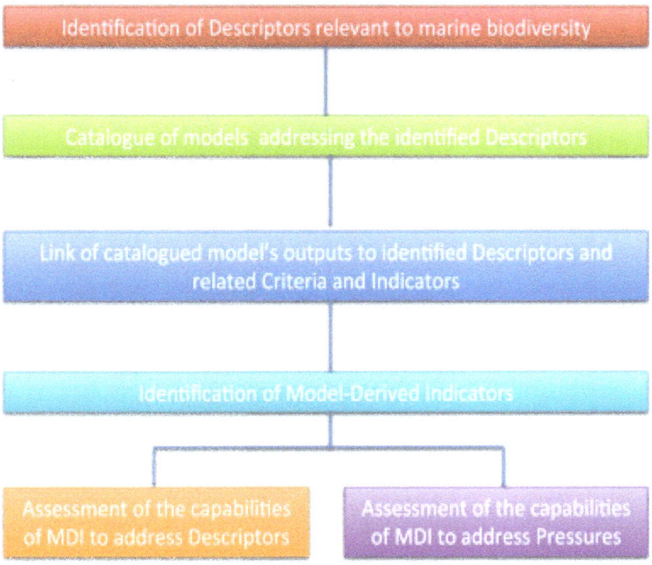

Figure 1: The work flow required for the assessment of the capabilities of the models to address environmental status for marine biodiversity.

Table 3: *Pressures* identified by the Marine Strategy Framework Directive.

	Pressure
1	Physical disturbance (sum)
2	Underwater noise
3	Marine litter
4	Interference with hydrological processes
5	Contamination (sum)
6	Contamination by radionucleids
7	Nutrient and organic matter enrichment
8	Introduction of microbial pathogens
9	Non-indigenous species
10	Extraction of living resources (sum)
11	Marine acidification

At European scale 44 ecological models were analysed for their capabilities to inform on the biodiversity-related *descriptors* [11, see Table 1]. The models are either operational i.e. tested and validated (24), or under development i.e. not yet validated (18), or conceptual (2). The type of models were grouped into 7 categories:

- biogeochemical: represents the dynamics and cycling of biogeochemical compounds of the lower trophic levels of the food web (1 model)
- meta-community: describe specific mechanistic processes to predict empirical community patterns, i.e. species composition and abundance (1)
- individual-based: represents individual organisms in a population having specific state characteristics, such as age, size, developmental stage, and physiological conditions (3)
- bio-optical: analyse and predict the optical properties of biological materials (6)
- food web: represent networks formed by feeding interactions between species to understand trophic patterns, population dynamics and implication for system stability and substance/energy flows (9)
- species distribution/habitat suitability: combine observations of species occurrence or abundance with environmental variables to predict distribution across selected habitats/spatial predictions on the suitability of locations for a target species, community or biodiversity (9)
- coupled hydrodynamic-biogeochemical: capture global scale patterns in physical-chemical components affecting lower trophic levels of the food web (15). When coupled to food web models they are known as end-to-end models.

From the 44 models, 201 MDI were identified, of which 129 are operational, 66 under development and 6 conceptual ones i.e. an indicator idea without practical measure/metric yet available. MDI were grouped into 7 major categories, based on what the indicators inform on:

- Biomass (115 MDI)
- Diversity (26)

- Physical, hydrological and chemical characteristics (24)
- Primary/secondary production (18)
- Spatial distribution (12)
- Ecological Network Analysis (4)
- Species life-history (2).

Among the biodiversity-related components of the MSFD such as microbes, phytoplankton up to fish, birds and marine mammals, the less frequently used food web models (9) were the ones that assessed most of the biodiversity-related components, while the more frequently applied coupled hydrodynamic-biogeochemical models (15) were the least inclusive in terms of number of components evaluated.

The majority of indicators (136) were derived from the Ecopath with Ecosim (Ewe, http://www.ecopath.org) software applications, which was also the only model applied to every regional sea. Of the 11 *descriptors* of the MSFD (Table 1), the evaluated models were able to address 8 of them but not contaminants (D8), litter (D10) and energy and noise (D11). Of the biodiversity-related GEnS *descriptors* models especially addressed biological diversity (D1) and food webs (D4), while non-indigenous species (D2) and sea-floor integrity (D6) were poorly addressed. In total 27 MDI were identified under these 4*descriptors*, of which the modelling capabilities to address them were highly heterogeneous, from 1 single model addressing only once a single *indicator* (e.g. 'Parameters describing the characteristics of the size spectrum of the benthic community' within D6) to all of the 44 models addressing 108 times a single *indicator*(e.g. 'Interactions between structural components' within D1). Considering their spatial coverage, the majority of MDI (53) related to the Mediterranean Sea and Bay of Biscay, followed by the North-East Atlantic Ocean (46), the Black Sea (29), the Baltic Sea (22) and other non-EU seas (11).

Among the predominant habitat types of the MSFD (water column, seabed and ice-associated habitats) the water column - and in particular the marine shelf - was the most comprehensively evaluated habitat by the models.

41 out of 44 models are in use to address collectively all the pressure impacts outlined in the MSFD (Annex III). The most addressed *pressure* is 'Inputs of nutrients and organic material' (44 models), followed by 'Marine Acidification' and 'Interference with the hydrological regime' (25). The least addressed *pressures* (1) were 'Introduction of microbial pathogens' and 'Contamination from radionucleids'.

The gap analysis of the models importantly showed that among the 4 biodiversity-related *descriptors* (D1, D2, D4 and D6), some of the related *indicators* were not addressed by any of the MDI. Also, the geographical coverage of the models was very heterogeneous, with several regional seas needing improvements in terms of number of models, model types, and capabilities to address MDI, as for example the Baltic Sea, which is the focus of the rest of this review.

Baltic Sea, Biodiversity and Human Pressures

The Baltic Sea (Fig. 2) features brackish waters and, consequently, both freshwater and marine species cohabit and are distributed according to the water salinity patterns that characterizes different areas. The classical knowledge of the Baltic Sea inhabited by relatively few species, i.e. having low inter-specific biodiversity, decreasing from more saline to more fresh waters, have been recently challenged. Ojaveer et al. [14] showed that not only does the Baltic Sea hosts some 6000 species, but furthermore that phytoplankton and zooplankton biodiversity is very high with more than 4000 taxa. Only the diversity of bottom dwelling zoobenthos and macroalgae is still considered comparably low.

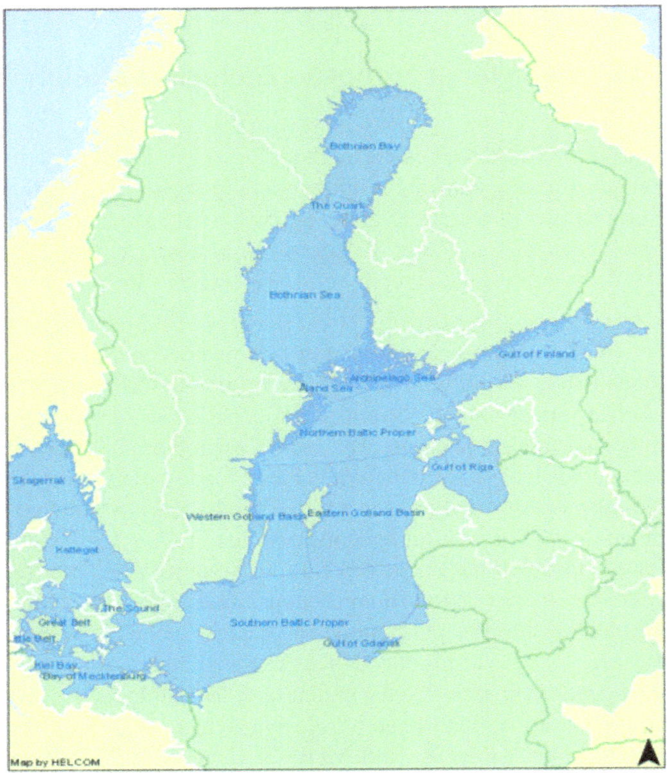

Figure 2: The Baltic Sea and its partition in subasins (Credits: HELCOM).

The main threats to the biodiversity of the Baltic Sea includes: fisheries, maritime activities (including shipping), physical damage and disturbance, recreational activities, eutrophication, hazardous substances, alien species, noise pollution, hunting, and climate change [15]. Overfishing, eutrophication, and drastic decline of marine mammals have been the most prominent changes in the Baltic Sea during the twentieth century [16].

Since the 1800 s, the Baltic Sea has changed from an oligotrophic clear-water sea to a eutrophic marine environment e.g. [17]. High nutrient concentrations stimulate growth of algae, which leads to imbalanced functioning of the system and extreme events that can cause: excess of filamentous algae and phytoplankton blooms, altered communities of fauna and flora, production of excess organic matter, increase in oxygen consumption, oxygen depletion and mortality of benthic organisms e.g. [18] and [19].

The spatial geographical, oceanographic, and climatological characteristics of the sea render the Baltic ecosystem highly susceptible to the environmental impacts of human activities at sea and in its catchment area [20]. The environmental status of the Baltic Sea is generally impaired [20]. The overall ecosystem health is degraded when the eutrophication and biodiversity status as well as the status of hazardous substances are evaluated together. In 2007 the Baltic Sea Action Plan (BSAP) [21], which aims to reach a healthy Baltic Sea by 2021, has been signed by the nine countries that surround it. The novelty of the BSAP is that the status of the ecosystem as wanted in the future is at the center, defining the management decisions that directly link abatement measures to the status of the Baltic Sea. Furthermore, a multi-model approach to characterize the nutrient loads, the retentions that occur between these sources and the sea, and the effects of various management strategies to reduce loads has been specifically developed to support the BSAP, providing more robust insights into patterns of loading and response when models agree, and priorities for additional research when models disagree [22].

In a recent assessment of the eutrophication status of the Baltic Sea it was found that the spatial extent of the eutrophication problem is expanding and now considered unacceptable for all the 17 evaluated open sea sub-basins [12], while earlier Bothnian Bay was classified unaffected by eutrophication [20]. Nutrient levels, rates of primary production (in terms of chlorophyll-a), Secchi depths and oxygen levels were used as indicators of direct and indirect effects of eutrophication of the sea. Even though a straight comparison with the previous eutrophication assessment [20] and [23] cannot be done as the assessment methodology and targets were different, it is a remarkable finding considering that the current management strategies for the Baltic Sea [13] do not require any nutrient reduction measure to the Bothnian Bay.

The expected combination of a rise in temperature and a decrease in salinity will result in a decrease in abundance and habitat occupied by marine species of fishes in the Baltic Sea. In contrast, habitats of freshwater species of fishes, particularly those whose growth or survival are enhanced by warmer temperatures, will increase. These changes in the fish community will affect fisheries and may require modifications to existing fisheries management policies. Long-term predictions of fish stock development are highly uncertain due to the uncertainties arises because salinity and temperature will change simultaneously and because these changes will have counteracting effects on biological phenomena. Furthermore, changes in salinity balance will affect other key hydrographic characteristics, such as inflows into the Baltic and thus oxygen concentrations in deeper water areas [16].

Other relevant human-induced pressures on the Baltic Sea include those related to shipping and contaminants e.g. [24]. Some of the shipping routes in the Baltic Sea are the busiest in the world. Potential negative environmental impacts related to the shipping are: the introduction of non-indigenous, potentially invasive species; emissions of nitrates, sulphates and carbon dioxide and a risk of accidental oil or chemical pollution. Contamination of the Baltic Sea certain fish species with dibenzo-p-dioxins and dibenzofurans has lead to recommendations to restrict the Baltic herring consumption e.g. [25], [26] and [27]. Model simulations indicate that contamination in the water column will continue to decrease especially if supported by further reduced emissions measures [28]. Tributyltin in paint used on ships and maritime installations contaminated the seabed, and although banned since 2008, concentrations remain high in certain areas, especially near dockyards e.g. [29] and [30].

Baltic Sea Models in Brief

Nine Baltic Sea ecosystem models were analysed for their potential of addressing one or more of the biodiversity-related *descriptors* (Table 4). Of those, seven are operational and two under development. The operational models include five coupled hydrodynamic-biogeochemical models, one food web model and one end-to-end model. Each model covers the entire Baltic Sea domain, except BaltProWeb that simulates the Baltic Proper basin only (see Fig. 2 for location). The still under development models include one coupled hydrodynamic-biogeochemical model and one end-to-end model.

Table 4: The Baltic Sea models and related number of Model-Derived Indicators (MDI) and living components in different habitats.

Name	Status	Type	# of MDI	Pelagic LTL	Pelagic HTL	Sea Ice	Benthic
BaltProWeb	oper.	food web	10	8	4	–	2
BALTSEM	oper.	hydr-bgc	7	4	–	–	–
ECOSMO	oper.	hydr-bgc	7	5	–	–	–
ECOSMO-SMS	un. dev.	end-to-end	+3	–	3	–	–
ERGOM-MOM	oper.	hydr-bgc	7	4	–	–	–
ERGOM-MOM-fish	oper.	end-to-end	+3	–	3	–	–
NEMO-BFM	un. dev.	hydr-bgc	14	12	–	4	7
RCO-SCOBI	oper.	hydr-bgc	7	4	–	–	–
SPBEM	oper.	hydr-bgc	7	4	–	–	–

The type and number of living components and marine habitats within the models are shown in Table 4 and Fig. 3, while the main features of the models are described extensively in the Supplemental material. Generally, the five operational coupled hydrodynamic-biogeochemical models are similar in that they describe the main functional groups of the pelagic ecological domain of the Baltic Sea. Main differences among them refer to their level of complexity (i.e. number of functional groups, see Fig. 3), vertical resolution, and sediment parameterisations. They all describe at least the dynamics of nitrogen, oxygen, phosphorous, phytoplankton, zooplankton and dead organic matter. Only three models are also Higher Trophic Level (HTL, i.e. include larger size classes than zooplankton up to top predators), of which one is under development. Finally, the under development coupled hydrodynamic-biogeochemical model NEMO-BFM adds to this framework a considerable level of complexity in terms of number of state variables (i.e. non-Redfield model), number of functional groups (higher diversity for cyanobacteria, spring blooms composition and zooplankton) and number of habitats (pelagic, benthic and sea ice).

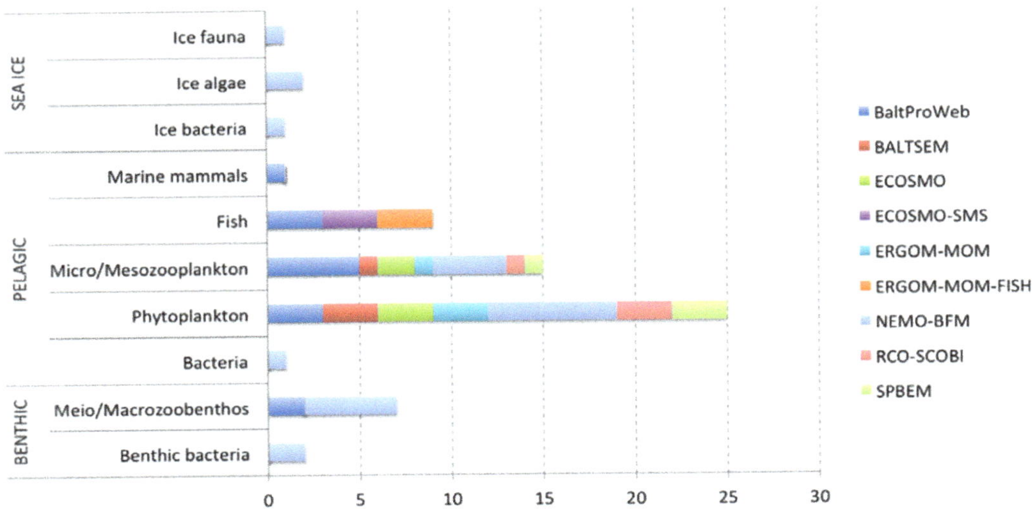

Figure 3: Type (y-axis) and number (x-axis) of living components of the Baltic Sea models in different marine habitats.

Criteria for the Identification of MDIS and Assessment of Models' Capabilities

The testing of an indicator-based approach to assess Baltic marine biodiversity has been conducted on a set of 22 national case studies and an overall assessment of the Baltic Proper sub-basin [15]. The main criteria used for assessing the status of marine biodiversity within a given area were:

- A sufficient number of indicators that describe a sufficiently broad array of biodiversity components for a given site or area;
- The definition of a desirable state for the selected indicators, including both a quantitative reference status (or target), as well as an acceptable deviation from this reference (the highest possible value being 50% from the reference value);
- The assessment of the overall status of the biodiversity by using the indicators and their reference levels and acceptable deviations as components in an overall assessment matrix termed BEAT (the HELCOM Biodiversity Assessment Tool).

Indicators were grouped under 4 categories according to the biodiversity segment of the HELCOM BSAP, and an additional category for supportive features:

- Category I, Marine Landscapes: Area-based habitat indicators (all types) and large geographic features
- Category II, Communities: Community indicators on structure and function of phytoplankton, zooplankton, zoobenthos, macrophytes, fish community, bird community, endangered habitats and biotopes
- Category III, Species: Single-species indicators of high profile species mainly fish, birds and mammals as well as indicators on endangered and alien species
- Category IV, Supporting features: Indicators of environmental parameters including e.g., water clarity, water temperature, oxygen concentrations, nutrients.

Similar criteria are used here while identifying the MDI, taking in consideration all possible diversity biological components (from LTL to HTL) and habitats (from benthic to pelagic to ice-associated habitats) and the possibility to have/set targets and/or reference values. Each MDI is then liked to the related MSFD criteria among the biodiverisity-related *descriptors*. Finally, depending on their capabilities with respect to each *indicator*, the models are classified as:

- currently able to address ✓
- potentially able to address with some changes to the model ✓
- not able to address and requiring different/new modelling tools/techniques ✗.

The criteria of classification is based on the minimum requirements, i.e. if a MDI already exists and addresses a certain *indicator*, the related model is classify as ✓, even though modifications for increasing the capabilities of the model to address biodiversity are anyhow encouraged. Similarly, if a MDI already exists and potentially can address a certain *indicator*, the related model is classified as ✓, even though such *indicator* might be addressed only to a certain extent and the usage/development of new features/model might be advisable. Finally, if a certain type of model is not able to address a certain*indicator* even after modifications, the model is classified as ✗ and some recommendations are provided, indicating suitable types of models with reference to that specific *indicator*.

RESULTS AND DISCUSSION

Model-Derived Indicators vs. Descriptors and Pressures

Of the 201 indicators derived from the full set of models, only 22 MDI are addressed by the Baltic Sea ecosystem models relevant to one or more *descriptors* (Table 5 and Fig. 4). Of the analysed biodiversity-related *descriptors* (biodiversity (D1), non-indigenous species (D2), food webs (D4) and sea-floor integrity (D6)) the Baltic models do not currently address *non-indigenous species* at all. Of the remaining *descriptors*, the Baltic models are able to also address commercial fish (D3), human-induced eutrophication (D5) and hydrological alteration (D7), but not contaminants (D8), contaminants in food (D9), litter (D10) and energy and noise (D11).

Table 5: Model-Derived Indicators (MDI) from MSFD descriptors and number of Baltic Sea models addressing them.

MDI	D1 Biological diversity	D4 Food webs	D6 Sea-floor integrity	D3 Commercial fish	D5 Human-induced eutrophication	D7 Hydrological alteration
Temperature	6	–	–	–	–	6
Salinity	6	–	–	–	–	6
pH	2	–	–	–	–	2
Oxygen penetration depth	7	–	7	–	7	7
Denitrification layer depth	1	–	1	–	1	1
Bacteria biomass	1	1	–	–	–	–
Phytoplankton biomass	7	7	–	–	7	–
Zooplankton biomass	7	7	–	–	–	–
Chlorophyll-a concentration	–	–	–	–	1	–
Primary production	–	7	–	–	7	–
Secondary production	–	9	–	–	–	–
Bacteria distributional range	1	–	–	–	–	–
Phytoplankton distributional range	6	–	–	–	6	–
Zooplankton distributional range	6	–	–	–	–	–
Fish: Herring biomass	3	3	–	3	–	–
Fish: Cod Biomass	3	3	–	3	–	–
Fish: Sprat Biomass	3	3	–	3	–	–
Marine mammal biomass (seals)	1	1	–	–	–	–
Zoobenthos biomass	2	2	2	–	–	–
Sea-ice algae biomass	1	1	–	–	–	–
Sea-ice bacteria biomass	1	1	–	–	–	–
Sea-ice fauna biomass	1	1	–	–	–	–
Total	69	46	10	9	29	22

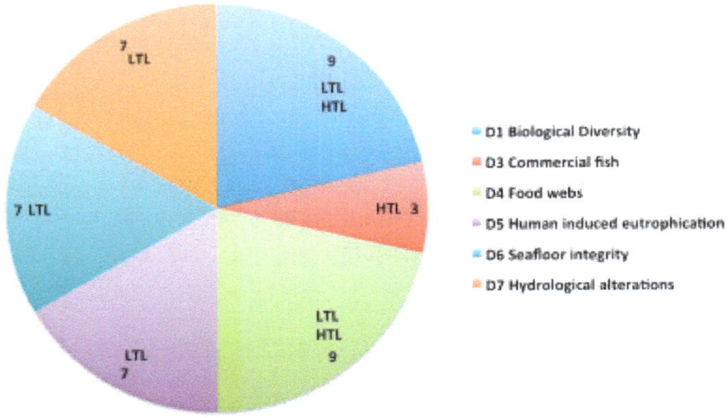

Figure 4: Baltic Sea models and addressed MSFD *descriptors*.

Biodiversity MDIs (19) are the most frequently addressed by the set of Baltic models (69 times), although mainly indirectly, linking MDIs mostly to species distributional ranges (*indicator* 1.1.1), to population biomasses (1.2.1), and to physical, hydrological and chemical conditions of the habitats (1.6.3). The second most addressed (46 times)*indicator* is *Food webs* by 13 MDIs (Table 5). Even though not the main targets of this study, *human-induced eutrophication* and *hydrological alteration* are both discretely addressed (22 times the former and 29 times the latter) by 5 and 6 MDIs, respectively. Of the remaining addressed *descriptors*, the least addressed (9 and 10 times, respectively) are *commercial fish* and *sea-floor integrity*, both by 3 MDI (Table 5).

The analysed Baltic models are capable of addressing biodiversity and food webs (D1 and D4) to some extent, while commercial fish (D3) is addressed only by the HTL models and human-induced eutrophication (D5), sea-floor integrity (D6) and hydrological alteration (D7) only by the Lower Trophic Level (LTL) models (Fig. 4).

All of the Baltic Sea LTL models provide indicators such as phytoplankton/zooplankton biomass and distributional ranges, relevant to biodiversity and food webs (D1 and D4,Table 5). While all of the LTL models provide information on the oxygen level in water and sediments, relevant to *biodiversity* and sea-floor integrity (D6), zoobenthos biomass is provided only by BaltProWeb and NEMO-BFM and is relevant also for D6, besides D1 and D4. Only BaltProWeb, ERGOM-MOM-Fish and the under development ECOSMO-SMS provide information on fish biomass (herring, sprat and cod), relevant to *descriptors*D1, D3 (commercial fish) and D4. Finally, BaltProWeb provides information on marine mammal biomass (i.e. grey seals), relevant also to D1, D3 and D4. The physical and chemical conditions, such as temperature, salinity, pH and oxygen penetration depth are important indicators relevant also to hydrological alteration (D7) and are addressed by all the LTL models, while denitrification layer depth only by the under development NEMO-BFM.

The analysed Baltic Sea models are totally able to address to some extent 8 of the 11*pressures* defined by the MSFD (Fig. 5 and Table 6), but not 'contaminants in radionuclides', 'introduction of microbial pathogens' and, most importantly, 'non-indigenous species'. Of the addressed *pressures*, 'interference with the hydrological regime', such as thermal regime changes, is addressed by each LTL model and by each MDI up to 66 times (Table 6). Also 'nutrients and organic matter enrichment' and 'marine acidification' are highly addressed by each of the by LTL models and by several MDI (13 and 11, respectively) several times (42 and 34, respectively, Table 6). The least addressed *pressures* are 'underwater noise' and 'marine litter', both in relation to marine mammals and addressed by the BaltProWeb model only, followed by "contamination" and "extraction of living resources" addressed only 9 and 10 times, respectively.

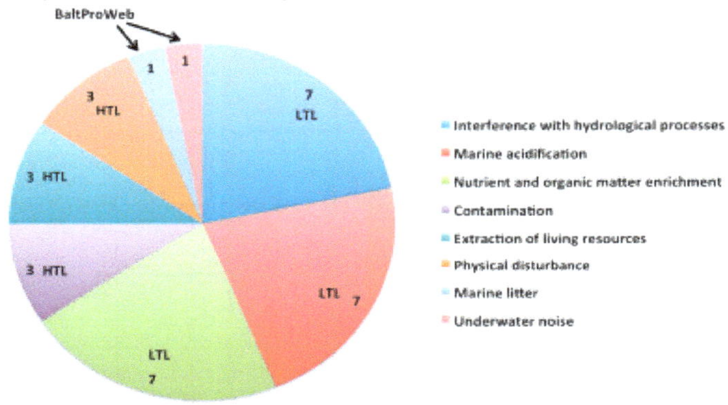

Figure 5: Baltic Sea models and addressed MSFD *pressures*.

Table 6: List of Model Derived Indicators (MDI) and corresponding MSFD *pressures* and number of Baltic Sea models addressing them.

MDI	Physical disturbance	Underwater noise	Marine litter	Interference with hydrological processes	Contamination	Nutrient and organic matter enrichment	Extraction of living resources	Marine acidification
Temperature	–	–	–	6	–	–	–	–
Salinity	–	–	–	6	–	–	–	–
pH	–	–	–	2	–	–	–	2
Oxygen penetration depth	7	–	–	7	–	7	–	–
Denitrification layer depth	1	–	–	1	–	1	–	–

Bacteria biomass	–	–	–	1	–	1	–	1
Phytoplankton biomass	–	–	–	7	–	7	–	7
Zooplankton biomass	–	–	–	7	–	7	–	7
Chlorophyll-a concentration	–	–	–	1	–	1	–	1
Primary production	–	–	–	7	–	7	–	7
Secondary production	–	–	–	9	–	9	–	9
Bacteria distributional range	–	–	–	1	–	1	–	1
Phytoplankton distributional range	–	–	–	6	–	6	–	6
Zooplankton distributional range	–	–	–	6	–	6	–	6
Fish: Herring biomass	3	–	–	3	3	–	3	–
Fish: Cod Biomass	3	–	–	3	3	–	3	–
Fish: Sprat Biomass	3	–	–	3	3	–	3	–
Marine mammal biomass (grey seals)	–	1	1	1	–	–	1	–
Zoobenthos biomass	2	–	–	2	–	2	–	–
Sea-ice algae biomass	1	–	–	1	–	1	–	1
Sea-ice bacteria biomass	1	–	–	1	–	1	–	1
Sea-ice fauna biomass	1	–	–	1	–	1	–	1
Total	22	1	1	82	9	58	10	50

Capability to Address the Biodiversity-Related MSFD Indicators

In this section the capabilities of the evaluated Baltic Sea models to address the biodiversity-related *indicators*, i.e. with reference to biodiversity (D1), non-indigenous species (D2), food webs (D4) and sea-floor integrity (D6), according to the criteria described in Section 2.4, are analysed in detail.

In general, the capabilities of the Baltic Sea models to address the MSFD *indicators* is quite heterogeneous (Table 7), from every model able to address several *indicators*linked to 'Phytoplankton distributional range' and 'Zooplankton distributional range', to none of the models producing or capable to produce any MDI addressing *indicators* such as 'Population genetic structure' within D1. The MDI 'Phytoplankton biomass' is referred to be capable of addressing *indicator* 1.2.1 'Population biomass' and 4.3.1 'Abundance trends of functionally important selected groups/species' by each Baltic model. However, depending on the model complexity, the number of Plankton Functional Types (PFTs) could be increased with the aim of increasing biodiversity at the population/community levels. For example, the reduction in the duration of the coastal sea-ice cover during the past century and the associated changes in vertical mixing are considered to have contributed to the increasing dominance of dinoflagellates over diatoms during the spring bloom in some parts of the Baltic Sea [31]. The type of blooming PFT has large consequences on the Baltic food web: while diatoms are usually large cells that sink rapidly, enriching the seabed of organic matter that can lead to anoxic conditions, dinoflagellates tend to be mostly consumed and recycled in the upper water column, feeding a more efficient microbial loop [32]. Besides the under-development NEMO-BFM model, none of the Baltic Sea LTL models currently simulates the abundance and distributional range of the spring dinoflagellates. It is suggested that all of the Baltic Sea LTL models would take such unique feature of the Baltic Sea vernal bloom into consideration for a better representation of e.g. the carbon paths/dynamics including the oxic, suboxic and anoxic conditions of the sea-floor within different areas of the Baltic Sea.

Table 7: Capabilities of the Baltic Sea models to address the biodiversity-related *indicators*: ✓ if the model is currently capable; (✓) if the model can be capable after some model's modifications; ✗ if the model is not capable and other model types are required.

Indicator	MDI	BaltProWeb	BALTSEM	ECOSMO	ERGOM	NEMO-BFM	RCO-SCOBI	SPBEM	Developments
1.1.1 Species Distributional range	Phytoplankton distrib. range	✓	✓	✓	✓	✓	✓	✓	
	Zooplankton distrib. range	✓	✓	✓	✓	✓	✓	✓	
	Bacterioplankton distrib. range	(✓)	(✓)	(✓)	(✓)	✓	(✓)	(✓)	e.g. new group/PFT
1.1.2 Species Distributional pattern within range	-	✗	✗	✗	✗	✗	✗	✗	e.g. habitat suitability models
1.1.3 Area covered by the species	-	(✓)	(✓)	(✓)	(✓)	(✓)	(✓)	(✓)	e.g. new group/PFT representing benthic species
1.2.1 Population abundance and/or biomass	Phytoplankton biomass	✓	✓	✓	✓	✓	✓	✓	
	Zooplankton biomass	✓	✓	✓	✓	✓	✓	✓	
	Bacterioplankton biomass	(✓)	(✓)	(✓)	(✓)	✓	(✓)	(✓)	e.g. new group/PFT
	Zoobenthos biomass	(✓)	(✓)	(✓)	(✓)	✓	(✓)	(✓)	e.g. new group/PFT
	Sea-ice algae biomass	(✓)	(✓)	(✓)	(✓)	✓	(✓)	(✓)	e.g. new group/PFT
	Sea-ice fauna biomass	(✓)	(✓)	(✓)	(✓)	✓	(✓)	(✓)	e.g. new group/PFT
	Sea-ice bacteria biomass	(✓)	(✓)	(✓)	(✓)	✓	(✓)	(✓)	e.g. new group/PFT
	Herring biomass	✓	(✓)	✓	✓	(✓)	(✓)	(✓)	e.g. coupling to HTL model
	Sprat biomass	✓	(✓)	✓	✓	(✓)	(✓)	(✓)	e.g. coupling to HTL model
	Cod biomass	✓	(✓)	✓	✓	(✓)	(✓)	(✓)	e.g. coupling to HTL model
	Grey seal biomass	✓	(✓)	(✓)	(✓)	(✓)	(✓)	(✓)	e.g. coupling to HTL model
1.3.1 Population demographic characteristics	Herring biomass (body size)	✓	(✓)	✓	✓	(✓)	(✓)	(✓)	e.g. coupling to HTL model
	Sprat biomass (body size)	✓	(✓)	✓	✓	(✓)	(✓)	(✓)	e.g. coupling to HTL model
	Cod biomass (body size)	✓	(✓)	✓	✓	(✓)	(✓)	(✓)	e.g. coupling to HTL model
	Grey seal biomass (body size)	✓	(✓)	(✓)	(✓)	(✓)	(✓)	(✓)	e.g. coupling to HTL model
1.3.2 Population genetic structure	-	✗	✗	✗	✗	✗	✗	✗	e.g models of population genetics structure
1.4.1 Habitat distributional range	-	✗	✗	✗	✗	✗	✗	✗	e.g. habitat suitability models
1.4.2 Habitat distributional pattern	-	✗	✗	✗	✗	✗	✗	✗	e.g. habitat suitability models
1.5.1 Habitat area	-	✗	✗	✗	✗	✗	✗	✗	e.g. habitat suitability models
1.5.2 Habitat extent	-	✗	✗	✗	✗	✗	✗	✗	e.g. habitat suitability models
1.6.1 Habitat: Condition of the typical species and communities	-	✗	✗	✗	✗	✗	✗	✗	e.g. habitat suitability models
1.6.2 Habitat: Relative abundance and/or biomass	-	✗	✗	✗	✗	✗	✗	✗	e.g. habitat suitability models
1.6.3 Habitat: physical, hydrological and chemical conditions	Temperature	✗	✓	✓	✓	✓	✓	✓	e.g. coupling to hydrodynamic model
	Salinity	✗	✓	✓	✓	✓	✓	✓	e.g. coupling to hydrodynamic model
	pH	✗	(✓)	✓	(✓)	✓	(✓)	(✓)	e.g. adding carbonate system to the model
	Oxygen penetration depth	✓	✓	✓	✓	✓	✓	✓	
	Denitrification layer depth	(✓)	(✓)	(✓)	(✓)	✓	(✓)	(✓)	
1.7.1 Composition and relative proportion of ecosystem components	-	✗	✗	✗	✗	✗	✗	✗	
2.1.1 Trends in abundance, temporal occurrence and spatial distribution of invasive non-indigenous species	-	(✓)	(✓)	(✓)	(✓)	(✓)	(✓)	(✓)	new group/PFT representing a single invasive species
2.2.1 Ratio between invasive non-indigenous species and native species	-	(✓)	(✓)	(✓)	(✓)	(✓)	(✓)	(✓)	new group/PFT representing a single invasive species
2.2.2. Impacts of non-indigenous invasive species	-	(✓)	(✓)	(✓)	(✓)	(✓)	(✓)	(✓)	new group/PFT representing a single invasive species
4.1.1. Performance of key predator species or trophic groups	Primary production	✓	✓	✓	✓	✓	✓	✓	
	Secondary production	✓	✓	✓	✓	✓	✓	✓	
4.2.1 Proportion of large fish at the top of food webs	Cod biomass	✓	(✓)	✓	✓	(✓)	(✓)	(✓)	e.g. coupling to HTL model

4.3.1 Abundance trends of functionally important selected groups/species	Phytoplankton biomass	✓	✓	✓	✓	✓	✓	✓	
	Zooplankton biomass	✓	✓	✓	✓	✓	✓	✓	
	Bacterioplankton biomass	✓	✓	✓	✓	✓	✓	✓	e.g. new group/PFT
	Zoobenthos biomass	✓	✓	✓	✓	✓	✓	✓	e.g. new group/PFT
	Sea-ice algae biomass	✓	✓	✓	✓	✓	✓	✓	e.g. new group/PFT
	Sea-ice fauna biomass	✓	✓	✓	✓	✓	✓	✓	e.g. new group/PFT
	Sea-ice bacteria biomass	✓	✓	✓	✓	✓	✓	✓	e.g. new group/PFT
	Herring biomass	✓	✓	✓	✓	✓	✓	✓	e.g. coupling to HTL model
	Sprat biomass	✓	✓	✓	✓	✓	✓	✓	e.g. coupling to HTL model
	Cod biomass	✓	✓	✓	✓	✓	✓	✓	e.g. coupling to HTL model
	Grey seals biomass	✓	✓	✓	✓	✓	✓	✓	e.g. coupling to HTL model
6.1.1. Type, abundance, biomass and areal extent of relevant biogenic substrate	-	✗	✗	✗	✗	✗	✗	✗	e.g. habitat suitability models
6.1.2. Extent of the seabed significantly affected by human activities	Oxygen penetration depth	✓	✓	✓	✓	✓	✓	✓	
	Denitrification layer depth	✓	✓	✓	✓	✓	✓	✓	e.g. new state variable
6.2.1. Presence of particularly sensitive and/or tolerant species	-	✓	✓	✓	✓	✓	✓	✓	e.g. new group(s)/PFT(s) representing (a) single species
6.2.2. Multi-metric indexes assessing benthic community condition and functionality	-	✗	✗	✗	✗	✗	✗	✗	e.g. habitat suitability models
6.2.3. Proportion of biomass or number of individuals in the macrobenthos	Zoobenthos biomass	✓	✓	✓	✓	✓	✓	✓	e.g. new group/PFT
6.2.4. Parameters describing the characteristics of the size spectrum of the benthic community	-	✗	✗	✗	✗	✗	✗	✗	e.g. habitat suitability models

Considering the different levels in the ecosystem structure defined in *descriptor*biological diversity (D1), biodiversity is especially addressed at the level of communities (for phytoplankton and zooplankton) and populations (for fish and marine mammals) in several indicators within D1. However, *indicators* representing biodiversity strictly at species and habitat levels are missing. For example, all of the Baltic LTL models simulate at least one generic group of cyanobacteria. However, the cyanobacteria bloom in the Baltic Sea is mainly composed of two dominant species: *Nodularia spumigena* and*Aphanizomenon flos-aquae*, which especially differ for the fact that the former is toxic, while the latter is not e.g. [33]. This is addressed by the 2 groups of cyanobacteria represented in the under-development NEMO-BFM model, but should also be included in other models. This would enable a better understanding of the abundance and distributional range of both species within the Baltic Sea, and would provide valuable information for improving management and mitigation measures. Besides the existing biogeochemical models, other types of model might also extensively address such*indicator* as for example individual-based models.

None of the *descriptor* non-indigenous species (D2) related *indicators* (2.1.1 to 2.2.2) are addressed by any of the Baltic models, but could potentially be after some modifications to models. For example, the distribution and abundance in the Baltic Sea of the invasive and potentially toxic dinoflagellate *Prorocentrum minimum* e.g. [34] could be potentially modelled by any of the Baltic LTL models after the implementation of a new PFT that would represent the target species.

Many of the habitat-related *indicators* within D1, such as 'Habitat distributional range', 'Habitat distributional pattern', 'Habitat area' and 'Habitat volume', cannot currently be addressed by the evaluated models and the development/usage of other types of tools such as habitat suitability models are thus suggested. Within the set of models analysed, descriptor food webs (D4) is discretely represented. In particular, *indicator* 4.3.1 linked to phytoplankton and zooplankton biomass and abundance is addressed by all of the models, while those linked to other trophic levels and living components, such as bacteria and fishes in *indicators* 4.1.1 and 4.2.1, are addressed only by some models, and an increase in the number of PFT is instead required for the rest of the models.

Several gaps are found when relating the analysed Baltic Sea models to the *indicators* of*descriptor* sea-floor integrity (D6) 6.1.1 to 6.2.4. In particular, the living components of the benthic habitat, except for the under-development NEMO-BFM, are poorly represented by the current set of models and required either an increase in complexity of the models, or the usage or development of new type of software such as habitat suitability models, which could not only describe both the biotic and abiotic parts of the system, but could also map those areas impacted by human activities such as fishing, drilling, enriching in organic matter and polluting.

The indicated suggestions are obviously not meant to be exhaustive among all the type of existing models, since newly developed models specifically targeted for the requirements of the MSFD should also be taken into consideration.

CONCLUSIONS

The conceptual workflow that allowed the identification of the biodiversity-related*descriptors*, the available models and their Model-Derived Indicators (MDI) (Fig. 1) has been presented. After a short review of the main characteristic of the Baltic Sea in relation to biodiversity and human pressures (Section 2.2), and a summary of the main features of the selected nine Baltic Sea models (Section 2.3), the analysis of the MDIs in relation to the number of models (Table 5), the type of living components (Fig. 3), and the related *descriptors* and *pressures* (Section 3.1) addressing them has been been discussed. Finally, a more detailed analysis that links the biodiversity-related *indicators*to each Baltic model (Section 3.2) has highlighted the current and potential capabilities of the available Baltic Sea models, together with some suggested developments of new features and/or modelling techniques.

22 MDI and 9 ecosystem models were found to be able to inform on and support the MSFD, which are 11% and 20%, respectively, of the total number found by Piroddi et al.[11] at European level. Despite the capability of the Baltic Sea ecosystem models to address several *pressures* and *descriptors*, several gaps still remain before the set of models would be entirely capable of assessing the environmental status of the Baltic Sea as requested by the MSFD. The gap analysis of the Baltic Sea study shows that some MSFD *descriptors*, such as food webs (D4), are described by the Baltic models to some extent, while others, such as non-indigenous species (D2) are not addressed at all, although there is potential for all of the models to address such descriptor. Sea-floor integrity (D6) is currently poorly addressed despite the tight coupling between the pelagic and benthic habitats in the shallow Baltic Sea, thus the usage and development of tools such as habitat suitability models is encouraged for the Baltic Sea. Most of the MDIs refer to biomass, while diversity in a more strict sense is poorly addressed apart from the under development NEMO-BFM, which adds considerable ecosystem complexity to the LTL Baltic Sea food web. Concerning habitat types, ice-associated habitats are described only by one model under development, despite sea ice covering the northern Baltic Sea every year. While all of the LTL models include some form of parameterisation of benthic chemical processes, only 2 models describe the living component of the Baltic seabed, which remains poorly represented.

Finally, the detailed analysis of the MSFD *indicators* in relation to each Baltic model inSection 3.2, where each model is classified as capable, potentially capable or not capable of addressing the biodiversity-related *indicators*, aims to provide support for both the modelling community, suggesting research areas where more effort is needed, but also for policy makers and stakeholders to enable a better understanding and direction for funding requirements.

The analysis here described is biased by the high representation of coupled hydrodynamic-biogeochemical models versus other type of models that are not described here. However, there is no doubt that, when describing the marine food web from microbes to top predators, the increasing interest is towards end-to-end models, which combine LTL and HTL models in a single quantitative tool that can be used for ecosystem-based management since capable of handling multiple impacts, such as those expected under climate change [35]. The main challenge is to fill the gaps between the biomass-based representation of the widely developed and used LTL biomass-based models and the *descriptors*, *criteria*, *indicators* and *pressures* defined in the MSFD, which only partly correlate to biomass. While the development of new indicators, better linked to the type of outputs that current available models generate, is highly encouraged, new or revised models specifically targeted to inform on MSFD are to be implemented. The first step towards better addressing marine biodiversity is certainly increasing the ecosystem complexity in terms of functional groups to a meaningful level. One approach is the use of a cost function to assess among differently complex models (e.g. in terms of functional groups) the one that score the highest skills, and that not necessarily is the most complex model e.g. [36]. A further step is to move from PFT diversity to species diversity. An example is the one of Artioli et al. [37] that modifies a LTL model with the addition of phytoplankton diversity to simulate invasion from phytoplankton, and thus directly addressing non-indigenous species (D2) of the MSFD, which none of the Baltic models is currently doing. With such a toll, the assessment of the likelihood of success of an invasion and the estimate of the potential impact on ecosystem structure would become possible. In general, a tighter link between the political ground of the MSFD and the quantitative marine ecosystem models is urgently needed and the development of innovative tools in terms of models and indicators that can be used to that end is strongly encouraged.

ACKNOWLEDGMENTS

This manuscript has resulted from the DEVOTES (DEVelopment Of innovative Tools for understanding marine biodiversity and assessing good Environmental Status) project funded by the European Union under the 7th Framework Programme, The Ocean of Tomorrow Theme (grant agreement no. 308392), www.devotes-project.eu. All of the MDIs can be found in the DEVOTool software as part of the DEVOTES FP7 project (http://www.devotes-project.eu/devotool/). DEVOTool enables browsing, ranking and extracting marine ecological indicators using various criteria and is continuously maintained and updated. The authors are grateful to Heliana Texeira for her valuable comments and to the anonymous reviewers for thoughtful suggestions that have improved the quality of this work.

REFERENCES

1. European Commission. Directive 2008/56/EC of the European Parliament and of the Council establishing a framework for community action in the field of marine environmental policy (Marine Strategy Framework Directive), Official Journal of the European Union L164, 2008, pp. 19–40.
2. L.D. Mee, R.L. Jefferson, D.d.A. Laffoley, M. Elliott, How good is good? Human values and Europe's proposed Marine Strategy Directive, Mar. Pollut. Bull. 56 (2008) 187–204.
3. European Commission. Directive 2000/60/EC of the European Parliament and of the Council of 23 October 2000 establishing a framework for community action in the field of water policy, Official Journal of the European Union L327, 2000, pp. 1–72.
4. European Commission. Commission Decision of 1 September 2010 on criteria and methodological standards on good environmental status of marine waters (notified under document C(2010) 5956)(2010/477/EU), Official Journal of the European Union L232, 2010, pp. 12–24.
5. A. Borja, M. Elliott, J.H. Andersen, A.C. Cardoso, J. Carstensen, J.C. Ferreira, A.- S. Heiskanen, J.C. Marques,

J.M. Neto, H. Teixeira, L. Uusitalo, M.C. Uyarra, N. Zampoukas, Good environmental status of marine ecosystems: What is it and how do we know when we have attained it? Mar. Pollut. Bull. 76 (1–2) (2013) 16–27.

6. P. Tett, R.J. Gowen, S.J. Painting, M. Elliott, R. Forster, D.K. Mills, E. Bresnan, E. Capuzzo, T.F. Fernandes, J. Foden, R.J. Geider, L.C. Gilpin, M. Huxham, A. L. McQuatters-Gollop, S.J. Malcolm, S. Saux-Picart, T. Platt, M.F. Racault, S. Sathyendranath, J. van der Molen, M. Wilkinson, Framework for understanding marine ecosystem health, Mar. Ecol. Prog. Ser. 494 (2013) 1–27.

7. A. Borja, D.M. Dauer, A. Gremare, The importance of setting targets and reference conditions in assessing marine ecosystem quality, Ecol. Indic. 12 (2012) 1–7.

8. I. Rombouts, G. Beaugrand, X. Fizzala, F. Gaill, S.P.R. Greenstreet, S. Lamare, F. L. Loch, A. McQuatters-Gollop, B. Mialet, N. Niquil, J. Percelay, F. Renaud, A. G. Rossberg, J.P. Feral, Food web indicators under the Marine Strategy Framework Directive: from complexity to simplicity? Ecol. Indic. 29 (2013) 246–254.

9. A. Borja, Grand Challenges in Marine Ecosystems Ecology, Frontiers in Marine Science, 2014, 1 (1).

10. E.A. Fulton, Approaches to end-to-end ecosystem models, J. Mar. Syst. 81 (2010) 171–183.

11. C. Piroddi, H. Teixeira, C. Lynam, C. Smith, M. Alvarez, K. Mazik, E. Andonegi, T. Churilova, L. Tedesco, M. Chifflet, et al., Using ecological models to assess ecosystem status in support of the European Marine Strategy Framework Directive, Ecol. Indic. 58 (2015) 175–191.

12. V. Fleming-Lehtinen, J.H. Andersen, J. Carstensen, E. Lysiak-Pastuszak, C. Murray, M. Pyhala, M. Laamanen, Recent developments in assessment methodology reveal that the Baltic Sea eutrophication problem is expanding, Ecol. Indic. 48 (2015) 380–388.

13. HELCOM, HELCOM Copenhagen Ministerial Declaration: taking further action to implement the Baltic Sea Action Plan reaching good environmental status for a healthy Baltic Sea, Tech. Rep., 2013.

14. H. Ojaveer, A. Jaanus, B.R. MacKenzie, G. Martin, S. Olenin, T. Radziejewska, I. Telesh, M.L. Zettler, A. Zaiko, Status of biodiversity in the Baltic Sea, PLoS One 5 (9) (2010) e12467.

15. HELCOM, Biodiversity in the Baltic Sea – An integrated thematic assessment on biodiversity and nature conservation in the Baltic Sea, Tech. Rep. No 116B, 2009.

16. B. MacKenzie, J. Alheit, C.D.J.P. Holm, C. Kinze, Ecological hypotheses for a historical reconstruction of upper trophic level biomass in the Baltic Sea and Skagerrak, Can. J. Fish. Aquat. Sci. 59 (2002) 173–190.

17. N. Kautsky, H. Kautsky, U. Kautsky, M. Waern, Decreased depth penetration of Fucus vesiculosus (L.) since the 1940s indicates eutrophication of the Baltic Sea, Mar. Ecol. Prog. Ser. 28 (1) (1986) 1–8.

18. B.B. Jorgensen, Seasonal oxygen depletion in the bottom waters of a danish fjord and its effect on the benthic community, Oikos 34 (1) (1980) 68–76.

19. H. Cederwall, R. Elmgren, Biological effects of eutrophication in the Baltic Sea, particularly the coastal zone, AMBIO 19 (1990) 109–112.

20. HELCOM, Ecosystem Health of the Baltic Sea 2003 – 2007: HELCOM Initial Holistic Assessment, Tech. Rep. No 122, 2010a.

21. HELCOM, Baltic Sea Action Plan. HELCOM Ministerial Meeting. Adopted in Krakow, Poland on 15th of November 2007, Tech. Rep., 2007.

22. F. Wulff, C. Humborg, H.E. Andersen, G. Blicher-Mathiesen, M. Czajkowski, K. Elofsson, A. Fonnesbech-Wulff, B. Hasler, B. Hong, V. Jansons, et al., Reduction of Baltic Sea nutrient inputs and allocation of abatement costs within the Baltic Sea catchment, AMBIO 43 (1) (2014) 11–25.

23. K. Eilola, J.L.S. Hansen, H.E.M. Meier, M.S. Molchanov, V. Ryabchenko, M. D. Skogen, Eutrophication Status Report of the North Sea, Skagerrak, Kattegat and the Baltic Sea: A model study. Present and future climate, Mar. Pollut. Bull. 115 (2013) 38.

24. A.L. Pikkarainen, Polycyclic aromatic hydrocarbons in Baltic Sea sediments, Polycycl. Aromat. Compd. 24 (4–5) (2004) 667–679.

25. P. Isosaari, A. Hallikainen, H. Kiviranta, P.J. Vuorinen, R. Parmanne, J. Koistinen, T. Vartiainen, Polychlorinated dibenzo-p-dioxins, dibenzofurans, biphenyls, naphthalenes and polybrominated diphenyl ethers in the edible fish caught from the Baltic Sea and lakes in Finland, Environ. Pollut. 141 (2) (2006) 213–225.

26. H. Karl, U. Ruoff, Dioxins, dioxin-like PCBs and chloroorganic contaminants in herring, clupea harengus, from different fishing grounds of the Baltic Sea, Chemosphere 67 (9) (2007) S90–S95.

27. K. Wiberg, A.T. Assefa, K.L. Sundqvist, I.T. Cousins, J. Johansson, M.S. McLachlan, A. Sobek, G. Cornelissen, A. Miller, J. Hedman, A. Bignert, H. Peltonen, M. Kiljunen, V. Shatalov, I. Cato, Managing the dioxin problem in the Baltic region with focus on sources to air and fish, Tech. Rep. ISBN 978-91-620-8652-7, ISSN 0282–7298, The Swedish Environmental Protection Agency, Sweden, 2013.

28. J.M. Armitage, M.S. McLachlan, K. Wiberg, P. Jonsson, A model assessment of polychlorinated dibenzo-p-dioxin and dibenzofuran sources and fate in the Baltic Sea, Sci. Total Environ. 407 (2009) 3784–3792.

29. K. Kannan, J. Falandysz, Butyltin residues in sediment, fish, fish-eating birds, harbour porpoise and human tissues from the polish coast of the Baltic Sea, Mar. Pollut. Bull. 34 (3) (1997) 203–207.

30. HELCOM, Hazardous substances in the Baltic Sea An integrated thematic assessment of hazardous substances in the Baltic Sea, Tech. Rep. No 120B, 2010b.

31. R. Klais, T. Tamminen, A. Kremp, K. Spilling, K. Olli, Decadal-scale changes of dinoflagellates and diatoms in the anomalous Baltic Sea spring bloom, PLoS One 6 (6) (2011) e21567.

32. K. Spilling, A. Kremp, R. Klais, K. Olli, T. Tamminen, Spring bloom community change modifies carbon pathways and C:N:P:Chl a stoichiometry of coastal material fluxes, Biogeosciences 11 (2014) 7275–7289.

33. M. Kahru, U. Horstmann, O. Rud, Satellite detection of increased cyanobacteria blooms in the Baltic Sea: Natural fluctuation or ecosystem change? Ambio 23 (1994) 469–472.

34. S. Hajdu, S. Pertola, H. Kuosa, Prorocentrum minimum (dinophyceae) in the Baltic sea: Morphology, occurrence - a review, Harmful Algae 4 (3) (2005) 471–480.

35. K.A. Rose, J.I. Allen, Y. Artioli, M. Barange, J. Blackford, F. Carlotti, R. Cropp, U. Daewel, K. Edwards, K. Flynn, et al., End-To-End models for the analysis of marine ecosystems: Challenges, issues, and next steps, Mar. Coast. Fish. 2 (2010) 115–130.

36. Y. Xiao, M.A.M. Friedrichs, Using biogeochemical data assimilation to assess the relative skill of multiple ecosystem models in the mid-atlantic bight: effects of increasing the complexity of the planktonic food web, Biogeosciences 1 (2014) 3015–3030.

37. Y. Artioli, C. Galienne, J. Holt, S. Wakelin, M. Butenschn, C. Schrum, U. Daewel, D. Pushpadas, H. Cannaby, B. Salihoglu, et al., Invasive species: an increasing threat to marine ecosystems under climate change?, Geophysical Research Abstracts, EGU General Assembly,15, EGU2013–13075, 2013.

CHAPTER 17

ASSESSING THE UTILITY OF GEOSPATIAL TECHNOLOGIES TO INVESTIGATE ENVIRONMENTAL CHANGE WITHIN LAKE SYSTEMS

Eirini Politi, John S. Rowan, Mark E.J. Cutler

Keywords: Lake change, Ecosystem services, remote sensing, Geospatial technology, Catchment pressures

INTRODUCTION

The world's freshwater ecosystems and the biodiversity they support are vital components of the global biosphere, yet remain fragile and vulnerable to anthropogenic disturbance and climate change (Dokulil, 2013, Schindler and Donahue, 2006,Vörösmarty et al., 2000 and Williamson et al., 2009). Whilst recognition grows of their role in the carbon cycle at regional and continental scales (e.g. Knoll et al., 2013,McDonald et al., 2013, Renwick et al., 2008 and Yang et al., 2008), understanding their global significance to biogeochemical cycling and ecosystem services has been hampered by uncertainties about absolute numbers and the distributions of different types, as well as limited information on their ecological, chemical and physical condition (Downing et al., 2006 and Verpoorter et al., 2012). Verpoorter et al. (2014) enumerated a global inventory of 117 million lakes (> 0.002 km^2), with a combined surface area of about 5 × 10^6 km^2, equating to 3.7% of the Earth's non-glaciated land mass. In Europe alone there are over 1.5 × 10^6 lakes with surface areas greater than 0.01 km^2 and at least 500,000 natural lakes larger than 0.1 km^2 (Kristiansen and Hansen, 1994). These systems span a continuum of size, depth, form, altitude, geology, climatic and hydrological regime. Importantly, lakes also sit within a wider landscape, catchment and hydrological network that means one needs to look beyond the lake itself to determine the drivers of lake water behaviour and any changes in that. For example, the Danube Basin is some 801,463 km^2 in size and according to the Global Lakes and Wetlands Database (Lehner and Döll, 2004) includes 592 lakes > 0.1 km^2. The large numbers involved and the relative remoteness of many lake districts, especially within large transnational catchments such as the Danube, presents major challenges in terms of managing the standing water resource and emphasises the need to be able to generalise about behaviour based on limited field observations.

The role that aquatic systems play in underpinning society's activities, health and well-being is increasingly being understood through the concept of ecosystem services, which links nature to an economic value to society (TEEB, 2010). Particularly since the launch of the United Nations Millennium Ecosystem Assessment (MEA) in 2005, there has been an increase in the awareness of the ecosystem services provided by aquatic and terrestrial ecosystems amongst policy makers (de Groot et al., 2010), matched by a significant increase in the number of scientific publications on the topic (de Araujo Barbosa et al., 2015, Large and Gilvear, 2014 and Seppelt et al., 2011). Remote sensing has the potential to provide an important source of information for quantifying and mapping dynamic terrestrial and aquatic ecosystem services and hydrological processes, including approaches to hydrologic modelling as reviewed by Xu et al. (2014). In many cases, given the large catchment area of many of the most important river and lake systems in the world (particularly those with large populations dependent upon them and their attendant ecosystem services), the synoptic, wide area coverage and frequent observations provided by satellite-based remote sensing are an important

Citation: Eirini Politi, John S. Rowan, Mark E.J. Cutler, Assessing the utility of geospatial technologies to investigate environmental change within lake systems, Science of The Total Environment, Volume 543, Part A, 1 February 2016, Pages 791-806, ISSN 0048-9697, http://dx.doi.org/10.1016/j.scitotenv.2015.09.136.

source of standardised information that is not prey to variability in national and regional in situ data collection methodologies and standards. As a result, remote sensing has become an important geospatial technology for deriving information about lakes and their catchments, which alongside advances in in situ sensor systems to capture spatial information in addition to measurements across time (Crawford et al., 2014) and analysis with Geographic Information Systems (GIS), provides an important step forward in our ability to model lake and catchment status and behaviour.

From both passive (e.g. multispectral and hyperspectral) and active (LiDAR and RADAR) remote sensing systems it is possible to retrieve the state of important lake and catchment variables that may vary across space at particular points in time. Recently, either directly or in combination with other data, remote sensing has been successfully used within the Danube Basin to map variations in biomass (Kovàcs, 2007), carbon stocks in riparian forests (Suchenwirth et al., 2014), habitat change (Kollár et al., 2011), historical land cover change (Crăciunescu et al., 2010), evapotranspiration (Rodell et al., 2011) and soil moisture (dall'Amico et al., 2012), to name but a few examples. However, it is important to recognise that whilst some catchment properties may be retrieved directly from remotely sensed data (e.g. canopy height with LiDAR or leaf area index from optical data) some catchment properties, such as land use, may only be inferred from remotely sensed data through contextual associations across, whilst others, such as raw material production, will require remotely sensed data to be coupled with other ancillary datasets, for example in situ measurements and modelled data (de Araujo Barbosa et al., 2015).

Following decades of fundamental research and the launch of several new satellite programmes a report by London Economics to the UK Space Agency suggests that we stand at the dawn of a 'New Space Age' (London Economics, 2015, p. 104), one in which there will be substantial growth in applications of remote sensing data. Indeed the future of European (and global) remote sensing for catchment, ecosystem and lake-based studies looks very promising, particularly with respect to data availability and the new generation Copernicus satellites of the European Space Agency (ESA). The Copernicus satellites build on the success and capabilities of other currently operational (e.g. Landsat series) and non-operational (e.g. Envisat MERIS and AATSR) instruments and include the Sentinel-2 Multispectral Imager (MSI) (launched June 2015), Sentinel-3 Ocean and Land Colour Instrument (OLCI) and Sentinel-3 Sea and Land Surface Temperature Radiometer (SLSTR), the first of which is due for launch late 2015. Sentinel-2 MSI has 13 wavebands in the visible and NIR at 10–60 m spatial resolution and a 5-day revisit time. Sentinel-3 OLCI at a 300 m spatial resolution will employ 21 wavebands in the visible and NIR. Sentinel-3 SLSTR will employ 9 wavebands, with a nominal 500 m spatial resolution in the visible and NIR and 1 km at the TIR. In addition, Sentinel-3 will have a revisit time of 1–2 days. These optical and thermal sensors combine short revisit times, fine spatial resolutions and multiple spectral channels with a free data access policy. Here we signpost some of the new opportunities for catchment and lake studies provided by these new systems by collating experiences of previous research and how these and the plethora of new data sources will contribute to future lake and catchment management.

Several authors have provided comprehensive reviews of the potential and current status of remote sensing to map and quantify a range of ecosystem services (e.g.Andrew et al., 2014 and de Araujo Barbosa et al., 2015). This paper builds upon those comprehensive reviews and others by exploring the utility of remote sensing to capture spatial and temporal patterns of change in key catchment processes and the use of these data to inform policy and management decisions. Particular reference is made to examples from the Danube Basin highlighting information derived from remote sensing data relevant for catchment water balance estimation and modelling, as well as land cover change and other critical ecosystem functions. By focussing on the catchment we evaluate how to derive information on drivers of change in lake behaviour rather than explicitly exploring the use of remote sensing to detect change in the lake water quality itself, which is the subject a separate paper in this Special Issue (Tyler et al., in this volume).

LAKES, LANDSCAPE LIMNOLOGY AND REMOTE SENSING

Soranno et al. (2010) introduced landscape limnology as a new conceptual framework for understanding lake behaviour based on a series of important comparative and regional limnological studies (e.g. Riera et al., 2000, Webster et al., 2000 and Soranno et al., 2009). Landscape limnology is the spatially-explicit study of lakes within freshwater, terrestrial and human landscapes to determine the effects of pattern on ecosystem processes over multiple temporal and spatial scales (Fig. 1). It highlights that landscape settings influence lake functions through the hydromorphological setting mediated by the extent of human modification, superimposed upon which is the effect of climate change.

An extension to this framework is shown in Table 1, which uses four hierarchical structuring elements consistent with the water body classification attributes used in the EU Water Framework Directive (EU WFD, 2000). The first element is the ecoregion which sets out the broadest climatic, physiographic and dominant biomes. Landscape setting establishes the specific characteristics in terms of hydro-climatic factors such as continentality, the elevation of the lake and its catchment (e.g. whether montane, piedmont or lowland) and biodiversity characteristics. The lake-catchment relation captures the relative importance of 'catchment' versus 'lake' processes in relation to water, sediment and nutrient budgets, including the critical role of autochthonous versus allochthonous carbon fluxes (cf. Tranvik et al., 2009 and Kutser et al., 2015). Finally, lake morphometry (size, shape, orientation, water depth, etc.) governs lake processes and characteristics such as mixing regime, residence time, shore zone energetics and riparian habitats (Hutchinson, 1957).

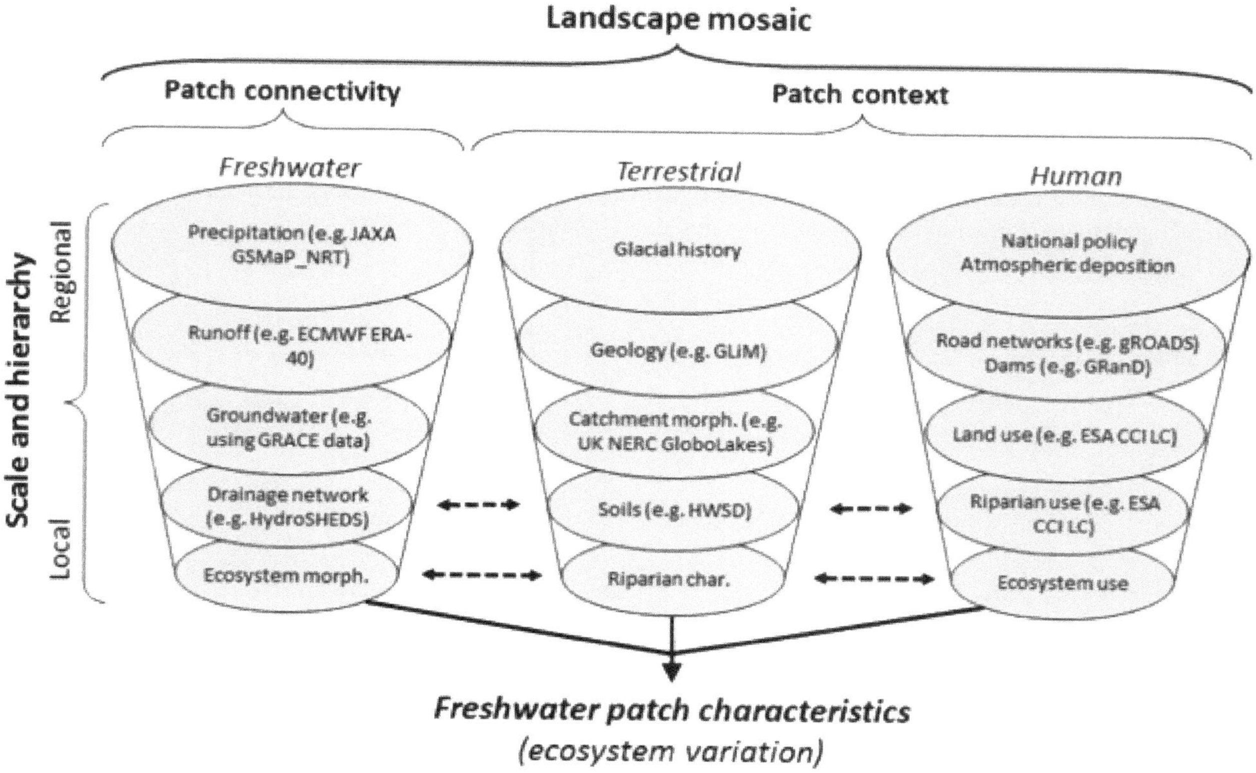

Figure 1: Landscape limnology framework connecting local and regional lake characteristics with examples of globally available datasets and EO products (after Soranno et al., 2009). JAXA GSMaP_NRT (Japan Aerospace Exploration Agency Global Satellite Mapping of Precipitation in Near-Real-Time); ECMWF ERA-40 (European Centre for Medium-Range Weather Forecasts 40-year re-analysis model data); GRACE (NASA/DLR's Gravity Recovery and Climate Experiment); HydroSHEDS (World Wildlife Fund Hydrological data and maps based on SHuttle Elevation Derivatives at multiple Scales); GLiM (Global Lithological Map database); UK NERC GloboLakes (United Kingdom Natural Environment Research Council GloboLakes project); HWSD (Harmonised World Soil Database); gROADS (Global Roads Open Access Data Set); GRanD (Global Reservoir and Dam database); ESA CCI LC (European Space Agency Climate Change Initiative Land Cover).

Table 1: WFD compliant eco-geomorphic framework for contextualising lake behaviour.

	Scale	Hydro-climatic	Terrestrial	Biotic	Human
↑ Regional influences	Ecoregion	• Latitude/Longitude • Solar radiation regime • Temperature • Precipitation • Wind regime	• Geological structure and lithology • Physiography • Soils	• Regional biomes • Biomass • Phenology	• Airborne pollution • Climate change
	Landscape setting	• Continentality • Orographic effects • Environ. lapse rates • Runoff quantity and timing • Ice cover	• Elevation • Topographic variation • Drainage network and density • Glacial history	• Habitat diversity • Species dispersal • Flux of humic substances	• Economy • Demographics • Land cover and management • Water resource management
SCALE	Catchment-lake relation	• Hydrological regime • Hydrological pathways • Surface:groundwater contribution	• Hydrogeology • Groundwater and solute flux • Sediment budget • Salinity	• Biological continuity • Nutrient cycling • Acidity	• Impoundments • Water transfers • Drainage • Land management • Riparian land cover
Increasingly site specific	Lake hydromorphology (water body scale)	• Size and shape • Stratification and mixing depth • Water residence time • Resuspension and internal loading	• Shoreline complexity • Littoral:pelagic ratio • Riparian habitat structure	• Appropriate species composition • Integrity of trophic function • Alien species	• Water level manipulation • Cumulative pressures and disturbance
↓	Meso-habitat patches	• Microclimate • Exposure • Erosion/sedimentation	• Littoral slope • Substrate • Wave energy	• Local carbon flux • Pollution • Habitat patches	• Local pressures to shore and lake bed

Untangling the effects of multiple pressures on lake structure and function, and determining the environmental pressures that will have the largest effect on a given lake or lake type, is a challenge that requires multi-scale observations of a suite of functionally relevant and reliable indicators of lake ecosystem condition for a larger number of lakes than has been possible with traditional monitoring. Remote sensing provides a plethora of datasets and products that, in combination with ancillary data (e.g. in situ measurements and observations, modelled data), can be used to improve our understanding of lake system response to climate and non-climate pressures within a catchment. For example, land cover/use (and consequently trends in the riparian development of a lakeshore), soil moisture and elevation are, amongst others, standard remote sensing products, whilst water balance modelling techniques have recently assimilated remotely sensed information on climatic variables (e.g. evapotranspiration, precipitation and land surface temperature), soil moisture and vegetation properties. Taking the catchment of Lake Balaton (the largest freshwater body within the Danube Basin) as an example, we hereby present examples of global datasets that are available from various sources at varying temporal frequencies, time periods and spatial resolutions (Fig. 2, Table 2), and which are often used as catchment indicators of lake change. The next sections review the applicability of remote sensing in deriving some of these products in the realm of limnological research at catchment-wide scales across the globe, and specifically in the Danube Basin.

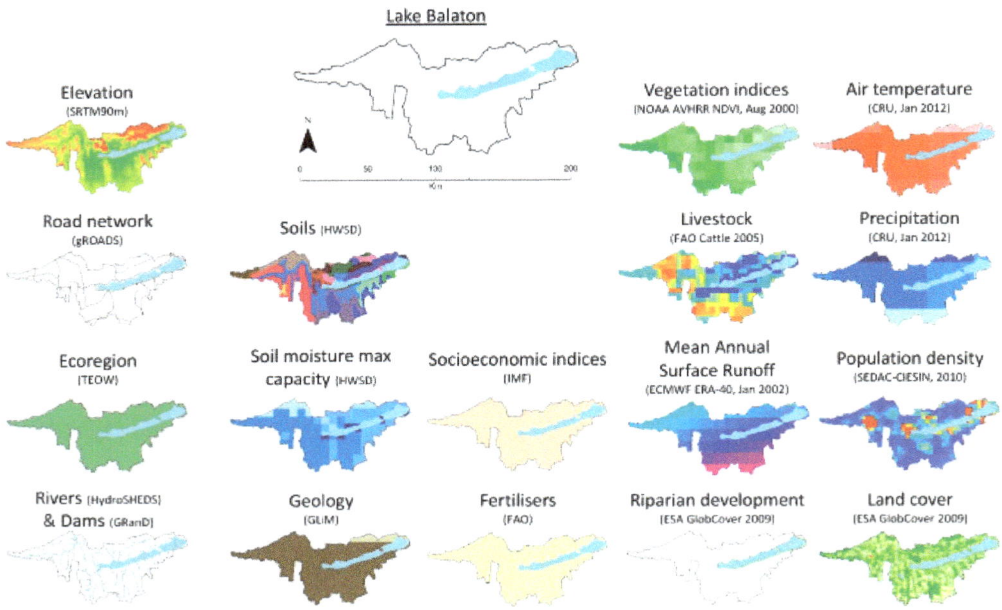

Figure 2: Examples of climate and non-climate datasets depicting lake pressure attributes in the catchment of Lake Balaton. SRTM90m (Shuttle Radar Topography Mission at 90 m spatial resolution); gROADS (Global Roads Open Access Data Set); TEOW (Terrestrial Ecoregions of the World); GRanD (Global Reservoir and Dam database); HWSD (Harmonised World Soil Database); GLiM (Global Lithological Map database); IMF (International Monetary Fund); FAO (Food and Agriculture Organisation); NOAA (National Oceanic and Atmospheric Administration); AVHRR (Advanced Very High Resolution Radiometer); NDVI (Normalised Difference Vegetation Index); ECMWF ERA-40 (European Centre for Medium-Range Weather Forecasts 40-year re-analysis model data); ESA GlobCover (European Space Agency Global Land Cover product); CRU (Climatic Research Unit); SEDAC (Socioeconomic Data & Applications Center); CIESIN (Center for International Earth Science Information Network).

Table 2: Examples of global data sources for deriving catchment attributes (period(s) of projected values are shown in square brackets).

Dataset/database	Product parameter(s)	Time period	Format	Units	Spatial resolution
Harmonised World Soil Database (HWSD)	Soil type	2008	Raster	SMU	1 km
Global Lithological Map Database (GLiM)	Geology	2012	Raster	–	62.5 km
Global Reservoir and Dam Database v.1.1 (GRanD)	Location and age of dams; spatial extent of reservoirs	2011	Vector	–	–
Terrestrial Ecoregions of the World (TEOW)	16 ecoregions	2001	Polygons	–	–
HydroSHEDS (USGS)	River network	2008	Vector	–	1 km
Gridded Population of the World v.3 (GPWv3)	Population density	1990, 1995, 2000, [2005], [2010], [2015]	Raster	Persons per km^2	5 km
Shuttle Radar Topography Mission v.4.1 (SRTM90m)	DEM (elevation)	2008	Raster	m	90 m

Climatic Research Unit Time Series v.3.21 (CRU TS 3.21)	Precipitation	1901–2012; monthly	NetCDF	mm	62.5 km
	Potential evapotranspiration			mm	
	Temperature			10 × °C	
European Centre for Medium-Range Weather Forecasts (ECMWF) ERA-40	Runoff	Sep 1957–Aug. 2002; monthly	NetCDF	m/day	~ 15 km
Global Roads Open Access Data Set (gROADSv1)	Road network	1980–2010	Vector	–	–
European Space Agency (ESA) Climate Change Initiative (CCI) Land Cover (LC)	Land cover	1998–2002, 2003–2007, 2008–2012	Raster	–	300 m
Gridded Livestock of the World (FAO, Food and Agriculture Organisation)	Cattle, poultry, buffalo, goat, pig, sheep	2000, 2005	Raster	Animals per km^2	5 km
Fertilisers (FAO)	Nitrogen & phosphate	2002–2010; annual	csv	Tonnes per 1000 ha	–
Irrigated land (actual & potential) (FAO)	Agricultural area irrigated; Total area equipped for irrigation	1961–2011; annual	csv	1000 ha	–
Gross Domestic Product (GDP) per capita, International Monetary Fund (IMF)	GDP per capita	1980–[2018]; annual	csv	$ (USD)	–

CATCHMENT WATER BALANCE

In recent decades the combined effects of economic growth and rapidly increasing global human population has led to increased demand for freshwater, which in turn has resulted in the intensification of water withdrawal from surface freshwater bodies and groundwater pools. Human activities have caused global water scarcity, particularly in arid and semi-arid regions, where overuse of natural water resources has taken place (Wada et al., 2011). At the global scale agriculture is the main water consumer with over 75% of abstractions, far outweighing the amounts used for industrial (20%) or domestic purposes (5%) (FAO, 2014 and UNEP, 2008). However, significant regional variations also exist reflecting local climate and physiographic features (Shiklomanov, 2000), imprinting an often complex picture of change when trying to assess the strong impact that global climate change has on freshwater supplies (WWAP, 2012). Determining the extent and severity of impacts of human activities and climate change on water resources across the globe, particularly in ungauged basins, has hitherto been a major issue. This has often been addressed through water balance modelling that can generate spatial and temporal variations of water availability in lake catchments and river basins, and thus contribute towards the management of water resources from local to global scales (Mitsova, 2014).

In principle, water balance models take into account the amount of water input through precipitation and snowmelt, and the amount of water output through evaporation, transpiration, surface (including river) runoff, groundwater discharge and percolation. There exist numerous water balance models, one of the earlier and more simplistic being the Thornthwaite and Mather (TM) model (Thornthwaite and Mather, 1955 and Thornthwaite and Mather, 1957), a revision of an earlier model by Thornthwaite (1948), which requires precipitation, air temperature and latitude to produce estimates of evapotranspiration, soil moisture storage, snow storage, surplus and runoff (McCabe and Markstrom, 2007). Many studies have since developed new or modified existing water balance models for use in different applications and at various spatio-temporal scales and the reader is referred to Xu and Singh, 1998 and Xu and Singh, 2004 andBoughton (2005) for comprehensive reviews of various water balance modelling techniques.

A key requirement of large hydrological systems, such as the Danube Basin, is that these water balance techniques and models must be applicable to regional or even continental scales, which is potentially where remote sensing can have a part to play. At a continental scale applicable water balance models include FAO AQUASTAT (2001) that was developed to map agricultural water use in Africa, WaterGAP that simulates global water availability and use in large drainage basins (Döll et al., 2003 and Alcamo et al., 2003) and GWAVA that models global water availability with demonstrable applications in Africa, South America and Europe (Meigh et al., 1999). In addition, the European Union Water and Global Change (WATCH) project created a multi-model ensemble bringing together nine large-scale hydrological models to compare their ability to capture mean annual runoff and quantify the uncertainty related to the modelling outputs. It was found that even though the performance of each individual WATCH model differed, their ensemble mean produced reliable estimates of runoff in European catchments (Gudmundsson et al., 2012a and Gudmundsson et al., 2012b).

The application of water balance modelling specifically within the Danube Basin in the past has tended to make use of historic weather station measurements and simulation data as input parameters (e.g. Mauser and Bach, 2009, Petrovič et al., 2010, Klein et al., 2011 and Kling et al., 2012) rather than making direct use of remote sensing observations. However, the GLOWA project (Global Change of the water cycle; funded by the German Ministry of Research and Education (BMBF)) is an example of one such initiative that exemplified the potential of remote sensing to address the impact of global change on regional water resources. Within this initiative is included the GLOWA-Danube project (dedicated to the Upper Danube watershed) that aims to investigate the sustainability of future water use using modelling techniques (Ludwig et al., 2003a). Within the GLOWA-Danube framework satellite-retrieved land cover information was used in a physically-based water balance model to provide spatio-temporal estimates of evapotranspiration, soil moisture, snow cover and runoff (Ludwig et al., 2003b). The study combined multi-temporal coarse spatial resolution NOAA AVHRR images and spectral unmixing techniques to produce sub-pixel land cover information at the fine scales required for the detailed description of various hydrological processes, and showed that the approach produced improved model results compared to the use of conventionally used land cover data (e.g. CORINE maps).

Role of Remote Sensing in Water Balance Studies

Water balance models require a variety of data ranging from climatic variables to information on soil and vegetation properties. Remote sensing has the potential to produce some of this information at large enough spatial and temporal scales to fulfil the need for continuous water resource monitoring applications at scales varying from local to global (Xu et al., 2014) and it is not uncommon for a combination of remote sensing sensors to be used in hydrological modelling and water balance applications (e.g.Awange et al., 2008 and Milzow et al., 2011). Even though remotely sensed observations complement in situ observations in providing data inputs for water balance modelling techniques, its potential was realised more than four decades ago. For example, evapotranspiration estimations based on remotely sensed canopy temperatures date back to the 1970s (e.g. Stone and Horton, 1974, Blad and Rosenberg, 1976, Heilman et al., 1976 and Jensen and Chery, 1980), as does the estimation of remotely sensed soil water content, also known as soil moisture (Wang and Qu, 2009) and the mapping of snow cover (e.g. Kyle et al., 1978). On the other hand, precipitation can be estimated using remotely sensed data and various global products already exist, but their use in water balance modelling has only recently been realised (e.g. Pereira-Cardenal et al., 2011). Finally, there are no remote sensing products for air temperature, however Deus et al. (2013) propose the use of remotely sensed land surface temperature (LST) as a substitute for air temperature in water balance modelling, and several methods incorporate LST to estimate evapotranspiration (Verstraeten et al., 2008). Whilst Xu et al. (2014) and Karimi and Bastiaanssen (2015) provide excellent reviews on the integration of remotely sensed observations to hydrologic modelling in general, here we briefly review the status of remote sensing as a sources of functional information to understand the role of the catchment in changing lake behaviours.

Remote Sensing Sensors, Scales and Products

Evapotranspiration (ET)

Remote sensing is a promising tool for the representation of the spatial distribution of evapotranspiration (ET) across large spatial scales, which is required for distributed water balance modelling (Glenn et al., 2007). Several approaches have been proposed in the literature, commonly using optical and thermal data, and less frequently satellite microwave data. There are two types of methods that are used to estimate ET from remote sensing data; empirical and physical models. Empirical or statistical relationships make use of remotely sensed vegetation indices, such as the normalised difference vegetation index (NDVI) or the soil-adjusted vegetation index (SAVI) (Glenn et al., 2007), which are based on the fraction of the ground area covered or shaded by vegetation (Pereira et al., 2015). However, these methods are insensitive to conditions of water or salinity stress, which is not true for physical models (Pereira et al., 2015). Physical or analytical models solve the surface energy balance (SEB) equation using land surface temperature (LST) estimates based on remote sensing observations. The advantage of SEB models is that they can be used to estimate ET in areas with diverse vegetation types and heterogeneous vegetation cover (e.g. Minacapilli et al., 2009 and Pôças et al., 2013). The reader can refer to, for example, Glenn et al. (2007), Verstraeten et al. (2008),Kalma et al. (2008), Pereira et al. (2015) and Karimi and Bastiaanssen (2015) for detailed reviews of ET estimation methods based on remote sensing data.

The utility of remote sensing in water balance modelling, and particularly in the estimation of ET, depends on the specifications of the sensor and the cloud conditions (Kalma et al. 2008). Medium spatial resolution sensors, such as the Advanced Spaceborne Thermal Emission and Reflection Radiometer (ASTER) on-board the Terra satellite and the Thematic Mapper (TM), Enhanced TM (ETM +) and Operational Land Imager (OLI) on-board the Landsat series have a 16-day revisit capability, which means temporal interpolation techniques are needed to provide the daily inputs often required for hydrological models. However, ASTER has a spatial resolution of 15 m in the visible and NIR wavebands and 90 m in the TIR wavebands, while OLI offers a pixel size of 30 m in the visible and NIR and 100 m in the TIR, which in many cases around the globe and particularly within the Danube Basin, makes them suitable for mapping individual agricultural fields and smaller land parcels (Pereira et al., 2015). On the other hand, the Terra/Aqua Moderate Resolution Imaging Spectrometer (MODIS) and NOAA Advanced Very High Resolution Radiometer (AVHRR) provide daily observations depending on cloud cover at a spatial resolution of 1 km. Both MODIS and NOAA AVHRR can provide LST estimates with an accuracy of 1.0 °C at 1 km or 5 km spatial resolution (Glenn et al., 2007) and less than 1.3 °C at 4.4 km (Pinheiro et al., 2006), respectively. Finally, geostationary satellites at relatively coarse (1–5 km) spatial resolutions in the visible and TIR have very frequent revisit times (15–30 min) and could be used to derive LST, however their constant viewing angle introduces problems at high latitudes which limits their global applicability (Diak et al., 2004).

Overall, whilst the above illustrate the potential of remote sensing to provide reliable information on ET, its practical application in water balance modelling has often been hindered by concerns over accuracy. As Karimi and Bastiaanssen (2015) note, the reliability of remotely-sensed derived ET is often case and location specific, but the ensemble mean ET products currently under development may be a significant step towards addressing these concerns.

Soil Moisture

The most common method for the retrieval of soil moisture is the use of microwave remote sensing sensors, both active and passive, and dates back to the 1970s (e.g.Eagleman and Ulaby, 1975, Schmugge et al., 1974, Njoku and Kong, 1977 and Mo et al., 1982). Remotely sensed soil moisture can only be estimated in the uppermost layer of the soil surface (i.e. top few centimetres), but assimilation of these observations within Soil-Vegetation-Atmosphere Transfer (SVAT) models helps retrieve soil moisture in the root zone (Wigneron et al., 2003), which is a key variable in hydrological modelling (e.g.Wilker et al., 2006 and Seneviratne et al., 2010).

There are many different surface soil moisture retrieval techniques that have been proposed in the literature, some used to derive global products, and these can be both empirical and physically-based (de Jeu et al., 2008 and Vereecken et al., 2008). While passive microwave sensors (radiometers) detect microwave radiation that is naturally emitted by the Earth's surface, active microwave sensors (RADAR) transmit an electromagnetic pulse and measure the scattered microwave energy back from the Earth surface. The success of microwave sensors in measuring soil moisture lies in the fact that changes in the soil dielectric constant due to changes in water content are detectable, particularly at low-frequency microwave regions (1–10 GHz) (e.g. Njoku and Kong, 1977, Wagner et al., 2007 and de Jeu et al., 2008). A major advantage of microwave remote sensing is that can it be used day and night regardless of cloud cover. A limitation is that dense vegetation cover and soil roughness introduce noise to soil moisture measurements (e.g. Wagner et al., 2007, de Jeu et al., 2008 and Seneviratne et al., 2010), but low-frequency microwave bands have been found to reduce this effect (Schmugge et al., 2002 and Njoku et al., 2003). In addition, a recent study in the Upper Danube basin demonstrated that soil moisture retrieval errors can be spatially diverse due to different factors (e.g. land surface heterogeneity) that introduce noise in the remote sensing observations, and that the retrieval accuracy strongly depends on the scale of observation (Loew, 2008).

Microwave instruments used in soil moisture retrieval include the non-operational Nimbus 7 Scanning Multichannel Microwave Radiometer (SMMR) (27–148 km spatial resolution), and Aqua Advanced Microwave Scanning Radiometer-EOS (AMSR-E) (38–56 km) for which global soil moisture products exist (e.g. Njoku et al., 2003 and Owe et al., 2008). The F-series Special Sensor Microwave Imager (SSM/I) and Special Sensor Microwave Imager Sounder (SSMIS) lack low-frequency wavebands that are desired for soil moisture retrieval. Nevertheless, soil moisture retrieval methods based on SSM/I data have been proposed in the literature (e.g. Jackson et al., 2002 and Wen et al., 2005). The relatively high-frequency TRMM TMI and WindSat radiometer data have also been used for the retrieval of soil moisture datasets (e.g. Bindlish et al., 2003, Gao et al., 2006 and Parinussa et al., 2012). ESA's Soil Moisture and Oceanic Salinity (SMOS) (30–50 km spatial resolution) was launched in 2009 and employs an L-band, which offers deeper vegetation penetration making it potentially more reliable over densely vegetated regions (Seneviratne et al., 2010). Within the upper Danube region initial results on the validation of soil moisture products suggested that radio-frequency interference can be a problem (dall'Amico et al., 2012) but that the spatial pattern of soil moisture variability was similar to in situ measurements, even if SMOS tended to generally underestimate soil moisture (Schlenz et al., 2012).

Radar scatterometers can also be suitable for the retrieval of soil moisture, depending on the frequency they operate in (Seneviratne et al., 2010). These include ESA's ERS-1/2 Advanced Microwave Instrument Scatterometer (AMI-SCAT) (50 km spatial resolution) and the METOP-A/B/C Advanced Scatterometer (ASCAT) (50 km) (Wagner et al., 2003 and Bartalis et al., 2007). Finally, NASA's Soil Moisture Active and Passive (SMAP) mission carries on-board active and passive L-band sensors and was launched on the 31st January 2015. The mission is designed to collect global observations of soil moisture at a spatial resolution of 9 km, with a revisit time of 2–3 days.

Soil moisture itself is a Global Climate Observing System (GCOS) Essential Climate Variable (ECV) and ESA's Climate Change Initiative (CCI) programme has funded the creation of a more than 30-year long soil moisture dataset by merging active (ERS AMI-SCAT and METOP ASCAT) and passive (SMMR, SSM/I, TMI, and AMSR-E) soil moisture retrievals (Wagner et al., 2012). This product at 0.25° spatial resolution aims to address the discrepancies that have been observed (e.g. Reichle et al., 2007 and Rüdiger et al., 2009) between different products developed for different (or the same) microwave sensors.

Other sensors that have been used in the retrieval of soil moisture include gravity recovery instruments (Seneviratne et al., 2010) and even global positioning system (GPS) receivers (Larson et al., 2008). Seasonal variations in terrestrial water content (TWC), including soil moisture, groundwater, snow and surface water, result in variations of the Earth's gravity field, which can be accurately measured with NASA/DLR's Gravity Recovery and Climate Experiment (GRACE) (e.g. Tapley et al., 2004, Wahr et al., 2004,Andersen et al., 2005, Swenson et al., 2008 and Rodell et al., 2009). However, separation of the soil moisture contribution to the total observed terrestrial water content requires knowledge of the other TWC components, making this technique complex and reliant on additional data being available. In addition, any estimates can only be done at much coarser spatial resolutions (at best 400 km) than those offered by microwave remote sensing. GPS satellite receivers use L-band frequencies, similar to the SMOS and SMAP sensors, and have shown the potential to estimate soil moisture at spatial resolutions of circa 300 m, but the effects of ground roughness and of different vegetation and soil types, requires further investigation (Larson et al., 2008). The reader is referred to Schmugge et al. (2002), Wigneron et al. (2003), de Jeu et al. (2008) and Seneviratne et al. (2010) for excellent reviews on the remote sensing of soil moisture.

Precipitation

Precipitation estimates with remote sensing can be generated using both in situ radar systems and space-borne sensors. Michaelides et al. (2009) and Tapiador et al. (2012)provide excellent reviews of remote sensing techniques for precipitation measurements. Typically, single polarisation weather radar systems have been used to derive rainfall rate, but limitations such as calibration issues and contamination by ground returns amongst others make them less than ideal for precipitation mapping. Recent technological advances have led to the development of more suitable polarimetric radar systems (also referred to as dual-polarisation radars) that overcome some of the limitations attached to single polarisation weather radars (Michaelides et al., 2009). For about three decades, rainfall has been retrieved from satellite instruments employing the visible, infrared and microwave regions of the spectrum. However, most of the satellite observations are discarded due to cloud contamination and issues with the variability of emissivity over land-water areas and especially in coastal areas. Another limitation is the coarse spatial resolution of the satellite sensors that ranges from 10 km^2 over land to 50 km^2 over the ocean, and the common assumption that rainfall is horizontally and vertically homogeneous.

According to Stephens and Kummerow (2007) the optimal approach to retrieve precipitation from space is by combining data from passive and active sensors. A major step towards this direction was made with the launch in 1997 of the space-borne Tropical Rainfall Measuring Mission (TRMM) that carries on-board two primary rainfall sensors; the 13.8 GHz Precipitation Radar (PR) and the TRMM Microwave Imager (TMI). Numerous algorithms for the retrieval of rainfall from TRMM data have since been developed (e.g., Iguchi et al., 2000 and Haddad et al., 1997), leading to the TRMM PR 2A25 and TRMM TMI 2A12 rainfall products. However, these two products exhibit spatial and seasonal differences that have yet to be explained (Wang et al., 2009) and are geographically limited to latitudes between 35°N–35°S. A third sensor on-board TRMM, the Visible and Infrared Radiometer (VIRS) can only provide indirect estimations of rainfall by means of precipitation indices.

Even though most available precipitation retrieval methods focus on rainfall rather than snowfall, high-frequency passive microwave wavebands (e.g. on AMSU) can provide snow detection and frozen precipitation estimations (Kongoli et al., 2004, Vila et al., 2007 and Michaelides et al., 2009). In fact, the Microwave Surface and Precipitation Products System (MSPPS) (now Microwave Integrated Retrieval System (MIRS)) is a suite of operational global products derived from the NOAA Advanced Microwave Sounding Unit (AMSU), including precipitation rate, total precipitable water and snow cover (Ferraro et al., 2005 and Vila et al., 2007).

The need for uniformly calibrated precipitation products that are independent of variation caused by different sensors is to be addressed by the international Global Precipitation Measurement (GPM) mission, which employs a constellation of passive microwave sensors. The GPM aims to provide high resolution satellite precipitation products every 2–3 h for use in various applications, such as hydrology and climatic studies. Building on the capabilities of the TRMM PR and TMI, the GPM-Core Observatory, which was launched in February 2014, covers latitudes between 65°N–65°S and contributes to the GPM constellation of sensors that combined cover the entire globe with datasets (Levels 0–3) freely available online. A global dataset for satellite-derived rainfall is the Global Satellite Mapping of Precipitation in Near-Real-Time (GSMaP_NRT) by the Japan Aerospace Exploration Agency (JAXA) Global Rainfall Watch System (Ver. 3.0), which provides global (60°N–60°S) rainfall rate (mm/h) at hourly intervals by combining various remote sensing data, including the GPM-Core Microwave Imager (GMI), TRMM TMI, DMSP (Defence Meteorological Satellite Programme) SSMIS, NOAA AMSU and METOP AMSU.

Snow Cover

Both optical and microwave remote sensing are used for snow cover mapping. Even though optical remote sensing is limited to mapping snow extent, much more information can be derived from microwave data; this includes snow water equivalent (SWE; mass of water per unit area), snow depth (SD) (a relevant measurement because of its relationship to SWE), snow extent, and snow state (wet/dry). The use of optical data is based on the fact that snow reflects radiation strongly in the visible and very poorly in the NIR regions of the EM spectrum. Dozier (1989) mapped snow using the ratio of Landsat TM reflectance in the visible and NIR, a technique that had been previously used in the 1970s (e.g. Kyle et al., 1978) and was later adapted to develop the normalised-difference snow index (NDSI) for MODIS (Hall et al., 2002 and Hall and Riggs, 2007) and Landsat ETM + (Salomonson and Appel, 2004). NOAA AVHRR has also been used to map snow cover (e.g. Gesell, 1989, Voigt et al., 1999, Akyürek and Şorman, 2002, Matikainen et al., 2002 and Hüsler et al., 2012) despite its coarse spatial resolution (1 km), because it has a long data archive and a very frequent revisit time (twice daily). A limitation of using optical data to map snow cover is that it is often difficult to separate clouds from snow, because snow may show a similar spectral response to clouds in the visible and TIR (Akyürek and Şorman, 2002 and Miller et al., 2005). Another challenge is that the reflectance of snow decreases with impurity and age (seasonal and even daily ageing) (König et al., 2001 and Dietz et al., 2012) and vegetation (resulting in surface heterogeneity) might also affect the reflected signal (Nolin, 2010).

Snow attenuates microwave radiation, which is positively correlated with the amount of snow in the snowpack (Chang et al., 1987, König et al., 2001 and Clifford, 2010). In a similar manner to soils, the dielectric constants of ice and water are very different, which enables the use of microwave remote sensing to map snow. The depth of snow that can be mapped depends on the microwave wavelength; ranging from a minimum of around 2 cm (Dietz et al., 2012) to a maximum of 100 times the microwave wavelength (Clifford, 2010). In microwave remote sensing of snow, the size and shape of the snow crystals can affect the estimations (Foster et al., 1999), as well as depth, temperature, snow state and density (Schmugge et al., 2002). In addition, the presence of dense vegetation and liquid water in the snowpack can lead to underestimation of SD and SWE (Dietz et al., 2012), when passive microwave sensors are used. Using early morning satellite overpasses (local time) can minimise the impact of wet snow on snow mapping (Schmugge et al., 2002). Alternatively, active microwave remote sensing is more suitable for mapping wet snow (e.g. Baghdadi et al., 1997 and König et al., 2001), with the added bonus of finer spatial resolutions than passive microwave satellite sensors.

In the 1980s, the US National Weather Service (NWS) developed operational remote sensing products for snow hydrology, using NOAA AVHRR and GOES to produce periodic river basin snow cover extent maps (Schmugge et al., 2002) and since 2002 the Aqua AMSR-E has been used to produce a global SWE Level 3 product. MODIS data have also been used to derive a suite of snow products at various spatial (500 m–30 km) and temporal (daily-monthly) resolutions. In addition, the snowmelt runoff model (SRM) was adapted to use remotely-sensed snow cover information at basin-scales (Martinec et al., 2008). The reader can refer to, for example, König et al. (2001), Schmugge et al. (2002), Clifford (2010), Nolin (2010) and Dietz et al. (2012) for more detailed reviews on the remote sensing of snow cover.

Land Surface Temperature (LST)

Atmospherically corrected brightness temperature retrieved using remote sensing thermal sensors are used in land surface temperature retrieval. There are three techniques for the retrieval of LST from remote sensing data; single channel, generalised split window (GSW) and dual angle (or dual algorithm) (DA) approach. The single channel method requires a comprehensive radiative transfer model and atmospheric profiles. The split window approach, combines two thermal wavebands and accounts for atmospheric effects based on the differential absorption in adjacent infrared bands. However, because of the large emissivity differences between thermal wavebands over land due to vegetation, topography and soil, this method can be problematic. Land surface emissivity studies have been conducted to overcome this problem and various techniques have been proposed in the literature for NOAA AVHRR and Terra/Aqua MODIS (e.g. Wan and Dozier, 1996, Becker and Li, 1990, Mao et al., 2005 and Pinheiro et al., 2006), (A)ATSR (e.g. Sòria and Sobrino, 2007 and Galve et al., 2009) and MSG Spinning Enhanced Visible and Infrared Imager (SEVIRI) (e.g. Jiang and Li, 2008, Qian et al., 2013 and Freitas et al., 2013). The Terra/Aqua MOD11 LST product uses a split-window approach that utilises thermal bands 31 and 32. The dual angle approach is based on the differential absorption of a single waveband due to different viewing angles. In order to apply this technique two scenes simultaneously acquired by different satellites, e.g. the geostationary MSG and polar-orbiting Television Infrared Observation Satellites (TIROS-series) (Chédin et al., 1982) or a single scene from a satellite with multi-angle acquisition capability, such as the (Advanced) Along-Track Scanning Radiometer ((A)ATSR) on-board Envisat, can be used. The second option is more commonly employed using (A)ATSR data (e.g. Kustas and Norman, 1997, Sobrino et al., 1996, Sobrino et al., 2004 and Sòria and Sobrino, 2007) or GOES and Multifunction Transport Satellite (MTSAT) data (Freitas et al., 2013). Finally, a robust empirical technique has been developed specifically for the Terra Advanced Spaceborne Thermal Emission and Reflection Radiometer (ASTER); the Temperature Emissivity Separation (TES) algorithm (Gillespie et al., 1998) that uses ASTER TIR wavebands to estimate LST. The two main advantages of using Terra ASTER for LST retrievals are its fine spatial resolution (90 m) and more frequent revisit capability (twice daily) compared to other thermal sensors (e.g. NOAA AVHRR, (A) ATSR, MODIS, etc.).

Even though mostly thermal sensors are used in LST-retrievals, there exist a few studies where passive microwave sensors were employed (Tomlinson et al., 2011), which have the advantage of providing measurements even on cloudy days. Despite the coarse spatial resolution of passive microwave sensors (5–70 km depending on frequency), the AMSR-E (Chen et al., 2011) and more commonly the SSM/I (e.g. McFarland et al., 1990,Basist et al., 1998, Peterson et al., 2000, Williams et al., 2000 and Fily et al., 2003) have been used to estimate LST. For examples of in-depth reviews on the remote sensing of land surface temperature, the reader can refer to Prata et al. (1995), Dash et al. (2001),Schmugge et al. (2002), Tomlinson et al. (2011) and Li et al. (2013).

MAPPING CATCHMENT LAND COVER AND OTHER DRIVERS OF LAKE BEHAVIOUR

There are multiple catchment drivers of lake change due to intensification of human activities and climate change. Lakes across the globe can be affected by a combination of different stressors and pressures, based on vicinity to human establishments, type and intensity of human activities within the catchment, type and ratio of natural over altered landscape in the catchment, protection status and geographical location. Human activities such as agriculture, deforestation, urbanisation and industrial activities can be grouped under the umbrella of land cover/land use change (LULC), which may have various direct and indirect effects on lake water quality, for example increased nutrient loading and eutrophication (e.g. Huang et al., 2013 and Keatley et al., 2011), pollution (e.g. IJC, 1980), acidification (e.g. Nghiem et al., 2011), anoxic conditions (e.g. Valero-Garcés et al., 2000). The impact of other human activities such as tourism and commercial fishing on lake ecosystem status and ecosystem services is not possible to map and quantify with remote sensing unless their impact is measured indirectly. For example, by mapping forest area reduction due to tourism-related development (Chaplin and Babyn, 2013) or estimating epipelagic fish distribution and abundance in shallow waters (Churnside et al., 2003).

Despite the potential of remote sensing to provide direct estimation of a wide variety of ecosystem functions and services (cf. those described above) both Andrew et al. (2014) and de Araujo Barbosa et al. (2015) note that the majority of published studies that have attempted to map ecosystem functions and services (and by inference related catchment processes) do so using indirect proxies, such as land cover and land use data, derived mainly from remotely sensed sources (Table 3). It is pertinent, therefore, to reflect briefly on how land cover information is derived, and more importantly, the limitations and considerations to its application for monitoring change in ecosystem functions and services of large areas.

Table 3: Linkage between selected catchment/landscape features, fluvial processes, ecosystems services and their detection with remote sensing. (builds on Andrew et al., 2014 and Large and Gilvear, 2014). *Indicative*means the list is not exhaustive.

Catchment feature, or land cover	Inferred processes and characteristics	Ecosystem functions	Ecosystem services (*indicative*)	RS products (*required*)	Remote sensing source (*indicative*)
Catchment delineation	Catchment area, morphometry and hydrodynamic behaviour	Range of freshwater and terrestrial habitats	Water quantity and quality; seasonality	Topography	DEM derived from varied sources, including LiDAR, RADAR, UAV-based photography and aerial photography
Slope	Low slopes reduce energy gradient for transfer of water, sediment & nutrients, promoting storage and biogeochemical processing; reworking of sediment in active reaches	Hydraulic diversity; channel dynamism; habitat creation; sediment storage; habitat heterogeneity; increased wetted perimeter	Flood mitigation; water quality	Topography	DEM derived from varied sources, including LiDAR, RADAR, UAV-based photography and aerial photography
Catchment water balance and water level fluctuations	Water dynamics, quantity, quality and provenance of water and sediment runoff controlling biogeochemical processes including nutrient flux; snow accumulation; runoff volumes and timings; groundwater recharge and connectivity within alluvial flatlands; seasonal variations in water levels; movement of migratory species; crop phenology	Provisioning; supporting and regulating; cultural climate regulation; removal of pollutants; fish production	Water supply; biodiversity; carbon sequestration; flood mitigation and regulation; migratory species; tourism	Evapotranspiration Precipitation Soil moisture Water, snow/ice extent and duration Water level Groundwater	Thermal RS, VIs, climate data RADAR, PM RADAR (e.g. SMAP) Optical, RADAR, PM RADAR altimetry Gravity surveys, subsidence
Hyporheic zones and groundwater	Links to water flow and water quality in surface waters	Groundwater dependent terrestrial ecosystems including wetlands	Water supply; Flood mitigation and regulation; Biogeochemical filtration; biodiversity	Land cover map Species map, spectral diversity	Multispectral and multi-temporal RS, HS. Range or variability of biochemistry, spectral indices or reflectance variation
Hydromorphological alteration	Elimination of flood inundation; loss of channel dynamism; altered supply of sediment & nutrients	Loss of natural land cover; hydrological alteration; habitat change	None	Land cover Feature extraction (e.g. embankment length, presence of dams etc)	Multispectral RS, HS

Riparian/river bank woodland	Shading, allochthonous leaf litter and woody debris input	Habitat creation and hydraulic diversity; cooling of water, food source	Biodiversity; fisheries	Spectral diversity Proxy indicators (e.g. canopy structure, productivity, gap fraction)	Range or variability of biochemistry, spectral indices or reflectance variation Varied e.g. LiDAR, Multispectral, HS, HY, multi-temporal RS
Floodplain physical habitat mosaic	Hydromorphological heterogeneity and channel dynamism; varied land use patterns	Range of freshwater-terrestrial habitats; ecotone creation	Biodiversity; fisheries	Spectral diversity, species map, land cover map, LAI, biomass	Range or variability of biochemistry, spectral indices or reflectance variation, LiDAR, image texture, Multispectral, HS, HY
Catchment disturbance	Potential for changing hydrological regime, channel dynamism, nutrient cycling, water quality, biodiversity	Potential loss/degradation of natural land cover; hydrological alteration	Changed natural ecosystem services	Change in biomass/plant traits Plant stress Fire detection Drought monitoring	Multi-temporal RS VIs, HY Thermal anomalies, fire products (e.g. MODIS) Water content, surface temperature & ET
Wetlands	Plant and animal succession processes; enhanced nutrient cycling and storage; semi-aquatic habitats	Carbon sequestration, phosphorous uptake and denitrification; habitat heterogeneity; flow attenuation; refugia	Water supply; water quality; biodiversity	Land cover map Species map; spectral diversity	Multispectral and multi-temporal RS, HS. Range or variability of biochemistry, spectral indices or reflectance variation
Floodplain forest	Substrate stabilisation; enhanced hydraulic processes	Flow attenuation, enhanced nutrient cycling and storage; habitat heterogeneity	Carbon sequestration; flood mitigation; biodiversity; water quality	Land cover map; biomass	Multispectral and multi-temporal RS, HS
Floodplain lakes	Water storage; nutrient cycling	Refugia; habitat heterogeneity	Water supply; carbon sequestration; water quality; fisheries; biodiversity	Land cover map Water colour (phytoplankton, suspended matter, CDOM) and water clarity, LWST, ice cover	Multispectral and multi-temporal RS, HS. Spectral analysis, multispectral RS. Thermal RS

Agriculture	Potential for increased runoff; enhanced sediment input; water quality deterioration	Loss of natural land cover; hydrological alteration	Natural ecosystem services reduced with increased crop production	Land cover map, LAI	Multispectral and multitemporal RS, HS, HY, LiDAR
Woodland plantation	Substrate stabilisation; enhanced hydraulic roughness	Flow attenuation; biomass increase	Timber production; flood mitigation	Land cover map, biomass	Multispectral and multitemporal RS, HS, HY, LiDAR
Urban areas	Potential for increased runoff and water quality deterioration	Loss of natural land cover; hydrological alteration	None	Land cover map	Multispectral and multi-temporal RS, HS

LiDAR (Light Detection and Ranging); MS (Multispectral, e.g. Landsat Thematic Mapper); HS (High spatial resolution imagery (e.g. WorldView-3, Google Earth); HY (hyperspectral imagery/data); PM (passive microwave); VIs (vegetation indices); ET (evapotranspiration); CDOM (coloured dissolved organic matter); LWST (lake water surface temperature).

Table 4: Example global land cover datasets suitable for large area land cover analysis.

Product	Sensor	Satellite	Spatial resolution	Temporal resolution
GLC2000[1]	VGT	SPOT	1 km	Single date: 2000
MCD12Q1[2]	MODIS	Aqua, Terra	500 m	Annual: 2001–2007
ESA CCI-LC[3]	MERIS	ENVISAT	300 m	Three epochs: 2008–2012; 2003–2007; 1998–2002
ESA CCI LC Seasonality (Three products: vegetation, snow cover & burned area)	MERIS	ENVISAT	300 m	Annual

1 European Commission (EC) Global Vegetation Monitoring (GVM) Unit Global Land Cover (GLC) 2000 product.

2 NASA Land Processes Distributed Active Archive Center (LP DAAC) Land Cover Type product.

3 European Space Agency (ESA) Climate Change Initiative (CCI) Land Cover (LC) product.

Perhaps the most common and one of the most basic forms of land cover analysis using remotely sensed data is land cover classification. This relates spectral and textural information within an image to specific land cover classes (as defined by the user) to produce a land cover map, from which land use may be inferred. There are many methods of image classification, from 'traditional' statistically-based per-pixel methods that assign individual pixels to land cover classes based upon their spectral response, to object-based approaches which segment the landscape into parcels based upon spectral response, shape, texture and other object characteristics. Additionally, a classification may be 'hard' where pixels are assigned to a single class, or 'soft' where pixels are assigned multiple class memberships. The latter is particularly relevant, when information relating to small scale changes in land cover are important and where measuring change requires greater precision. More often it is used when the spatial resolution of the instrument does not match the intrinsic scale of land cover heterogeneity within an area of interest. For example, Probeck et al. (2005) were able to derive subscale land cover information from relatively coarse (1 km) NOAA AVHRR imagery of the upper Danube by using a fuzzy classification approach, from which results were produced that compared favourably with independent datasets. A similar approach was used by Ludwig et al. (2003b). In both cases the fuzzy classification method used was spectral unmixing but alternatives include the use of artificial neural networks (e.g.Aitkenhead et al., 2008) and other types of machine learning.

The methods used to classify remotely sensed data are a field of research in themselves, and whilst helpful reviews of the current state of land cover extraction from remote sensing are available (e.g. Cihlar, 2000, Aplin, 2004 and Li et al., 2014), the myriad of methods and approaches leads to variability in the land cover map outputs derived by individual researchers and groups, limiting their wider applicability. To address this, some studies have employed the use of standardised LULC products (e.g. Table 4). These products tend to be produced from a single satellite sensor, removing variability associated with multiple sensor characteristics. Additionally, products may be single-year or multi-year products, enabling change to be detected using a standardised set of data characteristics and land cover classes. Thus, the use of a LULC product may reduce some of the uncertainty associated with individuals undertaking a classification exercise in isolation. However, the uptake in use of LULC products has been slow, put down partly to a lack of awareness but also a lack of confidence in spatial accuracy (Pfeifer et al., 2012 and Congalton et al., 2014). Most operational products are still restricted to coarse spatial resolution data (Table 4), which whilst providing frequent temporal updates may be a deterrent towards their use for some ecosystem studies.

An additional limitation of LULC analysis is determining a biophysical basis for linking particular LULC classes to ecosystem functions and services. For individuals and groups undertaking their own classification there is often a tension when defining appropriate LULC classes to map between those features and classes that can be extracted from remotely sensed data, i.e. those that are spectrally separable, and those features and classes that are directly correlated with ecosystem function and/or catchment processes. The resulting class definitions are, therefore, often a pragmatic solution using features and LULC classes that can be detected remotely and using these as proxy indicators of the environmental variable of interest (e.g. Table 3). These proxies, however, may have limited relevance, being often the result of hypothesised but largely untested relationships (Andrew et al., 2014 and Seppelt et al., 2011) leading to unreliable outputs when compared to field observations (e.g. Eigenbrod et al., 2010). This potentially becomes even more problematic when using standardised LULC products, where there are often inconsistencies in class definitions between products (Congalton et al., 2014) and where different products are produced at different spatial resolutions, thus limiting their use as proxy indicators.

Also, by assuming that the same ecosystem value can be applied to the whole of a particular class ignores any scale dependency in ecosystem processes and leads to inaccuracies when compared with independently observed data (Eigenbrod et al., 2010). Thus ecosystem services derived from land cover information will be highly dependent upon the efficacy of linking a land cover class to a particular ecosystem service/function, the spatial resolution of the data and the accuracy of image classification (de Araujo Barbosa et al., 2015).

The accuracy of image classification is an important, but often ignored factor in determining the reliability of areal change in land cover class(es) derived from a classified map. Change in key indicator classes, such as woodland or urban areas, are important in understanding changes in lake water quality, habitat and other critical ecosystem services. However, as Olofsson et al. (2013) effectively argue, information relating to the accuracy of land cover outputs and products, including commonly used measures such as overall accuracy and the kappa coefficient, fail to provide enough information to effectively judge the effect that classification error has on estimates of areal change. To address this they suggest mapped areas should be adjusted using a stratified estimator approach (Olofsson et al., 2013). At the very least, the uncertainty and error associated with the mapping of ecosystem properties should be addressed and reported, with Rocchini et al. (2013) providing a useful summary of good practise for this very purpose. Even when using globally available LULC products the stated accuracy is not always acceptable for the purpose of mapping change (Congalton et al., 2014).Karimi and Bastiaanssen (2015) noted that the overall accuracies reported for global land cover maps varied between 69 and 87%, suggesting that such products should be used with caution in water accounting applications.

All of the limitations above suggest that whilst LULC information derived from remotely sensed sources has the potential for quantifying and mapping selected catchment and ecosystem functions and services across large areas, the limitations in terms of applicability (biophysical basis of land cover classes), data quality (accuracy) and data properties (spatial and temporal resolution) should be considered carefully when deciding how such data should be used. Fuller et al. (2003) provide a useful commentary on experiences of mapping the UK with remotely sensed data highlighting the limitations of comparing between products, whilst Seppelt et al. (2011) provide a useful review of critical questions that should be raised when reviewing ecosystem service assessments at a regional scale, which are directly relevant to remote sensing-based studies.

REMOTE SENSING OF LAKE LEVEL AND VOLUME

Finally, whilst this review has thus far focussed on catchment functions and characteristics, it is also pertinent to briefly assess the potential of remote sensing to provide information regarding lakes themselves. The remote sensing techniques and sensors used to map lake water quality are reviewed by Tyler et al., in this Special Issue. Remote sensing can also have a role to play in detecting water bodies (e.g. Verpoorter et al., 2014), and estimating lake water level change and volume.

Both water level and volume are important with respect to lake ecology, being determinants of water residence time, flushing rate and mixing. They are also highly influenced by human activities, with the construction of dams and impoundments, and associated freshwater withdrawal for domestic, agricultural and industrial use responsible for significantly altering water level (and, thus, volume) in some lakes and reservoirs globally (in Lake Chad, for example). However, not all lake water level change is caused by direct human intervention. Climate change and natural climatic fluctuations are also important drivers of water level change/fluctuation in freshwater bodies, particularly in closed lake basins in endorheic regions (Hutchinson, 1957).

Remote sensing of lake water level

Water level change can be retrieved from remote sensing altimetry data (e.g. Birkett, 1995, Crétaux and Birkett, 2006, Hwang et al., 2011, Mercier et al., 2002, Ponchaut and Cazenave, 1998 and Swenson and Wahr, 2009) with very high accuracy (e.g. 3–33 cm;Birkett and Beckley, 2010; 25–53 cm; Frappart et al., 2006). Altimeters can be radar (microwave) or laser (visible and IR), but the second are sensitive to the presence of clouds. They both transmit pulses and receive the reflected signal, so the theoretical basis for measurements is similar (Duan and Bastiaanssen, 2013). Lake surface water level is estimated by calculating the difference between satellite height in respect to a reference surface (e.g. Earth's centre) and the satellite-to-surface range (calculated by the time taken for the pulse to return to the sensor since transmission), correcting at the same time for instrumental, atmospheric and other effects (Duan and Bastiaanssen, 2013). The disadvantage of altimeters is that they can only return measurements from along their track, which does not cover the globe. As a result, only specific water bodies (that fall into the satellite's track) can be detected. Laser altimeters such as the Geoscience Laser Altimeter System (GLAS) on-board ICESat (Ice, Cloud, and land Elevation Satellite) are more suitable for relatively small water bodies due narrower footprint size (~ 100 m) compared to radar altimeters (several kilometres).

Several radar altimeters were until recently or currently operational, including ERS Radar Altimeter (RA) and Envisat RA-2, the Poseidon sensors on-board TOPEX/Poseidon, Jason-1 and Jason-2 (or Ocean Surface Topography Mission, OSTM), and GeoSat Fellow On (GFO) Radar Altimeter. These sensors were used to create three distinguished databases that include lake water level estimates: (1) ESA's River and Lake project, (2) US Department of Agriculture (USDA) Global Reservoir and Lake Monitoring (GRLM) programme, and (3) Laboratoire d'Etudes en Géophysique et Océanographie Spatiales (LEGOS) and Geodesy, Oceanography and Hydrology from Space (GOHS) Hydroweb. A fourth database is based on ICESat laser altimetry; ICESat-GLAS level 2 Global Land Surface Altimetry data (ICESat-GLAS). For an excellent overview of these databases see Duan and Bastiaanssen (2013).

Remote Sensing of Lake Water Quantity (Volume)

Lake water quantity (volume) cannot be measured directly from remote sensing. Water level information is combined with bathymetric information of the water body in order to produce estimates of lake volume. Apart from using traditional sonar techniques, bathymetric maps can also be produced using GPS and laser transit survey data (e.g.Wilcox and Los Huertos, 2005), and in fact the notion of a laser bathymeter is not new (Muirhead and Cracknell, 1986). To replace the need for bathymetric predictions, new techniques that make use of visible and IR-based lake surface area estimations have been developed for the retrieval of lake water volume (e.g. Duan and Bastiaanssen, 2013 and Lu et al., 2013). Finally, satellite altimetry-derived water levels of lakes can provide estimates of lake water volume change, when combined with satellite or otherwise-derived surface water area (e.g. Becker et al., 2010) or detailed bathymetric maps (Crétaux and Birkett, 2006).

OBSERVATIONS AND CONCLUSIONS

We began this review by presenting a suite of datasets (Fig. 2), some derived from remote sensing, that have the potential to capture both status and change in lake catchments. This is presented as a precursor to understanding both the causes and drivers of lake change, and which can complement direct observations of lake behaviour by satellite remote sensing (Tyler et al., in this issue). Whilst by no means an exhaustive review of the capabilities of remote sensing and geospatial technologies in this field, this paper points towards some of the key catchment variables that can be estimated, both directly and indirectly, as well as the limitations and challenges currently faced.

From the work that is reviewed it is clear that remote sensing is an important tool for mapping ecosystem functions and aiding our understanding of catchment drivers of lake change, with promising results frequently reported. However, it is also clear that many of the results are highly specific to a particular time and/or place, and thus a manifestation of the 'one time one place' approach identified by Woodcock (2002) as a common cause of uncertainty in the application of remotely sensed data for other applications. Time and again various reviews point to continued issues with respect to multiple scales, sensors, reliability and other methodological constraints that limit the operational application of remote sensing for extracting critical ecosystem functions. Many suggest further work is needed to develop techniques that combine multi-source information to provide disaggregated products for catchment monitoring and modelling.

In part a response to these issues, several reviews refer to increasing efforts to integrate information across a variety of sensors and scales, and the development of many standardised remotely sensed products (Andrew et al., 2014 and Pfeifer et al., 2012). Issues of spatial and biophysical relevancy persist, but with the advent of higher spatial resolution global datasets becoming available (e.g. Hansen and Loveland, 2012) and the introduction of new complementary in situ sensor systems capable of measurements at high temporal and spatial -frequencies (Crawford et al., 2014) at least some of these concerns will continue to be addressed.

As stated in our introduction, the future of remote sensing for catchment, ecosystem and lake-based studies looks very promising. ESA's Copernicus programme will not only include Sentinel's 2 and 3 (which themselves will include multiple versions, e.g., Sentinel 3A, 3B and 3C, thereby providing a constellation of 'observers') but also includes a C-band SAR sensor on-board Sentinel-1 (launched April 2014) to provide observations of ice, oil spills and land surface monitoring. Additionally, future missions such as ESA's BIOMASS (planned for 2020) will provide enhanced information on above-ground biomass dynamics at 50 m spatial resolution, whilst ESA's FLEX (Fluorescence Explorer), Germany's EnMAP (30 m spatial resolution) and NASA's HYSPIRI (60 m) satellite sensors will allow regional to global assessments of plant biophysical and biochemical status, biogeochemical cycling, erosion monitoring and ecosystem fragmentation and disturbance. Thus the range of spaceborne sensors for deriving quantitative estimates of critical ecosystem and catchment variables appears to be in good shape looking forward.

In addition to new optical and SAR sensors described above and elsewhere in this review, there is clearly much progress in developing geospatial technologies which will have the potential to revolutionise the mapping of ecosystem functions. These include the development of UAV platforms for low cost, high resolution data gathering, as well as new hyperspectral sensor technologies and multispectral LiDAR that have the potential for direct estimation of biophysical variables (Woodhouse et al., 2011) and plant traits, removing the need for reliance on simple proxies. Furthermore, the existence of platforms that facilitate the use of extensive datasets and long remote sensing archives, such as the Google Earth Engine (GEE) for environmental data analysis, can and will provide, in part, the necessary infrastructure that the era of 'Big Data' asks for.

To maximise the exploitation of new technologies for ecosystem service assessmentAndrew et al. (2014) note the need for close collaboration between remote sensing scientists, ecologists and social scientists. Clearly, for this application the same is true of remote sensing scientists and limnologists, hydrologists, ecologists and many more to truly understand the drivers of lake change and behaviour across the globe, with the potential for much of this to be informed by remote sensing observations. Multidisciplinary collaboration is a necessity for the integration and assimilation of geospatial data and new technologies in hydrological modelling and traditional limnological studies, and in order to be able to understand and interpret both the uncertainties of the data used and the research outputs. The inclusion of remote sensing scientists in limnological research is therefore considered prudent and often an essential step towards understanding remote sensing data and the specific uncertainties, interpretation and limitations related to remote sensing products before their use. In this respect, projects such as the UK Natural Environment Research Council-funded GloboLakes project (Global Observatory of Lake Responses to Environmental Change) is one of several initiatives to address issues of lake change in just such a way, and which will lead to paradigm shifting developments in the characterisation of the planet's freshwater resources and responses to environmental change.

ACKNOWLEDGEMENTS

This work was supported by NERC Consortium Grant NE/J024279/1 — GloboLakes (Global Observatory of Lake Responses to Environmental Change): www.globolakes.ac.uk The Authors also wish to thank the constructive feedback from the anonymous reviewers.

REFERENCES

1. Aitkenhead, M.J., Flaherty, S., Cutler, M.E.J., 2008. Evaluating neural networks and evidence pooling for land cover mapping. Photogramm. Eng. Remote. Sens. 74, 1019–1032.
2. Akyürek, Z., Şorman, A.Ü., 2002. Monitoring snow-covered areas using NOAA-AVHRR data in the eastern part of Turkey. Hydrol. Sci. J. 47, 243–252. http://dx.doi.org/10. 1080/02626660209492927.
3. Alcamo, J., Döll, P., Henrichs, T., Kaspar, F., Lehner, B., Rösch, T., Siebert, S., 2003. Development and testing of the WaterGAP 2 global model of water use and availability. Hydrol. Sci. J. 48, 317–337. http://dx.doi.org/10.1623/hysj.48.3.317.45290.
4. Andersen, O.B., Seneviratne, S.I., Hinderer, J., Viterbo, P., 2005. GRACE-derived terrestrial water storage depletion associated with the 2003 European heat wave. Geophys. Res. Lett. 32, L18405. http://dx.doi.org/10.1029/2005GL023574.
5. Andrew, M.E., Wulder, M.A., Nelson, T.A., 2014. Potential contributions of remote sensing to ecosystem service assessments. Prog. Phys. Geogr. 38, 328–353.
6. Aplin, P., 2004. Remote sensing: land cover. Prog. Phys. Geogr. 28, 283–293.
7. Awange, J.L., Sharifi, M.A., Ogonda, G., Wickert, J., Grafarend, E.W., Omulo, M.A., 2008. The falling Lake Victoria water level: GRACE, TRIMM and CHAMP satellite analysis of the lake basin. Water Resour. Manag. 22, 775–796.
8. Baghdadi, N., Gauthier, Y., Bernier, M., 1997. Capability of multitemporal ERS-1 SAR data for wet-snow mapping. Remote Sens. Environ. 60, 174–186.
9. Bartalis, Z., Wagner, W., Naeimi, V., Hasenauer, S., Scipal, K., Bonekamp, H., Figa, J., Anderson, C., 2007. Initial soil moisture retrievals from the METOP-A Advanced Scatterometer (ASCAT). Geophys. Res. Lett. 34, L20401. http://dx.doi.org/10.1029/ 2007GL031088.
10. Basist, A., Grody, N.C., Peterson, T.C., Williams, C.N., 1998. Using the Special Sensor Microwave/Imager to monitor land surface temperatures, wetness and snow cover. J. Appl. Meteorol. 37, 888–911.
11. Becker, F., Li, Z.-L., 1990. Towards a local split window method over land surfaces. Int. J. Remote Sens. 11, 369–393. http://dx.doi.org/10.1080/01431169008955028.
12. Becker, M., LLovel, W., Cazenave, A., Güntner, A., Crétaux, J.-F., 2010. Recent hydrological behavior of the East African great lakes region inferred from GRACE, satellite altimetry and rainfall observations. Compt. Rendus Geosci. 342, 223–233.
13. Bindlish, R., Jackson, T.J., Wood, E., Gao, H., Starks, P., Bosch, D., Lakshmi, V., 2003. Soil moisture estimates from TRMM Microwave Imager observations over the Southern United States. Remote Sens. Environ. 85, 507–515.
14. Birkett, C., 1995. The contribution of TOPEX/POSEIDON to the global monitoring of climatically sensitive lakes. J. Geophys. Res. 100, 25179–25204.
15. Birkett, C.M., Beckley, B., 2010. Investigating the performance of the Jason-2/OSTM radar altimeter over lakes and reservoirs. Mar. Geod. 33, 204–238.
16. Blad, B.L., Rosenberg, N.J., 1976. Evaluation of resistance and mass transport evapotranspiration models requiring canopy temperature data. Agron. J. 68, 764–769.
17. Boughton, W., 2005. Catchment water balance modelling in Australia 1960–2004. Agric. Water Manag. 71, 91–116.
18. Chang, A.T.C., Foster, J.L., Hall, D.K., 1987. Nimbus-7 SMMR derived global snow cover parameters. Ann. Glaciol. 9, 39–44.
19. Chaplin, J., Babyn, L., 2013. Using remote sensing and GIS to investigate the impacts of tourism on forest cover in the Annapurna Conservation Area, Nepal. Appl. Geogr. 43, 159–168.
20. Chédin, A., Scott, N.A., Berroir, A., 1982. A single-channel, double-viewing angle method for sea surface temperature determination from coincident METEOSAT and TIROSN radiometric measurements. J. Appl. Meteorol. 21, 613–618.
21. Chen, S., Chen, X., Chen, W., Su, Y., Li, D., 2011. A simple retrieval method of land surface temperature from AMSR-E passive microwave data — a case study over Southern
22. China during the strong snow disaster of 2008. Int. J. Appl. Earth Obs. Geoinf. 13, 140–151.
23. Churnside, J.H., Demer, D.A., Mahmoudi, B., 2003. A comparison of Lidar and echosounder measurements of fish schools in the Gulf of Mexico. ICES J. Mar. Sci. 60, 147–154. http://dx.doi.org/10.1006/jmsc.2002.1327.

24. Cihlar, J., 2000. Land cover mapping of large areas from satellites: status and research priorities. Int. J. Remote Sens. 21, 1093–1114.

25. Clifford, D., 2010. Global estimates of snow water equivalent from passive microwave instruments: history, challenges and future developments. Int. J. Remote Sens. 31, 3707–3726. http://dx.doi.org/10.1080/01431161.2010.483482.

26. Congalton, R.G., Jianyu, G., Yadav, K., Thenkabail, P., Ozdogan, M., 2014. Global land cover mapping: a review and uncertainty analysis. Remote Sens. 6, 12070–12093.

27. Crăciunescu, V., Flueraru, C., Stăncălie, G., 2010. The usage of the historical cartographic datasets and the remote sensing data for the better understanding and mapping of the 2006 Danube floods in Romania. Acta Geod. Geophys. Hung. 45, 112–119.

28. Crawford, J.T., Loken, L.C., Casson, N.J., Smith, C., Stone, A.G., Winslow, L.A., 2014. Highspeed limnology: using advanced sensors to investigate spatial variability in biogeochemistry and hydrology. Environ. Sci. Technol. 49, 442–450.

29. Crétaux, J.-F., Birkett, C., 2006. Lake studies from satellite radar altimetry. Compt. Rendus Geosci. 338, 1098–1112.

30. dall'Amico, J.T., Schlenz, F., Loew, A., Mauser, W., 2012. First results of SMOS soil moisture validation in the Upper Danube Catchment. IEEE Trans. Geosci. Remote Sens. 50, 1507–1516.

31. Dash, P., Göttsche, F.-M., Olesen, F., Fischer, H., 2001. Retrieval of land surface temperature and emissivity from satellite data: physics, theoretical limitations and current methods. J. Indian Soc. Remote Sens. 29, 23–30.

32. de Araujo Barbosa, C.C., Atkinson, P.M., Dearing, J.A., 2015. Remote sensing of ecosystem services: a systematic review. Ecol. Indic. 52, 430–443.

33. de Groot, R.S., Alkemade, R., Braat, L., Hein, L., Willemen, L., 2010. Challenges in integrating the concept of ecosystem services and values in landscape planning, management and decision making. Ecol. Complex. 7, 260–272.

34. de Jeu, R.A.M., Wagner, W., Holmes, T.R.H., Dolman, A.J., van de Giesen, N.C., Friesen, J., 2008. Global soil moisture patterns observed by space borne microwave radiometers and scatterometers. Surv. Geophys. 29, 399–420. http://dx.doi.org/10.1007/s10712- 008-9044-0.

35. Deus, D., Gloaguen, R., Krause, P., 2013. Water balance modeling in a semi-arid environment with limited in situ data using remote sensing in Lake Manyara, East African Rift, Tanzania. Remote Sens. 5, 1651–1680. http://dx.doi.org/10.3390/rs5041651.

36. Diak, G.R., Mecikalski, J.R., Anderson, M.C., Norman, J.M., Kustas, W.P., Torn, R.D., DeWolf, R.L., 2004. Estimating land surface energy budgets from space; review and current efforts at the University of Wisconsin-Madison and USDA-ARS. Am. Meteorol. Soc. http://dx.doi.org/10.1175/BAMS-85-1-65.

37. Dietz, A.J., Kuenzer, C., Gessner, U., Dech, S., 2012. Remote sensing of snow — a review of available methods. Int. J. Remote Sens. 33, 4094–4134.

38. Dokulil, M.T., 2013. Impact of climate warming on European inland waters. Inland Waters 4, 27–40. http://dx.doi.org/10.5268/IW-4.1.705.

39. Döll, P., Kaspar, F., Lehner, B., 2003. A global hydrological model for deriving water availability indicators: model tuning and validation. J. Hydrol. 270, 105–134.

40. Downing, J.A., Prairie, Y.T., Cole, J.J., Duarte, C.M., Tranvik, L.J., Striegl, R.G., McDowell, W.H., Kortelainen, P., Caraco, N.F., Melack, J.M., Middelburg, J.J., 2006. The global abundance and size distribution of lakes, ponds, and impoundments. Limnol. Oceanogr. 51, 2388–2397.

41. Dozier, J., 1989. Spectral signature of alpine snow cover from the Landsat Thematic Mapper. Remote Sens. Environ. 28, 9–22.

42. Duan, Z., Bastiaanssen, W.G.M., 2013. Estimating water volume variations in lakes and reservoirs from four operational satellite altimetry databases and satellite imagery data. Remote Sens. Environ. 134, 403–416.

43. Eagleman, J.R., Ulaby, F.T., 1975. Remote sensing of soil moisture by Skylab radiometer and scatterometers sensors. J. Astronaut. Sci. 23, 147–159.

44. Economics, L., 2015. The Case for Space. Report to the UK Space Agency Available at http://www.ukspace.org/wp-content/uploads/2015/07/LE-Case-for-Space-2015-Full-Report.pdf.

45. Eigenbrod, F., Armsworth, P.R., Anderson, B.J., Heinemayer, A., Gillings, S., Roy, D.B.,

46. Thomas, C.D., Gaston, K.J., 2010. The impact of proxy-based methods on mapping the distribution of ecosystem services. J. Appl. Ecol. 47, 377–385.

47. FAO, 2014. Coping with water scarcity: An action framework for agriculture and food security. FAO Water Report 38. Food and Agriculture Organization of the United Nations, Rome.

48. FAO AQUASTAT, 2001. Available at http://www.fao.org/nr/water/aquastat/main/index.stm.

49. Ferraro, R.R., Weng, F., Grody, N.C., Zhao, L., Meng, H., Kongoli, C., Pellegrino, P., Qiu, S., Dean, C., 2005. NOAA operational hydrological products derived from the Advanced Microwave Sounding Unit. IEEE Trans. Geosci. Remote Sens. 43, 1036–1049.

50. Fily, M., Royer, A., Goïta, K., Prigent, C., 2003. A simple retrieval method for land surface temperature and fraction of water surface determination from satellite microwave brightness temperatures in sub-Arctic areas. Remote Sens. Environ. 85, 328–338.

51. Foster, J.L., Hall, D.K., Chang, A.T.C., Rango, A., Wergin, W., Erbe, E., 1999. Effects of snow crystal shape on the scattering of passive microwave radiation. IEEE Trans. Geosci. Remote Sens. 37, 1165–1168.

52. Frappart, F., Calmant, S., Cauhopé, M., Seyler, F., Cazenave, A., 2006. Preliminary results of ENVISAT RA-2-derived water levels validation over the Amazon Basin. Remote Sens. Environ. 100, 252–264.

53. Freitas, S.C., Trigo, I.F., Macedo, J., Barroso, C., Silva,

R., Perdigão, R., 2013. Land surface temperature from multiple geostationary satellites. Int. J. Remote Sens. 34, 3051–3068. http://dx.doi.org/10.1080/01431161.2012.716925.

54. Fuller, R.M., Smith, G.M., Devereux, B.J., 2003. The characterisation and measurement of land cover change through remote sensing: problems in operational applications? Int. J. Appl. Earth Obs. Geoinf. 4, 243–253.

55. Galve, J.M., Coll, C., Prata, F., 2009. Monthly land surface temperature maps over European zone using advanced along track scanning radiometer data for 2007. Proceedings of the Geoscience and Remote Sensing Symposium. 2009 IEEE international, IGARSS 2009 4, pp. 292–295.

56. Gao, H., Wood, E.F., Jackson, T.J., Drusch, M., Bindlish, R., 2006. Using TRMM/TMI to retrieve surface soil moisture over the Southern United States from 1998 to 2002. J. Hydrometeorol. 7, 23–38.

57. Gesell, G., 1989. An algorithm for snow and ice detection using AVHRR data, an extension to the APOLLO software package. Int. J. Remote Sens. 10, 897–905. http://dx.doi.org/ 10.1080/01431168908903929.

58. Gillespie, A., Rokugawa, S., Matsunaga, T., Cothern, J.S., Hook, S., Kahle, A.B., 1998. A temperature and emissivity separation algorithm for advanced spaceborne thermal emission and reflection radiometer (ASTER) images. IEEE Trans. Geosci. Remote Sens. 36, 1113–1126.

59. Glenn, E.P., Huete, A.R., Nagler, P.L., Hirschboeck, K.K., Brown, P., 2007. Integrating remote sensing and ground methods to estimate evapotranspiration. CRC Crit. Rev. Plant Sci. 26, 139–168.

60. Gudmundsson, L., Tallaksen, L.M., Stahl, K., Clark, D.B., Dumont, E., Hagemann, S., Bertrand, N., Gerten, D., Heinke, J., Hanasaki, N., Voss, F., Koirala, S., 2012a. Comparing large-scale hydrological model simulations to observed runoff percentiles in Europe. J. Hydrometeorol. 13, 604–620.

61. Gudmundsson, L., Wagener, T., Tallaksen, L.M., Engeland, K., 2012b. Evaluation of nine large-scale hydrological models with respect to the seasonal runoff climatology in Europe. Water Resour. Res. 48, W11504. http://dx.doi.org/10.1029/2011WR010911.

62. Haddad, Z., Smith, E.A., Kummerow, C.D., Iguchi, T., Farrar, M., Darden, S., Alves, M., Olson, W., 1997. The TRMM 'Day-1' radar/radiometer combined rain-profiling algorithm. J. Meteorol. Soc. Jpn. 75, 799–809.

63. Hall, D.K., Riggs, G.A., 2007. Accuracy assessment of the MODIS snow-cover products. Hydrol. Process. 21, 1534–1547.

64. Hall, D.K., Riggs, G.A., Salomonson, V.V., DiGirolamo, N., Bayr, K.J., 2002. MODIS snowcover products. Remote Sens. Environ. 83, 181–194.

65. Hansen, M.C., Loveland, T.R., 2012. A review of large area monitoring of land cover change using Landsat data. Remote Sens. Environ. 122, 66–74.

66. Heilman, J.L., Kanemasu, E.T., Rosenberg, N.J., Blad, B.L., 1976. Thermal scanner measurement of canopy temperatures to estimate evapotranspiration. Remote Sens. Environ. 5, 137–145.

67. Huang, J., Zhan, J., Yan, H., Wu, F., Deng, X., 2013. Evaluation of the impacts of land use on water quality: a case study in the Chaohu Lake Basin. Sci. World J. 2013, 329187. http://dx.doi.org/10.1155/2013/329187.

68. Hüsler, F., Jonas, T., Wunderle, S., Albrecht, S., 2012. Validation of a modified snow cover retrieval algorithm from historical 1-km AVHRR data over the European Alps. Remote Sens. Environ. 121, 497–515.

69. Hutchinson, G.E., 1957. A treatise on limnology, Volume I, Geography, physics and chemistry. Wiley, New York.

70. Hwang, C., Kao, Y., Tangdamrongsub, N., 2011. A preliminary analysis of lake level and water storage changes over lakes Baikal and Balkhash from satellite altimetry and gravimetry. Terr. Atmos. Ocean. Sci. 22, 97–108.

71. Iguchi, T., Kozu, T., Meneghini, R., Awaka, J., Okamoto, K., 2000. Rain-profiling algorithm for the TRMM precipitation radar. J. Appl. Meteorol. 39, 2038–2052.

72. IJC, 1980. Pollution in the Great Lakes basin from land use activities. Report, International Joint Commission.

73. Jackson, T.J., Hsu, A.Y., Neill, P.E.O., 2002. Surface soil moisture retrieval and mapping using high-frequency microwave satellite observations in the Southern Great Plains. J. Hydrometeorol. 3, 688–699.

74. Jensen, J.R., Chery Jr., D.L., 1980. Landsat crop identification for watershed water balance determinations. Int. J. Remote Sens. 1, 345–359. http://dx.doi.org/10.1080/ 01431168008948247.

75. Jiang, G.-M., Li, Z.-L., 2008. Split-window algorithm for land surface temperature estimation from MSG1-SEVIRI data. Int. J. Remote Sens. 29, 6067–6074. http://dx.doi.org/10. 1080/01431160802235860.

76. Kalma, J.D., McVicar, T.R., McCabe, M.F., 2008. Estimating land surface evaporation: a review of methods using remotely sensed surface temperature data. Surv. Geophys. 29, 421–469.

77. Karimi, P., Bastiaanssen, W.G.M., 2015. Spatial evapotranspiration, rainfall and land use data in water accounting — part 1: review of the accuracy of the remote sensing data. Hydrol. Earth Syst. Sci. 19, 507–532.

78. Keatley, B.E., Bennett, E.M., MacDonald, G.K., Taranu, Z.E., Gregory-Eaves, I., 2011. LandUse Legacies Are Important Determinants of Lake Eutrophication in the Anthropocene. PLoS ONE 6, e15913. http://dx.doi.org/10.1371/journal.pone.0015913.

79. Klein, B., Krahe, P., Lingemann, I., Nilson, E., Kling, H., Fuchs, M., 2011. Assessing climate change impacts on water balance in the Upper Danube Basin based on a 23 member RCM Ensemble. Proceedings of the XXVth Conference of the Danube Countries, Hungary.

80. Kling, H., Fuchs, M., Paulin, M., 2012. Runoff conditions in the upper Danube basin under an ensemble of climate change scenarios. J. Hydrol. 424–425, 264–277.

81. Knoll, L.B., Vanni, M.J., Renwick, W.H., Dittman, E.K., Gephart, J.A., 2013. Temperate reservoirs are large

carbon sinks and small CO2 sources: Results from high-resolution carbon budgets. Glob. Biogeochem. Cycles 27, 52–64. http://dx.doi.org/10.1002/gbc. 20020.

82. Kollár, S., Vekerdy, Z., Márkus, B., 2011. Forest habitat change dynamics in a riparian woodland. Spat. Stat. 7, 371–376.

83. Kongoli, C., Grody, N.C., Ferraro, R.R., 2004. Interpretation of AMSU microwave measurements for the retrievals of snow water equivalent and snow depth. J. Geophys. Res. - Atmos. 109, D24111.

84. König, M., Winther, J.-G., Isaksson, E., 2001. Measuring snow and glacier ice properties from satellite. Rev. Geophys. 39, 1–27.

85. Kovàcs, F., 2007. Assessment of regional variations in biomass production using satellite image analysis between 1992 and 2004. Trans. GIS 11, 911–926.

86. Kristiansen, P., Hansen, H.O. (Eds.), 1994. European rivers and lakes: Assessment of their environmental stateEEA Environmental Monographs 1. Denmark, European Environment Agency.

87. Kustas, W.P., Norman, J.M., 1997. A two-source approach for estimating turbulent fluxes using multiple angle thermal infrared observations. Water Resour. Res. 33, 1495–1508.

88. Kutser, T., Verpoorter, C., Paavel, B., Tranvik, L.J., 2015. Estimating lake carbon fractions from remote sensing data. Remote Sens. Environ. 157, 138–146.

89. Kyle, H.L., Curran, R.J., Barnes, W.L., Escoe, D., 1978. A cloud physics radiometer. Proceedings of Third Conference on Atmospheric Radiation, Davis, CA, pp. 107–109.

90. Large, A.R.G., Gilvear, D.J., 2014. Using Google Earth, a virtual globe-imaging platform, for ecosystem services-based river assessment. River Res. Appl. http://dx.doi.org/10. 1002/rra2798.

91. Larson, K.M., Small, E.E., Gutmann, E.D., Bilich, A.L., Braun, J.J., Zavorotny, V.U., 2008. Use of GPS receivers as a soil moisture network for water cycle studies. Geophys. Res. Lett. 35, L24405. http://dx.doi.org/10.1029/2008GL036013.

92. Lehner, B., Döll, P., 2004. Development and validation of a global database of lakes, reservoirs and wetlands. J. Hydrol. 296, 1–22.

93. Li, Z.-L., Tang, B.-H., Wu, H., Ren, H., Yan, G., Wan, Z., Trigo, I.F., Sobrino, J.A., 2013. Satellitederived land surface temperature: current status and perspectives. Remote Sens. Environ. 131, 14–37.

94. Li, M., Zang, S.Y., Zhang, B., Li, S.S., Wu, C.S., 2014. A review of remote sensing image classification techniques: the role of spatio-contextual information. Eur. J. Remote Sens. 47, 389–411.

95. Loew, A., 2008. Impact of surface heterogeneity on surface soil moisture retrievals from passive microwave data at the regional scale: the Upper Danube case. Remote Sens. Environ. 112, 231–248.

96. Lu, S., Ouyang, N., Wu, B., Wei, Y., Tesemma, Z., 2013. Lake water volume calculation with time series remote-sensing images. Int. J. Remote Sens. 34, 7962–7973. http://dx.doi. org/10.1080/01431161.2013.827814.

97. Ludwig, R., Mauser, W., Niemeyer, S., Colgan, A., Stolz, R., Escher-Vetter, H., Kuhn, M., Reichstein, M., Tenhunen, J., Kraus, A., Ludwig, M., Barth, M., Hennicker, R., 2003a. Web-based modelling of energy, water and matter fluxes to support decision making in mesoscale catchments — the integrative perspective of GLOWA-Danube. Phys. Chem. Earth 28, 621–634.

98. Ludwig, R., Probeck, M., Mauser, W., 2003b. Mesoscale water balance modelling in the Upper Danube watershed using sub-scale land cover information derived from NOAA-AVHRR imagery and GIS-techniques. Phys. Chem. Earth 28, 1351–1364.

99. Mao, K., Qin, Z., Shi, J., Gong, P., 2005. A practical split-window algorithm for retrieving land-surface temperature from MODIS data. Int. J. Remote Sens. 26, 3181–3204. http://dx.doi.org/10.1080/01431160500044713.

100. Martinec, J., Rango, A., Roberts, R., 2008. Snowmelt Runoff Model (SRM) User's Manual. New Mexico State University Press, New Mexico.

101. Matikainen, L., Kuittinen, R., Vepsäläinen, J., 2002. Estimating drainage area-based snowcover percentages from NOAA AVHRR images. Int. J. Remote Sens. 23, 2971–2988. http://dx.doi. org/10.1080/01431160110071923.

102. Mauser, W., Bach, H., 2009. PROMET — large scale distributed hydrological modelling to study the impact of climate change on the water flows of mountain watersheds. J. Hydrol. 376, 362–377.

103. McCabe, G.J., Markstrom, S.L., 2007. A monthly water-balance model driven by a graphical user interface. U.S. Geological Survey Open-File report 2007–1088 (6 pp.).

104. McDonald, C.P., Stets, E.G., Striegl, R.G., Butman, D., 2013. Inorganic carbon loading as a primary driver of dissolved carbon dioxide concentrations in the lakes and reservoirs of the contiguous United States. Glob. Biogeochem. Cycles 27, 1–11.

105. McFarland, M.J., Miller, R.L., Neale, C.M.U., 1990. Land surface temperature derived from the SSM/I passive microwave brightness temperatures. IEEE Trans. Geosci. Remote Sens. 28, 839–845.

106. Meigh, J.R., McKenzie, A.A., Sene, K.J., 1999. A grid-based approach to water scarcity, estimates for Eastern and Southern Africa. Water Resour. Manag. 13, 85–115.

107. Mercier, F., Cazenave, A., Maheu, C., 2002. Interannual lake level fluctuations (1993–1999) in Africa from Topex/Poseidon: connections with ocean–atmosphere interactions over the Indian Ocean. Glob. Planet. Chang. 32, 141–163.

108. Michaelides, S., Levizzani, V., Anagnostou, E., Bauer, P., Kasparis, T., Lane, J.E., 2009. Precipitation: measurement remote sensing, climatology and modelling. Atmos. Res. 94, 512–533.

109. Miller, S.D., Lee, T.F., Fennimore, R.L., 2005. Satellite-based imagery techniques for daytime cloud/snow delineation from MODIS. J. Appl. Meteorol. 44, 987–997.

110. Milzow, C., Krogh, P.E., Bauer-Gottwein, P., 2011.

Combining satellite radar altimetry, SAR surface soil moisture and GRACE total storage changes for hydrological model calibration in a large poorly gauged catchment. Hydrol. Earth Syst. Sci. 15, 1729–1743.

111. Minacapilli, M., Agnese, C., Blanda, F., Cammalleri, C., Ciraolo, G., D'Urso, G., Iovino, M., Pumo, D., Provenzano, G., Rallo, G., 2009. Estimation of actual evapotranspiration of

112. Mediterranean perennial crops by means of remote-sensing based surface energy balance models. Hydrol. Earth Syst. Sci. 13, 1061–1074. http://dx.doi.org/10.5194/ hess-13-1061-2009.

113. Mitsova, D., 2014. Coupling land use change modelling with climate projections to estimate seasonal variability in runoff from an urbanizing catchment near Cincinnati, Ohio. Int. J. Geo-Inf. 3, 1256–1277.

114. Mo, T., Choudhury, B.J., Schmugge, T.J., Wang, J.R., Jackson, T.J., 1982. A model for microwave emission from vegetation-covered fields. J. Geophys. Res. 87, 11,229–11,237. http://dx.doi.org/10.1029/ JC087iC13p11229.

115. Muirhead, K., Cracknell, A.P., 1986. Airborne lidar bathymetry. Int. J. Remote Sens. 7, 597–614. http:// dx.doi.org/10.1080/01431168608954714.

116. Nghiem, V.T., Nedjai, R., Messaoud, N.N., 2011. The consequences of changes in forest land cover in the Alpine and Jurassic massifs on the physico-chemical status of deep lake waters. J. Alp. Res. 99–3.

117. Njoku, E.G., Kong, J.A., 1977. Theory for passive microwave remote sensing of near-surface soil moisture. J. Geophys. Res. 82, 3108–3118.

118. Njoku, E.G., Jackson, T.J., Lakshmi, V., Chan, T.K., Nghiem, S.V., 2003. Soil moisture retrieval from AMSR-E. IEEE Trans. Geosci. Remote Sens. 41, 215–229.

119. Nolin, A.W., 2010. Recent advances in remote sensing of seasonal snow. J. Glaciol. 56, 1141–1150.

120. Olofsson, P., Foody, G.M., Stehman, S.V., Woodcock, C.E., 2013. Making better use of accuracy data in land change studies: estimating accuracy and area and quantifying uncertainty using stratified estimation. Remote Sens. Environ. 129, 122–131.

121. Owe, M., de Jeu, R., Holmes, T., 2008. Multisensor historical climatology of satellite-derived global land surface moisture. J. Geophys. Res. 113, F01002. http:// dx.doi.org/ 10.1029/2007JF000769.

122. Parinussa, R.M., Holmes, T.R.H., de Jeu, R.A.M., 2012. Soil moisture retrievals from the WindSat spaceborne polarimetric microwave radiometer. IEEE Trans. Geosci. Remote Sens. 50, 2683–2694.

123. Pereira, L.S., Allen, R.G., Smith, M., Raes, D., 2015. Crop evapotranspiration estimation with FAO56: past and future. Agric. Water Manag. 147, 4–20. http://dx.doi.org/10.1016/j. agwat.2014.07.031.

124. Pereira-Cardenal, S.J., Riegels, N.D., Berry, P.A.M., Smith, R.G., Yakovlev, A., Siegfried, T.U., Bauer-Gottwein, P., 2011. Real-time remote sensing driven river basin modelling using radar altimetry. Hydrol. Earth Syst. Sci. 15, 241–254.

125. Peterson, T.C., Basist, A.N., Williams, C.N., Grody, N.C., 2000. A blended satellite-in situ near-global surface temperature dataset. Bull. Am. Meteorol. Soc. 81, 2157–2164.

126. Petrovič, P., Mravcová, K., Holko, L., Kostka, Z., Miklánek, P., 2010. Basin-wide water balance in the Danube River Basin. In: Brilly, M. (Ed.), Hydrological Processes of the Danube River Basin. Springer Science + Business Media http://dx.doi.org/10.1007/978- 90-481-3423-6_7.

127. Pfeifer, M., Disney, M., Quaife, T., Marchant, R., 2012. Terrestrial ecosystems from space: a review of earth observation products for macroecology applications. Glob. Ecol. Biogeogr. 21, 603–624.

128. Pinheiro, A.C.T., Mahoney, R., Privette, J.L., Tucker, C.J., 2006. Development of a daily long term record of NOAA-14 AVHRR land surface temperature over Africa. Remote Sens. Environ. 103, 153–164.

129. Pôças, I., Cunha, M., Pereira, L.S., Allen, R.G., 2013. Using remote sensing energy balance and evapotranspiration to characterize montane landscape vegetation with focus on grass and pasture lands. Int. J Appl. Earth Obs. Geoinf. 21, 159–172.

130. Ponchaut, F., Cazenave, A., 1998. Continental lake level variations from Topex/Poseidon (1993–1996). C. R. Acad. Sci. Ser. IIa 326, 13–20.

131. Prata, A.J., Caselles, V., Coll, C., Sobrino, J.A., Ottlé, C., 1995. Thermal remote sensing of land surface temperature from satellites: current status and future prospects. Remote Sens. Rev. 12, 175–224. http://dx.doi.org/10.1080/02757259509532285.

132. Probeck, M., Ludwig, R., Mauser, W., 2005. Fusion of NOAA AVHRR imagery and geographical information system techniques to derive subscale land cover information for the upper Danube watershed. Hydrol. Process. 19, 2407–2418.

133. Qian, Y.-G., Li, Z.-L., Nerry, F., 2013. Evaluation of land surface temperature and emissivities retrieved from MSG/ SEVIRI data with MODIS land surface temperature and emissivity products. Int. J. Remote Sens. 34, 3140–3152. http://dx.doi.org/10.1080/ 01431161.2012.716538.

134. Reichle, R.H., Koster, R.D., Liu, P., 2007. Comparison and assimilation of global soil moisture retrievals from the Advanced Microwave Scanning Radiometer for the Earth Observing System (AMSR-E) and the Scanning Multichannel Microwave Radiometer (SMMR). J. Geophys. Res. 112, D09108.

135. Renwick, W.H., Vanni, M.J., Zhang, Q., Patton, J., 2008. Water quality trends and changing agricultural practices in a Midwest US watershed, 1994–2006. J. Environ. Qual. 37, 1862–1874.

136. Riera, J.L., Magnuson, J.J., Kratz, T.K., Webster, K.E., 2000. A geomorphic template for the analysis of lake districts applied to the Northern Highland lake district, Wisconsin, USA. Freshw. Biol. 43, 301–318.

137. Rocchini, D., Foody, G.M., Nagendra, H., Ricotta, C., Anand, M., He, K.S., Amici, V., Kleinschmit, B., Förster,

M., Schmidtlein, S., Feilhauser, H., Ghisla, A., Metz, M., Neteler, M., 2013. Uncertainty in ecosystem mapping by remote sensing. Comput. Geosci. 50, 128–135.

138. Rodell, M., Velicogna, I., Famiglietti, J.S., 2009. Satellite-based estimates of groundwater depletion in India. Nature 460, 999–1002. http://dx.doi.org/10.1038/nature08238.

139. Rodell, M., McWilliams, E.B., Famiglietti, J.S., Beaudoing, H.K., Nigro, J., 2011. Estimating evapotranspiration using an observation based terrestrial water budget. Hydrol. Process. 25, 4082–4092.

140. Rowan, J.S., 2010. Developing a lake hydromorphology typology for the UK. SNIFFER Research Report WFD104 (50 pp.).

141. Rüdiger, C., Calvet, J.-C., Gruhier, C., Holmes, T.R.H., de Jeu, R.A.M., Wagner, W., 2009. An intercomparison of ERS-Scat and AMSR-E soil moisture observations with model simulations over France. J. Hydrometeorol. 10, 431–447.

142. Salomonson, V.V., Appel, I., 2004. Estimating fractional snow cover from MODIS using the normalized difference snow index. Remote Sens. Environ. 89, 351–360.

143. Schindler, D.W., Donahue, W.F., 2006. An impending water crisis in Canada's western prairie provinces. Proc. Natl. Acad. Sci. U. S. A. 103, 7210–7216.

144. Schlenz, F., dall'Amico, J.T., Loew, A., Mauser, W., 2012. Uncertainty assessment of the SMOS validation in the Upper Danube Catchment. IEEE Trans. Geosci. Remote Sens. 50, 1517–1529.

145. Schmugge, T., Gloersen, P., Wilheit, T.T., Geiger, F., 1974. Remote sensing of soil moisture with microwave radiometers. J. Geophys. Res. 79, 317–323.

146. Schmugge, T.J., Kustas, W.P., Ritchie, J.C., Jackson, T.J., Rango, A., 2002. Remote sensing in hydrology. Adv. Water Resour. 25, 1367–1385.

147. Seneviratne, S.I., Corti, T., Davin, E.L., Hirschi, M., Jaeger, E.B., Lehner, I., Orlowsky, B., Teuling, A.J., 2010. Investigating soil moisture–climate interactions in a changing climate: a review. Earth Sci. Rev. 99, 125–161.

148. Seppelt, R., Dormann, C.F., Eppink, F.V., Lautenbach, S., Schmidt, S., 2011. A quantitative review of ecosystem service studies: approaches, shortcomings and the road ahead. J. Appl. Ecol. 48, 630–636.

149. Shiklomanov, I.A., 2000. Appraisal and assessment of world water resources. Water Int. 25, 11–32. http://dx.doi.org/10.1080/02508060008686794.

150. Sobrino, J.A., Li, Z.-L., Stoll, M.P., Becker, F., 1996. Multi-channel and multi-angle algorithms for estimating sea and land surface temperature with ATSR data. Int. J. Remote Sens. 17, 2089–2114. http://dx.doi.org/10.1080/01431169608948760.

151. Sobrino, J.A., Sòria, G., Prata, A.J., 2004. Surface temperature retrieval from Along Track Scanning Radiometer 2 data: algorithms and validation. J. Geophys. Res. 109, D11101. http://dx.doi.org/10.1029/2003JD004212.

152. Soranno, P.A., Webster, K.E., Cheruvelil, K.S., Bremigan, M.T., 2009. The lake landscapecontext framework: linking aquatic connections, terrestrial features and human effects at multiple spatial scales. Verh. Int. Ver. Limnol. 30, 695–700.

153. Soranno, P.A., Cheruvelil, K.S., Webster, K.E., Bremigan, M.T., Wagner, T., Stow, C.A., 2010. Using landscape limnology to classify freshwater ecosystems for multi-ecosystem management and conservation. Bioscience 60, 440–454.

154. Sòria, G., Sobrino, J.A., 2007. ENVISAT/AATSR derived land surface temperature over a heterogeneous region. Remote Sens. Environ. 111, 409–422.

155. Stephens, G.L., Kummerow, C.D., 2007. The remote sensing of clouds and precipitation from space: a review. J. Atmos. Sci. 64, 3742–3765.

156. Stone, L.R., Horton, M.L., 1974. Estimating evapotranspiration using canopy temperatures: field evaluation. Agron. J. 66, 450–454.

157. Suchenwirth, L., Stümer, W., Schmidt, T., Förster, M., Kleinschmit, B., 2014. Large-scale mapping of carbon stocks in riparian forests with self-organising maps and the knearest neighbor algorithm. Forests 5, 1635–1652.

158. Swenson, S.C., Wahr, J., 2009. Monitoring the water balance of Lake Victoria, East Africa, from space. J. Hydrol. 370, 163–176.

159. Swenson, S., Famiglietti, J., Basara, J., Wahr, J., 2008. Estimating profile soil moisture and groundwater variations using GRACE and Oklahoma Mesonet soil moisture data. Water Resour. Res. 44, W01413.

160. Tapiador, F.J., Turk, F.J., Petersen, W., Hou, A.Y., García-Ortega, E., Machado, L.A.T., Angelis, C.F., Salio, P., Kidd, C., Huffman, G.J., de Castro, M., 2012. Global precipitation measurement: methods, datasets and applications. Atmos. Res. 104, 70–97.

161. Tapley, B.D., Bettadpur, S., Ries, J.C., Thompson, P.F., Watkins, M.M., 2004. GRACE measurements of mass variability in the earth system. Science 305, 503–505. http://dx. doi.org/10.1126/science.1099192.

162. TEEB, 2010. The economics of ecosystems and biodiversity available at http://www. teebweb.org/resources/ecosystem-services/.

163. Thornthwaite, C.W., 1948. An approach toward a rational classification of climate. Geogr. Rev. 38, 55–94.

164. Thornthwaite, C.W., Mather, R.J., 1955. The water balance. Publications in Climatology 8. DIT, Laboratory of Climatology, Centerton, New Jersey, USA, pp. 1–86.

165. Thornthwaite, C.W., Mather, R.J., 1957. Publ. no. 10. Laboratory of Climatology, Centerton, New Jersey, USA.

166. Tomlinson, C.J., Chapman, L., Thornes, J.E., Baker, C., 2011. Remote sensing land surface temperature for meteorology and climatology: a review. Meteorol. Appl. 18, 296–306. http://dx.doi.org/10.1002/met.287.

167. Tranvik, L.J., Downing, J.A., Cotner, J.B., Loiselle, S.A., Striegl, R.G., Ballatore, T.J., Dillon, P., Finlay, K., Fortino, K., Knoll, L.B., Kortelainen, P.L., Kutser, T., Larsen, S., Laurion, I., Leech, D.M., Leigh McCallister, S., McKnight, D.M., Melack, J.M., Overholt, E., Porter,

J.A., Prairie, Y., Renwick, W.H., Roland, F., Sherman, B.S., Schindler, D.W., Sobek, S.,

168. Tremblay, A., Vanni, M.J., Verschoor, A.M., Von Wachenfeldt, E., Weyhenmeyer, G.A., 2009. Lakes and reservoirs as regulators of carbon cycling and climate. Limnol. Oceanogr. 54, 2298–2314.
169. Tyler, A.N., Hunter, P.D., Spyrakos, E., Groom, S., Constaninescua, A., Kitchen, J., Paun, A., 2015. Developments in Earth observation for the assessment and monitoring of inland, transitional, coastal and shelf-sea waters. Sci. Total Environ. (in this issue).
170. UNEP, 2008. Vital water graphics — an overview of the state of the world's fresh and marine waters. 2nd edition. UNEP, Nairobi, Kenya 92-807-2236-0.
171. Valero-Garcés, B.L., Navas, A., Machin, J., Stevenson, T., Davis, B., 2000. Responses of a saline lake ecosystem in a semiarid region to irrigation and climate variability. AMBIO: J. Hum. Environ. 29, 344–350.
172. Vereecken, H., Huisman, J.A., Bogena, H., Vanderborght, J., Vrugt, J.A., Hopmans, J.W., 2008. On the value of soil moisture measurements in vadose zone hydrology: a review. Water Resour. Res. 44. http://dx.doi.org/10.1029/2008WR006829.
173. Verpoorter, C., Kutser, T., Tranvik, L., 2012. Automated mapping of water bodies using Landsat multispectral data. Limnol. Oceanogr. Methods 10, 1037–1050. http://dx. doi.org/10.4319/lom.2012.10.1037.
174. Verpoorter, C., Kuster, T., Seekell, D.A., Travnik, L.J., 2014. A global inventory of lakes based on high-resolution satellite imagery. Geophys. Res. Lett. 41. http://dx.doi.org/10. 1002/2014GL060641.
175. Verstraeten, W.W., Veroustraete, F., Feyen, J., 2008. Assessment of evapotranspiration and soil moisture content across different scales of observation. Sensors 8, 70–117.
176. Vila, D., Ferraro, R., Joyce, R., 2007. Evaluation and improvement of AMSU precipitation retrievals. J. Geophys. Res. 112, D20119. http://dx.doi.org/10.1029/2007JD008617.
177. Voigt, S., Koch, M., Baumgartner, M.F., 1999. A multichannel threshold technique for NOAA AVHRR data to monitor the extent of snow cover in the Swiss Alps. Proceedings of IUGG99 Symposium HS2, interactions between the cryosphere, climate and greenhouse gases, Birmingham. IAHS Publ. no. 256.
178. Vörösmarty, C.J., Green, P., Salisbury, J., Lammers, R.B., 2000. Global water resources: vulnerability from climate change and population growth. Science 289, 284–288.
179. Wada, Y., van Beek, L.P.H., Bierkens, M.F.P., 2011. Modelling global water stress of the recent past: on the relative importance of trends in water demand and climate variability. Hydrol. Earth Syst. Sci. 15, 3785–3808.
180. Wagner, W., Scipal, K., Pathe, C., Gerten, D., Lucht, W., Rudolf, B., 2003. Evaluation of the agreement between the first global remotely sensed soil moisture data with model and precipitation data. J. Geophys. Res. 108, 4611. http://dx.doi.org/10.1029/ 2003JD003663.
181. Wagner, W., Blöschl, G., Pampaloni, P., Calvet, J.-C., Bizzarri, B., Wigneron, J.-P., Kerr, Y., 2007. Operational readiness of microwave remote sensing of soil moisture for hydrologic applications. Nord. Hydrol. 38, 1–20.
182. Wagner, W., Dorigo, W., de Jeu, R., Fernandez, D., Benveniste, J., Haas, E., Ertl, M., 2012. Fusion of active and passive microwave observations to create and essential climate variable data record on soil moisture. ISPRS Annals of the Photogrammetry, Remote Sensing and Spatial Information Sciences, I-7, eXXII ISPRS Congress, Melbourne, Australia.
183. Wahr, J., Swenson, S., Zlotnicki, V., Velicogna, I., 2004. Time-variable gravity from GRACE: First results. Geophys. Res. Lett. 31, L11501. http://dx.doi.org/10.1029/ 2004GL019779.
184. Wan, Z., Dozier, J.A., 1996. Generalized split-window algorithm for retrieving land-surface temperature for space. IEEE Trans. Geosci. Remote Sens. 34, 892–905.
185. Wang, L., Qu, J.J., 2009. Satellite remote sensing applications for surface soil moisture monitoring: a review. Front. Earth Sci. China 3, 237–247. http://dx.doi.org/10.1007/ s11707-009-0023-7.
186. Wang, N.-Y., Liu, C., Ferraro, R., Wolff, D., Zipser, E., Kummerow, C., 2009. TRMM 2A12 Land precipitation product — status and future plans. J. Meteorol. Soc. Jpn. 87, 237–253.
187. Webster, K.E., Soranno, P.A., Baines, S.P., Kratz, T.K., Bowser, C.J., Dillon, P.J., Campbell, P., Fee, E.J., Hecky, R.E., 2000. Structuring features of lake districts: landscape controls on lake chemical responses to drought. Freshw. Biol. 43, 499–515.
188. Wen, J., Jackson, T.J., Bindlish, R., Hsu, A.Y., 2005. Retrieval of soil moisture and vegetation water content using SSM/I data over a corn and soybean region. J. Hydrometeorol. Spec. Sect. 6, 854–863.
189. Wigneron, J.-P., Calvet, J.-C., Pellarin, T., Van de Griend, A.A., Berger, M., Ferrazzoli, P., 2003. Retrieving near-surface soil moisture from microwave radiometric observations: current status and future plans. Remote Sens. Environ. 85, 489–506.
190. Wilcox, C., Los Huertos, M., 2005. A simple, rapid method for mapping bathymetry of small wetland basins. J. Hydrol. 301, 29–36.
191. Wilker, H., Drusch, M., Seuffert, G., Simmer, C., 2006. Effects of the near-surface soil moisture profile on the assimilation of L-band microwave brightness temperature. J. Hydrometeorol. 7, 433–442.
192. Williams, C.N., Basist, A., Peterson, T.C., Grody, N.C., 2000. Calibration and validation of land surface temperature anomalies derived from the SSM/I. Bull. Am. Meteorol. Soc. 81, 2141–2156.
193. Williamson, C.E., Saros, J.E., Vincent, W.F., Smol, J.P., 2009. Lakes and reservoirs as sentinels, integrators, and regulators of climate change. Limnol. Oceanogr. 54, 2273–2282.
194. Woodcock, C.E., 2002. Uncertainty in remote sensing. In: Foody, G.M., Atkinson, P.M. (Eds.), Uncertainty in

Remote Sensing and GIS. Wiley, Chichester.

195. Woodhouse, I., Nichol, C., Sinclair, P., Jack, J., Morsdorf, F., Malthus, T.J., Patenaude, G., 2011. A multispectral canopy LiDAR demonstrator project. IEEE Geosci. Remote Sens. Lett. 8, 839–843.

196. WWAP, 2012. The United Nations World Water Development Report 4: Managing Water under Uncertainty and Risk. World Water Assessment Programme, Paris, UNESCO.

197. Xu, C.-Y., Singh, V.P., 1998. A review on monthly water balance models for water resources investigations. Water Resour. Manag. 12, 31–50.

198. Xu, C.-Y., Singh, V.P., 2004. Review on regional water resources assessment models under stationary and changing climate. Water Resour. Manag. 18, 591–612.

199. Xu, X., Li, J., Tolson, B.A., 2014. Progress in integrating remote sensing data and hydrologic modelling. Prog. Phys. Geogr. 38, 464–498.

200. Yang, H., Xing, Y., Xie, P., Ni, L., Rong, K., 2008. Carbon source/sink function of a subtropical, eutrophic lake determined from an overall mass balance and a gas exchange and carbon burial balance. Environ. Pollut. 151, 559–568.

CHAPTER
18

CURRENT MICROBIAL WATER QUALITY RISK ASSESSMENT

Gemma Dunn, Leila Harris, Christina Cook, Natalie Prystajecky

Keywords: Microbial pathogens, Risk assessment and management, Canada, British Columbia, Ontario

INTRODUCTION

Microbial risk assessment and management of water quality is an important concern and focus of governmental regulation and scientific inquiry (WHO, 2004). Microorganisms (bacteria, protozoa and viruses) are ubiquitous in aquatic environments. While most microorganisms are benign and serve essential ecosystem functions, some can be (or produce substances that are) harmful to ecosystem and human health. Microbially-contaminated drinking water has long been implicated in human illness and historically, attention has focused on finished (end-product or tap) water quality. However, these strategies are increasingly regarded as insufficient to prevent disease outbreaks and more attention has been paid to preventing illness from source-to-tap (Byleveld et al., 2008 and Summerscales and McBean, 2011). Furthermore, it has been increasingly noted that poor microbial water quality can be harmful to aquatic organisms and ecosystem function (Gozlan et al., 2006, Miller et al., 2002, Miller et al., 2011 and Weitz and Wilhelm, 2012). Human health and ecosystem health are implicitly related, particularly when considering microbial risk along the source-to-tap spectrum (Serveiss and Ohlson, 2007 and Davies and Mazumder, 2003).

A comprehensive understanding of the risks (both existing and potential) to water quality can be achieved through evaluation of the entire water supply system. This concept is known as 'source-to-tap', whereby a high quality source, in combination with effective treatment and safe distribution, supported by legislation and ongoing water quality testing, will yield water that is safe for human consumption. As a cornerstone of drinking water quality risk assessment (Hrudey, 2011 and Krewski et al., 2002), the source-to-tap framework has, to date, been mobilized with a focus on human health. Comprehensive risk assessment and management approaches that include the entire source-to-tap gradient are more likely to successfully address not only point-source pollution but also the more complex impact of non-point sources of water contaminants within a given watershed (Phillips, 1988). Sound risk assessment and management is crucial for both drinking water provision (Byleveld et al., 2008, Hamilton et al., 2006 and Jayarante, 2008) as well as broader ecosystem protection (Serveiss and Ohlson, 2007). Indeed, risk management in a source to tap framework is thought to be critical given the recognition that hazards are innumerable and resources to deal with them are limited (Dominguez-Chicas and Scrimshaw, 2010, Gelting, 2009, Hamilton et al., 2006, Hrudey, 2004 and Hrudey, 2009).

In Canada, some aspects of microbial testing for water quality are mandated across all jurisdictions (*Escherichia coli* and total coliform monitoring at tap for drinking water). However, beyond general guidelines recommending the adoption of a multi-barrier approach [1] by the Canadian Council of Ministers of the Environment Water Quality Task Group and the Federal–Provincial–Territorial Committee (CCME FPTC) on Drinking Water, there is no overarching Canadian framework specific to microbial risk assessment and management, anywhere along the source to tap framework (CCME, 2004). In the absence of a mandated national framework, considerable diversity in risk assessment and management of microbial water contamination exists across the country.

Little is known about the gap (or variance) between the theory and practical application of risk assessment and management. This study aims to address that gap by examining

Citation: Gemma Dunn, Leila Harris, Christina Cook, Natalie Prystajecky, Current microbial water quality risk assessment, Canada, Science of The Total Environment, Volumes 468–469, 15 January 2014, Pages 544-552, ISSN 0048-9697, http://dx.doi.org/10.1016/j.scitotenv.2013.08.004.

the ways different Canadian agencies and practitioners apply concepts and tools related to microbial assessment and management. Our objective is to describe current practices in two Canadian provinces, British Columbia (BC) and Ontario, and to situate them with respect to the literature and best practices. Our evaluation extends across the source-to-tap gradient and is attentive to both ecosystem and human health concerns. We do not attempt to compare or analyze the wide range of formalized risk assessment and management tools currently available. Rather we ascertain which ones, if any, practitioners are currently using. While we recognize the importance of chemical or other contamination risks, these considerations are beyond the scope of this study.

Our research included a literature review of current practices (see Section 2) and interviews with practitioners from selected agencies we expect are engaged in microbial risk assessment and management, from source-to-tap. Section 3 outlines the methods used. In the results (Section 4) we highlight key water quality monitoring issues as well as the considerable variability in risk assessment and management approaches between provinces and among agencies. In Section 5, we discuss current practices in light of best practices derived from the literature and find considerable divergence in on the ground practices. In Section 6, we consider the implications of these findings, particularly in light of recommendations in the literature for preventative, multi-barrier approaches that consider both ecosystem health and human health concerns. Looking forward we consider the outcomes of this research within the context of new developments in microbial water quality monitoring such as tests derived from genomics and metagenomics based research.

RISK ASSESSMENT AND MANAGEMENT

The terms risk assessment and risk management are intrinsically linked and often conflated; however, their meanings are quite distinct. Risk assessment is a scientifically based process involving four key steps: hazard identification/assessment/measurement; hazard characterization (e.g. dose–response analysis); exposure assessment; and risk characterization (Hunter et al., 2003). Risk assessment undertaken within the context of a source-to-tap framework should offer an improved, integrated understanding of the various components of the water supply system, their strengths and weaknesses, and the existing and potential threats to water quality so that informed decisions can be made. Risk management refers to the control options, the legal considerations and risk management decisions (including economic and social factors) to reduce or mitigate risk. This includes the task of managing the assessed risks in the face of uncertainty, balancing consideration of potential hazards with available treatment and mitigation strategies as well as resources (Hamilton et al., 2006).

In many sectors risk assessment and management are central to operations and protocols. Although methods may vary, risk assessment and management practices have been utilized for decades in the energy utility sector, in industries such as automotive and food, and among High Reliability Organizations (HROs) such as aviation and nuclear power plants. Comparatively speaking, formalized, explicit and routine risk assessment practices in the water sector are relatively newer and less widespread (Egerton, 1996 and Pollard et al., 2004).

In 2004, The World Health Organization (WHO) published their first set of guidelines emphasizing risk-based management of water (WHO, 2004). In doing so, the WHO advocated the application of broader, more comprehensive approaches to manage water quality challenges. This is indicative of a move away from a reactive approach of focusing on treated water (narrowly focusing on end-product testing at the tap), which can only highlight a potential health problem after the water has been consumed, toward a preventative risk management approach that looks comprehensively at the entire water system from the source-to-tap (Hrudey, 2003). Increasingly, preventative approaches are considered to be more reliable and cost effective to protect public health (Byleveld et al., 2008, Dominguez-Chicas and Scrimshaw, 2010 and Hamilton et al., 2006). Coupled with this has been a transition from implicit (implied but not formally expressed) to explicit (formalized and routine) frameworks for risk assessment and management, particularly within the international water utility sector (Hrudey et al., 2006, Pollard et al., 2004 and Summerill et al., 2010a). This is characterized by the introduction of procedures such as Hazard Analysis and Critical Control Points (HACCP), Qualitative Microbial Risk Assessment (QMRA) and Water Safety Plans (WSP). Table 1 provides a summary, with key citations, of several risk assessment tools currently available and in use.

Table 1: Selection of risk assessment & management tools identified in the literature.

Risk assessment tool	Brief description	Reference
Failure Modes and Effects (Criticality) Analysis (FME(C)A)	A systematic process to identify potential failure modes (causes and effects) based on experience with similar processes. (Widely used in manufacturing industries)	Dominguez-Chicas and Scrimshaw (2010), Hamilton et al. (2006), Pollard et al. (2004)
Critical Control Points (CCP)	A point, step or procedure (i.e. a critical failure area) that can be identified in a system. (Derived from FMEA)	Hamilton et al. (2006), Summerill et al. (2010a), Yokoi et al. (2006)

Hazard Analysis and Critical Control Points (HACCP)	A preventative risk management system in which a point, step or procedure (i.e. a critical failure area) can be identified in a system to which corrective actions can be applied so that a potential hazard can be prevented, eliminated, or reduced to an acceptable level. Also includes verification procedures to ensure the plan is adequate.(Builds on CCP. Originally developed in the food industry initially for NASA. Now widely used in food and pharmaceutical industries)	Davison et al. (2005), Dominguez-Chicas and Scrimshaw (2010), Hamilton et al. (2006), Miller et al. (2005), Pollard et al. (2004), Yokoi et al. (2006), Jayarante (2008)
Water Safety Plans (WSP)	Comprehensive risk assessment and management plan to identify and prioritize potential threats to water quality at each step in a specific system's water supply chain (from source to tap) implementing best practices to mitigate threats to drinking water.(Derived from HACCP and the multi-barrier approach)	Ashbolt (2004), Davison et al. (2005),Byleveld et al. (2008), Hamilton et al. (2006), Schijven et al. (2011), Smeets et al. (2010), Summerill et al., 2010a and Summerill et al., 2010b, Vieria (2007), Yokoi et al. (2006), Jayarante (2008), Miller et al. (2005), Gelting (2009), Austin et al. (2012), Hrudey (2011), Bartram et al. (2009),Gunnarsdottir et al. (2012)
Total Quality Management (TQM)	A holistic, integrated management approach whereby all members of an organization participate in maintaining and improving processes, products, services. Also described as a cultural initiative that fosters a collaborative environment between departments within an organization to improve overall organizational quality.	O' Connor (2002), Hamilton et al. (2006),Hrudey, 2003 and Hrudey, 2004
Quantitative Microbial Risk Assessment (QMRA)	A systematic quantitative assessment process to estimate the risks of human exposure to an array of microorganisms that can cause infectious disease outbreaks. Combines dose response information for the infectious agent with information on the distribution of exposures.	Haas et al. (1999), Ashbolt (2004),Ashbolt et al. (2010), Cool et al. (2010),Schijven et al. (2011), Signor and Ashbolt (2006), Smeets et al. (2010), Goss and Richards (2008), Benke and Hamilton (2008)
International Organization for Standardization (ISO) 14001 & 9001[a]	An internationally recognized family of certified standards approved through independent assessment. 14001 — Environmental Management: a systematic approach to minimize negative impacts, increase operational efficiency and identify cost savings, underpinned by continuous improvement.9001 — Quality Management: a framework for an organization to focus on customer and product requirements, process performance and effectiveness in the systems delivery with a key focus on continuous improvement and objective measurement.	Serveiss and Ohlson (2007), Jayarante (2008), Miller et al. (2005), Summerill et al. (2010a), Vieria (2007)
Ecological Risk Assessment & Regional Risk Model	A process to evaluate the potential adverse effects of human activities on the ecological health of an ecosystem at a particular site. An explicit expression of the environmental value (species, ecological resource, or habitat type) that is to be protected.	Serveiss and Ohlson (2007)
Catchment Risk Management	Risk analysis of water supply at a catchment-scale. Considers a multitude of possible sources of hazardous events, caused by natural or human factors such as wild animals erosion, land use, industry, traffic, and recreational activities.	Miller et al. (2005)

a In addition, ISO standard for Risk Management (ISO 31000:2009) provides principles for effective risk management and corporate governance. However, unlike ISO 14001 and 9001, ISO 31000 cannot be used for certification purposes, although it does provide guidance for internal or external audit programs.

Perhaps most notable of all these approaches is the WSP (based on HACCP and advocated by the WHO), which is a risk-based preventative approach to managing drinking water safety from catchment to consumer (source-to-tap). A WSP is an iterative process whereby the threats to the system; the capacity of the system to cope with threats; the ability to respond if barriers fail; and measures to improve the system are all characterized and incorporated into planning (Bartram et al., 2009, Gelting, 2009 and Hrudey, 2011). As such, WSPs involve an explicit risk assessment and management philosophy (Hrudey et al., 2006, Pollard et al., 2004 and Summerill et al., 2010a).

The uptake of explicit, formalized and routine risk assessment and management approaches in the water sector is gaining momentum around the world with a number of jurisdictions introducing policy and legal requirements. For example, the Dutch Drinking Water Act (2001) created a legal requirement for water utilities to use QMRA for drinking water from surface and vulnerable groundwater (Schijven et al., 2011). Water Safety Plans in particular are gaining popularity and are widely used across Europe, Africa and the Americas (Gelting, 2009). Iceland legislated drinking water utilities use Water Safety Plans in 1995 (Gunnarsdottir et al., 2012). The Australian Drinking Water Guidelines promote the implementation of Water Safety Plans in states and territories. For example, New South Wales has recommended water suppliers implement WSPs, and Victoria has regulated utilities to implement risk-based management plans through the Water Safety Act 2003 (Byleveld et al., 2008).

In Canada, it has been increasingly emphasized that a transition from implicit to explicit approaches in the area of water safety is necessary, particularly since the Walkerton crisis of 2000 and the report of the Walkerton Inquiry in 2002. Recommended approaches for securing drinking water include risk assessment and management, the source-to-tap approach, and WSPs (Ivey et al., 2006, Hrudey, 2011 and O' Connor, 2002).

METHODS

Rationale

To date, explicit material which links the theory of microbial assessment and management to actual practice is not readily available. In Canada, each province has a unique approach to water governance in general, and microbial risk assessment and management in particular. Key differences include regulatory requirements and provincial guidelines on risk assessment and management (Cook et al., in press and Ngueng Feze et al., under review). Furthermore, within a single watershed multiple stakeholders engage in microbial risk assessment and management including watershed authorities, water utilities and the health authorities. Because each agency has a distinct mandate, priorities and interests in water quality, we hypothesize that microbial risk management approaches will differ. We predict that health authorities and water utilities may prioritize risk assessment for human health (including drinking or recreational water), while watershed authorities will stress broader ecosystem health considerations in source water. We also expect important differences across these entities based on their mandates, data availability, and monitoring and assessment capacity. It is precisely these types of differences that have not been previously documented and that our study aims to uncover.

Selection of Interviewees

The research employed a case study approach using semi-structured interviews to gather qualitative data (Yin, 2003, Summerill et al., 2010b and Taylor et al., 2013). In the two Canadian provinces studied (BC and Ontario) interviews were conducted with experienced personnel from select agencies to provide insights into "on-the-ground" risk assessment and management practices along the source-to-tap spectrum. Three case study watersheds were selected in each of the two provinces (Abbotsford, Kelowna and Victoria in BC and Toronto, Kitchener–Waterloo and Ottawa in Ontario). In each of the six watersheds, we examined the same three types of agencies that we anticipated would be actively engaged in microbial risk management (from source-to-tap): watershed authorities, water utilities and health authorities. Specific criteria for case study selection included type of: watershed (urban or mixed urban–rural); source water (lake, reservoir or river; areas served exclusively by groundwater sources were excluded); drinking water treatment (e.g. chlorination, UV and/or filtration); water purveyor management (e.g. municipal government, private utility); population served (e.g. medium (> 10,000) to large drinking water systems (> 90, 000)); and historical water quality or quantity challenges (Table 2).

Table 2: Case study selection.

	Watershed authority	Water utility	Health authority
BC	Fraser Valley Regional District	Abbotsford	Fraser Health Authority
	Okanagan Basin Water Board	Kelowna	Interior Health Authority
	Capital Regional District Watershed	Victoria	Vancouver Island Health Authority
ON	Toronto Conservation Authority	Toronto	Toronto Regional Health Authority
	Grand River Conservation Authority	Kitchener–Waterloo	Waterloo Health Authority
	Rideau Valley Conservation Authority	Ottawa	Ottawa Public Health

Collection of Data

Narrative data was collected through eighteen semi-structured interviews (conducted by telephone, lasting approximately 1 h) with practitioners employed in these agencies including water systems operators, municipal and provincial employees. This social science research approach enabled sufficiently open discussion with experienced practitioners to reveal nuances of microbial risk practices while ensuring interview discussions did not stray too far from the research objectives (Summerill et al., 2010a and Taylor et al., 2013). We sought to interview those with expertise in water management and policy, specifically individuals who would be most familiar with the day-to-day microbial risk assessment and management approaches engaged by these agencies. The narrative data presented here reflects the knowledge and opinions of the interviewees. Evidence of risk assessment and management practices and policies (such as HACCP or TQM reports) was not requested.

Twenty-three interview questions addressed four key themes: 1) how microbial risk is measured and assessed by the organization (including what data is collected, types of risk assessment tools used, and interpretation of results); 2) how the organization manages microbial risk (including risk management plans); 3) how risk is communicated (including sharing of information internally and externally with policy makers and other stakeholders); and 4) how policy and legislation affects 'on-the-ground' microbial risk management practices. The interviews were recorded, transcribed and qualitatively analyzed using theme codes (overarching issues) based on the questions asked of participants and sub-theme codes (specific points that fit within an overarching theme) derived from the content of the discussions. Free and informed consent of the participants was obtained; the study protocol was approved by the University of British Columbia, Vancouver, Canada, Behavioral Research Ethics Board (H12-01626, July 2012).

RESULTS

Four core themes emerged from the interviews: issues pertaining to water quality monitoring; limited and variable application of risk assessment tools; limited use of risk management plans; and significant differences in risk assessment practices across the two study provinces.

Water Quality Monitoring

Water quality monitoring is the first step toward microbial risk assessment and management (characterization and measurement of the hazard). In both provinces water utilities are the primary collectors of microbial water quality data, often (but not always) sharing the test results with health authorities and watershed authorities. Some health authorities and watershed authorities conduct/enforce microbial sampling, but the extent and drivers for undertaking their own sampling are highly varied.

Microbial Monitoring

E. coli and total coliforms are the most commonly measured microbial water quality indicators in accordance with legislated requirements (BC Reg. 200/2003 Drinking Water Protection Regulation; O. Reg. 169/03 Ontario Drinking Water Quality Standards), followed by indirect measures (primarily water treatment process indicators) such as turbidity and chlorine residual. All six water utilities test for *E. coli* and total coliforms and they are legally required to provide these test results to the health authorities. (Health authorities in turn may undertake their own additional microbial sampling). Sampling type and frequency vary between types of agency interviewed and between provinces (on provinces see Cook et al., in press). Unlike Ontario, BC's drinking water quality regulation does not prescribe regularity or sample types (see Table 3). Five of the 18 agencies interviewed (from both BC and Ontario) monitor for specific pathogens, typically testing for *Giardia* and *Cryptosporidium*. Across both provinces a few agencies (usually water utilities) occasionally undertake additional testing such as Microscopic Particulate Analysis (MPA) (as a surrogate of protozoa), enterococci bacteria, *Bacteroides* and bacteriophage, but this testing is not performed routinely.

Table 3: Comparison on microbial risk assessment and management approaches at the provincial scale.

	British Columbia	Ontario
Microbial water quality standards	No total coliforms, *E. coli*, or fecal coliforms (B.C. Reg. 200/2003 Drinking Water Protection Regulation)	No Total coliforms or *E. coli* (heterotrophic plate counts) (O. Reg 169/03 Ontario Drinking Water Quality Standards)
Legislated sampling frequency	Population-based; no regularity prescribed; sample type not specified	Population-based; regularity prescribed; sample type specified
Centralized data collection	No — there is no centralized or provincial water quality database.	Yes — a centralized system for province-wide water quality data is collected and maintained by the Ministry of Environment.
Multi-barrier approach	Voluntary approach — provincial guidelines for the multi-barrier approach have been established.	The multi-barrier approach is regulated through several provincial laws, incl. Safe Drinking Water Act and the Ontario Water Resources Act, Clean Water Act.
Source water protection	Part 5 of the BC Drinking Water Protection Act (BC DWPA) and Regulation (BC DWPR) enables Drinking Water Protection Plans to be developed, but these are not compulsory. Other regulatory tools with powers to protect source waters include: Water Act, Forest and Range Practices Act, Environmental Management Act and Land Act.	Clean Water Act, 2006, S.O., 2006, c.22 mandates source protection planning for municipal water sources (untreated surface or groundwater).

Risk assessment	• No legislative imperative to conduct drinking water quality risk assessment, unless a Drinking Water Officer orders a source water or system assessment (DWPA 2001s.18). • No consistent risk assessment and risk abatement methodologies for drinking water quality or ecosystem health.	Formal risk assessment (of drinking water sources are required in accordance with the province's Safe Water Act 2002). Safe Drinking Water Act 2002s.15(1).
Risk management	• No legislative imperative to conduct drinking water risk management. • No consistent risk management methodologies. • Emergency Response procedures are required under (DWPA 2001s.10).	Required by legislation Clean Water Act, 2006 287/07, 54–60 (this includes Emergency Response Procedures).

Human Health Focus

Of the three agency types interviewed watershed authorities are the only ones to explicitly include both the protection of human and ecosystem health as part of their mandate. However, not all watershed authorities engage in microbial water quality monitoring in line with this mandate. One Ontario watershed authority interviewee commented that since the Walkerton tragedy watershed authorities are somewhat behooved to include *E. coli* in their examination of water quality but cautioned how this information can be used, interpreted and shared. Microbial data have limited utility when collected through infrequent grab samples; without routine year-round sampling, this data cannot be incorporated into most microbial risk assessment tools. Since water quality monitoring for human health falls under the jurisdiction of other agencies (water utilities and health authorities), three interviewees indicated the futility of collecting microbial water quality information, particularly when information is collected more rigorously by other agencies. As such, watershed authorities may only use microbial information to better understand the overall state of the watershed, rather than inclusion in formalized microbial risk management. In the absence of this information some watershed authorities indicated that they tend to focus on chemical contamination, which they felt they had a better capacity to monitor.

Limitations of Current Testing Approaches

The availability of suitable tools to characterize microbial risk is essential for successful risk assessment and management. Interviewees shared five key underlying concerns related to microbial data collection. First, they indicated concern about their limited ability to detect specific pathogens, particularly the absence of tools to detect viruses in water. The limitations of using *E. coli* as a surrogate indicator for all pathogens were specifically highlighted, since strong empirical evidence indicates that *E. coli* is not always predictive of pathogens occurrence, particularly for viruses and protozoa. I.e. the absence of *E. coli* is not a guarantee that a water sample is pathogen free (Harwood et al., 2005 and Leclerc et al., 2001). Second, interviewees expressed concern related to the inability to identify the source of contamination with certainty. Currently, practitioners can target only the most likely species that may be a cause for concern with inadequate resolution, which interviewees indicated is an ineffective use of limited resources. Third, is concern regarding the time delay between collecting a sample and receiving the test results, highlighting the challenge to protect public health when water advisories depend on culture-based testing results that take 18–24 h to generate (without consideration for transport time to the laboratory). The time lag was identified in both recreational and drinking water quality contexts and is also highlighted in the literature (Hrudey, 2004,Hrudey, 2011, Hamilton et al., 2006, Vieria, 2007 and Gelting, 2009). The fourth concern is the inability to establish whether a particular microorganism will cause disease in humans (i.e. some methods cannot distinguish between pathogenic and non-pathogenic strains). Fifth, interviewees noted the inherent variability of sampling and testing methods between agencies, which may impede data sharing.

Together, these five concerns emphasize ongoing challenges for agencies engaged in characterizing microbial risk. It is clear across the range of agencies interviewed that legislated indicators alone are insufficient for assessment and microbial risk management needs and as such, additional microbial tests are being applied. Interviewees emphasized a desire for microbial tests with quicker turn-around-times and that answered specific risk associated questions (e.g. whether a microorganism is pathogenic to humans or whether groundwater is vulnerable to pathogen intrusion).

Limited and Variable Use of Risk Assessment

Risk assessment practices across the 18 agencies interviewed are limited and variable in terms of type of methods used, scope of application and frequency of use.

Types of Risk Assessment Methods Used

Among our case study agencies there appears to be limited uptake of the established formalized methods as identified in Table 1. In some cases, interviewees reported little awareness of such tools. Of the methods and tools identified in Table 1, HACCP & CCP were the most commonly identified or familiar to our interviewees; however, the full implementation of these methods is limited. Instead, interviewees indicated they typically use existing tools informally; drawing on elements of these tools (particularly HACCP) to develop their own tailor-made risk assessment better suited to their needs or resource capacities (financial and staff). Interviewees in both provinces identified a number of constraints to their ability to practice microbial risk assessment and management. These include: insufficient data collection to "do justice to a risk assessment"; financial and staff capacity (to collect more data more frequently, conduct specific microbial risk assessment projects or to introduce an industry standardized tools such as HACCP or ISO certification); limited regulatory requirements (e.g. there are no regulated standards for *Cryptosporidium*); and lack of tools (e.g. to assess microbial risks to groundwater). Only two of the 18 agencies (both in Ontario) use QMRA routinely. One application is by a health authority for beach (recreational) water quality purposes and the other is by a water utility. Both agencies acknowledged the technical support provided by Health Canada in their implementation of QMRA. One Ontario water utility interviewee indicated that the absence of provincial funding limited the capacity of practitioners to use QMRA; instead collaboration with academic researchers was driving its use in municipalities. Two watershed agencies (both in BC) conduct ecological or environmental risk assessments: one includes microbial water quality data and the other focuses only on chemical risk. Challenges to conduct ecological risk assessments on a watershed scale identified by the interviewees (mentioned in 4.1.2 and 4.1.3) are found in the literature (Serveiss and Ohlson, 2007).

Scope and Frequency of Application

Risk assessments that are formalized, routine and applied system-wide have not been widely adopted by the agencies interviewed, other than the Ontario water utilities, which conduct these from source-to-tap. The most common applications of risk assessments are short-term or one-off assessments. For example, prior to design and construction of a water treatment plant or a project specific assessment might be undertaken to examine one particular component of the entire water system (e.g. recreational water quality at a certain beach, or until engineering controls for a specific contamination source could be established). No one reported using any of these models explicitly for ecosystem health.

Limited Use of Risk Management Plans

Ten interviewees noted that their agency does not have any formal risk management plan in place for human health or ecosystem purposes, including all six watershed authorities in BC and Ontario. As one BC water utility interviewee observed, "...from the point the water enters the intake, down to the customer's tap, we do not have a risk management process. Not an explicit one certainly." Of the agencies that stated they do have a risk management plan the extent of these were varied and often project specific (e.g. requirement for filtration deferral). Only one agency identified having ever used a specific microbial risk management plan, for groundwater under direct influence of surface water; however, the interviewee noted that they no longer use this method.

While many of the interviewees acknowledged their agency does not have a formalized risk management plan for microbial risk, all water utilities and health authorities have a formal Emergency Response Procedure (ERP), which is a regulatory requirement in both BC and Ontario. These procedures are activated once an actual (e.g. *E. coli* present in treated water) or perceived (e.g. elevated total coliform count or turbidity measurement) microbial contamination event occurs and typically includes a communication strategy. In line with this finding, a few interviewees suggested that their current approach to risk assessment or management is reactive.

Interestingly, despite the absence of microbial risk assessment practices, there is general confidence among the agencies interviewed (13 out of the 18 interviewees) that they are well positioned to handle a microbial contamination event. This confidence could be attributed to a variety of factors including: i) having formalized response and communication procedures in place (some municipalities have designated crisis management teams); ii) routine monitoring at least at the tap (continuous monitoring in the water treatment plant or compliance monitoring as required by legislation); iii) land ownership around source waters (particularly the ability to exert control over source water protection); iv) system knowledge (such as lessons learned from previous incidents); and v) having a contingency plan in place to augment their water supply.

Differing Provincial Approaches to Risk Assessment and Risk Management

Ontario and BC have fundamentally different governance approaches to the application risk assessment and management (see Table 3).

Ontario legislation requires that operators of municipal water systems conduct microbial risk assessments and all three of the Ontario water utilities interviewed confirmed they conduct formal microbial risk assessment in accordance with the Ontario's Safe Drinking Water Act 2002s.15(1). This legislation requires municipal drinking water systems operators to prepare operational plans according to the Director's directions. The directions prescribe a Drinking Water Quality Management System (DWQMS) specifying minimum requirements for risk assessment and risk management in operational plans (Schedule A). As one Ontario water utility interviewee described, "It's a pretty legislated regulatory environment that we function in." Indeed, penalties for non-compliance are onerous; as of January 2013 water system owners and operators in Ontario bear personal responsibility (financial penalties and/or imprisonment). In contrast, BC has no legislative imperative; a drinking water officer may require a BC water utility to undertake but this is discretionary. Only one of the three BC water utilities interviewed acknowledged using a formalized risk assessment and management approach, which was a requirement to obtain a filtration deferral for water treatment. None of the BC water utilities undertakes formal microbial risk assessment or management on a routine basis. One interviewee observed, "the expectations [in BC] are not as high [in comparison with Ontario], so they are not difficult to meet. In terms of drinking water health, in most other jurisdictions... there is more stringent management within the regulation."

Given the small sample size, we can only suggest that indeed the regulatory differences between the two provinces appear to have an important impact in terms of the microbial risk practices being undertaken by these agencies. We discuss these key provincial differences in Section 5.1.

DISCUSSION

There are several axes of variability that emerged as important in the analysis: 1) approaches to microbial risk assessment and management are variable between provinces; 2) water utilities, health authorities and watershed authorities approach, and may be required to approach, risk assessment and management differently; 3) risk assessment and management practices (particularly in BC) diverge from the literature on best practices in ways that are potentially significant; 4) there is limited, if any, focus on risk assessment for ecosystem health from the microbial perspective and 5) testing method limitations.

Varying Approaches between Provinces

British Columbia and Ontario have notably different water governance frameworks. In response to recommendations in the Walkerton Inquiry (O' Connor, 2002), Ontario overhauled their water governance approach. A strong legislative and regulatory framework was introduced with a source-to-tap focus, and includes risk management protocols to enforce and harmonize approaches across the province (Jayarante, 2008 and O' Connor, 2002). Specifically, Ontario is the first province in Canada to introduce a legislated, semi-quantitative risk assessment framework for source water protection. As one Ontario water utility interviewee described, "We have a multi-barrier approach and in fact it is now enshrined in legislation".

In contrast, legislated requirements in BC are narrower in scope; the BC approach to source protection might best be described as voluntary (see part 5 of the BC DWPA). Moreover, the BC DWPA and its regulations (see Table 3) are outcome-based in that they specify the outcomes to be achieved, not how to achieve them. BC does have provincial guidelines for the multi-barrier approach, but none explicitly for risk assessment and risk management. Some BC interviewees attributed confidence in their water quality to their ability to protect their source water (either through limited land use activity or fully protected watersheds). Indeed, BC's topography lends itself more easily to source protection with approximately 60% of the province's population receiving water from protected watersheds (including the cities of Vancouver and Victoria). However, despite the fact that fully protected watersheds serve more than half the population, BC has the highest number of boil water advisories per capita (Eggerston, 2008). Furthermore, the province has also had the highest rate of enteric (gastrointestinal) infections of all the Canadian provinces, a proportion of which may be attributable to waterborne transmission (Isaac-Renton et al., 2003). Historically, BC water supplies have not always been safe and highlight that a perception of pristine water province-wide may actually undermine risk management to the extent that this perception may contribute to the lack of preventative practices.

Varying Approaches between Watersheds and Similar Agencies

Few clear patterns emerge in terms of type, methods, scope and frequency of risk assessment and management between agencies with similar mandates. Our results show that in our two case provinces, between watersheds and between agencies (even of the same type) risk assessment and management practices vary widely. While watershed authorities typically focus on overall watershed characterization and assessment (using environmental, ecological or regional risk assessment tools, but not always including microbial pathogens), health authorities and water utilities focus primarily on drinking water supply using TQM based tools, which incorporate elements of HACCP and CCP. Not only does the scope and frequency of microbial risk assessment vary but so do the methods applied. No single tool has been adopted widely. Agencies that do practice risk assessment often prefer to develop their own tools. None of the agencies studied have pursued formal certification (e.g. ISO 9001 and 14001), citing limited capacity (financial and staff) as an impediment.

Limited Focus on Ecosystem Health

There is limited evidence in the literature that ecosystem-focused risk assessments are routinely incorporated into water risk assessment and management activities (Pollard and Huxham, 1998). Serveiss and Ohlson (2007) argue that in Canada water quality standards are human-centric and watershed planning is oriented toward drinking water quality protection, focusing primarily on chemical and physical measures, omitting a broader ecosystem perspective. A truly integrated source-to-tap risk management approach should emphasize both human and ecosystem health. The concept of integrating human and ecological risk assessment has been advocated in the environmental literature (Harvey et al., 1995, Sekizawa and Tanabe, 2005 and Scott et al., 2005) but only applied infrequently to aquatic environments (Moiseenko et al., 2006, Orme-Zavaleta and Munns, 2008 and Wang, 2006). Moreover, in the literature, ecological risk assessments for aquatic environments have focused largely on chemical risk assessment, rather than microbial risk assessment (Brock, 2013, Ellis, 2000, González-Pleiter et al., 2013 and Leeuwangh et al., 1993). An integrated risk management approach is consistent with the "One Health" approach advocated by organizations such as the WHO and the World Organization for Animal Health: that environments, animals and humans health are interrelated in a complex system with many common stressors. An integrated approach allows for a broader, system-wide approach to managing microbial risks and creates efficiencies by using common sets of tools for risk characterization, assessment and management. Consider the example of an algal bloom: algal toxins are harmful to mammals (human, aquatic and terrestrial mammals) and the bloom can also change nutrient and gas cycling in the ecosystem resulting in adverse outcomes at multiple trophic levels in the ecosystem. Here, ecosystem and human health are clearly connected and could benefit from an integrated approach. In our case study entities, we find ecosystem risk assessments are largely absent and, when present, are not integrated with human health risk assessments. Even in large water utilities, which have both watershed and drinking water departments, we found little integration of ecosystem and human health assessments, and minimal emphasis on the microbial dimensions of ecosystem health in general. Even within the same organization, different tools are being used in ways that can serve as an obstacle to integration. One interviewee from BC postulated that ecosystem health is on the whole poorly understood. He observed, "one of the biggest problems I think we have is the assumption that we can treat something, therefore, we don't have to worry so much about the quality of the source. I think there needs to be a strong connection between the quality of the source and what's coming out the pipe. I don't think that we want to promote the notion that we can — we should allow the degradation of the source simply by increasing the treatment. So an emphasis on watershed planning, on controlling non-point sources, on people knowing how their actions affect the quality of the source and how that relays into cost savings in terms of treatment."

Divergence from Literature on Best Practice

On the whole, our research also indicates that current approaches to microbial risk assessment and management in BC case study sites, and to a lesser degree in Ontario, remain largely reactive. Formal microbial risk assessment and management practices are not widely adopted across agencies; they are not typically carried out on a continuous basis (but rather on an ad hoc, as needed basis); and they are not carried out across the full water system from source to tap, rather components of the system (e.g. focusing only on the treatment dimensions, rather than incorporating quality of source water as well). Evidencing a reactive risk management pattern, agencies indicated that risk assessments are typically instigated after a contamination event. While ERPs have an essential role, these are best thought of as reactive approaches. A broader proactive approach as advocated by the WHO would require microbial hazard identification and efforts to prevent and remediate contamination events. We find that in BC in particular microbial risk assessment and management approaches for water quality have not yet achieved the vision of the WHO. Internationally and in Canada, despite the emphasis on source protection, ecosystem health appears to be a lower priority. Economic and political factors; assumptions made is risk assessment scenarios; and other factors can enable or constrain microbial risk assessment and management practices (see Reimann and Banks, 2004). Further work on why microbial risk might not be pursued, or complex tradeoffs with other human and ecosystem health considerations would be a fruitful avenue for future exploration.

We postulate that the regulatory context in Ontario is having some impact on actual practices of risk assessment and management. Although not as extensively as we might expect to be able to more adequately integrate human and ecosystem approaches, or to share data across agencies. In terms of other important trajectories toward the adoption of more integrated and explicit approaches, it is also worth noting that the Canadian province of Alberta is introducing a legal requirement for drinking water safety plans by December 2013; but again, this will likely focus on human health rather than an integrated risk management approach.

Limitations of Current Testing Methodologies

The success of microbial risk assessment and management is only as good as the available risk characterization or measurement tools. In addition to identifying gaps in current approaches to risk assessment and management, our research highlights key limitations to current microbial testing methods, a critical first step in risk assessment and management. As discussed in Section 4.1.3, interviewees were most concerned with the turn-around-times and suitability of current tools to enable comprehensive and routine microbial risk assessment. While all microbial classes (bacteria, protozoa and viruses) are harmful to ecosystem and human health, most of the available tools are narrow in range, targeting only bacterial pathogens (Davison et al., 2005). As noted, some interviewees identified the need for additional testing parameters to enable thorough risk assessment and risk management.

Developments in genomics and biotechnology may resolve some of the methodological limitations to risk characterization identified by interviewees. For example, water metagenomics studies may identify new targets for microbial testing, which may allow risk characterization along a wider, and more representative, set of microbial parameters (Thomas et al., 2012 and Languille et al., 2012). Current risk characterization approaches rely on detection of microbial indicators, which do not accurately predict the occurrence of all bacterial, viral and protozoan pathogen, while pathogen specific testing is cost prohibitive and lack the sensitivity required for adequate risk characterization. Novel markers that could better predict the occurrence of bacterial, viral and protozoan pathogens would improve risk assessment and management activities.

Furthermore, biotechnology advancements in testing platforms will likely allow more sensitive assays (lower limits of detection), faster turn-around-times, identification of contamination source, and are expected to be available in the near future at equal or lower costs. For example, quantitative real-time polymerase chain reaction (qPCR), which detects the nucleic acid of a microorganism rather than culturing organisms, provides the benefit of being able to detect even organisms that cannot easily be cultured such as viruses, protozoa and some difficult to culture bacteria. This platform can test nearly any microbial target, is flexible (new assays can easily be developed) and is widely applied in clinical diagnostics. We see considerable potential for new markers on new qPCR platforms to be integrated into current risk assessment and management approaches. Novel tests that are more sensitive and specific have the potential to significantly improve microbial risk assessment and management capacities. It should be noted, however, that nucleic acid based approaches like qPCR do have methodological limitations, such as challenges with compounds that may inhibit the assay or detection of dead cells. Technical advancements have addressed many of these technical challenges and are detailed by others (Girones et al., 2010 and Aw and Rose, 2012).

CONCLUSIONS

In sum, our research identifies considerable variability in risk assessment and management frameworks currently applied in BC and Ontario. We find that the variability between and within provinces, as well as between agencies, is indicative of an overall lack of a uniform approach. We find that the most common applications of risk assessments are short-term or one-off assessments. Risk assessments that are formalized, routine (or continuous) and applied system-wide (i.e. the entire water supply system — source-to-tap) have not been adopted widely. Our findings of varied approaches between and within provinces are broadly consistent with the literature on water governance in Canada, which indicates considerable fragmentation, and that water management is often ad hoc (Bakker and Cook, 2011, Hill et al., 2008, Hrudey, 2011 and Weibust, 2009). We also observe that ecological risk assessment is largely absent among the entities interviewed. Our research suggests that current approaches to microbial water quality are still largely implicit and reactive, diverging from the literature on best practices in the water sector.

The literature suggests the need for preventative and explicit risk management for the protection of both human and ecosystem health, but our case study provinces fail to meet these best practices in a comprehensive manner. In terms of significance, this work highlights the need for improved microbial risk assessment and management frameworks in Canada. At a minimum, data needs to be comparable and integrated, and resources and expertise need to be shared in order to overcome some of the issues identified by interviewees. Adoption of improved microbial testing, based on genomics and biotechnological advances, would provide richer data, faster, and improve risk assessment and management. As well, we see considerable scope to more adequately evaluate the importance of different regulatory frameworks, either in terms of actual risk management practices, or better still, linkages to human and ecosystem health outcomes. In the absence of considerable progress toward these ends, human and ecosystem health may continue to be compromised.

ACKNOWLEDGMENTS

This research was conducted as a part of a three-year (2011–2014) project Applied Metagenomics of the Watershed Microbiome. The authors would like to thank Genome British Columbia and Genome Canada for financially supporting this research. Thank you also to the Canadian Water Network and the Alberman Fellowship at the Hebrew University of Jerusalem for additional financial support. For more information about this project, please visit our website www.watersheddiscovery.ca. The authors would particularly like to acknowledge the contributions from research team members Judy Isaac-Renton, Karen Bakker, Ida Ngueng Feze, Beverly Holmes and Maggie Low.

REFERENCES

1. Ashbolt NJ. Risk analysis of drinking water microbial contamination versus disinfection by-products (DBP's). Toxicology 2004;198(1–3):255–62.
2. Ashbolt NJ, Schoen ME, Soller JA, Roser DJ. Predicting pathogen risks to aid beach management: the real value of quantitative microbial risk assessment (QMRA). Water Res 2010;44:4692–703.
3. Austin Z, Alcock RE, Christley RM, Haygarth PM, Heathwaite AL, Latham SM, et al. Policy, practice and decision making for zoonotic disease management: water and Cryptosporidium. Environ Int 2012;40:70–8.
4. Aw TG, Rose JB. Detection of pathogens in water: from phylochips to qPCR to pyrosequencing. Curr Opin Biotechnol 2012;23(3):422–30.
5. Bakker K, Cook C. Water governance in Canada: innovation in the context of fragmentation. Int J Water Resour Dev 2011;27(2):275–89.
6. Bartram J, Corrales L, Davison A, Deere D, Drury D, Gordon B, et al. Water safety plan manual: step-by-step risk management for drinking-water suppliers. Geneva: World Health Organization; 2009.
7. Benke KK, Hamilton AJ. Quantitative microbial risk assessment: uncertainty and measures of central tendency for skewed distributions. Stoch Environ Res Risk Assess 2008;22(4):533–9.
8. British Columbia Drinking Water Protection Act (BC DWPA), Drinking Water Protection Regulation B.C. Reg. 200/2003. O.C. 508/2003.
9. Brock TC. Priorities to improve the ecological risk assessment and management for pesticides in surface water. Integr Environ Assess Manag 2013;9(3):e64–74. [Apr 22].
10. Byleveld PM, Deere D, Davison A. Water safety plans: planning for adverse events and communicating with consumers. J Water Health 2008;6:1–9.
11. Canadian Council of Ministers of the Environment (CCME). From source to tap: guidance on the multi-barrier approach to safe drinking water. Ottawa, Ontario: Canadian
12. Council of Ministers of the Environment Water Quality Task Group and the Federal–Provincial–Territorial Committee on Drinking Water; 2004.
13. Cook C, Prystajecky N, Ngueng Feze I, Joly Y, Dunn G, Kirby E, et al. A comparative analysis of microbial drinking water quality testing in three Canadian provinces. Can Water Resour J 2013. http://dx.doi.org/10.1080/07011784.2013.822186 [in press, accepted July].
14. Cool G, Rodriguez MJ, Bouchard C, Levallois P, Joerin F. Evaluation of the vulnerability to contamination of drinking water systems for rural regions in Quebec, Canada. J Environ Plan Manag 2010;53(5):615–38.
15. Davies JM, Mazumder A. Health and environmental policy issues in Canada: the role of watershed management in sustaining clean drinking water quality at surface sources. J Environ Manage 2003;68:273–86.
16. Davison A, Howard G, Stevens M, Callan P, Fewtrell L, Deere D, et al. Water safety plans: managing drinking-water quality from catchment to consumer. Prepared for theGeneva: World Health Organisation; 2005 [WHO/SDE/WSH/05.06].
17. Dominguez-Chicas A, Scrimshaw MD. Hazard and risk assessment for indirect potable reuse schemes: an approach for use in developing water safety plans. Water Res 2010;44(2):6115–23.
18. Egerton AJ. Achieving reliable and cost effective water treatment. Water Sci Technol 1996;33(2):143–9.
19. Eggerston L. Investigative report: 1766 boil-water advisories now in place across Canada. Can Med Assoc J 2008;178(10):1261–3.
20. Ellis JB. Risk assessment approaches for ecosystem responses to transient pollution events in urban receiving waters. Chemosphere 2000;41(1–2):85–91.
21. Gelting R. Water safety plans: CDC's role. J Environ Health 2009;72(4):44–5.
22. Girones R, Ferrús MA, Alonso JL, Rodriguez-Manzano J, Calgua B, Corrêa Ade A, et al. Molecular detection of pathogens in water—the pros and cons of molecular techniques. Water Res 2010;44(15):4325–39. [Aug].
23. González-Pleiter M, Gonzalo S, Rodea-Palomares I, Leganés F, Rosal R, Boltes K, et al. Toxicity of five antibiotics and their mixtures towards photosynthetic aquatic organisms: implications for environmental risk assessment. Water Res 2013;47(6):2050–64.
24. Goss M, Richards C. Development of a risk-based index for source water protection planning, which supports the reduction of pathogens from agricultural activity entering water resources. J Environ Manage 2008;87(4):623–32.

25. Gozlan RE, Peeler EJ, Longshaw M, St-Hilaire S, Feist SW. Effect of microbial pathogens on the diversity of aquatic populations, notably in Europe. Microbes Infect 2006;8(5): 1358–64.

26. Gunnarsdottir MJ, Gardarsson SM, Elliott M, Sigmundsdottir G, Bartram J. Benefits of water safety plans: microbiology, compliance, and public health. Environ Sci Technol 2012;46(14):7782–9.

27. Haas CN, Rose JB, Gerba CP. Quantitative microbial risk assessment. New York: Wiley; 1999464.

28. Hamilton P, Gale P, Pollard SJT. A commentary on recent water safety initiatives in the context of water utility risk management. Environ Int 2006;32:958–66.

29. Harvey T, Mahaffey KR, Velazquez S, Dourson M. Holistic risk assessment: an emergingprocess for environmental decisions. Reg Toxicol Pharmacol 1995;22:110–7.

30. Harwood VJ, Levine AD, Scott TM, Chivukula V, Lukasik J, Farrah SR, et al. Validity of the indicator organism paradigm for pathogen reduction in reclaimed water and public health protection. Appl Environ Microbiol 2005;71:3163–70.

31. Hill C, Furlong K, Bakker K, Cohen C. Harmonization versus subsidiarity in water governance: a review of water governance and legislation in the Canadian provinces and territories. Can Water Resour J 2008;33(4):1–18.

32. Hrudey SE. Drinking water risk management principles for a total quality management framework. In: Hrudey SE, editor. Drinking water safety — a total quality management approach. NERAM, Institute for Risk Research, University of Waterloo; 2003. p. 3–14.

33. Hrudey SE. Drinking water risk management principles for a total quality management framework. J Toxicol Environ Health A 2004;67:1555–66.

34. Hrudey SE. Chlorination disinfection by-products, public health risk tradeoffs and me. Water Res 2009;43:2057–92.

35. Hrudey SE. Safe drinking water policy in Canada — turning hindsight into foresight. C.D Howe Institute commentary: the water series no. 323Toronto: CD Howe Institute; 2011.

36. Hrudey SE, Hrudey EJ, Pollard SJT. Risk management for assuring safe drinking water. Environ Int 2006;32:948–57.

37. Hunter PR, Payment P, Ashbolt N, Bartram J. Chapter 3 — assessment of risk. Assessing microbial safety of drinking water: improving approaches and methods. Published on behalf of World Health Organization and organization for economic co-operation and development. London: IWA Publishing; 2003. p. 79–109.

38. Isaac-Renton JIR, Peck S, Kendal P. Health drinking water in British Columbia: a physicians' update. B C Med J 2003;45(9):436.

39. Ivey JL, de Loë RC, Kreutzwiser RD. Planning for source water protection in Ontario. Appl Geogr 2006;26:192–209.

40. Jayarante A. Application of a risk management system to improve drinking water safety. J Water Health 2008;6(4):547–57.

41. Krewski D, Balbus J, Butler-Jones D, Haas C, Isaac-Renton JIR, Roberts KJ, et al. Managing health risks from drinking water—a report to the Walkerton inquiry. J Toxicol Environ Health A 2002;65(21):1635–823.

42. Languille MGI, Laird MR, Hsiao WWL, Chiu TA, Eisen JA, Brinkman FSL. MicrobeDB: a locally maintainable database of microbial genomic sequences. Bioinformatics 2012;28(14): 1947–8.

43. Leclerc H, Mossel DAA, Edberg SC, Struijk CB. Advances in the bacteriology of the coliform group: their suitability as markers of microbial water safety. Annu Rev Microbiol 2001;55:201–34.

44. Leeuwangh P, Brock TC, Kersting K. An evaluation of four types of freshwater model ecosystem for assessing the hazard of pesticides. Hum Exp Toxicol 1993;13(12):888–99.

45. Miller MA, Gardner IA, Kreuder C, Paradies DM, Worcester KR, Jessup DA, et al. Coastal freshwater runoff is a risk factor for Toxoplasma gondii infection of southern sea otters (Enhydra lutris nereis). Int J Parasitol 2002;32(8):997–1006.

46. Miller R, Whitehill B, Deere D. A national approach to risk assessment for drinking water catchments in Australia. 4th World Water Congress. Water Supply Water Serv Manag 2005;5(2):123–34.

47. Miller KM, Li S, Kaukinen KH, Ginther N, Hammill E, Curtis JM, et al. Genomic signatures predict migration and spawning failure in wild Canadian salmon. Science 2011;331(6014): 214–7.

48. Moiseenko TI, Voinov AA, Megorsky VV, Gashkina NA, Kudriavtseva LP, Vandish OI, et al. Ecosystem and human health assessment to define environmental management strategies: the case of long-term human impacts on an Arctic lake. Sci Total Environ 2006;369(1–3):1–20.

49. Ngueng Feze I, Kirby E, Prystajecky N, Cook C, Knoppers BM, Ozdenmir V, et al. Legal analysis of water quality assessment biotechnologies: is Canada ready to ride the next wave of metagenomics testing? Ott Law Rev 2013w. [under review, submitted March].

50. O' Connor DR. Part two. Report of the Walkerton Inquiry. A strategy for safe water. Ottawa: Ministry of the Attorney General; 2002. p. 582.

51. Orme-Zavaleta J, Munns Jr WR. Integrating human and ecological risk assessment: application to the cyanobacterial harmful algal bloom problem. Adv Exp Med Biol 2008;619:867–83.

52. Phillips JD. Nonpoint source pollution and spatial aspects of risk assessment. Ann Assoc Am Geogr 1988;78(4):611–23.

53. Pollard P, Huxham M. The European Water Framework Directive: a new era in the management of aquatic ecosystem health? Aquat Conserv 1998;8(6):773–92.

54. Pollard SJT, Strutt JE, MacGillivary BH, Hamilton PD, Hrudey SE. Risk analysis in the water utility sector: a review of drivers, tools and techniques. Process Saf Environ Prot 2004;82(B6):453–62.

55. Reimann C, Banks D. Setting action levels for drinking water: are we protecting our health or our economy (or our backs!)? Sci Total Environ 2004;322:13–21.

56. Schijven JF, Teunis RF, Rutjes SA, Bouwknegt M, de Roda Husman AM. QMRAspot: a tool for quantitative microbial risk assessment from surface water to potable water. Water Res 2011;45(17):5564–76.

57. Scott MJ, Brandt CA, Bunn AL, Engel DW, Eslinger PW, Miley TB, et al. Modeling long-term risk to environmental and human systems at the Hanford Nuclear Reservation: scope and findings from the initial model. Environ Manage 2005;35(1):84–98.

58. Sekizawa J, Tanabe S. A comparison between integrated risk assessment and classical health/environmental assessment: emerging beneficial properties. Toxicol Appl Pharmacol 2005;207(2 Suppl.):617–22.

59. Serveiss VB, Ohlson DW. Using ecological risk assessment principles in a Source Water Protection Assessment. Hum Ecol Risk Assess 2007;13:402–17.

60. Signor RS, Ashbolt NJ. Pathogen monitoring offers questionable protection against drinking water risks: a QMRA (quantitative microbial risk analysis) approach to assess management strategies. Water Sci Technol 2006;54(3):261–8.

61. Smeets PW, Rietveld LC, van Dijk JC, Medema GJ. Practical applications of quantitative microbial risk assessment (QMRA) for water safety plans.Water Sci Technol 2010;61(6): 1561–658.

62. Summerill C, Smith J, Webster J, Pollard SJT. An international review of the challenges associated with securing buy-in for water safety plans within providers of drinking water supplies. J Water Health 2010a;8(2):387–98. [http://www.iwaponline.com/ jwh/008/jwh0080387.htm].

63. Summerill C, Pollard SJT, Smith JA. The role of organizational culture and leadership in water safety plan implementation for improved risk management. Sci Total Environ 2010b:4319–27.

64. Summerscales IM, McBean EA. Application of risk assessment tools to small drinking water systems in British Columbia. Water Qual Res J Can 2011;46(4):332–44.

65. Taylor CM, Pollard SJT, Angus AJ, Rocks SA. Better by design: rethinking interventions for better environmental regulation. Sci Total Environ 2013;408(20):488–99.

66. Thomas T, Gilbert J, Meyer F. Metagenomics — a guide from sampling to data analysis. Microb Inform Exp 2012;2(1):3.

67. Vieria JMP. Water safety plans: methodologies for risk assessment and risk management in drinking water systems. Water in Celtic countries: quantity, quality and climate variability 2007;310:57–67.

68. Wang M. Optimal environmental management strategy and implementation for groundwater contamination prevention and restoration. Environ Manage 2006;37(4): 553–66.

69. Weibust I. The green leviathan: the case for a federal role in environmental policy. Aldershot: Ashgate Publishing; 2009.

70. Weitz JS, Wilhelm SW. Ocean viruses and their effects on microbial communities and biogeochemical cycles. F1000 Biol Rep 2012;4:17.

71. World Health Organization. Guidelines for drinking-water quality. Recommendations, 3rd ed., vol. 1. Geneva: World Health Organization; 2004.

72. Yin RK. Applications of case study research. 2nd ed. Sage Publications; 2003.

73. Yokoi H, Embutsu I, Yoda M, Waseda K. Study on the introduction of hazard analysis and critical control point (HACCP) concept of the water quality management in water supply systems. Water Sci Technol 2006;53(4):483–92.

CHAPTER 19 ASSESSING WATER SCARCITY BY SIMULTANEOUSLY CONSIDERING ENVIRONMENTAL FLOW REQUIREMENTS, WATER QUANTITY, AND WATER QUALITY

Junguo Liu, Qingying Liu, Hong Yang

Keywords: Environmental flow requirement, Water quality, Water quantity, Water scarcity, Huangqihai River Basin

INTRODUCTION

Freshwater is a fundamental resource for human well-being and the natural environment; it is regarded as the most essential natural resource in the world (Gleick, 1993). Over the past few decades, climate change and human socioeconomic development have greatly changed global hydrological cycles, threatening human water security, the health of aquatic environments and river biodiversity (Vörösmarty et al., 2010, Jacobsen et al., 2012 and van Vliet et al., 2013). Given this situation, increasing attention has been paid to assessing the environmental flow requirement (EFR) of rivers and water scarcity (Vörösmarty et al., 2010 and Kirby et al., 2014).

EFR is defined as the quantity, timing, and quality of the water flows required to sustain freshwater and estuarine ecosystems and the human livelihoods and well-being that depend upon these ecosystems (Brisbane Declaration, 2007). More than 200 methods are being used worldwide to calculate EFR that is needed to maintain healthy rivers (Tharme, 2003). These methods can be grouped into four categories: hydrological approach, hydraulic rating, habitat simulation, and holistic methods. The selection of an appropriate method is primarily constrained by the availability of data for a region, as well as by local limitations in terms of time, funding, expertise, and logistical support.

Citation: Junguo Liu, Qingying Liu, Hong Yang, Assessing water scarcity by simultaneously considering environmental flow requirements, water quantity, and water quality, Ecological Indicators, Volume 60, January 2016, Pages 434-441, ISSN 1470-160X, http://dx.doi.org/10.1016/j.ecolind.2015.07.019.

The main approaches used to assess water scarcity include the Falkenmark water stress indicator (Falkenmark et al., 1989), the IWMI indicator (Seckler et al., 1998), the criticality ratio (Alcamo et al., 2000), and the water poverty index (Sullivan et al., 2003). These approaches all focused on water quantity, but did not account for water quality for water scarcity assessment. Zeng et al. (2013) developed a simple indicator that combines quantity with quality in an easily understood way. However, it did not include a realistic approach to quantifying EFR. It is worth noting that there is an increasing awareness of explicitly considering EFR in the assessment of water scarcity in the hydrology community. Hoekstra et al. (2012) assumed EFR to be 80% of the total water resources in the assessment of global water quantity scarcity. This assumption was too simplistic, as it did not consider the complexity of EFR in a river regime. Hence, there is a need for a water scarcity assessment approach that can consider both water quantity and water quality, while also permitting a realistic consideration of EFR. Such a combined approach can provide more complete information on water scarcity.

The objective of this paper was to improve the water scarcity assessment approach by incorporating the water volume needed for EFR in the assessment. This new holistic water scarcity assessment approach provides an indicator that combines the status of quantity, quality and EFR (QQE indicator). The data used in this approach is easy to obtain, and can be quickly applied for a region. To demonstrate the application of the improved approach, we used it to assess the water scarcity in an arid and semi-arid region, the Huangqihai River Basin in Inner Mongolia, China.

STUDY AREA AND METHODS

Case Study Area

The Huangqihai River Basin is located in the central part of Inner Mongolia Autonomous Region in China (Fig. 1). The average annual temperature of the basin is 4.6 °C, with mean monthly temperatures ranging from a minimum of −18 °C in January to a maximum of 26 °C in July and a freeze up period of more than 150 days. Annual precipitation ranges from an average of 270 mm in the southeast to 300 mm in the northwest. As an arid and semi-arid region, precipitation is unevenly distributed within a year, with 80% falling between June and September (Yu et al., 2013).

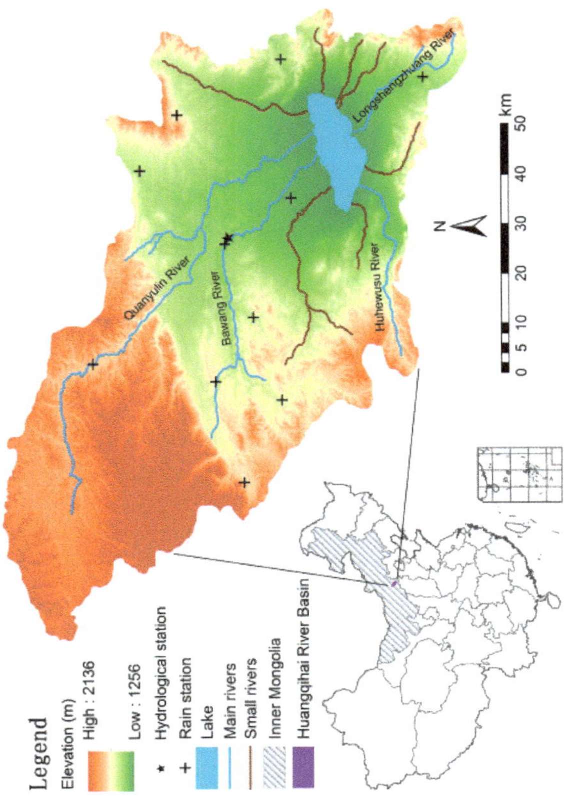

Figure 1: Location of the Huangqihai River Basin in Inner Mongolia, China.

There are 11 primary rivers in the basin (Fig. 1). Only 4 of them have year-round recharge (the Bawang, Quanyulin, Huhewusu, and Longshengzhuang rivers); the others are seasonal rivers or have dried up completely (Li et al., 2013). The Bawang River and the Quanyulin River are the main water sources for the Huangqihai Lake, contributing more than 77% of the total annual river flows. In recent years, several reservoirs were constructed in upstream regions. They intercepted water flowing into the downstream reaches, thereby changing the natural runoff patterns. The downstream rivers often ran dry and underwent siltation, leading to a rapid shrinking of the Huangqihai Lake (Ma et al., 2002).

The basin is currently facing the problems of water shortage and poor water quality, as well as deterioration of ecosystem quality. The water availability per capita is only 985 m^3/year. With the increasing demand for water from the domestic and industrial sectors, conflicts for water use between agriculture and the other sectors have become more acute. Environmental water use has been deprived. The water quality in many river sections is below Grade III, the minimum quantity standard of water for direct usage. The poor quality of water further intensifies the water shortage problem because it reduces the usable water in the basin.

A Quantity–Quality-EFR (QQE) Water Scarcity Indicator

The following equations are used to construct the QQE water scarcity indicator:

$$S_{qqe}=S_{quantity}(P)|S_{quality} \quad (1)$$

$$S_{quantity}=BWF/BWA=W\times R/(BWR-EFR) \quad (2)$$

$$S_{quality}=\frac{GWF}{BWR} \quad (3)$$

where *Sqqe* is the overall water scarcity index, which is a comprehensive indicator to reflect water scarcity by considering water quantity, water quality and EFR. $S_{quantity}$ is the index of water quantity scarcity; $S_{quality}$ is an index that quantifies the pollution-based water scarcity; P is the percentage of EFR in total blue water resources (BWR), and if not specifically mentioned, it is associated with the EFR for maintaining a level of "good" habitat quality. BWF (m^3) is the blue water footprint; BWA (m^3) is the blue water availability, which equals BWR (m^3) minus EFR (m^3); W (m^3) is the blue water withdrawal; R is the water consumption ratio; and GWF is the gray water footprint (m^3). GWF is used to quantify the amount of water required to dilute pollutants in wastewater sufficiently to meet environmental water quality standards (Hoekstra et al., 2011). When $S_{quantity}$ > 1.0, the available fresh water is not enough to meet the water consumption in the basin; at ≤1.0, there is sufficient water to meet consumption needs. If $S_{quality}$ > 1.0, the freshwater in this basin is insufficient to dilute the polluted water to the desired level specified in the water quality standard, and the region is experiencing water stress with respect to water quality; at ≤1.0, there is enough water available for dilution (Hoekstra et al., 2011).

Simulation of Hydrological Processes and Blue Water Resources

We used the Agricultural Policy/Environmental eXtender (APEX) model to simulate the basin's hydrological processes. The APEX model (Williams and Izaurralde, 2006) was developed by the Blackland Research and Extension Center (Temple, Texas, USA). It is a flexible, dynamic, and physically based distributed model that can be used to simulate the impacts of land use and management on watersheds. The applicability of the APEX model has been tested for the Huangqihai River Basin in previous studies (Liu et al., 2014 and Wang et al., 2014). The results show that the APEX model is appropriate to simulate the hydrological processes throughout the Huangqihai River Basin.

The basin's BWR is the sum of the surface runoff, lateral subsurface flow, and percolation below the root zone. The APEX model simulates these flows for each sub-basin and for the entire basin on a daily basis. Surface runoff volume was simulated using a modified Soil Conservation Service (SCS) curve number technique described by Williams (1995), and the lateral subsurface flows and percolation were computed using the storage routing and pipe flow equations of Gassman et al. (2010). We used the average BWR within the simulated period (1982–2011) in this study.

The APEX model required a digital elevation model (DEM), soil data, land cover data, and climate data. The DEM data were obtained at a resolution of 90 m (http://www.gscloud.cn). Soil data were obtained at a resolution of 1 km from the Harmonized World Soil Database (http://webarchive.iiasa.ac.at/Research/LUC/External-World-soil-database/HTML/index.html?sb=1). The land cover data were based on the 1993 Landsat Thematic Mapper (TM) data at a resolution of 30 m (Chen and Liu, 2014). Daily maximum and minimum air temperatures, wind speed data, and relative humidity data from 1980 to 2011 were recorded at the Jining Weather Station (Fig. 1) (http://cdc.cma.gov.cn/choiceStation.do). Daily precipitation data during this period were obtained from 10 rainfall stations in the Huangqihai River Basin (Fig. 1). All of the data were rescaled to the 90-m DEM resolution.

Environmental Flow Requirement (EFR)

The EFR of the Huangqihai River Basin equals the sum of the in-stream flow requirements for all the rivers in the basin. Due to the lack of historical data on river flow, we used simulated flow data for the calculation. The APEX model simulated the long-term (1980–2011) natural flows in the absence of human disturbance for the basin's main and small rivers. The results of the first two years were used as the warm-up period for the model, and the river discharge for the rest of the study period (from 1982 to 2011) was used to calculate the EFR of the rivers based on the Tennant (1976) method. We used the following equations:

$$\text{EFR} = \sum_{i=1}^{12} e_{ij} \quad (4)$$

$$e_{ij} = 3600 \times 24 \times n_i \times Q_i \times P_{ij} \quad (5)$$

where EFR*j* (m^3) is the annual environmental flow requirement of a river at fish-habitat quality level *j* (*j* = Maximum, Optimum, Outstanding, Excellent, Good, Moderately degraded, and Severely degraded). There is a temporal variation in the proportion of the mean annual flow (MAF) to maintain different levels of fish-habitat conditions (Table 1). *eij* (m^3) is EFR in month *i* at habitat quality level *j*, *ni* is the number of days in month *i*, *Qi* (m^3/s) is the mean daily flow in month *i* (calculated by the APEX model), and *Pij* (%) is the percentage of the mean annual flow in month *i* at habitat quality level *j* (Table 1).

Table 1: Temporal variation in the range of proportions of the mean annual flow (MAF) that must be allocated to maintain various levels of habitat quality.

Flow category or habitat quality	Recommended flow (% of MAF)	
	October to March	April to September
Maximum	200	200
Optimum	60–100	60–100
Outstanding	40	60
Excellent	30	50
Good	20	40
Moderately degraded	10	30
Highly to severely degraded	≤10	≤10

Sources: Tennant (1976) and Arthington (2012).

The Tennant (1976) method (also known as the "Montana" method) relies solely on the recorded or estimated river flow regimes to calculate EFR (Jowett, 1997 and Arthington, 2012). The method is rapid, inexpensive, and easy to use. It is the most frequently used hydrological method to assess the habitat quality and the associated EFR in different areas of the world (Ubertini et al., 1996, Tharme, 2003, Smakhtin et al., 2006, Kumara and Srikantaswamy, 2011, Arthington, 2012 and Pastor et al., 2014). In China, the Tennant method and the recommended EFRs for different levels of habitat qualities have been found also applicable in many regions, including in the arid and semi-arid regions (Xia et al., 2006, Jiang et al., 2006, Wang et al., 2013 and Men et al., 2014). In this study, we quantified different levels of EFR to maintain the respective levels of habitat quality defined in the Tennant method. We paid particular attention to the proportion of BWR required to maintain a "good" level of habitat quality as the acceptable EFR. Below it, the habitat quality is in degraded status.

Blue Water and Gray Water Footprints

The BWF represents the consumptive use of surface and ground water (Hoekstra et al., 2011). To calculate BWF for the Huangqihai River Basin, we first calculated the BWF of each sector in this region by multiplying the water withdrawal by a water consumption ratio (i.e., the proportion of water consumption to total water withdrawal). This is because that part of the water withdrawal is return flow; it is not really "consumed" but returns to the river systems and can be reused by downstream users (Cai et al., 2003, Anisfeld, 2010 and Hoekstra et al., 2012). No specific data were available for the water consumption ratios for the basin. We therefore used the overall water consumption ratios of three sectors for Inner Mongolia in the present study: 63% for agriculture, 62% for industry, and 69% for domestic sector. These ratios were obtained from the Inner Mongolia Water Resources Bulletin (http://www.nmgslw.gov.cn/info/infoView.jsp?idcontent=18405).

GWF can be calculated using a general approach proposed by Hoekstra et al. (2011):

$$GWF = L/(C_{max} - C_{nat}) \quad (6)$$

where GWF (m^3) is the gray water footprint, *L* (kg/yr) is the pollutant load, C_{max} (mg/L) is the maximum acceptable concentration of the ambient water quality, and C_{nat} (mg/L) is the natural pollutant concentration in the receiving water body. When natural concentrations are not known, one can assume $C_{nat} = 0$ (Hoekstra et al., 2011). In this study, we assumed $C_{nat} = 0$ due to data lacking. Using a value of 0 represents a conservative approach, since it increases the likelihood that sufficiently large environmental flows will be released to dilute the target pollutants to acceptable levels.

We calculated GWF for the farming, livestock, domestic rural consumption, domestic urban consumption, and industrial sectors. Chemical oxygen demand (COD) is the largest pollution factor in wastewater from the domestic and industrial sectors, and nitrogen is the major pollutant in the farming and livestock sectors (Zeng and Liu, 2013). To account for GWF, we selected COD as the main domestic and industrial pollutant, and total nitrogen as the main farming and livestock pollutant in the basin's water bodies. Based on the China's *Environmental Quality Standards for Surface Water* (MEPC, 2002), water quality is categorized into five grades, in which Grade III indicates that the water is suitable for fishing, aquaculture, and swimming; higher grades (IV and V) indicate poor water quality. Grade III was used to calculate GWF with a *Cmax* of 20 mg/L for COD and a *Cmax* of 1 mg/L for total nitrogen, following Zeng et al. (2013).

Data for water withdrawals and pollutant loads in 2010 for the five sectors considered were obtained from the planning report "Protection and Recovery Plan of Huangqihai Wetland" (Li et al., 2013), published by the Chinese Research Academy of Environmental Sciences.

RESULTS

BWR and EFR

The annual BWR in the Huangqihai basin totaled 144 × 10^6 m^3/yr. The river flows were concentrated between April and September, accounting for 92.6% of the total annual flow. From October to March, there were almost no flows in many rivers. The annual flow of the Bawang and Quanyulin rivers located in the northern basin was more than 10 times the flow of the Huhewusu and Longshengzhuang rivers located in the southern basin (Fig. 1).

Our analysis showed that EFR in the basin ranged from 170 × 10^6 m^3/yr to 8 × 10^6 m^3/yr in association with the maximum to severely degraded habitat conditions defined in the Tennant method (Fig. 2). To keep the rivers' habitat health at a "good" level, at least 37 × 10^6 m^3 of water should be left in the rivers, which was equivalent to 26% of the total BWR. This means that, on average, over one quarter of BWR is required to sustain a healthy habitat of the rivers within the basin.

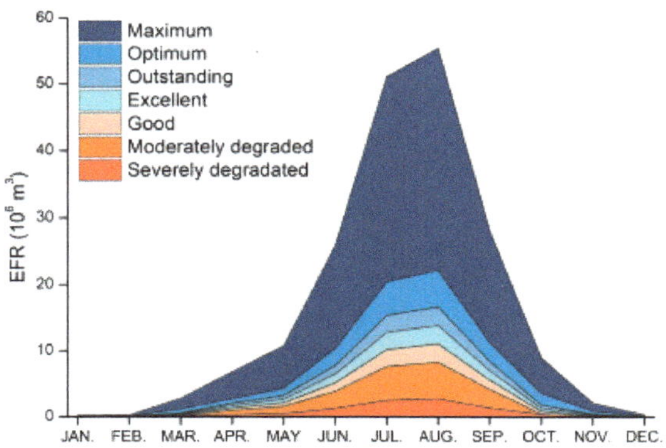

Figure 2: Different levels of total environmental flow requirement (EFR) in the *Huangqihai River Basin*. Habitat quality levels correspond to the flows shown in Table 1.

We also quantified the monthly mean flows and EFR at a "good" habitat health level for the four main rivers in the basin (Fig. 3). The total annual EFR at a "good" level was 10 × 10^6 m^3/yr for the Bawang River, 15 × 10^6 m^3/yr for the Quanyulin River, 1 × 10^6 m^3/yr for the Huhewusu River, and 0.9 × 10^6 m^3/yr for the Longshengzhuang River.

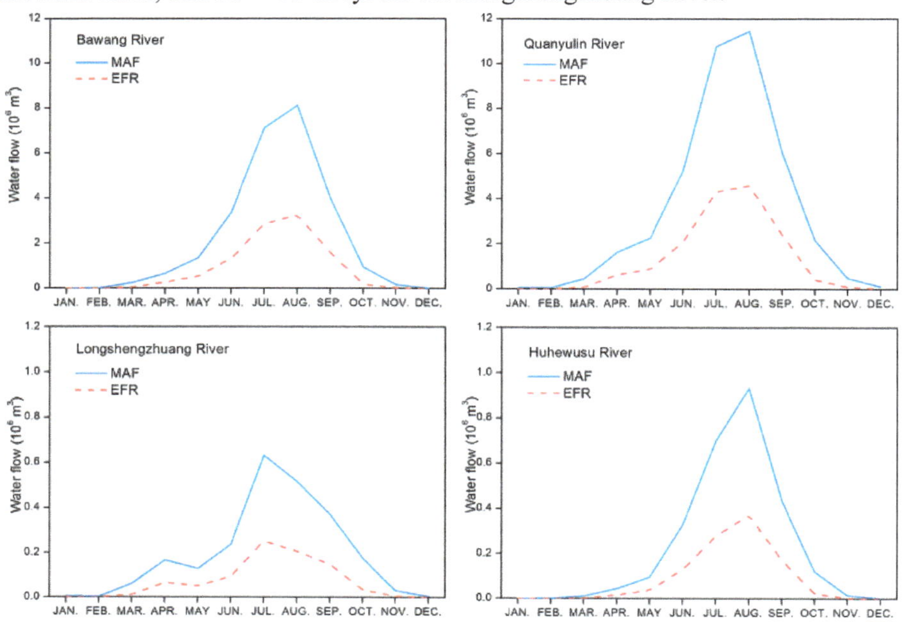

Figure 3: Environmental flow requirement (EFR) and mean annual flow (MAF) for the four main rivers in the *Huangqihai River*

Basin; EFR represents the recommended flow for the "good" level defined in Table 1.

BWF and GWF

The BWF of the Huangqihai River Basin was 135 × 10^6 m^3/yr in 2010, which was equivalent to about 94% of the BWR. Among the socioeconomic sectors, the BWF of the farming and livestock sectors accounted for 85% of the total (116 × 10^6 m^3/yr). The BWF of the industry sector was 9 × 10^6 m^3/yr (7% of the total), and rural and urban domestic sectors account for 11 × 10^6 m^3/yr (8% of the total).

The total GWF in the basin was 2.05 × 10^9 m^3/yr in 2010, which was 14.2 times the total BWR (Fig. 4). The highest GWF was from livestock, 713 × 10^6 m^3/yr, followed by industry and domestic urban consumption, 525 × 10^6 m^3/yr and 523 × 10^6 m^3/yr, respectively. The GWF of the three sectors accounted for 86% of the total GWF.

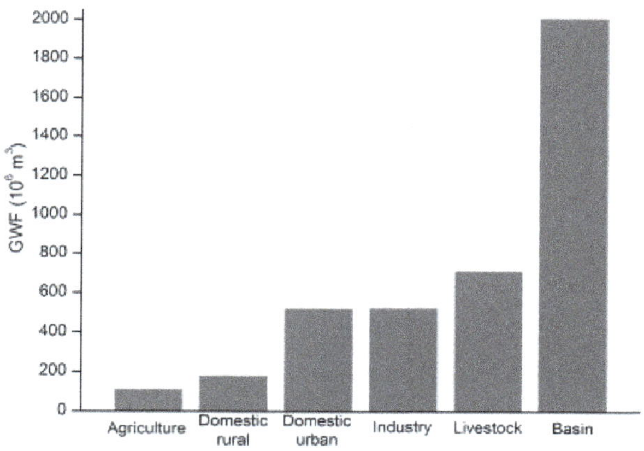

Figure 4: The gray water footprint (GWF) in the *Huangqihai River Basin.*

Water Scarcity Assessment

The $S_{quantity}$ and $S_{quality}$ for the Huangqihai River Basin were estimated at 1.3 and 14.2, respectively. Both were above the threshold of 1.0, suggesting that the basin suffered from both water quantity scarcity and water quality scarcity. The overall water scarcity indicator, *Sqqe*, is thus 1.3(26%)|14.2.

We compared the BWF, BWR, and different levels of blue water availability on a monthly basis (Fig. 5) to identify quantity-induced water scarcity in each month. During a year, the basin suffers from quantity-induced water scarcity problem mainly in three months, i.e., May, June, and July (in all the three months BWF is even higher than the BWR). Although precipitation and runoff are high in the growing season, water resources cannot meet water demand of the economic sectors, in particular the irrigation water demand. In other months, $S_{quantity}$ is mostly below 1.0 when the EFR is met to maintain a "good" level of habitat quality. The basin does not have water quantity induced water scarcity in these months.

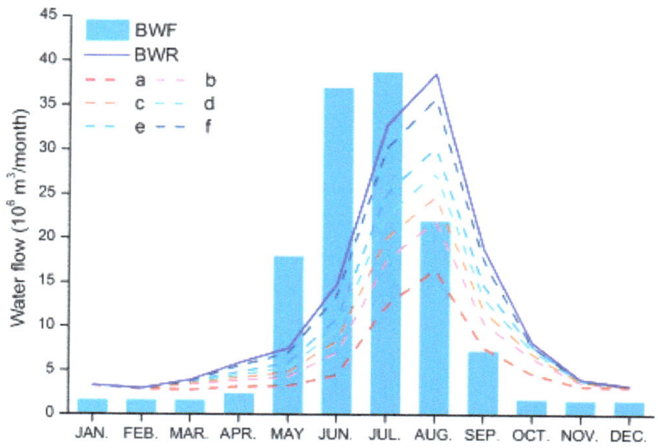

Figure 5: The blue water resource, blue water footprint and blue water available at different habitat level in the Huangqihai River Basin. The bar shows the blue water footprint, the solid line is the blue water resources, the dotted lines are blue water available for human use after accounting for EFR at different habitat level: (a) Optimum, (b) Outstanding, (c) Excellent, (d) Good, (e) Moderately degraded, (f) Severely degraded. (For interpretation of the references to color in this figure legend, the reader is referred to the web version of this article.)

DISCUSSION

Advantages and Limitations of the QQE Water Scarcity Indicator

The QQE water scarcity indicator developed in this study provides an easy to obtain and to understand measurement that contains the information of water quantity and quality status, as well as EFR. Although the case study is conducted in an arid and semi-arid area in China, the procedure can be adapted to any other areas in the world to provide a comprehensive assessment on water scarcity.

For the Huangqihai River Basin, the QQE water scarcity is 1.3(26%)|14.2. By default, the value in the bracket indicates that 26% of the total BWR of the basin is required, i.e., EFR, for maintaining the ecological habitat status at the 'good' level. By specification, one can also use the percentage of EFR to indicate any other levels of ecological habitat status.

Introducing the WF concept in the water scarcity assessment allows for accounting of the actual water consumption of local water resources as well as the ability to control water pollution. The $S_{quantity}$ of 1.3 in the basin indicates that the water consumptive use in the economic sectors amounts to 1.3 times of blue water availability of the basin when the EFR is met for maintaining the good ecological habitat condition. The basin is therefore enduring severer quantity induced water scarcity. The $S_{quality}$ of 14.2 indicates that 14.2 times of BWR would be needed to dilute the polluted water to meet the water quality standard for general use. With such a high value of $S_{quality}$, it can be expected that in many areas in the basin, the water is too polluted to be used for any purpose. A mere reduction of human water uses would not immediately mean a reduction of the water scarcity in the basin due to the remaining water pollution effects.

Despite the advantages specified above, the QQE indicator does have some limitations. It remains a rule of thumb indicator, as the other existing indicators, instead of a precise measurement. It is more suitable to be used for a quick and approximate assessment of water scarcity status on large scales and for obtaining general information on variations across regions. For detailed assessment, more specific local factors, such as technology, rainfall patterns and economic structure, have to be taken into consideration.

Unable to incorporate green water resources is a major limitation of the QQE indicator. This is also the problem in all the existing water scarcity indicators. Incorporating green water resources in water scarcity assessment is a challenge faced with the relevant scientific communities (Hoekstra et al., 2012). Another demerit of the QQE indicator is related to its hybrid nature, which contains three individual, through linked, components. The indicator is not as straight forward as the existing indicators, which use a single value to indicate the status of water scarcity. It requires some professional knowledge to understand the indicator and interpret the information contained. Furthermore, as the value of $S_{quantity}$ is linked with the EFR, which is tied with the level of ecosystem habitat health, a comparison of QQE across regions would not be appropriate when the habitat health is set at different levels.

Environmental FLOW REQUIREMENT

Like all other models, uncertainties are inevitable in the results. In the simulation of the hydrological processes with the APEX model, the calibrated model differed from the observed flows by up to 11%. It would be reasonable to increase the estimate of 26% by 11% (i.e., to 29%) to provide a safety margin that would ensure the needs of the environment are met. This finding is similar to the results of Smakhtin et al. (2004), who calculated the EFR in northern China and obtained values of 25–30% of the total BWR. Both this study and Smakhtin et al. (2004) showed that the EFR is generally no more than 30% of the total BWR in northern China.

The EFR calculated in this study is much lower than the assumption of 80% proposed byHoekstra et al. (2012) and adopted by Zeng et al. (2012). If we had used the percentage of Hoekstra et al. (2012) for EFR, the basin would have faced a much higher level of$S_{quantity}$ with all months having serious water scarcity, especially in May ($S_{quantity}$ = 11.9) and June ($S_{quantity}$ = 12.4). It is likely that the assumption of 80% overestimates EFR, and will lead to an overestimate of the quantity-based water scarcity problem.

Setting different levels of habitat quality will influence the percentage of EFR in total BWR, as well as the value of water scarcity indicator. As shown in our study, when the ecosystem health is prioritized and river habitat is required to be maintained at an "optimum" level, 40% of BWR is required to meet EFR (Fig. 5). Consequently, the $S_{quantity}$will be over 2.0(40%). When the river habitat is at a "severely degraded" level, no more than 7% of BWR will be allocated to river ecosystems. The available water resources in the basin can almost meet human's economic use ($S_{quantity}$ = 1.0(7%)). Suppose the river ecosystem health is completely ignored, and the river runs dry, the $S_{quantity}$ will be 0.94(0%). In this case, there will be no water scarcity for humans, but this will result in losing almost all of natural habitat and biological assemblages. The basic ecosystem functions will be destroyed (King et al., 2000). This will eventually make the region inhabitable for humans.

Reducing the Water Scarcity in the Huangqihai River Basin

With the ongoing urbanization and economic development in the region, the domestic and industrial water supply has been increasing continuously. The increased water demand has been met mainly with overexploitation of groundwater. It has resulted in a serious decline in the groundwater table, and some areas surrounding the Huangqihai Lake have developed cones of depression (Li and Wang, 2008). Such a trend cannot last for long because the overexploitation will eventually lead to exhaustion of available groundwater resources.

Given the fact that the irrigation accounts for 61% of the water use, while the contribution of the farming sector to the regional GDP is only 10%, a reduction of irrigation water use would be one of the options that the region should pursue. Based on this study, suppose a reduction of 20% of irrigation water use, 36 million m^3 of water would be released. The reduction would allow for the meet of the water uses for other economic sectors as well as the EFR. The reduction of GDP would be only about 2%.

Water pollution is another important cause of water scarcity. As the livestock and industrial sectors are major sources of water pollution in the basin, stringent control on the pollution discharge in these two sectors can significantly reduce the pollution load and consequently ease the water quality induced water scarcity. This is particularly so for the livestock sector whose pollution is typically diffused and hence difficult to be dealt with by centralized wastewater treatment plants. Limiting the expansion of livestock production in the basin is therefore important. Currently, the over-grazing and degradation of pastoral land are serious, which impose negative effects on the BWR availability in the basin (Li et al., 2013). Reducing the pressure on pastoral land is therefore also conducive for water conservation and making more BWR available for use.

Shortcomings of this Study

We would like to point out some shortcomings relating to the lack of data for the water scarcity assessment in the Huangqihai River Basin. The Tennant method was used to calculate the EFR mainly because it can be applied with only hydrological data. The Tennant method may have not been able to capture all the hydrological and ecological characteristic of the river system, leading to uncertainties in the estimation of the EFRs for the different levels of habitat qualities in this study.

When calculating BWF, we used the water consumption ratios of Inner Mongolia due to the lack of specific data in the studied area. This may lead to errors. But when local water managers were interviewed, they believed the ratios in the Huangqihai basin were not much different from the average in Inner Mongolia.

We only present results for one specific year mainly because water quality data were only available for one year (2010). This does not allow us to provide an analysis for multiple years and observe the changes. It is clear that the value of the QQE water scarcity indicator changes with the fluctuation of rainfall (and hence BWR each year). An average over a few years or for a year with normal rainfall would be more appropriate to represent a general status of water scarcity.

For this reason, in calculating the QQE water scarcity indicator for the Huangqihai basin, we actually used the average BWR and EFR for the period 1982–2011. This caused a slight time inconstancy. We calculated the QQE with BWR and EFR for the year 2010 and found only minor differences with the QQE values reported in this study (because the BWR in 2010 was close to the average BWR during the period 1982–2011).

Local data on natural background concentration are not available. When calculating GWF, we assumed natural pollutant concentration to be zero. This may not be realistic because nitrogen exists in the natural water bodies. The assumption may overestimate the assimilation capacity for chemical substance, and underestimate GWF, but such an underestimation is believed to be marginal and will not largely influence the assessment results (Hoekstra et al., 2011). For example, Franke et al. (2013) suggested using background concentration of Ammonium N–NH_4 of 0.015 mg/L following Chapman (1996). By using this background concentration in our study, the re-calculated GWF is only 1.5% higher than the current estimate. Hence, assuming a value of zero for natural background concentration will only lead to very small errors.

CONCLUSION

In this study, we proposed an approach to assessing water scarcity by considering both water quantity and water quality, while also explicitly accounting for the EFR. We used this method in the Huangqihai River Basin in China. We demonstrated that the QQE indicator can provide a more comprehensive picture of water scarcity status for a basin under the given levels of water uses in economic sectors, EFR and water pollution. For the Huangqihai River Basin, the results show that the basin is currently suffering from both quantity- and quality-induced water scarcity problems. The EFR for maintaining the good ecological status cannot be met in some months during a year. The very high value of quality related water scarcity index suggests that water pollution plays a big role in causing the water scarcity in the river basin. This implies that only focusing on reducing water use may not help significantly alleviate the water scarcity situation. Instead, a special attention has to be paid to reducing water pollution and encouraging water conservation. The QQE approach provides a general method to assess water scarcity, and it could be applied in other regions where both the hydrological and ecological data are scarce.

ACKNOWLEDGEMENTS

The present work was partially developed within the framework of the Panta Rhei Research Initiative of the International Association of Hydrological Sciences (IAHS) by the working group “Water Scarcity Assessment: Methodology and Application”. This study was supported by the Projects of International Cooperation and Exchanges of the National Natural Science Foundation of China (41161140353), the Natural Science Foundation of Beijing, China (8151002), the National Natural Science Foundation of China(91325302, 91425303), the International S&T Cooperation Program from the Ministry of Science and Technology of China (2012DFA91350), the National Program for Support of Top-notch Young Professionals, the Fundamental Research Funds for the Central Universities (TD-JC-2013-2). We purchased the historical hydrological and climate data from the local Hydrological Bureau of Wulanchabu City in Inner Mongolia. This data cannot be shared without permit from the Bureau. Other data can be obtained upon request from the authors.

REFERENCES

1. Alcamo, J., Henrichs, T., Rösch, T.,2000. World water in 2025: global modeling and scenario analysis for the world commission on water for the 21st century. In: Report A0002. Center for Environmental Systems Research, University of Kassel, Kassel, Germany.
2. Anisfeld, S.C., 2010. Water Resources. Island Press, Connecticut Ave., NW, Washington.
3. Arthington, A.H., 2012. Environmental Flow: Saving Rivers in the Third Millennium. University of California Press, Berkeley and Los Angeles, California, USA.
4. Brisbane Declaration, 2007. The Brisbane Declaration: Environmental Flows are Essential for Freshwater Ecosystem Health and Human Well-Being, September 3–6, Brisbane, Australia, Available at: http://www.eflownet.org/viewinfo. cfm?linkcategoryid=4&linkid=13&siteid=1&FuseAction=display
5. Cai, X., Rosegrant, M.W., Ringler, C., 2003. Physical and economic efficiency of water use in the river basin: implications for efficient water management. Water Resour. Res. 1, 1–12.
6. Chen, M., Liu, J., 2014. Landscape dynamics in the Huangqihai Lake Basin from 1976 to 2010. J. Northwest For. Univ. 29, 165–170 (in Chinese).
7. Chapman, D. (Ed.), 1996. Water Quality Assessments: A Guide to the Use of Biota, Sediments and Water in Environmental Monitoring. , second ed. UNESCO/WHO/UNEP.
8. Falkenmark, M., Lundqvist, J., Widstrand, C., 1989. Macro-scale water scarcity requires micro-scale approaches. Nat. Resour. Forum 13, 258–267.
9. Franke, N.A., Boyacioglu, H., Hoekstra, A.Y., 2013. Grey water footprint accounting: Tier 1 supporting guidelines. Value of Water Research Report Series No. 65. UNESCO-IHE, Delft, the Netherlands.
10. Gleick, P.H., 1993. Water and conflict: fresh water resources and international security. Int. Secur. 18, 79–112.
11. Hoekstra, A.Y., Chapagain, A.K., Aldaya, M.M., Mekonnen, M.M., 2011. The Water Footprint Assessment Manual: Setting the Global Standard. Earthscan Press, London, UK.
12. Hoekstra, A.Y., Mekonnen, M.M., Chapagain, A.K., Mathews, R.E., Richter, B.D., 2012. Global monthly water scarcity: blue water footprints versus blue water availability. PLOS ONE 7, e32688.
13. Jacobsen, D., Milner, A., Brown, L., Dangles, O., 2012. Biodiversity under threat in glacier-fed river systems. Nat. Clim. Change 2, 361–364.
14. Jiang, D., Wang, H., Li, L., 2006. Progress in ecological and environmental Mwater requirements research and applications in China. Water Int. 31 (2), 145–156.
15. Jowett, I.G., 1997. Instream flow methods: a comparison

of approaches. Regul. Rivers: Res. Manag. 13, 115–127.

16. King, J.M., Tharme, R.E., De Villiers, M., 2000. Environmental flow assessments for rivers:manualfor the building blockmethodology.In: DeVilliers,M. (Ed.),Water Research Commission, Pretoria, South Africa.
17. Kirby, J.M., Connor, J., Ahmad, M.D., Gao, L., Mainuddin, M., 2014. Climate change and environmental water reallocation in the Murray–Darling Basin: impacts on flows, diversions and economic returns to irrigation. J. Hydrol., http://dx.doi.org/ 10.1016/j. jhydrol.2014.01.024 (in press).
18. Kumara, B.K.H., Srikantaswamy, S., 2011. Environmental flow requirements in Tungabhadra River, Karnataka, India. Nat. Resour. Res. 20 (3), 193–205.
19. Li, J., Xiao, N., Quan, Z., Luo, Z., 2013. Protection and Recovery Plan of Huangqihai Wetland. China Environmental Science Research Institute, Wulanchabu City of Inner Mongolia Autonomous Region People's Government of ChaYouQianQi, Inner Mongolia, China (in Chinese).
20. Li, Q., Wang, R., 2008. Wulanchabu city water resources development and utilization problems and countermeasures. Inner Mongolia Water Conserv. 5, 66–67 (in Chinese).
21. Liu, Q., Liu, J., Wang, X., 2014. Applicability of the APEX model for Huangqihai River Basin. J. Arid Land Resour. Environ. 28, 185–190 (in Chinese).
22. Ma,X.,Xin,W.,Peng,Y.,Wen,L.,2002.The environmental problem of ChaYouQianQi, and its comprehensive control countermeasures. Inner Mongolia For. Invest. Des. 25 (Supplement), 6–8 (in Chinese).
23. Men, B., Yu, T., Kong, F., Yin, H., 2014. Study on the minimum and appropriate instream ecological flow in Yitong River based on Tennant method. Nat. Environ. Pollut. Technol. 13 (3), 541–546.
24. MEPC, 2002. Environmental Quality Standards for Surface Water. [GB3838-2002], Ministry of Environment Protection, People's Republic of China. Standards Press of China, Beijing, China (in Chinese).
25. Pastor, A.V., Ludwig, F., Biemans, H., Hoff, H., Kabat, P., 2014. Accounting for environmental flow requirements in global water assessments. Hydrol. Earth Syst. Sci. 18 (12), 5041–5059.
26. Seckler, D., Amarasinghe, U., Molden, D., 1998. World Water Demand and Supply, 1990 to 2025: Scenarios and Issues. Research Report 19. International Water Management Institute, Colombo, Sri Lanka.
27. Smakhtin, V., Revenga, C., Döll, P., 2004. A pilot global assessment of environmental water requirements and scarcity. Water Int. 29, 307–317.
28. Smakhtin, V.U., Shilpakar, R.L., Hughes, D.A., 2006. Hydrology-based assessment of environmental flows: an example from Nepal. Hydrol. Sci. J. 51 (2), 207–222.
29. Sullivan, C.A., Meigh, J.R., Giacomello, A.M., 2003. The water poverty index: development and application at the community scale. Nat. Resour. Forum 27, 189–199.
30. Tennant, D.L., 1976. Instream flow regimens for fish, wildlife, recreation and related environmental resources. Fisheries 1, 6–10.
31. Tharme, R.E., 2003. A global perspective on environmental flow assessment: emerging trends in the development and application of environmental flow methodologies for rivers. River Res. Appl. 19, 397–441.
32. Ubertini, L., Manciola, P., Casadei, S., 1996. Evaluation ofthe minimum instream flow of the Tiber river basin. Environ. Monit. Assess. 41 (2), 125–136.
33. van Vliet, M.T.H., Franssen, W.H.P., Yearsley, J.R., Ludwig, F., Haddeland, I., Lettenmaier, D.P., Kabat, P., 2013. Global river discharge and water temperature under climate change. Glob. Environ. Change 23, 450–464.
34. Vörösmarty, C.J., McIntyre, P.B., Gessner, M.O., Dudgeon, D., Prusevich, A., Green, P., Glidden, S., Bunn, S.E., Sullivan, C.A., Liermann, C.R., Davies, P.M., 2010. Global threats to human water security and river biodiversity. Nature 467, 555–561.
35. Wang, Q., Liuang, G., He, B., Zhou, H., 2013. Analysis and measures of ecological water demand in lower reaches of Biliuhe reservior. South-to-North Water Transf.
36. Water Sci. Technol. 11, 1–5 (in Chinese).
37. Wang, X., Yen, H., Liu, Q., Liu, J., 2014. An auto-calibration tool for the agricultural policy environmental extender model. Trans. ASABE 4, 1087–1098.
38. Williams, J.R., 1995. The EPIC model. In: Singh, V.P.(Ed.), Computer Models ofWatershed Hydrology. Water Resources Publications, Highlands Ranch, Colorado, pp. 909–1000.
39. Williams, J.R., Izaurralde, R.C., 2006. The APEX Model. CRC Press, Taylor and Francis Group, Boca Raton, FL.
40. Xia, J., Feng, H.-L., Zhan, C.-S., Niu, G.-W., 2006. Determination of a reasonable percentage for ecological water-use in the Haihe River Basin, China. Pedosphere 16 (1), 33–42.
41. Yu, L., Qin, Y., Liu, J., 2013. Analysis water resources features of Huangqihai Basin. Inner Mongolia Water Resour. 2, 36–38 (in Chinese).
42. Zeng, Z., Liu, J., 2013. Historical trend of grey water footprint of Beijing, China. J. Nat. Res. 28, 1169–1178 (in Chinese).
43. Zeng, Z., Liu, J., Koeneman, P.H., Zarate, E., Hoekstra, P.D.I.A., 2012. Assessing water footprint at river basin level: a case study for the Heihe River Basin in Northwest China. Hydrol. Earth Syst. Sci. Discuss. 9, 5779–5808.
44. Zeng, Z., Liu, J., Savenije, H.H.G., 2013. A simple approach to assess water scarcity integrating water quantity and quality. Ecol. Indic. 34, 441–449.

CHAPTER 20 OPEN SOURCE SOFTWARE FOR VISUALIZATION AND QUALITY CONTROL OF CONTINUOUS HYDROLOGIC AND WATER QUALITY SENSOR DATA

Jeffery S. Horsburgh, Stephanie L. Reeder, Amber Spackman Jones, Jacob Meline

Keywords: Data management, Open source software, Sensor data, Python, Hydrologic observations, Quality control

INTRODUCTION

Environmental monitoring with *in situ* environmental sensors presents many challenges for data management, particularly for large-scale networks consisting of multiple sites, sensors, and personnel. Over the past decade, there has been a drastic increase in the use of automated data collection in scientific research. The high frequency, extended duration, and spatial distribution of data collection efforts require cyberinfrastructure to support and facilitate research using sensor data streams. Researchers and practitioners need tools for data import and storage as well as data access and management. In addition to addressing the challenges presented by managing the sheer quantity of data, monitoring network managers need practices to ensure high data quality, including standard procedures and software tools for data post processing and quality control.

In this paper we describe a workflow for scripted quality control editing of continuous, *in situ* time series datasets and the architecture and functionality of an open source software tool called ODM Tools Python that implements this workflow. ODM Tools Python enables users to query and export, visualize, and edit time series observations stored in an Observations Data Model (ODM) database (Horsburgh et al., 2008). ODM was developed as a standard data model in which to organize, store, and describe point observations (e.g., observations made at fixed point monitoring sites such as streamflow, water quality, and weather monitoring stations) with sufficient metadata for observations to be unambiguously interpreted by multiple users. ODM is implemented in relational database software to permit flexibility in querying and data retrieval. ODM Tools Python is a modernized and advanced version of the original ODM Tools software, which, along with ODM, was developed as part of the HydroServer software stack (Horsburgh et al., 2010) within the Consortium of Universities for the Advancement of Hydrologic Science, Inc. (CUAHSI) Hydrologic Information System (HIS) (Horsburgh et al., 2009 and Tarboton et al., 2009).

Previous versions of ODM Tools (Horsburgh et al., 2011) were developed in Microsoft Visual Studio .NET and only supported ODM databases implemented in Microsoft SQL Server, limiting deployment to Microsoft Windows-based computers. Functionality consisted of exporting data series and associated metadata, plotting and summary of single data series, generation of derivative data series, and editing data series using a set of simple tools implemented within the ODM Tools graphical user interface (GUI). However, there were no capabilities for recording or storing the sequence of edits made to a data series in the quality control process. Furthermore, there were not adequate capabilities to support the full quality control workflow, especially for long time series that were difficult to process in a single editing session or time series continually being updated as new data were collected. These were major limitations that affected the usefulness of ODM Tools where recording the provenance of data edits was desired or required. Irrespective of the tool being used to perform quality control, it remains common for some scientists to perform these steps at the same time as data visualization

Citation: Jeffery S. Horsburgh, Stephanie L. Reeder, Amber Spackman Jones, Jacob Meline, Open source software for visualization and quality control of continuous hydrologic and water quality sensor data, Environmental Modelling & Software, Volume 70, August 2015, Pages 32-44, ISSN 1364-8152, http://dx.doi.org/10.1016/j.envsoft.2015.04.002.

and analysis without preserving the provenance of the quality control process in a way that it is accessible to others or for future reference. We sought to address this deficiency with the software described in this paper.

The new, Python-based version of ODM Tools described here adds a modernized GUI with dockable components, multiple platform support (Windows, Linux, and Mac), support for multiple relational database management systems (RDBMS) (Microsoft SQL Server, PostgreSQL, and MySQL), enhanced plotting and visualization, and automated scripting of quality control edits performed on data series through an integrated Python script editor and console. Additional improvements include enhanced queries for data selection and export, interactive data selection directly on a time series plot, additional options for how edits to data series are versioned and saved, and the ability to plot multiple data series simultaneously with various plot types. The contribution of this paper is in the description the quality control editing workflow for continuous sensor data streams and in the architecture of the ODM Tools Python software that implements the workflow and enables the features listed above. We anticipate that the functionality that we have developed in ODM Tools Python will be widely applicable to managers of continuous sensor data and that the architecture, software development approach, and deployment approach that we have used for the ODM Tools Python software will be informative for researchers developing similar tools.

In Section 2 we provide further background on quality control for hydrologic and water quality sensor data streams and associated challenges. Section 3 describes our workflow for supporting the quality control editing process. Section 4 describes our approach to address these challenges and the architecture and software implementation for ODM Tools Python. Section 5 describes the deployment approach we selected for the ODM Tools Python software, and Section 6 provides an example of using ODM Tools Python for quality control editing and post processing of a continuous water quality time series. Finally, we conclude with a summary of the research results.

BACKGROUND

Quality Control of Sensor Data

Errors in continuous environmental datasets primarily occur from fouling of sensors and from sensor calibration shift, although anomalies and erroneous data values can occur for many other reasons, including failures of sensors, recorders, transmission systems, or unforeseen environmental conditions that adversely affect sensor readings (Wagner et al., 2006). Producing high quality, continuous data streams from raw sensor output requires application of data corrections to mitigate for instrument calibration drift, fouling, and other errors (Mourad and Bertrand-Krajewski, 2002 and Campbell et al., 2013).Fig. 1 shows common types of errors in raw sensor data streams and illustrates the associated types of corrections needed in the quality control process. These are also listed in Table 1, which provides a more extensive list of common error types along with examples and description of the most common procedures for correction. The errors listed in Table 1 are illustrative of those commonly encountered within raw environmental sensor datasets, and the software described in this paper was developed to advance data from raw to quality controlled versions using the corrections described in Table 1.

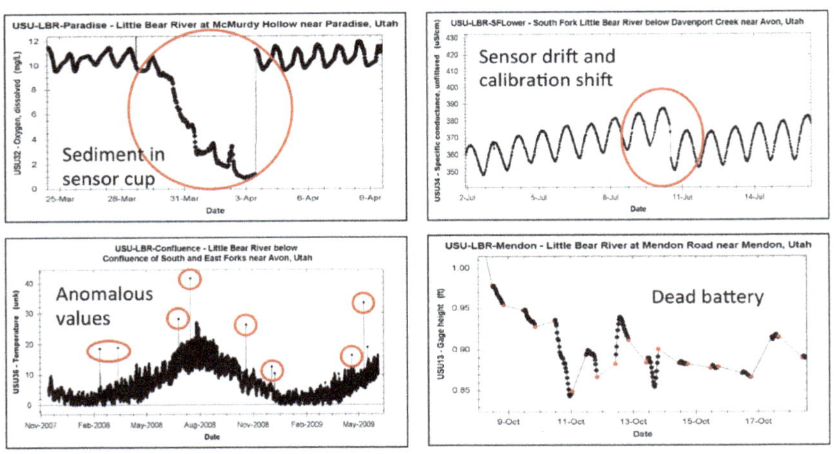

Figure 1: Example anomalous sensor data.

Table 1: Common types of error in hydrologic and water quality sensor data streams.

Type of error	Description and examples	Typical correction method
Sensor drift	Error in sensor readings due to electronic drift in the sensor reading away from the instrument's calibration during the period between calibrations.	Variable data correction
Sensor fouling	Error in sensor readings due to biological growth or other residue buildup on the sensor between maintenance visits.	Variable data correction

Out of range values	Data values that are beyond the range of plausible values for the particular phenomenon being measured.	Deletion or interpolation and flagging
Data value persistence	Constant data values that are recorded when a sensor becomes stuck in a single position (e.g., in the case of a wind direction sensor) or when a sensor fails and the datalogger repeatedly records the last measured value.	Deletion
Failed sensor or logger	Erroneous data values recorded by a sensor or datalogger that has failed (e.g., a dissolved oxygen sensor fails and the datalogger records dissolved oxygen concentrations that drop to 0 mg/L).	Deletion
Power failure	A battery or power supply fails to supply required levels of power to a sensor or datalogger resulting in suspect measurements.	Deletion or flagging
Adverse site conditions	Conditions immediately surrounding the sensor are not representative of the site (e.g., sediment buildup in a sensor cup, ice buildup around a sensor).	Deletion
Incorrect offset or calibration	Data values that are in error by a constant value due to a datalogger program applying an incorrect offset or an error during sensor calibration.	Add or subtract a constant
Truncation	Data values recorded at the reporting limit for a sensor because its maximum or minimum recording level has been exceeded.	Deletion or flagging
Skipped or no-data values	Gaps in data caused by datalogger errors or skipped scans.	Insertion and/or interpolation and flagging

Correction procedures listed in Table 1 include simple methods such as: deletion of erroneous values, insertion of individual values to fill brief no-data gaps, and interpolation of erroneous values using prior and subsequent values that are known to be good. More sophisticated corrections include variable data corrections applied to correct for sensor drift and fouling. It is beyond the scope of this paper to describe the circumstances under which each of these corrections should be applied. Rather, we describe a workflow and software that enable application of any of these corrections according to the judgment of the data analyst in a way that the provenance of the changes can be saved. Wagner et al. (2006) provide excellent guidance and discussion on which correction procedures to use for various data collection situations encountered in hydrologic and water quality monitoring. Similarly, Fiebrich et al. (2010) review checks and correction procedures for common meteorological variables.

Another important distinction is that of "automated" quality control versus "interactive" quality control editing performed by an analyst. Most automated quality control procedures are focused on automatically flagging raw data values that do not meet one of several plausibility tests (e.g., Sheldon, 2008, Lerner et al., 2011 and Taylor and Loescher, 2013). These include programmatic checks aimed at identifying instances of the types of errors described in Table 1 in raw data streams. Algorithms for identifying these errors range from simple range checks to more complicated techniques using statistical models or machine learning methods (e.g., Moatar et al., 2001, Hill et al., 2009,Fiebrich et al., 2010, White et al., 2010 and Dereszynski and Dietterich, 2012). While these techniques can be quite good at identifying and flagging potentially erroneous values, they do not excel at choosing and applying appropriate data corrections to resolve the error for which the data were flagged, steps that generally require the attention and expertise of field or data technicians (Fiebrich et al., 2010 and White et al., 2010). This paper, the workflow we present, and the ODM Tools Python software focus on this secondary level of quality control (or post processing) and provide a set of tools for applying the common data corrections listed in Table 1.

The Role of Visualization in Data Quality Control

Quality control of sensor data is often an iterative process that requires visualization of raw data to identify anomalies, to inspect modified data after application of quality control procedures to ensure corrections were correctly applied, and to examine additional datasets that provide context for an editing session. For example, in deciding whether anomalous values in a sensor dataset need to be edited, it is common to inspect data for the same or other variables at the same monitoring site (i.e., internal consistency) or to examine data for the same variable collected at a nearby monitoring site (i.e., external consistency) to determine whether corrections should be made (Moatar et al., 2001 and Campbell et al., 2013). The iterative process of quality control editing and visualization can be tedious and challenging, especially when the tools used for each are not the same or where action is required to update a visualization each time an edit is made.

Challenges in Integrating Tools for Visualization and Quality Control of Sensor Data

While most scientists use the general techniques described above to visualize and perform quality control of sensor data, there is large variety in the ways in which this is currently done. This heterogeneity arises because: 1) scientists work across computing platforms (e.g., Windows, Linux, Macintosh); 2) scientists have a spectrum of technical expertise with respect to automation of tasks like data quality control via computer programming; and 3) scientists use the software tools that they are most comfortable with and that meet their combination of technical expertise and computing platform. For example, some scientists open data and perform adjustments in Microsoft Excel using built in visualization capabilities to plot and verify the changes they have made. Some script their data edits using R, Python, or Matlab and use the native or imported plotting capabilities of these languages for visualization. Examples in this category include the sensorQC R package developed by the United States Geological Survey (https://github.com/USGS-R/sensorQC) and the Georgia Coastal Ecosystems (GCE) Data Toolbox for Matlab (https://gce-svn.marsci.uga.edu/trac/GCE_Toolbox/). Some scientists use available commercial software systems such as the Aquarius software package from Aquatic Informatics (http://aquaticinformatics.com/products/aquarius-time-series/) or the Kisters WISKI software (http://www.kisters.net/wiski.html), which have good functionality but can cost enough that they are out of reach for many small research groups. Some commercial products are also tied to specific instrument/data collection equipment manufacturers, limiting their general applicability. There are currently few software products designed specifically for quality control editing of continuous environmental time series data with integrated visualization capabilities. Meeting this need is challenging given the diversity of computing platforms and technical expertise among potential users.

Maintaining the provenance of data edits is another major technical challenge for data managers. It is common for data managers to maintain "raw" and "quality controlled" versions of their data, but it is less common for data managers to maintain a complete provenance record of how they moved from "raw" to "quality controlled" versions (Lerner et al., 2011, Campbell et al., 2013 and Mason et al., 2014). Data edits made manually can produce quality controlled versions of data series, but since the edits are manual and not recorded, they can be impossible to reproduce. In contrast, scripting of quality control edits provides a written record of the various steps involved in the quality control process, effectively encoding data corrections in a way that they can be easily revised and re-executed at any time. Indeed, scripting of data management and analysis tasks has been recommended as a best practice (e.g., Borer et al., 2009), and some groups have begun implementing scripting to post process environmental data (Mason et al., 2014). Scripting is, however, difficult for many scientists that have no computer science background or expertise and subsequently lack the requisite computer programming skills.

Python as a Platform for Data Quality Control Editing

Although we have pointed out that many engineers and scientists do not yet use scripting in their day-to-day work, the number of those that do is growing. For example, the use of Python for scientific computing on diverse computing platforms has increased significantly in recent years (Millman and Aivazis, 2011). Python provides a high-level, object-oriented, interpreted programming language that has been used extensively for exploratory, interactive, and computational scientific research. Major advantages of Python include its platform independence, the rich collection of functionality that can be added via extension packages, and the relative ease with which Python can be integrated with software written in other languages (Perez et al., 2011). For example, the Python data analysis library pandas (http://pandas.pydata.org) includes several easy to use and highly optimized data structures and analysis tools suitable for the types of analyses and functionality needed for performing quality control on continuous environmental data streams. Another advantage of Python is the ability to both work in scripting/command line mode and to develop complex software programs with GUIs. These characteristics made Python a good candidate for the development work described in this paper.

One difficulty in using Python for scientific computing and in software development is getting a complete set of Python tools needed for a particular project installed on a computer. This challenge remains, although new Python distributions such as Enthought Canopy (https://www.enthought.com/products/canopy/) and the Anaconda Python distribution (https://store.continuum.io/cshop/anaconda/) provide an entire Python ecosystem with many of the most popular Python packages for science, math, engineering, and data analysis within a Python environment that can be easily installed. In fact, many scientists maintain multiple Python environments on their computers for different purposes. Using Python for the software development effort described in this paper presented two deployment-related challenges that had to be overcome: 1) deploying the software we developed and all required Python components to computers running one of multiple operating system platforms; and 2) deploying the software in a way that it did not overwrite or conflict with a user's existing Python environments.

DATA QUALITY CONTROL EDITING WORKFLOW

Quality control of continuous sensor datasets is often an ongoing process (Mason et al., 2014). Some data collection efforts may continue for months or years, and it is not always feasible to wait until the end of data collection activities to perform quality control to produce finalized versions of collected data for use in scientific analyses. Furthermore, quality control activities typically lag behind the collection of data as scientists need time to review data within the context of environmental conditions (Campbell et al., 2013). Because of these disparities in timeframe between raw and quality controlled data, a quality control editing workflow is needed that supports creation of quality controlled versions of the data while data collection activities continue. Implementing data edits within a script enables regeneration of a processed, quality controlled version of the data from the raw data on demand, and the script effectively becomes both the definition of the quality controlled version of the data and the record of the provenance of data edits. Subsequent edits to the data can be added to the script as needed (e.g., as new data are collected and added to the raw time series).

Fig. 2 shows the data quality control workflow that we developed and implemented for generating and maintaining scripted quality control edits. To begin the workflow, a user first selects a raw time series for editing and plots the data to visually inspect it and determine whether edits are needed. The user then begins an editing session, which creates a memory copy of the dataset on which any edits are performed. If the user is editing a time series for the first time, a new script is created to record the sequence of data edits performed within the editing session. The user then performs edits, each of which is recorded as one or more lines of code in the script. The user can add descriptive comments to the script at any time. At the end of the editing session, the user creates code to delete any unreviewed data from the time series (e.g., in the case where the user does not review all the way to the end of the raw data record during the editing session) and saves the script to disk. The user can then create code to save the edited data series back to the database as a new, quality controlled version of the data. The workflow implements best practices in that all edits are made on a copy of the raw data (Campbell et al., 2013), and the raw data are maintained in the database in their original condition.

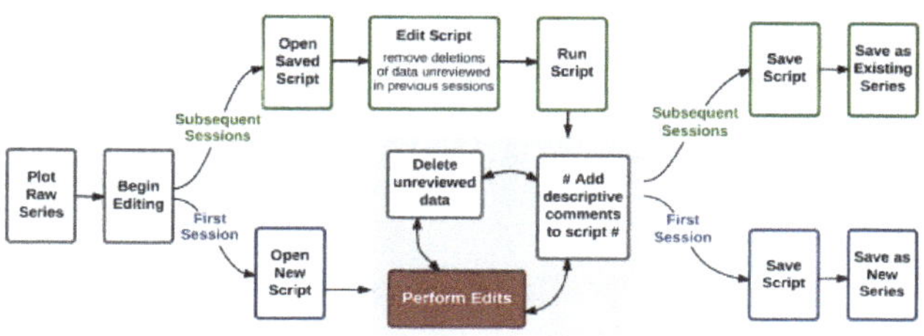

Figure 2: Workflow for scripting quality control editing of continuous time series data.

When new raw data have been added to a previously edited time series or when a user wants to return and complete quality control editing for a time series they have reviewed only partially, a subsequent editing session begins in much the same way. The user again selects the raw time series for editing and begins an editing session. However, instead of creating a new script, the user opens the script that they saved during their previous editing session. The user then removes from the end of the script any lines that deleted unreviewed data and executes the script on the raw data. The result is a memory copy of the time series that contains quality controlled data up to the point that the user left off during their previous editing session, with any new or unreviewed raw data appended to end. The user can then continue the same process described above – perform additional edits as needed, add descriptive comments, and then delete any unreviewed data. These steps are appended to the original script. At the end of the current editing session, the user saves the modified script to disk and has a choice of whether to save the resulting edited time series back to the database as a completely new version or to save it back to the database by overwriting the previous quality controlled time series.

This workflow is generic, and does not prescribe a specific coding language or implementation. However, the process can be drastically simplified by selecting a simple language for capturing the scripts, implementing convenient data editing functions that can be used within the scripts, and developing user interface tools for generating the code. In the following sections we describe a software implementation of the above workflow using Python scripting within a cross-platform, open-source, desktop software program.

SOFTWARE IMPLEMENTATION

One of the key goals of this research was to investigate how data visualization and quality control editing could be coupled within a single software environment while preserving the provenance of the data editing process. This goal and the challenges described in the sections above inspired the following design requirements for the software we developed:

1. Multi-platform support for Windows, Linux, and Macintosh
2. Multi-database support for Microsoft SQL Server, MySQL, and PostgreSQL
3. Preserving the provenance of all data edits through scripting
4. Coupled graphical and scripting-based editing interface for users
5. Linked visualization and editing capabilities for immediate user feedback
6. Simple software deployment on multiple platforms

For developing the GUI for ODM Tools Python, we chose to adopt an approach similar to that of Williams (2009), who developed a GUI-based software program called Rattle for performing data mining using the statistical computing environment R. Within Rattle, R code is created and executed based on a user's actions within the GUI – e.g., clicking on a button within the GUI generates the R code for the function tied to that button. There are several advantages to this approach when considering the list of requirements above. First, it provides convenient access to the most relevant code functions via buttons in the GUI, but translates each user action into executable code that can be captured in a script. The resulting script can serve as a record of all of the user actions that were performed. Second, it provides a way to support novice users (who will focus on functionality within the GUI) and advanced users (who may go directly to scripting and skip much of the GUI interaction). Finally, implementing scripting within the GUI enables immediate visualization of results via graph and table-based views of the data. We successfully used a similar approach to integrate scripting of data visualization and analysis using R within the HydroDesktop client application for the CUAHSI HIS (Horsburgh and Reeder, 2014).

Architecture

The ODM Tools Python software was developed in Python as a desktop program with a GUI. It is a client application for a local or remote ODM database implemented within an RDBMS. Time series data are retrieved from an ODM database into local memory for data visualization and editing within ODM Tools and then the results can be saved back to the ODM database. The architecture of the ODM Tools Python software consists of four layers (Fig. 3). In the following sections we describe each of these layers and highlight their main functionality.

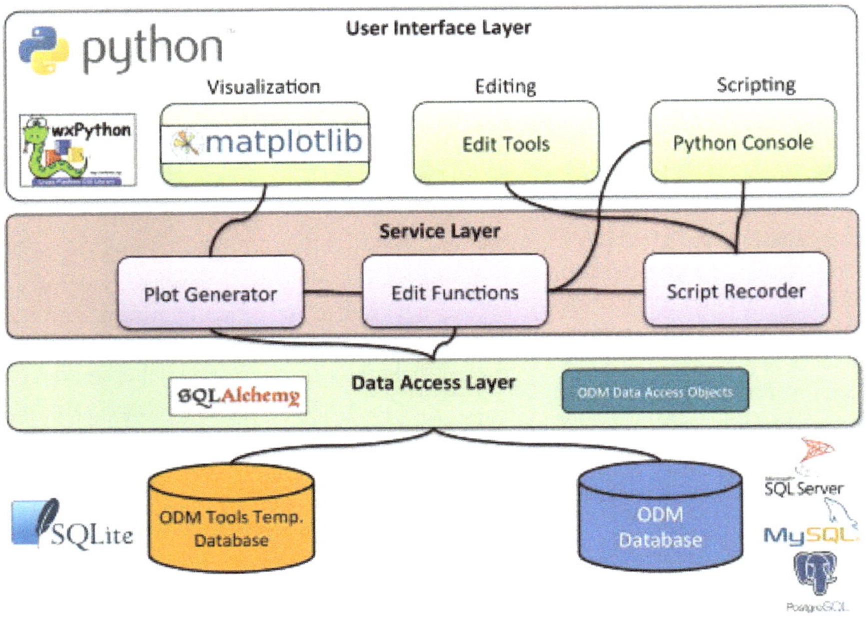

Figure 3: ODM Tools Python software architecture.

Data Storage Layer

In the CUAHSI HIS, an ODM database implemented within an RDBMS is the underlying data storage mechanism that supports publication of observational data via standardized web services that query data from the ODM database in response to user requests and then return data in a standard XML schema called Water Markup Language (WaterML) (Zaslavsky et al., 2007 and Taylor, 2012). Regardless of whether users intend to publish their data using the CUAHSI HIS, ODM provides a convenient and robust structure within which to store, manage, and manipulate sensor data. User access control for ODM databases is handled by the RDBMS within which they are deployed. Using ODM Tools Python requires users to have an appropriate user account to access an ODM database. ODM's existing capability for storing environmental time series data and the query and data manipulation capabilities of the RDBMS made them a suitable platform on which to build the software we envisioned. ODM is described by Horsburgh et al. (2008), and additional documentation including instructions on implementing ODM can be found on the CUAHSI HIS website (http://his.cuahsi.org/odmdatabases.html). At the time of this writing, ODM Tools Python is fully functional with ODM 1.1.1 databases, and we are working on compatibility with a soon-to-be released revision of ODM called ODM2. Sample ODM databases for testing the software are available on the ODM Tools Python GitHub website.

Data Access Layer

ODM Tools Python uses a SQLAlchemy-based data access layer (http://www.sqlalchemy.org/). This layer serves to abstract data from the ODM database and provides a set of programmable objects that facilitate data management within the ODM Tools software application – rather than repeatedly programming Structured Query Language (SQL) queries directly against the ODM database. This is a significant improvement over earlier versions of ODM Tools, which used SQL queries throughout the code that were difficult to maintain. As a new feature, the data access layer also provides support for connections to ODM databases within multiple RDBMS (currently Microsoft SQL Server, MySQL, and PostgreSQL), whereas older versions worked only with Microsoft SQL Server. Multiple database support is facilitated by SQLAlchemy, which includes SQL dialect implementations for the various RDBMS. An additional benefit of the data access layer is that it insulates the code of the service layer and the user interface layer from the data storage layer. Changes to the data storage layer, for example updating ODM 1.1.1 to ODM2, only affect the data access layer, minimizing the amount of code that has to be refactored as the storage layer evolves.

Additionally, ODM Tools Python uses SQLite as an in-memory database to temporarily store and manipulate time series data objects during data editing sessions. SQLite is a software library that implements a file-based, serverless, transactional SQL database engine (http://www.sqlite.org). The performance of the in-memory SQLite database is dependent upon the amount of available memory on the user's computer. In our testing, loading a time series with approximately 250,000 data values representing 15-min sampling frequency for more than 7 years required approximately 300 MB of available memory.

Service Layer

The service layer consists of a set of Python-based services containing the core functionality of the software application. A service is an object that creates and receives data objects defined in the data access layer. The GUI interacts with the data access layer through the service layer. For example, a Plot Generator service provides functions for creating the visualizations within the GUI, the Edit Functions service provides functions for performing data quality control editing on time series data, and the Script Recorder service manages the automated recording of Python function calls during data editing sessions. The Edit Functions service was designed as a Python module that is installed automatically on a user's computer with the ODM Tools Python software deployment. It serves as the underlying code for the data editing tools on the ODM Tools Python editing toolbar, and, because it is independent of the ODM Tools Python GUI, it can be added to and used within any Python script. This means that a data editing Python script created within the ODM Tools software and then saved to a file can be executed independent of the ODM Tools software as long as the Edit Functions Python module is present. Table 2 lists the data editing functions available in the ODM Tools Python Edit Functions Python module and provides examples of how they are used.

Table 2: ODM Tools Python data editing functions in the data editing service.

Function Name	Description
add_points()	Given a list of data values, inserts values from the list into the data series Example: points = **[[**(u'8.5', **None,** datetime.datetime(2015, 1, 6, 0, 0),'-7', datetime.datetime(2015, 1, 6, 7, 0)**, None, None,** u'nc', **None, None,** 1, 35, 21, 1, 0)**]][**0**]** edit_service.add_points(points)
change_value()	Modifies selected data values by adding, subtracting, multiplying by, or setting equal to a constant value Example: edit_service.change_value(10, '+')
create_method()	Creates a new method in the underlying ODM database Example: edit_service.create_method('Quality controlled data created using ODM Tools Python'**), None)**
create_qualifier()	Creates a new data value qualifier in the underlying ODM database Example: edit_service.create_qualifier('I', 'Quality controlled data created using ODM Tools Python')
create_qcl()	Creates a new data processing/quality control level in the underlying ODM database Example: edit_service.create_qcl('Raw', 'Raw data', 'Raw data that have not undergone quality control.')
create_variable()	Creates a new variable in the underlying ODM database Example: edit_service.create_variable('DO', 'Dissolved Oxygen', 'Not Applicable', 'mg/L', 'Surface Water', 'Field Observation', **False,** 30, 'min', 'Average', 'Water Quality', -9999)
delete_points()	Deletes selected data values from the data series Example: edit_service.delete_points()
drift_correction()	Applies a linear drift correction on the selected data values Example: edit_service.drift_correction(-0.15)
filter_value()	Selects data values falling within a specific data value range Example: edit_service.filter_value(8.25, '>')
filter_date()	Selects data values falling within a specific date range Example: edit_service.filter_date(datetime.datetime(2007, 9, 30, 23, 59,59), datetime.datetime(2007, 9, 1, 0, 0))
data_gaps()	Selects data values on either side of data gaps of a given time duration Example: edit_service.data_gaps(30.0, 'minute')
flag()	Adds a data qualifying comment to selected values Example: edit_service.flag(1)
interpolate()	Linearly interpolates the selected data values in the data series using the previous and next data values Example: edit_service.interpolate()

reset_filter()	Removes all applied filters for data selection Example: edit_service.reset_filter()
restore()	Discards any changes made and replaces the in-memory database with a new copy of the data series Example: edit_service.restore()
get_filtered_points()	Gets the data values that meet the criteria of the currently applied filter Example: edit_service.get_filtered_points()
get_series_points()	Gets the data values in the currently selected data series Example: edit_service.get_series_points()
filter_from_previous()	Alerts ODM Tools to select data values from the full set or the existing selected set of data values Example: edit_service.filter_from_previous(**True**)
value_change_threshold()	Selects all data values where the change from one value to the next is greater than or less than a threshold Example: edit_service.value_change_threshold(0.01,'>')
save()	Saves the modified data series to the database, overwriting the series from which it was created Example: edit_service.save()
save_as()	Saves the modified data series to the database as a new data series Example: new_variable = series_service.get_variable_by_id(35) new_qcl = series_service.get_qcl_by_id(1) \new_method = series_service.get_method_by_id(62) edit_service.save_as(new_variable, new_method, new_qcl)
save_exsiting()	Saves the modified data series to the database by overwriting an existing data series in the database Example: new_variable = series_service.get_variable_by_id(35) new_qcl = series_service.get_qcl_by_id(1) new_method = series_service.get_method_by_id(62) edit_service.save_existing(new_variable, new_method, new_qcl)
select_points()	Given a list of date/time values, selects the corresponding data value points associated with those dates Example: points = [[datetime.datetime(2007, 9, 14, 8, 30)]][0] edit_service.select_points([], points)

User Interface Layer

The user interface layer provides the GUI within which users can visualize and export data, generate summary statistics, and perform data quality control editing. The GUI was designed and implemented using wxPython (http://www.wxpython.org/), which is a toolkit for Python that provides programmers with components for building interactive GUIs. The matplotlib plotting library (http://matplotlib.org/) was used for creating the data visualizations and provides interactive zooming and panning of the plot. We added customized mouse hover functionality to display individual data values and a plot-based data selection tool to aid in data editing. The user interface layer also includes an integrated Python script editor and console that were implemented using the PyCrust component of wxPython. These libraries are cross-platform, which was necessary for meeting our design objectives and enabled us to reuse the same code on Windows, Linux, and Macintosh platforms.

The ODM Tools Python GUI consists of a ribbon control, a data visualization panel, a time series selection panel, a table viewer panel, a Python script editor, and an integrated Python console. Fig. 4 shows the standard configuration of these components upon opening the software (not all components are shown). Components can be toggled for visibility from the View menu at the top of the window. Several components of the GUI (e.g., the time series selection panel, table view panel, Python script editor, and Python console) are dockable, which enables users to arrange and configure their window(s) according to their preference.

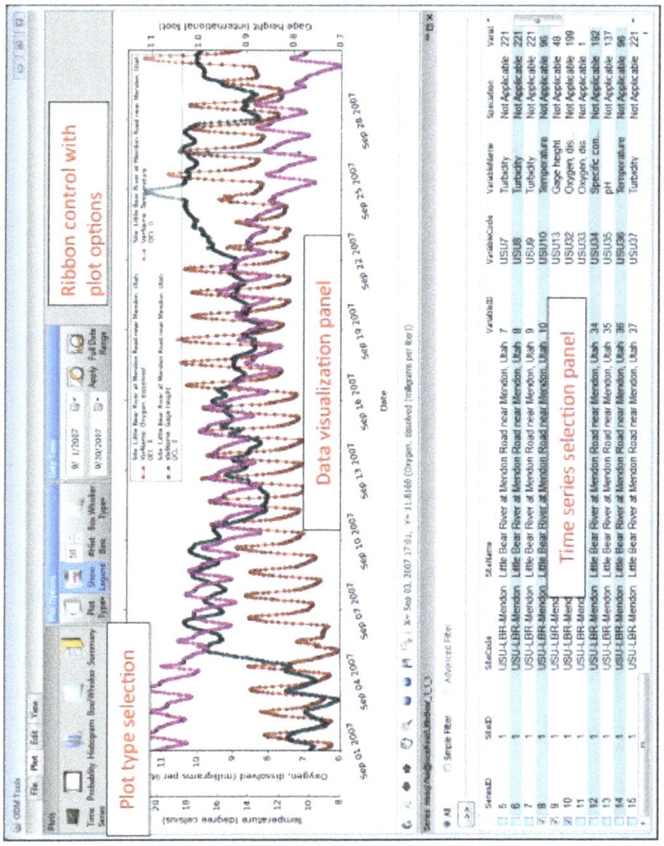

Figure 4: ODM Tools Python graphical user interface.

ODM Tools Python provides several data editing buttons on the main ribbon toolbar, including value filtering (i.e., selection of data values based on user input criteria), insertion or deletion of values, linear interpolation, linear drift correction (a form of variable data correction described by Wagner et al. (2006)), value adjustment (i.e., set equal to, multiply by, or offset by), and value flagging. Flags are an important component of sensor data quality control and serve as indicators of suspect data, failures of quality control checks, or methods used to process or estimate data values (Campbell et al., 2013).

The ODM Tools Python scripting interface automatically records the user's actions as they click these buttons to perform corrections and adjustments on data series in the quality control process. Each button on the data editing toolbar is mapped to a data editing function in the ODM Tools Python Edit Functions service. When a button is clicked, it executes the underlying data editing function on the local memory copy of the data, it fires an equivalent line of code to the Python script editor (Fig. 5), and updates the display of data in the GUI (plot and table) to reflect the change. The sequence of edits automatically recorded in a Python script can be saved as a file, ensuring that the editing steps are traceable and reproducible.

Within ODM Tools Python, all edits are made on a copy of the raw data in order to preserve the original raw data. When a data series is selected for editing, a copy of the data series is created within an in-memory SQLite database. All subsequent edits and modifications are made to the memory copy of the time series. When editing is finished, the user can choose whether to save the edited series. Saving writes the time series stored in memory to the underlying ODM database as a new version, and the memory copy is destroyed. Users can choose whether to create an entirely new version when saving or to overwrite an existing quality controlled version (e.g., in the case of a secondary editing session). These options preserve the metadata associated with the data series and permit the user to specify the version by modifying the method, variable, or quality control level. Using SQLite as the in-memory storage for data series being edited provides a robust and convenient structure that supports querying and filtering in a memory object that does not have to be persisted on disk.

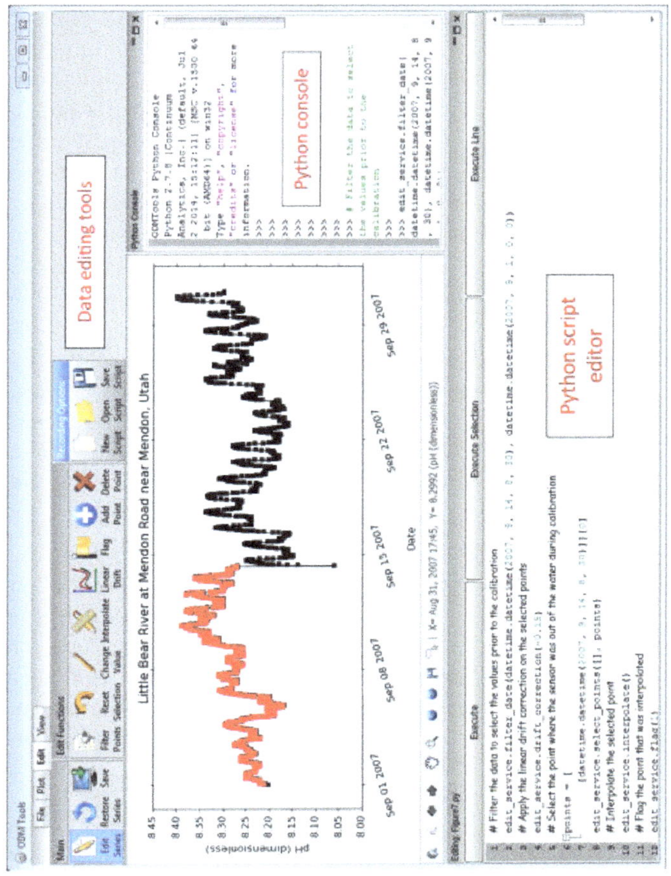

Figure 5: ODM Tools Python graphical user interface in editing mode with integrated Python script editor and console.

The time series selection panel at the bottom of the window enables users to select which data series are shown in the data visualizations or which series is selected for editing. Simple filters can be applied to the list of data series stored within the attached database to identify a subset of data series that meet particular criteria. This is especially useful where the underlying ODM database contains many data series. Users can right click on a data series listing in the time series selection panel and choose an option in the context menu to export the data or the metadata, providing a simple method for exporting data from the underlying database to a file on the user's computer.

Enhanced visualizations within ODM Tools Python include the ability to plot multiple time series simultaneously (Fig. 4). This is a significant improvement over previous versions of ODM Tools and is particularly useful in data quality control editing. Anomalous values in the time series being edited can easily be compared to other time series at the same site or a time series for the same variable from another nearby site to determine whether the anomalous values should be removed, interpolated, left alone, etc. Additional plot types supported by ODM Tools Python include probability plots, histograms, and box-and-whisker plots. Users can switch between plot types by selecting the plot type on the ribbon control.

SOFTWARE DEPLOYMENT

ODM Tools Python users currently have multiple choices for running the software. The first option is to download the source code for a particular release of ODM Tools Python from the GitHub repository and then run ODM Tools directly from the source code. This requires that a user first configure a Python environment on his/her computer that has all of the requisite Python modules and components already installed. Documentation on the ODM Tools Python source code repository website describes the Python modules that are required for running ODM Tools Python from source code. Prior to posting the source code for a release, it is tested on Windows, Mac, and Linux platforms to ensure that it will work for users on these platforms. This approach may be preferable for experienced Python users who want flexibility in how Python and the various components are installed. Potential developers who want to modify the source code of ODM Tools can either download the source code for a particular release or they can clone the GitHub repository to get the latest development code. Running from source code is currently the only option available for Linux users.

Knowing that there are many within the scientific community that might want to use the ODM Tools Python software but may lack the software expertise for setting up a development environment for running from source code (or that have an established Python environment on their computer that they do not want to change), we designed two additional ODM Tools Python software deployments: 1) a no-install zip file; and 2) executable software installers. Both of these use a Python virtual environment within which the ODM Tools Python software runs. The virtual environment is an isolated working copy of Python that is deployed with the software and contains the correct version of all of the requisite Python modules. It does not require installing anything in the user's existing Python environments and, therefore, will not conflict with any existing Python environments on the user's computer.

For the no-install option, a zip file containing the Python virtual environment and the latest source code for ODM Tools is downloaded and extracted on a user's computer. This enables users to run ODM Tools from source code without an install, but within the encapsulated Python environment delivered within the zip file. The no-install option requires launching the main ODM Tools Python form from the Python console. This approach is preferable for users who may not have administrative access to run software installers on their computers. This option works on both Windows and Mac platforms and does not have any pre-requisites, as the Python environment delivered within the zip file contains Python and all of the required software libraries. The last option is to use one of the executable installers provided via an ODM Tools Python software release. Users can currently download executable installers for both Windows and Mac OS X from the ODM Tools GitHub website, and, like the no-install option, there are no prerequisites.

EXAMPLE APPLICATION

An example is presented in this section to illustrate the use of ODM Tools Python for performing quality control edits and to highlight the benefits of the ODM Tools Python scripting functionality. Fig. 6 illustrates the ODM Tools Python data editing workflow for a time series of pH data measured in the Little Bear River of northern Utah, USA. These data were recorded by an *in situ* pH sensor every 30 min. The time period between sensor cleaning and calibration checks at this site is approximately 2 weeks, during which some sensor fouling and instrument drift occurs. In panel (a) of Fig. 6, the pH data series was plotted by selecting it in the series selection panel and then placed in editing mode using the "Edit Series" button on the ribbon toolbar. The calibration date and resulting shift in the data can be seen near the middle of the plot. There is also a single anomalous value that resulted from the sensor being out of the water during the calibration process.

In panel (b), the pH data values prior to the sensor calibration were selected using a date filter in the "Filter Points" tool (ODM Tools highlights selected data values in red). Points can also be selected using other filters in the Filter Points tool (e.g., value threshold, value change threshold, data gaps, etc.), by drawing a selection polygon around them on the plot using a Lasso tool, or by clicking on them in the table view. In panel (c), a linear drift correction was applied to remove the sensor drift between the beginning of the data series and the calibration date. For this example, a correction value of −0.15 pH units was applied based on calibration check values measured in the field and recorded in the field notes. In panel (d), the single remaining anomalous data value that resulted when the sensor was out of the water for calibration was selected and interpolated using the "Interpolate" tool on the toolbar. Panel (e) shows the dialog for applying data qualifying comments accessed by clicking on the "Flag" button on the editing toolbar. ODM Tools applies the selected qualifying comment to any data values that are selected; in this case, the single data value that was interpolated was flagged with a qualifying comment indicating that it was interpolated. Finally, panel (f) shows both the original and edited pH data series after the edited data series was saved to the database as a new version.

The Python script generated by this editing workflow is shown in Fig. 7. Each button click and/or action in the graphical user interface using the editing tools on the toolbar was recorded in the Python script editor as one or more lines of code that call a function in the ODM Tools Python edit service. The script in Fig. 7 is shown within the ODM Tools Python script editor, which includes usability features such as code folding, syntax highlighting, and execution buttons to make it easier for users to work with the code. Within the script editor, users can add comments to the code to indicate why they added particular edits or performed particular actions. The final Python script can then be saved to the user's hard drive as a record of the edits that were performed and for later execution, modification, or review by other scientists or data analysts. This may help address inconsistencies and human bias that can occur with manual data quality control (Campbell et al., 2013).

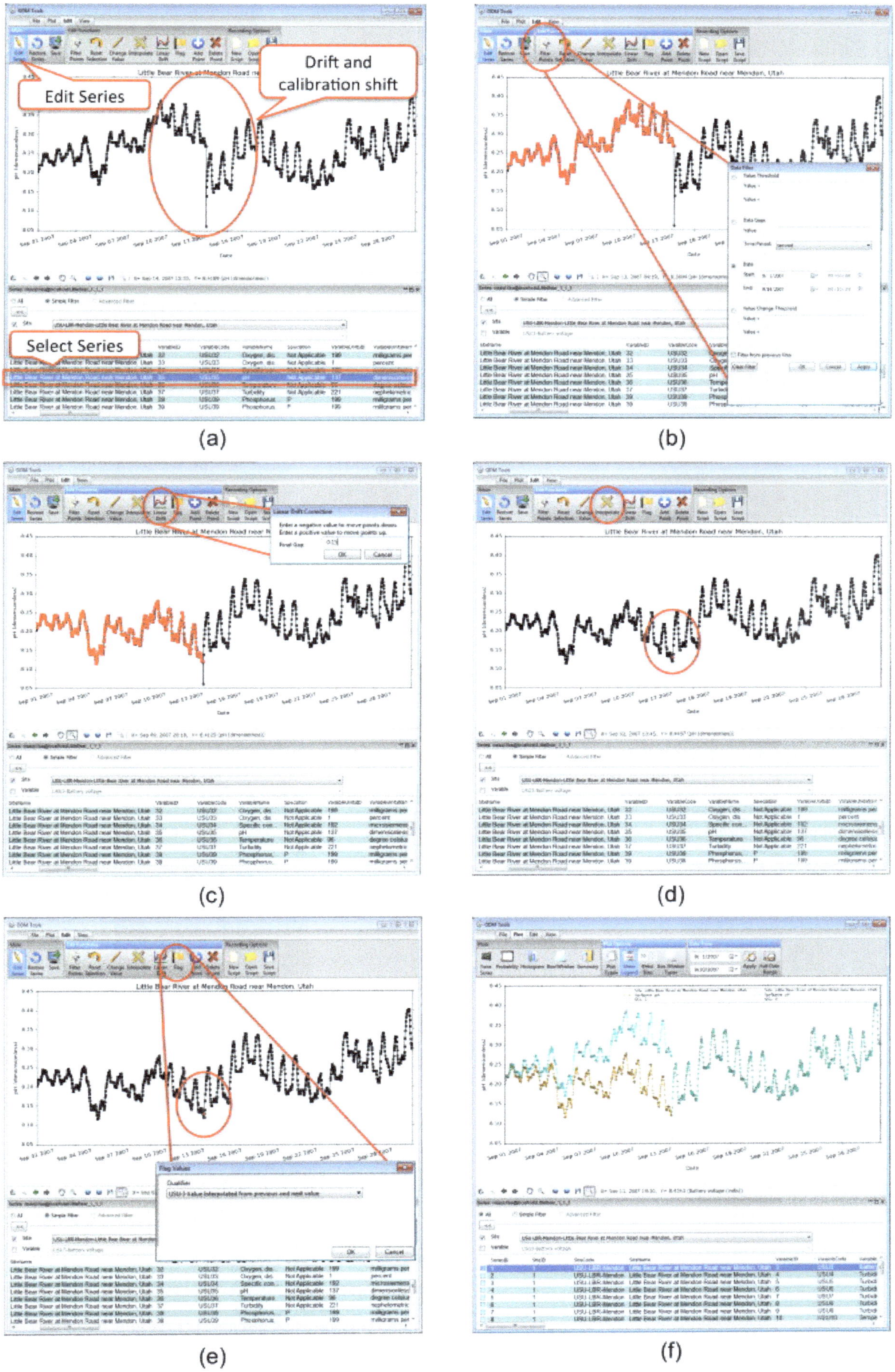

Figure 6: Example of ODM Tools Python data editing workflow.

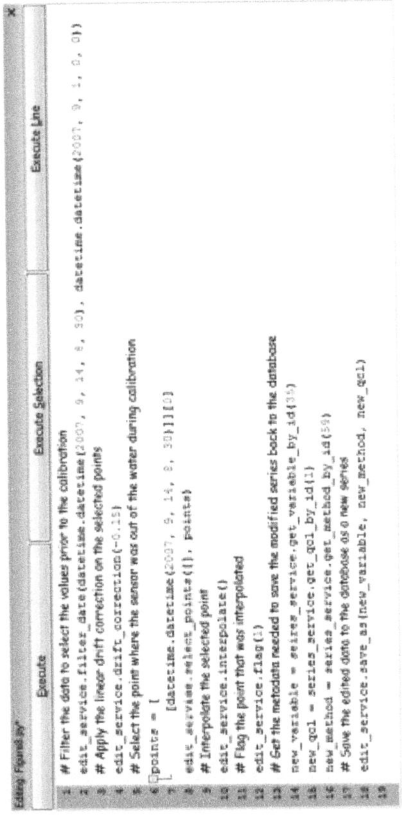

Figure 7: Listing of the Python code generated by the data editing steps in the example. The code is shown in the ODM Tools Python integrated Python script editor.

SUMMARY AND FUTURE WORK

We have developed a new version of ODM Tools in Python that provides multiple platform support, multiple database support, and support for automated scripting of quality control edits performed on data series – each of which were major limitations in the original versions of ODM Tools. Overcoming the platform and database limitations provides flexibility for users, regardless of their chosen platform or RDBMS. The automated scripting capabilities provide an effective method for tracking the quality control edits performed during an editing session, preserving the provenance of how quality controlled datasets were created, and enabling subsequent editing sessions and regeneration of the quality controlled data at any time.

Several new features were included in modernizing the ODM Tools GUI. The integrated ribbon control provides flexibility in implementing toolbar buttons and menus and is more consistent with common desktop software with which users will be familiar. Dockable window components enable users to configure their window(s) to suit their needs. The ability to plot multiple data series simultaneously with various plot types is particularly useful in the quality control editing process as users can examine an edited series versus the raw data to make sure data edits have been applied correctly, or they can examine a series that is being edited alongside other time series collected at the same or nearby sites to investigate anomalous data and determine whether edits need to be made. The code of the GUI also serves as an example of how interactive user interfaces can be created using Python, which can be challenging in practice.

The workflow we developed for scripting time series data quality control editing is a simple but effective procedure for creating and maintaining scripts that encode data edits, and it is facilitated by the redesigned ODM Tools software. ODM Tools combines editing functions exposed via buttons on the main GUI toolbar with automatically generated Python code. The strength of this approach is that it exposes the ability to perform data editing actions through the GUI, but translates any actions within the GUI into executable Python code. The resulting script serves as a record of the edits and can be re-executed at any time to generate the same result. Users can create new quality control editing scripts based on their visual editing through the ODM Tools GUI or they can open and execute the scripts resulting from previous editing sessions. This approach supports both novice and more experienced users.

Last, we believe that the deployment approach we chose for ODM Tools Python may be instructive for how Python-based software can be more easily distributed without requiring users to set up their own Python environment with all of the prerequisite components. Deploying ODM Tools Python with its own Python virtual environment enables it to run on computers that do not have Python installed, providing a quick and easy way to get up and running. For users that already have one or more Python environments on their computer, it isolates the Python components used with ODM Tools Python and ensures that the required versions of Python and modules run regardless of whether other versions exist in other Python environments. This avoids version conflicts that could lead to errors in ODM Tools or other Python programs the user has created. Finally, more advanced users or potential developers can access the source code from the GitHub repository and can run from source code within a Python environment that they have set up. These multiple options provide flexibility in reaching a broad range of potential users. Indeed, we anticipate that ODM Tools Python will be useful for data managers needing procedures and software tools for data post processing and quality control.

Future work will be aimed at adding additional editing tools to the data editing service and the data editing toolbar. These may include two point variable data corrections, smoothing algorithms, and data aggregation/derivation algorithms. Visualization of data qualifying comments with plotted data is another area we are seeking to improve. We are also working to better support the ability for users to define and use custom, user-defined functions within the Python scripting environment for data quality control. We have recently completed a round of modifications to improve the performance of the software, and we are currently in the process of making ODM Tools compatible with ODM2, which is the next version of ODM. ODM2 adds capability for integrating both sensor and sample-based observational data within a single database and enhances the ability to describe and manage both types of environmental observations. This capability will make ODM2 and ODM Tools Python more versatile tools for data managers working with multiple types of environmental observations data.

ACKNOWLEDGMENTS

This work was supported by the National Science Foundation

under EPSCoR grant IIA 1208732 awarded to Utah State University, as part of the State of Utah EPSCoR Research Infrastructure Improvement Award. Additional support was provided by National Science Foundation grants EAR 1224638 and ACI 1339834. Support for previous versions of the ODM Tools software was provided by National Science Foundation grant EAR 0622374. Any opinions, findings, and conclusions or recommendations expressed in this material are those of the authors and do not necessarily reflect the views of the National Science Foundation. We gratefully acknowledge Chris Cox, Dave Eiriksson, and Joe Crawford who provided valuable testing and feedback on the ODM Tools Python software.

REFERENCES

1. Borer, E.T., Seabloom, E.W., Jones, M.B., Schildhauer, M., 2009. Some simple guidelines for effective data management. ESA Bull. 90 (2), 205e214. http:// dx.doi.org/10.1890/0012-9623-90.2.205.
2. Campbell, J.L., Rustad, L.E., Porter, J.H., Taylor, J.R., Ethan, W., Shanley, J.B., Gries, C., Henshaw, D.L., Martin, M.E., Wade, M., Boose, E.R., Dereszynski, E.W., 2013. Quantity is nothing without quality. Bioscience 63 (7), 574e585. http:// dx.doi.org/10.1525/bio.2013.63.7.10.
3. Dereszynski, E.W., Dietterich, T.G., 2012. Probabilistic models for anomaly detection in remote sensor data streams. In: Proceedings of the Twenty-third Conference on Uncertainty in Artificial Intelligence. http://arXiv:1206.5250.
4. Fiebrich, C.A., Morgan, C.R., McCombs, A.G., Hall Jr., P.K., McPherson, R.A., 2010. Quality assurance procedures for mesoscale meteorological data. J. Atmos. Ocean. Technol. 27 (10), 1565e1582. http://dx.doi.org/10.1175/ 2010JTECHA1433.1.
5. Hill, D.J., Minsker, B.S., Amir, E., 2009. Real-time Bayesian anomaly detection in streaming environmental data. Water Resour. Res. 45 (4) http://dx.doi.org/10.1029/2008WR006956.
6. Horsburgh, J.S., Reeder, S., 2014. Data visualization and analysis within a hydrologic information system: integrating with the R statistical computing environment.
7. Environ. Model. Softw. 52, 51e61. http://dx.doi.org/10.1016/ j.envsoft.2013.10.016.
8. Horsburgh, J.S., Tarboton, D.G., Maidment, D.R., Zaslavsky, I., 2008. A relational model for environmental and water resources data. Water Resour. Res. 44, W05406. http://dx.doi.org/10.1029/2007WR006392.
9. Horsburgh, J.S., Tarboton, D.G., Maidment, D.R., Zaslavsky, I., 2011. Components of an environmental observatory information system. Comput. Geosci. 37, 207e218. http://dx.doi.org/10.1016/j.cageo.2010.07.003.
10. Horsburgh, J.S., Tarboton, D.G., Schreuders, K.A.T., Maidment, D.R., Zaslavsky, I., Valentine, D., 2010. Hydroserver: a platform for publishing space-time hydrologic datasets. In: Proceedings 2010 American Water Resources Association Spring Specialty Conference Geographic Information Systems (GIS) and Water Resources VI, Orlando, Florida. American Water Resources Association, Middleburg, Virginia, TPS, pp. 10e11.
11. Horsburgh, J.S., Tarboton, D.G., Piasecki, M., Maidment, D.R., Zaslavsky, I., Valentine, D., Whitenack, T., 2009. An integrated system for publishing environmental observations data. Environ. Model. Softw. 24 (8), 879e888. http:// dx.doi.org/10.1016/j.envsoft.2009.01.002.
12. Lerner, B., Boose, E., Osterweil, L., Ellison, A., Clarke, L., 2011. Provenance and quality control in sensor networks. In: Jones, M., Gries, C. (Eds.), Proceedings of the Environmental Information Management Conference, September 2011, pp. 98e103. http://dx.doi.org/10.5060/D2NC5Z4X.
13. Mason, S.J.K., Cleveland, S.B., Llovet, P., Izurieta, C., Poole, G.C., 2014. A centralized tool for managing, archiving, and serving point-in-time data in ecological research laboratories. Environ. Model. Softw. 51, 59e69. http://dx.doi.org/ 10.1016/j.envsoft.2013.09.008.
14. Millman, K.J., Aivazis, M., 2011. Python for Scientists and Engineers. Comput. Sci. Eng. 13 (2), 9e12. http://dx.doi.org/10.1109/MCSE.2011.36.
15. Moatar, F., Miquel, J., Poirel, A., 2001. A quality-control method for physical and chemical monitoring data. Application to dissolved oxygen levels in the river Loire (France). J. Hydrol. 252 (1e4), 25e36. http://dx.doi.org/10.1016/S0022- 1694(01)00439-5.
16. Mourad, M., Bertrand-Krajewski, J.L., 2002. A method for automatic validation of long time series of data in urban hydrology. Water Sci. Technol. 45 (4e5), 263e270.
17. Perez, F., Granger, B.E., Hunter, J.D., 2011. Python: an ecosystem for scientific computing. Comput. Sci. Eng. 13 (2), 13e21. http://dx.doi.org/10.1109/ MCSE.2010.119.
18. Sheldon Jr., W.M., 2008. Dynamic, rule-based quality control framework for realtime sensor data. In: Gries, C., Jones, M.B. (Eds.), Proceedings of the Environmental Information Management Conference 2008: Sensor Networks, September 2008, pp. 145e150. http://gce-lter.marsci.uga.edu/public/files/pubs/ wsheldon_dynamic_qc_eimc2008_final.pdf.
19. Tarboton, D.G., Horsburgh, J.S., Maidment, D.R., Whiteaker, T., Zaslavsky, I., Piasecki, M., Goodall, J., Valentine, D., Whitenack, T., 2009. Development of a community hydrologic information system. In: Anderssen, R.S., Braddock, R.D., Newham, L.T.H. (Eds.), 18th World IMACS Congress and MODSIM09 International Congress on Modelling and Simulation, Modelling and Simulation Society of Australia and New Zealand and International Association for Mathematics and Computers in Simulation, July 2009, ISBN 978-0-9758400-7-8, pp. 988e994.
20. Taylor, J.R., Loescher, H.L., 2013. Automated quality control methods for sensor data: a novel observatory approach. Biogeosciences 10, 4957e4971. http://dx.doi.org/ 10.5194/bg-10-4957-2013.
21. Taylor, P. (Ed.), 2012. OGC WaterML 2.0: Part 1 e Timeseries, OGC Candidate Standard OGC 10-126r2, Version 2.0.0. https://portal.opengeospatial.org/files/?artifact_id¼48531 (accessed 27.12.14.).
22. Wagner, R.J., Boulger, R.W., Oblinger, C.J., Smith,

B.A., 2006. Guidelines and Standard Procedures for Continuous Water-quality Monitors: Station Operation, Record Computation, and Data Reporting. U.S. Geological Survey Techniques and Methods 1-D3, 51 pp. þ 8 attachments. http://pubs.water.usgs.gov/tm1d3.

23. White, D.L., Sharp, J.L., Eidson, G., Parab, S., Ali, F., Esswein, S., 2010. Real-time quality control (qc) processing, notification, and visualization services, supporting data management of the Intelligent River©. In: Proceedings of the 2010 South Carolina Water Resources Conference, October 2010.

24. Williams, G.J., 2009. Rattle: a data mining GUI for R. R J. 1 (2), 45e55. http://journal. r-project.org/archive/2009-2/RJournal_2009-2_Williams.pdf.

25. Zaslavsky, I., Valentine, D., Whiteaker, T. (Eds.), 2007. CUAHSI WaterML. OGC Discussion Paper OGC 07-041r1. Version 0.3.0. http://portal.opengeospatial.org/files/?artifact_id¼21743 (accessed 27.12.14.).

INDEX